Molecular Evolution and

Molecular Evolution and Adaptive Radiation

Edited by

Thomas J. Givnish and Kenneth J. Sytsma
University of Wisconsin – Madison

PUBLISHED BY THE PRESS SYNDICATE OF THE UNIVERSITY OF CAMBRIDGE
The Pitt Building, Trumpington Street, Cambridge, United Kingdom

CAMBRIDGE UNIVERSITY PRESS
The Edinburgh Building, Cambridge CB2 2RU, UK   http://www.cup.cam.ac.uk
40 West 20th Street, New York, NY 10011-4211, USA   http://www.cup.org
10 Stamford Road, Oakleigh, Melbourne 3166, Australia
Ruiz de Alarcón 13, 28014 Madrid, Spain

© Cambridge University Press 1997

This book is in copyright. Subject to statutory exception and
to the provisions of relevant collective licensing agreements,
no reproduction of any part may take place without
the written permission of Cambridge University Press.

First published 1997
First paperback edition 2000

Printed in the United States of America

Typeset in Palatino

A catalog record for this book is available from the British Library

Library of Congress Cataloging in Publication data
Molecular evolution and adaptive radiation / edited by Thomas J. Givnish, Kenneth J. Sytsma
    p. cm.
1. Biology – Classification – Molecular aspects. 2. Adaptation (Biology)
I. Givnish, Thomas J. II. Sytsma, Kenneth Jay.
QH83.M664 1997
578.4–dc21

ISBN 0 521 57329 7 hardback
ISBN 0 521 77929 4 paperback

As buds give rise by growth to fresh buds,
and these, if vigorous, branch out and overtop
on all sides many a feebler branch, so by generation
I believe it has been with the great Tree of Life,
which fills with its dead and broken branches
the crust of the earth, and covers the surface with
its ever branching and beautiful ramifications.

> Charles Darwin
> *On the Origin of Species*

Contents

Preface   xiii

List of Authors   x

I. Introduction

1. Adaptive radiation and molecular systematics: issues and approaches   1
    Thomas J. Givnish

2. Homoplasy in molecular vs. morphological data: the likelihood of correct phylogenetic inference   55
    Thomas J. Givnish and Kenneth J. Sytsma

II. Integrative studies

3. Adaptive radiation of the Hawaiian silversword alliance: congruence and conflict of phylogenetic evidence from molecular and non-molecular investigations   103
    Bruce G. Baldwin

4. The chronicle of marsupial evolution   129
    Mark S. Springer, John A. W. Kirsch, and Judd A. Chase

5. Evolutionary origins of phenotypic diversity in *Daphnia*   163
    John K. Colbourne, Paul D. N. Hebert, and Derek J. Taylor

6. Evolutionary trends in the ecology of New World monkeys inferred from a combined phylogenetic analysis of nuclear, mitochondrial, and morphological data   189
    Inés Horovitz and Axel Meyer

7. Adaptive radiation in the aquatic plant family Pontederiaceae: insights from phylogenetic analysis   225
    Spencer C. H. Barrett and Sean W. Graham

8. Molecular evolution and adaptive radiation in *Brocchinia* (Bromeliaceae: Pitcairnioideae) atop tepuis of the Guayana Shield   259
    Thomas J. Givnish, Kenneth J. Sytsma, James F. Smith, William J. Hahn, David H. Benzing, and Elizabeth M. Burkhardt

III. Convergence

9. You aren't (always) what you eat: evolution of nectar-feeding among Old World fruitbats (Megachiraptera: Pteropodidae) — 313
   John A. W. Kirsch and François-Joseph Lapointe

10. Leapfrog radiation in floral and vegetative traits among twig epiphytes in the orchid subtribe Oncidiinae — 331
    Mark W. Chase and Jeffrey D. Palmer

11. Adaptation, cladogenesis, and the evolution of habitat association in North American tiger beetles: a phylogenetic perspective — 353
    Alfried P. Vogler and Paul Z. Goldstein

IV. Rapid radiations

12. Molecular phylogenetic tests of speciation models in Lake Malawi cichlid fishes — 375
    Peter N. Reinthal and Axel Meyer

13. Rapid radiation due to a key innovation in columbines (Ranunculaceae: *Aquilegia*) — 391
    Scott A. Hodges

14. Origin and evolution of *Argyranthemum* (Asteraceae: Anthemideae) in Macaronesia — 407
    Javier Francisco-Ortega, Daniel J. Crawford, Arnoldo Santos-Guerra, and Robert K. Jansen

V. Reproductive strategies

15. Plant-pollinator interactions and floral radiation in *Platanthera* (Orchidaceae) — 433
    Jeffrey R. Hapeman and Ken Inoue

16. Phylogenetic perspectives on the evolution of dioecy: adaptive radiation in the endemic Hawaiian genera *Schiedea* and *Alsinodendron* (Caryophyllaceae: Alsinoideae) — 455
    Ann K. Sakai, Stephen G. Weller, Warren L. Wagner, Pamela S. Soltis, and Douglas E. Soltis

17. Ecological and reproductive shifts in the diversification of the endemic Hawaiian *Drosophila* — 475
    Michael P. Kambysellis and Elysse M. Craddock

VI. Character divergence and community assembly

18. History of ecological selection in sticklebacks: uniting experimental and phylogenetic approaches 511
    Eric B. Taylor, John Donald McPhail, and Dolph Schluter

19. Phylogenetic studies of convergent adaptive radiations in Caribbean *Anolis* lizards 535
    Todd Jackman, Jonathan B. Losos, Allan Larson, and Kevin de Queiros

VII. Macroevolutionary patterns

20. Molecular and morphological evolution during the post-Palaeozoic diversification of echinoids 559
    Andrew B. Smith and D. T. J. Littlewood

21. How fast is speciation? Molecular, geological, and phylogenetic evidence from adaptive radiation of fishes 585
    Amy R. McCune

Index 611

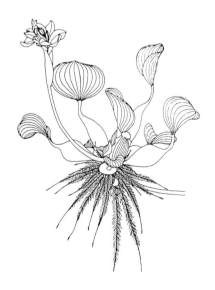

Authors

Bruce G. Baldwin
Jepson Herbarium
  and Department of
  Integrative Biology
1001 Valley Life Sciences
  Building 2465
University of California
Berkeley, California 94720-2465

Spencer C. H. Barrett
Department of Botany
University of Toronto
Toronto, Ontario M5S 3B2
Canada

David H. Benzing
Department of Biology
Oberlin College
Oberlin, Ohio 44074

Elizabeth M. Burkhardt
Department of Organismic and
  Evolutionary Biology
Harvard University
Cambridge, Massachusetts
  02138

Judd A. Chase
Department of Biology
University of California
Riverside, California 92521

Mark W. Chase
Molecular Systematics Section
Royal Botanic Gardens, Kew
Richmond, Surrey, TW9 3DS
United Kingdom

John K. Colbourne
Department of Zoology
University of Guelph, Ontario
N1G 2W1
Canada

Elysse M. Craddock
Division of Natural Sciences
Purchase College
State University of New York
Purchase, New York 10577-1400

Daniel J. Crawford
Department of Plant Biology
The Ohio State University
Columbus, Ohio 43210-1293

Javier Francisco-Ortega
Department of Botany
University of Texas
Austin, Texas 78713-7640

Thomas J. Givnish
Department of Botany
University of Wisconsin
Madison, Wisconsin 53706

Paul Z. Goldstein
Dept. of Ecology &
  Evolutionary Biology
University of Connecticut
Storrs, Connecticut 06269

Sean W. Graham
Department of Botany
University of Toronto
Toronto, Ontario M5S 3B2
Canada

William J. Hahn
Department of Earth and
  Environmental Sciences
Columbia University
New York, New York 10027

Jeffrey R. Hapeman
Department of Botany
University of Wisconsin
Madison, Wisconsin 53706

Paul D. N. Hebert
Department of Zoology
University of Guelph, Ontario
N1G 2W1
Canada

Scott A. Hodges
Department of Ecology,
  Evolution and Marine
  Biology
University of California
Santa Barbara, California 93106

Inés Horovitz
Department of Ecology and
  Evolution
State University of New York at
  Stony Brook
Stony Brook, New York
  11794-5245

Ken Inoue
Biological Institute and
  Herbarium
Faculty of Science, Shinshu
  University
Matsumoto 390
Japan

Todd Jackman
Department of Biology
Campus Box 1137
Washington University
St. Louis, Missouri 63130-4899

Robert K. Jansen
Department of Botany
University of Texas
Austin, Texas 78713-7640

Michael P. Kambysellis
Department of Biology
New York University
New York, New York 10003

John A. W. Kirsch
Director, Zoological Museum
University of Wisconsin
Zoological Museum
Madison, Wisconsin 53706

François-Joseph Lapointe
Departement de Sciences
  Biologiques
Universite de Montréal
Montréal, Quebec H3C 3J7
Canada

Allan Larson
Department of Biology
Campus Box 1137
Washington University
St. Louis, Missouri 63130-4899

Axel Meyer
Department of Ecology and
 Evolution
State University of New York at
 Stony Brook
Stony Brook, New York 11794-
 5245

Jeffrey D. Palmer
Department of Biology
Indiana University
Bloomington, Indiana 47405

Kevin de Queiroz
Division of Amphibians and
 Reptiles
National Museum of Natural
 History
Smithsonian Institution
Washington, D. C., 20560 USA

Peter N. Reinthal
Department of Biology
Eastern Michigan University
Ypsilanti, Michigan 48197

Ann K. Sakai
Department of Ecology and
 Evolutionary Biology
University of California
Irvine, California 92717

Arnoldo Santos-Guerra
Jardín de Aclimatación de La
 Orotava
Puerto de La Cruz
E-38400, Tenerife
Canary Islands, Spain

Dolph Schluter
Department of Zoology and
 Centre for Biodiversity
 Research
University of British Columbia
Vancouver, British Columbia
 V6T 1Z4
Canada

Andrew B. Smith
Department of Palaeontology
The Natural History Museum
Cromwell Road
London SW7 5BD
United Kingdom

Douglas E. Soltis
Department of Botany
Washington State University
Pullman, Washington
 99164-4238

Pamela S. Soltis
Department of Botany
Washington State University
Pullman, Washington
 99164-4238

Mark S. Springer
Department of Biology
University of California
Riverside, California 92521

Kenneth J. Sytsma
Department of Botany
University of Wisconsin
Madison, Wisconsin 53706

Derek J. Taylor
Department of Biology
University of Michigan
Ann Arbor, Michigan
 48109-1048

Eric B. Taylor
Department of Zoology and
 Centre for Biodiversity
 Research
University of British Columbia
Vancouver, British Columbia
 V6T 1Z4
Canada

Alfried P. Vogler
Department of Entomology
The Natural History Museum
London, SW7 5BD
 and
Department of Biology
Imperial College at Silwood Park
Ascot, Berkshire, SL5 7PY
United Kingdom

Warren L. Wagner
Department of Botany
Smithsonian Institution
Washington, D. C. 20560

Stephen G. Weller
Department of Ecology and
 Evolutionary Biology
University of California
Irvine, California 92717

Preface

Adaptive radiation – the rise of a diversity of ecological roles and attendant adaptations in different species within a lineage – has been central to evolutionary biology since Charles Darwin first observed the remarkable finches of the Galápagos. It is one of the most important processes bridging ecology and evolution, and studies of its operation in groups like the African rift-lake cichlids and Hawaiian silverswords have helped shape our modern understanding of adaptation and diversity in the biological world. Yet few of the classic cases of adaptive radiation have been investigated rigorously. The fundamental problem is that the very traits undergoing radiation (e.g., beak size and shape) were also often used to help classify the organisms in question. This approach can easily become circular, with traits being traced down evolutionary pathways determined, at least in part, by the traits themselves.

The emergence of molecular systematics over the past 15 years has provided powerful tools with which to infer phylogeny independent of phenotype, and thus made the time ripe for a revolution in the study of adaptive radiation. Over the past seven years, adaptive radiation has become the focus of many cladistic studies and molecular systematic analyses, and has sparked studies in a number of related areas. Several key questions at the interface of ecology, evolutionary biology, systematics, and biogeography are now being addressed satisfactorily for the first time.

This book is the first attempt to synthesize the recent explosion of research on adaptive radiation based on molecular systematics. It represents the proceedings of an international symposium held at McGill University on July 7-8, 1995, and attended by more than 300 faculty and students. We invited a distinguished group of investigators, working on a variety of plant and animal groups, to address a number of fundamental questions that arise in the study of adaptive radiation. Their research exemplifies a variety of analytical approaches, organisms, and geographical settings, and often involves the use of molecular data and modern analytical techniques in conjunction with more traditional approaches, including comparative morphology, cytology, and paleontology. Their contributions synthesize recent research on many classic cases of adaptive radiation; several address fascinating but previously unstudied groups.

We included a wide range of organisms – vertebrates and invertebrates, plants and animals – in order to encompass a diversity of evolutionary patterns and processes, and to make available to a broad audience studies that would ordinarily appear in journals with largely non-overlapping readerships. Interesting research on some groups (e.g., birds, whales, corals, bacteria, algae) could not be included due to limits on space or the availability of manuscripts as this book went to press. We deliberately chose to emphasize studies on flowering plants, to counteract a certain bias in the classical literature and to highlight the strengths of several recent studies on angiosperms. The size and structural conservatism of the chloroplast genome, inherited maternally, allows for sophisticated studies of phylogeny, hybridization, and reticulate evolution. The latter may play a momentous role not merely in plant evolution, but also – through primary and secondary endosymbiosis, and the rampant swapping of genetic material among prokaryotes – in the grand radiation of life on earth.

Overview

Givnish (Chapter 1) provides a broad conceptual framework for the study of adaptive radiation, reviewing recent developments in the field (with special emphasis on the contributions of molecular systematics), and summarizing important research issues involving phylogeny, adaptation, historical ecology, genetics, development, biogeography, tempo, and predictability. Givnish and Sytsma (Chapter 2) demonstrate that, at least for plants, molecular studies generally provide more numerous and more highly consistent characters than morphological studies. Computer simulations show that, as a result, molecular data are far more likely to yield correct phylogenies; equations are provided for the minimum number of characters at a given level of consistency needed for a particular level of confidence in the resulting phylogeny.

Chapters 3 through 8 are integrative studies that report research on several different aspects of adaptive radiation; read as a whole, they could be taken as an alternative introduction to the field. Baldwin (Chapter 3) provides a magisterial account of the phylogenetic, biosystematic, physiological, and biogeographical studies on the Hawaiian silversword alliance, a group of plants that underwent a spectacular radiation in habit and habitat following an initial colonization from California less than six million years ago. Springer et al. (Chapter 4) show that marsupials have undergone rampant convergence not only with placental mammals but with themselves as well. Gliding, arboreality, and burrowing evolved three times; syndactyly and diprotodonty, twice. Backward-oriented pouches arose four times, mainly in connection with burrowing; their retention in leaf-eating koalas may facilitate the vital transfer of gut flora to the young via coprophagy. Unexpectedly, the great radiation of Australian marsupials includes *Dromiciops* from South America. Based on their research on the freshwater zooplankton genus *Daphnia*, Colbourne et al. (Chapter 5) argue that melanism, headshields and spines, and plumose setae have evolved repeatedly in response to exposure to high levels of ultraviolet radiation, predation, and water turbidity, respectively.

Horovitz and Meyer (Chapter 6) use molecular and morphological evidence to analyze relationships and ecological innovations in New World monkeys. They infer a basal split into two major clades: the mainly herbivorous Atelidae, and heavily insectivorous/nectivorous Cebidae. Body size itself may be a key innovation in this radiation, with the atelids generally being large and the cebids generally being small. Spider monkeys and their relatives have prehensile tails and bimanual locomotion, providing access to fruit and leaves near the tips of branches; uakaris and their relatives have sharp canines and teeth with low cusp relief, which may allow use of hard-husked fruits and seeds. Barrett and Graham (Chapter 7) reconstruct evolution in the water hyacinths, an important family of aquatic plants in the New World. Surprisingly, they found that perennials arose several times from an annual ancestor, and that floral self-incompatibility evolved before tristyly, calling into question some leading evolutionary theories. Givnish et al. (Chapter 8) document the pattern of adaptive radiation in *Brocchinia,* a bromeliad genus endemic to the tepuis and sandplains of the Guayana Shield. The impounding habit and absorptive trichomes in this group are key innovations that allowed the rise of carnivory, ant-fed myrmecophily, nitrogen fixation, and tank epiphytism, entrained concerted convergence in several morphological traits, and led to ecological dominance and widespread distribution in nutrient-poor habitats. The "key landscape" of the barren eastern tepuis may have played a large role in stimulating this adaptive radiation. It is interesting, however, that radiation resulted in less speciation than in the remaining taxa, which are generally short-statured, endemic to individual mountains or adjacent lowlands, and perhaps poorly dispersed.

Chapters 9 through 11 focus primarily on evolutionary convergence. Kirsch and Lapointe (Chapter 9) use phylogenetically structured and unstructured analyses to examine the evolution of nectar-feeding in the Old World fruitbats. Six genera (previously classified as macroglossines) share a brushy tongue, reduced dentition, and a narrow muzzle. DNA-DNA hybridization data, however, show that these traits have undergone repeated convergence, and that the macroglossine bats are thus polyphyletic. Chase and Palmer (Chapter 10) describe shifts in the vegetative form, lifecycle, and floral morphology of oncidioid twig epiphytes, a phylogenetic "twig" of the orchid family composed of short-lived species that inhabit the smallest, outermost branches of their host trees. In this group, certain features of vegetative morphology – generally subject to convergence and parallelism in other plants – are a reliable guide to phylogeny, while several aspects of flower form – generally used to classify species and genera – show repeated convergence in response to pollination by wasps, halictid bees, euglossine bees, birds, and butterflies. Neoteny and other forms of heterochrony appear to have played an important role in vegetative evolution, allowing some species to reproduce in as little as four months. Vogler and Goldstein (Chapter 11) show that tiger beetles exhibit repeated, convergent shifts in the seasonal timing of adult activity in connection with the invasion of higher latitudes, providing tests of the taxon-cycle and taxon-pulse hypotheses.

Chapters 12 through 14 address the phenomenon of rapid radiation, in which recently derived species differ much morphologically but little genetically, as might be expected at the outset of many radiations. Reinthal and Meyer (Chapter 12) provide remarkable evidence of very recent, parallel radiations with little genetic change in a group of closely related cichlids from Lake Malawi. While their data cannot differentiate between sympatric and microallopatric modes of speciation, it is clear that speciation and ecological divergence can occur over very small spatial scales – a finding with enormous implications for within-lake diversification in the African cichlids. Hodges (Chapter 13) suggests that the evolution of nectar spurs was a key innovation "spurring" greatly increased rates of speciation and floral diversification in columbines and perhaps in other plant groups. Francisco-Ortega et al. (Chapter 14) demonstrate that the radiation of *Argyranthemum* in Macaronesia has generally involved what might be termed "horizontal evolution", with species adapted to windward (relatively moist) or leeward (relatively dry) habitats forming two clades, with habitats on a given island being colonized mainly by species inhabiting similar habitats on other islands.

Chapters 15 through 17 examine the evolution of reproductive strategies. Hapeman and Inoue (Chapter 15) report a fascinating investigation of floral evolution in *Platanthera*, the largest temperate genus of orchids. Ancestral pollination by nocturnal settling moths has led to pollination by nocturnal and diurnal hawkmoths, butterflies, bees, empidid flies, and mosquitoes. Closely related species often place their pollinia on different parts of the pollinator's body; a fringed labellum has evolved repeatedly in association with hawkmoth pollination. Sakai et al. (Chapter 16) document the repeated evolution of wind pollination, dioecy, and narrow leaves in connection with the invasion of dry, windy habitats in Hawaiian species of *Alsinodendron* and *Schiedea*. Kambysellis and Craddock (Chapter 17) trace the evolution of oviposition sites and related female reproductive adaptations in the Hawaiian drosophiloids. Their report of a tendency toward oviposition on richer, more clumped resources may provide a key to understanding the rise of lek mating systems and the strong pressures exerted by sexual selection in the evolution of this remarkable group.

Chapters 18 and 19 investigate the basis for character divergence and for regularity in the assembly of ecological communities, two key issues in the study of adaptive radiation. Taylor et al. (Chapter 18) provide molecular data suggesting that benthic and limnetic species of sticklebacks speciated and diverged ecologically in sympatry, a remarkable finding that accords with the conclusions presented by Reinthal and Meyer (Chapter 12). They summarize the experimental evidence that competition in sticklebacks results in divergent selection on trophic structures, providing unique support for the actual operation of character displacement in adaptive radiation. Such divergent selection, coupled with sexual selection based on trophic structures, could easily lead to ecological speciation *in situ*. Jackman et al. (Chapter 19) show that, on each of the four islands of the Greater Antilles, *Anolis* lizards have diversified into similar sets of disparate ecological roles, with the roles having evolved in essentially the same sequence. This fundamental result belies the claim by S. J. Gould (in *Wonderful Life*) that a "replay of the tape would lead evolution down a pathway radically different from the road actually taken."

The last two chapters investigate phenomena over macroevolutionary timescales. Smith and Littlewood (Chapter 20) show that sea urchins evolved the infaunal habit (i.e., became sand dollars) roughly 200 million years ago, about the same time that specialized grazing forms with rasping lanterns also appeared. Transitions from planktotrophic to lecithotrophic larvae have occurred on at least five occasions. Molecular divergence and morphological divergence are both correlated with time since taxonomic divergence, but are unrelated to each other; changes in larval and adult morphology have been essentially independent. Finally, McCune (Chapter 21) uses molecular, geological, and paleontological data to demonstrate that lacustrine fishes show exceptionally high rates of speciation, at least relative to birds or arthropods endemic to oceanic islands. This surprising result dovetails with those reported by Reinthal and Meyer (Chapter 12) and Taylor et al. (Chapter 18).

Acknowledgments

We would like to thank the authors for their fine efforts, and express our gratitude to the reviewers who helped ensure that the chapters met the highest standards. Helpful comments were provided by Joseph Felsenstein, Dana Geary, John Kirsch, Karen Strier, and John Wenzel; special thanks go to the Adaptive Radiation group in the Department of Botany at the University of Wisconsin, including Jeff Hapeman, Austin Mast, Alison Mahoney, Molly Nepokroeff, Tom Patterson, Chris Pires, Aaron Rodriguez, and Michelle Zjhra. David Green, Marty Lechowitz, and Jean-Francois Noel provided key logistical support for the McGill symposium. The artistry, technical virtuosity, and unbounded enthusiasm of Kandis Elliot and James Jaeger in designing and typesetting the book and redrafting the figures are most deeply appreciated.

Thomas J. Givnish and Kenneth J. Sytsma
University of Wisconsin, Madison

We wish to thank Princeton University Press for permission to adapt drawings of the heads of Darwin's finches and Hawaiian honeycreepers for use in the following figures:

Figures 1.1 and 1.2: Grant, P. *Ecology and evolution of Darwin's finches.*
Copyright © 1986 by Princeton University Press. Reprinted by permission of Princeton University Press.

Figure 1.3: Pratt, H. D., Bruner, P. L., and Berett, D. G. *Birds of Hawaii and the tropical Pacific.*
Copyright © 1987 by Princeton University Press. Reprinted by permission of Princeton University Press.

1 Adaptive radiation and molecular systematics: issues and approaches

Thomas J. Givnish

Adaptive radiation – the rise of a diversity of ecological roles and attendant adaptations in different species within a lineage (see Osborn 1902; Huxley 1942; Simpson 1953; Mayr 1970; Futuyma 1986; Givnish et al. 1994, 1995; Schluter 1996a) – is one of the most important processes bridging ecology and evolution. As exemplified by Darwin's finches (Grant 1986), adaptive radiation has profound implications for the origin of adaptations, the genesis of biological diversity, and the coexistence of related species. It also poses several fundamental questions involving the study of phylogeny and its relationship to ecology, morphology, physiology, behavior, and biogeography (Table 1.1).

Divergence among species in a radiation may reflect selection to avoid competition with closely related, ecologically similar taxa, or simply differences in the environments facing each species (Lack 1947; Bowman 1963). Selection for ecological divergence may facilitate the coexistence of species generated by other mechanisms, or may itself drive speciation (Brown and Wilson 1956; Grant 1975; Schluter 1988a, 1996b). Radiations are often most apparent on oceanic islands or similarly isolated habitats (e.g., lakes, mountaintops); few groups are able to colonize such remote areas, eliminating competition from lineages long specialized on particular resources while increasing the importance of competition with close relatives. The short life of most islands also helps ensure that ecologically divergent but related species can still be recognized as a monophyletic lineage (the set of descendants of a shared ancestor). Fossils demonstrate, however, that adaptive radiation occurs frequently on continents as well, as illustrated by the progressive invasion of savannas by horses and their allies (MacFadden 1992) and the functional diversification of ungulate herbivores (Jernvall et al. 1996).

History of the concept

Darwin (1859) used the concept of adaptive radiation extensively in *The Origin of Species*, but the term itself was coined by the palaeontologist Henry Fairfield Osborn in 1902. Osborn linked adaptive radiation with its converse, evolutionary convergence, noting that mammals on different continents diverged into forms adapted for running, digging, flying, and swimming, with each form in a lineage being convergent in many respects with the analogous forms emerging from other lineages. Of the founders of the Modern Synthesis in the 1930s and 40s, Julian Huxley (1942) and George Gaylord Simpson (1944, 1953) viewed adaptive radiation as a central evolutionary concept. Simpson (Osborn's successor at the American Museum) discussed adaptive radiation at length from a palaeontological perspective, using fossils as evidence for both phylogenetic relationships and the ecological divergence of organisms into different "adaptive zones"; he advanced the idea that a "key innovation" (e.g., wings) could accelerate speciation in a lineage (e.g., birds) by opening access to an entirely new range of resources. Julian Huxley (grandson of Thomas Henry Huxley, Darwin's great advocate) played an important and perhaps underappreciated role in the Modern Synthesis, contributing an ecological perspective in his illuminating discussion of adaptive radiation and related topics. Although the founders of the Modern Synthesis – Fisher, Haldane, Dobzhansky, Wright, Huxley, Stebbins, Mayr, and Simpson – brought to bear their respective strengths in genetics, development, cytology, systematics, biogeography, and palaeontology, none was truly an ecologist. But the developmentalist Huxley probably came closest, having mentored Charles Elton (the eminent population biologist), E. B. Ford (the founder of ecological genetics), and – most important in our context – David Lack.

Lack (1940, 1947) conducted extensive studies on Darwin's finches (Geospizidae) to address some of the issues that Darwin began to perceive after returning from his 1835 visit to the Galápagos and receiving key insights into the apparent relationships of the birds and plants of that archipelago from John Gould and Joseph Hooker (Sulloway 1984; Grant 1986). When Lack began his research in the 1930s he believed – as did many at that time – that most of the characteristics that define species were of

Table 1.1. Some key conceptual issues involved in the study of adaptive radiation.

Phylogeny – Do phenotypic similarities among species in a lineage reflect ecological similarities, close relationships, or both? Do phylogenies based on molecular data materially change our interpretation of relationships and evolutionary trends within radiations? What is the relative reliability of phylogenies inferred from morphological vs. molecular characters?

Adaptation – What is the evidence that differences among species are adaptive and fit them to different ecological roles? Were competitive pressures or divergent environments more important in driving species divergence?

Historical ecology – What are the evolutionary pathways leading to various adaptations? To what extent does an organism's present-day ecology reflect its phylogenetic history? What environmental factors or species traits appear to have been important in favoring the evolution of specific traits?

Speciation – Can selection for ecological divergence result in speciation? Do "key innovations" accelerate speciation? What traits *are* associated with high rates of speciation or genetic divergence?

Genetics – Do lineages undergoing adaptive radiation display unusually high levels of genetic divergence among populations? What is the relationship between genetic and morphological divergence?

Development – What are the developmental and genetic bases for the lability of the phenotypic characters that underlie an adaptive radiation? How much of the morphological diversity seen across species can be generated by heterochronic or homeotic shifts in just a few aspects of development?

Biogeography – What is the relationship of phylogeny to geography within a radiation? Can adaptive radiation occur sympatrically? Are there "key landscapes" – either in terms of geography or the constellation of species and environmental conditions facing a lineage – that can favor the invasion of new adaptive zones and/or spur speciation?

Tempo – What is the tempo of adaptive radiation and what is its role in macroevolution? Does the pace of phenotypic divergence slow through time?

Predictability – Will similar adaptive radiations occur when different lineages invade similar ecological landscapes? What is the relationship of adaptive radiation to ecological community structure? How many closely related species can coexist? Are radiations more likely on oceanic and ecological islands?

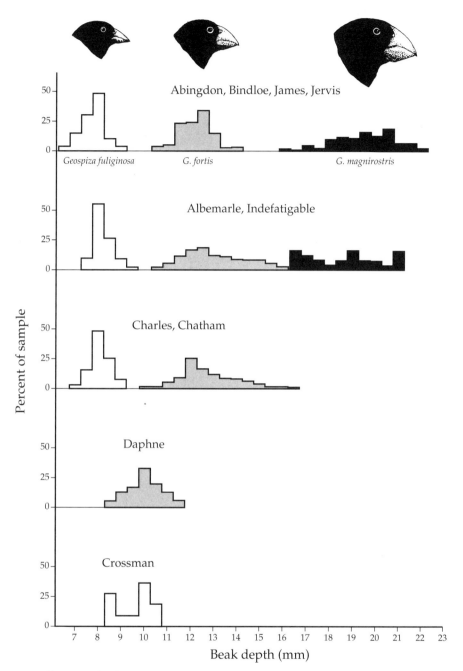

Figure 1.1. Character displacement in beak depth in Darwin's finches from the Galápagos (Lack 1947; drawings after Grant 1986). The small-beaked ground finch (*Geospiza fuliginosa*) is absent from Daphne, where its closest potential competitor – the medium-beaked ground finch (*G. fortis*) – has evolved a somewhat smaller beak, substantially overlapping the beak depth (and hence, the seed diet) of *G. fuliginosa* elsewhere. Similarly, where *G. fortis* is absent, *G. fuliginosa* has evolved larger beaks than where it co-occurs with its larger-beaked competitor. After the large-beaked ground finch (*G. magnirostris*) began to breed on Daphne after the El Niño event in 1983, large seeds became relatively rare during the subsequent drought and selection favored *G. fortis* with smaller beaks (Grant 1994).

little selective or ecological value. But his viewpoint changed radically in the early 40s, and his pivotal 1947 book probably did more to bring adaptive radiation – and ecological thinking generally – into the Modern Synthesis than any other single work (e.g., see Huxley 1942; Muller 1949; Wright 1949; Simpson 1953; Mayr 1963). Lack observed that different finch species have different diets, and that their beaks were apparently adapted to those diets, as Darwin had suspected long ago. The seed-eating ground finches have thick, heavy beaks; the fruit-eating cactus finches, conical beaks; and the insect-eating tree finches and warbler finches, slender beaks. Lack's most remarkable finding was that, when certain species of seed-eating finches co-occurred on the same island, they were more divergent in beak depth than when they occurred alone; species with very similar beaks rarely co-occurred (Figure 1.1). This suggested that competition for food selected for divergence in diet and bill shape, and that it prevented similar species from coexisting; that is, ecology shaped evolution and biogeography. Presumably, the absence of competitors from most other lineages (a common feature on isolated islands) helped drive this process of divergence. Lack drew a provisional family tree for Darwin's finches based on their different beak sizes and shapes, placing taxa with similar bills and diets on the same branch, with the least specialized forms close to the base of the tree (Figure 1.2). This tree encapsulated not only the presumed relationships among different species, but also the history of the ecology of different forms, the possible directions of beak evolution, and (to a much lesser extent) the pattern of ancestral migration from island to island.

Lack's research, and the seminal books by Huxley and Simpson, laid the foundation for future research on such crucial topics as character displacement, competitive release on islands, resource partitioning, and the minimum amount of divergence needed for species coexistence, and helped inspire the great flowering of evolutionary ecology initiated by Hutchinson and MacArthur in the 1950s and 60s. Lack's writings directly inspired research on the mechanisms that underlie divergence in Darwin's finches (Bowman 1963; Schluter and Grant 1982; Grant and Grant 1993) and, more generally, on the role of competition in structuring ecological communities (e.g., Terborgh 1971; Diamond 1973; Connor and Simberloff 1978; Strong et al. 1979, 1984; Connell 1980; Harvey et al. 1983; Colwell and Winkler 1984; Grant and Schluter 1984; Schluter 1984, 1988a) and of natural selection in sculpting character displacement between close competitors (e.g., Boag and Grant 1981, 1984a,b; Schluter and Grant 1984; Schluter et al. 1985; Grant 1986; Grant 1994; Schluter 1994). These classic studies on Darwin's finches, and widely cited accounts summarizing other famous cases of adaptive radiation – African rift-lake cichlids, Australian marsupials, Hawaiian honeycreepers, silverswords, and lobeliads, Lake Baikal gammarids and cottoids, New World phloxes – have helped shape the modern conceptual framework of ecology and evolutionary biology (see Gulick 1932; Berg 1935; Amadon 1950; Grant 1963; Mayr 1963, 1970; Carlquist 1965, 1970, 1974; Grant and Grant 1965; Carson et al. 1970; Fryer and Iles 1972; Greenwood 1974, 1978; Stebbins 1974; Grant 1981, 1986; Smith and Todd 1984; Schluter et al. 1985; Futuyma 1986; Schaeffer and Lauder 1986; Carr et al. 1989; Liem 1990; Wilson 1992; Fryer 1996; Novacek 1996).

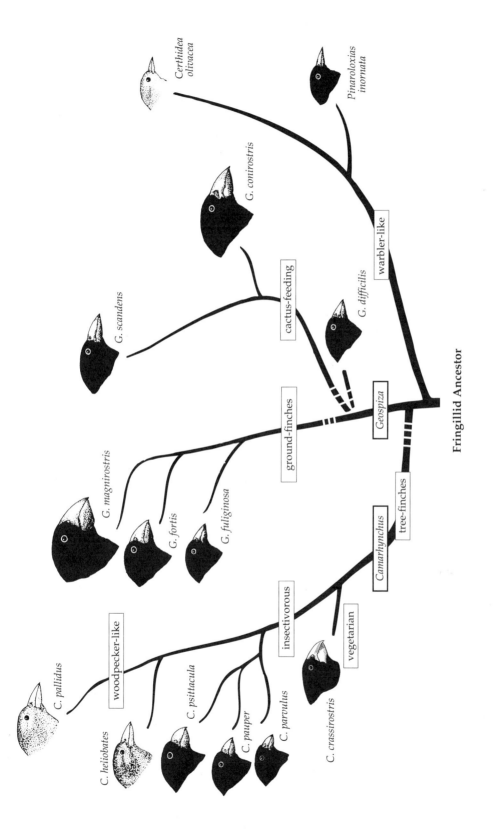

Figure 1.2. Beak size and shape in Darwin's finches (profiles after Grant 1986), superimposed on Lack's (1947) morphology-based phylogeny.

Adaptive radiation: paradigms in need of re-examination

Recently, however, it has become clear that few of these classic cases – the crown jewels of evolutionary ecology – had been studied rigorously, at least from a phylogenetic viewpoint (see Givnish 1987; Skinner 1988; Carr et al. 1989; Baldwin et al. 1990; Meyer et al. 1990, 1991; Baum and Larson 1991; Sytsma et al. 1991; Albert et al. 1992; Baldwin 1992; Chase and Palmer 1992; Givnish et al. in Armbruster 1992; Farrell and Mitter 1993, 1994; Knox et al. 1993; Kocher et al. 1993; Baldwin and Robichaux 1995; Givnish et al. 1995). The fundamental problem is that, in almost every case, the very characters whose radiation was under study (e.g., beak size and shape) were also used to help classify the organisms in question (e.g., Darwin's finches as studied by Lack 1947). This exercise can easily become circular, with traits being traced down evolutionary pathways determined, at least in part, by the traits themselves (Givnish et al. 1995).

Difficulties can arise even when the characters under study are excluded from phylogenetic analysis, if the remaining traits are subject to high levels of convergence or chance recurrence, or are linked genetically or developmentally to the characters undergoing radiation. A particularly insidious problem can arise if several, seemingly unrelated characters undergo "concerted convergence" (Givnish and Sytsma 1997; see Chapters 2 and 8), in which characters converge independently as a result of selection imposed on several traits simultaneously by a shared environment. As a consequence of this process, distantly related but ecologically similar organisms might converge in several of the morphological characters used for classification.[1] Such homoplasy can seriously skew the inference of phylogeny. This problem may be especially severe for island radiations, in which much of the phenotypic variation among species can be concentrated in the relatively few characters that underlie each radiation (see Grant 1986; Baldwin and Robichaux 1995; Givnish et al. 1995).

Potential contribution of molecular systematics

Given these fundamental problems with past approaches to adaptive radiation, several central findings in ecology and evolutionary biology may need re-examination. The primary challenge is systematic: **any rigorous, non-circular study of adaptive radiation must be based on a phylogeny that has been derived independently of the traits involved in that radiation.**[2] The emergence of molecular systematics over the last decade has provided the basis for deriving such phylogenies, and has made the time ripe for a revolution in the analysis of adaptive radiation.

---

[1] For example, in Hawaiian members of the carnation family (Chapter 16), dry windy environments favor narrow leaves, compact growth forms, wind pollination, and dioecy; rain forest understories favor broader leaves, tall woody growth forms, animal pollination, and hermaphroditic flowers. Culling floral characters (e.g., petal size and shape, pollen/ovule ratios) from an analysis of the evolution of breeding systems would not necessarily eliminate circularity caused by concerted convergence under similar environmental conditions.

[2] Brooks and McLennan (1994) argue that circularity is of less concern than the degree of congruence among characters involved vs. uninvolved in the radiation (see also de Queiroz 1996). This argument, while not without merit, may underestimate the problems introduced by pleiotropy, epistasis, and concerted convergence. These phenomena tend to blur the distinction between characters "involved" vs. "uninvolved" in a radiation, and amplify the potential for problems of circular reasoning.

Several studies have now shown that sequence and restriction-site variation in nuclear and organellar DNA provide a powerful tool for inferring relationships among species, genera, families, and higher taxonomic categories of various groups of organisms (e.g., see Hillis 1987, 1996; Sytsma 1990; Miyamoto and Cracraft 1991; Clegg and Zurawski 1992; Hamby and Zimmer 1992; Smith 1992; Olsen et al. 1992; Soltis et al. 1992; Baldauf and Palmer 1993; Chase et al. 1993; Clegg 1993; Doyle 1993; Hillis et al. 1993, 1994; Kocher et al. 1993; Olmstead et al. 1993; Schneider et al. 1993; Avise 1994; Brunsfield et al. 1994; Doyle et al. 1994; Liaud et al. 1994; Manhart 1994; Mishler et al. 1994; Sytsma and Hahn 1994, 1996; Baldwin et al. 1995; Bharathan and Zimmer 1995; Bogler et al. 1995; Chase et al. 1995; Clark et al. 1995; Halanych et al. 1995; Meyer 1995; Olmstead and Reeves 1995; Arnason and Gullberg 1996; Brower 1996; Lai et al. 1996; Ledje and Arnason 1996; Moran et al. 1996; Romano and Palumbi 1996; Shaw 1996; Sytsma and Baum 1996; Xu et al. 1996). Givnish and Sytsma (Chapter 2) present data showing that, for plants, phylogenetic analyses based on DNA sequences or chloroplast DNA restriction sites show significantly higher values of the consistency index (CI) – and thus, significantly less homoplasy – than those based on morphology. In addition, both sequence and restriction-site studies involve a much greater number of informative characters than morphological studies involving the same number of taxa. Based on computer simulations, Givnish and Sytsma (1997, Chapter 2) show that the probability of correct phylogenetic inference for trees with a given number of taxa increases with the consistency and number of variable/informative characters. The large number of independent characters sampled in chloroplast DNA analyses, their low level of homoplasy, and the well-defined nature of the universe of characters provide detailed, precise data for inferring phylogenetic relationships. Analyses based on mitochondrial DNA, while somewhat less powerful, show similar advantages for phylogenetic analyses of animal groups (e.g., see Avise 1994). Comparing phylogenies based on organellar DNA (often inherited uniparentally) with those based on nuclear DNA (inherited biparentally) can provide a direct means of screening for hybridization and/or introgression, and for identifying the parental taxa involved (Hamby and Zimmer 1992; Rieseberg and Brunsfeld 1992; Arnold 1993; Bachmann 1995; Haufler et al. 1995; McDade 1995; Rieseberg and Morefield 1995; Delwiche and Palmer 1996).

During the past seven years, several groups of researchers – studying organisms that display a rich diversity of adaptations in several geographical contexts – have begun to use molecular systematics as a basis for studies of adaptive radiation (e.g., Meyer et al. 1990, 1991; Springer and Kirsch 1991; Sytsma et al. 1991; Albert et al. 1992; Baldwin 1992; Chase and Palmer 1992; Smith 1992; Sturmbauer and Meyer 1992; Farrell and Mitter 1993, 1994; Kocher et al. 1993; Knox et al. 1993; Moran et al. 1993; Givnish et al. 1994, 1995; Hodges and Arnold 1994a,b; Milinkovitch et al. 1994; Baldwin and Robichaux 1995; Gillespie et al. 1995; Alvesgormes et al. 1995; Graham and Barrett 1995; Kambysellis et al. 1995; Meyer et al. 1996a; Tarr and Fleischer 1995; Barraclough et al. 1996; Berenbaum et al. 1996; Brower 1996; Francisco-Ortega et al. 1996; Monson 1996; Ricklefs 1996; Bleiweiss et al. 1997). Several groups of plants, fish, birds, mammals, lizards, insects, echinoderms, and other invertebrates have now been investigated in this way, on oceanic islands, in lakes and other islandlike environments, and across continents. These investigations have been immeasurably aided by the rapid development and evaluation of a variety of algorithms for (i) deriving phylogenies from molecular and/or morphological data (e.g., maximum parsimony, maximum likelihood, neighbor-joining [Farris 1972; Felsenstein 1978; Maddison et al. 1984; Saitou 1988; Albert et al. 1993; Bull et al. 1993; Dixon and Hillis 1993; Swofford 1993, 1996; Hillis 1996]); (ii) measuring the relative strength of support for different branches within a phylogeny (Felsenstein 1985, 1988; Farris 1989a,b; Bremer 1988; Archie 1989, 1996; Hillis et al. 1994; Hillis 1995, 1996; Farris et al. 1996); and (iii) inferring phenotypic character-states in ancestral taxa from a phylogeny and the distribution of character-states in the terminal (i.e., present-day) taxa (Lauder 1981; Felsenstein 1985; Gittelman and Kot 1990; Harvey and Pagel 1991; Maddison and Maddison 1992; Garland et al. 1993; Brooks and McLennan 1994; Losos and Miles 1994; Page 1994; Wenzel and Carpenter 1994; Martins and Hansen 1996).

Prompted by this explosion of research, Ken Sytsma and I organized an international symposium on Molecular Evolution and Adaptive Radiation, held at McGill University in Montreal during June 1995, to assess progress and alternative approaches in this exciting new field of research. We believe that the study of adaptive radiation based on molecular systematics provides a powerful new approach to answering several key questions at the interface of ecology, evolutionary biology, systematics, and biogeography (Table 1.1); many of these questions are addressed by one or more chapters in this volume.

Definition of adaptive radiation – scope of issues

The definition of adaptive radiation proposed here – **the evolution of a diversity of ecological roles and attendant adaptations in different species within a lineage** – mirrors an essential aspect of the explicit definitions given by many authors, including Osborn (1902), Huxley (1942), Simpson (1953), Grant (1963), Mayr (1970), Futuyma (1986), Grant (1986), Wilson (1992), and Schluter (1996a) (see Table 1.2). Simpson's definition seems to include an element of what might be called explosive diversification, a rapid and nearly contemporaneous divergence into several different ecological roles. However, Simpson (1953) made it clear that a similar pattern of ecological divergence into different adaptive zones, driven by similar processes of competition and natural selection, might proceed gradually as well. This makes sense: the Australian marsupials would be an adaptive radiation whether they diverged in ten thousand or ten million years; our ignorance of the rate of divergence in a lineage should not prevent us from recognizing it as a radiation. It is interesting that several influential publications on adaptive radiation – such as Lack (1947), Amadon (1950), and Carlquist (1965) – gave no explicit definition of this concept; apparently, its usage was so well accepted that the authors did not feel obliged to provide one. Osborn (1902) himself believed that he had defined adaptive radiation much earlier, in a discussion of the "functional radiation" of different groups of mammals (Osborn 1893).

Recently, Guyer and Slowinski (1993) advocated a different definition of adaptive radiation, based on an increase in the rate of speciation; Stanley (1979) had used a similar notion in discussing macroevolutionary trends (Table 1.2). Guyer and Slowinski's goal was to test Simpson's (1953) hypothesis that the evolution of a "key innovation" (e.g., wings) that permit the invasion of a new adaptive zone (e.g., consumption of flying prey) should increase the rate of speciation within a lineage that later underwent adaptive radiation in that zone. Slowinski and Guyer (1989) and Guyer and Slowinski (1993) advanced a series of statistical tests of this hypothesis, based solely on phylogenies for groups that evolved such key innovations and for sister-groups that did not. Several systematists (Sanderson and Donoghue 1994, 1996; Slowinski and Guyer 1994; Barraclough et al. 1996; Rohde 1996; Sanderson and Wojciechowski 1996) have followed their lead, using phylogenies to test whether particular traits accelerated speciation and thus acted as "key innovations."

Defining adaptive radiation in this way is inappropriate and, I believe, ultimately unproductive. It ignores the traditional usage and conceptual framework

Table 1.2. Representative definitions of adaptive radiation.

Author	Definition
Osborn (1902)	"Differentiation in habit in several directions from a primitive type"
Huxley (1942)	"Invasion of different regions of the environment by different lines within a group, and secondarily their exploitation of different modes of life"
Simpson (1953)	"More or less simultaneous divergence of numerous lines all from much the same ancestral adaptive type into different, also diverging adaptive zones"
Grant (1963)	"Diversification of a group of organisms in relation to the ways of making a living and reproducing successfully"
Mayr (1970)	"Evolutionary divergence of members of a single phyletic line into a series of different niches or adaptive zones"
Stanley (1979)	"The rapid proliferation of new taxa from a single ancestral group"
Futuyma (1986)	"Diversification into different ecological niches by species derived from a common ancestor"
Grant (1986)	"The evolutionary diversification of a single lineage into a variety of species with different adaptive properties"
Wilson (1992)	"The spread of species of common ancestry into different niches"
Guyer and Slowinski (1993)	"Some organisms have features that allow them to speciate more prolifically or become extinct less frequently than organisms without these features"
Skelton (1993)	"An episode of significantly sustained excess of cladogenesis over extinction, with adaptive divergence cued by the appearance of some form of ecological stimulus"
Schluter (1996a)	"A proliferation of species within a single clade accompanied by significant interspecific divergence in the kinds of resources exploited and in the morphological and physiological traits used to exploit these resources"
Givnish (1997)	"The origin of a diversity of ecological roles and attendant adaptations in different species within a lineage"

established for adaptive radiation, dating back more than a century, at least to Osborn (1893) if not Darwin (1859). It focuses on species number (perhaps a preoccupation of some systematists) to the exclusion of all other evolutionary phenomena. Most important, it conflates "adaptive radiation" with mere speciation. As Huxley (1942) recognized and Mayr (1942, 1963, 1970) argued so convincingly as a central evolutionary thesis, speciation often occurs simply because gene flow is interrupted by geographic barriers or by intrinsic biological factors (e.g., ploidy, premating barriers, philopatry), allowing populations to diverge genetically to such an extent that their members cannot interbreed successfully (or, at least, not as well as crosses within populations) when the populations contact each other again. Clearly, such speciation may have nothing to do with adaptive radiation. The flightless Hawaiian swordtail crickets (*Laupala*) seem to fall into this category of "non-adaptive radiation": different species on the same island differ in their songs, but differ little in ecology; much of the speciation in this genus is geographic, involving interisland dispersal events (Shaw 1995). *Cyanea* (Campanulaceae) – the largest genus of flowering plants endemic to Hawai`i – has undergone spectacular divergence in leaf form, flower shape, and plant height, but much of the speciation in this group may reflect restricted gene flow, caused by dispersal of its fruits by sedentary forest-interior birds (Givnish et al. 1995). Tepui-dwelling *Brocchinia* has evolved more mechanisms of nutrient capture than any other bromeliad genus, but has far fewer species than closely related *Navia* (Chapter 8). Many *Navia* appear to be ecologically equivalent, but different species occur on almost every tepui, perhaps because *Navia* and its relatives have naked seeds with no means of long-distance dispersal. To argue, based on Guyer and Slowinski's (1993) definition and the extent of speciation in each group, that *Navia* has undergone a more massive adaptive radiation than *Brocchinia* would be to turn the real state of affairs upside down.

It *is* reasonable to hypothesize, as did Simpson (1953), that adaptive radiation and invasion of new adaptive zones should spur speciation. However, Guyer and Slowinski's definition prevents a logical test of this hypothesis by conflating the traditional concept of adaptive radiation with speciation. Including increased speciation as part of the definition of adaptive radiation – as done by Skelton (1993) and Schluter (1996a; see Chapter 18) – also seems counterproductive. In most instances, adaptive radiation will surely be accompanied by an increase in species number within a lineage. But including such an increase in the definition of adaptive radiation would interfere with testing the very idea that radiation should increase speciation. As Graham and Barrett (Chapter 7) argue, adaptive radiation can increase, leave unchanged, or even decrease speciation.

Ecological and adaptive divergence can occur without speciation. Character displacement – often viewed as an important aspect of adaptive radiation based on resource partitioning – may arise within a species. For example, males and females of single species of *Anolis* lizards isolated on islands in the Caribbean often diverge in body size and perch height (Schoener 1974; see Chapter 19). Even more remarkably, Davidson (1978) has shown that seed-harvesting ants show greater within-colony variation in the size of individual workers where they co-occur with fewer competing species. In general, the rates of speciation and adaptive divergence within radiations can be decoupled (Foote 1996; Fortey et al. 1996).

The term "adaptive radiation" has been used so widely, and with so little reference to its historical usage and conceptual connotations, that it might be useful to distinguish a few related patterns and/or processes, even though they intergrade with each other to some degree and two or more may apply to the same clade.

NON-ADAPTIVE RADIATION – This term might be used to describe speciation without appreciable ecological divergence and evolution of corresponding adaptations (or involve adaptations that are only coincidentally related to such divergence). Allopatric speciation caused by geographic barriers and/or restricted dispersal in sedentary organisms can easily result in non-adaptive radiation (Carlquist 1974; Diamond et al. 1976; Gould and Calloway 1980; Brooks and McLennan 1993, 1994; Givnish et al. 1994, 1995; Chapter 8), as has been most recently argued for land snails on Crete (Gittenberger 1991) and Madeira (Cameron et al. 1996) and for brooding Antarctic sea urchins (Poullin and Feral 1996). Groups of burrowing mammals (e.g., gophers, moles, or mole rats) that speciate extensively based on restricted dispersal, but whose populations evolve adaptations (e.g., coat color, size of forelimbs) to the soils of specific regions, might provide examples that blur the distinction between non-adaptive and adaptive radiation.

DEVELOPMENTAL RADIATION – The first great flowering of morphological body-plans in the Precambrian, captured so beautifully in the Burgess Shale, may not have been primarily an adaptive radiation, at least initially (Gould 1989). It may have been just then that metazoans began to assemble cassettelike kits of homeotic genes that governed fundamental aspects of the development of multi-celled organisms – differentiation relative to inside vs. outside, front vs. back, top vs. bottom; segmentation; and differentiation of segments (see Carroll 1995; Valentine et al. 1996; Erwin et al. 1997). Once such cassettes evolved, they could rapidly be duplicated, arranged, activated, and interrelated in different combinations in the genome, with very different consequences for morphology and for the developmental linkages among morphological and physiological traits. These linkages may now underlie the Baupläne of our modern phyla, most of which first appeared in the Burgess Shale. As a diversity of bodyplans evolved, competition and divergent selection may have allowed several to invade distinct adaptive zones – reef-building by corals and sponges, burrowing by polychaete worms, deposit-feeding by sea cucumbers, pelagic feeding by comb jellies. Selection almost surely also played a role in the initial blossoming of multi-cellular life: increased levels of atmospheric oxygen may have made larger, more efficient forms possible, and an emerging arms race between predators and prey may have favored more complex morphologies and behaviors (Erwin et al. 1997).

The developmental radiation of multi-cellular life in the Precambrian is mirrored in every adaptive radiation, in the genetic and developmental machinery underlying the divergence of beak shape in finches, of jaw morphology in African rift-lake cichlids, of spines in sea urchins, of growth-forms in bromeliads, of flowers in water hyacinths. Such developmental radiations are of interest in their own right. When extensive morphological divergence has occurred within a lineage, but there is no evidence that such variation is adaptive, it might be parsimonious to refer to that lineage simply as a developmental radiation.

SEXUAL RADIATION – The striking multiplication of drosophilid flies in the Hawaiian Islands has sparked research by many leading evolutionary biologists (e.g., Carson et al. 1967, 1970; Heed 1968, 1971; Spieth 1968, 1982; Carson 1983, 1992; Beverly and Wilson 1985; Kaneshiro 1983, 1993; DeSalle et al. 1987; Thomas and Hunt 1991; DeSalle 1995; Kambysellis et al. 1995; Kaneshiro et al. 1995; Chapter 15). Males in many of the 800 or so species in this group have evolved remarkable behaviors and morphologies to attract mates: intricate dances, leks on specific host plants, black wing blotches in the large "picture-winged" group, highly modified mouthparts, eyes on long stalks, and spoon-shaped, split, or bristly tarsi. The detailed structure of the male genitalia separates many species as well (Kaneshiro et al. 1995). It seems likely that most of this divergence has been driven not by adaptation to different environments or ecological roles, but by the unpredictable vagaries of mate choice. Once females with a preference for a particular male form or behavior appear in a population, they select for such males, and may indirectly favor female offspring with such a preference, whether or not the preferred male phenotype is correlated in a given population with male genetic quality. Just as adaptive radiation into a variety of ecological roles can increase diversity within a lineage, so too can a sexual radiation, based on interpopulational divergence in the characteristics that act as premating barriers (Mayr 1970). Unlike adaptive radiation, however, there may be little predictability in the direction sexual selection may take evolution. The consistent presence of particular resources (e.g., fruit, insects, flowers, leaves) and resource spectra (e.g., fruits or insects of different sizes, chemical compositions, handling times) on different continents and islands favors the predictable evolution of frugivores, insectivores, pollinators, and herbivores in those regions, and of resource partitioning by species in each of these guilds. But sexual selection might favor mating based on leg size or shape in some populations, while in other groups it might favor a distinctive plumage, song, scent, or lock-and-key fit between genitalia. Sexual selection can act as a species-intrinsic factor promoting speciation more or less independently of adaptation to the environment, in much the same way as limited dispersal and/or geographic barriers can act as extrinsic factors promoting such non-adaptive radiation. In plants, radiations based on divergence in floral morphology may be sexual as well as ecological in nature. For example, the evolution of divergent flowers in columbines adapted to different pollinators (Chapter 9) may reflect speciation in response to the evolution of premating isolating barriers (i.e., sexual or reproductive differentiation) as much as it does divergence in adaptations to attract different pollinators (i.e., ecological differentiation).

COEVOLUTIONARY RADIATION – Coevolution can spur speciation through arms races with predators/herbivores, co-adaptation with mutualists, or co-speciation of hosts and commensal parasites or mutualists (see Ehrlich and Raven 1964; Grant and Grant 1965; Berenbaum 1973; Janzen 1974; Stebbins 1974; Gilbert 1980; Pierce 1987; Farrell and Mitter 1990, 1993; Farrell et al. 1991; Mitter et al. 1991; Moran and Baumann 1994; Thompson 1994; Bogler et al. 1995; Brower 1996; Lai et al. 1996; Moran 1996; Moran et al. 1996; Page 1996; Pellmyr et al. 1996). A coevolutionary radiation by, say, a plant group in pollinators would be directly analogous (or indeed, equivalent) to an adaptive radiation when that radiation is essentially one-sided – for example, when the pollinators predate the plant group in question, and are evolutionarily little influenced by them. However, when evolutionary interactions between two or more groups create new "adaptive zones" for all – for example, when some plant species evolve novel chemical defenses, new animal species evolve novel countermeasures, and so on – it seems unlikely that we could predict the directions that coevolution would take, at least not as well as we can envision how competition might lead to a partitioning of pre-existing resources. To the extent that coevolution between mutualists favors philopatry and/or reduced dispersal by both parties (Cushman and Murphy 1993a,b), mutualism may become associated with high speciation rates that reflect limited gene flow more than adaptive radiation.

In some plant groups, floral characters account for 40% of the morphological traits used to classify species pollinated by birds or specialized insects, 15% of those used to classify species pollinated by unspecialized insects, and 4% of those used to classify species pollinated by wind or water (Grant 1949). In such cases, the question naturally arises as to whether coevolution with specialized pollinators or other mutualists actually accelerates speciation, or whether it merely results in phenotypic differences that are more likely to be detected and used by human taxonomists (Crisp 1994).

Research issues in adaptive radiation

Of the questions imminent in any adaptive radiation, perhaps the most salient (Table 1.1) involve certain aspects of phylogeny, adaptation, historical ecology, speciation, genetics, development, biogeography, tempo, and predictability. Space limitations prohibit a detailed discussion of each of these issues here, but below I discuss some of the most important issues to provide a conceptual framework for studies of adaptive radiation. The remaining chapters in this volume provide detailed analyses and alternative perspectives on many of the questions addressed.

Phylogeny

Several studies have now shown that phenotypic similarities among species in a lineage can reflect ecological similarities, close relationships – or both. In Darwin's finches, the available data on protein polymorphism (Yang and Patton 1981) yields a distance-based phylogeny largely consistent with those based on morphology, using either qualitative (Lack 1947; Figure 1.2) or quantitative approaches (Schluter 1984; Grant 1986). Clearly, phenotypic similarity reflects both ecological similarity and close relationship in Darwin's finches, although incongruities detected in a recent re-analysis of the allozyme data (Stern and Grant 1996) show that it would be desirable to use more powerful evidence. The exemplary analysis of salamander relationships by Titus and Larson (1995) also demonstrated a high degree of congruence (97.2%) between morphology and mtDNA sequence data.

In the extant Hawaiian honeycreepers (Drepanidinae), the pattern is less straightforward, suggesting that morphology reflects partly ancestry, partly ecology. Recent analyses by Tarr and Fleischer (1995) of restriction-site variation in mtDNA suggest that nectarivores with reddish plumage and tubular tongues – the i'iwi (*Vestiaria coccinea*), `apapane (*Himatione sanguinea*), and `akohekohe (*Palmeria dolei*) – do indeed form a monophyletic clade[3] (Figure 1.3). This clade appears to have emerged from a paraphyletic group of insectivores with slender bills and green, yellow, or orange plumage – the foliage-gleaning `amakihi (*Hemignathus virens*) and `anianiau (*H. parvus*), the koa pod-cracking `akepa (*Loxops caeruleirostris, L. coccinea*), and the bark-probing Kaua'i creeper (*Oreomystis bairdii*), with the warblerlike Maui creeper (*Paroreomyza montana*) at the base of the tree (ecological descriptions from Scott et al. [1986] and Tarr and Fleischer [1995]). Remarkably, however, the slender-billed Kaua'i creeper appears to be the closest relative of the heavy-billed, finch- or parrotlike seed-eaters of the Psittarostrini, represented by the Laysan finch (*Telespiza cantans*). While this last relationship can be supported by some interpretations of the morphological data (cf. Pratt 1979, 1992), the resolution of the insectivores into three clades was unexpected, as were the identities of the clades which were the closest relatives to the nectarivores and the granivores.

The cichlids of the East African Great Lakes provide a striking illustration of morphology reflecting ecology much more than phylogeny (Figure 1.4). These fish have undergone independent, spectacular adaptive radiations and explosive speciations in Lakes Tanganyika, Victoria, and Malawi, involving several hundred species in all (Fryer and Iles 1972; Greenwood 1984; Meyer et al. 1994; Fryer 1996). A key innovation underlying these radiations is the pharyngeal jaw, a bony plate with toothlike projections in the roof of the cichlid mouth. The pharyngeal jaw acts as a secondary food processor, freeing the outer jaw to become highly specialized for the capture of particular prey (Liem 1973; Galis and Drucker 1996). Functional types include algae grazers, deposit feeders, planktivores, molluscivores, ambush predators feeding on other fish, egg-feeders, and scale- and tail-nippers. Kocher et al. (1993) analyzed mtDNA sequence variation to argue that ecologically similar, morphologically convergent species evolved independently in Lake Tanganyika and Lake Malawi (Figure 1.4). Extensive research by Axel Meyer and his colleagues suggests that (i) the largest group of cichlids, in Lake Tanganyika, may be polyphyletic; (ii) the species flocks in Lake Malawi and Lake Victoria are each monophyletic, sister to each other, and derived from the haplochromine cichlids of Lake Tanganyika; and (iii) a few non-endemic haplochromine cichlids (e.g., *Astatoreochromis, Astatotilapia*) are sister to the Malawi-Victoria flocks, and are also derived from a Tanganyikan ancestor (Meyer et al. 1990, 1991, 1994, 1996b; Nishida 1991; Sturmbauer and Meyer 1992, 1993; Meyer 1993; Sültmann et al. 1995).

In several groups discussed in this volume, molecular phylogenies provide evidence that convergence, parallelism, and/or recurrence have occurred many times,

---

[3] A monophyletic group, or clade, consists of a set of all taxa descended from a single common ancestor. A group is polyphyletic if it contains some or all of the descendants of at least two unrelated ancestors. A group is paraphyletic if some of the descendants of the common ancestor are not themselves included.

and that such homoplasy seriously skews the inference of relationships from morphological characters. Such cases include the Hawaiian silversword alliance (Chapter 3), Australian and South American marsupials (Chapter 4), water hyacinths and their allies (Chapter 7), bromeliads (Chapter 8), Old World fruitbats (Chapter 9), epiphytic and terrestrial orchids (Chapters 10 and 15), African rift-lake cichlids (Chapter 11),

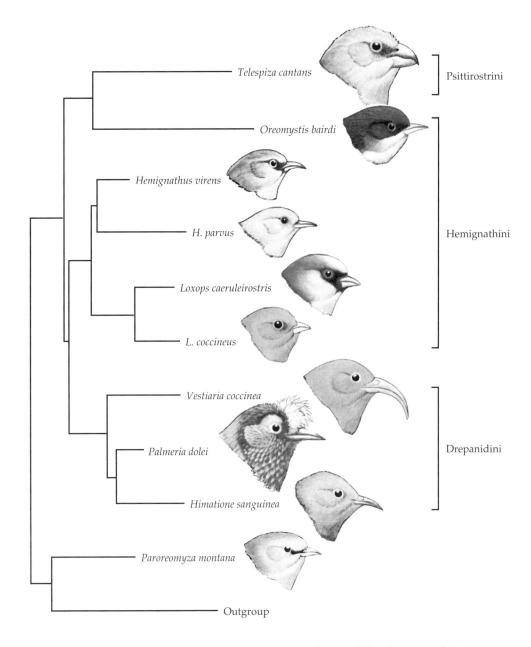

Figure 1.3. Molecular phylogeny of Hawaiian honeycreepers (Tarr and Fleischer 1995), showing pattern of beak evolution. Profiles adapted from Pratt et al. (1987).

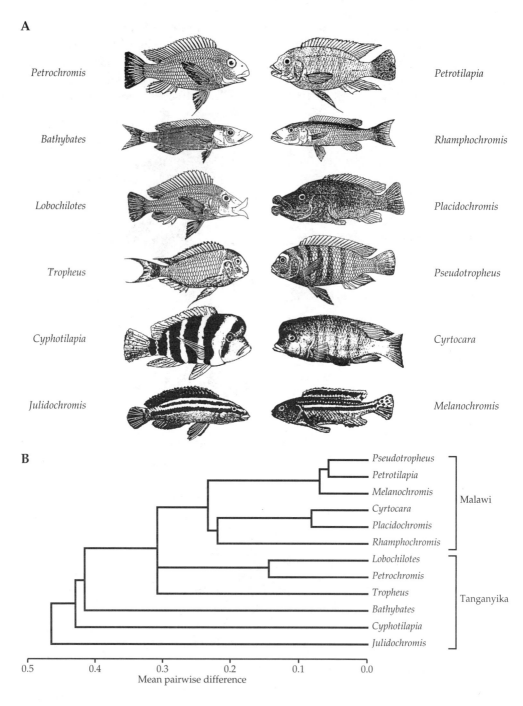

Figure 1.4. Convergent evolution of selected cichlids from Lake Malawi and Lake Tanganyika (after Kocher et al. 1993). (**A**) Six pairs of morphologically similar cichlids; Tanganyikan fish are on the left. The specific features shared are: (1) rasping jaw morphology; (2) fusiform body, associated with piscivory; (3) fleshy lips; (4) mbuna habit, associated with algal grazing on rocky substrates; (5) nuchal hump; and (6) horizontal striping. (**B**) An unrooted mtDNA phylogeny shows that similar forms have evolved independently in each lake.

Canadian sticklebacks (Chapter 18), and Caribbean *Anolis* (Chapter 19). The relative reliability and differential utility of molecular vs. morphological data are discussed at length in Chapter 2.

Adaptation

DEMONSTRATION – To show that a particular character-state (e.g., a specific leaf size in plants or jaw morphology in fish) is adaptive requires one of three approaches, involving comparative techniques, functional analyses, or populational studies (see Givnish 1987). When unrelated but ecologically similar species repeatedly share certain features of their form, physiology, or behavior, such evolutionary convergence provides some of the most compelling evidence that such features are adaptive – that is, that they increase the relative fitness of their bearer in the ecological context in question[4] (e.g., Merçot et al. 1994). However, while such comparative data provide powerful evidence that a particular trait is adaptive, they cannot directly provide any insight into *why* it is adaptive.

Studies that address the functional significance of traits – for example, the effect of leaf thickness, N content, and stomatal conductance on photosynthesis and water loss in different light environments – can illuminate why certain traits are adaptive in certain circumstances but not in others, by showing how variation in those traits affects certain key aspects of organismal performance and, by inference, competitive ability. Experimental manipulations of phenotypic traits (e.g., trimming crossbill mandibles [Benkham and Lindholm 1991] or stonefly wings [Marden and Kramer 1994]) is an especially powerful tactic to break down genetic and developmental correlations. When the insights obtained through functional analyses are combined with energetic or other constraints to produce optimality models, quantitative predictions can be produced that permit powerful tests of adaptation (e.g., Horn 1979; Wilson 1980; Givnish 1982a, 1986a,b, 1988, 1995; Cowan 1986; Parkhurst 1986; Seger and Stubblefield 1996). Neither functional analyses nor optimality models, however, can demonstrate that variation in particular traits actually affects fitness.

Populational studies can link phenotypic variation to differences in survivorship and reproduction, and thereby directly demonstrate the role of natural selection in maintaining or changing the distribution of phenotypes in a population (Kettlewell 1958; McNeilly and Antonovics 1968; Ford 1971; Jones et al. 1977; Endler 1986; Schluter 1988b; Freriksen et al. 1994; Schluter 1994; Reznick and Travis 1996). Such studies must address genetic and developmental correlations among traits if the selective signifi-

---

[4] Gould and Vrba (1982), Baum and Larson (1991), and Lauder (1996) argue that "adaptation" should be reserved to describe cases in which a trait was initially favored by selection to fulfill the same "purpose" it now serves. Traits that increase the fitness of their bearers today but arose for different reasons are termed aptations by Gould and Vrba (1982); traits that were favored for one reason but later became co-opted for another role (i.e., preadaptations) are termed exaptations. While heuristic, the aptation/exaptation terminology does not seem particularly useful. From an operational viewpoint alone, it is often quite difficult to establish "why" a particular trait in a population is favored by natural selection today; it is usually impossible to demonstrate the action of natural selection in the past, let alone "why" particular variants were favored. Furthermore, it seems arbitrary to exclude traits as adaptations simply because we do not know the history of their function (see also Fisher 1985; Endler 1986; Reeve and Sherman 1993; Amundson 1996; Vermeij 1996).

cance of variation in a specific trait is to be assessed (Lande and Arnold 1983; Mitchell-Olds and Shaw 1987; Arnold 1994; Sinervo and Basolo 1996). Grant and Grant (1995) provide a valuable illustration of the kind of careful approach needed: although the 1976–77 drought in the Galápagos favored an increase in bill length and depth but a decrease in bill width in *Geospiza fortis*, bill width actually increased through time as a consequence of the strong genetic correlations among all three bill dimensions and the very strong selection for greater bill length and depth. While such populational studies can provide definitive proof that natural selection favors a particular trait in a given context, they cannot specify why such a trait is adaptive. In this respect, populational studies provide insights somewhat like those afforded by comparative analyses. Populational studies, however, are generally labor-intensive and difficult to pursue in more than a few populations/species; comparative analyses are inherently less definitive, but can be pursued over a much wider range of taxa, potentially encompassing a much broader range of ecological factors and selection pressures.

AVOIDING FALSE DICHOTOMIES IN THE COMPARATIVE APPROACH – In recent years, important advances have been made in the use of phylogenies in comparative analyses (see Felsenstein 1985; Greene 1986; Coddington 1988, 1990; Baum and Larson 1991; Harvey and Pagel 1991; Maddison and Maddison 1992; Garland et al. 1993; Page 1994; Larson and Losos 1996; Martins and Hansen 1996; Nee et al. 1996). These techniques are a major addition to the armamentarium of evolutionary biology, and one or more are employed by almost every chapter in this volume (see section below on **Historical ecology** as well). Frumhoff and Reeve (1994) outline some of the limits of these techniques for inferring adaption. In addition, certain aspects of the approach recommended by Harvey and Pagel (1991) – and defended vociferously by Harvey et al. (1995a,b) – seem biased and should be abandoned. Harvey and Pagel (1991) assumed that phenotypic similarity may reflect shared ancestry or similar ecology; they argued that, if such similarity can be explained in terms of shared ancestry, it was parsimonious to account for it *in those terms alone*. This is clearly wrong and biases the results of comparative analyses toward phylogenetic explanations and against explanations based on shared selection pressures (Westoby et al. 1995a,b).

How might we use phylogenies and the comparative approach to identify an adaptation? Consider a derived phenotypic trait shared by two (or preferably, more) taxa, which are themselves ecologically similar but distantly related. These circumstances provide *a priori* evidence of convergence and suggest that the trait is an adaption in the ecological context in question (Figure 1.5). Conversely, if a derived trait is shared only by taxa which belong to one lineage but are themselves ecologically dissimilar, it seems reasonable to identify such a trait as a *phylogenetic constraint* (*sensu* Gould and Lewontin 1979) – a reflection of some shared genetic heritage, developmental program, or architectural Bauplan peculiar to a lineage (Figure 1.5). Such phylogenetic constraints should be contrasted with the energetic, mechanical, or functional constraints on organismal function imposed by physics, chemistry, or biomechanics and which are likely to apply across lineages. Gould and Lewontin (1979) and Gould (1983, 1993) labeled as "Panglossian" and "adaptationist" evolutionary biologists who seek explanations for organismal form, physiology, and behavior only in

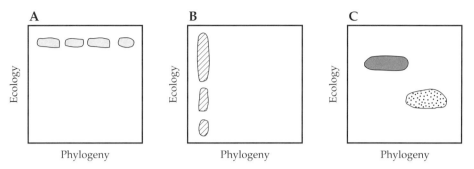

Figure 1.5. Schematic showing how to identify whether similar morphologies are a result of (**A**) convergence, (**B**) phylogenetic constraints, or (**C**) both. Clouds represent species with different ecologies and/or phylogenetic affinities; species with the same, ecologically significant character-state (e.g., wing shape, mechanism of seed dispersal) are shaded the same way.

terms of natural selection, and called for approaches that incorporate phylogenetic constraints as well. It is ironic that while Gould and Lewontin argued against advancing unjustified, *ad hoc* arguments that specific traits are adaptations, they provided no methodology for identifying phylogenetic constraints – thus opening the door for a multiplication of uncritical, *ad hoc* arguments that specific traits are instead phylogenetic constraints. The comparative technique just outlined provides one means for identifying phylogenetic constraints *at least under certain circumstances*, rescuing us from an equally Panglossian "constraintist" approach.

However, the fundamental limitation of this technique is that, over time and perhaps as a result of adaptive radiation on a grand scale, many lineages become ecologically and morphologically specialized. Individual lineages may also become associated with a particular suite of competitors or predators that preclude evolution in certain directions. Consequently, when we find that specific traits differ between such lineages (Figure 1.5), it may be impossible to determine whether such phenotypic differences reflect differences in ecology (i.e., adaptation), differences in ancestry (i.e., phylogenetic constraints) – or both. Westoby et al. (1995a,b), in an analysis of ecological patterns in seed mass in the Australian flora, has persuasively argued that the comparative approach championed by Harvey and Pagel (1991) fails to take the last contingency into account and instead simply asserts that such cases reflect shared ancestry alone. The rejoinders by Harvey et al. (1995a,b) are, at least to this author, unconvincing.

Consider gymnosperms, in which there is an overwhelming correlation – at the levels of individual species, genera, and families – between dioecy and fleshy, animal-dispersed propagules and between monoecy and wind-dispersed propagules (Givnish 1980, 1982b). This association accords with a simple model based on sexual selection (Bawa 1980; Givnish 1980). Thomson and Barrett (1981) correctly warned, however, that this analysis (like other early studies of the ecological correlates of breeding systems) failed to take phylogeny into account, and that the observed pattern might reflect phylogenetic constraints on breeding systems and mode of seed dispersal at the familial level. Donoghue (1989) used a family-level phylogeny of gymnosperms to test whether the distribution of breeding systems and dispersal syndromes showed statistically significant support for dioecy evolving after or concurrently with fleshy fruits; he found support for the latter but not the former (*contra* Harvey and Pagel 1991). Yet Donoghue's test has the potential to throw the baby out with the bathwater. Selection might have overwhelmingly favored dioecy in lineages with fleshy fruits and yet *appear* insignificant if (i) there are few shifts between dispersal syndromes and (ii) it is assumed that persistence of a particular breeding system or mode of seed dispersal is a phylogenetic constraint, not an adaptation. The likelihood of error is especially great when the test is fundamentally flawed (as in Donoghue 1989 and Harvey and Pagel 1991) by ignoring variability in breeding systems within individuals, species, genera, and families – such variability constitutes *prima facie* evidence that a trait assumed to be a legacy (i.e., dioecy in Cupressaceae, Podocarpaceae, Taxaceae) is actually evolutionarily labile. As Donoghue and Ackerley (1996) argue, when there is evidence that a trait is evolutionarily labile and may track ecological selection pressures closely, it is important to conduct phylogenetically structured and unstructured tests; when lability is high (i.e., when ancestry is not an important determinant of phenotype), standard contingency analyses or nested analyses of variance (like those originally used) may be appropriate. It should always be recognized that the phylogenetic distribution of a trait is not the only evidence bearing on its adaptive value; experimental, populational, or functional data can often be adduced to bolster an argument that a trait is adaptive. Similarly, developmental and genetic data might be adduced to help argue that a trait is a phylogenetic constraint. Phylogenetic techniques are a rather inefficient means of inferring whether a trait is adaptive, based solely on its distribution among a limited set of close relatives.

PREADAPTATION – Preadaptations (exaptations *sensu* Gould and Vrba 1982) are thought to play an important role in adaptive radiation (Simpson 1953; Mayr 1970). A trait that simultaneously performs two functions – or better, two traits or structures that perform the same function – can act as a vital evolutionary "bridge," allowing an organism to elaborate a novel function while retaining its ancestral functions, and permitting the progressive perfection of complex organs that fulfill subtly different roles at different points in evolution. Classic examples of preadaptations include (i) feathers, which may have originally played a role in thermoregulation in dinosaurs and later facilitated the evolution of flight in birds (Bakker 1975; Ostrom 1979; Gould and Vrba 1982; Novacek 1996); (ii) insect wings, which may have originally aided thermoregulation (Kingsolver and Koehl 1994) or water-surface skimming (Marden and Kramer 1994) and later permitted flight; (iii) rudimentary eye-spots, which initially permitted discrimination of light vs. dark and then repeatedly evolved into the eyes of crayfish, snails, octopi, and vertebrates, permitting the increasingly sophisticated resolution and processing of visual images (Nilsson and Pelger 1994; Dawkins 1996; Ridley 1996); and (iv) the oral and pharyngeal jaws of cichlids, in which the latter duplicates the food-processing function of the former, freeing the oral jaw to become increasingly specialized in its additional function of prey capture (Liem 1973; Galis and Drucker 1996). The nature of preadaptations is such that the duplication of structures or functions may be obvious only in retrospect, often after critical transitional taxa have become extinct. As a consequence, field studies of preadaptations and their role in adaptive radiation are sorely needed; the impounding habit and absorbent trichomes of bromeliads (Chapter 8) provide promising material for such research.

CHARACTER DISPLACEMENT – To demonstrate that interspecific competition has played a leading role in causing divergence between closely related species, one must employ carefully controlled comparative observations (Grant 1975, 1986; Grant and Grant 1993; Grant 1994) or – ideally – experimental manipulations (Schluter 1994; Chapter 18). Such studies are lamentably uncommon. Nevertheless, it is heartening to see natural selection operate in the expected direction in these cases: hybrid sticklebacks diverging from the morphology of an introduced benthic or limnetic form, and decreased bill depth of medium-beaked ground finches on Daphne after the invasion of the large-beaked ground finches in 1983. More studies like these are needed to show whether competition actually favors the observed divergence of flower shapes in columbines (Chapter 13), or the distribution of large- and small-bodied geckos on different islands of the Seychelles (Radtkey 1996).

COMPETITIVE RELEASE AND KEY INNOVATIONS – The absence/removal of competitors – or the evolution of an adaptation that allows invasion of a new adaptive zone – should greatly increase the likelihood that a lineage will undergo adaptive radiation (Mayr 1942; Simpson 1953; Skelton 1993). Studies of adaptive radiation should address whether one or the other of these factors has played a significant role. Clark and Johnston (1996) make an especially convincing case that *both* factors triggered the adaptive radiation of the notothenioid fishes, a bizarre group dominating Antarctic waters at shallow and moderate depths. The Antarctic ice sheet extended across the continental shelf in the Eocene and early Oligocene, making the extinction of the previous fauna inevitable; nototheniids subsequently radiated into this spectrum of empty ecological zones. A molecular phylogeny (Bargelloni et al. 1994) indicates that "antifreeze" glycoproteins – a key innovation required for survival in waters cold enough to freeze body fluids – evolved at the base of the suborder Notothenioidei, after its divergence from the Bovichidae and contemporaneous with the invasion of Antarctic waters. Notothenioids are descended from a benthic ancestor and lack a swim-bladder; in spite of this constraint, several species have invaded the water column, buoyed by extensive fat deposits and reduced ossification of skeleton and scales (Clark and Johnston 1996). Antarctic surface-water temperatures dropped dramatically beginning about 15 to 20 million years ago; one of the latest-diverging lineages, the ice-fish (Channichthyidae), are cryopelagic specialists that are unique in having lost all of their red blood cells. The high oxygen-holding capacity of plasma and sea water under extremely cold conditions, in combination with very low metabolic rates, apparently make erythrocytes unnecessary (Eastman 1993).

NATURE VS. NURTURE – Given that different species in many adaptive radiations occur in different habitats, the question naturally arises as to how much of the phenotypic variation among species is genetically determined, and how much does it instead reflect the differences in the habitats occupied by those species? In some organisms (e.g., birds), it appears unlikely that the environment has a major effect on adaptively important trophic structures (e.g., beaks). But in other organisms – particularly plants, which are notoriously plastic in their development – care must be taken to determine the genetic basis for phenotypic differences between species. In the Hawaiian silversword alliance, common-garden studies show that many of the differences between species in leaf size, venation, and growth form persist in cultivation (Chapter 3). On the other hand, diet has been shown to affect certain aspects of jaw morphology in sticklebacks (Chapter 18) and other fish. Common-garden studies (in the classic tradition of Clausen et al. 1941) and other investigations of phenotypic plasticity should be conducted as a routine part of the analysis of any adaptive radiation.

Historical ecology

The value of the insights that can be obtained by inferring the form, physiology, behavior, ecology, and geographic distribution of ancestors from a phylogeny and the characteristics of present-day taxa is now well established and is one of the most exciting features of modern evolutionary biology (Lauder 1981; Felsenstein 1985; Harvey and Pagel 1991; Maddison and Maddison 1992; Garland et al. 1993; Brooks and McLennan 1994; Page 1994; Wenzel and Carpenter 1994; Harvey et al. 1996; Martins and Hansen 1996; Silvertown and Dodd 1996). This approach is central to almost all the analyses presented in this volume, and detailed arguments regarding the utility and weaknesses of particular techniques and strategies will not be recapitulated here. However, a few of the most important limitations of the "historical ecology" approach (Brooks and McLennan 1994) must be mentioned. First, the branching topology of a phylogeny, its resolution, and the distribution of character-states among the terminal (present-day) taxa can strongly limit the inference of ancestral phenotype, distribution, and ecology. For example, based on the most recent molecular data, lungfishes and the coelacanth appear to be sister to each other, and together are sister to all tetrapods (Zardoya and Meyer 1996). As a consequence of this tree topology, it is essentially impossible to determine which morphological traits – among those that differ between lungfishes and the coelacanth – characterized the earliest land vertebrates.

Second, because different tree topologies often result in different inferences about the pattern of evolution, consideration should be given to the degree of confidence in individual nodes, and to the evolutionary implications if an inferred relationship is incorrect; sensitivity analyses should be conducted (Donoghue and Ackerley 1996; Sytsma and Baum 1996). Different techniques (e.g., maximum parsimony or maximum likelihood) can yield somewhat different phylogenies from the same matrix of taxa and character-states. Even if only parsimony is employed, there is a substantial chance that *any* phylogeny – whether based on morphological or molecular data – is incorrect, at least in part, given the current state of the art (see Chapter 2; Givnish and Sytsma 1997).

Third, the topology of a phylogeny (and the validity of conclusions drawn from it) can depend critically on which taxa are included; relationships among a subset of ingroup taxa can be altered if another taxon is added to the analysis, even if it is not sister to any of the other taxa in question. Sytsma and Baum (1996) graphically illustrate this principle, showing that relationships among broad groups of angiosperms are sensitive to the number and relationships of the outgroups used, and to the number and relationships (i.e., the density of taxon sampling) of the ingroups employed.

This is especially troubling because it suggests a harsh limit to the credibility of any inferred phylogeny. Extinction is part of evolutionary history, and can remove the possibility of including certain taxa in an analysis; this, in turn, may limit our ability to infer relationships among extant forms.

Fourth, inference of ancestral character-states often depends on the assumptions made about evolution (e.g., accelerated or delayed transformation) (Maddison and Maddison 1992). To minimize the appearance of convergence as an artifact, most chapters in this volume assume accelerated transformation. It may, however, be more conservative to consider only those changes that are insensitive to whether character-state transformation is assumed to be accelerated or delayed. Finally, ancestral phenotypes, distributions, or ecologies can only be inferred, not demonstrated, in the absence of conclusive fossil evidence. We can use parsimony, maximum likelihood, or other distance-related techniques to infer phylogenies and ancestral character-states in a rigorous, repeatable, plausible fashion. But we must always remember that evolution need not proceed in the most parsimonious, most likely, or most efficient fashion.

Speciation

ORIGIN OF DIVERGENCE – The paradigm of allopatric speciation (Mayr 1942, 1963, 1970) holds that partial or complete reproductive isolation is most likely to arise when gene flow between conspecific populations is interrupted, as a result of a new ecological barrier (e.g., a mountain range, desert, or river) or the colonization of a remote site. Interruption of gene flow would allow allopatric populations to diverge based on differences between areas in natural selection (Fisher 1930; Dobzhansky 1937; Mayr 1942), different trajectories of sexual selection (Lande 1981; Iwasa and Pomiankowski 1995), or genetic drift in the presence (Mayr 1963, 1970) or absence (Wright 1940) of population bottlenecks. If reproductive isolation between such populations is incomplete when they later come into secondary contact, reduced viability or fecundity of hybrids can select for additional pre- and postmating isolating mechanisms (Liou and Price 1994).

If speciation is allopatric, then selection for ecological divergence would be based – at least initially – on environmental differences, not competition with close relatives; such selection would play only an indirect role in speciation. However, speciation can occur sympatrically as well, at least under certain circumstances (e.g., see Liou and Price 1994; Orr 1995; Turelli and Orr 1995; Coyne 1996a; Mertz and Palumbi 1996; Rieseberg et al. 1996). Potential mechanisms of sympatric speciation include (i) polyploidy and chromosomal speciation (Stebbins 1971; White 1978); (ii) host race formation (Feder et al. 1990a,b); (iii) simple genetic changes at loci affecting pre- or postmating isolation (Bradshaw et al. 1995; Coyne 1996b; Davis and Wu 1996; Mertz and Palumbi 1996; True et al. 1996); (iv) infectious agents that reduce the viability or fecundity of hybrids (Werren et al. 1995)[5]; (v) evolution of parasites from hosts, as in

---

[5] Especially intriguing is the discovery of two airborne species of *Herpesvirus* that are pandemic in squirrel and spider monkeys (Albrecht and Fleckenstein 1992; Fleckenstein and Desrosiers 1992). In their respective hosts, these viruses are essentially asymptomatic; in other primates, however, they can rapidly prove fatal. Such herpesviruses may have profound importance as mediators of ecological competition, and might serve to isolate populations reproductively as well.

the adelphoparasitic red algae (Goff et al. 1996) and parasitic social insects (Hölldobler and Wilson 1990); (vi) hybridization followed by genetic coadaptation (Rieseberg et al. 1996); and (vii) ecological speciation (Schluter 1996b).

Any of these mechanisms would permit competition with close relatives to play a direct role, from the outset, in shaping differences between species in an adaptive radiation; the last mechanism would permit competition to help drive speciation as well. Schluter (1996b) argues that, in at least a dozen fish lineages, sympatric speciation has occurred postglacially as a result of divergent ecological selection pressures and rapid reproductive isolation. He asserts that these lineages all share (i) the rapid evolution of assortative mating based on trophic traits; (ii) the persistence of sympatric populations despite hybridization; (iii) substantial niche differentiation between sympatric congeners, often involving species feeding on plankton vs. benthos; and (iv) relatively high viability and fecundity of hybrids. Competition for prey leads to strong selection pressures for morphological divergence, even in a hybrid swarm (see Chapter 18); when coupled with assortative mating based on the very traits that allow efficient harvesting of different resources, then reproductive isolation, speciation, and ecological and reproductive character displacement can rapidly ensue. Does such a process underlie the very high rate of speciation seen in many groups of freshwater fishes (see Chapter 21)? More generally, how important is sexual selection on ecologically significant traits as a process driving adaptive radiation? Beak size and shape in Darwin's finches affects not only diet and ecological coexistence, but mate choice as well (Grant 1986); how important is the latter as an accelerator of divergence favored initially by resource competition?

Grant and Grant (1996a,b) argue that hybridization can play a crucial (if counterintuitive) role in adaptive radiation. In the Galápagos, the cactus finch (*Geospiza scandens*) and the medium-beaked ground finch (*G. fortis*) occasionally hybridize, with the offspring being intermediate in diet and beak shape. During a long drought beginning in 1977, the small soft seeds preferred by *G. scandens* became rare and the finch nearly became extinct. In the meantime, *G. fortis* and *G. fortis* × *G. scandens* hybrids survived better because the larger, tougher seeds they ate did not decline as much in abundance. With the advent of heavy rainfall with an El Niño in 1983, the scorched Galápagos became carpeted by showy meadows of annuals, which produce the small seeds preferred by *G. scandens*; with large numbers of tough and soft seeds, the hybrid became ecologically disadvantaged. Grant and Grant (1996a,b) suggest that, were it not for the preservation of *G. scandens* genes in hybrids and backcrosses, distinctive alleles and genetic combinations might have disappeared. Hybridization might thus play a creative role in adaptive radiation, preserving novelties during times of stress, and creating new genetic combinations on which selection might act. Carr (1995) has made a similar argument regarding the creative role of hybridization in the Hawaiian silversword alliance.

KEY INNOVATIONS – Guyer and Slowinski (1993) made the important point that any synapomorphy for a group that has undergone extensive speciation could, in principle, be advanced as a "key innovation" *sensu* Simpson (1953). They argued convincingly that – *if attention is restricted to phylogenetic information alone* – the only way to test whether a particular trait (e.g., wings) accelerated speciation was to conduct statistical analyses of the rates of diversification in several pairs of sister-groups in which the trait is or is not present (see also Heard and Hauser 1995). The strength of the inference that a specific trait accelerated speciation should be greater the wider the taxonomic range of groups investigated, to the extent that broader sampling breaks down the correlated occurrence of that trait with any others that might accelerate speciation.

However, this approach – while used by many systematists (Sanderson and Donoghue 1994, 1996; Slowinski and Guyer 1994; Barraclough et al. 1996; Rohde 1996; Sanderson and Wojciechowski 1996) to test whether traits significantly accelerate speciation – does not necessarily identify "key innovations." Simpson envisioned a key innovation as an adaptation (a microevolutionary or population/species-level phenomenon) that permitted access to a new range of resources, resulting in a wave of subsequent speciation (a macroevolutionary or clade-level phenomenon). But traits that increase the likelihood of speciation with little or no effect on adaptation – notably, limited dispersal (Mayr 1942, 1963, 1970; Diamond et al. 1976; Givnish et al. 1995; Cameron et al. 1996; Poullin and Feral 1996; Givnish 1997) – would be wrongly labeled a "key innovation" by the Guyer-Slowinski test. Moderately low rates of dispersal – low enough to interrupt gene flow, but high enough to allow the occasional colonization of new habitats (cf. Vermeij 1987; Bleiweiss 1990) – are likely to result in the highest rates of speciation, but should not be viewed as adaptations or key innovations.

While adaptive radiation may tend to accelerate speciation by permitting the invasion of new adaptive zones, I would argue that diversification in some classic cases of adaptive radiation has, in fact, often been non-adaptive and related to limited dispersal (see **Definition of adaptive radiation**). Consider the case of the African rift-lake cichlids, involving more than 1,000 species (Meyer et al. 1990, 1991; Fryer 1996). These fish have radiated spectacularly into different ecological roles, but surely embrace no more than 15 or 20 major trophic strategies. Even if we allow for some partitioning by depth (see Chapter 12; Meyer et al. 1996b) or particle size, it is clear that geographic speciation and narrow endemism among ecologically similar species must have played a major role. Dispersal between the three largest lakes – Victoria, Tanganyika, and Malawi – is minimal and largely independent radiations and speciations have occurred in each (Meyer et al. 1990, 1991, 1996b; Kocher et al. 1993). In cichlids, extensive intralacustrine speciation is favored by (i) mouth brooding and limited dispersal of young; (ii) use of rocks for shelter from predatory species, reflected in philopatry among adults in several groups (e.g., mbuna); (iii) the insular occurrence of rock outcrops suitable for many species scattered around the periphery of each lake; and (iv) periodic draw-downs of lake level and dissection of lakes into separate basins (see Mayr 1970; Fryer and Iles 1972; Greenwood 1974, 1978; Meyer et al. 1990, 1996b; Fryer 1996; Johnson et al. 1996; Verheyen et al. 1996; Chapters 12 and 21).

Limited dispersal in short, low-elevation species of *Brocchinia* is associated with more speciation than in clades that have undergone extensive radiation in mechanisms of nutrient capture (Chapter 8). *Brocchinia* embraces far fewer species than the related genus *Navia*, an ecologically stereotyped but poorly dispersing group. More generally,

Givnish (1997) argues that, in the understories of closed tropical forests, the sedentary nature of avian frugivores leads to high speciation rates in plants with fleshy fruits; many of the largest angiosperm genera (e.g., *Chamaedorea, Dipsis, Geonoma, Psychotria, Solanum*) bear fleshy fruits and are native to wet forest understories. The much greater dispersal capacity – and presumably, much lower speciation rate (Givnish et al. 1995) – of fleshy-fruited plants dependent on more vagile birds from forest edges or open forests may account for the lack of a gross relationship between plant species richness at the family level and mode of seed dispersal (see Herrera 1989; Ricklefs and Renner 1994).

DIVERSITY IN RELATION TO BODY SIZE – Differences in dispersal ability may interact with other factors to generate size-related patterns in species richness. Stanley (1979) argued that the species richness of mammal clades decreases with body size because larger species had lower population densities and hence greater probabilities of extinction, while smaller species had higher rates of speciation, based on limited dispersal and partitioning of the environment in a more coarse-grained fashion. Presumably, these broad principles help explain why rodents are exceptionally diverse among mammals (cf. Purvis 1996). The high diversity of bats – paradoxical given their power to fly – may actually reflect the rather sedentary behavior of many rain forest species (J. Kirsch, pers. comm.). In more vagile groups (e.g., many birds) or less interrupted terrains, adaptive radiation *per se* may play a much greater role in promoting speciation than mere geographic isolation; such appears to be the case in Darwin's finches. In both vagile and sedentary groups, however, adaptive diversification should play a key role in determining whether closely related species can co-occur (MacArthur 1972).

In plant lineages that colonize oceanic islands and undergo adaptive radiation, the humpbacked relationship of species richness to plant height may reflect both adaptation and the effects of dispersal on speciation. For example, in *Cyanea* (Lobelioideae), the largest angiosperm genus (ca. 65 spp.) native to Hawai'i, most species have a maximum stature of 2–4 m, with a few shorter and a few taller species (Figure 1.6). Among plants colonizing oceanic islands, there is selection for woodiness and increased

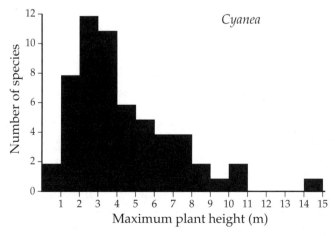

Figure 1.6. Histogram of maximum plant height in *Cyanea* (Campanulaceae: Lobelioideae) in the Hawaiian Islands, based on data tabulated by Lammers 1990.

stature, given that (i) many woody plants are unable to disperse long distances based on their large seed size (Carlquist 1965, 1974); (ii) herbaceous plants with lighter seeds are disproportionately represented; and (iii) competition for light favors taller individuals in productive, crowded environments (Darwin 1859; Givnish 1982a, 1988, 1997; Tilman 1988; see opposing argument by Carlquist 1974). Tall woody plants have repeatedly evolved from herbaceous ancestors on oceanic islands (Carlquist 1974), evidence via convergence of selection for increased stature, which would tend to extend the upper end of the species-height histogram (Figure 1.6). Indeed, *Cyanea* species are mainly rain forest trees and treelets; some (e.g., *C. leptostegia, C. hamatiflora*) are the tallest trees to have evolved on Hawai`i from non-tree ancestors (see Carlquist 1974; Wagner et al. 1990; Givnish et al. 1995). On the other hand, reduced dispersal in short species might have led to high speciation rates, creating a "hump" in the species-height histogram at low statures (Figure 1.6). Indeed, several short-statured species in the same clade often coexist on a given island, while generally only one tall species per clade is found (Givnish et al. 1995 and unpubl. data). This example illustrates how quasi-deterministic processes – competition for light and the effect of reduced dispersal on speciation – might generate a humpbacked relationship of diversity to body size. By contrast, Gould (1996) has argued that random processes (and a homeostatic constraint on minimum body size in bacteria) might be the predominant factors generating a similar relationship across living organisms.

MUTUALISM – In some groups, dispersal, adaptation, and speciation may be interrelated in interesting ways. For example, the superfamily Lycaenoidea is the largest group of butterflies in the world, comprising roughly 32% (ca. 5,400) of all described species (Taylor et al. 1993). The lycaenoids are characterized by at least three striking ecological features which might be related to their great diversity: (i) small size; (ii) remarkable dietary breadth, including carnivorous and parasitic taxa, as well as herbivorous species that feed on a large number of plant families; and (iii) a high incidence of ant associations, many of them mutualistic (see Pierce 1985, 1987; Pierce and Young 1986; Pierce et al. 1987). Mutualism should select for reduced dispersal; indeed, many lycaenoids have unusually low dispersal distances (ca. 10–100 m) between the site of their birth and where they mate. Cushman and Murphy (1993a,b) argue that, as a consequence of mutualism and low dispersal, the number of lycaenoid species is very large and most species have narrow ranges. Speciation might be similarly accelerated in other radiations involving mutualists.

SEXUAL SELECTION – As indicated above, sexual selection and competition for ecological resources may interact to accelerate speciation and adaptive radiation, based on assortative mating on trophic traits. In addition, certain patterns of resource use might increase the intensity of sexual selection, and thereby lead to an auto-catalytic acceleration of both speciation and adaptive radiation. Such a process may underlie, in part, the remarkable diversity (>800 spp.) of the Hawaiian drosophilids. Kambysellis and Craddock (Chapter 14) have found an evolutionary trend in these flies, from species which oviposit on common but low-quality resource patches (e.g., rotting leaves) to those which use rarer but high-quality patches (e.g., rotting bark, fluxes of tree exudates); the latter species-rich group includes the modified-mouthparts clade and (especially) the picture-winged clade. I suggest that the use of highly aggregated, high-quality patches by females would itself favor the evolution of leks and other forms of intense male-male competition for access to females; such sexual selection, in turn, has been implicated as a runaway process driving speciation in the Hawaiian drosophilids (Carson 1986; Kaneshiro 1987, 1989). Repeated speciation and isolation of small populations might themselves facilitate the evolution of new ecological preferences within a clade – by increasing the number of independent demes and the variation in the resource environments they face – and thereby accelerate adaptive radiation. Kambysellis and Craddock (Chapter 14) develop the first half of this argument in detail based on their extensive research, discussing the implications of female resource use, male sexual competition, and the larval environment for the evolution of Hawaiian drosophilids.

Genetics

Although the morphological divergence seen in many adaptive radiations suggests that overall genetic divergence between species might be substantial, in many cases the reverse is true, presumably reflecting (i) the recent origin of a radiation and (ii) the large effect of a few genes affecting ecologically significant traits and reproductive isolation. For example, many insular plant groups show little genetic divergence between species (Crawford 1990; Baldwin et al. 1991; Baldwin 1992; Crawford et al. 1992; Givnish et al. 1995; Böhle et al. 1996; Kim et al. 1996; Francisco-Ortega 1996; Chapters 3, 14, and 16 in this volume). While this pattern partly reflects low rates of molecular evolution in Asteraceae (Kim and Jansen 1995), it extends to other families as well; it has caused many investigators to rely on sequencing rapidly evolving regions, such as the ITS region of nuclear ribosomal DNA or the *trn*L-*trn*F region of cpDNA. As expected, there can be little sequence or restriction-site variation coupled with dramatic changes in morphology and ecology. Examples include major shifts of flower length and leaf shape in *Cyanea*, of growth form in *Echium*, of growth form, leaf shape, and physiology in the silversword alliance, and of habitat in the silversword alliance, *Alsinodendron-Schiedea*, *Argyranthemum*, *Cyanea*, *Echium*, and Macaronesian tree lettuces.

Hybridization appears to have played a major role in speciation within *Cyrtandra* on Hawai`i (Smith et al. 1996) and *Argyranthemum* in Macaronesia (Francisco-Ortega et al. 1996), and a less prominent role in the Hawaiian silversword alliance (Baldwin and Robichaux 1995; Carr 1995). Shifts in chromosome number and arrangement, where investigated, are generally minor or non-existent; an exception is the silversword alliance, where reciprocal translocations and aneuploidy are common and involved in partial reproductive isolation (Carr and Kyhos 1986; Chapter 3).

It must be recognized that some adaptations can be inherited independently of the bearer's nuclear genome. For example, bacterial symbionts in the guts of plant-sucking Homoptera are transmitted from mother to offspring, and provide vital amino acids and possibly vitamins to their insect hosts. Such symbionts – which almost surely are a "key innovation" that permit life based on a nutrient-deficient diet of phloem sap – appear to have arisen independently several times: once in the mealy-bugs, twice in white-flies, and twice in aphids (Moran and Baumann 1994). Horizontal transfers of adaptively crucial traits are rampant today in many groups of prokaryotes, as alarmingly illustrated by the rapid evolution of resistance to antibiotics in disease organisms (e.g., *Staphylococcus, Streptococcus*) via the exchange of plasmids (Udo and Grubb 1990; Cohen 1992; Neu 1992). Horizontal transfer played a crucial role in eukaryote evolution as well, with the capture and taming of various bacteria as mitochondria, chloroplasts, and (possibly) flagellae conferring aerobic respiration, photosynthesis, and efficient motion to their bearers (Margulis 1981; Khakina 1992). Indeed, the early history of life seems as much marked by anastomosis and a rich amalgamation of adaptive features as by strict evolutionary branching and divergence (Cavalier-Smith et al. 1994; Delwiche et al. 1995; McFadden and Gilson 1995; Barns et al. 1996; Delwiche and Palmer 1996; Palmer and Delwiche 1996). The acquisition of chloroplasts by the great clades of red, brown, golden, and green algae, the capture of aerobic bacteria as mitochondria by all eukaryotes, and the rampant swapping of genetic material among bacteria and archaebacteria (many with the most bizarre ecological tastes – methane-, sulfur-, and hydrogen-eaters, halophiles in saltpans, thermophiles in hot springs) call into question the validity of an evolutionary "tree" as a model for the early diversification of life. Our current cladistic tools will almost surely be inadequate for analyzing such evolutionary trends; when an "organism" can have as many as seven genomes (not counting plasmids or transposons) and can inherit several in different ways, new approaches and perspectives will be needed to understand the true nature of "organismal evolution."

Development

Two quite different kinds of explanations can be sought for any phenotypic trait, involving "how" it arises through development and "why" it might be favored by selection. Such explanations – involving structuralist and Darwinian perspectives – are complementary (e.g., Wake and Larson 1987; Roth et al. 1994; Amundson 1996; Klingenberg and Ekau 1996; Novacek 1996; Wake and Hansen 1996), even though they are sometimes seen as antagonistic alternatives (e.g., see Gould and Lewontin 1979; Lauder 1996). Just as the diversity of available ecological resources facing a lineage may partly explain "why" its species evolve different adaptations, so too the nature and lability of the developmental and genetic programs present in a lineage may partly explain "how" such adaptations arise and "why" some lineages undergo massive radiations while others do not (Givnish 1986c).

Understanding the basis for the developmental lability underlying specific cases of adaptive radiation promises to be an extremely productive area of inquiry over the next ten years (see Jacobs et al. 1995; Alberch and Blanco 1996; Gilbert et al. 1996; Shubin and Wake 1996). A relatively small number of homeobox-containing genes (e.g., *Hox* gene clusters) are now known to control many aspects of development in metazoans; the genetic sequence, position, and function of such genes are highly conserved, with more advanced phyla having a greater number of homeotic genes and gene clusters (Figure 1.7; Carroll 1995; Erwin et al. 1997). Developmental novelty within a group can arise through gene duplications within *Hox* gene clusters, changes in the function of individual homeobox genes, and altered interactions among such genes (Holland 1992;

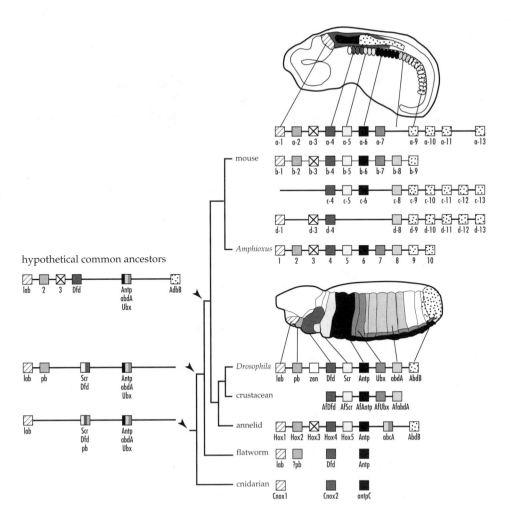

Figure 1.7. Phylogenetic distribution of *Hox* gene clusters in major phyla (redrawn from Carroll 1995 and Erwin et al. 1997). The position of *Hox* genes usually parallels the physical arrangement of the domains they affect, as indicated by the shading of different body parts in the mouse embryo (above) and *Drosophila* larva (below). Through evolution, *Hox* genes have undergone expansion, duplication, and changes in sequence, allowing the development of more complex body plans.

Halder et al. 1995; Holland and Garcia-Fernandez 1996; Meyer et al. 1995; Meyer 1996; Muller and Wagner 1996). The identities and pattern of symmetry of floral organs in angiosperms are controlled by a limited set of genes, mostly members of the MADS box family, analogous to the homeobox genes in animals and yeast (Coen and Meyerowitz 1991; Jack et al. 1994; Tandre et al. 1995; Theißen et al. 1996; Albert et al. 1997).

Heterochrony – simple shifts in the initiation, prolongation, or rate of a developmental process – can generate substantial morphological diversity within a lineage, based on changes at one or a few loci (Gould 1977; Alberch et al. 1979; Guerrant 1982; Wake and Larson 1987; McNamara 1995). Klingenberg and Ekau (1996), for example, show that repeated evolutionary shifts from benthic to pelagic habit in Antarctic notothenioid fish involve simple shifts of a shared growth trajectory during the larval stage; such shifts mainly affect fins and other swimming structures, and are manifested in bursts of morphological change associated with the invasion of open water. Genetic changes involving one or a few loci can have far-reaching effects, as shown by a recent study on the effects of thyroid hormone on salamander development. Rose (1996) found that the thyroxin sensitivity of several traits in *Eurycea* corresponded to their range of occurrence and developmental sequence in non-plethodontid salamanders, suggesting that simple changes in thyroxin level might profoundly influence many characteristics simultaneously. Similarly, Roth et al. (1994) showed that simple increases in neuron cell size – associated with higher ploidy – decreased the morphological complexity of brain tissue in frogs and salamanders. Changes at just four loci can increase the lifespan of *Caenorhabditis* five-fold (Lakowski and Hekimi 1996). A simple biomechanical decoupling of two jaw structures appears to accelerate speciation and functional diversification in the loricarioid catfishes (Schaeffer and Lauder 1996).

The genetic and developmental bases for traits may exert far-reaching constraints on the nature and direction of change within a given radiation. Jacobs et al. (1995) showed that the progressive disabling of the biochemical pathways for the synthesis of eumelanin and pheomelanin in tamarins (*Saguinus*) led to a unidirectional trend toward brighter pelage in this group of South American monkeys. More profoundly, Schluter (1996c) argues that – despite strong selection pressures – evolution in sticklebacks (Chapter 18) has largely been "along the genetic lines of least resistance," determined by the pattern of genetic covariance among several key traits. If this hypothesis proves correct and is generalizable, it would have substantial importance for the interpretation of adaptive radiations. However, as Schluter (1996c) himself notes, the pattern of genetic covariance across species may be determined partly by gene flow between populations; if selection helps drive differentiation among populations, then the apparently strong developmental control of adaptive radiation may prove to be a mirage, a result of the interaction of selection and migration.

Some lineages may have potentialities that others lack. For example, many herbaceous members of the Asteraceae produce some rudimentary woody tissue; this ability and their superb means of long-distance dispersal may explain why the treelike habit has evolved so frequently in the Asteraceae (Carlquist 1974; Givnish 1997). The late development of the keel bone and associated muscles in rails and allied birds may help explain why they have frequently lost the power of flight on islands and become the avian equivalent of small mammals (Olson 1973). Richard Lewontin (pers. comm.) suggests that direct tests of the importance of developmental lability will soon become possible, as transgenic organisms are included in laboratory selection experiments. Does developmental lability determine whether some lineages (like Darwin's finches on the Galápagos and lobeliads on Hawai`i) undergo massive radiation while others (like Galápagos mockingbirds and Hawaiian *Metrosideros*) do not? It hardly seems coincidental that beak dimensions in Darwin's finches vary 10 times more than in many mainland sparrow populations (Grant 1986).

Biogeography

The use of phylogenies to infer vicariance or dispersal events has a rich history (Brundin 1966; Carson 1970, 1981; Rosen 1978; Nelson and Platnick 1981; Beverly and Wilson 1985; Cracraft 1988; Wiley 1988; Carr et al. 1989; Funk and Brooks 1990; Baldwin et al. 1991; Crisci et al. 1991; Page 1991, 1994; Sytsma et al. 1991; Thorpe et al. 1991; Bremer 1992; Moritz et al. 1992; Martin and Dowd 1993; Avise 1994; Brooks and McLennan 1994; Givnish et al. 1994, 1995; Sperling and Harrison 1994; Baldwin and Robichaux 1995; DeSalle 1995; Funk and Wagner 1995; Gillespie and Croom 1995; Lammers 1995; Lowrey 1995; Morrone and Crisci 1995; Böhle et al. 1996; Francisco-Ortega et al. 1996; Malhotra et al. 1996). Insights obtained include the role of continental drift in the diversification of *Nothofagus* and ratites on the southern continents, in the global radiation of gigantic carnivorous dinosaurs (*Carcharodontosaurus* in Africa, *Acrocanthosaurus* in North America, *Gigantosaurus* in South America) 100 million years ago (Serreno et al. 1996), and in the parallel radiations of various mammalian groups following the breakup of Pangaea. Several issues are still hotly debated, including how to atomize terrains, how to code widespread species, how to distinguish the roles of dispersal vs. vicariance, and which algorithm (e.g., component analysis, Brooks parsimony) to employ in reconstructing historical biogeography (Morrone and Crisci 1995). Four questions are of special interest in the context of adaptive radiation:

First, what is the relationship between species distribution and adaptive divergence? A central tenet of evolutionary ecology is that species must diverge by some minimum amount in diet, habitat, or temporal activity in order to coexist; the development of statistical tests for non-random species distributions and ecological divergences has been of central importance in studies of character displacement and community structure. The interplay of ecology and geography in a phylogenetic context is a major objective of studies of adaptive radiation (e.g., see Grant 1986).

Second, does adaptive radiation ever proceed sympatrically? Speciation (i.e., reproductive isolation) is generally thought to transpire in allopatric populations, with character displacement and other forms of divergent selection occurring after such populations come into secondary contact (either through gradual directional selection or differential survival of colonists). Phylogenetic reconstructions – coupled, where possible, with experiments on competition and interbreeding (see Chapter 18) – can provide important insights into whether speciation and adaptive divergence can occur *in situ*, as it appears to do in African rift-lake cichlids (Chapter 12) and other groups of fishes (Chapters 18 and 21) (see section above on **Speciation**).

Third, is there evidence that taxon cycles occur on islands and archipelagoes, with more recently divergent taxa in a radiation occupying higher elevations or more inland or closed sites? In discussing the evolution of Melanesian ant faunas, Wilson (1961) argued that coastal habitats would be repeatedly invaded by new colonists, exerting pressure on previous arrivals to invade less crowded environments inland. Over time, this process would lead to progressive specialization on higher elevations, speciation, and (ultimately, as superior competitors arose or erosion eliminated appropriate habitats) extinction. Molecular phylogenies provide a direct means of testing this hypothesis in a variety of contexts.

Finally, are there "key landscapes" that favor the rise of innovations that, in turn, facilitate the invasion of new adaptive zones? For example, it seems likely that wet tropical forests are key landscapes for the evolution of epiphytism, given that only the soil is a reliable source of moisture and nutrients in habitats receiving less frequent rainfall, and given the tremendous advantage in light capture achieved by using other plants as supports. Givnish et al. (Chapter 8) argue that a geographic set of tepuis with a relatively large area of naked sandstone favored the rise of carnivory, ant-fed myrmecophily, and other unusual mechanisms of nutrient capture in the bromeliad genus *Brocchinia*; they support this idea by showing not only that these tepuis were the cradle of such key innovations in *Brocchinia*, they also appear to be the ancestral landscape of *Heliamphora* (Sarraceniaceae), the carnivorous South American sun-pitchers.

Tempo

The pace of speciation and adaptive diversification can be inferred from fossils (e.g., MacFadden 1992; Foote 1996; Jernvall et al. 1996; Novacek 1996; Chapters 4, 5, and 20), from geological dates for islands or filled lake basins (e.g., Carson 1983; Meyer et al. 1990; Givnish et al. 1994, 1995; Baldwin and Robichaux 1995; Tarr and Fleischer 1995; Wagner and Funk 1995; Chapters 12 and 21), or from molecular clocks (e.g., Sibley and Ahlquist 1982; Thomas and Hunt 1991; DeSalle 1992; Böhle et al. 1996; Kim et al. 1996; Chapters 3, 14, and 21). The resulting data provide insights into the rates of taxonomic and ecological divergence within a lineage, and thereby put important limits on the tempo of biological diversification. Inclusion of a time-scale also permits the correlation of particular features of an adaptive radiation with known environmental changes, as with the shift from browsing to grazing under drier climates in horses and their allies in the Northern hemisphere (Figure 1.8) and in large diprotodontid marsupials in Australia (Chapter 4).

Measurement of evolutionary tempo is also fundamental to testing what might be termed the "macroevolutionary hypothesis," the idea that characteristics of clades which increase speciation and decrease extinction may play a large role – perhaps greater than that of natural selection and other forces operating within populations – in determining the nature of the species that populate the world (e.g., see Jackson 1974; Stanley 1975, 1979; Williams 1975; Sepkoski 1984; Gould 1985; Vrba and Gould 1986; Jablonski 1987; Vermeij 1987, 1996; Valentine et al. 1991; Novacek 1996). Vermeij (1987, 1996) argues that, in many marine clades, there is little evidence of net adaptive progress at the level of individual organisms over geological time, but that clades have shown improvement in terms of increased speciation and decreased extinction. If this pattern were to prove generally true, would it mean that adaptive radiation is an epiphenomenon, or that it is an important macroevolutionary process enhancing the representation of labile, fast-speciating clades?

One of the most remarkable findings reported in this volume is that freshwater fish appear to have a much higher rate of speciation and adaptive diversification than other groups of terrestrial vertebrates and invertebrates (Chapter 21). Schluter (1996c) argues that sympatric speciation may occur rapidly in fish if selection pressures on trophic strategies are strongly divergent, and if assortative mating occurs and is based on trophic traits. Circumstantial support for divergent selection in freshwater fish is indeed widespread (Robinson and Wilson 1994). I suggest that one other factor be considered as well. The trophic structures of several fishes are developmentally plastic and can be altered by the diet they consume. So, if (i) a dietary shift occurs in a fish population as a result of invading a new lake or a different ecological zone within a lake already occupied, (ii) a novel trophic form arises, and (iii) some females begin to choose to mate with males having this novel form, then (iv) selection for a geneti-

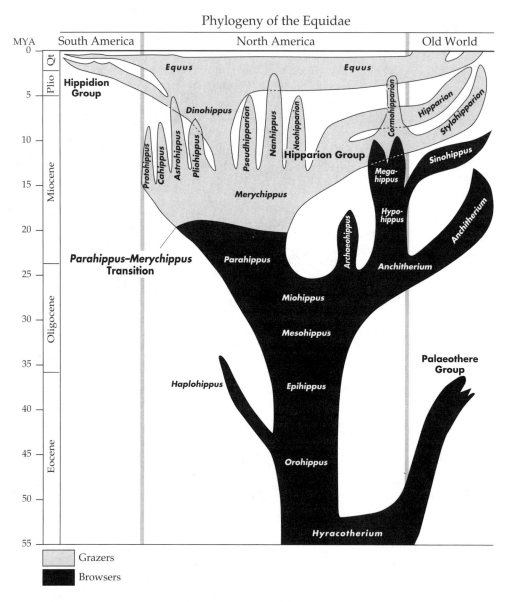

Figure 1.8. Adaptive radiation in horses and their allies, based on the fossil record (redrawn from MacFadden 1992).

cally more fixed expression of the novel trophic form will arise. In other words, developmental plasticity in response to changes in diet might bootstrap the ecological speciation process (Schluter 1996c) if it is coupled with sexual selection.

Predictability

One of the important theoretical differences between an adaptive radiation (based on natural selection and competition for ecological resources) and a sexual radiation (based on sexual selection and competition for mates) is that the pattern of adaptive radiation should be more or less predictable. Intra- and interspecific competition for similar resources should lead to convergence: kangaroos, wombats, jerboa marsupials, sugar gliders, native cats, Tasmanian devils, marsupial wolves, and saber-tooths are, in several important regards, functional equivalents to placental antelopes, groundhogs, kangaroo mice, flying squirrels, ocelots, wolverines, wolves, and saber-toothed tigers (Carlquist 1965; see Chapter 4). The woody habit and arborescent growth form repeatedly evolve in herbaceous lineages colonizing remote oceanic islands (Carlquist 1974; Givnish 1997), and flightlessness repeatedly arises in birds colonizing such islands. Similar climates on different continents favor similar distributions of plant growth forms (e.g., Raunkiaer 1934; Orians and Solbrig 1977; Mooney 1977) and leaf morphologies (see review by Givnish 1988). In contrast, the precise path taken by sexual selection is inherently unpredictable (Fisher 1930; Lande 1981; Iwasa and Pomiankowski 1995) – in some lineages it might lead to bigger or more complex horns and antlers, while in others it might lead to reddish fur, blue noses, brightly colored bills, long tails, or spatulate tarsi. While certain ecological conditions might favor more intense patterns of sexual selection, and perhaps result in a predictably greater exaggeration of secondary sexual characteristics, the exact nature of those characteristics remains unpredictable.

Yet it is clear that the genetic and developmental programs available within a lineage can constrain the direction taken by evolution (see preceding section on **Development**). Which lineage is present, with whatever phylogenetic constraints, might strongly affect the evolutionary trajectory of an adaptive radiation; so too might the particular competitors, predators, and mutualists with which members of the radiating lineage must interact at various stages of history. Such considerations led Gould (1989) to assert that historical contingency plays such a dominant role in evolutionary diversification that a "replay of the tape [of evolutionary diversification] would lead evolution down a pathway radically different from the road actually taken."

Is convergence or historical contingency the dominant process in adaptive radiation? While I would agree with Gould that genetic and developmental constraints can severely limit the direction evolution can take a particular lineage, I have argued elsewhere (Givnish 1986c) that this misses the point: natural selection operates on many populations and species simultaneously, favoring organisms with near-optimal traits that are favorably controlled (i.e., not linked to deleterious traits by the genetics or developmental pathway idiosyncratic to a particular lineage). We cannot predict which lineage of herbaceous plants invading an oceanic island will evolve the treelike habit; we can predict, however, that arborescent plants will evolve from some lineage on that island, provided that the climate supports productive vegetation (Givnish 1997).

The exemplary research by Jonathan Losos and his colleagues on Caribbean *Anolis* (see Chapter 19) shows that, despite historical contingencies and ecological differences among the four islands of the Greater Antilles, morphologically convergent species of these lizards have evolved on each island that fill the same set of ecological roles; furthermore, species filling different roles appear to have evolved in the same sequence on each island. As Jackman et al. (Chapter 19) note, "at least for anoles of the Greater Antilles, the tape has been replayed, four times, and the outcome has been substantially the same each time." Similarly, Foote (1996) shows that comparable amounts of morphological (and presumed ecological) divergence occurred at comparable rates in crinoids during their initial radiation in the early Paleozoic and their secondary radiation in the Triassic after the Permian mass extinction. This occurred despite the fact that most of the major adaptive zones occupied by marine animals during the Paleozoic were not completely vacated after the late Permian (Raup 1979; Bambach 1983, 1985; Erwin et al. 1987). Thus, even though crinoids experienced greater competition from other lineages in the Mesozoic than they did in the early Paleozoic, they radiated a comparable amount at roughly the same rate each time.

In spite of these suggestive findings, we should not conclude that historical contingency is unimportant in adaptive radiation, or in evolution more generally. Additional research is needed to study why the outcome of anoline or crinoid evolution appears to be so predictable, and why other radiations seem so unpredictable. Furthermore, we should not downplay historical contingency too much when we consider that some of the most striking examples of convergent radiation – anoles, crinoids, marsupial and placental mammals, arborescent Asteraceae or Lobelioideae on oceanic islands – often involve diversification based on very similar genetic and developmental programs, in the same lineage or closely related lineages. Contingency (in the form of which competing lineages are present) almost surely plays an important role in determining the extent (or lack) of radiation in many groups, including mammalian and avian lineages on different continents. How else can we account for the extraordinary diversification of browsers and grazers among marsupials in Australia, cervids and bovids in Laurasia, and the extinct notoungulates and horselike litopterns in South America? Or for the mass extinction of most of the South American megafauna after the Panamanian isthmus emerged, allowing the invasion of the North American megafauna? Much additional work is needed on how ecological forces, developmental constraints, and historical contingency interact to determine whether the course of adaptive radiation is repeatable and predictable.

Conclusions

Adaptive radiation is the incarnation of speciation and adaptation, our two great cosmological concerns as evolutionary biologists. It comes as no surprise, then, that the study of adaptive radiation is intimately intertwined with a broad range of questions involving phylogeny, adaptation, speciation, genetics, development, biogeography, and evolutionary tempo and predictability (Table 1.1). Many of these – notably those involving adaptation and phylogenetic inference, the link between ecology and speciation, and the control of speciation and adaptation by genetics and development – represent the unfinished business of the Modern Synthesis. Molecular systematics is a powerful new tool with which to attack several of these issues; it provides the crucial means of assessing evolutionary relationships in a precise fashion independent of the form, physiology, behavior, ecology, distribution, and development of the organisms under study.

The breadth of the issues involved in any adaptive radiation, and the diverse skills and immense amount of effort required to resolve them, trumpet the need for collaborative research. No one person is likely to have the training, insight, time, and physical resources needed to conduct incisive studies of the molecular systematics, ecology, biogeography, comparative morphology, genetics, development, and selection regimes of any clade. It is no accident that 17 of 21 chapters in this volume are co-authored. Ken Sytsma and I consider ourselves fortunate to have each other as collaborators, given our shared interest in almost all aspects of adaptive radiation, and our complementary skills in systematics and phylogenetic reconstruction and in ecology and adaptation, respectively. Similarly complementary skills and interests are brought to bear by many of the other teams represented in this volume, and by others active in the field.

Finally, it must be noted that while studies of adaptive radiation may provide us with many exciting scientific insights, several of the most remarkable examples of adaptive radiation are in grave peril. By their nature, many adaptive radiations are found on islands or islandlike habitats. The very lack of competitors and predators that helped evoke a great flowering of adaptive divergence in many island groups now makes them vulnerable when exotic organisms are introduced; the small size of islands and other insular environments makes the species that inhabit them especially sensitive to habitat destruction caused by economic or population pressures. And so it is that the introduction of the Nile perch threatens to extirpate the glittering shoals of cichlids in East Africa; the arrival of avian pox, avian malaria, mosquitoes, and exotic birds have virtually eliminated the Hawaiian honeycreepers at low elevations; introduced rabbits and feral cats, together with habitat destruction, threaten many Australian marsupials and have already caused some to vanish; the brown tree snake has annihilated the native Guam avifauna (at least in the wild); introduced predatory snails are destroying dozens of species of narrowly endemic land snails on Hawai`i and other island chains in the Pacific; and the ravening maws of exotic pigs, goats, cattle, axis deer, and rats – together with the loss of avian pollinators and seed dispersers – has severely threatened many species of *Cyanea* on Hawai`i, and driven several others to extinction. *Cyanea,* one of the most striking examples of adaptive radiation in flowering plants, now has more species on the U. S. lists of endangered and threatened taxa than any other genus of plants or animals. Unfortunately, some of the very factors that promote high speciation and adaptive divergence in *Cyanea* – limited dispersal, narrow elevational and geographic ranges, specialized pollination biology – also make its species especially susceptible to extinction (Givnish et al. 1995); perhaps the same is true of other island plant groups as well. We have an obligation, as students of adaptive radiation and as sentient beings, to use whatever means are necessary to ensure that such vivid illustrations of evolution – some of the most important keys to our biological cosmology – do not perish from the earth.

Acknowledgments

I would like to express my deep gratitude to Ken Sytsma for helpful discussions of several issues raised in this paper, and for suggesting a number of key references. My thoughts on adaptive radiation have diversified and become immeasurably enriched as a result of our long-term collaborations on the ecology and evolution of bromeliads and rapateads from the Guayana Shield, lobeliads from Hawai`i, and commelinoids worldwide, as well as our students' studies on a wide variety of other plant groups. I also would like to thank Tom Patterson for helping track down some early papers on adaptive radiation. David Ackerley, Victor Albert, Bruce Baldwin, Robert Bleiweiss, Elysse Craddock, Jorge Crisci, Donna Fernandez, Javier Francisco-Ortega, Morris Goodman, Linda Graham, George Lauder, Jonathan Losos, Michael Kambysellis, John Kirsch, Tom Kocher, Axel Meyer, Christopher Pires, Dolph Schluter, and John Wenzel provided helpful comments or access to useful reprints or preprints.

References

Alberch, P., and Blanco, M. J. 1996. Evolutionary patterns in ontogenetic transformation: from laws to regularities. International Journal of Developmental Biology 40:845–858.

Alberch, P., Gould, S. J., Oster, G. E., and Wake, D. B. 1979. Size and shape in ontogeny and phylogeny. Paleobiology 5:296–317.

Albert, V. A., Chase, M. W., and Mishler, B. D. 1993. Character-state weighting for cladistic analysis of protein-coding DNA sequences. Annals of the Missouri Botanical Garden 80:752–766.

Albert, V. A., Gustafsson, M. H. G., and Di Laurenzio, L. 1997. Ontogenetic systematics, molecular developmental genetics, and the angiosperm petal. In press in *Molecular systematics of plants, II*, P. S. Soltis, D. E. Soltis, and J. J. Doyle, eds. New York, NY: Chapman and Hall.

Albert, V. A.,Williams, S. E., and Chase, M. W. 1992. Carnivorous plants: phylogeny and structural evolution. Science 257:1491–1495.

Albrecht, J. C., and Fleckenstein, B. 1992. Primary structure of the *Herpesvirus saimiri* genome. Journal of Virology 66:5047–5048.

Alvesgormes, J. A., G. Orti, M. Haygood, W. Heiligenberg, and A. Meyer. 1995. Phylogeny of the South American electric fishes (Order Gymnotiformes) and the evolution of their electrogenic system – a synthesis based on morphology, electrophysiology, and mitochondrial DNA sequences. Molecular Biology and Evolution 12:298–318.

Amadon, D. 1950. The Hawaiian honeycreepers (Aves, Drepaniidae). Bulletin of the American Museum of Natural History 95:151–262.

Amundson, R. 1996. Historical development of the concept of adaptation. Pp. 11–53 in *Adaptation*, M. R. Rose and G. V. Lauder, eds. New York, NY: Academic Press.

Archie, J. W. 1989. Homoplasy excess ratios: new indices for measuring levels of homoplasy in phylogenetic systematics and a critique of the consistency index. Systematic Zoology 38:253–269.

Archie, J. W. 1996. Measures of homoplasy. Pp. 153–188 in *Homoplasy: the recurrence of similarity in evolution*, M. J. Sanderson and L. Hufford, eds. San Diego, CA: Academic Press.

Armbruster, W. S. 1992. Phylogeny and the evolution of plant-animal interactions. BioScience 42:12–20.

Arnason, U., and Gullberg, A. 1996. Cytochrome-b nucleotide sequences and the identity of five primary lineages of extant cetaceans. Molecular Biology and Evolution 13:407–417.

Arnold, M. L. 1993. *Iris nelsonii* (Iridaceae): origin and genetic composition of a homoploid species. American Journal of Botany 80:577–583.

Arnold, S. J. 1994. Is there a unifying concept of sexual selection that applies to both plants and animals? American Naturalist 144:S1–S12.

Avise, J. C. 1994. *Molecular markers, natural history and evolution*. New York, NY: Chapman and Hall.

Bachmann, K. 1995. Nuclear DNA markers for the evolution of *Microseris* (Asteraceae). Pp. 15–25 in *Experimental and molecular approaches to plant biosystematics*, P. C. Hoch and A. G. Stephenson, eds. Monographs in Systematic Botany 25. St. Louis, MO: Missouri Botanical Garden.

Bakker, R. T. 1975. Dinosaur renaissance. Scientific American 232:58–78.
Baldauf, S. L., and Palmer, J. D. 1993. Animals and fungi are each other's closest relatives: congruent evidence from multiple proteins. Proceedings of the National Academy of Sciences, USA 90:11558–11562.
Baldwin, B. G. 1992. Phylogenetic utility of the internal transcribed spacers of nuclear ribosomal DNA in plants: an example from the Compositae. Molecular Phylogenetics and Evolution 1:3–16.
Baldwin, B. G., Kyhos, D. W., and Dvořák, J. 1990. Chloroplast DNA evolution and adaptive radiation in the Hawaiian silversword alliance (Madiinae, Asteraceae). Annals of the Missouri Botanical Garden 77:96–109.
Baldwin, B. G., Kyhos, D. W., Dvořák, J., and Carr, G. D. 1991. Chloroplast DNA evidence for a North American origin of the Hawaiian silversword alliance (Asteraceae). Proceedings of the National Academy of Sciences, USA 88:1840–1843.
Baldwin, B. G., and Robichaux, R. H. 1995. Historical biogeography and ecology of the Hawaiian silversword alliance (Asteraceae): new molecular phylogenetic perspectives. Pp. 259–287 in *Hawaiian biogeography: evolution on a hot spot archipelago*, W. L. Wagner and V. A. Funk, eds. Washington, D. C.: Smithsonian Institution Press.
Baldwin, B. G., Sanderson, M. J., Porter, J. M., Wojciechowski, M. F., Campbell, C. S., and Donoghue, M. J. 1995. The ITS region of nuclear ribosomal DNA: a valuable source of evidence on angiosperm phylogeny. Annals of the Missouri Botanical Garden 82:247–277.
Bambach, R. K. 1983. Ecospace utilization and guilds in marine communities through the Phanerozoic. Pp. 719–746 in *Biotic interactions in recent and fossil communities*, M. J. S. Tevesz and P. L. McCall, eds. New York, NY: Plenum.
Bambach, R. K. 1985. Clades and adaptive variety: the ecology of diversification in marine faunas through the Phanerozoic. Pp. 191–253 in *Phanerozoic diversity patterns*, J. W. Valentine, ed. Princeton, NJ: Princeton University Press.
Bargelloni, L., Ritchie, P. A., Patarnello, T., Battaglia, B., Lambert, D. M., and Meyer, A. 1994. Molecular evolution at subzero temperatures: mitochondrial and nuclear phylogenies of fishes from Antarctica (Suborder Notothenioidei) and the evolution of antifreeze glycopeptides. Molecular Biology and Evolution 11:854–863.
Barns, S. M., Delwiche, C. F., Palmer, J. D., and Pace, N. R. 1996. Perspectives on archaeal diversity, thermophily and monophyly from environmental ribosomal RNA sequences. Proceedings of the National Academy of Sciences, USA 93:9188–9193.
Barraclough, T. G., Harvey, P. H., and Nee, S. 1996. Rate of *rbc*L sequence evolution and species diversification in flowering plants (angiosperms). Proceedings of the Royal Society of London, Series B 263:589–591.
Baum, D. A. 1995. The comparative pollination biology of *Adansonia* (Bombacaceae). Annals of the Missouri Botanical Garden 82:440–470.
Baum, D. A., and Larson, A. 1991. Adaptation reviewed: a phylogenetic methodology for studying character macroevolution. Systematic Zoology 40:1–18.
Bawa, K. S. 1980. Evolution of dioecy in flowering plants. Annual Review of Ecology and Systematics 11:15–40.
Benkham, C. W., and Lindholm, A. K. 1991. The advantages and evolution of morphological novelty. Nature 349:519–520.
Berenbaum, M. R. 1973. Coumarins and caterpillars: a case for coevolution. Evolution 37:163–179.
Berenbaum, M. R., Favet, C., and Schuler, M. A. 1966. On defining key innovations in an adaptive radiation-cytochrome P450's and Papilionidae. American Naturalist 148:S139-S155.
Berg, L. S. 1935. Über die vermeintlichen marinen Elemente in der Fauna und Flora des Baikalsees. Zoogeographica 2:455–470.
Beverly, S. M., and Wilson, A. C. 1985. Ancient origin for Hawaiian Drosophilidae inferred from protein comparisons. Proceedings of the National Academy of Sciences, USA 82:4753–4757.
Bharathan, G., and Zimmer, E. A. 1995. Early branching events in monocotyledons - partial 18S ribosomal DNA sequence analysis. Pp. 81–107 in *Monocotyledons: systematics and evolution*, P. Rudall, P. J. Cribb, D. F. Cutler, and C. J. Humphries, eds. Kew, England: Royal Botanic Gardens.
Bleiweiss, R. 1990. Ecological causes of clade diversity in hummingbirds: a neontological perspective on the generation of diversity. Pp. 354–380 in *Causes of evolution: a paleontological perspective*, R. M. Ross and W. R. Allmon, eds. Chicago, IL: University of Chicago Press.

Bleiweiss, R., Kirsch, J. A. W., and Mathews, J. C. 1997. DNA hybridization evidence for the principal lineages of hummingbirds (Aves: Trochilidae). Molecular Biology and Evolution 14:325–343.
Boag, P. T., and Grant, P. R. 1981. Intense natural selection in a population of Darwin's finches (Geospizinae) in the Galápagos. Science 214:82–85.
Boag, P. T., and Grant, P. R. 1984a. The classical case of character release: Darwin's finches (*Geospiza*) on Isla Daphne Major, Galápagos. Biological Journal of the Linnean Society 22:243–287.
Boag, P. T., and Grant, P. R. 1984b. Darwin's finches (Geospiza) on Isla Daphne Major, Galápagos: breeding and feeding ecology in a climatically variable environment. Ecological Monographs 54:463–489.
Bogler, D. J., Neff, J. L., and Simpson, B. B. 1995. Multiple origins of the yucca-yucca moth association. Proceedings of the National Academy of Sciences, USA 92:6864–6867.
Böhle, U.-R., Hilger, H. H., and Martin, W. F. 1996. Island colonization and evolution of the insular woody habit in *Echium* L. (Boraginaceae). Proceedings of the National Academy of Sciences, USA 93:11740–11745.
Bowman, R. I. 1963. Evolutionary patterns in Darwin's finches. Occasional Papers of the California Academy of Sciences 44:107–140.
Bradshaw, H. D., Wilbert, S. M., Otto, K. G., and Schemske, D. W. 1995. Genetic mapping of floral traits associated with reproductive isolation in monkeyflowers (*Mimulus*). Nature 376:762–765.
Bremer, K. 1988. The limits of amino-acid sequence data in angiosperm phylogenetic reconstruction. Evolution 42:795–803.
Bremer, K. 1992. Ancestral areas: a cladistic reinterpretation of the center of origin concept. Systematic Biology 4:436–445.
Brooks, D. R., and McLennan, D. A. 1993. Comparative study of adaptive radiations with an example using parasitic flatworms (Platyhelminthes, *Cercomeria*). American Naturalist 142:755–772.
Brooks, D. R., and McLennan, D. A. 1994. Historical ecology as a research programme: scope, limitations and the future. Pp. 1–27 in *Phylogenetics and ecology*, P. Eggleton and R. Vane-Wright, eds. London, England: Academic Press.
Brower, A. V. Z. 1996. Parapatric race formation and the evolution of mimicry in *Heliconius* butterflies – a phylogenetic hypothesis from mitochondrial DNA sequences. Evolution 50:195–221.
Brown, W. L., Jr., and Wilson, E. O. 1956. Character displacement. Systematic Zoology 5:49–64.
Brundin, L. 1966. Transantarctic relationships and their significance. Kungl. Svens. Vetenskapakad. Handl. 11:1–472.
Brunsfeld, S. J., Soltis, P. S., Soltis, D. E., Gadek, P. A., Quinn, C. J., Strenge, D. D., and Ranker, T. A. 1994. Phylogenetic relationships among the genera of Taxodiaceae and Cupressaceae: evidence from *rbc*L sequences. Systematic Botany 19:253–262.
Bull, J. J., Huelsenbeck, J. P., Cunningham, C. W., Swofford, D. L., and Waddell, P. J. 1993. Partitioning and combining data in phylogenetic analysis. Systematic Biology 42:384–397.
Cameron, R. A. D., Cook, L. M., and Hallows, J. D. 1996. Land snails on Porto Santo – adaptive and non-adaptive radiation. Proceedings of the Royal Society of London, Series B. 351:309–327.
Carlquist, S. 1965. *Island life.* New York, NY: Natural History Press.
Carlquist, S. 1970. *Hawaii: a natural history.* New York, NY: Natural History Press.
Carlquist, S. 1974. *Island biology.* New York, NY: Columbia University Press.
Carr, G. D. 1995. A fully fertile intergeneric hybrid derivative from *Argyroxiphium sandwicense* ssp. *macrocephalum* × *Dubautia menziesii* (Asteraceae) and its relationship to plant evolution in the Hawaiian Islands. American Journal of Botany 82:1574–1581.
Carr, G. D., and Kyhos, D. W. 1986. Adaptive radiation in the Hawaiian silversword alliance (Compositae-Madiinae). II. Cytogenetics of artificial and natural hybrids. Evolution 40:969–976.
Carr, G. D., Robichaux, R. H., Witter, M. S., and Kyhos, D. W. 1989. Adaptive radiation of the Hawaiian silversword alliance (Compositae-Madiinae): a comparison with Hawaiian picture-winged *Drosophila*. Pp. 79–97 in *Genetics, speciation, and the founder principle*, L. V. Giddings, K. Y. Kaneshiro, and W. W. Anderson, eds. New York, NY: Oxford University Press.
Carroll, S. 1995. Homeotic genes and the evolution of arthropods and chordates. Nature 376:479–485.
Carson, H. L. 1970. Chromosome tracers of the origin of species. Science 168:1414–1418.
Carson, H. L. 1981. Chromosomal tracing of evolution in a phylad of species related to *Drosophila hawaiiensis*. Pp. 286–297 in *Evolution and speciation: essays in honor of M. J. D.White*, W. R. Atchley and D. S. Woodruff, eds. Cambridge, England: Cambridge University Press.

Carson, H. L. 1983. Chromosomal sequences and interisland colonizations in the Hawaiian *Drosophila*. Genetics 103:465–482.

Carson, H. L. 1986. Sexual selection and speciation. Pp. 391–409 in *Evolutionary processes and theory*, S. Karlin and E. Nevo, eds. London, England: Academic Press.

Carson, H. L. 1992. Inversions in Hawaiian *Drosophila*. Pp. 407–439 in Drosophila *inversion polymorphism*, C. B. Krimbas and J. R. Powell, eds. Boca Raton, FL: CRC Press.

Carson, H. L., Clayton, E. E., and Stalker, H. D. 1967. Karyotypic stability and speciation in Hawaiian *Drosophila*. Proceedings of the National Academy of Sciences, USA 57:1280–1283.

Carson, H. L., Hardy, D. E., Spieth, H. T., and Stone, W. S. 1970. The evolutionary biology of the Hawaiian Drosophilidae. Pp. 437–543 in *Essays in evolution and genetics in honor of Theodosius Dobzhansky*, M. K. Hecht and W. C. Steere, eds. Amsterdam, The Netherlands: N-Holland.

Cavalier-Smith, T., Allsopp, M. T. E. P., and Chao, E. E. 1994. Chimeric conundra: are nucleomorphs and chromists monophyletic or polyphyletic? Proceedings of the National Academy of Sciences, USA 91:11368–11372.

Chase, M. W., Duvall, M. R., Hills, H. G., Conran, J. G., Cox, A. V., Eguiarte, L. E., Hartwell, J., Fay, M. F., Caddick, L. R., Cameron, K. M., and Hoot, S. 1995. Molecular phylogenetics of Lilianae. Pp. 109–137 in *Monocotyledons: systematics and evolution*, P. Rudall, P. J. Cribb, D. F. Cutler, and C. J. Humphries, eds. Kew, England: Royal Botanic Gardens.

Chase, M. W., and Palmer, J. D. 1992. Floral morphology and chromosome number in the subtribe Oncidiinae (Orchidaceae): evolutionary insights from a phylogenetic analysis of chloroplast DNA restriction site variation. Pp. 324–337 in *Molecular systematics of plants*, P. S. Soltis, D. E. Soltis, and J. J. Doyle, eds. New York, NY: Chapman and Hall.

Chase, M. W., Soltis, D. E., Morgan, D., Les, D. H., Duvall, M. R., Price, R., Hills, H. G., Qiu, Y., Kron, K. A., Rettig, J. H., Conti, E., Palmer, J. D., Clegg, M. T., Manhart, J. R., Sytsma, K. J., Michaels, H. J., Kress, W. J., Donoghue, M. J., Clark, W. D., Hedren, M., Gaut, B. S., Jansen, R. K., Kim, K., Wimpee, C. F., Smith, J. F., Furnier, G. R., Strauss, S. H., Xiang, Q., Plunkett, G. M., Soltis, P. M., Eguiarte, L. E., Learn, G. H., Graham, S., and Albert, V. A. 1993. Phylogenetics of seed plants: an analysis of nucleotide sequences from the plastid gene *rbc*L. Annals of the Missouri Botanical Garden 80:528–580.

Clark, A., and Johnston, I. A. 1996. Evolution and adaptive radiation of Antarctic fishes. Trends in Ecology and Evolution 11:212–218.

Clark, L. G., Zhang, W., and Wendel, J. F. 1995. A phylogeny of the grass family (Poaceae) based on *ndh*F sequence data. Systematic Botany 20:436–460.

Clausen, J., Keck, D. D., and Hiesey, W. M. 1941. *Experimental studies on the nature of species. IV. Genetic structure of ecological races*. Washington, D. C.: Carnegie Institute of Washington Publication 242.

Clegg, M. T. 1993. Chloroplast gene sequences and the study of plant evolution. Proceedings of the National Academy of Sciences, USA 90:363–367.

Clegg, M. T., and Zurawski, G. 1992. Chloroplast DNA and the study of plant phylogeny: present status and future prospects. Pp. 1–13 in *Plant molecular systematics*, D. E. Soltis, P. S. Soltis, and J. J. Doyle, eds. New York, NY: Chapman and Hall.

Coddington, J. A. 1988. Cladistic tests of adaptationist hypotheses. Cladistics 4:3–22.

Coddington, J. A. 1990. Bridges between evolutionary pattern and process. Cladistics 6:379–386.

Coen, E. S., and Meyerowitz, E. M. 1991. The war of the whorls: genetic interactions controlling flower development. Nature 353:31–37.

Cohen, M. L. 1992. Epidemiology of drug resistance: implications for a post-antimicrobial era. Science 257:1050–1055.

Colwell, R. K., and Winkler, D. 1984. A null model for null models in evolutionary ecology. Pp. 344–359 in *Ecological communities: conceptual issues and the evidence*, D. R. Strong, Jr., D. Simberloff, L. G. Abele, and A. B. Thistle, eds. Princeton, NJ: Princeton University Press.

Connell, J. H. 1980. Diversity and the coevolution of competition, or the ghost of competition past. Oikos 535:131–138.

Connor, E. F., and Simberloff, D. S. 1978. Species number and compositional similarity of the Galápagos flora and avifauna. Ecological Monographs 48:219–248.

Cowan, I. R. 1986. Economics of carbon fixation in higher plants. Pp. 133–170 in *On the economy of plant form and function*, T. J. Givnish, ed. New York, NY: Cambridge University Press.

Coyne, J. 1996a. Speciation in action. Science 272:700–701.

Coyne, J. A. 1996b. Genetics of differences in pheromonal hydrocarbons between *Drosophila melanogaster* and *Drosophila simulans*. Genetics 143:353–366.

Cracraft, J. 1988. Deep-history biogeography: retrieving the historical pattern of evolving continental biotas. Systematic Zoology 37:221–236.
Crawford, D. J. 1990. *Plant molecular systematics: macromolecular approaches.* New York, NY: John Wiley and Sons.
Crawford, D. J., Stuessy, T. F., Cosner, M. B., Haines, D. W., Silva, M., and Baeza, M. 1992. Evolution of the genus *Dendroseris* (Asteraceae, Lactuceae) on the Juan Fernandez Islands – evidence from chloroplast and ribosomal DNA. Systematic Botany 17:676–682.
Crisci, J. V., Cigliano, M. M., Morrone, J. J., and Roig Juñent, S. 1991. Historical biogeography of southern South America. Systematic Zoology 40:152–171.
Crisp, M. D. 1994. Evolution of bird-pollination in some Australian legumes (Fabaceae). Pp. 281–309 in *Phylogenetics and ecology*, P. Eggleton and R. Vane-Wright, eds. London, England: Academic Press.
Cushman, J. H., and Murphy, D. D. 1993a. Conservation of North American Lycaenids – an overview. Pp. 37–44 in Conservation biology of Lycaenidae (butterflies), T. W. New, ed. Gland, Switzerland: International Union for the Conservation of Nature.
Cushman, J. H., and Murphy, D. D. 1993b. Susceptibility of lycaenid butterflies to endangerment: implications for invertebrate conservation. Wings (Xerces Society) 17:16–21.
Darwin, C. 1859. *On the origin of species by means of natural selection.* London, England: John Murray.
Davidson, D. W. 1978. Size variability in the worker caste of a social insect (*Veromessor pergandei* Mayr) as a function of the competitive environment. American Naturalist 112:523–532.
Davis, A. W., and Wu, C. I. 1996. The broom of the sorcerer's apprentice: the fine structure of a chromosomal region causing reproductive isolation between two sibling species of *Drosophila*. Genetics 143:1287–1298.
Dawkins, R. 1996. *Climbing Mount Improbable.* New York, NY: W. W. Norton and Company.
Delwiche, C. F., Kuhsel, M., and Palmer, J. D. 1995. Phylogenetic analysis of *tuf*A sequences indicates a cyanobacterial origin of all plastids. Molecular Phylogenetics and Evolution 4:110–128.
Delwiche, C. F., and Palmer, J. D. 1996. Rampant horizontal transfer and duplication of RuBisCo genes in eubacteria and plastids. Molecular Biology and Evolution 13:873–882.
de Queiroz, K. 1996. Including the characters of interest during tree reconstruction and the problems of circularity and bias in studies of character evolution. American Naturalist 148:100–108.
DeSalle, R. 1992. The origin and possible time of divergence of the Hawaiian Drosophilidae: evidence from DNA sequences. Molecular Biology and Evolution 9:905–916.
DeSalle, R. 1995. Molecular approaches to biogeographic analysis of Hawaiian Drosophilidae. Pp. 72–89 in *Hawaiian biogeography: evolution on a hot spot archipelago*, W. L. Wagner and V. A. Funk, eds. Washington, DC: Smithsonian Institution Press.
DeSalle, R., Friedman, T., Prager, E. M., and Wilson, A. C. 1987. Tempo and mode of sequence evolution in mitochondrial DNA of Hawaiian *Drosophila*. Journal of Molecular Evolution 26:137–164.
Diamond, J. M. 1973. Distributional ecology of New Guinea birds. Science 179:759–769.
Diamond, J. M., Gilpin, M. E., and Mayr, E. 1976. Species-distance relation for birds of the Solomon archipelago, and the paradox of the great speciators. Proceedings of the National Academy of Sciences, USA 73:2160–2164.
Dixon, M. T., and Hillis, D. M. 1993. Ribosomal-RNA secondary structure: compensatory mutations and implications for phylogenetic analysis. Molecular Biology and Evolution 10:256–267.
Dobzhansky, T. 1937. *Genetics and the origin of species.* New York, NY: Columbia University Press.
Donoghue, M. J. 1989. Phylogenies and the analysis of evolutionary sequences, with examples from seed plants. Evolution 43:1137–1156.
Donoghue, M. J., and Ackerley, D. D. 1996. Phylogenetic uncertainties and sensitivity analyses in comparative biology. Philosophical Transactions of the Royal Society of London, Series B 351:1241–1249.
Doyle, J. A., Donoghue, M. J., and Zimmer, E. A. 1994. Integration of morphological and ribosomal RNA data on the origin of angiosperms. Annals of the Missouri Botanical Garden 81:419–450.
Doyle, J. J. 1993. DNA, phylogeny, and the flowering of plant systematics. BioScience 43: 380–389.
Eastman, J. T. 1993. *Antarctic fish biology: evolution in a unique environment.* New York, NY: Academic Press.
Ehrlich, P. H., and Raven, P. H. 1964. Butterflies and plants: a study in coevolution. Evolution 18:586–608.
Endler, J. A. 1986. *Natural selection in the wild.* Princeton, NJ: Princeton University Press.
Erwin, D. H., Valentine, J. W., and Sepkoski, J. J., Jr.. 1987. A comparative study of diversification events – the early Paleozoic versus the Mesozoic. Evolution 41:1177–1186.

Erwin, D., Valentine, J., and Jablonski, D. 1997. The origin of animal body plans. American Scientist 85:126–137.

Farrell, B. D., and Mitter, C. 1990. Phylogenesis of insect/plant interactions: have *Phyllobrotica* leaf-beetles (Chrysomelidae) and the Lamiales diversified in parallel? Evolution 44:1389–1403.

Farrell, B. D., and Mitter, C. 1993. Phylogenetic determinants of insect/plant community assembly. Pp. 253–266 in *Species diversity in ecological communities,* R. Ricklefs and D. Schluter, eds. Chicago, IL: University of Chicago Press.

Farrell, B. D., and Mitter, C. 1994. Adaptive radiations in insects and plants: time and opportunity. American Zoologist 34:57–69.

Farrell, B. D., Dussourd, D., and Mitter, C. 1991. Escalation of plant defense: do latex/resin canals spur plant diversification? American Naturalist 138:881–900.

Farris, J. S. 1972. Estimating phylogenetic trees from distance matrices. American Naturalist 106:645–668.

Farris, J. S. 1989a. The retention index and the rescaled consistency index. Cladistics 5:417–419.

Farris, J. S. 1989b. The retention index and homoplasy excess. Systematic Zoology 38:406–407.

Farris, J. S., Albert, V. A., Källersjö, M., Lipscomb, D., and Kluge, A. G. 1996. Parsimony jackknifing outperforms neighbor-joining. Cladistics 12:99–124.

Feder, J. L., Chilcote, C. A., and Bush, G. L. 1990a. The geographic pattern of genetic differentiation between host associated populations of *Rhagoletis pomonella* (Diptera: Tephritidae) in the eastern United States and Canada. Evolution 44:570–594.

Feder, J. L., Chilcote, C. A., and Bush, G. L. 1990b. Regional, local and microgeographic allele frequency variation between apple and hawthorn populations of *Rhagoletis pomonella* in western Michigan. Evolution 44:595–608.

Felsenstein, J. 1978. Cases in which parsimony or compatibility methods will be positively misleading. Systematic Zoology 27:401–410.

Felsenstein, J. 1985. Confidence limits on phylogenies: an approach using the bootstrap. Evolution 39:783–791.

Felsenstein, J. 1988. Phylogenies from molecular sequences: inference and reliability. Annual Reviews of Genetics 22: 521–565.

Fisher, D. C. 1985. Evolutionary morphology: beyond the analogous, the anecdotal, and the ad hoc. Paleobiology 11:120–138.

Fisher, R. A. 1930. *The genetical theory of natural selection.* Oxford, England: Clarendon Press.

Fleckenstein, B., and Desrosiers, R. C. 1982. *Herpesvirus saimiri* and *Herpesvirus ateles.* Pp.253–332 in *The herpesviruses, vol. 1,* B. Roizman, ed. New York, NY: Plenum Press.

Foote, M. 1996. Ecological controls on the evolutionary recovery of post-Paleozoic crinoids. Science 274:1492–1495.

Ford, E. B. 1971. *Ecological genetics, 3rd ed.* London, England: Chapman and Hall.

Fortey, R. A., Briggs, D. E. G., and Mills, M. A. 1996. The Cambrian evolutionary explosion – decoupling cladogenesis from morphological disparity. Biological Journal of the Linnean Society 51:13–33.

Francisco-Ortega, J., Jansen, R. K., and Santos-Guerra, A. 1996. Chloroplast DNA evidence of colonization, adaptive radiation, and hybridization in the evolution of the Macaronesian flora. Proceedings of the National Academy of Sciences, USA 93:4085–4090.

Freriksen, A., De Ruiter, b. L. A., Groenenberg, H.-J., Scharloo, W., and Heinstra, P. W. H. 1994. A multilevel approach to the significance of genetic variation in alcohol dehydrogenase of *Drosophila.* Evolution 48:781–790.

Frumhoff, P. C., and Reeve, H. K. 1994. Using phylogenies to test hypotheses of adaptation: a critique of some current proposals. Evolution 48:172–180.

Fryer, G. 1996. Endemism, speciation and adaptive radiation in great lakes. Environmental Biology of Fishes 45:109–131.

Fryer, G., and Iles, T. D. 1972. *The cichlid fishes of the great lakes of Africa.* Neptune City, NJ: T. F. H. Publications.

Funk, V. A., and D. R. Brooks. 1990. *Phylogenetic systematics as the basis of comparative biology.* Washington, D. C.: Smithsonian Institution Press.

Funk, V. A., and W. L. Wagner. 1995. Biogeography of seven ancient Hawaiian plant lineages. Pp. 160–194 in *Hawaiian biogeography: evolution on a hot spot archipelago,* W. L. Wagner and V. A. Funk, eds. Washington, D. C.: Smithsonian Institution Press.

Futuyma, D. J. 1986. *Evolutionary biology.* Sunderland, MA: Sinauer Associates.

Galis, F., and Drucker, E. G. 1996. Pharyngeal biting mechanics in centrarchid and cichlid fishes: insights into a key evolutionary innovation. Journal of Evolutionary Biology 9:641–671.

Garland, T., Dickerman, A. W., Janis, C. M., and Jones, J. A. 1993. Phylogenetic analysis of covariance by computer simulation. Systematic Biology 42:265–292.

Gilbert, L. E. 1980. Food web organization and conservation of neotropical diversity. Pp. 11–33 in *Conservation biology: an evolutionary-ecological perspective*, M. E. Soulé and B. A. Wilcox, eds. Sunderland, MA: Sinauer Associates.

Gilbert, S. F., Optiz, J. M., and Raff, R. A. 1996. Resynthesizing evolutionary and developmental biology. Developmental Biology 173:357–372.

Gillespie, R. G., and Croom, H. B. 1995. Comparison of speciation mechanisms in web-building and non-web-building groups within a lineage of spiders. Pp. 121–146 in *Hawaiian biogeography: evolution on a hot spot archipelago*, W. L. Wagner and V. Funk, eds. Washington, D. C.: Smithsonian Institution Press.

Gillespie, R. G., Croom, H. B., and Palumbi, S. R. 1994. Multiple origins of a spider radiation in Hawai`i. Proceedings of the National Academy of Sciences, USA 91:2290–2294.

Gittelman, J. L., and Kot, M. 1990. Adaptation: statistics and a nullmodel for estimating phylogenetic effects. Systematic Zoology 39:227–241.

Gittenberger, E. 1991. What about non-adaptive radiation? Biological Journal of the Linnean Society 43:263–272.

Givnish, T. J. 1980. Ecological constraints on the evolution of breeding systems in seed plants: dioecy and dispersal in gymnosperms. Evolution 34:959–972.

Givnish, T. J. 1982a. On the adaptive significance of leaf height in forest herbs. American Naturalist 120:353–381.

Givnish, T. J. 1982b. Outcrossing vs. ecological constraints in the evolution of dioecy. American Naturalist 119:849–865.

Givnish, T. J. 1986a. Optimal stomatal conductance, allocation of energy between leaves and roots, and the marginal cost of transpiration. Pp. 171–213 in *On the economy of plant form and function*, T. J. Givnish, ed. Cambridge, England: Cambridge University Press.

Givnish, T. J. 1986b. Biomechanical constraints on canopy geometry in forest herbs. Pp. 525–583 in *On the economy of plant form and function*, T. J. Givnish, ed. Cambridge, England: Cambridge University Press.

Givnish, T. J. 1986c. On the use of optimality arguments. Pp. 3–9 in *On the economy of plant form and function*, T. J. Givnish, ed. Cambridge, England: Cambridge University Press.

Givnish, T. J. 1987. Comparative studies of leaf form: assessing the relative roles of selective pressures and phylogenetic constraints. New Phytologist 106(Suppl.):131–160.

Givnish, T. J. 1988. Adaptation to sun vs. shade: a whole-plant perspective. Australian Journal of Plant Physiology 15:63–92.

Givnish, T. J. 1995. Plant stems: biomechanical adaptation for energy capture and influence on species distributions. Pp. 3–49 in *Plant stems: physiology and functional morphology*, B. L. Gartner, ed. New York, NY: Chapman and Hall.

Givnish, T. J. 1997. Adaptive plant evolution on islands: classical patterns, molecular data, new insights. In press in *Evolution on islands*, P. R. Grant, ed. Oxford, England: Oxford University Press.

Givnish, T. J., and Sytsma, K. J. 1997. Consistency, characters, and the likelihood of correct phylogenetic inference. Molecular Phylogenetics and Evolution 7:320–333.

Givnish, T. J., Sytsma, K. J., Smith, J. F., and Hahn, W. J. 1994. Thorn-like prickles and heterophylly in *Cyanea*: adaptations to extinct avian browsers on Hawaii? Proceedings of the National Academy of Sciences, USA 91:2810–2814.

Givnish, T. J., Sytsma, K. J., Smith, J. F., and Hahn, W. S. 1995. Molecular evolution, adaptive radiation, and geographic speciation in *Cyanea* (Campanulaceae, Lobelioideae). Pp. 288–337 in *Hawaiian biogeography: evolution on a hot spot archipelago*, W. L. Wagner and V. Funk, eds. Washington, D. C.: Smithsonian Institution Press.

Goff, L. J., Moon, D. A., Nyvall, P., Stache, B., Mangin, K., and Zuccarello, G. 1996. The evolution of parasitism in the red algae – molecular comparisons of adelphoparasites and their hosts. Journal of Phycology 32:297–312.

Gould, S. J. 1977. *Ontogeny and phylogeny*. Cambridge, MA: Harvard University Press.

Gould, S. J. 1983. The hardening of the modern synthesis. Pp. 71–93 in *Dimensions of Darwinism*, M. Greene, ed. Cambridge, England: Cambridge University Press.

Gould, S. J. 1985. The paradox of the first tier: an agenda for paleobiology. Paleobiology 11:2–12.

Gould, S. J. 1989. *Wonderful life.* New York, NY: Norton.
Gould, S. J. 1993. Fulfilling the spandrels of world and mind. Pp. 310–336 in *Understanding scientific prose,* J. Selzer, ed. Madison, WI: University of Wisconsin Press.
Gould, S. J. 1996. *Full house: the spread of excellence from Plato to Darwin.* New York, NY: Harmony Press.
Gould, S. J., and Calloway, C. B. 1980. Clams and brachiopods – ships that pass in the night. Paleobiology 6:383–396.
Gould, S. J., and Lewontin, R. C. 1979. The spandrels of San Marco and the Panglossian paradigm: a critique of the adaptationist programme. Proceedings of the Royal Society of London, Series B 205:581–598.
Gould, S. J., and Vrba, E. S. 1982. Exaptation – a missing term in the science of form. Paleobiology 8:4–15.
Graham, S. W., and Barrett, S. C. H. 1995. Phylogenetic systematics of Pontederiales: implications for breeding-system evolution. Pp. 415–441 in *Monocotyledons: systematics and evolution,* P. Rudall, P. J. Cribb, D. F. Cutler, and C. J. Humphries, eds. Kew, England: Royal Botanic Gardens.
Grant, B. R., and Grant, P. R. 1993. Evolution of Darwin's finches caused by a rare climatic event. Proceedings of the Royal Society of London, Series B 251:111–117.
Grant, B. R., and Grant, P. R. 1996a. High survival of Darwin's finch hybrids: effects of beak morphology. Ecology 77:500–509.
Grant, P. R. 1975. The classical case of character displacement. Evolutionary Biology 8:237–337.
Grant, P. R. 1981. Speciation and the adaptive radiation of Darwin's finches. American Scientist 69:653–663.
Grant, P. R. 1986. *Ecology and evolution of Darwin's finches.* Princeton, NJ: Princeton University Press.
Grant, P. R. 1994. Ecological character displacement. Science 266:746–747.
Grant, P. R., and Grant, B. R. 1995. Predicting microevolutionary responses to directional selection on heritable variation. Evolution 49:241–251.
Grant, P. R., and Grant, B. R. 1996b. Speciation and hybridization in island birds. Philosophical Transactions of the Royal Society of London, Series B 351:765–772.
Grant, P. R., and Schluter, D. 1984. Interspecific competition inferred from patterns of guild structure. Pp. 201–233 in *Ecological communities: conceptual issues and the evidence,* D. R. Strong, Jr., D. Simberloff, L. G. Abele, and A. B. Thistle, eds. Princeton, NJ: Princeton University Press.
Grant, V. 1949. Pollination systems as isolating mechanisms in angiosperms. Evolution 3:82–97.
Grant, V. 1963. *The origin of adaptations.* New York, NY: Columbia University Press.
Grant, V., and Grant, K. A. 1965. *Flower pollination in the phlox family.* New York, NY: Columbia University Press.
Greene, H. W. 1986. Diet and arboreality in the Emerald Monitor, *Varanus prasinus,* with comments on the study of adaptation. Fieldiana (Zoology, New Series) 31:1–12.
Greenwood, P. H. 1974. Cichlid fishes of Lake Victoria, East Africa: the biology and evolution of a species flock. Bulletin of the British Museum of Natural History (Zoology), Suppl. 6:1–134.
Greenwood, P. H. 1978. A review of the pharyngeal apophysis and its significance in the classification of African cichlid fishes. Bulletin of the British Museum of Natural History (Zoology) 33:297–323.
Greenwood, P. H. 1984. African cichlids and evolutionary theories. Pp. 141–154 in *Evolution of fish species flocks,* A. A. Echelle and I. Kornfeld, eds. Orono, ME: University of Maine Press.
Guerrant, E. O., Jr. 1982. Neotenic evolution of *Delphinium nudicaule* (Ranunculaceae): a hummingbird-pollinated larkspur. Evolution 36:699–712.
Gulick, A. 1932. Biological peculiarities of oceanic islands. Quarterly Review of Biology 7:405–445.
Guyer, C., and Slowinski, J. B. 1993. Adaptive radiation and the topology of large phylogenies. Evolution 47:253–263.
Halanych, K. M., Bacheller, J. D., Aguinaldo, A. M. A., Liva, S. M., Hillis, D. M., and Lake, J. A. 1995. Evidence from 18S DNA that the lophophorates are protostome animals. Science 267:1641–1643.
Halder, G., Callaerts, P., and Gehring, W. J. 1995. Induction of ectopic eyes by targeted expression of the *eyeless* gene in *Drosophila.* Science 267:1788–1792.
Hamby, R. K., and Zimmer, E. A. 1992. Ribosomal RNA as a phylogenetic tool in plant systematics. Pp. 50–91 in *Plant molecular systematics,* D. E. Soltis, P. S. Soltis, and J. J. Doyle, eds. New York, NY: Chapman and Hall.
Harvey, P. H., Colwell, R. K., Silvertown, J. W., and May, R. M. 1983. Null models in ecology. Annual Review of Ecology and Systematics 14:189–211.
Harvey, P. H., Leigh Brown, A. J., Maynard Smith, J., and Nee, S. (eds.). 1996. *New uses for new phylogenies.* Oxford, England: Oxford University Press.

Harvey, P. H., and Pagel, M. D. 1991. *The comparative method in evolutionary biology.* Oxford, England: Oxford University Press.

Harvey, P. H. H., Read, A. F., and Nee, S. 1995a. Why ecologists need to be phylogenetically challenged. Journal of Ecology 83:533–536.

Harvey, P. H. H., Read, A. F., and Nee, S. 1995b. Further remarks on the role of phylogeny in comparative ecology. Journal of Ecology 83:733–734.

Haufler, C. H., Soltis, D. E., and Soltis, P. S. 1995. Phylogeny of the *Polypodium vulgare* complex: insights from chloroplast DNA restriction site data. Systematic Botany 20:110–119.

Heard, S. B., and Hauser, D. L. 1995. Key evolutionary innovations and their ecological mechanisms. History of Biology 10:151–173.

Heed, W. B. 1968. Ecology of the Hawaiian Drosophilidae. University of Texas Publications 6818:387–419.

Heed, W. B. 1971. Host plant specificity and speciation in Hawaiian *Drosophila.* Taxon 20:115–121.

Herrera, C. M. 1989. Seed dispersal by animals: a role in angiosperm diversification? American Naturalist 133:309–322.

Hillis, D. M. 1987. Molecular versus morphological approaches to systematics. Annual Review of Ecology and Systematics 18:23–42.

Hillis, D. M. 1995. Approaches for assessing phylogenetic accuracy. Systematic Biology 44:3–16.

Hillis, D. M. 1996. Inferring complex phylogenies. Nature 383:130–131.

Hillis, D. M., Allard, M. W., and Miyamoto, M. M. 1993. Analysis of DNA sequence data: phylogenetic inference. Methods in Enzymology 224: 456–487.

Hillis, D. M., Huelsenbeck, J. P., and Cunningham, C. W. 1994. Application and accuracy of molecular phylogenies. Science 264: 671–676.

Hodges, S. A., and Arnold, M. L. 1994a. Floral and ecological isolation between *Aquilegia formosa* and *Aquilegia pubescens.* Proceedings of the National Academy of Sciences, USA 91:2493–2496.

Hodges, S. A., and Arnold, M. L. 1994b. Columbines: a geographically widespread species flock. Proceedings of the National Academy of Sciences, USA 91:5129–5132.

Holland, H. H. 1992. *Adaptation in natural and artificial systems.* Cambridge, MA: MIT Press.

Holland, P. W. H., and Garcia-Fernandez, J. 1996. *Hox* genes and chordate evolution. Developmental Biology 173:1382–1395.

Hölldobler, B., and Wilson, E. O. 1990. *The ants.* Cambridge, MA: Belknap Press.

Horn, H. S. 1979. Adaptation from the perspective of optimality. Pp. 48–61 in *Topics in plant population biology,* O. T. Solbrig, S. Jain, G. B. Johnson, and P. H. Raven, eds. New York, NY: Columbia University Press.

Huxley, J. 1942. *Evolution: the modern synthesis.* New York, NY: Harper and Brothers.

Iwasa, Y., and Pomiankowski, A. 1995. Continual change in mate preferences. Nature 377:420–422.

Jablonski, D. 1987. Heritability at the species level: analysis of geographical ranges of Cretaceous mollusks. Science 283:360–363.

Jack, T., Fox, G. L., and Meyerowitz, E. M. 1994. *Arabidopsis* homeotic gene APETALA3 ectopic expression: transcriptional and post-transcriptional regulation determine floral organ identity. Cell 76:703–716.

Jackson, J. B. C. 1974. Biogeographic consequences of eurytopy and stenotopy among marine bivalves and their evolutionary significance. American Naturalist 108:541–560.

Jacobs, S. C., Larson, A., and Cheverud, J. M. 1995. Phylogenetic reconstructions and orthogenetic evolution of coat color among tamarins (genus *Saguinus*). Systematic Biology 44:515–532.

Janzen, D. H. 1974. Tropical blackwater rivers, animals, and mast fruiting in the Dipterocarpaceae. Biotropica 6:69–105.

Jernvall, J., Hunter, J. P., and Fortelius, M. 1996. Molar tooth diversity, disparity, and ecology in Cenozoic ungulate radiations. Science 274:1489–1492.

Johnson, T. C., Scholz, C. A., Talbot, M. R., Kelts, K., Ricketts, R. D., Nogobi, G., Beuning, K., Ssemanda, I., and Mcgill, J. W. 1996. Late Pleistocene desiccation of Lake Victoria and rapid evolution of cichlid fishes. Science 273:1091–1093.

Jones, J. S., Leith, B. H., and Rawlings, P. 1977. Polymorphism in *Cepaea:* a problem with too many solutions? Annual Review of Ecology and Systematics 8:109–143.

Kambysellis, M. P., Ho, K.-F., Craddock, E. M., Piano, F., Parisi, M., and Cohen, J. 1995. Pattern of ecological shifts in the diversification of Hawaiian *Drosophila* inferred from a molecular phylogeny. Current Biology 5:1129–1139.

Kaneshiro, K. Y. 1983. Sexual selection and direction of evolution in the biosystematics of Hawaiian Drosophilidae. Annual Review of Entomology 28:161–178.

Kaneshiro, K. Y. 1987. The dynamics of sexual selection and its pleiotropic effects. Behavior Genetics 17:559–569.

Kaneshiro, K. Y. 1989. The dynamics of sexual selection and founder effects in species formation. Pp. 279–296 in *Genetics, speciation and the founder principle,* L. V. Giddings, K. Y. Kaneshiro, and W. W. Anderson, eds. New York, NY: Oxford University Press.

Kaneshiro, K. Y. 1993. Habitat-related variation and evolution by sexual selection. Pp. 89–101 in *Evolution of insect pests,* K. C. Kim and B. A. McPheron, eds. New York, NY: Wiley and Sons.

Kaneshiro, K. Y., Gillespie, R. G., and Carson, H. L. 1995. Chromosomes and male genitalia of Hawaiian *Drosophila:* tools for interpreting phylogeny and geography. Pp. 57–71 in *Hawaiian biogeography: evolution on a hot spot archipelago,* W. L. Wagner and V. A. Funk, eds. Washington, DC: Smithsonian Institution Press.

Kettlewell, H. B. D. 1958. A survey of the frequencies of *Biston betularia* (L) (Lep.) and its melanic forms in Great Britain. Heredity 12:51–72.

Khakina, L. N. 1992. *Concepts of symbiogenesis: a historical and critical study of the research of Russian botanists.* New Haven, CN: Yale University Press.

Kim, K. J., and Jansen, R. K. 1995. *ndh*F sequence evolution and the major clades in the sunflower family. Proceedings of the National Academy of Sciences, USA 92:10379–10383.

Kim, S.-C., Crawford, D. J., Francisco-Ortega, J., and Santos-Guerra, A. 1996. A common origin for woody *Sonchus* and five related genera in the Macaronesian Islands: molecular evidence for extensive radiation. Proceedings of the National Academy of Sciences, USA 93:7743–7748.

Kingsolver, J. G., and Koehl, M. A. R. (eds.). 1994. *Selective factors in the evolution of insect wings.* Palo Alto, CA: Annual Reviews, Inc.

Klingenberg, C. P., and Ekau, W. 1996. A combined morphometric and phylogenetic analysis of an ecomorphological trend – pelagization in Antarctic fishes (Perciformes, Notothenidae). Biological Journal of the Linnean Society 59:143–177.

Knox, E., Downie, S. R., and Palmer, J. D. 1993. Chloroplast genome rearrangements and the evolution of giant lobelias from herbaceous ancestors. Molecular Biology and Evolution 10:414–430.

Kocher, T. D., Conroy, J. A., McKaye, K. R., and Stauffer, J. R. 1993. Similar morphologies of cichlid fish in Lakes Tanganyika and Malawi are due to convergence. Molecular Phylogenetics and Evolution 2:158–165.

Lack, D. 1940. Evolution of the Galápagos finches. Nature 146:324–327.

Lack, D. 1947. *Darwin's finches.* Cambridge, England: Cambridge University Press.

Lai, C. Y., Baumann, P., and Moran, N. 1996. The endosymbiont (*Buchnera* sp.) of the aphid *Diuraphis noxia* contains plasmids consisting of Trpeg and tandem repeats of pseudogenes. Applied and Environmental Microbiology 67:332–339.

Lakowski, B., and Hekimi, S. 1996. Determination of life-span in *Caenorhabditis elegans* by four clock genes. Science 272:1010–1013.

Lammers, T. G. 1990. Campanulaceae. Pp. 420–489 in *Manual of flowering plants of Hawai`i,* W. L. Wagner, D. R. Herbst, and S. H. Sohmer, eds. Honolulu, HI: Bishop Museum Publications.

Lammers, T. G. 1995. Patterns of speciation and biogeography in *Clermontia* (Campanulaceae, Lobelioideae). Pp. 338–362 in *Hawaiian biogeography: evolution on a hot spot archipelago,* W. L. Wagner and V. A. Funk, eds. Washington, DC: Smithsonian Institution Press.

Lande, R. 1981. Models of speciation by sexual selection on polygenic characters. Proceedings of the National Academy of Sciences, USA 78:3721–3725.

Lande, R., and Arnold, S. J. 1983. The measurement of selection on correlated characters. Evolution 37:1210–1226.

Larson, A., and Losos, J. B. 1996. Phylogenetic systematics of adaptation. Pp. 187–220 in *Adaptation,* M. R. Rose and G. V. Lauder, eds. New York, NY: Academic Press.

Lauder, G. V. 1981. Form and function: structural analysis in evolutionary morphology. Paleobiology 7:430–442.

Lauder, G. V. 1996. The argument from design. Pp. 55–91 in *Adaptation,* M. R. Rose and G. V. Lauder, eds. New York, NY: Academic Press.

Ledje, C., and Arnason, U. 1996. Phylogenetic analyses of complete cytochrome-b genes of the order Carnivora with particular emphasis on the Caniformia. Journal of Molecular Evolution 42:135–144.

Liaud, M.-F., Valentin, C., Martin, W., Bouget, F.-Y., Kloareg, B., and Cerff, R. 1994. The evolutionary origin of red algae as deduced from the nuclear genes encoding cytosolic and chloroplast glyceraldehyde-3-phosphate dehydrogenase from *Chondrus crispus.* Journal of Molecular Evolution 38:319–327.

Liem, K. F. 1973. Evolutionary strategies and morphological innovations: cichlid pharyngeal jaws. Systematic Zoology 22:425–441.

Liem, K. 1990. Key evolutionary innovations, differential diversity, and symecomorphosis. Pp. 147–170 in *Evolutionary innovations*, M. H. Nitecki, ed. Chicago, IL: University of Chicago Press.

Liou, L. W., and Price, T. D. 1994. Speciation by reinforcement of premating isolation. Evolution 48:1415–1459.

Losos, J. B., and Miles, D. B. 1994. Adaptation, constraint, and the comparative method: phylogenetic issues and methods. Pp. 60–98 in *Ecological morphology: integrative organismal biology*, P. C. Wainwright and S. M. Reilly, eds. Chicago, IL: University of Chicago Press.

Lowrey, T. K. 1995. Phylogeny, adaptive radiation, and biogeography of Hawaiian Tetramelopium (Asteraceae: Astereae). Pp. 195–220 in *Hawaiian biogeography: evolution on a hot spot archipelago*, W. L. Wagner and V. A. Funk, eds. Washington, D. C.: Smithsonian Institution Press.

MacArthur, R. H. 1972. *Geographical ecology.* New York, NY: Harper and Row.

MacFadden, B. J. 1992. Interpreting extinctions from the fossil record: methods, assumptions, and case examples using horses (family Equidae). Pp. 17–45 in *Extinction and phylogeny*, M. J. Novacek and Q. D. Wheeler, eds. New York, NY: Columbia University Press.

Maddison, W. P., and Maddison, D. R. 1992. *MacClade: analysis of phylogeny and character evolution, vers. 3.0.* Sunderland, MA: Sinauer Associates.

Maddison, W. P., Donoghue, M. J., and Maddison, D. R. 1984. Outgroup analysis and parsimony. Systematic Zoology 33: 83–103.

Malhotra, A., Thorpe, R. S., Black, H., Daltry, J. C., and Wüster, W. 1996. Relating geographic patterns to phylogenetic processes. Pp. 187–202 in *New uses for new phylogenies*, P. H. Harvey, A. J. Leigh Brown, J. Maynard Smith, and S. Nee, eds. Oxford, England: Oxford University Press.

Manhart, J. R. 1994. Phylogenetic analysis of green plant *rbcL* sequences. Molecular Phylogenetics and Evolution 3:114–127.

Marden, J. H., and Kramer, M. G. 1994. Surface-skimming stoneflies: a possible intermediate stage in insect flight evolution. Science 266:427–430.

Margulis, L. 1981. *Symbiosis in cell evolution: life and its environment on the early earth.* San Francisco, CA: W. H. Freeman.

Martin, P. G., and Dowd, J. M. 1993. Using sequences of *rbcL* to study phylogeny and biogeography of *Nothofagus* species. Australian Systematic Botany 6:441–447.

Martins, E. P., and Hansen, T. F. 1996. A microevolutionary link between phylogenies and comparative data. Pp. 273–288 in *New uses for new phylogenies*, P. H. Harvey, A. J. Leigh Brown, J. Maynard Smith, and S. Nee, eds. Oxford, England: Oxford University Press.

Mayr, E. 1942. *Systematics and the origin of species.* New York, NY: Columbia University Press.

Mayr, E. 1963. *Animal species and evolution.* Cambridge, MA: Harvard University Press.

Mayr, E. 1970. *Populations, species, and evolution.* Cambridge, MA: Belknap Press.

McDade, L. A. 1995. Hybridization and phylogenetics. Pp. 305–331 in *Experimental and molecular approaches to plant biosystematics*, P. C. Hoch and A. G. Stephenson, eds. Monographs in Botany 53. St. Louis, MO: Missouri Botanical Garden.

McFadden, G., and Gilson, P. 1995. Something borrowed, something green: lateral transfer of chloroplasts by secondary endosymbiosis. Trends in Ecology and Evolution 10:12–17.

McNamara, K. J. 1995. *Evolutionary change and heterochrony.* New York, NY: John Wiley and Sons.

McNeilly, T., and Antonovics, J. 1968. Evolution in closely adjacent plant populations. IV. Barriers to gene flow. Heredity 23:205–218.

Merçot, H., Defaye, D., Capy, P., Pla, E., and David, J. R. 1994. Alcohol tolerance, ADH activity, and ecological niche of *Drosophila* species. Evolution 48:756–757.

Mertz, E. C., and Palumbi, S. R. 1996. Positive selection and sequence rearrangements generate extensive polymorphism in the gamete recognition protein binding. Molecular Biology and Evolution 13:397–406.

Meyer, A. 1993. Phylogenetic relationships and evolutionary processes in East African cichlid fishes. Trends in Ecology and Evolution 8:279–284.

Meyer, A. 1995. Molecular evidence on the origin of tetrapods and the relationships of the coelacanth. Trends in Ecology and Evolution 10:111–116.

Meyer, A. 1996. The evolution of body plans: HOM/*Hox* cluster evolution, model systems, and the importance of phylogeny. Pp. 322–340 in *New uses for new phylogenies*, P. H. Harvey, A. J. Leigh Brown, J. Maynard Smith, and S. Nee, eds. Oxford, England: Oxford University Press.

Meyer, A., Knowles, L., and Verheyen, E. 1996a. Widespread geographic distribution of mitochondrial haplotypes in Lake Tanganyika rock-dwelling cichlid fishes. Molecular Ecology 5:341–350.

Meyer, A., Kocher, T. D., Basasibwaki, P., and Wilson, A. C. 1990. Monophyletic origin of Lake Victoria cichlid fishes suggested by mitochondrial DNA sequences. Nature 347:550–553.

Meyer, A., Kocher, T. D., and Wilson, A. C. 1991. African fishes. Nature 350:467–468.

Meyer, A., Montero, C., and Spreinat, A. 1994. Evolutionary history of the cichlid fish species flocks of the East African great lakes inferred from molecular phylogenetic data. Advances in Limnology 44:409–425.

Meyer, A., Montero, C., and Spreinat, A. 1996b. Molecular phylogenetic inferences about the evolutionary history of the East African cichlid fish radiations. Pp. 303–323 in *IDEAL (International decade of East African lakes): the limnology, climatology and paleoclimatology of the East African lakes,* T. Johnson and E. Odada, eds. London, England: Gordon and Breach Scientific Publishers.

Meyer, A., Ritche, P. A., and Witte, K. E. 1995. Predicting developmental processes from evolutionary patterns: a molecular phylogeny of the zebrafish (*Danio rerio*) and its relatives. Philosophical Transactions of the Royal Society of London, Series B 394:103–111.

Milinkovitch, M. C., Meyer, A., and Powell, J. R. 1994. Phylogeny of all major groups of cetaceans based on DNA sequences from 3 mitochondrial genes. Molecular Biology and Evolution 11:939–948.

Mishler, B. D., Lewis, L. A., Buchheim, M. A., Renzaglia, K. S., Garbary, D. J., Delwiche, C. F., Zechman, F. W., Kantz, T. S., and Chapman, R. L. 1994. Phylogenetic relationships of the "green algae" and "bryophytes." Annals of the Missouri Botanical Garden 81:451–583.

Mitchell-Olds, T., and Shaw, R. G. 1987. Regression analysis of natural selection: statistical and biological interpretation. Evolution 41:1149–1161.

Mitter, C., Farrell, B. D., and Futuyma, D. J. 1991. Phylogenetic studies of insect/plant interactions: insights into the genesis of diversity. Trends in Ecology and Evolution 6:290–293.

Miyamoto, M. M., and Cracraft, J. (eds.). 1991. *Phylogenetic analysis of DNA sequences.* New York, NY: Oxford University Press.

Mooney, H. A. (ed.). 1977. *Convergent evolution in Chile and California: Mediterranean climate ecosystems.* Stroudsburg, PA: Dowden, Hutchinson, and Ross.

Monson, R. K. 1996. The use of phylogenetic perspective in comparative plant physiology and developmental biology. Annals of the Missouri Botanical Garden 83:3–16.

Moran, N. 1996. Accelerated evolution and Muller's ratchet in endosymbiotic bacteria. Proceedings of the National Academy of Sciences, USA 93:2873–2878.

Moran, N., and Baumann, P. 1994. Phylogenetics of cytoplasmically inherited microorganisms of arthropods. Trends in Ecology and Evolution 9:115–120.

Moran, N. A., Vondohlen, O. D., and Baumann, P. 1996. Faster evolutionary rates in endosymbiotic bacteria than in cospeciating insect hosts. Journal of Molecular Ecology 41:727–731.

Moran, P., Kornfield, I., and Reinthal, P. 1993. Molecular systematics and radiation of the haplochromine cichlids (Teleostei: Perciformes) of Lake Malawi. Copeia 1994:274–288.

Moritz, C., Schneider, C. J., and Wake, D. B. 1992. Evolutionary relationships within the *Ensatina eschscholtzii* complex confirm the ring species interpretation. Systematic Biology 41:273–291.

Morrone, J. J., and Crisci, J. V. 1995. Historical biogeography – introduction to methods. Annual Review of Ecology and Systematics 26:373–401.

Muller, G. E., and Wagner, G. P. 1996. Homology, *Hox* genes, and developmental integration. American Zoologist 36:4–13.

Muller, H. J. 1949. Redintegration of the symposium on genetics, paleontology, and evolution. Pp. 421–455 in *Genetics, paleontology, and evolution,* G. L. Jepsen, E. Mayr, and G. G. Simpson, eds. Princeton, NJ: Princeton University Press.

Nee, S., Holmes, E. C., Rambaut, A., and Harvey, P. H. 1996. Inferring population history from molecular phylogenies. Pp. 66–80 in *New uses for new phylogenies,* P. H. Harvey, A. J. Leigh Brown, J. Maynard Smith, and S. Nee, eds. Oxford, England: Oxford University Press.

Nelson, G., and Platnick. N. I. 1981. *Systematics and biogeography: cladistics and vicariance.* New York, NY: Columbia University Press.

Neu, H. C. 1992. The crisis in antibiotic resistance. Science 257:1064–1073.

Nilsson, D.-E., and Pelger, S. 1994. A pessimistic estimate of the time required for an eye to evolve. Proceedings of the Royal Society of London, Series B 256:53–58.

Nishida, M. 1991. Lake Tanganyika as an evolutionary reservoir of old lineages of East African cichlid fishes: inferences from allozyme data. Experientia 47:974–979.

Novacek, M. J. 1996. Paleontological data and the study of adaptation. Pp. 311–359 in *Adaptation*, M. R. Rose and G. V. Lauder, eds. New York, NY: Academic Press.

Olmstead, R. G., and Reeves, P. A. 1995. Evidence for the polyphyly of the Scrophulariaceae based on chloroplast *rbc*L and *ndh*F sequences. Annals of the Missouri Botanical Garden 82:176–193.

Olmstead, R. G., Bremer, B., Scott, K. M., and Palmer, J. D. 1993. A parsimony analysis of the Asteridae sensu lato based on *rbc*L sequences. Annals of the Missouri Botanical Garden 80: 700–722.

Olsen, G. J., Overbeek, R., Larsen, N., Marsh, T. L., Mccaughey, M. J., Maciukenas, M. A., Kuan, W.-M., Macke, T. J., Xing, Y., and Woese, C. R. 1992. The ribosomal database project. Nucleic Acids Research 20(Suppl.): 2199–2200.

Olson, S. L. 1973. Evolution of the rails of the South Atlantic Islands (Aves: Rallidae). Smithsonian Contributions to Zoology, 152:1–53.

Orians, G. H., and Solbrig, O. T. 1977. *Convergent evolution in warm deserts*. Stroudsburg, PA: Dowden, Hutchinson, and Ross.

Orr, H. A. 1995. The population genetics of speciation – the evolution of hybrid incompatibilities. Genetics 139:1805–1813.

Osborn, H. F. 1893. The rise of the Mammalia in North America. Proceedings of the American Association for the Advancement of Science 42:187–227.

Osborn, H. F. 1902. The law of adaptive radiation. American Naturalist 36:353–363.

Ostrom, J. H. 1979. Bird flight: how did it begin? American Scientist 67:46–56.

Page, R. D. M. 1991. Clocks, clades, and cospeciation: computing rates of evolution and timing of cospeciation rates in host-parasite assemblages. Systematic Zoology 40:188–198.

Page, R. D. M. 1994. Maps between trees and cladistic analysis of historical associations among genes, organisms, and areas. Systematic Biology 43:58–77.

Page, R. D. M. 1996. Temporal congruence revisited – comparison of mitochondrial DNA sequence divergence in cospeciating pocket gophers and their chewing lice. Systematic Biology 45:151–167.

Palmer, J. D., and Delwiche, C. F. 1996. 2nd-hand chloroplasts and the case of the disappearing nucleus. Proceedings of the National Academy of Sciences, USA 93:7432–7435.

Parkhurst, D. F. 1986. Internal leaf structure: a three-dimensional perspective. Pp. 215–249 in *On the economy of plant form and function*, T. J. Givnish, ed. New York, NY: Cambridge University Press.

Pellmyr, O., Thompson, J. N., Brown, J. M., and Harrison, R. G. 1996. Evolution of pollination and mutualism in the yucca moth lineage. American Naturalist 148:827–847.

Pierce, N. E. 1985. Lycaenid butterflies and ants: selection for nitrogen fixing and other protein rich food plants. American Naturalist 125:888–895.

Pierce, N. E. 1987. The evolution and biogeography of associations between lycaenid butterflies and ants. Oxford Surveys in Evolutionary Biology 4:89–116.

Pierce, N. E., and Young, W. R. 1986. Lycaenid butterflies and ants: two-species stable equilibria in mutualistic, commensal, and parasitic interactions. American Naturalist 128:216–227.

Pierce, N. E., Kitching, R. L., Buckley, R. C., Taylor, M. F. J., and Benbow, K. 1987. Costs and benfits of cooperation between the Australian lycaenid butterfly, *Jalmenus evagoras*, and its attendant ants. Behavioral Ecology and Sociobiology 21:237–248.

Poulin, E., and Feral, J. P. 1996. Why are there so many species of brooding Antarctic echinoids? Evolution 50:820–830.

Pratt, H. D. 1979. A systematic analysis of the endemic avifauna of the Hawaiian Islands. Ph.D. dissertation, Louisiana State University, Baton Rouge.

Pratt, H. D., 1992. Systematics of the Hawaiian "creepers" *Oreomystis* and *Paroreomyza*. Condor 94:836–846.

Pratt, H. D., Bruner, P. L., and Berrett, D. G. 1987. *The Birds of Hawaii and the Tropical Pacific*. Princeton, NJ: Princeton University Press.

Purvis, A. 1996. Using interspecies phylogenies to test macroevolutionary hypotheses. Pp. 153–168 in *New uses for new phylogenies*, P. H. Harvey, A. J. Leigh Brown, J. Maynard Smith, and S. Nee, eds. Oxford, England: Oxford University Press.

Radtkey, R. R. 1996. Adaptive radiation of day-geckos (*Phelsuma*) in the Seychelles archipelago – a phylogenetic analysis. Evolution 50:604–623.

Raunkiaer, C. 1934. *The life forms of plants and statistical plant geography*. Oxford, England: Clarendon Press.

Raup, D. M. 1979. Size of the Permo-Triassic bottleneck and its evolutionary implications. Science 206:217–218.

Reeve, H. K., and Sherman, P. W. 1993. Adaptation and the goals of evolutionary research. Quarterly Review of Biology 68:1–32.

Reznick, D., and Travis, J. 1996. The empirical study of adaptation in natural populations. Pp. 243–289 in *Adaptation,* M. R. Rose and G. V. Lauder, eds. New York, NY: Academic Press.

Ricklefs, R. E. 1996. Phylogeny and ecology. Trends in Ecology and Evolution 11:229–230.

Ricklefs, R. E., and Renner, S. S. 1994. Species richness within families of flowering plants. Evolution 48:1619–1636.

Ridley, M. 1996. *Evolution, 2nd ed.* Oxford, England: Blackwell Scientific Publications.

Rieseberg, L. H., and Brunsfeld, S. J. 1992. Molecular evidence and plant introgression. Pp. 151–176 in *Molecular systematics of plants,* P. S. Soltis, D. E. Soltis, and J. J. Doyle, eds. New York, NY: Chapman and Hall.

Rieseberg, L. H., and Morefield, J. D. 1995. Character expression, phylogenetic reconstruction, and the detection of reticulate evolution. Pp. 333–353 in *Experimental and molecular approaches to plant biosystematics,* P. C. Hoch and A. G. Stephenson, eds. Monographs in Botany 53. St. Louis, MO: Missouri Botanical Garden.

Rieseberg, L. H., Sinervo, B., Linder, C. R., Ungerer, M. C., and Arias, D. M. 1996. Role of gene interactions in hybrid speciation: evidence from ancient and experimental hybrids. Science 272:741–745.

Robinson, B. W., and Wilson, D. S. 1994. Character release and displacement in fishes: a neglected literature. American Naturalist 144:596–627.

Rohde, K. 1996. Robust phylogenies and adaptive radiations – a critical re-examination of methods used to identify key innovations. American Naturalist 148:481–500.

Romano, S. L., and Palumbi, S. R. 1996. Evolution of scleractinian corals inferred from molecular systematics. Science 271:640–642.

Rose, C. S. 1996. An endocrine-based model for developmental and morphogenetic diversification in metamorphic and paedomorphic urodeles. Journal of Zoology 239:253–284.

Rosen, D. E. 1978. Vicariant patterns and historical explanation in biogeography. Systematic Zoology 27:159–188.

Roth, G., Blanke, J., and Wake, D. B. 1994. Cell size predicts morphological complexity in the brains of frogs and salamanders. Proceedings of the National Academy of Sciences, USA 91:4796–4800.

Saitou, N. 1988. Property and efficiency of the maximum likelihood method for molecular phylogeny. Journal of Molecular Evolution 27:261–273.

Sanderson, M. J., and Donoghue, M. J. 1994. Shifts in diversification rate with the origin of angiosperms. Science 264:1590–1593.

Sanderson, M. J., and Donoghue, M. J. 1996. Reconstructing shifts in diversification rates on phylogenetic trees. Trends in Ecology and Evolution 11:15–20.

Sanderson, M. J., and Wojciechowski, M. F. 1996. Diversification rates in a temperate legume clade – are there so many species of *Astragalus* (Fabaceae)? American Journal of Botany 83:1488–1502.

Schaeffer, S. A., and Lauder, G. V. 1986. Historical transformation of functional design: evolutionary morphology of feeding mechanisms in loricarioid catfishes. Systematic Zoology 35:489–508.

Schaeffer, S. A., and Lauder, G. V. 1996. Testing historical hypotheses of morphological change – biomechanical decoupling in loricarioid catfishes. Evolution 50:1661–1675.

Schluter, D. 1984. Morphogenetic and phylogenetic relationships among the Darwin's finches. Evolution 38:921–930.

Schluter, D. 1988a. Character displacement and the adaptive divergence of finches on islands and continents. American Naturalist 131:799–824.

Schluter, D. 1988b. Estimating the form of natural selection on a quantitative trait. Evolution 42:849–861.

Schluter, D. 1994. Experimental evidence that competition promotes divergence in adaptive radiation. Science 266:798–801.

Schluter, D. 1996a. Ecological causes of adaptive radiation. American Naturalist 148:S40–S64.

Schluter, D. 1996b. Ecological speciation in post-glacial fishes. Philosophical Transactions of the Royal Society of London, Series B 351:807–814.

Schluter, D. 1996c. Adaptive radiation along genetic lines of least resistance. Evolution 50:1766–1774.

Schluter, D., and Grant, P. R. 1982. The distribution of *Geospiza difficilis* in relation to *G. fuliginosa* in the Galápagos Islands: tests of three hypotheses. Evolution 36:1213–1226.

Schluter, D., and Grant, P. R. 1984. Determinants of morphological patterns in communities of Darwin's finches. American Naturalist 123:175–196.

Schluter, D. T., Price, D., and Grant, P. R. 1985. Ecological character displacement in Darwin's finches. Science 227:1056–1059.

Schneider, H., Schneider, M. P. C., Sampaio, I., Harada, M. L., Stanhope, M., Czelusniak, J., and Goodman, M. 1993. Molecular phylogeny of the New World monkeys (Platyrrhini, Primates). Molecular Phylogenetics and Evolution 2:225–242.

Schoener, T. W. 1974. Resource partitioning in ecological communities. Science 185:27–39.

Scott, J. M., Mountainspring, S., Ramsey, F. L., and Kepler, C. B. 1986. *Forest bird communities of the Hawaiian Islands: their dynamics, ecology, and conservation.* Lawrence, KS: Allen Press.

Seger, J., and Stubblefield, J. W. 1996. Optimization and adaptation. Pp. 93–123 in *Adaptation,* M. R. Rose and G. V. Lauder, eds. New York, NY: Academic Press.

Sepkoski, J., Jr. 1984. A kinetic model of Phanerozoic taxonomic diversity. III. Post-Paleozoic families and mass extinctions. Paleobiology 14:322–330.

Serreno, P. C., Dutheil, D. B., Iarochene, M., Larsson, H. C. E., Lyon, G. H., Magwene, P. M., Sidor, C. A., Varicchio, D. J., and Wilson, J. A. 1996. Predatory dinosaurs from the Sahara and late Cretaceous faunal differentiation. Science 272:986–989.

Shaw, K. L. 1995. Biogeographic patterns of two independent Hawaiian cricket radiations (*Laupala* and *Prognathogryllus*). Pp. 39–56 in *Hawaiian biogeography: evolution on a hot spot archipelago,* W. L. Wagner and V. A. Funk, eds. Washington, D. C.: Smithsonian Institution Press.

Shaw, K. 1996. Sequential radiations and patterns of speciation in the Hawaiian cricket genus *Laupala* inferred from DNA sequences. Evolution 50:237–255.

Shubin, N., and Wake, D. B. 1996. Phylogeny, development, and morphological variation. American Zoologist 36:51–60.

Sibley, C. G., and Ahlquist, J. E. 1982. The relationships of the Hawaiian honeycreepers (Drepanidini) as indicated by DNA-DNA hybridization. Auk 99:130–140.

Silvertown, J., and Dodd, M. 1996. Comparing plants and connecting traits. Philosophical Transactions of the Royal Society of London, Series B 351:1233–1239.

Simpson, G. G. 1944. *Tempo and mode in evolution.* New York, NY: Columbia University Press.

Simpson, G. G. 1953. *The major features of evolution.* New York, NY: Columbia University Press.

Sinervo, B., and Basolo, A. L. 1996. Testing adaptation using phenotypic manipulations. Pp. 149–185 in *Adaptation,* M. R. Rose and G. V. Lauder, eds. New York, NY: Academic Press.

Skelton, P. W. 1993. Adaptive radiation: definition and diagnostic tests. Pp. 45–58 in *Evolutionary patterns and processes,* D. R. Lees and D. Edwards, eds. London, England: Academic Press.

Skinner, M. W. 1988. Comparative pollination ecology and floral evolution in Pacific Coast *Lilium*. Ph.D. dissertation, Harvard University.

Slowinski, J. B., and Guyer, C. 1989. Testing the stochasticity of patterns of organic diversity: an improved null model. American Naturalist 134:907–921.

Slowinski, J. B., and Guyer, C. 1994. Testing whether certain traits have caused amplified diversification: an improved method based on a model of random speciation and extinction. American Naturalist 142:1019–1024.

Smith, A. B. 1992. Echinoid phlogeny: molecules and morphology accord. Trends in Ecology and Evolution 7:224–229.

Smith, G. R., and Todd, T. N. 1984. Evolution of species flocks of fishes in north temperate lakes. Pp. 45–68 in *Evolution of fish species flocks,* A. A. Echelle and I. Kornfeld, eds. Orono, ME: University of Maine Press.

Smith, J. F., Burke, C. C, and Wagner, W. L. 1996. Interspecific hybridization in natural populations of *Cyrtandra* (Gesneriaceae) on the Hawaiian Islands – evidence from RAPD markers. Plant Systematics and Evolution 200:61–77.

Soltis, D. E. , P. S. Soltis, and J. J. Doyle (eds.). 1992. *Plant molecular systematics.* New York, NY: Chapman and Hall.

Sperling, F. A. H., and Harrison, R. G. 1994. Mitochondrial DNA variation within and between species of the *Papilio machaon* group of swallowtail butterflies. Evolution 48:408–422.

Spieth, H. T. 1968. Evolutionary implications of the mating behavior of the species of *Antopocerus* Drosophilidae in Hawaii. University of Texas Publications 6818:319–333.

Spieth, H. T. 1982. Behavioral biology and evolution of the Hawaiian picture-winged species group of *Drosophila*. Pp. 351–437 in *Evolutionary biology, vol. 14,* M.K. Hecht, B. Wallace and G.T. Prance, eds. New York, NY: Plenum Press.

Springer, M. S., and Kirsch, J. A. W. 1991. DNA hybridization, the compression effect, and the radiation of diprotodontian marsupials. Systematic Zoology 40:131–151.

Stanley, S. M. 1975. A theory of evolution above the species level. Proceedings of the National Academy

of Sciences, USA 72:646–650.
Stanley, S. M. 1979. *Macroevolution: pattern and process.* San Francisco, CA: W. H. Freeman.
Stebbins, G. L. 1971. *Chromosomal evolution in higher plants.* Reading, MA: Addison-Wesley.
Stebbins, G. L. 1974. *Flowering plants: evolution above the species level.* Cambridge, MA: Belknap Press.
Stern, D. L., and Grant, P. R. 1996. A phylogenetic re-analysis of allozyme variation among populations of Galápagos finches. Zoological Journal of the Linnean Society 118:119–134.
Strong, D. R., Jr., Simberloff, D., Abele, L. G., and Thistle, A. B. (eds.). 1984. *Ecological communities: conceptual issues and the evidence.* Princeton, NJ: Princeton University Press.
Strong, D. R., Jr., Szyska, L., and Simberloff, D. 1979. Tests of community-wide character displacement against null hypotheses. Evolution 35:897–913.
Sturmbauer, C., and Meyer, A. 1992. Genetic divergence, speciation, and morphological stasis in a lineage of African cichlid fishes. Nature 358:578–581.
Sturmbauer, C., and Meyer, A. 1993. Mitochondrial phylogeny of the endemic mouthbrooding lineages of cichlid fishes from Lake Tanganyika in Eastern Africa. Molecular Biology and Evolution 10:751–768.
Sulloway, F. J. 1984. Darwin and the Galápagos. Biological Journal of the Linnean Society 21:29–59.
Sültmann, H., Mayer, W. E., Figueroa, F., Tichy, H., and Klein, J. 1995. Phylogenetic analysis of cichlid fishes using nuclear DNA markers. Molecular Biology and Evolution 12:1033–1047.
Swofford, D. L. 1993. *PAUP: phylogenetic analysis using parsimony, vers. 3.1,* Champaign, IL: Illinois Natural History Survey.
Swofford, D. L. 1996. *PAUP*: phylogenetic analysis using parsimony, vers. 4.0.* Sunderland, MA: Sinauer Associates.
Sytsma, K. J. 1990. DNA and morphology: inferences of plant phylogeny. Trends in Ecology and Evolution 5:104–110.
Sytsma, K. J., and Baum, D. A. 1996. Molecular phylogenies and the diversification of the angiosperms. Pp. 314–340 in Flowering *plant origin, evolution, and phylogeny,* D. W. Taylor and L. J. Hickey, eds. New York, NY: Chapman and Hall.
Sytsma, K. J., and Hahn, W. J. 1994. Molecular systematics: 1991–1993. Progress in Botany 55:307–333.
Sytsma, K. J., and Hahn, W. J. 1996. Molecular systematics: 1994–1995. Progress in Botany 58:470–499.
Sytsma, K. J., Smith, J. F., and Berry, P. E. 1991. Biogeography and evolution of morphology, breeding systems, flavonoids, and chloroplast DNA in the four Old World species of *Fuchsia* (Onagraceae). Systematic Botany 16:257–269.
Tandre, K., Albert, V. A., Sundås, and Engström, P. 1995. Conifer homologues to genes that control floral development in angiosperms. Plant Molecular Biology 27:69–78.
Tarr, C. L., and Fleischer, R. C. 1995. Evolutionary relationships of the Hawaiian honeycreepers (Aves, Drepanidinae). Pp. 147–159 in *Hawaiian biogeography: evolution on a hot spot archipelago,* W. L. Wagner and V. A. Funk eds. Washington, D. C.: Smithsonian Institution Press.
Taylor, M. F. J., McKechnie, S. W., Pierce, N. E., and Kreitman, M. 1993. The lepidopteran mitochondrial control region: structure and evolution. Molecular Biology and Evolution 10:1259–1272.
Terborgh, J. 1971. Distribution on environmental gradients: theory and a preliminary interpretation of distributional pattern in the avifauna of the Cordillera Vilcabamba, Peru. Ecology 52:23–40.
Theißen, G., Kim, J. T., and Saedler, H. 1996. Classification and phylogeny of the MADS-box multigene family suggest defined roles of MADS-box gene subfamilies in the morphological evolution of the eukaryotes. Journal of Molecular Evolution 43:484–516.
Thomas, R. H., and Hunt, J. A. 1991. The molecular evolution of the alcohol dehydrogenase locus and the phylogeny of Hawaiian *Drosophila.* Molecular Biology and Evolution 8:687–702.
Thompson, J. N., 1994. *The coevolutionary process.* Chicago, IL: University of Chicago Press.
Thomson, J. D., and Barrett, S. C. H. 1981. Selection for outcrossing, sexual selection, and the evolution of dioecy in plants. American Naturalist 18:443–449.
Thorpe, R. S., Brown, R. P., Malhotra, A., and Wüster, W. 1991. Geographic variation and population systematics: distinguishing between ecogenetics and phylogenetics. Boll. Zool. 58:329–335.
Tilman, D. 1988. *Resource competition and community structure.* Princeton, NJ: Princeton University Press.
Titus, T. A., and Larson, A. 1995. A molecular phylogenetic perspective on the evolutionary radiation of the salamander family Salamandridae. Systematic Biology 44:125–151.
True, J., Mercer, J., and Laurie, C. 1996. A genome-wide survey of hybrid incompatibility factors by the introgression of marked segments of *Drosophila mauritiana* chromosomes into *Drosophila simulans.* Genetics 142:819–837.

Turelli, M., and Orr, H. A. 1995. The dominance theory of Haldane's Rule. Genetics 140:389–402.

Udo, E. E., and Grubb, W. B. 1990. Transfer of resistance determinants from a multi-resistant *Staphylococcus aureus* isolate. Journal of Medical Microbiology 35:72–79.

Valentine, J. W., Erwin, D. H., and Jablonski, D. 1996. Developmental evolution of metazoan body plans: the fossil evidence. Developmental Biology 173:373–381.

Valentine, J. W., Tiffney, B. H., and Sepkoski, J. J., Jr. 1991. Evolutionary dynamics of plants and animals: a comparative approach. Palaios 6:81–88.

Verheyen, E., Ruben, L., Snoeks, M., and Meyer, A. 1996. Mitochondrial phylogeography of rock-dwelling cichlid fishes reveals evolutionary influence of historical lake level fluctuations of Lake Tanganyika, Africa. Philosophical Transactions of the Royal Society of London, Series B 351:797–805.

Vermeij, G. J. 1987. *Evolution and escalation: an ecological history of life*. Princeton, NJ: Princeton University Press.

Vermeij, G. J. 1996. Adaptations of clades: resistance and response. Pp. 363–380 in *Adaptation*, M. R. Rose and G. V. Lauder, eds. New York, NY: Academic Press.

Vrba, E. S., and Gould, S. J. 1986. The hierarchical expansion of sorting and selection: sorting and selection cannot be equated. Paleobiology 12:217–228.

Wagner, W. L, and Funk, V. A. (eds.). 1995. *Hawaiian biogeography: evolution on a hot spot archipelago*. Washington, D. C.: Smithsonian Institution Press.

Wagner, W. L., Herbst, D. R., and Sohmer, S. H. (eds.). 1990. *Manual of the flowering plants of Hawai`i*. Honolulu, HI: Bishop Museum Publications.

Wagner, W. L., Weller, S. G., and Sakai, A. K. 1995. Phylogeny and biogeography in *Schiedea* and *Alsinodendron* (Caryophyllaceae). Pp. 221–258 in *Hawaiian biogeography: evolution on a hot spot archipelago*, W. L. Wagner and V. A. Funk, eds. Washington, D. C.: Smithsonian Institution Press.

Wake, D. B., and Hansen, J. 1996. Direct development in the lungless salamanders: what are the consequences for developmental biology, evolution and phylogenesis? International Journal of Developmental Biology 40:859–869.

Wake, D. B., and Larson, A. 1987. Multidimensional analysis of an evolving lineage. Science 238:42–48.

Wenzel, J. W., and Carpenter, J. M. 1994. Comparing methods: adaptive traits and tests of adaptation. Pp. 79–101 in *Phylogenetics and ecology*, P. Eggleton and R. Vane-Wright, eds. London, England: Linnean Society.

Werren, J. H., Windsor, D., and Guo, L. 1995. Distribution of *Wolbachia* among neotropical arthropods. Proceedings of the Royal Society of London, Series B 262:197–204.

Westoby, M., Leishman, M., and Lord, J. 1995a. On misinterpreting the phylogenetic correction. Journal of Ecology 83:531–534.

Westoby, M., Leishman, M., and Lord, J. 1995b. Further remarks on phylogenetic correction. Journal of Ecology 83:727–729.

White, M. J. D. 1978. *Modes of speciation*. San Francisco, CA: W. H. Freeman and Company.

Wiley, E. O. 1988. Vicariance biogeography. Annual Review of Ecology and Systematics 19:513–542.

Williams, G. C. 1975. *Sex and evolution*. Princeton, NJ: Princeton University Press.

Wilson, E. O. 1961. The nature of the taxon cycle in the Melanesian ant fauna. American Naturalist 95:169–193.

Wilson, E. O. 1980. Caste and the division of labor in leaf-cutter ants (Hymenoptera: Formicidae: Atta). II. The ergonomic optimization of leaf-cutting. Behavioural Ecology and Sociobiology 7:157–165.

Wilson, E. O. 1992. *The diversity of life*. Cambridge, MA: Belknap Press.

Wright, S. 1940. The statistical consequences of Mendelian heredity in relation to speciation. Pp. 161–183 in *The new systematics*, J. S. Huxley, ed. Oxford, England: Clarendon.

Wright, S. 1949. Adaptation and selection. Pp. 365–389 in *Genetics, paleontology, and evolution*, G. L. Jepsen, E. Mayr, and G. G. Simpson, eds. Princeton, NJ: Princeton University Press.

Xu, X. F., Janke, A., and Arnason, U. 1996. The complete mitochondrial DNA sequence of the Greater Indian Rhinoceras, *Rhinoceras unicornis*, and the phylogenetic relationship among Carnivora, Perissodactyla, and Artiodactyla. Molecular Biology and Evolution 13:1167–1173.

Yang, S. Y., and Patton, J. I. 1981. Genic variability and differentiation in the Galápagos finches. Auk 98:230–242.

Zardoya, R., and Meyer, A. 1996. Evolutionary relationships of the coelacanth, lungfishes, and tetrapods based on the 28S ribosomal RNA gene. Proceedings of the National Academy of Sciences, USA 93:5449–5454.

2 Homoplasy in molecular vs. morphological data: the likelihood of correct phylogenetic inference

Thomas J. Givnish and Kenneth J. Sytsma

There is no evidence to date that molecular data are less homoplastic than morphological data.

Sanderson and Donoghue 1989

If more reliable results are obtained when there is less homoplasy and if molecular data are less homoplastic than morphological data, then it follows that molecular data are superior. We consider the first part of this equation questionable; that is, the relationship between level of homoplasy and reliability or confidence is weak at best.

Donoghue and Sanderson 1992

Homoplasy – the independent evolutionary origin or loss of one or more traits in different organisms – can distort the inference of phylogenetic relationships, tying together similar but unrelated taxa. In recent years, a number of authors (working primarily with plants) have debated the relative extent of homoplasy in morphological and molecular data, and the possible implications for using such data to infer phylogeny (Sytsma and Gottlieb 1986b; Sibley and Ahlquist 1987; Gottlieb 1988; Mishler et al. 1988; Palmer et al. 1988; Patterson 1988; Archie 1989a,b; Sanderson and Donoghue 1989; Sytsma 1990; Sytsma et al. 1991a; Wake 1991; Donoghue and Sanderson 1992, 1994; Givnish and Sytsma 1992; Patterson et al. 1993; Donoghue 1994; Doyle et al. 1994; Hillis et al. 1994; Larson 1994; Mishler 1994; Donoghue and Sanderson 1994; Halanych et al. 1995; Hillis 1995, 1996; Miyamoto and Fitch 1995; Titus and Larson 1995; Doyle 1996). A resolution of this debate has proven elusive, partly because most studies of homoplasy have focused on pattern rather than process; partly because some statistical analyses have been incomplete or flawed; and partly because, as a result, no consensus has emerged on the mutually complementary (but different) roles that morphology and molecules can play in understanding phylogeny and adaptive evolution.

Homoplasy

Pattern vs. process

Following Hennig (1979), most discussions of homoplasy have focused on three patterns of character or character-state evolution in which it can be manifested: **convergence** (independent shifts of a' → a' and b' → a' in different lineages), **parallelism** (independent shifts of a' → a' in some but not all taxa in a lineage), and **reversal** (one or more shifts of a' → a within a lineage, after a shift of a → a' defining the origin of that lineage) (e.g., see Futuyma 1986; Meier et al. 1991; Donoghue and Sanderson 1992; Yeates 1992; McShea 1996). This terminology, unfortunately, has several shortcomings. These include the intergradation of certain phenomena (i.e., convergence and parallelism, convergence and reversal) depending on the range of taxa considered; possible confusion regarding the mechanism by which a given pattern might arise, based on widespread use of certain terms (i.e., convergence) in other contexts (i.e., response to similar selective pressures); and most important, the complete separation of the evolutionary pattern by which homoplasy might arise from the underlying process or mechanism (see Hennig 1979, p. 93). Although it would be hard to argue that morphological and molecular data differ in the extent to which they might display phyletic convergence, parallelism, or reversal, it seems quite likely that such data differ in their susceptibility to – and overall amount of – homoplasy caused by different underlying mechanisms.

Homoplasy can arise through four different mechanisms: **evolutionary convergence, recurrence, transference,** and **character misclassification**. Organisms that occupy similar ecological roles are often evolutionarily convergent, with phenotypic similarities that reflect similar selection pressures, not shared ancestry. Convergence is particularly likely to confuse analyses of lineages that have undergone a recent adaptive radiation (e.g., Darwin's finches). In such groups, the fundamental question is whether species that share an unusual character-state (e.g., some aspect of beak form) are closely related phylogenetically, ecologically, or both (Givnish 1987, 1997; Sytsma et al. 1991a; Armbruster 1992, 1996; Kocher et al. 1993; Givnish et al. 1994, 1995; Goldblatt et al. 1995; Johnson 1995). Some authors have argued that convergence may be less frequent among certain kinds of molecular characters (e.g., DNA restriction sites), to the extent that they are less directly exposed to selection than are many morphological characters (Sytsma and Gottlieb 1986a,b; Palmer et al. 1988; Givnish and Sytsma 1992).

Homoplasy can also arise in the absence of selection, through recurrent mutation (Sanderson and Donoghue 1989). If a particular character shows only a limited number of states (e.g., A, T, C, or G at a given position in a DNA sequence) then, as evolution proceeds in a lineage, analogous but non-homologous mutations can accumulate by chance in species that lack a recent common ancestor. It is perhaps intuitive that the more extensive has been evolution in a lineage of a given size (i.e., the greater the number of new mutations fixed), the greater should be the extent of recurrence homoplasy within that lineage (Felsenstein 1978). Other things being equal, recurrence should thus be greater within classes, orders, and families than it is, on average, within genera. Recurrence homoplasy should also increase with the number of taxa in a lineage: the greater the number of equally divergent species being compared, the greater the chance that two or more will share a mutation based on chance, not shared ancestry (Sanderson and Donoghue 1989). Recurrent mutation provides a potential cause for character reversal, as does evolutionary convergence. At one end of the spectrum of evolutionary parallelism – the repeated, independent rise of certain character-states among closely related organisms – the phenomenon can be seen simply in terms of recurrence: among taxa that share a similar genetic background and developmental program, it is likely that only a few (not necessarily identical) genetic changes can result in parallel phenotypic changes. At the other extreme, parallelism can grade into evolutionary convergence if it is based, at least in part, on similar selection pressures affecting some members of a clade.

What we term transference (related, in part, to horizontal gene transfer or xenology of Gray and Fitch [1983]) can generate homoplasy if a trait is propagated among taxa in a manner that differs from the usual vertical, biparental form of transmission in non-hybridizing taxa. For example, many distantly related, infectious microbes have acquired resistance to the same set of antibiotics through the spread and incorporation of plasmids carrying the genetic sequences for such resistance (Mitsuhashi and Krcmery 1984; Thomas 1989); such plasmid-mediated transfer of genetic material among bacteria and viruses is thought to be fairly widespread (see Dykhuizen and Green 1991; Maynard Smith et al. 1991; Médigue et al. 1991; Souza et al. 1992; Valdez and Piñero 1992). Similarly, genes can move from *Agrobacterium tumefaciens* into various dicotyledonous host plants via Ti-plasmids (Weising et al. 1988). Hybridization and/or introgression can also lead to tranference homoplasy, based on the differential propagation (including lineage sorting) of maternally transmitted organellar genomes or plasmids vs. the biparentally transmitted nuclear genome (Neigel and Avise 1986; Palmer et al. 1988; Neale and Sederoff 1989; Sytsma 1990; Rieseberg and Brunsfeld 1992). Finally, homoplasy can arise through the transfer of genetic sequences between nuclear and organellar genomes (Newton 1988; Lonsdale 1989; Baldauf and Palmer 1990; Palmer 1992a,b), or through gene duplication within a given genome (Odrzykoski and Gottlieb 1984; Soltis et al. 1987; Gottlieb 1988; Lonsdale 1989; Sytsma and Smith 1992). Although transference might affect both molecular and morphological traits, it seems more likely to affect molecular characters. However, well-known techniques exist for identifying some of these types of molecular transference and for analyzing them appropriately in a phylogenetic context (e.g., see Baldwin et al. 1990; Hamby and Zimmer 1992; Rieseberg and Brunsfeld 1992; Clark et al. 1994). In addition, Maddison (1995) suggests that the lack of coalescence among gene trees at or near the species level (giving rise to a conflict between "gene" vs. "species" phylogenies [Pamilo and Nei 1988]) may be as much a problem with morphological characters as it is with molecular characters.

Finally, homoplasy can arise from observer error in misclassifying characters and/or character-states. The homology of morphological character-states can often only be established by detailed comparative studies of organ development. Such studies (which are rarely conducted) are needed to exclude the possibilities that (i) similar phenotypes are merely convergent (analogous, not homologous); (ii) that putatively alternative states of the same character (e.g., red vs. yellow fruits) are actually states of different characters (presence/absence of anthocyanin, and presence/absence of carotenoids); or (iii) that putatively different characters (e.g., capitate or racemose inflorescences) are not actually alternative states of the same character (inflorescence axis contracted vs. elongate). Given the abundant positional data available to test the homology of many kinds of molecular characters (e.g., DNA restriction sites and sequences) and the ability to conduct straightforward tests to detect processes that might confuse naive analyses of such data (e.g., hybridization, DNA re-arrangements), and given the rampant problems of epistasis and pleiotropy that may affect any phenotypic trait, it seems more likely that character misclassification will affect morphological data. Considerable debate regarding the establishment of homology in morphological and molecular characters continues, however (Stevens 1984; Patterson 1988; Nelson 1989; de Pinna 1991; Williams 1993; Donoghue and Sanderson 1994; Hall 1994; Doyle 1996; Sanderson and Hufford 1996).

Previous statistical analyses

Sanderson and Donoghue (1989) attempted to test three hypotheses related to some of those just outlined, examining the extent of homoplasy in phylogenies that differ in (i) number of taxa, (ii) level of taxonomic divergence, and (iii) reliance on morphological vs. molecular variation. Their analysis focused on a study of 60 cladistic analyses based on parsimony published between 1980 and 1988, involving plant and animal groups that vary in size from 4 to 68 taxa, and that range in taxonomic rank from genus to class. They used $H = 1 - CI$ as a measure of overall homoplasy, where the consistency index $CI = m/s$ (Kluge and Farris 1969) of a phylogenetic tree is the ratio of its minimum possible length m (based on the number of variant characters) to its actual length s (using the symbol convention of Archie 1996). The consistency index of the most parsimonious (shortest) tree for a lineage is widely viewed as an inverse measure of overall homoplasy within that group, involving all components of homoplasy in our terminology; the value of CI depends, however, only on the distribution of changes in character states within a tree, not on inferences regarding the selective vs. neutral origin of homoplastic character states. H is an operational but minimal estimate of total homoplasy, based on conflicts among different characters; the total extent of actual homoplasy may be greater, due to concordance among homoplastic characters (e.g., concerted convergence – see below), or exclusion of homologous characters (e.g., nuclear DNA sequences) that would show conflict with those included in an analysis (e.g., organellar DNA sequences in a group that has undergone introgression [Rieseberg and Brunsfeld 1992]).

Sanderson and Donoghue (1989) discovered that CI declined with number of taxa: the more species in a lineage, the lower the consistency index tends to be, and hence, the greater is overall homoplasy; similar results have been obtained by Archie (1989a), Meier et al. (1991), and Donoghue and Sanderson (1992). Sanderson and Donoghue (1989) also found no relationship of homoplasy to the level of taxonomic divergence (genus vs. family or order) among the species involved. They concluded that studies based on molecular and morphological data showed indistinguishable levels of overall homoplasy when the number of taxa in each study was taken into account, and suggested that, in any event, there may be little relationship between the level of homoplasy and the likelihood of correct phylogenetic inference (see also Donoghue and Sanderson 1992). This analysis, however, was conducted before the recent explosion of plant phylogenetic analyses based on variation in chloroplast DNA (cpDNA) restriction sites and sequences. Only 4 studies based on cpDNA restriction-site data and 4 based on DNA sequences were examined, compared with 42 studies based on morphological data (including one on a group of imaginary organisms, the Dendrogramaceae [Churchill et al. 1984]).

Objectives

Our objectives here are, first, to determine whether Sanderson and Donoghue's conclusions remain valid when a more representative and statistically powerful group of molecular studies are examined – or whether, as we predict, homoplasy is greater for morphological data than for molecular data, and greater for variation within families and orders than within genera. Second, using simulations based on molecular data for three plant phylogenies, we test Sanderson and Donoghue's proposition (and a preliminary finding by Meier et al. 1991) that recurrent mutation can account for the observed decline in CI – and increase in homoplasy – with number of taxa. Third, we examine how the observed levels of homoplasy and numbers of variable characters affect the likelihood of correct phylogenetic inference, using equations derived from a simulation involving different rates of mutation in clades with a known phylogeny (Givnish and Sytsma 1997). Finally, we discuss the implications these results have for the relative degree of confidence to be placed in molecular vs. morphological phylogenies, and for the utility of such phylogenies for interpreting patterns of morphological evolution.

Methods

Homoplasy in molecular vs. morphological data

We compiled and analyzed the results of 40 cladistic studies based on cpDNA restriction-site variation in plants and 52 based on DNA sequence variation (including all those on plants published or known to us through early 1993 for restriction sites, and through 1995 for sequences), in addition to the 41 based on morphological variation in plant, animal, and fungal groups examined by Sanderson and Donoghue (1989), and one additional morphological analysis (Rodman 1991) that involves the same familial taxa as one of our sequence studies (Table 2.1). All studies were based on Wagner or Fitch parsimony. Following Sanderson and Donoghue (1989), modified consistency indices (designated CI' here) were calculated for each lineage after autapomorphies were excluded. This was done to ensure that the results from the molecular studies were comparable to those from morphological studies, which often do not report autapomorphies.

Although several indices have been proposed as alternative or superior measures of homoplasy (i.e., the **adjusted consistency index** [Klassen et al. 1991], **retention index** [Farris 1989a] or **homoplasy excess ratio** [Archie 1989a], **homoplasy slope ratio** [Meier et al. 1991], **D measure** [Brooks et al. 1986], and **F-ratio** [Farris 1972]; see recent review by Archie 1996), CI and CI' are the two most widely reported and consequently best suited for broad-scale comparative studies. Critiques of the consistency index as a measure of homoplasy (Archie 1989a,b; Farris 1989a,b; Faith and Cranston 1991) have themselves been countered by Goloboff (1991a,b), Klassen et al. (1991),

Table 2.1. Summary of data used for statistical comparisons of standardized consistency indices excluding autapomorphies (CI'), number of taxa, and equivalent number of binary-state characters in phylogenetic studies based on morphology, cpDNA restriction sites, or DNA sequences (see text).

Study*	Taxonomic level†	Number of taxa	CI'	Number of characters§	No. of characters / No. of taxa
Morphological variation:					
Jansen 1981	Genus	6	0.79	11	1.83
Funk 1982	"	25	0.63	39	1.56
Cane 1983	"	27	0.40	89	3.30
Grismer 1983	"	6	0.84	21	3.50
Ladiges and Humphries 1983	"	14	0.72	28	2.00
Collette and Russo 1985	"	20	0.68	73	3.65
Anderberg 1986	"	9	0.79	19	2.11
Campbell 1986	"	20	0.44	24	1.20
Cracraft 1986	"	9	0.81	30	3.33
Eckenwalder and Barrett 1986	"	37	0.39	62	1.68
Fink and Fink 1986	"	12	0.70	32	2.67
Guyer and Savage 1986	"	24	0.60	30	1.25
Herman 1986	"	35	0.43	65	1.86
Livezey 1986	"	4	1.00	13	3.25
Vilgalys 1986	"	11	0.53	20	1.82
Jamieson et al. 1987	"	11	0.64	42	3.82
Ladiges et al. 1987	"	29	0.43	62	2.14
Schuh and Polhemus 1980	Family	10	0.80	37	3.70
Baum 1983	"	29	0.31	126	4.34
Baum and Saville 1985	"	17	0.53	24	1.41
Cutler and Gibbs 1985	"	17	0.54	13	0.76
Fink 1985	"	27	0.49	245	9.07
Crother et al. 1986	"	5	0.72	23	4.60
Bremer 1987	"	29	0.56	47	1.62
Carpenter 1987	"	7	0.76	26	3.71
Kitching 1987	"	57	0.45	283	4.96
Kellogg and Campbell 1987	"	65	0.37	39	0.60
Wighton and Wilson 1987	"	16	0.50	15	0.94
P. Stevens unpubl.	"	68	0.32	94	1.38
Dahlgren and Rasmussen 1983	Order	15	0.60	44	2.93
Nelson 1984	"	22	0.63	70	3.18
Rodman et al. 1984	"	20	0.35	69	3.45
Brooks et al. 1985	"	9	0.93	28	3.11
Cracraft 1985	"	13	0.66	59	4.54
Crane 1985	"	20	0.62	31	1.55
Dahlgren and Bremer 1985	"	47	0.26	59	1.26
Lipscomb 1985	"	36	0.34	67	1.86
Mishler and Churchill 1985	"	11	0.82	28	2.55
Doyle and Donoghue 1986	"	20	0.50	62	3.10
Gauthier 1986	"	8	0.89	84	10.50
Gabrielson and Garbary 1987	"	15	0.55	31	2.07
Rodman 1991	"	26	0.34	107	4.12
Mean		**21.6**	**0.59**	**56.5**	**2.91**
Standard deviation		**15.3**	**0.19**	**54.4**	**1.94**
cpDNA restriction site variation:					
Palmer and Zamir 1982	Genus	10	0.93	14	1.40
Sytsma and Gottlieb 1986a	"	9	0.90	55	6.11
Jansen and Palmer 1988	"	12	0.65	123	10.25
Jansen and Palmer 1988	"	16	0.60	55	3.44
Sytsma and Smith 1988	"	15	0.77	10	0.67
Chase and Palmer 1989	"	13	0.84	66	5.08
Baldwin et al. 1990	"	24	0.68	33	1.38
Doebley 1990	"	17	0.94	32	1.88
Doyle et al. 1990	"	13	0.93	42	3.23
Duvall and Doebley 1990	"	23	0.86	37	1.61
Soreng 1990	"	46	0.84	73	1.59
Soreng et al. 1990	"	22	0.61	120	5.45
Suh and Simpson 1990	"	24	0.88	93	3.88
Sytsma et al. 1990	"	15	0.73	66	4.40
Wallace and Jansen 1990	"	25	1.00	78	3.12
Crawford et al. 1991	"	26	0.94	48	1.85
Philbrick and Jansen 1991	"	14	0.74	34	2.43
Rieseberg et al. 1991	"	24	0.93	26	1.08
Spooner et al. 1991	"	22	0.82	67	3.05
Sytsma et al. 1991a	"	9	0.97	192	21.33
Wendel et al. 1991	"	5	1.00	32	6.40
Chase and Palmer 1992 pers. comm.	"	52	0.75		
Kadereit and Sytsma 1992	"	23	0.68	164	7.13
Kellogg 1992	"	6	0.81	21	3.50
Olmstead et al. 1990	"	25	0.80	108	4.32
Givnish et al. 1994	"	32	0.95	78	2.44
Smith and Sytsma 1994	"	37	0.85	159	4.30
Wiegrefe et al. 1994	"	28	0.86	122	4.36
Givnish et al. in prep.	"	30	0.80	258	8.60
Jansen et al. 1990	Family	57	0.46	328	5.75

Study*	Taxonomic level†	Number of taxa	CI'	Number of characters§	No. of characters / No. of taxa
Ranker et al. 1990	Family	11	0.83	15	1.36
Soltis et al. 1990	"	13	0.94	31	2.38
Bremer and Jansen 1991	"	33	0.46	161	4.88
Lavin and Doyle 1991	"	20	0.85	52	2.60
Sytsma et al. 1991b	"	47	0.52	87	1.85
Olmstead and Palmer 1992	"	43	0.36	447	10.40
Soltis et al. 1993	"	44	0.76	311	7.07
Givnish et al. 1994	"	13	0.95	59	4.54
Givnish et al. in prep	"	46	0.64	105	2.28
Downie et al. 1992	Order	99	0.36	55	0.56
Mean		26.1	0.78	98.9	4.31
Standard deviation		17.9	0.17	95.4	3.73
DNA sequence variation:					
Baum et al. 1994	Genus	24	0.80	154	6.42
Kim and Jansen 1994	"	18	0.62		
Liden et al. 1995	"	15	0.61	129	8.60
Sang et al. 1994	"	16	0.92	34	2.13
Bain and Jansen 1995	"	31	0.90	166	5.35
Oxelman and Liden 1995a	"	64	0.36	396	6.18
Sang et al. 1995	"	13	0.92	37	2.85
Udovicic et al. 1995	"	25	0.69	212	8.48
Yuan et al. 1996	"	28	0.58		
Hamby and Zimmer 1988	Family	10	0.73	117	11.70
Kim et al. 1992	"	25	0.47	223	8.92
Xiang et al. 1993	"	46	0.47		
Conti et al. 1993	"	12	0.63	105	8.75
Soltis et al. 1993	"	42	0.40		
Brunsfeld et al. 1994	"	33	0.53	148	4.48
Johnson and Soltis 1994	"	31	0.57	127	4.10
Manen et al. 1994	"	41	0.75		
Morgan et al. 1994	"	43	0.46	279	6.49
Nadot et al. 1994	"	30	0.70	90	3.00
Shinwari et al. 1994	"	14	0.80		
Steele and Vilgalys 1994	"	24	0.70	258	10.80
Barker et al. 1995	"	34	0.43	197	5.78
Bremer et al. 1995	"	56	0.38	395	7.05
Campbell et al. 1995	"	23	0.55		
Clark et al. 1995	"	47	0.44		
Hedren et al. 1995	"	40	0.37		
Hoot et al. 1995	"	10	0.73		
Johnson and Soltis 1995	"	45	0.58		
Kim and Jansen 1995	"	94	0.38	1060	11.30
Natali et al. 1995	"	50	0.64		
Olmstead and Reeves 1995	"	32	0.46		
Scotland et al. 1995	"	24	0.53		
Soltis et al. 1995	"	46	0.56		
Bremer et al. 1987	Order	26	0.53	84	3.23
Donoghue et al. 1992	"	12	0.52	104	8.67
Duvall et al. 1993	"	122	0.27		
Price and Palmer 1993	"	48	0.38		
Qui et al. 1993	"	83	0.28		
Hamby and Zimmer 1992	"	60	0.39	729	12.15
Kron and Chase 1993	"	41	0.52		
Olmstead et al. 1993	"	37	0.31		
Olmstead et al. 1993	"	42	0.39		
Rodman et al. 1993	"	25	0.45	559	22.36
Bremer et al. 1994	"	63	0.30	514	8.16
Cosner et al. 1994	"	117	0.28		
Hasebe et al. 1994	"	64	0.25		
Wolf et al. 1994	"	45	0.33		
Bharathan & Zimmer 1995	"	37	0.45		
Hempel et al. 1995	"	57	0.32		
Hoot and Crane 1995	"	18	0.54		
Nickrent and Soltis 1995	"	62	0.28		
Oxelman and Liden 1995b	"	26	0.40	368	14.20
Conti et al. 1996	"	80	0.28	698	8.73
Mean		39.6	0.51	298.3	8.58
Standard deviation		23.5	0.14	281.6	4.90

* Citations for morphological studies are given by Sanderson and Donoghue (1990) and are not included in our bibliography unless cited elsewhere in this paper.
† Indicates studies based on variation within taxonomic groupings at the level of genus, family, or order and above.
§ Equivalent number of binary-state characters (Sokal and Shao 1985), based on a comparison of CI' and tree length. Values for morphological studies are those reported by Sanderson and Donoghue (1989); all others are calculated directly from data given in the original studies. Several recent sequence studies, however, do not give enough information to permit calculation of the equivalent number of binary-state characters.

and Meier et al. (1991). The last two papers state that no measure of homoplasy is, as desired, sensitive to internal data structure but insensitive to the number of taxa included. Clearly, however, any effect of number of taxa on homoplasy across lineages can be largely eliminated by treating it as a covariate.

Based on Sanderson and Donoghue's (1989) finding that CI' declines with N, the number of taxa in a lineage and outgroup, we calculated linear regressions between ln CI' and ln N across (i) all morphological studies, (ii) all cpDNA restriction-site studies, and (iii) all DNA sequence studies. Sanderson and Donoghue (1989) regressed ln CI' on N, but we found a better fit (measured by r^2) to a linear relationship between ln CI' and ln N in many cases; log-log analyses also reduced sensitivity to outliers, and increased the evenness of the distribution of studies along the N-axis. Differences between regression lines for morphological, restriction-site, and sequence data were evaluated using analyses of covariance (ANCOVA) and Fisher exact tests (see Snedecor and Cochran 1989; de Queiroz and Wimberger 1993), to determine whether molecular data show less homoplasy when the number of taxa in a phylogeny is taken into account. Where possible, we also compared regression lines for studies restricted to a given taxonomic rank (i.e., within-genus, within-family, or within-order and above) in order to eliminate any bias due to differences between classes of studies in the taxonomic breadth they embrace (Table 2.2). There are significant excesses of cpDNA restriction-site studies at the generic level, and of DNA sequence studies at the family level and the ordinal level and above ($\chi^2 = 32.6$, P < 0.0001 for 4 d.f.).

To determine whether – as is widely assumed – molecular studies entail effectively more characters than morphological studies, we compiled data on B, the equivalent number of informative binary-state characters (Sokal and Shao 1985) for as many of our tabulated studies as possible (Table 2.1). Several DNA sequence studies did not provide enough information (i.e., CI' and the length of the shortest tree excluding autapomorphies, **or** CI, CI', and the length of the shortest tree including autapomorphies) to permit calculation of B. Given that Sanderson and Donoghue (1989) found a significant increase in B with N across studies, we calculated separate regressions between ln B and ln N for morphological, restriction-site sequence studies, and then evaluated differences between these regression lines using ANCOVA.

Table 2.2. Number of cladistic studies included in survey (Table 2.1), tabulated by kind of data and taxonomic level.

Kind of data	Taxonomic level		
	Genus	Family	Order or above
Morphological	17	12	13
cpDNA restriction site	29	10	1
DNA sequence	9	24	20

$\chi^2 = 32.6$, P < 0.0001 for 4 d.f.

Homoplasy in relation to taxonomic rank

Studies based on cpDNA restriction sites and DNA sequences are ones for which comparable characters and character-states were scored for a substantial number (> 10) of lineages of different taxonomic rank. We calculated linear regressions of ln CI' on ln N at the generic and familial levels for restriction-site studies, and at the generic, familial, and ordinal levels for sequence studies. We then evaluated the difference between the two regressions for each kind of data using ANCOVA, to determine whether there is more homoplasy at higher taxonomic levels when the number of taxa is taken into account.

Homoplasy in relation to number of taxa

The observed decline in CI' with N might be (i) a consequence of the increased probability that two or more taxa in a larger pool of taxa will share a mutation based on chance alone (Sanderson and Donoghue 1989), or (ii) an artifact of larger pools of taxa (i.e., larger genera or families) representing the outcome of a greater overall amount of evolutionary divergence – creating greater possibilities for recurrence – than smaller pools. To test between these possibilities, we followed a suggestion of Sanderson and Donoghue (1989) and randomly sampled subsets of three clades of angiosperms for which extensive data on cpDNA restriction sites or sequences were available, and then calculated CI' as a function of subset size. Our methodology parallels a similar study conducted by Meier et al. (1991) on a single set of morphological data compiled for *Collybia* fungi by Vilgalys (1986).

Two clades chosen for study – *Brocchinia* (Bromeliaceae) (Chapter 8) and *Clarkia* (Onagraceae) (Sytsma et al. 1990) – have nearly equal levels of homoplasy (H = 0.10 vs. 0.16, including autapomorphies; H' = 1 – CI' = 0.22 vs. 0.27, excluding autapomorphies) and numbers of ingroup taxa (17 vs. 14), but have phylogenies with different topologies and distributions of cpDNA restriction-site mutations (Figure 2.1, top and middle). *Brocchinia* has four subclades that are well-differentiated relative to the amount of variation within each subclade, whereas *Clarkia* tends to show more variation across species within each of its subclades than it does between the subclades. The third clade chosen for study – representatives of the monophyletic group of the 14 families bearing glucosinolates, or mustard oils (Rodman et al. 1993) – was chosen because it had a much higher level of homoplasy (H = 0.30, H' = 0.44) than the other two clades and a different topology and distribution of mutations (Figure 2.1, bottom), and because it was based on sequence (*rbc*L) variation, not site variation.

From each of these clades, we randomly selected x ingroup species without replacement, repeating the process 10 times for each integer $x \in \{3, 4, 5, \ldots, n-1\}$, where n is the total number of species within the clade. For each of these subsets we conducted a cladistic analysis of relationships using PAUP 3.1 (Swofford 1993), applying parsimony to the molecular variation in the species in that subset and in the single outgroup species designated for the clade in question. For the most parsimonious tree(s) corresponding to each subset, PAUP 3.1 calculated CI and the retention index RI as measures of the degree of internal consistency within the data set. Means and standard deviations of these indices were calculated for the ten independently cho-

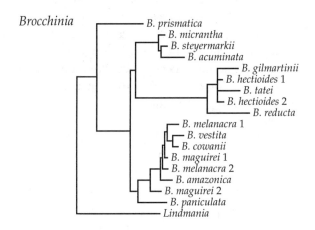

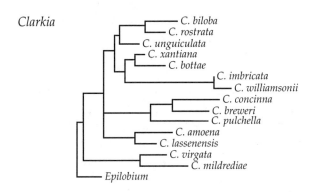

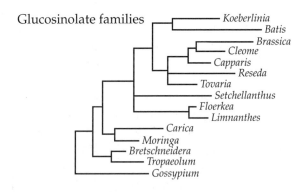

Figure 2.1 Phylograms of plant lineages used in the resampling analyses (see text). Within each diagram, the lengths of the horizontal branches are proportional to the amount of genetic change inferred along those branches using parsimony.

sen subsets of a given size within a clade, and related to the number of taxa characteristic of that group of subsets.

Homoplasy and the likelihood of correct phylogenetic inference

Givnish and Sytsma (1997) derive equations relating the likelihood of correct phylogenetic inference to CI (or CI'), the number of variable (or informative) characters, and the number of taxa, based on simulations involving the recurrent mutation of binary characters in the context of a dichotomous family tree with equal branch lengths. These simulations produce estimates of the **maximum** possible likelihood of correct phylogenetic inference for trees of a given size; shorter branches in less regular trees should be more difficult to resolve with a given number of characters (see Hillis et al. 1994) and level of homoplasy.

Probabilities of correct phylogenetic inference were calculated in two ways, based on a comparison of the topologies of the shortest tree(s) and the correct tree (Givnish and Sytsma 1997). Semi-strict correct inference occurred if the correct tree was among the shortest trees. Strict correct inference occurred only if the correct tree was the shortest tree; if the correct tree was among k shortest, topologically distinct trees, then strict correct inference had a probability of $1/k$. The means of strict and semi-strict correct inference were calculated for 50 replicate simulations involving each combination of mutation rate, number of characters, and number of taxa. Givnish and Sytsma (1997) also provided "pull-back" equations to eliminate the effect of number of taxa on CI and CI', providing standardized indices of consistency.

In this paper, we used the values of CI', number of informative characters, and number of taxa to calculate the maximum probabilities of strict and semi-strict correct inference for each of the morphological, restriction-site, and sequence studies tabulated here, using the equations derived by Givnish and Sytsma (1997). Inference probabilities were related to the number of taxa for each of the three groups of studies via linear regression, and the resulting relationships compared using ANCOVA.

Results

Homoplasy in molecular vs. morphological data

We found highly significant declines in ln CI' with ln N across phylogenies based on **morphology** ($r = -0.84$, $P < 0.0001$ for 40 d.f.), **cpDNA restriction sites** ($r = -0.57$, $P < 0.0001$ for 38 d.f.), and **DNA sequences** ($r = -0.76$, $P < 0.0001$ for 51 d.f.) (see Figure 2.2). These findings confirm the central conclusion of Sanderson and Donoghue (1989) that homoplasy tends to increase with the number of taxa in a phylogeny.

However, contrary to Sanderson and Donoghue (1989), we found that studies based on cpDNA restriction-site variation show significantly higher values of CI' – and hence, significantly lower levels of homoplasy – at a given number of taxa (Figure 2.2A). An analysis of covariance indicated that the slope of the decline in ln CI' with ln N is significantly greater ($P < 0.01$ for 1, 78 d.f.) for morphology-based phylogenies ($y = -0.414x + 0.593$) than for those based on cpDNA restriction sites ($y = -0.232x + 0.432$). In other words, CI' decreases – and homoplasy increases – more rapidly with the

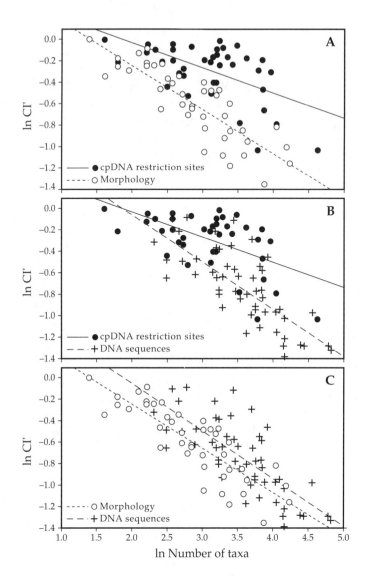

Figure 2.2 Relationships of the modified consistency index CI' to the number of taxa N for studies based on morphology, cpDNA restriction sites, or DNA sequences. Lines are least mean squares regressions for the logarithmically transformed variables. (**A**) Restriction-site studies have significantly higher levels of CI' than morphological studies with the same number of taxa, based on ANCOVA for intercept ($P < 0.01$) and Fisher exact tests ($P \ll 0.001$). (**B**) Restriction-site studies appear to show significantly higher levels of CI' than DNA sequence studies with the same number of taxa, based on a significant difference in the intercept (ANCOVA, $P < 0.001$ for 1, 91 d.f.) but not the slope of the regressions; this difference, however, appears to be an artifact of the taxonomic level at which restriction-site and sequence studies are conducted (see text). (**C**) Sequence studies have significantly higher levels of CI' than morphological studies involving the same number of taxa, based on a significant difference in intercept ($P < 0.01$, ANCOVA with 1,92 d.f.) but not slope.

number of taxa for morphological data than for cpDNA restriction-site data. The disparity in CI' between trees based on cpDNA restriction sites vs. morphology is least among trees of minimal size (ca. 3 spp.) and rises sharply with increasing tree size; it is particularly marked for lineages involving 10 or more taxa (Figure 2.2A).

The values of ln CI' for morphological studies lie above the regression line for cpDNA restriction-site studies in only 1 of 42 instances, compared with 22 of 40 restriction-site studies ($P \ll 0.001$, Fisher exact test). Conversely, the values of ln CI' based on cpDNA restriction-site studies lie below the morphological regression line in only 2 of 40 cases (Figure 2.2A), compared with 21 of 42 cases for morphological studies ($P \ll 0.001$, Fisher exact test). In other words, CI' is significantly higher – and homoplasy significantly lower – at a given number of taxa for studies based on cpDNA restriction sites than for those based on morphology (Figure 2.2A).

This finding is not an artifact of the difference in taxonomic rank shown by the two classes of studies (Table 2.2), with a significant excess of restriction-site studies at the generic level, and a significant excess of morphological studies at the familial level and above ($\chi^2 = 13.6$, $P < 0.002$ for 2 d.f.). If we restrict comparisons to the generic or familial level (there are not enough cases to allow a comparison at the ordinal level and above), CI' is still significantly higher at a given number of taxa for restriction-site studies (Figure 2.3), confirming our previous finding.

Studies based on DNA sequence variation display significantly lower values of ln CI' – and significantly higher levels of homoplasy – at a given number of taxa than do studies based on cpDNA restriction-site variation. The regressions differ significantly ($P < 0.001$ for 1, 91 d.f.) in intercept and slope (Figure 2.2B). However, this difference appears to be an artifact of the difference in taxonomic rank between the two groups (Table 2.2), with restriction-site studies focusing on genera and sequence studies focusing on families and higher taxonomic levels ($\chi^2 = 32.3$, $P < 0.0001$ for 2 d.f.). A comparison restricted to the generic level shows no significant difference in slope or intercept between sequence and restriction-site studies when the one outlier among the sequence studies is excluded (ANCOVA, $P > 0.05$). A comparison restricted to the familial level also shows no significant difference in slope or intercept.

Additional evidence that there are no intrinsic differences in the level of homoplasy exhibited by restriction-site vs. sequence data is provided by four groups of studies that present both kinds of data for the same taxa (Table 2.3). For 8 genera of Onagraceae, *rbc*L sequences and the sites recognized by 12 restriction enzymes throughout the chloroplast genome are known, the latter from a detailed analysis of double digests (Sytsma et al. 1991b; Conti et al. 1993). Both data sets generate the same shortest tree independently, and show nearly identical levels of homoplasy, with restriction sites showing slightly less homoplasy than *rbc*L sequences (Table 2.3). A similar pair of data sets for 17 genera of the Asteraceae (Jansen et al. 1992; Kim et al. 1992) showed a similar result, but with *rbc*L sequences showing slightly less homoplasy than cpDNA restriction sites. Olmstead et al. (1992), Olmstead and Sweere (1995), and R. Olmstead (pers. comm.) found slightly less homoplasy in sequence data for *rbc*L and *ndh*F than for cpDNA restriction-site data (Table 2.3). The *ndh*F sequence data generated 3 equally shortest trees; the restriction-site data, 44 shortest trees; and

Table 2.3. Comparison of CI, CI', and RI for trees based on cpDNA restriction sites vs. sequences, and involving the identical set of taxa. Unless otherwise noted, restriction-site data were scored for the entire chloroplast genome.

Group	No. of genera	Gene	No. of shortest trees (sites/sequence)	CI sites	CI sequence	CI' sites	CI' sequence	RI sites	RI sequence
Asteraceae[1]	17	rbcL	?	0.712	0.672	0.512	0.488	0.629	0.478
Onagraceae[2]	8	rbcL	1/1	0.846	0.814	0.649	0.625	0.635	0.618
Saxifragaceae[3]	19	rbcL	4/7	0.840	0.744	0.745	0.535	0.872	0.678
	19	matK	4/5	0.840	0.858	0.745	0.682	0.872	0.777
Solanaceae[4]	18	rbcL	44/67	0.688	0.731	0.427	0.444	0.381	0.444
	18	ndhF	44/3	0.688	0.853	0.427	0.591	0.381	0.497

[1] Data for taxa shared in studies by Kim et al. (1992), Jansen et al. (1992), and K.-J. Kim (pers. comm.).
[2] Data for taxa shared in studies by Sytsma et al. (1991b), Conti et al. (1993), and K. J. Sytsma (pers. comm.).
[3] Data for taxa shared in studies by Soltis et al. (1993), Johnson and Soltis (1994), and D. Soltis and L. Johnson (pers. comm.).
[4] Olmstead and Palmer (1992), Olmstead and Sweere (1995), and R. Olmstead (pers. comm.).

the *rbc*L sequence data, 67 shortest trees. Finally, Soltis et al. (1993) and Johnson and Soltis (1994) found roughly comparable levels of homoplasy in restriction-site data and in sequences for *rbc*L and *mat*K, and roughly comparable levels of taxonomic resolution as judged by the number of equally shortest trees. Across these studies, the average difference in CI between restriction-site and sequence data is -0.010 ± 0.090 ($P > 0.8$, 2-tailed t-test, 3 d.f.); the average corresponding difference in CI' is -0.023 ± 0.121 ($P > 0.6$, 2-tailed t-test, 3 d.f.).

Sequence studies yield a regression line that parallels that based on morphological studies but has a significantly greater intercept ($P < 0.01$, ANCOVA with 1, 92 d.f.), indicating that sequence studies entail significantly less homoplasy at a given number of taxa (Figure 2.2C). Comparisons restricted to the generic ($P < 0.005$ for 1, 23 d.f.) and familial levels ($P < 0.02$ for 1, 33 d.f.) show the same significant difference in intercept but not in slope, with sequence studies showing less homoplasy. The comparison restricted to the ordinal level and above showed a significant difference in slope ($P < 0.01$ for 1, 29 d.f.) but not in intercept. The ordinal regression line for sequence studies shows less homoplasy than that for morphological studies when the number of taxa exceeds 20; given that all but 2 of 22 sequence studies involve more than 20 taxa, this finding confirms the general tendency for DNA sequence studies to display less homoplasy than morphological studies.

The equivalent number of binary-state characters increases with the number of taxa in studies based on **morphology** ($\ln B = 0.69 \ln N + 1.78$, $P < 0.0001$ for 40 d.f.), **cpDNA restriction sites** ($\ln B = 0.74 \ln N + 1.96$, $P < 0.0005$ for 38 d.f.), and **DNA sequences** ($\ln B = 1.12 \ln N + 1.54$ for 23 d.f.). The relationships of $\ln B$ to $\ln N$ show no significant differences in slope (see Figure 2.4). Based on differences in their intercepts, however, sequence studies involve significantly more characters at a given number of taxa than restriction-site studies (ANCOVA, $P < 0.001$ for 1, 61 d.f.), and restriction-site studies in turn involve significantly more characters than morphological studies

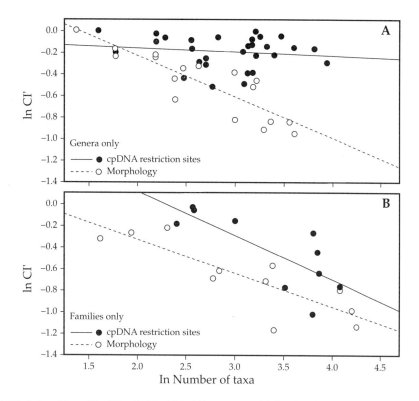

Figure 2.3 Relationships of ln CI' to ln N within (**A**) genera or (**B**) families for studies based on variation in cpDNA restriction sites vs. morphology. Lines are least mean squares regressions. For genera, the restriction-site regression has a significantly lower slope (ANCOVA, $P \ll 0.001$ for 1, 42 d.f.) and greater intercept ($P \ll 0.001$ for 1, 43 d.f.) than the morphological regression. For families, the restriction-site regression has a significantly ($P \ll 0.001$ for 1, 28 d.f.) greater intercept than the morphological regression, but does not differ significantly in slope from the morphological regression.

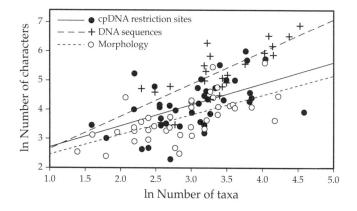

Figure 2.4 Equivalent number of binary-state characters (B) in relation to number of taxa in studies based on morphology (M), cpDNA restriction sites (R), and DNA sequences (S). Lines are least mean squares regressions between logarithmically transformed variables. Regressions do not differ significantly in slope, but the elevations of the sequence and restriction-site regressions are significantly greater than that of the morphological regression (ANCOVA, $P < 0.001$ for 1, 61 d.f., and $P < 0.025$ for 1, 78 d.f., respectively).

(ANCOVA, P < 0.025 for 1, 78 d.f.). Sequence studies involve significantly more characters than restriction-site studies as well (ANCOVA, P < 0.0001 for 1, 64 d.f.). At the mean number of taxa (N = 30.4) in our compilation, the expected number of informative characters is 62.6 for morphological studies, 88.8 for restriction-site studies, and 213.6 for sequence studies, representing advantages of 42.0% for restriction-site studies and 241.5% for sequence studies.

Homoplasy in relation to taxonomic rank

Among studies based on cpDNA restriction-site variation, the rate at which ln CI' declines – and homoplasy increases – with ln N is significantly lower (ANCOVA, P << 0.001 for 1, 35 d.f.) for genera than for families at a given number of taxa, resulting in lower values of CI' within families in studies involving large numbers of taxa (Figure 2.5). For DNA sequences, the intercept (but not the slope) of the regression for families is significantly greater (ANCOVA, P < 0.0001 for 1, 41 d.f.) than that for orders (Figure 2.6). The intercept for genera is greater than that for families (ANCOVA, P < 0.075 for 1, 30 d.f.), and significantly greater than that for orders (ANCOVA, P < 0.0001 for 1, 26 d.f.).

The tendency for CI' to be lower in molecular studies involving higher taxonomic categories is contrary to a conclusion by Sanderson and Donoghue (1989), who found no effect of taxonomic rank on the degree of homoplasy across all studies surveyed. It should be noted, however, that if morphological data are considered alone, there are indeed no significant differences in the intercepts of the regressions of ln CI' on ln N for studies conducted at the level of genus, family, or order (ANCOVA, P > 0.25 in all cases). There are two significant differences in slope but they show no consistent relationship to taxonomic rank, with the slope for familial studies being steeper than that for ordinal studies, and that for ordinal studies being steeper than that for generic studies.

Our findings for molecular studies are probably the result of greater recurrence in taxonomically broader (i.e., older or more rapidly evolving) lineages, in which

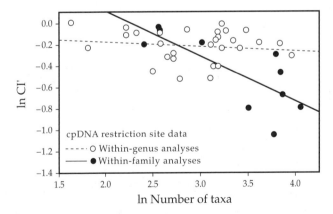

Figure 2.5 CI' in relation to the number of taxa in studies based on cpDNA restriction-site variation at the generic and familial levels.

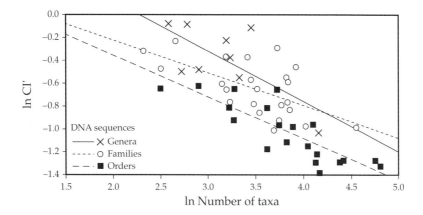

Figure 2.6. CI' in relation to the number of taxa in DNA sequence studies at the generic, familial, and ordinal levels.

greater numbers of mutations have become fixed along each line since the time of the founding ancestor (see section below on **Relation to likelihood of correct phylogenetic inference**). Phylogenies based on variation in cpDNA restriction sites or (in plants, at least currently) DNA sequences are unusual in that they are nearly all based on a fairly similar and comparable set of characters. In the former case, these involve 6-base restriction sites throughout the entire chloroplast genome recognized by 10 to 30 widely used endonucleases; in the latter, nucleotides of *rbc*L in most instances. It is true that restriction-site studies within certain large groups (e.g., Asteridae, monocotyledons) use slightly different cpDNA probes in order to handle efficiently the problems caused by inversions, deletions, and insertions, but such large-scale cpDNA rearrangements are relatively rare, and rigorous techniques exist for identifying and incorporating them in non-circular fashion in phylogenetic analyses (see Bowman et al. 1988; Palmer et al. 1988; Milligan et al. 1989; Downie and Palmer 1992; Knox et al. 1993; Givnish et al. 1994, 1995; Olmstead and Palmer 1994; Hahn et al. 1995).

In contrast, phylogenies based on morphology (and DNA sequences in some groups) entail different sets of characters that have been chosen by the investigator depending on the taxonomic breadth of the group being analyzed, to exclude characters that are judged on an *a priori* basis to evolve too fast and show too much homoplasy. Thus, shifts in homoplasy with taxonomic rank in morphological studies (and perhaps in DNA sequence studies) reflect changes in both recurrence **and** the inherent rate of evolution in the characters used for studies at different taxonomic levels.

Homoplasy in relation to number of taxa

Our studies on random subsets of the *Brocchinia, Clarkia*, and glucosinolate clades show that CI and CI' decrease more or less linearly with subset size (Figure 2.7), while RI increases or remains roughly constant (Figure 2.8). Of these indices, CI consistently shows the clearest trend and the least variation among replicates at a given number of taxa; RI often shows the most noise.

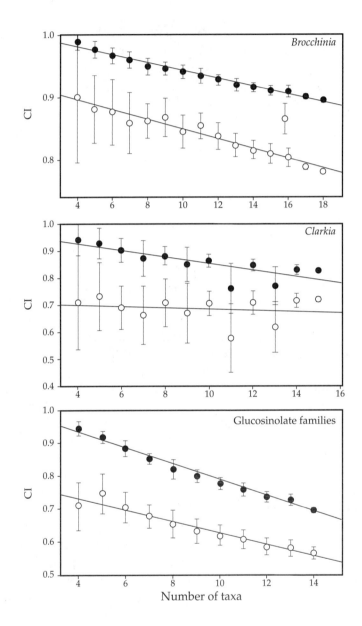

Figure 2.7. Relationships of the consistency indices CI (●, including autapomorphies) and CI' (○, excluding autapomorphies) to the number of taxa drawn randomly from *Brocchinia, Clarkia,* and the glucosinolate clade. Each replicate drawing includes n-1 ingroup taxa and the single outgroup taxon. Points represent means of ten replicates; bars indicate standard deviations. Lines represent the least mean squares regressions.

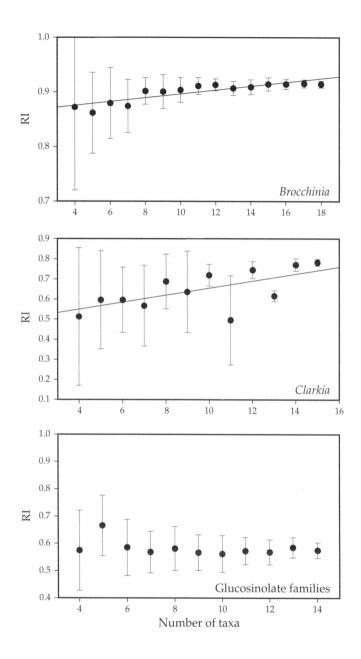

Figure 2.8. Retention index RI in relation to the number of taxa drawn randomly from *Brocchinia*, *Clarkia*, and the glucosinolate clade. Lines represent least mean squares regressions, where significant.

CI is always greater than CI' for subsets of a given size within a clade (Figure 2.7), as it must if any autapomorphies exist (Brooks et al. 1986). While the slope of CI' vs. N varies considerably across clades, the slope of CI is such that the predicted value of CI at N = 3 (including the outgroup) is roughly 1. This suggests a "pull-back" rule that would allow a comparison of consistency indices for two clades with different numbers of taxa, via a linear extrapolation of CI for subsets of the larger clade the same size as the small clade. If the entire data set were available, of course, our random subsampling approach could be applied directly to both clades, and the resulting functions CI(N) and CI'(N) plotted and compared statistically.

For all three clades, there is a strong relationship of homoplasy to the number of taxa (N) in randomly chosen subsets (Figure 2.9). We found the best approximation to a linear model with a new measure of homoplasy H*, defined as

$$H^* = (s - m) / m, \qquad (1)$$

where s = actual tree length, and m = minimum tree length. In a sense, s − m is a measure of the "noise" in a character set, while m is a measure of the "signal" conveyed by character state variation among taxa. Hence, H* is effectively a "noise/signal" ratio. Because CI = m/s (Kluge and Farris 1969),

$$H^* = 1/CI - 1 = H/(1 - H); \qquad (2)$$

thus, for CI close to 1,

$$H^* \approx 1 - CI = H. \qquad (3)$$

The value of H* varies from 0 to ∞ as CI varies from 1 to 0, compared with a range of 0 to 1 for H. The proposed index seems intuitively appealing, in that CI = 0.5 would imply a "noise/signal" ratio H* = 1, as compared with H = 0.5; a non-linear relationship between a homoplasy index and the consistency index seems appropriate, given that nearly unlimited amounts of recurrence and/or convergence must occur for CI to approach 0. It is important to note that, because H* and CI are inversely related (eq. 2), CI (or more precisely, CI/[1 − CI]) can be considered a "signal/noise" ratio.

In the glucosinolate clade, homoplasy increases linearly with number of taxa (H* = 0.035N − 0.079; r^2 = 0.994, P < 0.0001 for 9 d.f.). In *Brocchinia*, there is a slightly better fit to a quadratic model, with slightly decreasing increments to H with increasing numbers of taxa (H* = −0.588N^2 + 3.903N − 8.611; r^2 = 0.994, P < 0.0001 for each term for 12 d.f.), although a linear model is also highly significant (H* = 0.007N − 0.009; r^2 = 0.960, P < 0.0001 for 13 d.f.). In *Clarkia*, homoplasy increases linearly with number of taxa, although there is considerable scatter in the data (H* = 0.015N + 0.025; r^2 = 0.551, P < 0.01 for 10 d.f.).

Our results support Sanderson and Donoghue's (1989) proposition that the rise in homoplasy with number of taxa across lineages is an artifact of a statistical sampling process. That process is much like that underlying the famous "birthday paradox": as the number of people in a room approaches 47, the probability that any two of them – even if they are not twins – will share the same birthday approaches 95%. Similarly, as the number of taxa in a lineage increases, the probability that two or

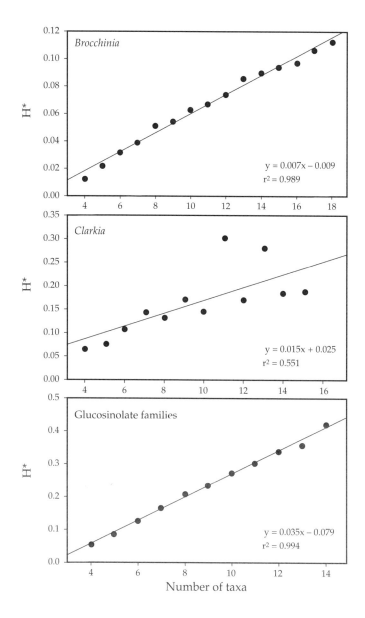

Figure 2.9. Modified homoplasy index (H*) in relation to the number of taxa drawn randomly from *Brocchinia*, *Clarkia*, and the glucosinolate clade. Lines represent least mean squares regressions.

more taxa will share a mutation by chance (i.e., recurrence or, in some deeper sense, convergence) increases, and with it, homoplasy. Note, however, that the slopes of H* on N differ markedly among the three clades studied, ranging from 0.007 for *Brocchinia* to 0.035 for the glucosinolate families, a five-fold difference. That is, particular character sets for particular lineages differ in the extent of homoplasy inherent in them, even after the purely statistical effects of the number of taxa involved in each

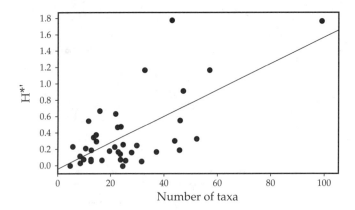

Figure 2.10. H*' in relation to the number of taxa involved in studies based on variation in cpDNA restriction sites. Line represents the least mean squares regression (H* = 0.016N - 0.057, P < 0.0001 for 38 d.f.).

has been removed – for example, by calculating the expected value of H* at a fixed value of N. Thus, although our data support Sanderson and Donoghue's view that the drop in CI' with N across lineages results partly from a purely statistical process, we cannot exclude the possibility that inherent differences between lineages in homoplasy which vary with their size (such as appears to be the case for studies of cpDNA restriction sites at the generic vs. familial level [see Figure 2.5]) are also partly responsible. The relationship across lineages of H* vs. N for cpDNA restriction-site data excluding autapomorphies (Figure 2.10) has a slope of 0.016, near the middle of the range of the values obtained by our random resampling studies. The scatter of points implies a two-fold difference among lineages/character sets in the inherent amount of homoplasy at a given number of taxa, based on a 95% confidence interval for the slope of 0.011 to 0.022.

Homoplasy in relation to the likelihood of correct phylogenetic inference

Based on simulations of binary character-state evolution involving dichotomous trees with equal branch lengths, Givnish and Sytsma (1997) concluded that the maximum probability S of semi-strict correct inference increases in a highly significant fashion with the standardized consistency $CI_9^{*'}$ and the number of informative binary-equivalent characters B, and a highly significant decline with the number of taxa N:

$$S = \text{trunc} (2.2575 \ln CI_9^{*'} + 0.3601 \ln B - 0.6117 \ln N + 1.5741) \qquad (4)$$

where trunc (x) = maximum (0, minimum (1, x)) ($r^2 = 0.880$, $P < 0.0001$ for 71 d.f.). Contours specifying the minimum numbers of informative characters and minimum values of $CI_9^{*'}$ required to ensure a given probability of semi-strict correct inference for 8, 16, 32, and 64 ingroup taxa are shown in Figure 2.11. The maximum probability of strict correct inference shows a qualitatively similar relationship to CI, B, and N:

$$S = \text{trunc} (2.0560 \ln CI_9^{*'} + 0.3811 \ln B - 0.6660 \ln N + 1.3902) \qquad (5)$$

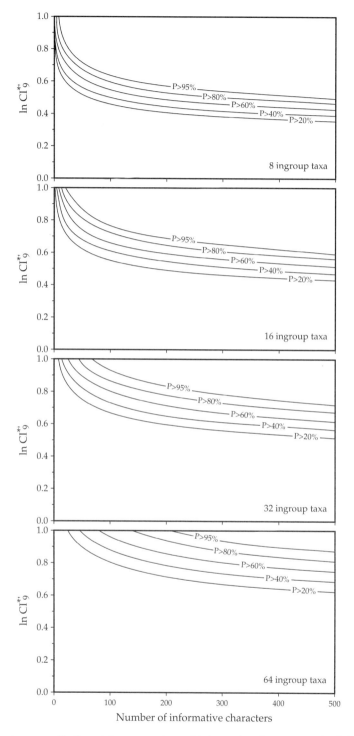

Figure 2.11. Contours specify the minimum numbers of informative characters and minimum values of $CI_9^{*\prime}$ required to ensure a given probability of semi-strict correct inference for 8, 16, 32, and 64 ingroup taxa and 1 outgroup. Based on equations devised by Givnish and Sytsma (1997).

($r^2 = 0.793$, $P < 0.0001$ for 71 d.f.). The standardized consistency index excluding autapomorphies, based on eq. (2) and the results of Givnish and Sytsma's (1997) simulation, is defined as:

$$CI_9^{*'} = 1/(1 + H_9^{*'}) = (N - 3)CI_N'/((N - 9)CI_N' + 6). \tag{6}$$

This index, and a corresponding one which includes autapomorphies, eliminates the spurious effect of the number of taxa on CI' and CI, respectively, and standardizes the consistency index to its expected value when 9 taxa are involved, including an outgroup.

Based on the observed values of $CI_9^{*'}$, the number of variable binary-equivalent characters, and the number of taxa for cpDNA restriction-site studies (Table 2.1), the calculated maximum probability of semi-strict correct inference (i.e., of the correct tree being among the shortest trees) is 0.88 ± 0.21 (n = 39); the corresponding value for strict inference (i.e., of the shortest tree being the correct tree) is 0.73 ± 0.26. These probabilities exceed by highly significant margins ($P < 0.0001$, 2-tailed t-test on $\sin^{-1}\sqrt{x}$ with 79 d.f.) those for current morphological studies, based on either semi-strict (0.53 ± 0.34) or strict correct inference (0.37 ± 0.33); the apparent advantage of restriction-site studies increases with the number of taxa (Figure 2.12). For plant DNA sequence studies, the average maximum probability of semi-strict correct inference is 0.84 ± 0.26 (n = 25); the corresponding value for strict correct inference is 0.74 ± 0.17. These probabilities are almost identical to those for restriction-site studies, and are much greater ($P < 0.0001$, 2-tailed t-test on $\sin^{-1}\sqrt{x}$ with 63 d.f.) than those for morphological studies, even though

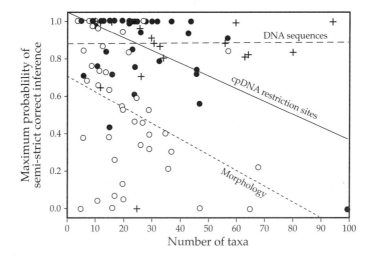

Figure 2.12. Maximum probability of semi-strict correct inference for actual phylogenetic analyses based on variation in DNA sequences (+), restriction sites (●), and morphology (O), shown as a function of number of taxa. Probabilities calculated using parameters for dichotomous trees drawn from Givnish and Sytsma (1997; Table 2). Lines indicate least mean squares regressions. The intercept of the line for cpDNA restriction sites is significantly greater than that of the line for morphology (ANCOVA, $P < 0.001$ for 1, 78 d.f.); there is no significant difference between the lines in slope.

the latter tend to be conducted at lower taxonomic levels. The maximum probability of semi-strict inference in morphological studies is significantly less than that in restriction-site or sequence studies involving the same number of taxa (P < 0.001, ANCOVA); the same is true for comparisons involving strict inference.

Discussion

Homoplasy in molecular vs. morphological data

Our data demonstrate that phylogenies based on cpDNA restriction sites show significantly less homoplasy than those based on morphology, when taxonomic rank and number of taxa are controlled. This finding contradicts earlier reports by Sanderson and Donoghue (1989) and Donoghue and Sanderson (1992), which were based on far fewer molecular studies, and mingled results based on chloroplast and mitochondrial restriction sites, DNA sequences, amino-acid sequences, and protein polymorphisms. Although we found DNA sequences to show roughly the same amount of homoplasy as morphological data – and hence, more homoplasy than restriction-site data – this appears to be a statistical artifact resulting from the tendency for sequence studies to be conducted at a higher taxonomic level than either restriction-site or morphological studies. Such systematic differences in taxonomic level between molecular and morphological studies were not considered or controlled in the analyses by Sanderson and Donoghue (1989) and Donoghue and Sanderson (1992).

These results are not an artifact of differences in homoplasy between plant and animal groups, or in the criteria used to recognize taxa (and hence, to determine N) in morphological and molecular studies. Animal studies include 21 of 42 morphological studies, but are absent or rare from both classes of molecular studies (Table 2.1). However, there is a very tight relationship (r = –0.84) between CI' and N across all morphological studies (Figure 2.2), and there are no significant differences in the slope (ANCOVA, P > 0.25 for 1, 37 d.f.) or intercept (P > 0.1 for 1, 38 d.f.) of the regressions for plant and animal groups. All studies – morphological and molecular – use morphological criteria for species recognition, so there are no obvious biases between groups of studies that could influence our results by systematically affecting estimates of N.

Why does variation in cpDNA restriction sites show less homoplasy than variation in morphology? We believe there are two principal reasons: the greater likelihood of homology in restriction sites, and their lower exposure to selection and convergence.

HOMOLOGY – The homology of morphological character-states can often only be established by detailed comparative studies of organ development. Such studies (which are rarely conducted) are needed to exclude the possibilities that (i) similar phenotypes are merely convergent; (ii) that putatively alternative states of the same character are not actually states corresponding to different characters; or (iii) that putatively different characters are not actually alternative states of the same character.

The last two potential problems involve the "atomization" of an organism into distinct characters and character-states. Given the existence of pleiotropy, epistasis, and other forms of genetic and developmental correlation, such an atomization may be difficult to accomplish, and even the most profound analyses may not be able to exclude interdependence among certain phenotypic characters. That is, phenotypic traits – including amino-acid sequences and electrophoretic protein polymorphisms – may not always be clearly separable in a genetic (and hence, evolutionary) sense.

By contrast, restriction sites are relatively well-defined genetic characters, involving the presence or absence of a particular 4-, 5-, or 6-base sequence at a particular location in the chloroplast genome. The architecture of the latter is highly conserved, relative to both the nuclear and mitochondrial genomes (Palmer 1987; Wolfe et al. 1987). Although re-arrangements are known to occur in a few families and orders, they are easily detected and incorporated in phylogenetic analyses (e.g., see Jansen and Palmer 1988; Knox et al. 1993; Givnish et al. 1995). Certainty regarding the homology of states of individual restriction sites is limited by the (i) possibility of convergent site gains and losses; (ii) failure to distinguish the distinct ways in which a site may be lost; and (iii) lack of positional resolution and consequent inability to distinguish nearby sites (see Knox 1992 for discussion). Positional resolution is limited proximally by the resolution of small differences in restriction fragment length, and ultimately by uncertainties introduced by insertion or deletion of DNA segments. The consequent inability to distinguish nearby sites is offset somewhat by the low probability of a 6-base site arising at random in a particular location.

Donoghue and Sanderson (1992) argue that convergence – the third potential problem in assessing homology – is not likely to be significant for most phylogenetic studies, in that systematists are sensitive to convergence and developmental plasticity, and are likely to exclude characters subject to them from analysis. While we agree with the basic thrust of their argument, it is somewhat simplistic for three reasons: (1) For many rare or poorly collected taxa, there are very few specimens from which to infer the level of habitat-specific convergence or plasticity that a population may show for states of certain characters. Species have been and will continue to be described, erroneously, on the basis of such character-states. (2) Systematists do, in fact, overlook ecological patterns of convergence in making classifications or conducting phylogenetic analyses. This is an easy error to make, even for extremely capable workers. For example, the presence of divided leaves has often been used to link several species of *Cyanea* (Campanulaceae) endemic to the Hawaiian Islands (Rock 1919; Lammers 1989, 1990). Such leaves are strongly associated with the possession of thorn-like prickles, otherwise absent in the family. Yet, on the basis of fossil evidence and a detailed molecular analysis, Givnish et al. (1994) inferred that divided leaves and thorn-like prickles both arose *at least three times independently*, as visual and mechanical defenses against browsing by recently extirpated flightless geese and goose-like ducks. (3) Most important, it is not clear how convergence and the phylogeny of closely related organisms can be detected simultaneously in rigorous fashion, if many of the characters determining the phylogeny are subject to convergence.

EXPOSURE TO SELECTION – In plants, many morphological characters and character-states are at least partly responsive to external selection pressures, as evidenced by repeated patterns of evolutionary convergence under similar ecological conditions (for floral characteristics, see Grant and Grant 1965; Faegri and van der Pijl 1971; Skinner 1988; Goldblatt et al. 1995; Johnson 1995; Armbruster 1996; Endress 1996; for vegetative traits, see Horn 1971; Mooney 1977; Orians and Solbrig 1977; Givnish 1979, 1984, 1986, 1987, 1988. 1995; Albert et al. 1992; Conran and Dowd 1993). Several morphological character-states do appear to have little or no functional significance (e.g., axile vs. basal placentation, 4-mery vs. 5-mery) and show little tendency toward convergence, but our impression is that many systematists (e.g., Cronquist 1981) downplay or disregard the possibility that some of the morphological traits they use to infer phylogeny may be adaptive and show convergence in certain circumstances.

Relative to morphological characters, cpDNA restriction sites probably have a low potential for convergence in response to external selection pressures. Such sites often involve at least one "silent" position of no phenotypic significance; they may occur in introns and in portions of exons that are deleted post-transcriptionally; and even within coding regions, the 4 to 6 bases of a site may not align with the codon reading frame. Further, a typical study localizes several hundred sites scattered more or less evenly throughout the chloroplast genome; it seems unlikely that even the strongest external selection pressure could affect more than a few of these sites – localized in a few genes with organellar function only – at any one time.

OTHER POTENTIAL FACTORS– Sanderson and Donoghue (1989) observe that, other things being equal, phylogenies based on unordered multi-state characters will show greater homoplasy than those based on binary characters. They then argue that if phylogenies based on morphology were to show more homoplasy, it might be partly because morphological characters are frequently multi-state and molecular characters binary. Our analysis – based on converting all characters to an equivalent number of binary-state characters – clearly refutes this view. When characters are expressed on this equalized and fair basis, sequence and restriction-site studies have many more characters than morphological studies involving the same number of taxa (Figure 2.4). It should be noted that many DNA sequence studies involve a high proportion of multi-state characters, and that such characters may sometimes account for a majority of the informative binary-state character-equivalents (see Table 2.4 for examples).

Homoplasy in relation to taxonomic rank

Our data show that homoplasy depends on the taxonomic rank of the taxa involved in phylogenies based on variation in cpDNA restriction sites, being lower in studies within genera than in studies within families when the effect of the number of taxa is controlled. A similar trend is seen in DNA sequence studies. Taken together, our findings contradict Sanderson and Donoghue (1989), whose conclusion regarding the absence of an effect of taxonomic rank was based largely on morphological data and on the mingling of those with molecular data.

We believe that the most likely reason that homoplasy shows no dependence on taxonomic rank in morphological studies is close to that suggested by Sanderson and Donoghue (1989): "Systematists undoubtedly choose their character sets to reflect an appropriate amount of variation for the rank of the terminal taxa used." However, this phrasing blurs a key underlying issue: in conducting phylogenetic studies, many systematists discard or never include particular morphological

Table 2.4. Distribution of binary vs. multi-state informative characters in four representative plant studies based on *rbc*L sequence variation.

	Study				Mean
	Conti et al. 1993	Rodman et al. 1993	Donoghue et al. 1992	Kim et al. 1992	
Number of sites:					
2-state	69	184	74	117	**111**
3-state	15	116	12	42	**46**
4-state	2	39	2	11	**14**
Proportion of sites:					
2-state	80%	54%	84%	69%	**72%**
3-state	18%	34%	14%	25%	**23%**
4-state	2%	12%	2%	6%	**6%**
Proportion of equivalent binary characters:					
2-state	66%	35%	73%	50%	**56%**
3-state	29%	44%	24%	36%	**33%**
4-state	6%	22%	6%	14%	**12%**

characters because those characters show too much "noise" or homoplasy relative to some notion – preconceived or emerging from unspoken assumptions as analysis proceeds – of what the actual phylogenetic relationships in a group are. In some ways, this is a sensible approach: why incorporate leaf width, for example, in a cladistic analysis when it is manifest that variation in this trait is subject to rampant evolutionary convergence and developmental plasticity? On the other hand, this approach can be abused and lead to problems of circularity, infinite regress, and selective use of data to justify views that are, consciously or unconsciously, preconceived. How is one to infer simultaneously the "correct" phylogeny for a group and which characters should be discarded (as too "noisy") to obtain that phylogeny? Is it not perhaps naive to suggest that "plant systematists are not really fooled by plasticity, rather, they have become adept at delimiting characters even in the face of considerable environmental variation" (Donoghue and Sanderson 1992, p. 345)?

We agree with Sanderson and Donoghue (1989) that this problem is not limited to morphological data. Different genetic sequences evolve at different rates, and each involves a finite number of characters; it is thus not surprising that plant molecular systematists find it efficient to use rapidly evolving sequences (e.g., *ndh*F and *atpb-rbc*L spacer in cpDNA, internal transcribed spacer (ITS) of nuclear ribosomal DNA) to infer relationships within genera, and more slowly evolving sequences (e.g., *rbc*L, 5S subunit gene of nrDNA) to infer relationships among families and orders.

However, three important points must be made. First, as they have been studied by molecular plant systematists, cpDNA restriction sites are characters that are fairly equivalent across taxa regardless of taxonomic rank. With only a few exceptions, the kinds of restriction-site characters assayed by systematists tend not to vary in a regular fashion with the taxonomic breadth of the organisms studied. Restriction-site surveys of some broad groups (e.g., the Asteridae [Downie and Palmer 1992] or the monocotyledons [Davis 1995; Hahn et al. 1995]) do rely on slower-evolving sites or cpDNA segments (i.e., the inverted repeat region) than those used in studying relationships within genera; the evolution over time of insertions and deletions within the large and small single copy regions rapidly precludes certainty at higher taxonomic levels regarding the homology of restriction fragments and sites therein. Limitations on the reliable inference of fragment and site homology have generally restricted the use of such data to within-genus and within-family questions, where they show demonstrably less homoplasy than morphological characters. Nevertheless, the fact that restriction-site characters are roughly comparable across taxonomic levels separates them qualitatively from morphological characters: the unspoken winnowing of characters perceived as aberrant are **not** likely to bias or distort restriction-site studies.

Second, there is no reason to believe that incorporating data on **both** slow- and fast-evolving sites or DNA sequences would strongly alter the phylogenetic inferences drawn, or at least not alter them more than such a practice would affect inferences from morphological data. For taxonomically narrow or slowly evolving groups, slow-evolving sites or sequences would show little variation and phylogeny would be inferred mainly from data from the faster-evolving segments. For broader or more rapidly evolving groups, phylogeny would be inferred mainly from the slow-evolving sites or sequences, with large amounts of recurrence homoplasy blurring or nullifying the "phylogenetic signal" detectable from faster-evolving segments.[1] However, we believe – *contra* Donoghue and Sanderson (1992) – that morphological characters differ in being more likely subject to convergence homoplasy than molecular characters (see above). The distinction between convergence and recurrence homoplasy is crucial, in that convergence, induced by shared ecological conditions, **can affect multiple traits simultaneously and send a sharp but erroneous "phylogenetic signal,"** whereas recurrence homoplasy (at least if averaged across a substantial number of sites) will generally send a blurred correct "signal" or a null signal (see Armbruster 1996 for example).

Finally, because DNA restriction sites and sequences are genetic and not merely phenotypic characters, they are less likely than morphological characters to be sources of homoplasy due to an incorrect atomization of traits or failure to take proper account of epistasis, pleiotropy, or genetic or developmental correlations. This is not to imply that DNA characters are not without difficulty: nearby sequence characters may be interdependent due to constraints on secondary structure (see Hixson and Brown 1986; Appels and Honeycutt 1986; Steele et al. 1988; Wheeler and Honeycutt 1988; Dixon and Hillis 1993); genome re-arrangements must be taken into account; and questions regarding homology by deletions and insertions in non-coding regions

---

[1]Difficulties can arise if different sites or sequences evolve at rates that differ only moderately. Bull et al. (1993) show that, if equal numbers of slow- and fast-evolving binary characters are considered, analyzing the combined data (as recommended by Miyamoto 1985; Kluge 1989; Barrett et al. 1991; Donoghue and Sanderson 1992) often leads to a less accurate phylogeny when there is an intermediate (i.e., 2- to 3-fold) level of disparity in evolutionary rates. This is a critical problem that must be addressed by advocates of the "total evidence" approach, but it would not affect molecular and morphological analyses differently.

must be resolved. Yet these problems, as well as those just discussed, seem less likely to corrupt a precise analysis of detailed relationships than those possibly associated with morphological characters.

In our view, morphological and molecular characters are mutually complementary. The former are undeniably superior in defining species, generating testable hypotheses on the composition and relationships of higher taxonomic groupings, and incorporating fossil evidence. The latter are more likely to be useful as a precise tool with which to test detailed phylogenetic hypotheses, given that they are more numerous, better defined, and show less homoplasy at a taxonomic level and given number of taxa.

Homoplasy in relation to the number of taxa

The results of our random resampling simulations tend to support Sanderson and Donoghue's (1989) conjecture that the observed decline in CI' with the number of taxa involved in a study is mainly a statistical artifact. As the number of taxa being compared increases, the chance that two or more will share a derived character-state as a result of chance, not ancestry, increases. Further, the extent of the observed increase in CI' (and decrease in homoplasy) with N in random resampling simulations within lineages approximates that seen across taxonomic groups (Figures 2.9 and 2.10).

Simulations by Givnish and Sytsma (1997) confirm and extend these conclusions, and provide a simple pull-back rule (eq. 6) that standardizes the consistency index so as to eliminate the effect of number of taxa. These simulations also imply that the decline in CI' with N should be greater in lineages with a greater total rate/amount of genetic divergence. It is not surprising, therefore, that the greatest decline in CI' with N was observed in the resampling of the glucosinolate genera, and the least in the resampling of the genera *Brocchinia* and *Clarkia*. The tendency for the decline in CI' to plateau at greater numbers of taxa in *Brocchinia* but not in the other clades may reflect the greater concentration of mutations deep in the *Brocchinia* phylogeny; at some fairly low number of taxa examined, most of the genetic variation between subclades is represented, and further declines in CI' will mainly reflect homoplasy among branch tips, excluding autapomorphies. The lower concentration of genetic change at the bases of the principal subclades in *Clarkia* and the glucosinolate genera makes them less subject to this plateauing of the CI' vs. N relationship.

Homoplasy in relationship to the likelihood of correct phylogenetic inference

Simulations by Givnish and Sytsma (1997) demonstrate that the likelihood of correct phylogenetic inference increases with the number and consistency of variable independent characters, and decreases with the number of taxa, when all three parameters are considered jointly (see Figure 2.11). This finding is, in many ways, intuitively appealing. Mutations at individual nucleotides or restriction sites create a "picture," much like a pointillist painting by Georges Seurat or his Impressionist colleagues, of the relationships among taxa within a lineage. The more "dots" (or mutations) that are placed, and the more accurately they are placed on a canvas of a given size, the more likely we are to perceive correctly the evolutionary scene they imply. Conversely, if there are few dots or mutations per unit area (i. e., a low ratio of

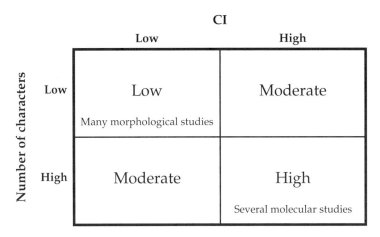

Figure 2.13. Proposed relationship of correct phylogenetic inference to CI and number of independent, variable characters.

variable characters to taxa), and/or if those present are misleadingly placed, we may imagine a scene – a set of phylogenetic relationships – totally at odds with the record of actual events set down by genetic evolution.

Our CLADOGENESIS results imply that, in general, we might expect a high frequency of correct inferences from large numbers of variable, independent characters that display a high standardized CI; a low frequency of correct inferences from a small number of characters that display a low standardized CI; and an intermediate frequency of correct inferences from studies involving either a small number of characters, or a low standardized CI, but not both (Figure 2.13). In general, DNA restriction-site studies involve far more informative binary-state character-equivalents than morphological studies involving the same number of taxa (Figure 2.4; overall means of B are 98.9 ± 95.4 vs. 56.5 ± 54.4, $P < 0.05$ for 2-tailed t-test with 81 d.f.), and a significantly higher degree of character consistency (Figure 2.3A: overall means for $CI_9^{*'}$ are 0.91 ± 0.08 vs. 0.77 ± 0.10, $P < 0.0001$ for 81 d.f.). In other words, most restriction-site studies fall into the lower right-hand cell in Figure 2.13, and many morphological studies fall into the upper left-hand cell, implying that restriction-site studies are likely to be a more reliable guide to phylogeny. Furthermore, given that DNA sequence studies generally involve even more characters, and appear to involve a higher degree of consistency than morphological studies if taxonomic rank is controlled (see discussion above on **Homoplasy in relation to molecular vs. morphological data**), it seems likely that DNA sequence variation would also prove more reliable than morphology for inferring detailed phylogenetic relationships.

It thus seems inescapable that DNA restriction sites and/or sequences – if controlled for the avoidable problems associated with genome re-arrangement (inversions, deletions, insertions) or transference (i.e., hybridization, introgression, horizontal transmission, or genetic exchange between nuclear and organellar genomes) – will generally provide a more reliable window on phylogenetic relationships than morphological characters (see Figure 2.12). It should be noted that our argument does not **directly** involve the greater likelihood of morphological homoplasy due to convergence or character misclassification; only the overall extent of homoplasy from all sources in molecular vs. morphological data is considered. To the extent that the greater amount of homoplasy in morphological data is due to convergence, and to the extent that selection in similar environments can lead to what we term **concerted convergence** (in which several, seemingly unrelated characters may converge in different organisms due to the selection pressures imposed on several traits simultaneously by a shared environment {research by Wake [1991], Crisp [1994], Titus and Larson [1995], Armbruster [1996] provides potential examples of this phenomenon}), morphological data may be even more misleading than their average level of homoplasy would indicate. In terms of our pointillist and signal analogies, concerted convergence would involve the simultaneous skewing of several points/signals in the same direction, whereas misclassification homoplasy would presumably involve independent, random skewing; the former seems more likely to alter our impression of relationships than the latter.

Synthesis

We have presented comparative data and the results of computer simulations to argue that certain kinds of molecular data are more likely to lead to correct phylogenetic inference than morphological data. Specifically, phylogenies based on cpDNA restriction-site variation show significantly less homoplasy than those based on morphology, when the effects of number of taxa and level of taxonomic analysis are controlled. When these same factors are controlled, phylogenies based on DNA sequences also appear to show less homoplasy than those based on morphology. Moreover, when both homoplasy **and** the number of variable characters is taken into account, the reliability of characters is strongly and inversely related to the degree of homoplasy they display: the more homoplasy a set of characters shows, the weaker are the inferences of phylogeny based on them. The probability of correct phylogenetic inference increases with the overall number of variable characters, and declines with the number of taxa under consideration.

These results contrast strongly with Donoghue and Sanderson's (1992) view (pp. 353–361) that morphological data lack serious disadvantages relative to molecular data and may aid phylogenetic inference by providing additional and possibly complementary characters, so that morphological and molecular characters should usually be combined and given equal weight in phylogenetic analyses. This view is based on four key assumptions: (1) the homology of morphological characters often seems more secure, based on their "complexity" (i.e., their detailed pattern of development and multi-faceted structure and position) (pp. 343–345); (2) convergence in morphological characters is no more important than in molecular characters (p. 343); (3) even if it were, there is no evidence that the reliability of characters for phylogenetic reconstruction is related to the degree of homoplasy they display (pp. 342–343); and (4) in any case, there is no evidence that morphological characters are indeed more homoplastic overall (pp. 346–347).

The validity of assumption (1) is debatable. The "complexity" of morphological characters is at once their greatest strength (Donoghue and Sanderson 1994) **and** their greatest weakness, a potentially potent support of homology, and an equally potent source of problems involving epistasis, pleiotropy, and the improper atomization of traits. Many DNA sequence or restriction-site characters have formidable support for their homology, based on literally scores of positional cues; the state of any individual character (a nucleotide or restriction site) can be more or less completely separated/atomized from these cues. Molecular analyses are not immune to problems of proper character atomization. In particular, DNA sequence studies are likely always to be plagued by difficulties caused by constraints on secondary structure (Appels and Honeycutt 1986; Steele et al. 1988; Wheeler and Honeycutt 1988) or by the interdependence of nucleotide mutations in small-scale insertions and deletions. Such subtleties are unlikely to affect restriction-site analyses, however, given their cruder resolution. Indeed, Hillis et al. (1994) found that phylogenies of P7 phages inferred from restriction sites deviated less from the actual phylogeny than those inferred from DNA sequence variation.

We believe that character misclassification is an important and overlooked source of homoplasy in morphological studies. Donoghue (1983) provided a phylogenetic analysis of a plant group based on morphology that is exceptional in terms of the effort devoted to assessing the homology of a large number of characters. Yet Donoghue and Sytsma (1993) and Donoghue and Baldwin (1993), after a detailed molecular systematic analysis showing (i) strong congruence between data sets based on variation in chloroplast DNA restriction sites and nuclear DNA ITS sequences, and (ii) a lack of congruence of either with morphology in the placement of one critical taxon, concluded that the "relationships imply homoplasy in conspicuous morphological characters." In particular, it appears that black fruit color and naked buds – crucial characters in the morphological analysis – have arisen independently in at least two lineages of *Viburnum*.

The lack of a one-to-one correspondence between phenotype and genotype makes the problem of character misclassification fundamental to any analysis based on morphology. Furthermore, there is no easy way to identify – except in hindsight – which morphological characters are "reliable" (i.e., homologous). Even a highly unusual, fairly complex character that may appear to have arisen only once – like the "hot-lips" inflorescence in certain species of *Cephaelis/Psychotria* (Rubiaceae) – can, on detailed analysis, be shown to be homoplastic, the likely result of convergence or recurrence (Nepokroeff and Sytsma 1997; see also Endress 1996). Similarly, a century of systematic thought based on morphology failed to identify the closest relative of the bizarre Hawaiian cliff succulent *Brighamia* (Campanulaceae), even though cpDNA restriction-site variation clearly links it to *Delissea* (Givnish et al. 1995).

Assumption (2) is arguable, but Donoghue and Sanderson (1992) fail to make the crucial distinction between **convergence** in molecular characters (a result of some but not all taxa in a lineage being exposed to a similar selection pressure), and the result of a strong but **uniform** pressure on **all** taxa. Darwin (1859) himself recognized that traits with essentially the same adaptive value in all environments could be as fertile a source of homologies as traits that were selectively neutral. In our context, the fact the cpDNA sequence for *rbc*L (the large subunit of RUBISCO, the enzyme responsible for CO_2 capture during photosynthesis and the most abundant enzyme on earth) is highly conservative and manifestly under strong selection pressure does **not** make it as subject to convergence as many morphological characters clearly are (e.g., see Grant and Grant 1965; Faegri and van der Pijl 1971; Givnish 1979, 1984, 1986, 1987, 1988, 1990, 1995; Ehleringer et al. 1981, 1986; Chase and Palmer 1992; Albert et al. 1992; Conran and Dowd 1993; Givnish et al. 1994, 1995; and several chapters in this volume). There is no evidence for any tendency for different forms of *rbc*L to be favored in different environments, and the very fact that *rbc*L is under strong, uniform selection pressures ensures that the residual variation – the only material on which phylogenetic analyses can be based – is, for all practical purposes, neutral. Indeed, analyses by Mark Chase (pers. comm.) show that more than 95% of the apparent mutations in the *rbc*L sequence during angiosperm evolution has been phenotypically synonymous or functionally equivalent, and hence, selectively neutral. This is not to say that particular sequences or restriction sites will not, in the future, be shown to be subject to true convergence (caused by selection pressures that vary between habitats); undoubtedly they will. However, in our judgment the preponderance of evidence suggests that many morphological characters are more subject to true convergence than are cpDNA restriction sites or DNA sequences. We do agree with Donoghue and Sanderson (1992) that the overall level of homoplasy shown by any character can be affected by factors other than its selective value – in our terminology, overall homoplasy has components based on recurrence, transference, and character misclassification as well.

Our results, summarized in the opening paragraph of this synthesis, demonstrate that assumptions (3) and (4) are clearly wrong. Indeed, after we provided them with our earlier results (Givnish and Sytsma 1992), Sanderson and Donoghue (1996) increased the sampling of molecular studies and found higher CI values for restriction-site data than for morphological data, but found about equal values for sequence and morphological data. Combining these CI results with the significantly greater number of characters for the molecular data in their sampling (means of 87 for restriction sites, 606 for sequences, and 34 for morphology) – and keeping in mind our finding of significant correlation of numbers of informative characters and reliability (Givnish and Sytsma 1997; see above) – the molecular data should be more reliable than morphological data. This finding is appealing as it is a restatement of a **general finding** (Felsenstein 1978; see Archie 1996 for review) **that it is more difficult to estimate trees accurately under high rates of evolution, and that levels of homoplasy correlate with evolutionary rates**.

Sanderson and Donoghue (1996) (see also Donoghue and Sanderson 1992, pp.342–343, 355–356; 1994, p. 401) attempt to weaken this link between level of homoplasy and reliability by invoking other factors that in their view are more important in determining reliability: (i) the degree of clade resolution, (ii) the level of clade support, and (iii) distribution of homoplasy across the tree. We certainly agree that these factors might interplay with overall homoplasy in determining reliability, but observe that their analysis suffers from common misconceptions and problems. First, a full dichotomous resolution of the tree should **not** be assumed to best represent the "true tree." As is clear in many chapters of this book and by the cases (*Clarkia*, *Fuchsia*, Oncidiinae orchids) cited by Donoghue and Sanderson (1992), rapid radiations will and should produce unresolved nodes (see also Sytsma and Baum 1996). Suggesting that "in all of these cases, the addition of even a few morphological characters could help resolve relationships" (Donoghue and Sanderson 1992, p.356), is highly speculative and indeed dangerous considering that this "extra" resolution can be achieved simply by the addition of one highly homoplaseous morphological character. A strong signal for a hard polytomy (no resolution) should not be diminished by invoking the old adage that "more resolution is better." Second, we suspect that clade resolution and bootstrap support are strongly linked to the kind of taxa sampling employed in molecular vs. morphological cladistic studies. Molecular studies typically employ a larger sampling than morphological studies as a consequence of the "individual" vs. "type" sampling, respectively (Table 2.1; Sanderson and Donoghue 1996, Table 2; see Doyle et al. 1994 and Sytsma and Baum 1996 for similar discussions of sampling problems in studies of angiosperm radiation). Multiple accessions per taxon used in the molecular vs. morphological studies will lead inevitably to decreased clade resolution and support.

How then should morphological and molecular characters be treated in phylogenetic analyses? While morphological and molecular data are mutually complementary and equally important in guiding the overall strategy of phylogenetic inference, we do not believe they should be used interchangeably, or that equal weight should necessarily be given to morphological characters, which appear to be generally more homoplaseous and more subject to convergence or character misclassification.

If morphological and molecular data separately generate quite incongruent phylogenetic relationships, we see little advantage and considerable danger in their indiscriminate combination, just as it is misleading to combine nuclear and chloroplast DNA when they are clearly tracking different evolutionary or genetic phenomena (Sytsma 1990). When Donoghue and Sanderson (1992) state that "in the worst scenario, morphological data may be so noisy that they lower the resolution of an otherwise highly resolved molecular data set" (p. 356), we believe that they do not perceive the worst case. In our opinion, the worst case involves the unopposed, incorrect resolution of mid-level or basal nodes in a tree by a few morphological characters that are nominally homoplastic (resulting in an apparent mid-level or basal synapomorphy), or that misleadingly show no homoplasy as a consequence of character misclassification. We note that, if a group of characters is subject to high levels of homoplasy, extreme caution must be exercised in using them to infer cladogenetic events that are not implied by groups of characters that systematically show less homoplasy (see also Bull et al. 1993; Miyamoto and Fitch 1995). This is especially true if the latter are numerous and strongly support many nodes. In that case, resolution of some of the remaining nodes by the more homoplaseous data set may be illusory

– the unresolved nodes may evolutionarily be truly unresolved, a result of multiple founder events from a single source population, or an explosive pattern of speciation (Sytsma 1990). Alternatively, concerted convergence among several seemingly unrelated characters could, if such characters were included in an analysis, overwhelm the phylogenetic "signal" implicit in less misleading characters at any particular node and result in an incorrect phylogeny. If, on the other hand, morphological and molecular data considered separately generate congruent phylogenies **and** involve roughly equal levels of consistency, then it seems appropriate to analyze relationships based on the combined data sets (Doyle et al. 1994; however, Bull et al. [1993] and Miyamoto and Fitch [1995] recommend not combining data sets to increase the significance of corroboration by independent sets of evidence). The suggestion of Donoghue and Sanderson (1992) that morphological characters may help resolve nodes in family trees that are left unresolved by molecular data deserves consideration. Different sets of characters may evolve at somewhat different rates at different times and in different portions of a lineage, so that there may be utility in using disparate data to resolve all branchings within a lineage. Yet Bull et al. (1993) demonstrate that even this course is fraught with danger. And, as we demonstrate above, the **lack** of molecular support for resolving a node in a rapidly evolving lineage might itself be a strong signal that should not be lost by the addition of homoplaseous morphological characters.

Recommended strategy

We agree with Donoghue and Sanderson's (1992) general view that morphological and molecular data have mutually complementary roles in the analysis of phylogenetic relationships, but disagree as to what those roles might be. They advocate combining both kinds of data in a formal cladistic analysis, placing equal weight and confidence in each. We propose instead that morphological data are best suited for recognizing species and suggesting broad patterns of relationships, and that molecular data – by dint of the greater number and higher consistency of characters involved – are a more precise guide to detailed phylogenetic relationships, barring any problems arising from hybridization, introgression, and lineage sorting.

Adopting our proposed strategy would have three added advantages. First, it would provide the basis for rigorous, non-circular studies of morphological evolution within a variety of groups, excluding characters whose radiation, convergence, or other significance is being investigated (Givnish 1987; Skinner 1988; Harvey and Purvis 1991; Sytsma et al. 1991; Albert et al. 1992; Baldwin 1992; Knox et al. 1993; Kocher et al. 1993; Givnish et al. 1994, 1995; Crisp 1994; Titus and Larson 1995).[2] We strongly believe that understanding patterns of morphological evolution is one of the most stimulating issues in evolutionary biology, bridging aspects of ecology, phy-

---

[2] Brooks and McLennan (1994) argue that circularity is of less concern than the degree of congruence among characters involved vs. uninvolved in a radiation. We believe that such arguments underestimate the problems introduced by pleiotropy, epistasis, and concerted evolution (see above); these phenomena tend to blur the distinction between characters "involved" vs. "uninvolved" in a radiation, and amplify the potential for problems of circular reasoning.

logeny, morphology, and natural selection. Molecular systematics provides an important tool for examining patterns of evolution in morphological characters – indeed, in every morphological character – and thus should be of great and continuing interest to all those whose main interest is morphological evolution. At the same time, molecular systematics has and will always depend on traditional morphological systematics for species definition and the generation of initial hypotheses regarding broad-scale relationships.

Second, molecular systematics can potentially provide an enormous wealth of independent data sets with which to test phylogenetic hypotheses. The validity of inferences based on one set of molecular data can be tested by comparison with other sets; if they produce congruent trees when analyzed separately and combined, then these data sets provide strong support for the relationships they indicate.

The last advantage of the molecular approach is probably the most fundamental and the least widely recognized: **the universe of characters is well defined and can be sampled exhaustively and without bias once the study parameters have set**. For example, if cpDNA restriction sites are examined, all variation across related species can be surveyed for a specified set of enzymes within the limits of resolution imposed by gel electrophoresis, Southern blotting, and the size and degree of similarity of heterologous probes. By contrast, the universe of morphological characters is manifestly not well defined, in terms of either the range of characters or their proper atomization. The well-defined nature of a sampling universe is crucial, in that it prevents *ad hoc, a priori*, or unconscious selection and removal of characters to support one phylogenetic hypothesis or another. It is interesting to note that in the history of ecology, our understanding of the compositional "relationships" among communities underwent a revolution once the need for explicit criteria specifying a sample universe was recognized and implemented (Whittaker 1956; Curtis 1959). The unconscious biases introduced in the sampling of ecological sites were so great that they produced a conclusion (supporting the super-organism hypothesis of Clements 1916) regarding the nature of compositional variation across communities exactly **opposite** to that which emerged (supporting the individualistic hypothesis of Gleason 1917) when unbiased sampling was adopted. It may not be unreasonable to expect that our views regarding relationships among organisms may undergo a similar revolution, at least in some cases, if we adopt an explicit sampling universe of characters for phylogenetic studies.

Conclusions

Homoplasy can arise through four mechanisms: convergence, recurrence, transference, and character misclassification. We argue that morphological characters are more likely to be subject to convergence and misclassification than are certain kinds of molecular characters (cpDNA restriction sites and DNA sequences), that recurrence should increase with a group's taxonomic breadth and number of taxa, and that the likelihood of correct phylogenetic inference should decrease with the extent of homoplasy displayed by a group of independent characters and increase with the number of such characters. An analysis of 104 phylogenetic studies based on parsimony demonstrates that cpDNA restriction-site studies show (i) significantly less overall homoplasy than those based on morphology, if the effects of number of taxa and taxonomic rank are controlled; (ii) no significant difference in homoplasy from DNA sequence studies, if the same factors are controlled; and (iii) a significant increase in homoplasy with the taxonomic breadth of the group involved, if the number of taxa is controlled. Restriction-site and sequence studies also involve significantly more binary-state character-equivalents than morphological studies involving the same number of taxa.

Resampling analyses of molecular data for random subsets of three clades suggest that the across-lineage increase in homoplasy with number of taxa is a statistical artifact, reflecting the increased number of species comparisons that can generate homoplasy via recurrence. The rate at which homoplasy increases with number of taxa rises with the total genetic divergence within a lineage, and is modulated by the distribution of divergence within phylogenies. The consistency index CI (including autapomorphies) shows a more regular pattern of change with number of taxa than does the modified consistency index CI' (excluding autapomorphies) or the retention index RI, suggesting that CI may be a better tool for comparative studies.

Given that studies based on restriction-site and sequence data generate more characters with a higher level of consistency than comparable studies based on morphology, and given that simulations (Givnish and Sytsma 1997) show that such advantages should lead to a higher likelihood of correct phylogenetic inference, molecular studies may often provide a more precise guide to detailed relationships than morphology. Morphology plays a vital complementary role in species recognition and in providing an initial, coarse-grained guide to phylogeny. Molecular systematics provides a basis for non-circular studies of adaptive radiation and convergence, and for powerful tests of phylogenetic hypotheses employing independent data sets with large numbers of well-defined, independent characters that show a relatively low degree of homoplasy. In addition, molecular characters are drawn exhaustively from a well-defined sample universe, eliminating possible distortions due to *ad hoc* or unconscious selection of characters.

Acknowledgments

We wish to thank K.-J. Kim, R. Olmstead, and D. Soltis for providing access to data on variation in cpDNA restriction sites and DNA sequences for matched taxa. D. Baum, E. Knox, R. Olmstead, M. Donoghue, and M. Chase provided useful comments and advice. This research was supported by NSF grants BSR-8806520, DEB-9020055, DEB-9007293, and DEB-9306943.

References

Albert, V. A., Williams, S. E., and Chase, M. W. 1992. Carnivorous plants: phylogeny and structural evolution. Science 257:1491–1495.

Appels, R., and Honeycutt, R. L. 1986. rDNA: evolution over a billion years. Pp. 81–135 in *DNA systematics, vol. II, plants*, S. K. Dutta, ed. Boca Raton, FL: CRC Press.

Archie, J. W. 1989a. Homoplasy excess ratios: new indices for measuring levels of homoplasy in phylogenetic systematics and a critique of the consistency index. Systematic Zoology 38:253–269.

Archie, J. W. 1989b. A randomization test for phylogenetic information in systematic data. Systematic Zoology 38:239–252.

Archie, J. W. 1996. Measures of homoplasy. Pp. 153–188 in *Homoplasy: the recurrence of similarity in evolution*, M. J. Sanderson and L. Hufford, eds. San Diego, CA: Academic Press.

Armbruster, W. S. 1992. Phylogeny and the evolution of plant-animal interactions. BioScience 42:12–20.

Armbruster, W. S. 1996. Exaptation, adaptation, and homoplasy: evolution of ecological traits in *Dalechampia* vines. Pp. 227–243 in *Homoplasy: the recurrence of similarity in evolution*, M. J. Sanderson and L. Hufford, eds. San Diego, CA: Academic Press.

Bain, J. F., and Jansen, R. K. 1995. A phylogenetic analysis of the aureoid *Senecio* (Asteraceae) complex based on ITS sequence data. Plant Systematics and Evolution 195:209–220.

Baldauf, S. L., and Palmer, J. D. 1990. Evolutionary transfer of the chloroplast *tuf*A gene to the nucleus. Nature 344:262–265.

Baldwin, B. G. 1992. Phylogenetic utility of the internal transcribed spacers of nuclear ribosomal DNA in plants: an example from the Compositae. Molecular Phylogenetics and Evolution 1:3–16.

Baldwin, B. G., Kyhos, D. W., and Dvorak, J. 1990. Chloroplast DNA evolution and adaptive radiation in the Hawaiian silversword alliance (Madiinae, Asteraceae). Annals of the Missouri Botanical Garden 77:96–109.

Barker, N. P., Linder, H. P., and Harley, E. H. 1995. Polyphyly of Arundinoideae (Poaceae): evidence from *rbc*L sequence data. Systematic Botany 20:423–435.

Barrett, M., Donoghue, M. J., and Sober, E. 1991. Against consensus. Systematic Zoology 40:486–493.

Baum, D. A., Sytsma, K. J., and Hoch, P. C. 1994. A phylogenetic analysis of *Epilobium* (Onagraceae) based on nuclear ribosomal DNA sequences. Systematic Botany 19:363–388.

Bharathan, G., and Zimmer, E. A. 1995. Early branching events in monocotyledons - partial 18S ribosomal DNA sequence analysis. Pp. 81–107 in *Monocotyledons: systematics and evolution*, P. Rudall, P. J. Cribb, D. F. Cutler, and C. J. Humphries, eds. Kew, England: Royal Botanic Gardens.

Bowman, C. M., Barker, R. F., and Dyer, T. A. 1988. In wheat ctDNA, segments of ribosomal protein genes are dispersed repeats, probably conserved by nonreciprocal recombination. Current Genetics 14:127–136.

Bremer, B., Andreasen, K., and Olsson, D. 1995. Subfamilial and tribal relationships in the Rubiaceae based on *rbc*L sequence data. Annals of the Missouri Botanical Garden 82:383–397.

Bremer, B. B., and Jansen, R. K. 1991. Comparative restriction site mapping of chloroplast DNA implies new phylogenetic relationships within Rubiaceae. American Journal of Botany 78:198–213.

Bremer, B., Olmstead, R. G., Struwe, L., and Sweere, J. A. 1994. *rbc*L sequences support exclusion of *Retzia, Desfontainia,* and *Nicodemia* from the Gentianales. Plant Systematics and Evolution 190:213–230.

Bremer, K., Humphries, C. J., Mishler, B. D., and Churchill, S. P. 1987. On cladistic relationships in green plants. Taxon 36:339–349.

Brooks, D. R., and McLennan, D. A. 1994. Historical ecology as a research programme: scope, limitations and the future. Pp. 1–27 in *Phylogenetics and ecology*, P. Eggleton and R. Vane-Wright, eds. London, England: Academic Press.

Brooks, D. R., O'Grady, R. T., and Wiley, E. O. 1986. A measure of the information content of phylogenetic trees, and its use as an optimality criterion. Systematic Zoology 35:571–581.

Brunsfeld, S. J., Soltis, P. S., Soltis, D. E., Gadek, P. A., Quinn, C. J., Strenge, D. D., and Ranker, T. A. 1994. Phylogenetic relationships among the genera of Taxodiaceae and Cupressaceae: evidence from *rbc*L sequences. Systematic Botany 19:253–262.

Bull, J. J., Huelsenbeck, J. P., Cunningham, C. W., Swofford, D. L., and Waddell, P. J. 1993. Partitioning and combining data in phylogenetic analysis. Systematic Biology 42:384–397.

Campbell, C. S., Donoghue, M. J., Baldwin, B. G., and Wojciechowski, M. F. 1995. Phylogenetic relationships in Maloideae (Rosaceae): evidence from sequences of the internal transcribed spacers of nuclear ribosomal DNA and its congruence with morphology. American Journal of Botany 82:903–918.

Chase, M. W., and Palmer, J. D. 1989. Chloroplast DNA systematics of Lilioid monocots: resources, feasibility, and an example from the Orchidaceae. American Journal of Botany 76:1720–1730.

Chase, M. W., and Palmer, J. D. 1992. Floral morphology and chromosome number in the subtribe Oncidiinae (Orchidaceae): evolutionary insights from a phylogenetic analysis of chloroplast DNA restriction site variation. Pp. 324–337 in *Molecular systematics of plants*, P. S. Soltis, D. E. Soltis, and J. J. Doyle, eds. New York, NY: Chapman and Hall.

Churchill, S. P., Wiley, E. O., and Haser, L. A. 1984. A critique of Wagner groundplan-divergence studies and a comparison with other methods of phylogenetic analysis. Taxon 33:212–232.

Clark, J. B., Maddison, W. P., and Kidwell, M. G. 1994. Phylogenetic analysis supports horizontal transfer of P-transposable elements. Molecular Biology and Evolution 11:40–50.

Clark, L. G., Zhang, W., and Wendel, J. F. 1995. A phylogeny of the grass family (Poaceae) based on *ndh*F sequence data. Systematic Botany 20:436–460.

Clements, F. E. 1916. Plant succession: an analysis of the development of vegetation. Carnegie Institute of Washington, Publication 242.

Conran, J. G., and Dowd, J. M. 1993. The phylogenetic relationships of *Byblis* and *Roridula* (Byblidaceae-Roridulaceae) inferred from partial 18S ribosomal RNA sequences. Plant Systematics and Evolution 188:73–86.

Conti, E., Fischbach, A., and Sytsma, K. J. 1993. Tribal relationships in the Onagraceae using *rbc*L sequence data. Annals of the Missouri Botanical Garden 80:672–685.

Conti, E., Litt, A., and Sytsma, K.J. 1996. Circumscription of Myrtales and their relationships to other rosids: evidence from *rbc*L sequence data. American Journal of Botany 83:221–233.

Cosner, M. E., Jansen, R. K., and Lammers, T. K. 1994. Phylogenetic relationships in the Campanulales based on *rbc*L sequences. Plant Systematics and Evolution 190:79–95.

Crawford, D. J., Palmer, J. D., and Kobayashi, M. 1991. Chloroplast DNA restriction site variation, phylogenetic relationships, and character evolution among sections of North American *Coreopsis* (Asteraceae). Systematic Botany 16:211–224.

Crisp, M. D. 1994. Evolution of bird-pollination in some Australian legumes (Fabaceae). Pp. 281–309 in *Phylogenetics and ecology*, P. Eggleton and R. Vane-Wright, eds. London, England: Academic Press.

Cronquist, A. 1981. *An integrated system of classification of flowering plants.* New York, NY: Columbia University Press.

Curtis, J. T. 1959. *The vegetation of Wisconsin.* Madison, WI: University of Wisconsin Press.

Darwin, C. 1859. *On the origin of species by means of natural selection.* London, England: John Murray.

Davis, J. I. 1995. A phylogenetic structure for the monocotyledons, as inferred from chloroplast DNA restriction site variation, and a comparison of measures of clade support. Systematic Botany 20:503–527.

de Pinna, M. C. C. 1991. Concepts and tests of homology in the cladistic paradigm. Cladistics 7:367–394.

de Queiroz, A., and Wimberger, P. H. 1993. The usefulness of behavior for phylogeny estimation: levels of homoplasy in behavioral and morphological characters. Evolution 47:46–60.

Dixon, M. T., and Hillis, D. M. 1993. Ribosomal-RNA secondary structure: compensatory mutations and implications for phylogenetic analysis. Molecular Biology and Evolution 10:256–267.

Doebley, J. 1990. Molecular systematics of *Zea* (Gramineae). Maydica 35:143–150.

Donoghue, M. J. 1983. The phylogenetic relationships of *Viburnum*. Pp. 143–166 in *Advances in cladistics, vol. 2*, N. I. Platnick and V. A. Funk, eds. New York, NY: Columbia University Press.

Donoghue, M. J. 1994. Progress and prospects in reconstructing plant phylogeny. Annals of the Missouri Botanical Garden 81:405–418.

Donoghue, M. J., and Baldwin, B. G. 1993. Phylogenetic analysis of *Viburnum* based on ribosomal DNA sequences from the internal transcribed spacer regions. American Journal of Botany 80(6):145.

Donoghue, M. J., Olmstead, R. G., Smith, J. F., and Palmer, J. D. 1992. Phylogenetic relationships of Dipsacales based on *rbc*L sequences. Annals of the Missouri Botanical Garden 79:333–345.

Donoghue, M. J., and Sanderson, M. J. 1992. The suitability of molecular and morphological evidence in reconstructing plant phylogeny. Pp. 340–368 in *Molecular systematics of plants*, P. S. Soltis, D. E. Soltis, and J. J. Doyle, eds. New York, NY: Chapman and Hall.

Donoghue, M. J., and M. J. Sanderson. 1994. Complexity and homology in plants. Pp. 393–421 in B. K. Hall, ed. *Homology: the hierarchical basis of comparative biology.* San Diego, CA: Academic Press.

Donoghue, M. J., and Sytsma, K. J. 1993. Phylogenetic analysis of *Viburnum* based on chloroplast DNA restriction site data. American Journal of Botany 80(6):146.

Downie, S. R., and Palmer, J. D. 1992. Restriction site mapping of the chloroplast DNA inverted repeat: a molecular phylogeny of the Asteridae. Annals of the Missouri Botanical Garden 79:266–283.

Doyle, J. A., Donoghue, M. J., and Zimmer, E. A. 1994. Integration of morphological and ribosomal RNA data on the origin of the angiosperms. Annals of the Missouri Botanical Garden 81:419–450.

Doyle, J. J. 1996. Homoplasy connections and disconnections: genes and species, molecules and morphology. Pp. 37–66 in *Homoplasy: the recurrence of similarity in evolution*, M. J. Sanderson and L. Hufford, eds. San Diego, CA: Academic Press.

Doyle, J. J., Doyle, J. L., and Brown, A. H. D. 1990. Chloroplast DNA phylogenetic affinities of newly described species in *Glycine* (Leguminosae: Phaseoleae). Systematic Botany 15:466–471.

Duvall, M. R., and Doebley, J. F. 1990. Restriction site variation in the chloroplast genome of *Sorghum* (Poaceae). Systematic Botany 15:472–480.

Duvall, M. R., Clegg, M. T., Chase, M. W., Clark, W. D., Kress, J. W., Hills, H. G., Equiarte, L. E, Smith, J. F., Gaut, B. S., Zimmer, E. A., and Learn, G. H., Jr. 1993. Phylogenetic hypotheses for the monocotyledons constructed from *rbc*L sequence data. Annals of the Missouri Botanical Garden 80:607–619.

Dykhuizen, D. E., and Green, L. 1991. Recombination in *Escherichia coli* and the definition of biological species. Journal of Bacteriology 173:7257–7268.

Ehleringer, J. R., Mooney, H. A., Gulmon S. L., and Rundel, P. W. 1981. Parallel evolution of leaf pubescence in *Encelia* in coastal deserts of North and South America. Oecologia 49:38–41.

Ehleringer, J. R., Ullmann, I., Lange, O., Farquhar, G. D., Cowan, Schulze, E.-D., and Ziegler, H. 1986. Mistletoes: a hypothesis concerning morphological and chemical avoidance of herbivory. Oecologia 70:234–237.

Endress, P. K. 1996. Homoplasy in angiosperm flowers. Pp. 303–325 in *Homoplasy: the recurrence of similarity in evolution*, M. J. Sanderson and L. Hufford, eds. San Diego, CA: Academic Press.

Faegri, K., and Pijl, L. van der. 1971. *The principles of pollination biology*, 2nd ed. New York, NY: Pergamon.

Faith, D. P., and Cranston, P. S. 1991. Could a cladogram this short have arisen by chance? On permutation tests for cladistic structure. Cladistics 7:1–28.

Farris, J. S. 1972. Estimating phylogenetic trees from distance matrices. American Naturalist 106:645–668.

Farris, J. S. 1989a The retention index and the rescaled consistency index. Cladistics 5:417–419.

Farris, J. S. 1989b. The retention index and homoplasy excess. Systematic Zoology 38:406–407.

Felsenstein, J. 1978. Cases in which parsimony or compatibility methods will be positively misleading. Systematic Zoology 27:401–410.

Futuyma, D. J. 1986. *Evolutionary biology*, 2nd ed. Sunderland, MA: Sinauer Associates.

Givnish, T. J. 1979. On the adaptive significance of leaf form. Pp. 375–407 in *Topics in plant population biology*, O. T. Solbrig, S. Jain, G. B. Johnson, and P. H. Raven, eds. New York, NY: Columbia University Press.

Givnish, T. J. 1984. Leaf and canopy adaptations in tropical forests. Pp. 51–84 in *Physiological ecology of plants of the wet tropics*, E. Medina, H. A. Mooney, and C. Vásquez-Yánes, eds. The Hague, Netherlands: Dr. Junk.

Givnish, T. J. 1986. Biomechanical constraints on canopy geometry in forest herbs. Pp. 525–583 in *On the economy of plant form and function*, T. J. Givnish, ed. Cambridge, England: Cambridge University Press.

Givnish, T. J. 1987. Comparative studies of leaf form: assessing the relative roles of selective pressures and phylogenetic constraints. New Phytologist 106(Suppl.):131–160.

Givnish, T. J. 1988. Adaptation to sun vs. shade: a whole-plant perspective. Australian Journal of Plant Physiology 15:63–92.

Givnish, T. J. 1990. Leaf mottling: relation to growth form and leaf phenology, and possible role as camouflage. Functional Ecology 6:463–474.

Givnish, T. J. 1995. Plant stems: biomechanical adaptation for energy capture and influence on species distributions. Pp. 3–49 in *Plant stems: physiology and functional morphology*, B. L. Gartner, ed. New York, NY: Chapman and Hall.

Givnish, T. J. 1997. Adaptive plant evolution on islands: classical patterns, molecular data, new insights. In press in *Evolution on islands*, P. R. Grant, ed. Oxford, England: Oxford University Press.

Givnish, T. J., and Sytsma, K. J. 1992. Chloroplast DNA restriction site data yield phylogenies with less homoplasy than analyses based on morphology or DNA sequences (abstract). American Journal of Botany 79 (Supplement):145.

Givnish, T. J., and Sytsma, K. J. 1997. Consistency, characters, and the likelihood of correct phylogenetic inference. Molecular Phylogenetics and Evolution 7:320–333.

Givnish, T. J., Sytsma, K. J., Smith, J. F., and Hahn, W. S. 1994. Thorn-like prickles and heterophylly in *Cyanea*: adaptations to extinct avian browsers on Hawaii? Proceedings of the National Academy of Sciences, USA 91:2810–2814.

Givnish, T. J., Sytsma, K. J., Smith, J. F., and Hahn, W. S. 1995. Molecular evolution, adaptive radiation, and geographic speciation in *Cyanea* (Campanulaceae). Pp. 288–337 in *Hawaiian biogeography: evolution on a hot spot archipelago*, W. L. Wagner and V. A. Funk, eds. Washington, DC: Smithsonian Institution Press.

Gleason, H. A. 1917. The structure and development of the plant association. Bulletin of the Torrey Botanical Club 43:463–481.

Goldblatt, P., Manning, J. C., and Bernhadt, P. 1995. Pollination biology of *Lapeirousia* subgenus *Lapeirousia* (Iridaceae) in Southern Africa; floral divergence and adaptation for long-tongued fly pollination. Annals of the Missouri Botanical Garden 82:517–534.

Goloboff, P. A. 1991a. Homoplasy and the choice among cladograms. Cladistics 7:215–232.

Goloboff, P. A. 1991b. Random data, homoplasy, and information. Cladistics 7: 395–406.

Gottlieb, L. D. 1988. Toward molecular genetics in *Clarkia*: gene duplication and molecular characterization of PGI genes. Annals of the Missouri Botanical Garden 75:1169–1179.

Grant, V., and Grant, K. A. 1965. *Flower pollination in the phlox family*. New York, NY: Columbia University Press.

Gray, G. S., and Fitch, W. M. 1983. Evolution of antibiotic resistance genes: the DNA sequence of a kanomycin resistance gene from *Staphylococcus aureus*. Molecular Biology and Evolution 1:57–66.

Hahn, W. J., Givnish, T. J., and Sytsma, K. J. 1995. Evolution of the monocot chloroplast inverted repeat: I. Evolution and phylogenetic implications of the ORF 2280 deletion. Pp. 579–587 in *Proceedings of the Kew monocot symposium*, P. Rudall, D. Cutler, and C. Humphries, eds. Kew, England: Royal Botanical Gardens.

Halanych, K. M., Bacheller, J. D., Aguinaldo, A. M. A., Liva, S. M., Hillis, D. M., and Lake, J. A. 1995. Evidence from 18s ribosomal DNA that the lophophorates are protostome animals inarticulate. Science 267:1641–1643.

Hall, B. K (ed.). 1994. *Homology: the hierarchical basis of comparative biology*. San Diego, CA: Academic Press.

Hamby, R. K., and Zimmer, E. A. 1988. Ribosomal RNA sequences for inferring phylogeny within the grass family (Poaceae). Plant Systematics and Evolution 34:393–400.

Hamby, R. K., and Zimmer, E. A. 1992. Ribosomal RNA sequences for inferring phylogeny within the grass family (Poaceae). Plant Systematics and Evolution 34:393–400.

Harvey, P. H., and Purvis, A. 1991. Comparative methods for explaining adaptations. Nature 351:619–623.

Hasebe, M., Omori, T., Nakazawa, M., Sano, T., Kato, M., and Iwatsuki, K. 1994. *rbc*L gene sequences provide evidence for the evolutionary lineages of leptosporangiate ferns. Proceedings of the National Academy of Sciences, USA 91:5730–5734.

Hedren, M., Chase, M. W., and Olmstead, R. G. 1995. Relationships in the Acanthaceae and related families as suggested by cladistic analysis of *rbc*L nucleotide sequences. Plant Systematics and Evolution 194:93–110.

Hempel, A. L., Reeves, P. A., Olmstead, R. G., and Jansen, R. K. 1995. Implications of *rbc*L sequence data for higher order relationships of the Loasaceae and the anomalous aquatic plant *Hydrostachys* (Hydrostachyaceae). Plant Systematics and Evolution 194:25–38.

Hennig, W. 1979. *Phylogenetic systematics*. Urbana, IL: University of Illinois Press.

Hillis, D. M. 1995. Approaches for assessing phylogenetic accuracy. Systematic Biology 44:3–16.

Hillis, D. M. 1996. Inferring complex phylogenies. Nature 383:130–131.

Hillis, D. M., Huelsenbeck, J. P., and Cunningham, C. W. 1994. Application and accuracy of molecular phylogenies. Science 264:671–677.

Hixson, J. E., and Brown, W. M. 1986. A comparison of the small mitochondrial RNA genes from the mitochondrial DNA of the great apes and humans: sequence, structure, evolution, and phylogenetic implications. Molecular Biology and Evolution 3:1–18.

Hoot, S. B., and Crane, P. 1995. Inter-familial relationships in the Ranunculidae based on molecular systematics. Plant Systematics and Evolution [Suppl.] 9:119–131.

Hoot, S. B., Culham, A., and Crane, P. R. 1995. The utility of *atp*B gene sequences in resolving phylogenetic relationships: comparison with *rbc*L and 18S ribosomal DNA sequences in the Lardizabalaceae. Annals of the Missouri Botanical Garden 82:194–207.

Horn, H. S. 1971. *The adaptive geometry of trees*. Princeton, NJ: Princeton University Press.

Jansen, R. K., and Palmer, J. D. 1988. Phylogenetic implications of chloroplast DNA restriction site variation in the Mutisieae (Asteraceae). American Journal of Botany 75:753–766.

Jansen, R. K., Holsinger, K. E., Michaels, H. J., and Palmer, J. D. 1990. Phylogenetic analysis of restriction site data at higher taxonomic levels: an example from the Asteraceae. Evolution 44:2089–2105.

Jansen, R. K., Michaels, H. J., Wallace, R. S., Kim, K. -J., Keeley, S. C., Watson, L. E., and Palmer, J. D. 1992. Chloroplast DNA variation in the Asteraceae: phylogenetic and evolutionary implications. Pp. 252–279 in *Molecular systematics of plants*, P. S. Soltis, D. E. Soltis, and J. J. Doyle, eds. New York, NY: Chapman and Hall.

Johnson, L. A., and Soltis, D. E. 1994. *mat*K DNA sequences and phylogenetic reconstruction in Saxifragaceae s. str. Systematic Botany 19:143–156.

Johnson, L. A., and Soltis, D. E. 1995. Phylogenetic inference in Saxifragaceae sensu stricto and *Gilia* (Polemoniaceae) using *mat*K sequences. Annals of the Missouri Botanical Garden 82:149–175.

Johnson, S. D. 1995. Pollination and the evolution of floral traits: selected studies in the Cape flora. Ph.D. dissertation, University of Cape Town, Cape Town, South Africa.

Kadereit, J., and Sytsma, K. J. 1992. Disassembling the genus *Papaver*: a restriction site analysis of chloroplast DNA. Nordic Journal of Botany 12:205–217.

Kellogg, E. A. 1992. Tools for studying the chloroplast genome in the Triticeae (Gramineae): an *Eco*R1 map, a diagnostic deletion, and support for *Bromus* as an outgroup. American Journal of Botany 79:186–197.

Kim, K.-J., and Jansen, R. K. 1994. Comparisons of phylogenetic hypotheses among different data sets in dwarf dandelions (*Krigia*, Asteraceae): additional information from internal transcribed spacer sequences of nuclear ribosomal DNA. Plant Systematics and Evolution 190:157–185.

Kim, K.-J., and Jansen, R. K. 1995. *ndh*F sequence evolution and the major clades in the sunflower family. Proceedings of the National Academy of Sciences, USA 92:10379–10383.

Kim, K.-J., Jansen, R. K., Wallace, R. S., Michaels, H. J., and Palmer, J. D. 1992. Phylogenetic implications of *rbc*L variation in the Asteraceae. Annals of the Missouri Botanical Garden 79:428–445.

Klassen, G. J., Mooi, R. D., and Locke, A. 1991. Consistency indices and random data. Systematic Zoology 40:446–457.

Kluge, A. G. 1989. A concern for evidence and a phylogenetic hypothesis of relationships among *Epicrates* (Boidae, Serpentes). Systematic Zoology 38:7–25.

Kluge, A. G., and Farris, J. S. 1969. Quantitative phyletics and the evolution of anurans. Systematic Zoology 18:1–32.

Knox, E. B. 1992. Evolution of the giant Senecios and giant Lobelias in Eastern Africa. Ph.D. dissertation, University of Michigan, Ann Arbor.

Knox, E., Downie, S. R., and Palmer, J. D. 1993. Chloroplast genome rearrangements and the evolution of giant lobelias from herbaceous ancestors. Molecular Biology and Evolution 10:414–430.

Kocher, T. D., Conroy, J. A., McKaye, K. R., and Stauffer, J. R. 1993. Similar morphologies of chichlid fish in Lakes Tanganyika and Malawi are due to convergence. Molecular Phylogenetics and Evolution 2:158–165.

Kron, K. A., and Chase, M. W. 1993. Systematics of the Ericaceae, Empetraceae, Epacridaceae and related taxa based upon *rbc*L sequence data. Annals of the Missouri Botanical Garden 80:735–741.

Lammers, T. G. 1989. Revision of *Brighamia* (Campanulaceae: Lobelioideae), a caudiciform succulent endemic to the Hawaiian Islands. Systematic Botany 14:133–138.

Lammers, T. G. 1990. Campanulaceae. Pp. 420–489 in *Manual of the flowering plants of Hawai'i*, W. L. Wagner, D. R. Herbst, and S. H. Sohmer, eds. Honolulu, HI: Bishop Museum Publications.

Larson, A. 1994. The comparison of morphological and molecular data in phylogenetic systematics. Pp. 371–390 in *Molecular ecology and evolution: approaches and applications*, B. Schierwater, G. P. Wagner, and R. DeSalle, eds. Basel, Switzerland: Birkhäuser Verlag.

Lavin, M., and Doyle, J. J. 1991. Tribal relationships of *Sphinctospermum* (Leguminosae): integration of traditional and chloroplast DNA data. Systematic Botany 16:162–172.

Lidén, M, Fukuhara, T., and Axberg, T. 1995. Phylogeny of *Corydalis*, ITS and morphology. Plant Systematics and Evolution [Suppl] 9:183–188.

Lonsdale, D. M. 1989. The plant mitochondrial genome. Pp. 230–295 in *The biochemistry of plants, volume 15: molecular biology*, P. K. Stumpf and E. E. Conn, eds. New York, NY: Academic Press.

Maddison, W. 1995. Phylogenetic histories within and among species. Pp. 273–287 in *Experimental and molecular approaches to plant biosystematics*. Monographs in Systematic Botany, Vol. 53, P. C. Hoch and A. G. Stephenson, eds. St. Louis, MO: Missouri Botanical Garden.

Manen, J.-F., Natali, A., and Ehrendorfer, F. 1994. Phylogeny of Rubiaceae-Rubieae inferred from the sequence of a cpDNA intergene region. Plant Systematics and Evolution 190:195–211.

Maynard Smith, J., Dowson, C. G., and Spratt, B. G. 1991. Localized sex in bacteria. Nature 349:29–31.

McShea, D. W. 1996. Complexity and homoplasy. Pp. 207–225 in *Homoplasy: the recurrence of similarity in evolution*, M. J. Sanderson and L. Hufford, eds. San Diego, CA: Academic Press.

Médigue, C., Rouxel, T., Vigier, P., Hénaut, A., and Danchin, A. 1991. Evidence for horizontal gene transfer in *Escherichia coli* speciation. Journal of Molecular Biology 222:851–856.

Meier, R., Kores, P., and Darwin, S. 1991. Homoplasy slope ratio: a better measurement of observed homoplasy in cladistic analyses. Systematic Zoology 40:74–88.

Milligan, B. G., Hampton, J. N., and Palmer, J. D. 1989. Dispersed repeats and structural reorganization in subclover chloroplast DNA. Molecular Biology and Evolution 6:355–368.

Mishler, B. D. 1994. The cladistic analysis of molecular and morphological data. American Joural of Physical Anthropology 94:143–156.

Mishler, B. D., Bremer, K., Humphries, C. J., and Churchill, S. P. 1988. The use of nucleic acid sequence data in phylogenetic reconstruction. Taxon 37:391–395.

Mitsuhashi, S., and Krcmery, V. (eds.). 1984. *Transferable antibiotic resistance: plasmids and gene manipulation*. New York, NY: Springer-Verlag.

Miyamoto, M. M. 1985. Consensus cladograms and general classifications. Cladistics 1:186–189.

Miyamoto, M. M., and Fitch, W. M. 1995. Testing species phylogenies and phylogenetic methods with congruence. Systematic Biology 44:64–76.

Mooney, H. A. (ed.). 1977. *Convergent evolution in Chile and California: Mediterranean climate ecosystems*. Stroudsburg, PA: Dowden, Hutchinson, and Ross.

Morgan, D. R., Soltis, D. E., and Robertson, K. R. 1994. Systematic and evolutionary implications of *rbc*L sequence variation in Rosaceae. American Journal of Botany 81:890–903.

Nadot, S., Bajon, R., and Lejeune, B. 1994. The chloroplast gene *rps*4 as a tool for the study of Poacee phylogeny. Plant Systematics and Evolution 191:27–38.

Natali, A., Manen, J. -F., and Ehrendorfer, F. 1995. Phylogeny of the Rubiaceae-Rubioideae, in particular the tribe Rubieae: evidence from a non-coding chloroplast DNA sequence. Annals of the Missouri Botanical Garden 82:428–439.

Neale, D. B., and Sederoff, R. R. 1989. Paternal inheritance of chloroplast DNA and maternal inheritance of mitochondrial DNA in loblolly pine. Theoretical and Applied Genetics 77:212–216.

Neigel, J. E., and Avise, J. C. 1986. Phylogenetic relationships of mitochondrial DNA under various demographic models of speciation. Pp. 515–534 in *Evolutionary processes and theory*, S. Karlin and E. Nev, eds. New York, NY: Academic Press.

Nelson, G. J. 1989. Cladistics and evolutionary models. Cladistics 5:275–289.

Nepokroeff, M., and Sytsma, K. J. 1997. Evolution in cloud forest members of *Psychotria* section *Notopleura*, Rubiaceae. In press in *Natural History of the Monteverde Cloud Forest Preserve*, N. Nadkarni and N. Wheelwright, eds. New York, NY: Oxford University Press.

Newton, K. J. 1988. Plant mitochondrial genomes: organization, expression and variation. Annual Review of Plant Physiology 39:503–532.

Nickrent, D. L., and Soltis, D. E. 1995. A comparison of angiosperm phylogenies from nuclear 18S rDNA and *rbc*L sequences. Annals of the Missouri Botanical Garden 82:208–234.

Odrzykoski, I. J., and Gottlieb, L. D. 1984. Duplications of genes coding 6-phosphogluconate dehydrogenase in *Clarkia* (Onagraceae) and their phylogenetic implications. Systematic Botany 9:479–489.

Olmstead, R. G., and Palmer, J. D. 1992. A chloroplast DNA phylogeny of the Solanaceae: subfamilial relationships and character evolution. Annals of the Missouri Botanical Garden 79:346–360.

Olmstead, R. G., and Palmer, J. D. 1994. Chloroplast DNA systematics – a review of methods and data analysis. American Journal of Botany 81:1205–1224.

Olmstead, R. G., and Reeves, P. A. 1995. Evidence for the polyphyly of the Scrophulariaceae based on chloroplast *rbc*L and *ndh*F sequences. Annals of the Missouri Botanical Garden 82:176–193.

Olmstead, R. G., and Sweere, J. A. 1995. Combining data in phylogenetic systematics: an empirical approach using 3 molecular data sets in the Solanaceae. Systematic Biology 43:467–481.

Olmstead, R. G., Bremer, B., Scott, K. M., and Palmer, J. D. 1993. A parsimony analysis of the Asteridae sensu lato based on *rbc*L sequences. Annals of the Missouri Botanical Garden 80:700–722.

Olmstead, R. G., Jansen, R. K., Michaels, H. J., Downie, S. R., and Palmer, J. D. 1990. Chloroplast DNA and phylogenetic studies in the Asteridae. Pp. 119–134 in *Biological approaches and evolutionary trends in plants*, S. Kawano, ed. San Diego, CA: Academic Press.

Olmstead, R. G., Michaels, H. J., Scott, K. M., and Palmer, J. D. 1992. Monophyly of the Asteridae and identification of their major lineages inferred from DNA sequences of *rbc*L. Annals of the Missouri Botanical Garden 79:249–265.

Orians, G. H., and Solbrig, O. T. 1977. A cost-income model of leaves and roots with special reference to arid and semiarid areas. American Naturalist 111:677–690.

Oxelman, B., and Lidén, M. 1995a. Generic boundaries in the tribe Sileneae (Caryophyllaceae) as inferred from nuclear rDNA sequences. Taxon 44:525–542.

Oxelman, B., and Lidén, M. 1995b. The position of *Circaeaster* – evidence from nuclear ribosomal DNA. Plant Systematics and Evolution (Suppl) 9:189–193.

Palmer, J. D. 1987. Chloroplast DNA evolution and biosystematic uses of chloroplast DNA variation. American Naturalist 130:S6–S29.

Palmer, J. D. 1992a. Comparison of chloroplast and mitochondrial genome evolution in plants. Pp. 99–133 in *Plant gene research, vol. 6*, R. G. Hermann, ed. Wien, Germany: Springer-Verlag.

Palmer, J. D. 1992b. Mitochondrial DNA in plant systematics: applications and limitations. Pp. 36–49 in *Molecular systematics of plants*, P. S. Soltis, D. E. Soltis, and J. J. Doyle, eds. New York, NY: Chapman and Hall.

Palmer, J. D., and Zamir, D. 1982. Chloroplast DNA evolution and phylogenetic relationships in *Lycopersicon*. Proceedings of the National Academy of Sciences, USA 79:5006–5010.

Palmer, J. D., Jansen, R. K., Michaels, H. J., Chase, M. W., and Manhart, J. R. 1988. Chloroplast DNA variation and plant phylogeny. Annals of the Missouri Botanical Garden 75:1180–1206.

Pamilo, P., and Nei, M. 1988. Relationships between gene trees and species trees. Molecular Biology and Evolution 5:568–583.

Patterson, C. 1988. Homology in classical and molecular biology. Molecular Biology and Evolution 5:603–625.

Patterson, C., Williams, D. M., and Humphries, C. J. 1993. Congruence between molecular and morphological phylogenies. Annual Review of Ecology and Systematics 24:153–188.

Philbrick, C. T., and Jansen, R. K. 1991. Phylogenetic studies of North American *Callitriche* (Callitrichaceae) using chloroplast DNA restriction fragment analysis. Systematic Botany 16:478–491.

Price, R. A., and Palmer, J. D. 1993. Phylogenetic relationships of the Geraniaceae and Geraniales from *rbc*L sequence comparisons. Annals of the Missouri Botanical Garden 80: 661–671.

Qui, Y. L., Chase, M. W., Les, D. H., and Parks, C. R. 1993. Molecular phylogenetics of the Magnoliidae: cladistic analyses of nucleotide sequences of the plastid gene *rbc*L. Annals of the Missouri Botanical Garden 80:587–606.

Ranker, T. A., Soltis, D. E., Soltis, P. S., and Gilmartin, A. J. 1990. Subfamilial phylogenetic relationships of the Bromeliaceae: evidence from chloroplast DNA restriction site variation. Systematic Botany 15:425–434.

Rieseberg, L. H., Beckstrom-Sternberg, S. M., Liston, A., and Arias, D. M. 1991. Phylogenetic and systematic inferences from chloroplast DNA and isozyme variation in *Helianthus* sect. *Helianthus* (Asteraceae). Systematic Botany 16:50–76.

Rieseberg, L. H., and Brunsfeld, S. J. 1992. Molecular evidence and plant introgression. Pp. 151–176 in *Molecular systematics of plants*, P. S. Soltis, D. E. Soltis, and J. J. Doyle, eds. New York, NY: Chapman and Hall.

Rock, J. F. 1919. A monographic study of the Hawaiian species of the tribe Lobelioideae, family Campanulaceae. Memoirs of the Bernice Bishop Museum 7:1–394.

Rodman, J. E. 1991. A taxonomic analysis of glucosinolate-producing plants. Part 2: cladistics. Systematic Botany 16:619–629.

Rodman, J., Price, R., Karol, K., Conti, E., Sytsma, K. J., and Palmer, J. D. 1993. Gene sequence evidence for monophyly of mustard oil families. Annals of the Missouri Botanical Garden 80:686–699.

Saitou, N. 1988. Property and efficiency of the maximum likelihood method for molecular phylogeny. Journal of Molecular Evolution 27:261–273.

Sanderson, M. J. 1991. In search of homoplastic tendencies: statistical inference of topological patterns in homoplasy. Evolution 45: 351–358.

Sanderson, M. J., and Donoghue, M. J. 1989. Patterns of variation in levels of homoplasy. Evolution 43:1781–1795.

Sanderson, M. J., and Donoghue, M. J. 1996. The relationship between homoplasy and confidence in a phylogenetic tree. Pp. 67–89 in *Homoplasy: the recurrence of similarity in evolution*, M. J. Sanderson and L. Hufford, eds. San Diego, CA: Academic Press.

Sanderson, M. J., and Hufford, L. (eds.). 1996. *Homoplasy: the recurrence of similarity in evolution*. San Diego, CA: Academic Press.

Sang, T., Crawford, D. J., Kim, S.-C., and Stuessy, T. F. 1994. Radiation of the endemic genus *Dendroseris* (Asteraceae) on the Juan Fernandez Islands: evidence from sequences of the ITS regions of nuclear ribosomal DNA. American Journal of Botany 81:1494–1501.

Sang, T., Crawford, D. J., Stuessy, T. F., and Silva O, M. 1995. ITS sequences and the phylogeny of the genus *Robinsonia* (Asteraceae). Systematic Botany 20:55–64.

Scotland, R. W., Sweere, J. A., Reeves, P. A., and Olmstead, R. G. 1995. Higher-level systematics of Acanthaceae determined by chloroplast DNA sequences. American Journal of Botany 82:266–275.

Shinwari, Z. K., Kato, H., Terauchi, R., and Kawano, S. 1994. Phylogenetic relationships among genera in the Liliaceae-Asparagoideae-Polygonatae s.l. inferred from *rbc*L gene sequence data. Plant Systematics and Evolution 192:263–277.

Sibley, C. G., and Ahlquist, J. E. 1987. Avian phylogeny reconstructed from comparisons of the genetic material, DNA. Pp. 95–121 in *Molecules and morphology in evolution: conflict or compromise?*, C. Patterson, ed. Cambridge, England: Cambridge University Press.

Skinner, M. W. 1988. Comparative pollination ecology and floral evolution in Pacific Coast *Lilium*. Ph.D. dissertation, Harvard University.

Smith, J. F., and Sytsma, K. J. 1994. Evolution in the Andean epiphytic genus *Columnea* (Gesneriaceae) Part II: Chloroplast DNA restriction site variation. Systematic Botany 19:317–336.

Snedecor, G. W., and Cochran, W. G. 1989. *Statistical methods. 8th ed*. Ames, IA: Iowa State University Press.

Sokal, R. R., and Shao, K.-T. 1985. Character stability in 39 data sets. Systematic Zoology 34:83–89.

Soltis, D. E., Morgan, D., Grabel, A., Soltis, P. S., and Kuzoff, R. 1993. Molecular systematics of Saxifragaceae. American Journal of Botany 80:1056–1081.

Soltis, D. E., Soltis, P. S., and Bothel, K. D. 1990. Chloroplast DNA evidence for the origins of the monotypic *Bensoniella* and *Conimitella* (Saxifragaceae). Systematic Botany 15:349–362.

Soltis, D. E., Xiang, Q.-Y., and Hufford, L. 1995. Relationships and evolution of Hydrangeaceae based on *rbc*L sequence data. American Journal of Botany 82:504–514.

Soltis, P. S., Soltis, D. E., and Gottlieb, L. D. 1987. Phosphoglucomutase gene duplications in *Clarkia* (Onagraceae) and their phylogenetic implications. Evolution 41:667–671.

Soreng, R. J. 1990. Chloroplast-DNA phylogenetics and biogeography in a reticulating group: study in *Poa* (Poaceae). American Journal of Botany 77:1383–1400.

Soreng, R. J., Davis, J. I., and Doyle, J. J. 1990. A phylogenetic analysis of chloroplast DNA restriction site variation in Poaceae subfam. Pooideae. Plant Systematics and Evolution 172:83–97.

Souza, V. T., Nguyen, T., Hudson, R. R., Piñero, D., and Lenski, R. E. 1992. Hierarchical analysis of linkage disequilibrium in *Rhizobium* populations: evidence for sex? Proceedings of the National Academy of Sciences, USA 89:8389–8393.

Spooner, D. M., Sytsma, K. J., and Conti, E. 1991. Chloroplast DNA evidence for genome differentiation in wild potatoes (*Solanum* sect. *Petota*: Solanaceae). American Journal of Botany 78:1354–1366.

Steele, K. P., Holsinger, K. E., Jansen, R. K., and Taylor, D. W. 1988. Phylogenetic relationships in green plants – a comment on the use of 5S ribosomal RNA sequences by Bremer et al. Taxon 37:135–138.

Steele, K. P., and Vilgalys, R. 1994. Phylogenetic analyses of Polemoniaceae using nucleotide sequences of the plastid gene *mat*K. Systematic Botany 19:126–142.

Stevens, P. F. 1984. Homology and phylogeny: morphology and systematics. Systematic Botany 9:395–409.

Suh, Y., and Simpson, B. B. 1990. Phylogenetic analysis of chloroplast DNA in North American *Gutierrezia* and related genera (Asteraceae: Astereae). Systematic Botany 15:660–670.

Swofford, D. L. 1993. *PAUP: phylogenetic analysis using parsimony, ver. 3.1*. Champaign, IL: Illinois Natural History Survey.

Sytsma, K. J. 1990. DNA and morphology: inferences of plant phylogeny. Trends in Ecology and Evolution 5:104–110.

Sytsma, K. J., and Baum, D. A. 1996. Molecular phylogenies and the diversification of angiosperms. Pp. 314–340 in *Flowering plant origin, evolution and phylogeny*, D. W. Taylor and L. J. Hickey, eds. New York, NY: Chapman and Hall.

Sytsma, K. J., and Gottlieb, L. D. 1986a. Chloroplast DNA evolution and phylogenetic relationships in *Clarkia* sect. *Peripetasma* (Onagraceae). Evolution 40:1248–1261.

Sytsma, K. J., and Gottlieb, L. D. 1986b. Chloroplast DNA evidence for the origin of the genus *Heterogaura* from a species of *Clarkia* (Onagraceae). Proceedings of the National Academy of Sciences, USA 83:5554–5557.

Sytsma, K. J., and Smith, J. F. 1988. DNA and morphology: comparisons in Onagraceae. Annals of the Missouri Botanical Garden 75:1217–1237.

Sytsma, K. J., and Smith, J. F. 1992. Molecular systematics of Onagraceae: examples from *Fuchsia* and *Clarkia*. Pp. 295–323 in *Molecular systematics of plants*, P. S. Soltis, D. E. Soltis, and J. J. Doyle, eds. New York, NY: Chapman and Hall.

Sytsma, K. J., Smith, J. F., and Gottlieb, L. D. 1990. Phylogenetics in *Clarkia* (Onagraceae): restriction site mapping of chloroplast DNA. Systematic Botany 15:280–295.

Sytsma, K. J., Smith, J. F., and Berry, P. E. 1991a. Biogeography and evolution of morphology, breeding systems, flavonoids, and chloroplast DNA in the four Old World species of *Fuchsia* (Onagraceae). Systematic Botany 16:257–269.

Sytsma, K. J., Smith, J. F., and Hoch, P. C. 1991b. A chloroplast DNA analysis of tribal and generic relationships within Onagraceae. American Journal of Botany 78:222.

Thomas, C. M. 1989. *Promiscuous plasmids of gram-negative bacteria.* New York, NY: Academic Press.

Titus, T. A., and Larson, A. 1995. A molecular phylogenetic perspective on the evolutionary radiation of the salamander family Salamandridae. Systematic Biology 44:125–151.

Udovicic, F., McFadden, G. I., and Ladiges, P. Y. 1995. Phylogeny of *Eucalyptus* and *Angophora* based on 5S rDNA spacer sequence data. Molecular Phylogenetics and Evolution 4:247–256.

Valdez, A. M., and Piñero, D. 1992. Phylogenetic estimation of plasmid exchange in bacteria. Evolution 46:641–656.

Vilgalys, R. 1986. Phenetic and cladistic relationships in *Collybia* sect. *Lepipedes* (Fungi: Basidiomycetes). Taxon 35:225–233.

Wake, D. B. 1991. Homoplasy: the result of natural selection, or evidence of design limitations? American Naturalist 138:543–567.

Wallace, R. S., and Jansen, R. K. 1990. Systematic implications of chloroplast DNA variation in the genus *Microseris* (Asteraceae: Lactuceae). Systematic Botany 15:606–616.

Weising, K., Schell, J., and Kahl, G. 1988. Foreign genes in plants: transfer, structure, expression, and applications. Annual Review of Genetics 22:421–477.

Wendel, J. F., Stewart, J. M., and Rettig, J. H. 1991. Molecular evidence for homoploid reticulate evolution among Australian species of *Gossypium*. Evolution 45:694–711.

Wheeler, W. C., and Honeycutt, R. L. 1988. Paired sequence difference in ribosomal RNAs: evolutionary and phylogenetic implications. Molecular Biology and Evolution 5:90–96.

Whittaker, R. H. 1956. Vegetation of the Great Smoky Mountains. Ecological Monographs 26:1–80.

Wiegrefe, S. J., Sytsma, K. J., and Guries, R. P. 1994. Phylogeny of elms (*Ulmus*, Ulmaceae): molecular evidence for a sectional classification. Systematic Botany 19:590–612.

Williams, D. M. 1993. A note on molecular homology: multiple patterns from single datasets. Cladistics 9:233–245.

Wolf, P. G., Soltis, P. S., and Soltis, D. E. 1994. Phylogenetic relationships of dennstaedtioid ferns: evidence from *rbc*L sequence variation. Molecular Phylogenetics and Evolution 3:383–392.

Wolfe, K. H., Li, W.-H., and Sharp, P. M. 1987. Rates of nucleotide substitution vary greatly among plant mitochondrial, chloroplast, and nuclear DNAs. Proceedings of the National Academy of Sciences, USA 84:9054–9058.

Xiang, Q.-Y., Soltis, D. E., Morgan, D. R., and Soltis, P. S. 1993. Phylogenetic relationships of *Cornus* L. sensu lato and putative relatives inferred from *rbc*L sequence data. Annals of the Missouri Botanical Garden 80:723–734.

Yeates, D. 1992. Why remove autapomorphies? Cladistics 8:387–389.

Yuan, Y.-M., Küpfer, P., and Doyle, J. J. 1996. Infrageneric phylogeny of the genus *Gentiana* (Gentianaceae) inferred from nucleotide sequences of the internal transcribed spacers (ITS) of nuclear ribosomal DNA. American Journal of Botany 83:641–652.

3 Adaptive radiation of the Hawaiian silversword alliance: congruence and conflict of phylogenetic evidence from molecular and non-molecular investigations

Bruce G. Baldwin

The Hawaiian silversword alliance (Compositae) is a lineage of 28 Hawaiian-endemic species in *Argyroxiphium*, *Dubautia*, and *Wilkesia* (Carr 1985) that has been regarded as "the most remarkable example of adaptive radiation in plants" (Raven et al. 1992). Evidence from multiple lines of systematic investigation discussed herein demonstrates that extreme and rapid divergence in morphological and physiological characters has occurred within the silversword alliance in association with minimal inter-island dispersal and major ecological shifts between strongly contrasting environments on Kaua`i, O`ahu, Maui Nui (i.e., the once-contiguous island group including Lana`i, Maui, and Moloka`i), and, perhaps, Hawai`i.

Morphological and ecological diversity

Diversity in the evergreen, woody and semi-woody growth forms represented in the silversword alliance includes acaulescent or short-stemmed, monocarpic and polycarpic rosette plants (*Argyroxiphium*, 5 species); long-stemmed, monocarpic and polycarpic rosette plants (*Wilkesia*, 2 species); and trees, shrubs, subshrubs, matplants, cushion plants, and lianas (*Dubautia*, 21 species) (Figures 3.1, 3.2; see Carr 1985). Climatic situations and vegetation types occupied by members of the silversword alliance span nearly the entire range of environmental conditions found in the Hawaiian Islands. Species occurrences have been recorded from 75 to 3,750 m in elevation in sites that can differ dramatically in annual precipitation, from less than 400 mm on leeward slopes of the youngest islands to over 12,000 mm in the mountains of central Kaua`i and West Maui, among the wettest sites known on earth (Carr 1985). Members of the silversword alliance can be found on dry shrublands, dry forests, subalpine shrublands, subalpine forests, alpine deserts, mesic forests, wet forests, and bogs and on young lava flows (Carr 1985).

Tissue osmotic and elastic properties that influence resistance to drought stress show marked variation across species (Robichaux 1985; Robichaux and Canfield 1985; Robichaux et al. 1990). In general, species restricted to dry habitats maintain leaf turgor at lower water potentials (i.e., under greater drought stress) than those of wet habitats, the contrasting responses apparently being attributable to different underlying cellular characteristics.

Extensive variation in vegetative morphology and anatomy among species of the silversword alliance (Carlquist 1957, 1959a,b; Carr 1985) appears, at least in part, to be correlated with variation in ecological setting. In general, species of *Dubautia* found in densely vegetated, wet environments have thin cuticles, thin leaves, and loosely organized mesophyll; those from exposed, dry environments tend to have thick cuticles, thick leaves, and compact mesophyll (Carlquist 1959a; Carr et al. 1989). In *Argyroxiphium*, copious extracellular mucilage in leaves and, in some species, a reflective, white indumentum on the leaf surface have been implicated in avoiding or resisting dessication and thermal damage in the open, high elevation environments to which the genus is restricted (Carlquist 1957, 1959a; Robichaux et al. 1990; Lipp et al. 1994; Melcher et al. 1994).

Leaf vasculature also varies greatly within the silversword alliance (Carlquist 1957, 1959a,b; Carr 1985), although the ecological or physiological significance of the variation is mostly uncertain. Venation patterns include a highly reticulate condition with a pronounced midvein – stereotypic of dicots – in the liana *Dubautia latifolia*, and a wide array of leaf types with multiple subparallel or longitudinally directed primary and secondary veins in many species. In *Wilkesia*, the venation is remarkably similar to that of many monocots, with strongly parallel veins and few cross-connecting veinlets.

Systematic background

Despite extensive phenotypic divergence within the silversword alliance, monophyly of the group was first recognized on the basis of morphological and anatomical studies (Keck 1936; Carlquist 1958a,b, 1959a,b). Carlquist (1959b) provided a detailed summary of anatomical characteristics common to members of the silversword alliance that are unusual in Compositae, except in subtribe Madiinae (see below). Carlquist's evidence of close similarity among members of *Argyroxiphium*, *Dubautia*, and *Wilkesia* was reinforced by later biosystematic and cytogenetic investigations by Carr and Kyhos (1981, 1986), who showed that hybrids of partial to full fertility could be produced in all interspecific and intergeneric crossing combinations attempted between species of the three genera.

Carr and Kyhos's (1981, 1986) cytogenetic work revealed that chromosomal evolution at the homoploid level has occurred within the silversword alliance, via reciprocal translocations of chromosome arms. Shared genomic arrangements and involvement of some chromosomes in more than one translocation event allowed Carr and Kyhos (1986, pers. comm.) to reconstruct a preliminary outline of unrooted relationships within much of the group. The morphological and cytogenetic data, supplemented with results from more recent investigations of the silversword alliance based on enzyme electrophoresis (Witter 1986, 1990; Witter and Carr 1988), flavonoid variation (Crins et al. 1988; Crins and Bohm 1990), and ecophysiological properties (Robichaux 1985; Robichaux and Canfield 1985; Robichaux et al. 1990) provide an abundance of evolutionary evidence that can be fruitfully compared and further interpreted with a molecular phylogenetic perspective.

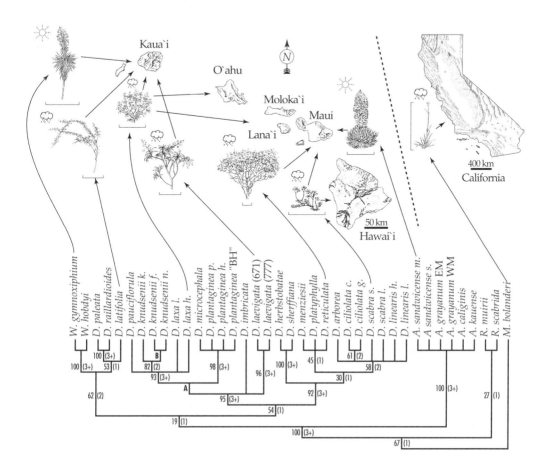

Figure 3.1. The semi-strict consensus of eight minimum-length trees from Fitch-parsimony analysis of 18–26S nuclear ribosomal DNA (nrDNA) internal transcribed spacer (ITS) sequences in representatives of 28 species of the Hawaiian silversword alliance and the American tarweed genera *Madia* and *Raillardiopsis*. Two outgroup species (not shown), *Adenothamnus validus* and *Raillardella pringlei*, were included in the analysis, conducted with PAUP 3.1 (Swofford 1993). Consistency index without uninformative characters (CI') = 0.67; retention index (RI) = 0.85. Bootstrap values, from 100 replicate analyses of the resampled data set (using the PAUP "heuristics" option and "closest" addition sequence of taxa), appear to left of phylogeny branches. Decay index values appear in parentheses to right of phylogeny branches. Generic abbreviations: *A.* = *Argyroxiphium*, *D.* = *Dubautia*, *M.* = *Madia*, *R.* = *Raillardiopsis*, *W.* = *Wilkesia*. EM = East Maui; WM = West Maui. Data, including collection and voucher information, are in Baldwin and Robichaux (1995), wherein only two minimum-length trees were recognized because of lack of unequivocal character support for branches A and B. Plant scale-bars = 1 meter. Arrows indicate island(s) or State (California) of occurrence, not necessarily the location within islands or California where the plants are found. Plant illustrations depict a variety of habits and habitat/island occurrences in the silversword alliance. Sun-symbol = restriction to dry habitats; rain-cloud symbol = restriction to wet habitats (including bogs). Wet habitats do not necessarily receive more annual precipitation than dry habitats, but do have more water available for plant uptake (see Baldwin and Robichaux 1995).

DNA-level phylogenetic studies

Investigations of the silversword alliance based on chloroplast DNA restriction-site variation (Baldwin et al. 1990, 1991) and sequences of the internal transcribed spacer (ITS) of nuclear ribosomal DNA (nrDNA) (Baldwin 1992; Baldwin and Robichaux 1995) yielded fine-scale resolution of relationships in the lineage. Inclusion of extra-Hawaiian outgroup species from subtribe Madiinae and subtribe Chaenactidinae of tribe Heliantheae s. lat. in the molecular studies allowed resolution of the phylogenetic root or ancestral node of the silversword alliance, thereby permitting a first estimate of the directionality of evolutionary change within the Hawaiian assemblage. It is still uncertain whether an in-depth phylogenetic analysis of the silversword alliance based on morphological and anatomical characters would yield comparable phylogenetic resolution and support.

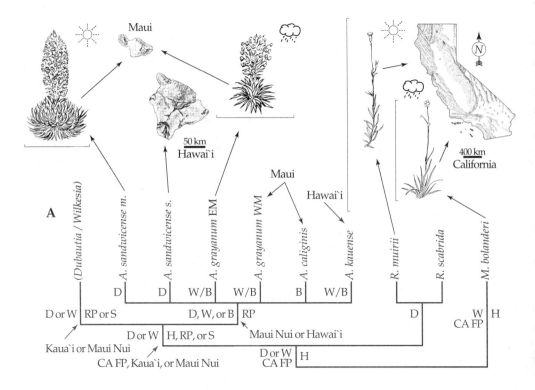

Figure 3.2. (**A**) *Argyroxiphium* and California tarweed ITS lineages. Ecological and biogeographic change in lineages of the silversword alliance estimated from nrDNA ITS sequences (from Figure 3.1). Habitat abbreviations (D or sun-symbol = dry, W = wet, B = bog, and rain-cloud symbol = wet or bog habitat), life-form abbreviations (H = herb, L = liana, M = mat plant, RP = rosette plant, S = shrub, and T = tree), and geographic areas (CA FP = California Floristic Province (see Raven and Axelrod 1978); Maui Nui = the once-contiguous island group including Lana`i, Maui, and Moloka`i) appear beside phylogeny branches where ancestral occurrence is unequivocal, based on MacClade 3.01 (Maddison and Maddison 1992) reconstructions; the actual shift to the indicated life-form, habitat, or area may precede such branches (e.g., with additional resolution of polytomies). Life-forms of species and distributional information conform with Carr (1985). Habitat designations are from Baldwin and Robichaux (1995).

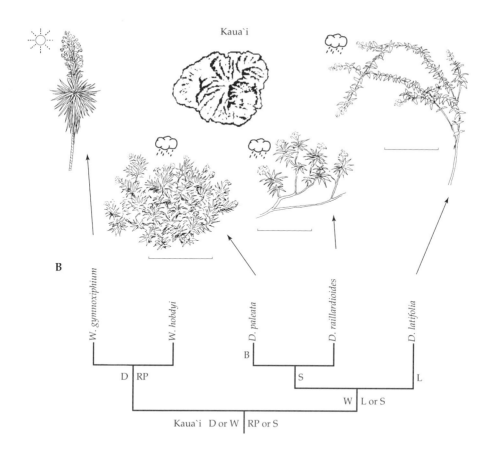

Figure 3.2. (B) *Wilkesia* and the ITS sister-lineage of Kaua`i *Dubautia*. See Figure 3.2A for discussion.

Monophyly and early radiation of the silversword alliance

Parsimony-based phylogenetic analyses of variation among cpDNA restriction sites (Baldwin et al. 1990, 1991) and nuclear rDNA ITS sequences (Baldwin and Robichaux 1995) from representatives of all but two or three extant species in the silversword alliance and multiple closely related outgroup taxa yielded trees with congruities pertinent to an understanding of the Hawaiian radiation. Both lines of evidence strongly support monophyly of the silversword alliance, as evidenced from high bootstrap and decay index values (Figures 3.1, 3.3, 3.4), in accord with the hypothesis of *in situ* diversification of the Hawaiian species from a single ancestral, insular taxon.

Rapid diversification of the silversword alliance into major lineages may be evidenced by the lack of resolution of relationships among the major cpDNA groups and weak support for the basal topology of the nrDNA tree (Figures 3.1, 3.3). Despite low levels of homoplasy, only a small minority of potentially informative characters in each data set shows variation with possible relevance to understanding basal relationships in the group. Similarly, only weak support exists for basal relationships

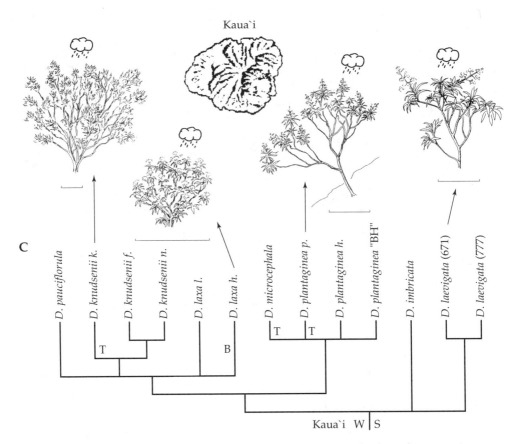

Figure 3.2. (**C**) The ITS lineage of Kaua`i species in *Dubautia* with DG1 and DG3 nuclear genomic arrangements. See Figure 3.2A for discussion.

among lineages of delphacid planthoppers that have apparently coevolved with the silversword alliance (G. K. Roderick, pers. comm.). Inability to reject an hypothesis of rate-constant evolution of nrDNA ITS sequences in the silversword alliance (Figure 3.5; B. G. Baldwin and M. J. Sanderson, in prep.) strengthens the case for "explosive" early radiation of the lineage, as might be expected following insular establishment of a founder species.

Continental origin and timing of the radiation

Molecular (Baldwin et al. 1991; Baldwin 1992) and biosystematic evidence (Kyhos et al. 1990; Baldwin et al. 1991; Carr et al. 1996) demonstrates that the silversword alliance descended from a North American tarweed ancestor in the *Madia/Raillardiopsis* group (Figures 3.1, 3.2A) of subtribe Madiinae (Heliantheae s. lat.) within the last 15 million years (Figure 3.5). Phylogenetic nesting of the silversword alliance within the tarweed radiation indicates that the Hawaiian group postdates considerable diversification of extant North American lineages in Madiinae (Baldwin 1996). Restriction of the vast majority of the ca. 86 species of American tarweeds to the

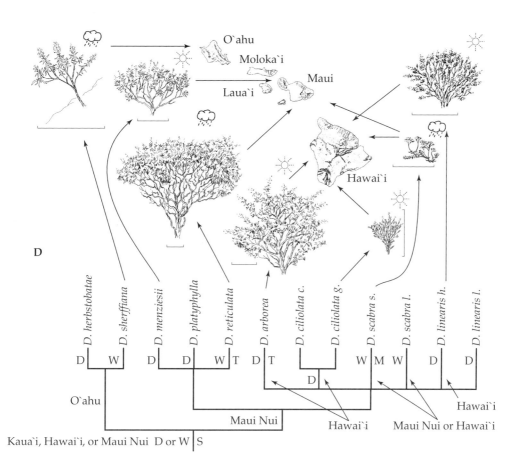

Figure 3.2. (D) *Dubautia* sect. *Railliardia* ITS lineage. See Figure 3.2A for discussion.

California Floristic Province (see Kyhos et al. 1990; Baldwin 1996) leaves little doubt that diversification of subtribe Madiinae was tied to development of a summer-dry climate in westernmost North America, a process that began ca. 15 million years ago (Mya) (Axelrod 1992; Flower and Kennett 1994). Dispersal of the silversword alliance ancestor to the Hawaiian Islands (almost certainly bird-mediated: see Baldwin and Robichaux 1995) must, therefore, have occurred within the last 15 My.

Reconstructions of lineage-branching times based on rate-constant nrDNA ITS evolution suggest that diversification of modern lineages of the silversword alliance has occurred within the last 6 My, approximately since the time of origin of the oldest modern high island, Kaua`i (Figure 3.5; B. G. Baldwin and M. J. Sanderson, in prep.). The radiations of Hawaiian lobelioids (Givnish et al. 1995, 1996), honeycreepers (Sibley and Ahlquist 1982), and fruit flies (Thomas and Hunt 1991) have been estimated to be 10 to 20 million years old, greatly predating the calculated age of the silversword alliance.

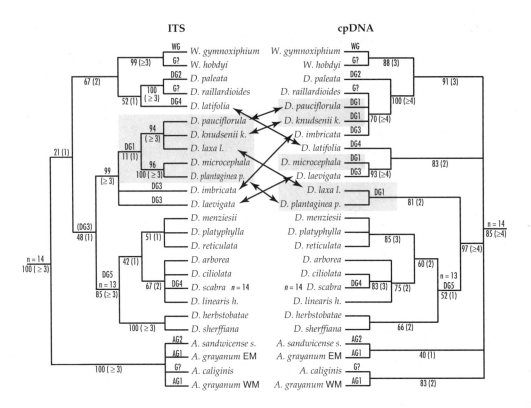

Figure 3.3. Comparison of phylogenetic relationships obtained from parsimony analyses (PAUP 3.1) of nrDNA ITS sequences (left; data in Baldwin and Robichaux 1995) and chloroplast DNA (cpDNA) restriction-site variation (right; data in Baldwin 1989; Baldwin et al. 1990). The two analyses involved the same set of DNA samples from representatives of 24 species of the Hawaiian silversword alliance and five perennial outgroup species in American Madiinae (*Adenothamnus validus*, *Madia bolanderi*, *Raillardella pringlei*, *Raillardiopsis muirii*, and *Raillardiopsis muirii*; not shown). The Hawaiian samples are the same set analyzed in detail in Baldwin et al. (1990); the continental samples are the same set analyzed in Baldwin and Robichaux (1995) minus *Arnica mollis* and *Hulsea algida*. The ITS tree is the semi-strict consensus of the four minimum-length trees (CI' = 0.75; RI = 0.86). The cpDNA tree is the semi-strict consensus of the seven minimum-length trees (consistency index, without uninformative characters = 0.76; retention index = 0.88). Bootstrap values, from 100 replicate analyses of the resampled data sets (using the PAUP "heuristics" option and "closest" addition sequence of taxa), appear below phylogeny branches. Generic abbreviations: *A.* = *Argyroxiphium*, *D.* = *Dubautia*, *W.* = *Wilkesia*. EM = East Maui; WM = West Maui. Nuclear genomic arrangement abbreviations and haploid chromosome numbers (from Carr and Kyhos 1986, pers. comm.) appear above branches where the occurrence of that arrangement is unequivocal, based on MacClade 3.01 (Maddison and Maddison 1992) reconstructions; the actual transformation may precede such branches with additional resolution of polytomies, genomic status of species, and interchange relationships in the silversword alliance. "G?" = unknown genomic arrangement. Placement of the parenthetical "DG3" in the ITS tree is based on a MacClade 3.01 reconstruction of chromosome evolution that involved undirected ordering of *Dubautia* and *Wilkesia* genomic arrangements. Shaded blocks delimit lineages that unequivocally possess the *Dubautia* Genome 1 arrangement (Carr and Kyhos 1986, pers. comm.). Arrows point to species that are not in directly opposing positions in the figure; conflicting placement of taxa in the ITS and cpDNA trees is not necessarily implied by arrows.

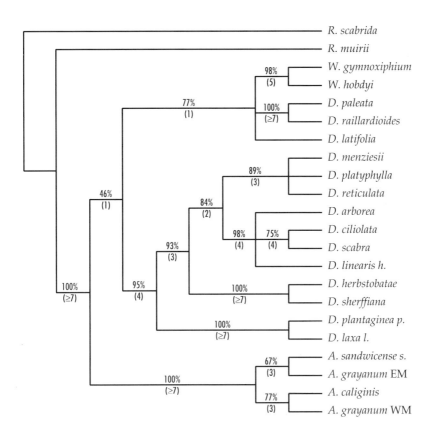

Figure 3.4. Strict consensus of the two minimum-length trees from a parsimony analysis of combined data from nrDNA ITS sequences (Baldwin and Robichaux 1995) plus cpDNA restriction sites (Baldwin et al. 1990) in 20 species of the Hawaiian silversword alliance and two California-tarweed outgroup taxa, *Raillardiopsis muirii* and *R. scabrida* (branch-and-bound analysis conducted with PAUP 3.1). The two minimum-length trees differed topologically only in placement of *Dubautia latifolia*, which was either sister to *Wilkesia/D. paleata/D. raillardioides* or sister to *D. paleata/D. raillardioides*. Bootstrap values (from 100 replicate "heuristic" analyses with "closest" addition sequence of taxa) appear above phylogeny branches. Decay index values (using the PAUP "heuristics" option and "closest" addition sequence of taxa, with semi-strict consensus of resulting trees) appear in parentheses below phylogeny branches. CI' = 0.76; RI = 0.88. Generic abbreviations: *A.* = *Argyroxiphium*, *D.* = *Dubautia*, *M.* = *Madia*, *R.* = *Raillardiopsis*, *W.* = *Wilkesia*. EM = East Maui; WM = West Maui. Five species of the silversword alliance (*D. imbricata, D. knudsenii, D. laevigata, D. microcephala, D. pauciflorula*) were not included in the analysis because ITS and cpDNA data from each of the five taxa were not combinable with data from the 22 species shown here ($p = 0.01$ in all analyses, using the combinability test (Farris et al. 1994, 1995) in a beta-version (d46) of PAUP* 4.0 provided by D. L. Swofford). ITS and cpDNA data of the 22 taxa included in the parsimony analysis were combinable ($p = 0.82$).

Note: Combinable data sets of 22 taxa can also be obtained by replacement of *D. laxa* and *D. plantaginea* with either *D. laevigata* and *D. microcephala* ($p = 0.47$) or *D. knudsenii* and *D. pauciflorula* ($p = 0.11$). Results from parsimony analyses of each of the last two 22 taxa data sets (not shown) place *D.* sect *Railliardia* (see Figure 3.2D) sister to other species in the silversword alliance and, with one data set, place *D. laevigata* + *D. microcephala* sister to the lineage comprising *Wilkesia, D. latifolia, D. paleata,* and *D. raillardioides* or, with the other data set, place *D. knudsenii* + *D. pauciflorula* sister to *D. paleata* + *D. raillardioides*. The three sets of trees each suggest a different phylogenetic position of the DG1/DG3 *Dubautia* lineage (see Figure 3.2C and section on ancient hybridization in Kaua`i *Dubautia*). The tree based on an analysis including *D. laxa* and *D. plantaginea* is shown because the placements of the two taxa in trees from separate analyses of ITS and cpDNA data are in closer agreement than the placements of either *D. laevigata* and *D. microcephala* or *D. knudsenii* and *D. pauciflorula* in the two sets of trees (see Figure 3.3).

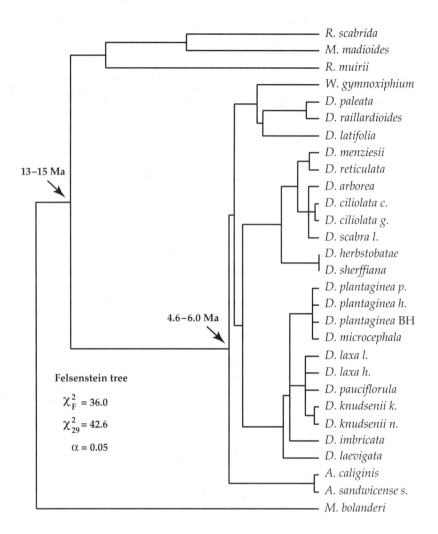

Figure 3.5. A rate-constant phylogeny of the Hawaiian silversword alliance and relatives based on nrDNA ITS sequences (B. G. Baldwin and M. J. Sanderson, in prep.). The tree topology was obtained by Fitch parsimony analysis of ITS sequences from representatives of 20 species in the Hawaiian silversword alliance and four perennial species in the American genera *Madia* and *Raillardiopsis*. Collection and voucher information for all taxa except *Madia madioides* (BGB 488, DAV) is in Baldwin and Robichaux (1995). Generic abbreviations: *A.* = *Argyroxiphium*, *D.* = *Dubautia*, *M.* = *Madia*, *R.* = *Raillardiopsis*, *W.* = *Wilkesia*. Taxa omitted from the tree that are present in Figures 3.1 or 3.2 are identical in ITS sequence to one of the taxa included here. Branch lengths were estimated by maximum likelihood with constraint to a rate-constant model of molecular evolution. Felsenstein's (1993) global test of rate constancy, a likelihood ratio test of unconstrained and clock-constrained trees, did not allow rejection of the null hypothesis of a constant rate of molecular evolution throughout the tree with 95% confidence. The sequence evolution model used was HKY 85: a two-parameter model that allows for unequal base frequencies. The transition/transversion ratio was estimated empirically using maximum likelihood. All analyses were conducted with a beta-version of PAUP*4.0 generously provided by D. L. Swofford. A maximum age of 13 to 15 My for the most recent common ancestor of the Hawaiian silversword alliance and California tarweeds (see text) dictates that the age of the most recent common ancestor of the silversword alliance is ≤ ca. 6 My, roughly the age of Kaua`i, the oldest modern high island of the Hawaiian archipelago.

Rates of molecular evolution and lineage diversification

Rate-constant evolution of nrDNA ITS sequences throughout the lineage that encompasses the Hawaiian silversword alliance and the Californian perennial species within *Madia/Raillardiopsis* (Figure 3.5; B. G. Baldwin and M. J. Sanderson, in prep.) does not accord with an overall acceleration of molecular evolution in the Hawaiian setting. In contrast, morphological and ecological disparity among members of the Hawaiian lineage appears much greater than among members of the continental *Madia/Raillardiopsis* group, a mostly herbaceous assemblage (Figures 3.1, 3.2A; see Baldwin 1996).

Measures of genetic similarity between species based on cpDNA (Baldwin et al. 1991) and nrDNA (Baldwin 1992; B. G. Baldwin, unpubl.) data indicate, in general, lower levels of divergence between taxa of the silversword alliance than between major lineages in *Madia/Raillardiopsis*, in keeping with the phylogenetic data. Striking morphological differences between members of the silversword alliance may be a consequence of evolutionary changes at a small number of loci, e.g., those involved in developmental regulation and physiological response. Interspecific divergence based on genes involved in morphological development, e.g., floral homeotic genes, may prove to be more closely correlated with morphological divergence within the *Madia/Raillardiopsis*/silversword alliance lineage than changes in cpDNA or nuclear rDNA sequences.

Evidence for accelerated diversification in the silversword alliance relative to the rate of lineage-branching within the continental *Madia/Raillardiopsis* sister-group is highly equivocal. Both cpDNA and nrDNA trees place members of a highly diverse *Madia/Raillardiopsis* sublineage, the group delimited by a base chromosome number of $n = 8$ (e.g., *R. muirii*, Figure 3.2A) in the sister-group of the silversword alliance (Baldwin 1993). The insignificant difference ($P > 0.05$) in species numbers between the Hawaiian silversword alliance and the $n = 8$ *Madia/Raillardiopsis* lineage (< 40:1 ratio of taxa in the sister-groups if $P = 2r/(r + s - 1)$, where r and s are numbers of lineages or species of the two groups [Slowinski and Guyer 1989; Sanderson and Donoghue 1996]), assuming a birth-and-death model of diversification with constant rates of lineage splitting, does not allow rejection of the null hypothesis of equal rates of diversification in the two lineages. A more sensitive test for differences in diversification rate between the Californian and Hawaiian sister-lineages could be conducted if information on timing of lineage branching throughout both groups was available. Unfortunately, rate-consistancy of ITS evolution is rejected in trees that include annual species of *Madia* (B. G. Baldwin and M. J. Sanderson, unpubl.), a result that may reflect a correlation between rate of molecular evolution and life history.

The possibility remains that the Hawaiian group has experienced higher rates of diversification than the continental sister-lineage. Absence of branching along a considerably greater length of the basal phylogeny stem of the silversword alliance relative to that of the continental sister-group in the rate-constant nrDNA ITS trees (Figure 3.5; B. G. Baldwin and M. J. Sanderson, in prep.) indicates that diversification of the modern species of the silversword alliance has occurred within a narrower time frame than diversification of the American sister-group. The possibility also exists that the silversword alliance has experienced higher rates of extinction than the continental tarweeds, given the more extreme dynamism of environmental conditions and spatial restriction of species distributions in the Hawaiian Islands compared to those in western North America. If diversification *and* extinction in the silversword alliance have been proceeding more rapidly than in *Madia/Raillardiopsis*, then an analysis of modern diversity alone may fail to detect any differences in rates of lineage branching.

A final consideration in comparing lineage diversities between North American and Hawaiian Madiinae is possible extinction of the closest tarweed relatives of the silversword alliance. The absence of polyploid lineages in *Madia/Raillardiopsis* that conceivably share a common genomic constitution with the tetraploid silversword alliance (see Baldwin and Robichaux, 1995; Baldwin 1996) raises the distinct possibility that the continental sister-group of the Hawaiian lineage within *Madia/Raillardiopsis* is extinct and may have been significantly less diverse than the Hawaiian clade. If the silversword alliance was of allopolyploid origin, the hypothesis of an extinct, continental, allopolyploid sister-group is more parsimonious than the alternative hypothesis of independent dispersal of different diploid tarweed species to the Hawaiian Islands, allopolyploid formation *in situ*, and subsequent extinction of the two diploid entities.

Acquired woodiness in the Hawaiian setting

The phylogenetic position of the silversword alliance within *Madia/Raillardiopsis* in both the cpDNA and nrDNA trees (Baldwin et al. 1991; Baldwin 1992) is consistent with evolution of extensive aboveground woodiness after dispersal of the silversword alliance ancestor to the Hawaiian Islands. All of the American members of the *Madia/Raillardiopsis* group are either annual or perennial herbs (sometimes with aerial stems proximally woody); the only tarweeds with well-developed aerial woodiness outside the Hawaiian Islands are in *Adenothamnus* and *Hemizonia*, genera that are more distantly related to the Hawaiian species than are *Madia* and *Raillardiopsis* (see Baldwin 1996).

An evolutionary tendency for woody life-forms to arise from herbaceous ancestors on islands has been long inferred by island biologists, including Darwin (1859), Wallace (1878), and Carlquist (1965, 1970). Possible explanations for the phenomenon include a greater propensity for long-distance dispersal among herbs relative to woody plants and the prevalence of conditions favorable to year-round plant growth on islands (see Carlquist 1965, 1970). Phylogenetic verification of derivation of the silversword alliance from a continental lineage comprised exclusively of herbaceous species is consistent with the hypothesis of acquired woodiness in island-dwelling plants and Carlquist's (1965) conclusion that the silversword alliance is an example of the phenomenon. In contrast, molecular phylogenetic data from the highly diversified Hawaiian lobelioids (Givnish et al. 1996) and Hawaiian *Psychotria* (Nepokroeff and Sytsma 1996) indicate that the colonizing ancestors of both groups were woody.

Resolution of life-form evolution within the silversword alliance is in part equivocal. Reconstructions based on the nrDNA ITS trees (Figure 3.2), using MacClade 3.0

(Maddison and Maddison 1992), indicate that the most recent common ancestor of all modern species in the Hawaiian lineage may have been an herb, a rosette plant, or a shrub. Two origins of the rosette plant life-form, one in *Argyroxiphium* and one in *Wilkesia*, and one origin of the shrub life-form are unequivocally reconstructed if either the herb or shrub habit is assumed to be the ancestral condition in the silversword alliance. Two origins of the shrub habit, once in the common ancestor of *D. paleata/D. raillardioides/D. latifolia* and once in the most recent common ancestor of the other members of *Dubautia*, is one of the maximally parsimonious reconstructions if the rosette plant habit is considered to be ancestral. The tree habit appears to have arisen independently at least four times, in *D. arborea, D. reticulata, D. knudsenii,* and *D. plantaginea/D. microcephala*: twice on Kaua`i, once on Maui, and once on Hawai`i.

Biogeographic history in the Hawaiian Islands

Basal phylogenetic resolution provided by the nrDNA ITS tree (Figures 3.1, 3.2) allowed reconstruction of the early biogeographic and ecological history of the silversword alliance (Baldwin and Robichaux 1995). Both the nrDNA ITS tree (Figures 3.1, 3.2) and minimum-length trees from a parsimony analysis of combined nrDNA and cpDNA data (Figure 3.4) indicate that *Argyroxiphium*, comprising the true silverswords and greenswords, is sister to the remaining species in the genera *Dubautia* and *Wilkesia* of the silversword alliance, albeit with weak support. In the *Dubautia/Wilkesia* lineage, three major nrDNA ITS sublineages were reconstructed: two that comprise all sampled species of the silversword alliance that occur at least in part on the oldest high island, Kaua`i (Figure 3.2B,C), and one phylogenetically nested lineage that corresponds to *Dubautia* sect. *Railliardia* (Figure 3.2D), a morphologically and chromosomally distinctive group of younger-island endemics, none of which has ever been documented from Kaua`i (see Carr 1985; Carr and Kyhos 1986).

The nrDNA ITS phylogeny of the silversword alliance (Figure 3.1) conforms to one of the simplest biogeographic models predicted for ancient Hawaiian groups (i.e., the "progression rule," *sensu* Funk and Wagner 1995) if the *Dubautia/Wilkesia* lineage is examined in isolation from *Argyroxiphium* (Baldwin and Robichaux 1995). The most parsimonious interpretation of biogeographic history in the nrDNA ITS lineage of *Dubautia/Wilkesia* is descent from a most recent common ancestor on Kaua`i (5.1 My), followed by dispersal to the younger high islands or island groups in the chain, which are oriented in a linear series of decreasing age from west to east: O`ahu (3.7 My), Maui Nui (1.9 My), and Hawai`i (0.4 My) (see Clague and Dalrymple 1987, Carson and Clague 1995 for data on island ages).

Completely congruent patterns of relationship within the cpDNA and nrDNA ITS lineages that correspond to *Dubautia* sect. *Railliardia* ("DG5" lineage in Figure 3.3, see Figure 3.2D) reinforce evidence from each data set that species endemic to each young island or island group radiated from a single founder species, with the possible exception of the Hawai`i endemics. The sequence of insular colonization in *Dubautia* sect. *Railliardia* is mostly equivocal based on the phylogenetic pattern alone. Lineage ages in the group, as estimated from the rate-constant nrDNA ITS tree

(Figure 3.5), are consistent with an hypothesis of progressive dispersal from Kaua`i to O`ahu to Maui Nui to Hawai`i.

The phylogenetic position of *Argyroxiphium* in the nrDNA tree (Figures 3.1, 3.2A) and the combined nrDNA and cpDNA tree (Figure 3.4) initially appears inconsistent with a Kaua`i ancestry of the silversword alliance (Baldwin and Robichaux 1995). If *Argyroxiphium* is truly sister to a group that comprises the remaining species of the silversword alliance, then the two lineages must be of equal age. Absence of species of *Argyroxiphium* from all but the two youngest islands, Maui and Hawai`i, is therefore enigmatic if the *Dubautia/Wilkesia* group indeed arose from a Kaua`i ancestor.

As discussed by Baldwin and Robichaux (1995; also see Carlquist 1995), restriction of *Argyroxiphium* to high elevation situations may explain the absence of any species of the genus on the old, highly eroded islands of Kaua`i and O`ahu, where such habitats have either disappeared or have become geographically restricted. Any species of *Argyroxiphium* on the older islands may have become extinct with loss of sufficient suitable habitat following successful dispersal of a founder species to Maui Nui or Hawai`i within the last two million years. Absence of *Argyroxiphium* from high-elevation bogs on Kaua`i (e.g., on Mt. Wai`ale`ale) that appear similar to bogs occupied by *A. caliginis* and *A. grayanum* on West Maui conforms to Carlquist's (1995) suggestion that the genus may have occupied prehistoric, dry, alpine sites on Kaua`i (those most similar climatically to habitats of montane perennials in Californian *Madia/Raillardiopsis*), with a shift to wet/bog habitats occurring only after colonization of the younger islands.

A hypothesis of ancient origin of the *Argyroxiphium* lineage and post-Pliocene radiation of modern species in the genus is supported by (i) extensive molecular change along the stem of the *Argyroxiphium* lineage in the rate-constant nrDNA ITS tree (Figure 3.5); (ii) an almost complete absence of nrDNA ITS variation among the extant species of silverswords and greenswords (Figure 3.5); and (iii) the small number of species of *Argyroxiphium* relative to the number of species in the apparent sister-group, *Dubautia/Wilkesia* (Figure 3.1). Based on the dating in Figure 3.5, the *Argyroxiphium* lineage is approximately as old as Kaua`i; the most recent common ancestor of modern species in *Argyroxiphium* is less than one million years in age. Evidence from allozymes of high genetic identities between taxa of *Argyroxiphium* led Witter (1986) to conclude similarly that the genus may be considerably older than the radiation of modern species.

Diversification of *Argyroxiphium* on Maui and Hawai`i

Evidence that diversification of the modern species of *Argyroxiphium* has occurred within only 1 My (Figure 3.5) belies the extensive divergence in morphological and ecological characteristics that has occurred between lineages in the genus (see Carr 1985). The most widely distributed species, *A. sandwicense*, comprises mostly monocarpic plants of dry cinder fields. These plants generally produce one massive terminal captulescence before dying. The other four species (one possibly extinct) are bog or wet-forest dwelling taxa that often branch at the base and flower repeatedly (polycarpic), but with smaller capitulescences than those of *A. sandwicense* (Figure 3.2A). The nearly glabrous leaves of the greensword, *A. grayanum*, contrast morphologically and anatomically with the densely white-hairy, light-reflective leaves of the true silverswords (Carlquist 1957, 1959b).

Resolution of phylogenetic relationships within *Argyroxiphium* has been a difficult challenge, in part because of a lack of genetic variation in cpDNA and nrDNA ITS sequences. In sharp contrast to the nrDNA ITS data, which strongly support monophyly of *Argyroxiphium* but offer no internal resolution of relationships (Figures 3.1, 3.2A; Baldwin and Robichaux 1995), cpDNA restriction-site data yield no evidence of monophyly of *Argyroxiphium* but do provide some intriguing phylogenetic resolution within the genus (Figure 3.3; Baldwin et al. 1990). Two restriction-site gains distinguish a cpDNA lineage that comprises the sampled individuals of *A. caliginis*, the West Maui silversword, and a West Maui sample of *A. grayanum*, the greensword. The other sample of *A. grayanum*, from an East Maui population, lacks the two restriction-site gains. The cpDNA identity of the sympatric West Maui silversword and greensword may indicate recent introgressive hybridization between the two species (Baldwin et al. 1990), which are known to form natural hybrids (Carr 1985). Insufficient cytogenetic (Carr and Kyhos 1986) or allozymic (Witter 1986; Witter and Carr 1988) data are available from *Argyroxiphium* to shed additional light on relationships in the genus.

Argyroxiphium plus *Wilkesia*: not a monophyletic group

Nuclear rDNA ITS and cpDNA data (Baldwin et al. 1990; Baldwin and Robichaux 1995) demonstrate that the rosette plants of the silversword alliance, i.e., all species of *Argyroxiphium* and *Wilkesia*, do not form a monophyletic group (Figures 3.1, 3.3). The two species of *Wilkesia*, *W. gymnoxiphium* (Figures 3.1, 3.2B) and *W. hobdyi*, both endemic to the dry western slopes of the oldest high island, Kaua`i, do show considerable superficial morphological similarity to species of *Argyroxiphium* (Figures 3.1, 3.2A). Like *A. sandwicense*, *W. gymnoxiphium* is a typically unbranched, monocarpic species of dry habitats that produces a massive terminal capitulescence from a large rosette of linear leaves (Figure 3.2B; see Carr 1985). The recently discovered *W. hobdyi*, in contrast, is polycarpic and extensively branched, with a low growing habit somewhat like a greensword and unlike the erect growth form of *W. gymnoxiphium*, which produces a pole-like stem up to five meters in height (Carr 1985). Keck (1936) submerged *Wilkesia* within *Argyroxiphium*, in recognition of a perceived close relationship between *W. gymnoxiphium* and the greensword, *A. grayanum*. Carlquist (1957) argued for maintenance of *Wilkesia* upon providing evidence of extensive anatomical dissimilarities between species of *Argyroxiphium* and *W. gymnoxiphium*. All subsequent authors have continued to recognize *Wilkesia* at the generic level (St. John 1971; Carr 1985, 1990). Molecular phylogenetic data (Figures 3.1, 3.3) reinforce Carlquist's (1957) argument that *Argyroxiphium* and *Wilkesia* are discrete groups, while leaving open the possibility that some characteristics shared exclusively by the two genera, such as the rosette plant habit, may be homologous, i.e., symplesiomorphic.

Morphological and ecological radiation of *Wilkesia* and *Dubautia* on Kaua`i

Both lines of molecular evidence support monophyly of *Wilkesia* (Figures 3.1, 3.3; Baldwin et al. 1990; Baldwin and Robichaux 1995). Genetic identity of *W. gymnoxiphium* and *W. hobdyi* in nrDNA ITS sequences (Baldwin and Robichaux 1995) accords with allozymic evidence of high genetic similarity between the two species (Witter and Carr 1988) and Witter's (1986) hypothesis that divergence within *Wilkesia* is recent, relative to the age of the genus. Based on the calibrated rate-constant ITS tree (Figure 3.5; B. G. Baldwin and M. J. Sanderson, in prep), the *Wilkesia* lineage appears to have diverged from its sister-group ca. 3 to 4 Mya.

Nuclear rDNA ITS and cpDNA data support an unsuspected sister-group relationship between *Wilkesia* and a sublineage of Kaua`i *Dubautia* that includes *D. paleata* and *D. raillardioides* (Figures 3.1, 3.2B, 3.3; Baldwin et al. 1990; Baldwin and Robichaux 1995). Close relationship between *D. paleata* and *D. raillardioides* (Figure 3.2B) was hypothesized by Carr (1985) on the basis of shared morphological characteristics (e.g., white corollas and large, purple peduncle-glands) otherwise unknown in Kaua`i *Dubautia*. Witter and Carr (1988) discovered a unique enzyme allele shared exclusively by the two species. Although experimental cytogenetic data from *D. raillardioides* and *W. hobdyi* are lacking, meiotic evidence from hybrids involving 19 other species in the silversword alliance led Carr and Kyhos (1986) to conclude that *W. gymnoxiphium* and *D. paleata* may share a unique genomic structural arrangement (i.e., DG2=WG in Figure 3.3), an interpretation consistent with more recent results (G. D. Carr and D. W. Kyhos, pers. comm.). Identical chromosomal arrangements in *W. gymnoxiphium* and *D. paleata*, if verified by subsequent analyses, may provide a structural genomic characteristic in further support of a *Wilkesia*/Kaua`i *Dubautia* lineage. Possible morphological evidence of a close relationship among the four species includes the shared possession of whorled leaves (polymorphic with opposite leaves in *D. paleata*), a characteristic of only one other species of Kaua`i *Dubautia* – *D. waialealae*, a bog-dwelling cushion plant of unknown genomic arrangement that remains to be sampled for molecular analysis. Intriguing similarities in growth form, leaf characteristics, and stem anatomy between *D. raillardioides* and *Wilkesia* (*W. hobdyi* in particular) (Carlquist 1958a; Carr 1985) may prove to be characteristics retained from a common ancestor.

Occurrence of *Wilkesia*, *Dubautia paleata*, and *D. raillardioides* within the same sublineage in cpDNA and nrDNA ITS trees (Figures 3.1, 3.2B, 3.3; Baldwin et al. 1990; Baldwin and Robichaux 1995) suggests that a great magnitude of morphological and ecological change occurred during evolution of the silversword alliance on Kaua`i. An exceptionally wide diversity of life-forms and habitat preferences is found among *Wilkesia* and the two species of *Dubautia* (Figure 3.2B; Carr 1985): monocarpic and polycarpic rosette plants of dry habitats (*W. gymnoxiphium*, *W. hobdyi*), widely branched, sprawling, wet-forest shrubs (*D. raillardioides*), and compact, bog-dwelling shrubs (*D. paleata*). No other island-endemic sublineage of the silversword alliance resolved from cpDNA or nrDNA ITS data encompasses such phenotypically disparate taxa. The disparity may be a consequence of greater antiquity of the *Wilkesia*/*Dubautia* radiation on Kaua`i compared to the timing of diversification of the other insular sublineages (Figure 3.5; B. G. Baldwin and M. J. Sanderson, in prep.). Clarification of genomic arrangements in the four species through continuing cytogenetic efforts by G. D. Carr and D. W. Kyhos (pers. comm.) may offer a third line of phylogenetic evidence to test relationships among *Wilkesia*, *D. paleata*, and *D. raillardioides*.

Placement of the mesic-forest liana, *Dubautia latifolia*, within a *Wilkesia*/*D. paleata*/*D. raillardioides* lineage, as seen in the nrDNA ITS tree (Figures 3.1, 3.2B) and the nrDNA + cpDNA tree (Figure 3.4), implies an even greater degree of morphological change within the Kaua`i *Dubautia* subgroup than indicated above. Leaf anatomical differences between *D. latifolia* and *Wilkesia* are extreme even from the perspective of variation across angiosperms in general (see section below on morphological and ecological diversity), not to mention the sharp contrasts in life-form and ecology between the two taxa. Lack of congruent placement of *D. latifolia* in the molecular trees as opposed to the cytogenetic groups (Figure 3.3; see section below on *Dubautia scabra*), however, leaves open the question of the relationships of *D. latifolia*, the sole representative of *D.* sect. *Venoso-reticulatae*, to other taxa in the silversword alliance. Allozyme data (Witter 1986; Witter and Carr 1988) parallel morphological indications of extreme divergence of *D. latifolia* from other members of the silversword alliance.

ANCIENT HYBRIDIZATION IN KAUA`I *DUBAUTIA* – Evidence for ancient hybridization in the silversword alliance is provided by strong conflicts between the cpDNA and nrDNA trees in phylogenetic placement of mesophytic shrub and tree species of Kaua`i *Dubautia* with genomic arrangements D1 (*D. knudsenii, D. laxa, D. microcephala, D. pauciflorula,* and *D. plantaginea*) and D3 (*D. laevigata* and, provisionally, *D. imbricata*) (Figure 3.3; see Carr and Kyhos 1986 for chromosomal data; G. D. Carr and D. W. Kyhos, unpubl. data). Robust support for placement of D1 species in three different lineages of the cpDNA tree (Figure 3.3) led Baldwin et al. (1990) to suspect that interspecific hybridization may account for discordant chloroplast and nuclear DNA relationships. Acceptance of the cpDNA topology as an accurate reflection of overall organismal phylogeny would require an hypothesis of multiple origins of similar chromosomal arrangements in Kaua`i *Dubautia*.

The nrDNA ITS tree (Figures 3.1, 3.2C) reinforces the cytogenetic perspective of nuclear DNA evolution; ITS relationships are perfectly consistent with a single origin of the D1 and D3 chromosomal arrangements, in conflict with the cpDNA patterns (Figure 3.3). On a finer scale of resolution than provided by the cytogenetic data, an allozyme-based phylogenetic estimate (Witter 1986) is congruent with the pattern of ITS relationships in the D1 *Dubautia* lineage, with support in both data sets for a *D. microcephala*/*D. plantaginea* clade. The allozyme data provide further resolution of a sister-group relationship between *D. knudsenii* and *D. pauciflorula* and monophyly of *D. laxa* (Witter 1986). The well-supported ITS lineage comprising *D. knudsenii, D. laxa,* and *D. pauciflorula* (Figure 3.1), a subgrouping of D1 species unresolved with allozyme data and in conflict with cpDNA data, corresponds precisely with the set of species possessing large, sessile glands on the disk corolla tubes, a characteristic otherwise unknown in the silversword alliance (Carr 1985).

Based on the foregoing molecular and cytogenetic comparisons within the silversword alliance, hybridization appears to have played at least a limited role in the evolution of the group, in support of Carr's (1995) conclusions. Homoplasy is an unlikely explanation for the incongruities between the nuclear and organellar data sets based on strong support in the molecular trees for conflicting relationships (Figure 3.3), strong rejection of combinability of the two data sets with inclusion of all DG1 and DG3 taxa (Figure 3.4), and the similarly low levels of divergence between taxa in cpDNA and nrDNA ITS sequences (Baldwin et al. 1990; Baldwin 1992; Baldwin and Robichaux 1995). Lineage sorting of cpDNAs (i.e., differential sorting through organismal lineages of divergent cpDNAs from a polymorphic ancestral species; see Pamilo and Nei 1988) appears untenable as an explanation for the conflicts, in view of the need for several different cpDNA types (each borne by different individuals) to persist through multiple speciation events involving island populations subject to repeated genetic bottlenecks. Patterns of allozyme variation in the silversword alliance appear inconsistent with long-term persistence of such high levels of genetic variation through diverging lineages (Witter 1986, 1990; Witter and Carr 1988). Hybridization remains the most likely explanation for the chloroplast and nuclear DNA conflicts, especially in light of high levels of interfertility of species, documented natural hybridization between species of Kaua`i *Dubautia*, and widespread sympatry among the ecologically similar, mesophytic shrubs and trees in question (see Carr 1985).

Data indicating speciation from an ancestor of hybrid constitution – powerful evidence of an evolutionary role for hybridization – may be provided by two taxa that possess the nuclear genomic arrangement D1, *Dubautia knudsenii* and *D. pauciflorula* (Figure 3.2C). As mentioned above, the two species are part of a well-supported monophyletic group with *D. laxa* in the nrDNA ITS trees, a relationship that is further reinforced by common possession of coarse, sessile corolla glands unknown in other members of the silversword alliance. Possession of a unique allozyme by *D. knudsenii* and *D. pauciflorula* (Witter and Carr 1988) provides evidence that the two species may be sister taxa. Placement of the two species in the cpDNA lineage of *D. raillardioides* (Figure 3.3) is consistent with the hypothesis that the cpDNAs of *D. knudsenii* and *D. pauciflorula* descend from a common ancestor that underwent introgressive hybridization with a member of the *D. raillardioides* lineage. Alternatively, these two species (like *D. imbricata*) may have independently acquired cpDNA from the *D. raillardioides* lineage. Additional sampling of cpDNAs within the D1 and D3 species may shed further light on the history of hybridization in Kaua`i *Dubautia*.

Despite molecular evidence that ancient hybridization has had a lasting effect on the genetic constitution of members of the silversword alliance, hybridization in the group has not been a sufficiently dominant evolutionary force to eliminate evidence of divergent organismal evolution. As noted above, even in Kaua`i *Dubautia* the nrDNA ITS lineage of D1 and D3 species is completely congruent with cytogenetic data and substantially consistent with allozyme results and morphological patterns despite conflicts with hypotheses of relationships reconstructed with cpDNA (Figure 3.3), a genome that appears to be prone to transfer across species lineages via introgressive hybridization in several plant groups (see Rieseberg 1995).

MOLECULAR INSIGHTS INTO CHROMOSOME EVOLUTION – Conformity of nrDNA ITS and cytogenetic data extends to an even broader scale to clarify directionality of chromosome evolution in the silversword alliance. Parsimony-based mapping of chromosome evolution on the ITS tree identifies the D3 genome, known at present only in the Kaua`i species *Dubautia laevigata* (Figure 3.2C) and (provisionally) *D. imbricata* (Carr 1985; Carr and Kyhos 1986), as the ancestral chromosome arrangement of the vast majority of species in the silversword alliance (Figure 3.3), i.e., those in the DG1/DG3 Kaua`i *Dubautia* lineage and the DG5/DG4 younger-island endemic sister-group, *Dubautia* sect. *Railliardia*. The DG3 arrangement is only one reciprocal translocation removed from the DG1 arrangement and is distinguished by a different (unequal) reciprocal translocation from the DG5 arrangement, common to all but one species of *Dubautia* sect. *Railliardia*, *D. scabra* (Carr and Kyhos 1986). Loss of a centric fragment following the unequal chromosomal interchange event in the ancestry of *Dubautia* sect. *Railliardia* would explain the shift to $n = 13$ from $n = 14$, the ancestral chromosome number common to all species of the silversword alliance outside *Dubautia* sect. *Railliardia* (Figure 3.3; see Carr 1985; Carr and Kyhos 1986). Further clarification of the pattern of chromosome evolution in the silversword alliance awaits refined understanding of chromosomal arrangements in the minority of species for which cytogenetic data are inconclusive (Carr and Kyhos 1986), e.g., in species of *Argyroxiphium* and *Wilkesia*, and better resolution of chromosomal relationships between Hawaiian and continental Madiinae.

Monophyly of the young-island endemic group of *Dubautia* species, i.e., *Dubautia* sect. *Railliardia* (Figure 3.2D), is among the best supported results from systematic investigations of the silversword alliance. Morphologically, the group is diagnosed by a pappus of flattened, plumose bristles and connate receptacular paleae (Carr 1985). Nine of the 10 species exclusively possess a unique chromosomal arrangement (DG5; Carr and Kyhos 1986), and chromosome number ($n = 13$; Carr 1985); the remaining species, *D. scabra*, bears an $n = 14$ genome (DG4; Carr and Kyhos 1986) that differs from the $n = 13$ D5 genome by a single unequal arm-interchange and presence of an additional centromere. Weak evidence for monophyly of *Dubautia* sect. *Railliardia* in the cpDNA (Figure 3.3; Baldwin et al. 1990) and allozymic (Witter 1986; Witter and Carr 1988) data sets is strongly reinforced by robust support of the lineage in the nrDNA ITS tree (Figures 3.1, 3.3) and the nrDNA ITS + cpDNA tree (Figure 3.4).

Ecological and morphological radiation in *Dubautia* following island colonization

The cpDNA and nrDNA ITS lineages corresponding to *Dubautia* sect. *Railliardia* ("DG5" lineage in Figure 3.3, see Figure 3.2D) provide completely congruent support for the occurrence of major ecological shifts accompanying diversification on the young islands or island groups east of Kaua`i. As pointed out by Baldwin and Robichaux (1995), each of the three major ITS sublineages within *Dubautia* sect. *Railliardia* [i.e., (i) *D. herbstobatae*/*D. sherffiana*, (ii) *D. menziesii*/*D. platyphylla*/*D. reticulata*, and (iii) *D. arborea*/*D. ciliolata*/*D. linearis*/*D. scabra*] includes species restricted to dry habitats and species restricted to wet habitats (Figure 3.2D). Consequently, the ITS trees yield evidence for shifts between dry and wet habitats concomitant with diversification on O`ahu, Maui Nui, and, possibly, Hawai`i. The semi-strict consensus of the maximally parsimonious cpDNA trees (Figure 3.3) reinforces Baldwin and

Robichaux's (1995) hypothesis by providing perfectly concordant resolution of relationships within *Dubautia* sect. *Railliardia*. As expected, parsimony analysis of the combined nrDNA ITS and cpDNA characters yielded a tree with increased resolution and enhanced internal support for relationships within the section (Figure 3.4). All relationships in the nrDNA+ cpDNA tree are congruent with those in the trees generated from independent analysis of each of the two molecular data sets.

The repeated association of habitat shifts with diversification in *Dubautia* sect. *Railliardia* (Figure 3.2D), evident from both lines of molecular phylogenetic data, suggests that major ecological changes played an important role in the evolution of new species following founder events (Baldwin and Robichaux 1995). A partial synopsis of the diversity in each of the three sublineages of *Dubautia* sect. *Railliardia* illustrates the results of ecological radiation on the young islands. The East Maui lineage includes the largest tree (up to 8 m in height) in the silversword alliance, *D. reticulata* (Figure 3.2D), an inhabitant of densely vegetated, wet forests, and a small to large shrub of dry scrub and cinder fields, *D. menziesii* (Figure 3.2D). Distributions of the two species and another member of the East Maui lineage, *D. platyphylla*, a large shrub of dry, deep ravines, abut on Mt. Haleakala, where steep climatic gradients allow close parapatry of ecologically disparate species. The O`ahu lineage comprises *D. herbstobatae*, a small shrub of dry ridges, and *D. sherffiana* (Figure 3.2D), a large shrub of mesic forest and wet ridges (Carr 1985). The two allopatric species, both endemic to the Wai'anae Mountains, are sufficiently divergent in leaf arrangement and venation patterns to obscure their sister-group relationship, first proposed by Witter and Carr (1988) on the basis of a shared, unique enzyme allele. The Hawai`i/Maui Nui lineage comprises (i) two endemics of dry scrub and lava flows on Hawai`i: *D. arborea* (Figure 3.2D), a large shrub to small tree, and *D. ciliolata* (Figure 3.2D), a small erect shrub, (ii) at least two subspecies of *D. linearis* (Figure 3.2D), diffusely branched shrubs of mostly dry situations on Hawai`i and Maui Nui, and (iii) *D. scabra* (Figure 3.2D), a mat-forming to suberect shrub of wet situations, including wet forests and recent lava flows, on Hawai`i and Maui Nui.

Directionalities of ecological shifts between wet and dry habitats in sublineages of *Dubautia* sect. *Railliardia* (Figure 3.2D) are equivocal if based on the molecular phylogenetic patterns alone (Baldwin and Robichaux 1995). Ecophysiological findings led Robichaux (1985) to conclude that the wet-forest tree *D. reticulata* (Figure 3.2D) on East Maui descended from an ancestor of dry habitats. Robichaux's hypothesis was based on the observation that tissue elastic properties of *D. reticulata* are intermediate between those of a species of *Dubautia* from mesic to wet habitats on Kaua`i (*D. knudsenii*; Figure 3.2C) and a species of *Dubautia* from dry habitats (*D. menziesii*; Figure 3.2D). Differences between tissue elastic properties of the Kaua`i species of *Dubautia* and *D. reticulata* were attributed in part to a phylogenetic constraint on *D. reticulata*, i.e., a limitation imposed by the xeric ancestry of *D. reticulata*.

DUBAUTIA SCABRA: HOMOPLASTIC GENOMIC ARRANGEMENT OR HYBRID HISTORY? – The only example of conflict between cytogenetic data and both lines of molecular data from the silversword alliance could indicate (i) an unusual instance of independent origin of similar genomic arrangements in different lineages or (ii) hybridization in the history of *Dubautia scabra*. Both cpDNA and nrDNA ITS trees (Figure 3.3) place *D. scabra* in a nested position within *Dubautia* sect. *Railliardia* ("DG5" lineage in Figure 3.3, see Figure 3.2D), in the Hawai`i/Maui Nui lineage with *D. arborea*, *D. ciliolata*, and *D. linearis*. If the two congruent molecular topologies accurately reflect all organismal relationships within *Dubautia* sect. *Railliardia*, then the most parsimonious hypothesis for origin of the $n = 14$ D4 genome in *D. scabra* is via ascending dysploidy from the $n = 13$ D5 genomic arrangement shared by all other species in the section. An hypothesis of reversed directionality of chromosome number evolution from $n = 13$ to $n = 14$ is complicated by the knowledge that the D4 genomic arrangement of *D. scabra* is shared by *D. latifolia* (G. D. Carr and D. W. Kyhos, pers. comm.), the Kaua`i vine that falls outside the lineage corresponding to *Dubautia* sect. *Railliardia* in the cpDNA and nrDNA ITS trees (Figure 3.3).

Independent evolution of the DG4 arrangement in *Dubautia scabra* and *D. latifolia* (Figure 3.3) is conceivable given cytogenetic evidence for exceptional lability of whole-arm interchange between the two chromosomes (designated "13" and "14" by Carr and Kyhos 1986) involved in the dysploid origin of the $n = 13$ D5 genome and the hypothetical dysploid origin of the *D. scabra* D4 genome. All three possible whole-arm interchanges have occurred between the two chromosomes; the different arrangements corresponding to the D3, D4, and D5 genomes (Carr and Kyhos 1986). From this perspective, the occurrence of a homoplastic whole-arm re-arrangement between the two chromosomes, with acquisition of a centromere, is a tenable hypothesis for origin of the *D. scabra* DG4 condition.

Congruence between a molecular phylogeny of planthoppers and their silversword-alliance hosts, at least within nested sublineages of both groups (G. K. Roderick, pers. comm.), is consistent with a divergent origin of *Dubautia scabra* within the Hawai`i/Maui Nui lineage of *Dubautia* sect. *Railliardia* (Figure 3.2D). With the provision of limited sampling, the same taxon of planthopper is apparently restricted to *D. ciliolata* and *D. scabra* (G. K. Roderick, pers. comm.), sister-species in the cpDNA (Figure 3.3) and cpDNA + ITS (Figure 3.4) trees. The planthopper taxon known only from *D. ciliolata* and *D. scabra* is nested phylogenetically within the planthopper lineage comprised of species restricted to members of *Dubautia* sect. *Railliardia* (G. K. Roderick, pers. comm.). A hybrid origin of *D. scabra*, involving *D. ciliolata* or a progenitor species, can be reconciled with the planthopper phylogeny if it is assumed that leaf characteristics important to planthopper specificity in *D. ciliolata* were inherited by *D. scabra*.

A reticulate origin of *D. scabra* involving an $n = 13$ species of *Dubautia* sect. *Railliardia* ("DG5" lineage in Figure 3.3, see Figure 3.2D) from the Hawai`i/Maui Nui molecular lineage and a species closely related to *D. latifolia*, which possesses the same chromosome arrangement as *D. scabra*, could explain the molecular and cytogenetic incongruence without invoking parallel chromosome evolution. Under this hybrid-origin scenario for *D. scabra*, the chromosome arrangement stabilized to the condition found in the paternal $n = 14$ parent and the nrDNA ITS sequences became homogeneous with those of the maternal $n = 13$ parent, the sole donor of cpDNA to the hybrid species. A hybrid origin of *D. scabra* could explain (i) the somewhat intermediate morphological (Carr 1985) characteristics of the species relative to other $n = 14$ species of *Dubautia* and $n = 13$ members of *Dubautia* sect. *Railliardia*, and (ii) the strong similarity in turgor maintenance characteristics between *D. scabra* and other $n = 14$ species of *Dubautia*. Absence of any species of *Dubautia* with an $n = 14$ D4 genome other than *D. scabra* on the young islands (i.e., outside Kaua`i) is a problematical aspect of the hybrid speciation hypothesis.

A divergent origin of *Dubautia scabra* from an ancestor closely related to *D. latifolia*, followed by recent, localized introgressive hybridization with one or more species having $n = 13$ in *Dubautia* sect. *Railliardia* from the Hawai`i/Maui Nui molecular lineage, is the least plausible hypothesis to explain the molecular and cytogenetic patterns (Figure 3.3). Although contemporary natural hybridization and possible introgression between *D. scabra* and other members of *Dubautia* sect. *Railliardia* is documented (Carr 1985; Crins et al. 1988), introgression that postdates wide dispersal of *D. scabra* cannot account for the molecular phylogenetic evidence. Identity of nrDNA ITS sequences from representatives of both subspecies of *D. scabra* on Hawai`i and on Maui (Baldwin and Robichaux 1995), where all but one other species (*D. linearis*) of the Hawai`i/Maui Nui molecular lineage comprising *D. arborea, D. ciliolata, D. linearis,* and *D. scabra* are absent, is inconsistent with recent introgression of ITS sequences from Hawai`i/Maui Nui *Dubautia* into *D. scabra*. The hypothetical origin of *D. scabra* cpDNA and ITS sequences via recent introgressive hybridization with a member of *Dubautia* sect. *Railliardia* (such as *D. ciliolata*) is also difficult to reconcile with the apparent absence of planthoppers on *D. scabra* apart from the taxon otherwise apparently restricted to *D. ciliolata*, based on limited sampling (G. K. Roderick, pers. comm.). Sampling of additional populations and genes of *D. scabra* and putative close relatives will be necessary to gain further insights into the history of this enigmatic species.

Conclusions

Molecular phylogenetic evidence from cpDNA restriction sites and nrDNA ITS sequences reinforce evidence from morphological, cytogenetic, and allozymic investigations that the endemic Hawaiian silversword alliance (28 species in *Argyroxiphium, Dubautia,* and *Wilkesia*) is a monophyletic group that rapidly radiated into extremely divergent life-forms and ecological situations. Both lines of molecular data indicate that the founder species of the silversword alliance was a western North American tarweed, a member of the herbaceous *Madia/Raillardiopsis* group (Compositae-Madiinae), that arrived in the Hawaiian Islands less than 15 Mya.

Diversification of the silversword alliance into the modern species of rosette plants, trees, shrubs, mat plants, cushion plants, and lianas has occurred within the last 6 My (i.e., within the history of the modern high Hawaiian Islands) based on estimates from nuclear rDNA ITS data. Radiation of the modern species of the alliance appears to have occurred more recently, and perhaps more rapidly, than diversification of the continental sister lineage. Accelerated molecular evolution in the silversword alliance is not supported by available DNA data.

Nuclear rDNA sequences and allozyme data support the hypothesis that *Argyroxiphium* is an ancient lineage that radiated relatively recently (< 2 Mya) on Maui and Hawai`i, with at least one major ecological shift between wet and dry habitats. Phylogenetic data from cpDNA and nrDNA are congruent with a model of minimal inter-island dispersal in *Wilkesia*/*Dubautia*, with diversification of all endemic species on each major island or island group from a single founder taxon. Diversification in each of the three young-island lineages of *Dubautia* sect. *Railliardia* was accompanied by at least one major ecological shift between dry and wet habitats. Placement of *Wilkesia*, *D. paleata*, and *D. raillardioides* in a common Kaua`i lineage in both molecular trees indicates that extreme morphological change accompanied diversification on the oldest modern high island, with ecological shifts among dry, wet, and bog habitats.

Strong conflict between cpDNA relationships and nrDNA/cytogenetic patterns indicates that ancient hybridization has had at least a minor evolutionary impact on species of Kaua`i *Dubautia*, which are widely sympatric and ecologically similar. Congruence of nrDNA ITS relationships and cytogenetic results in Kaua`i *Dubautia* demonstrates that (i) hybridization on Kaua`i has not been sufficient to eliminate genetic evidence of divergent evolution and (ii) genome D3, known from *D. laevigata* and (provisionally) *D. imbricata*, is the ancestral genomic arrangement of the vast majority of species in the silversword alliance. Nuclear rDNA and cpDNA trees are consistent with the cytogenetic interpretation of a single dysploid decrease to $n = 13$ from an ancestral chromosome number of $n = 14$ in the silversword alliance. Nesting of *D. scabra* ($n = 14$) within the $n = 13$ *Dubautia* lineage in both the cpDNA and nrDNA trees might be explained by either hybridization in the history of the species or dysploid increase and parallel origin of the $n = 14$ D4 genomic arrangement (also known from *D. latifolia* on Kaua`i) in *D. scabra*. Clarification of the complex evolutionary history of the silversword alliance by reference to multiple lines of evidence underscores the need for a broad-based approach to resolving relationships in groups that have undergone rapid radiation.

Acknowledgments

This work was supported by an award from the U. S. National Science Foundation (DEB-9458237). I thank Sue Bainbridge, Joan Canfield, Gerry Carr, Jan Dvořák, Betsy Gagné, Bob Hobdy, Don Kyhos, Art Medeiros, John Obata, Steve Perlman, Libby Powell, Rob Robichaux, Lars Walker, Marti Witter, and Ken Wood for collection assistance; Rob Robichaux, John Strother, Jonathan Wendel, and anonymous reviewers for helpful comments on the manuscript; Gerry Carr and Don Kyhos for sharing unpublished cytogenetic data; George Roderick for sharing unpublished phylogenetic data on Hawaiian planthoppers; Karen Klitz for drawing the plants in Figures 3.1 and 3.2; Tony Morosco for aiding in production of Figures 3.1 and 3.2; and Tom Givnish and Ken Sytsma for inviting me to participate in the Symposium on Molecular Evolution and Adaptive Radiation. I owe a special debt of gratitude to the late Lani Stemmermann for her generous field assistance and for sharing her contagious enthusiasm and love for the endangered flora of the Hawaiian Islands.

References

Axelrod, D. I. 1992. Miocene floristic change at 15 Ma, Nevada to Washington, USA The Palaeobotanist 41:234–239.

Baldwin, B. G. 1992. Phylogenetic utility of the internal transcribed spacers of nuclear ribosomal DNA in plants: an example from the Compositae. Molecular Phylogenetics and Evolution 1:3–16.

Baldwin, B. G. 1993. Molecular phylogenetics of *Madia* (Compositae-Madiinae) based on ITS sequences of 18–26S nuclear ribosomal DNA. American Journal of Botany 80 (Suppl.):130.

Baldwin, B. G. 1996. Phylogenetics of the California tarweeds and the Hawaiian silversword alliance (Madiinae; Heliantheae *sensu lato*). Pp. 377–391 in *Proceedings of the international Compositae conference, Kew, 1994, vol. 1. systematics*, D. J. N. Hind, H. Beentje, and G. V. Pope, eds. Kew, England: Royal Botanic Gardens.

Baldwin, B. G., and Robichaux, R. H. 1995. Historical biogeography and ecology of the Hawaiian silversword alliance (Asteraceae): new molecular phylogenetic perspectives. Pp. 259–287 in *Hawaiian biogeography: evolution on a hot spot archipelago*, W. L. Wagner and V. A. Funk, eds. Washington, DC: Smithsonian Institution Press.

Baldwin, B. G., Kyhos, D. W., and Dvořák, J. 1990. Chloroplast DNA evolution and adaptive radiation in the Hawaiian silversword alliance (Asteraceae-Madiinae). Annals of the Missouri Botanical Garden 77:96–109.

Baldwin, B. G., Kyhos, D. W., Dvořák, J., and Carr, G. D. 1991. Chloroplast DNA evidence for a North American origin of the Hawaiian silversword alliance. Proceedings of the National Academy of Sciences, USA 88:1840–1843.

Carlquist, S. 1957. Leaf anatomy and ontogeny in *Argyroxiphium* and *Wilkesia* (Compositae). American Journal of Botany 44:696–705.

Carlquist, S. 1958a. Wood anatomy of Heliantheae (Compositae). Tropical Woods 108:1–30

Carlquist, S. 1958b. Structure and ontogeny of glandular trichomes of Madinae (Compositae). American Journal of Botany 45:675–682.

Carlquist, S. 1959a. Vegetative anatomy of *Dubautia*, *Argyroxiphium*, and *Wilkesia* (Compositae). Pacific Science 13:195–210.

Carlquist, S. 1959b. Studies on Madiinae: anatomy, cytology, and evolutionary relationships. Aliso 4:171–236.

Carlquist, S. 1965. *Island Life*. New York, NY: Natural History Press.

Carlquist, S. 1970. *Island Biology*. New York, NY: Columbia University Press.

Carlquist, S. 1995. Introduction. Pp. 14–29 in *Hawaiian biogeography: evolution on a hot spot archipelago*, W. L. Wagner and V. A. Funk, eds. Washington, DC: Smithsonian Institution Press.

Carr, G. D. 1985. Monograph of the Hawaiian Madiinae (Asteraceae): *Argyroxiphium*, *Dubautia*, and *Wilkesia*. Allertonia 4:1–123.

Carr, G. D. 1990. *Wilkesia*. Pp. 375–376 in *Manual of the flowering plants of Hawai`i, vol. 1*, W. L. Wagner, D. R. Herbst, and S. H. Sohmer, eds. Honolulu, HI: University of Hawai'i Press and Bishop Museum Press.

Carr, G. D. 1995. A fully fertile intergeneric hybrid derivative from *Argyroxiphium sandwicense* ssp. *macrocephalum* and *Dubautia menziesii* (Asteraceae) and its relevance to plant evolution in the Hawaiian Islands. American Journal of Botany 82:1574–1581.

Carr, G. D., and Kyhos, D. W. 1981. Adaptive radiation in the Hawaiian silversword alliance (Compositae-Madiinae). I. Cytogenetics of spontaneous hybrids. Evolution 35:543–556.

Carr, G. D., and Kyhos, D. W. 1986. Adaptive radiation in the Hawaiian silversword alliance (Compositae-Madiinae). II. Cytogenetics of artificial and natural hybrids. Evolution 40:969–976.

Carr, G. D., Baldwin, B. G., and Kyhos, D. W. 1996. Cytogenetic implications of artificial hybrids between the Hawaiian silversword alliance and North American tarweeds (Asteraceae: Heliantheae-Madiinae). American Journal of Botany 83:653–660.

Carr, G. D., Robichaux, R. H., Witter, M. S., and Kyhos, D. W. 1989. Adaptive radiation of the Hawaiian silversword alliance (Compositae-Madiinae): a comparison with Hawaiian picture-winged *Drosophila*. Pp. 79–97 in *Genetics, speciation, and the founder principle*, L. V. Giddings, K. Y. Kaneshiro, and W. W. Anderson, eds. New York, NY: Oxford University Press.

Carson, H. L., and Clague, D. A. 1995. Geology and biogeography of the Hawaiian Islands. Pp. 14–29 in *Hawaiian biogeography: evolution on a hot spot archipelago*, W. L. Wagner and V. A. Funk, eds. Washington, DC: Smithsonian Institution Press.

Clague, D. A., and Dalrymple, G. B. 1987. The Hawaiian-Emperor volcanic chain, part I: geologic evolution. Pp. 5–54 in *Volcanism in Hawaii*, R. W. Decker, T. L. Wright, and P. H. Stauffer, eds. U. S. Geological Survey Professional Paper 1350. Washington, DC: U. S. Government Printing Office.

Crins, W. J., and Bohm, B. A. 1990. Flavonoid diversity in relation to systematics and evolution of the tarweeds. Annals of the Missouri Botanical Garden 77:73–83.

Crins, W. J., Bohm, B. A., and G. D. Carr. 1988. Flavonoids as indicators of hybridization in mixed population of lava-colonizing Hawaiian tarweeds (Asteraceae: Heliantheae: Madiinae). Systematic Botany 13:567–571.

Darwin, C. 1859. *On the origin of species by means of natural selection, or the preservation of favored races in the struggle for life.* London, England: Murray.

Farris, J. S., Källersjö, M., Kluge, A. G., and Bult, C. 1994. Testing significance of incongruence. Cladistics 10:315–319.

Farris, J. S., Källersjö, M., Kluge, A. G., and Bult, C. 1995. Constructing a significance test for incongruence. Systematic Biology 44:570–572.

Flower, B. P., and Kennett, J. P. 1994. The middle Miocene climatic transition: east Antarctic ice sheet development, deep ocean circulation and global carbon cycling. Palaeogeography, Palaeoclimatology, Palaeoecology 108:537–555.

Funk, V. A., and Wagner, W. L. 1995. Biogeographic patterns in the Hawaiian Islands. Pp. 259–287 in *Hawaiian biogeography: evolution on a hot spot archipelago*, W. L. Wagner and V. A. Funk, eds. Washington, DC: Smithsonian Institution Press.

Givnish, T. J., Sytsma, K. J., Smith, J. F., and Hahn, W. J. 1995. Molecular evolution, adaptive radiation, and geographic speciation in *Cyanea* (Campanulaceae, Lobelioideae). Pp. 288–337 in *Hawaiian biogeography: evolution on a hot spot archipelago*, W. L. Wagner and V. A. Funk, eds. Washington, DC: Smithsonian Institution Press.

Givnish, T. J., Knox, E., Patterson, T. B., Hapeman, J. R., Palmer, J. D., and Sytsma, K. J. 1996. The Hawaiian lobelioids are monophyletic and underwent a rapid initial radiation roughly 15 million years ago. American Journal of Botany 83 (Suppl.):159.

Keck, D. D. 1936. The Hawaiian silverswords: systematics, affinities, and phytogeographic problems of the genus *Argyroxiphium*. Occasional Papers Bernice P. Bishop Museum 11:1–38.

Kyhos, D. W., Carr, G. D., and Baldwin, B. G. 1990. Biodiversity and cytogenetics of the tarweeds (Asteraceae: Heliantheae-Madiinae). Annals of the Missouri Botanical Garden 77:84–95.

Lipp, C. C., Goldstein, G., Meinzer, F. C., and Niemczura, W. 1994. Freezing tolerance and avoidance in high-elevation Hawaiian plants. Plant, Cell and Environment 17: 1035–1044

Maddison, W. P., and Maddison, D. R. 1992. *MacClade, ver. 3: analysis of phylogeny and character evolution.* Sunderland, MA: Sinauer Associates.

Melcher, P. J., Goldstein, G., Meinzer, F. C., Minyard, B., Giambelluca, T. W., and Loope, L. L. 1994. Determinants of thermal balance in the Hawaiian giant rosette plant, *Argyroxiphium sandwicense*. Oecologia 98:412–418.

Nepokroeff, M., and Sytsma, K. J. 1996. Systematics and patterns of speciation and colonization in Hawaiian *Psychotria* and relatives based on phylogenetic analysis of ITS sequence data. American Journal of Botany 83 (suppl):181–182.

Pamilo, P., and Nei, M. 1988. Relationships between gene trees and species trees. Molecular Biology and Evolution 5:568–583.

Raven, P. H., and Axelrod, D. I. 1978. Origin and relationships of the California flora. University of California Publications in Botany 72:1–134.

Raven, P. H., Evert, R. F., and Eichhorn, S. E. 1992. *Biology of plants, 5th ed*. New York, NY: Worth Publishers.

Rieseberg, L. H. 1995. The role of hybridization in evolution: old wine in new skins. American Journal of Botany 82:944–953.

Robichaux, R. H. 1985. Tissue elastic properties of a mesic forest Hawaiian *Dubautia* species with 13 pairs of chromosomes. Pacific Science 39:191–194.

Robichaux, R. H., and Canfield, J. E. 1985. Tissue elastic properties of eight Hawaiian *Dubautia* species that differ in habitat and diploid chromosome number. Oecologia 66:77–80.

Robichaux, R. H., Carr, G. D., Liebman, M., and Pearcy, R. W. 1990. Adaptive radiation of the Hawaiian silversword alliance (Compositae-Madiinae): ecological, morphological, and physiological diversity. Annals of the Missouri Botanical Garden 77:64–72.

Sanderson, M. J., and Donoghue, M. J. 1996. Reconstructing shifts in diversification rates on phylogenetic trees. Trends in Ecology and Evolution 11:15–20.

Sibley, C. G., and Ahlquist, J. E. 1982. The relationships of the Hawaiian honeycreepers (Drepanidini) as indicated by DNA-DNA hybridization. Auk 99:130–140.

Slowinski, J. B., and Guyer, C. 1989. Testing the stochasticity of patterns of organismal diversity: an improved null model. American Naturalist 134:907–921.

St. John, H. 1971. The status of the genus *Wilkesia* (Compositae), and discovery of a second Hawaiian species. Hawaiian plant studies 34 (i.e., 38). Occasional Papers Bernice P. Bishop Museum 24:127–137.

Swofford, D. L. 1993. *PAUP: phylogenetic analysis using parsimony, ver. 3.1.* Champaign, IL: Illinois Natural History Survey.

Thomas, R. H., and Hunt, J. A. 1991. The molecular evolution of the alcohol dehydrogenase locus and the phylogeny of Hawaiian *Drosophila*. Molecular Biology and Evolution 8:687–702.

Wallace, A. R. 1878. *Tropical nature and other essays*. London, England: Macmillan and Co.

Witter, M. S. 1986. Genetic differentiation in the Hawaiian silversword alliance (Compositae: Madiinae). Ph.D. Dissertation, University of Hawai'i, Honolulu.

Witter, M. S. 1990. Evolution in the Madiinae: evidence from enzyme electrophoresis. Annals of the Missouri Botanical Garden 77:110–117.

Witter, M. S., and Carr, G. D. 1988. Adaptive radiation and genetic differentiation in the Hawaiian silversword alliance (Compositae: Madiinae). Evolution 42:1278–1287.

4 THE CHRONICLE OF MARSUPIAL EVOLUTION

Mark S. Springer, John A. W. Kirsch, and Judd A. Case

In *The Major Features of Evolution*, George Gaylord Simpson (1953) defined adaptive radiation as the "more or less simultaneous divergence of numerous lines all from the same ancestral adaptive type into different, also diverging adaptive zones." In Simpson's view, such radiations are promoted by the emergence of new adaptive types and the availability of ecologically open territory. His concept of adaptive radiation is wide-ranging, allowing for brief or prolonged periods of diversification, and for radiations limited in scope as well as those that "ramify into the most extraordinarily varied zones." Simpson contrasted adaptive radiation with progressive occupation, which "involves in any one period of time the change of only one or a few lines from one zone to another, with each transition involving a different ancestral type." For Simpson, all of the diversity of life could be explained by the intergrading processes of adaptive radiation and progressive occupation "plus the factor of geographic isolation which may permit essential duplication of adaptive types by different organisms in different regions" (Simpson 1953, p. 224). This duplication of adaptive types, presumably facilitated by similar environments and selective pressures, is manifested as convergent and/or parallel evolution. Simpson's definitions remain useful for understanding broad patterns of adaptive change (Price 1996), particularly those documented in the fossil record.

Within Class Mammalia, the most extensive adaptive radiation occurred among the placental mammals – the bats, rodents, ungulates, carnivores, whales, and their kin in infraclass Eutheria. Simpson (1953) argued that the great placental radiation (comprising more than 4,300 extant species [Wilson and Reeder 1993]) was due partly to the extinction of many reptilian groups at the end of the Mesozoic. The marsupials – the opossums, kangaroos, koalas, and their kin in infraclass Metatheria – are sister to the placental mammals but comprise an order of magnitude fewer species. However, the long and often isolated evolutionary history of marsupials (especially in Australia and South America during the Tertiary) resulted in a morphological and ecological diversification more nearly comparable to that seen in the placental mammals. This history involved episodes of adaptive radiation at several different taxonomic levels, progressive occupation of adaptive zones by constituent lineages, and convergent evolution both between marsupials and placentals and among marsupials themselves.

The parallel adaptive radiations of marsupials and placentals in time (but not always in space) provide part of the perennial fascination of the "alternative" metatherian mammals. However, such parallelisms are incomplete. Notably absent among marsupials are analogs of cetaceans and chiropterans, perhaps in the latter case because bats have probably been residents of Australia for at least as long as marsupials (Hand et al. 1994). Likewise, the nectar-feeding honey-possum (*Tarsipes*) has no real counterpart among placentals. Still, some of the most striking

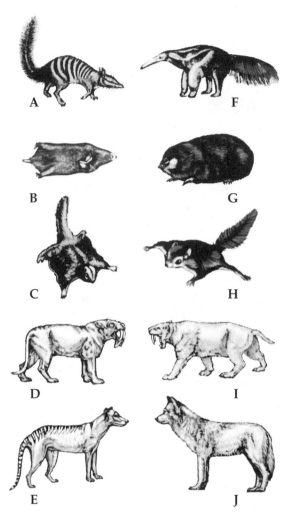

Figure 4.1. Examples of convergence involving marsupial (left) and placental (right) analogs. Animal sketches redrawn from sources listed parenthetically. (**A**) *Myrmecobius* (Ride 1970); (**B**) *Notoryctes* (Archer 1984); (**C**) *Petaurus* (Avers 1989); (**D**) *Thylacosmilus* (Ridley 1993); (**E**) *Thylacinus* (Ridley 1993); (**F**) *Myrmecophaga* (Avers 1989); (**G**) *Chrysochloris* (Bisacre et al. 1989); (**H**) *Glaucomys* (Bisacre et al. 1989); (**I**) *Smilodon* (Ridley 1993); (**J**) *Canis* (Ridley 1993).

examples of convergent evolution known involve pairs of marsupials and placentals (Figure 4.1), such as the numbat *Myrmecobius* and the anteater *Myrmecophaga*, the marsupial mole *Notoryctes* and the placental mole *Chrysochloris*, the sugar glider *Petaurus* and the flying squirrel *Glaucomys*, the marsupial sabertooth *Thylacosmilus* and the placental sabertooth *Smilodon*, and – perhaps most famously – the thylacine *Thylacinus* and the wolf *Canis*. In each of these cases, however, the convergence is obvious and limited to a few, mostly external features, and has not disguised other fundamental characteristics that reveal the true phylogenetic affinities of these taxa. For example, in spite of similar adaptations for gliding (e.g., the patagium) in *Petaurus* and *Glaucomys*, the sugar glider is clearly a marsupial based on numerous reproductive characters (e.g., the lateral positions of the paired vaginae with respect to the ureters, and the appearance of a pseudovaginal canal at the time of parturition) and osteological features (e.g., the presence of an alisphenoid contribution to the tympanic bulla). Furthermore, *Petaurus* is clearly a member of the marsupial order Diprotodontia by virtue of sharing numerous apomorphies with other members of that taxon. Similarly, the larger kangaroos and antelopes may seem at first to be close ecological analogs, yet their similarities are also incomplete and rather superficial. The kangaroo family – whose radiation postdates that of the antelopes – occupies a much broader range of niches, possibly including even carnivorous forms (Flannery 1989). Kangaroos themselves have lophodont molars, whereas ecologically analogous antelopes have selenodont grinders. Finally, kangaroos and antelopes differ profoundly in many non-dental characters, including several features of locomotion, reproduction, digestion, and physiology.

Far more interesting and taxonomically challenging are examples of possible convergence among marsupials themselves, for which it has proven difficult to disentangle synapomorphic and homoplastic similarity. For example, a longstanding controversy in marsupial systematics hinges on the phylogenetic affinities of Australian thylacines, which were morphologically very similar to the extinct South American borhyaenids. Either the thylacines (marsupial wolves) and borhyaenids inherited these common features from a recent common ancestor that entered the terrestrial-carnivore adaptive zone (implying a close relationship between these Australian and South American taxa), or these features evolved independently in the two groups. Similarly, there are strong similarities between the largely carnivorous American didelphid opossums and Australasian dasyurids, as well as between South American caluromyid opossums and Australasian folivorous ringtail possums, both of which are arboreal and have overlapping dietary preferences. Even within Australasia there are striking examples of convergence in certain characteristics: for example, gliding occurs in three possum genera (*Acrobates*, *Petauroides*, and *Petaurus*), each of which appears to have a nearest relative that is not volant.

With the application of molecular techniques to such cases in recent years, it has become clear that intra-marsupial convergence is extensive; indeed, in each of the instances just mentioned, molecular data provide strong support for the repeated, independent evolution of the characters in question. The marsupials thus illustrate the key role that molecular systematics can play in distinguishing anatomical homoplasy from true synapomorphy, in tracing the pattern of adaptive radiation within groups showing extensive homoplasy, and in examining the timing of morphological and ecological diversification. In this paper, we examine the impact of DNA/DNA hybridization and DNA sequence data on our understanding of marsupial evolution, one of the classic (and incompletely understood) cases of adaptive radiation. We focus on the following questions:

- How do phylogenies based on molecular data compare with those based on morphology?
- What implications do molecular phylogenies have for the detection of homoplasy (convergence or recurrence) among morphological characters?
- What is the time-frame over which the orders and families of marsupials diverged from each other? How does the timing of the marsupial radiation relate to plate-tectonic, climatic, ecologic, and floristic history?
- To what extent did marsupial evolution involve progressive occupation, adaptive radiation, or both? Has either phenomenon (and convergence) occurred more than once?

To provide background for addressing these questions, we first present an overview of marsupial paleontology, ecology, and traditional classification.

Marsupial natural history

Fossil record

Simpson (1945) placed all marsupials in one of six superfamilies in the order Marsupialia; more recent classifications (beginning with Ride 1964) have elevated these superfamilies and distribute living marsupials among as many as seven orders (Aplin and Archer 1987; Marshall et al. 1990), with additional orders for extinct taxa.

As recently as Marshall et al.'s (1990) paleontological review, it was thought that the oldest undisputed marsupial fossils were from the Cenomanian of Utah (Cifelli and Eaton 1987), approximately 90 million years ago (Mya), with a diversity of putatively Late Cretaceous taxa from both North and South America. Marshall et al. (1990) suggested that the Mesozoic diversification of marsupials in the Americas resulted in a single marsupial fauna (with regional endemism) that encompassed North and South America by the Maestrichtian; this fauna included supposed representatives of all American orders as well as a proto-Australian taxon (*Andinodelphys*) in South America. Even earlier estimates of the time of divergence between American and Australian marsupials were made based on globin amino-acid sequences (Richardson 1988; Hope et al. 1990). Marshall et al. (1990) suggested that, after the Cretaceous diversification of metatherians in the Americas, one lineage arrived in Australia before Australia separated from Antarctica 56 Mya and then diversified into the fossil and extant Australian marsupials.

Over the last few years, however, several discoveries have modified our understanding of early marsupial evolution. A probable marsupial was found in the Albian of Utah, from roughly 101 Mya (Cifelli 1993a). A more diverse array of late Cretaceous marsupial fossils has been found in North America (Cifelli 1990a,b, 1993a,b; Eaton 1993a,b). There are also marsupials from the late Cretaceous of Asia (Szalay 1994; Trofimov and Szalay 1994). Marsupials from supposedly late Cretaceous sediments of South America, in turn, are now regarded as early Paleocene (Bonaparte 1990; Gayet et al. 1991). Thus, the gap between the oldest North and South American marsupials approaches 35 to 40 My (Harland et al. 1989). Furthermore, this hiatus is not just a result of the absence of Cretaceous mammal faunas from South America, as Hauterivian and Campanian mammalian faunas are known from this continent. However, all of the Cretaceous South American mammals are non-tribosphenic (e. g., symmetrodonts, dryolestoids) and appear to represent an endemic Gondwanan group that evolved in isolation from Laurasian mammals for at least much of the Cretaceous (Bonaparte 1990). Therefore, there is no geologic or paleontologic evidence that marsupials were present in Gondwanaland prior to the beginning of the Tertiary. The entrance of marsupials into Australia has also been pushed back to the late Paleocene or early Eocene with the discovery of the marsupial-bearing Tingamarra Local Fauna (Godthelp et al. 1992; Archer et al. 1993). As we shall discuss below, these new discoveries – together with chronological inferences from DNA data – suggest a quite different scenario of early marsupial dispersal and diversification from that envisioned by Marshall et al. (1990).

Near the end of the Tertiary, several South American marsupials (including most of the large carnivorous borhyaenids) went extinct before the Great American

Interchange, which occurred at approximately 3 Mya with the emergence of a solid land route across the Panamanian isthmus (Marshall et al. 1982). Subsequent to the Interchange and the invasion of placental mammals, additional marsupial taxa such as the sabertooth *Thylacosmilus* and argyrolagids suffered extinction as well. In the case of the herbivorous, saltatorial argyrolagids, the role of competition in their demise is unclear because no real placental analogs replaced them.

The timing of early marsupial radiation in Australasia has been largely inaccessible from a paleontological perspective because of the incompleteness and discontinuity of the fossil record from pre-Oligocene to late Oligocene (Woodburne et al. 1985, 1993). But as far as we can infer from that record, only one major Australasian group – the enigmatic, possibly rabbit-like order Yalkaparidontia (Archer et al. 1988) – has become extinct in the interim.

Ecology of extant forms

Among the three extant orders of South American marsupials, the Didelphimorphia is the most diverse; it includes 63 species of small- to medium-sized opossums with diets ranging from omnivorous (e. g., *Caluromys*) to largely carnivorous (e. g., *Lestodelphys*, *Lutreolina*). The single living species of Microbiotheria (*Dromiciops gliroides*) resembles the smaller didelphids in its habitat (terrestrial or semi-arboreal) and mainly invertebrate diet. *Dromiciops* is restricted to the bamboo-beech forests of southern Chile and adjacent Argentina. Caenolestids (the only living taxa of Paucituberculata) are small-bodied marsupials that are shrewlike in appearance. They prey on invertebrates and small vertebrates and typically occur in scrub adjacent to the meadows of high, moist Andean paramos.

Of the four extant Australasian orders of marsupials, Notoryctemorphia is probably now monotypic and includes only the blind marsupial mole (*Notoryctes typhlops*), distributed over the deserts of central Australia. Marsupial bandicoots and bilbies (Order Peramelina) are medium-sized, efficient diggers; many if not all construct burrows. Bandicoot dentition is well suited to an eclectic diet of insects and other arthropods, small vertebrates, fruits, and soft tubers. Bandicoots are also noteworthy for their extremely accelerated ontogeny; Cockburn (1990) argues that this allows them to exploit temporally patchy environments. Dasyuromorphia includes insectivorous and carnivorous forms that range in size from mouse-sized antechinuses (*Antechinus* and related genera), planigales (*Planigale*), and dunnarts (*Sminthopsis*), to the dog-sized (and recently extinct) thylacines; the majority (such as the medium-sized *Dasyurus* and *Sarcophilus*) are terrestrial. Also included in this order is the termite-eating numbat (*Myrmecobius*). Diprotodontia, the most diverse Australasian order, includes mostly herbivorous forms (kangaroos, wallabies, wombats, koala, honey-possum, ringtail possums, cuscuses, pygmy possums) that occupy a wide variety of ecological roles. Some kangaroos (*Dendrolagus*) are arboreal, while the largest are saltatory grazers; wombats graze and construct burrows. The remaining diprotodontian families (excepting the koala) are collectively called "possums" by Australians and are arboreal browsers or nectarivores. In addition to these extant diprotodonts, there were extinct and often large forms such as the granivorous ektopodontids, rhinoceros-sized diprotodontids, and carnivorous thylacoleonids (marsupial lions).

Thus, three of the extant metatherian orders are found only in South America (together with a few recent immigrants to North America), while the remaining four occur only in Australasia (Table 4.1). The geographic disjunction of these two ordi-

Table 4.1. A classification of extant marsupial orders and families, using ordinal-level names adopted by Marshall et al. (1990). Current geographic distributions for orders and species-numbers for families [from authors in Wilson and Reeder (1993)] are also listed. Genera not mentioned in the text are bracketed.

Class Mammalia
    Subclass Theria
        Infraclass Metatheria
            Order **Didelphimorphia** (North and South America)
                Family **Didelphidae** (58 spp., 12 genera: *Didelphis, Philander,*
                    Lutreolina, Chironectes, Metachirus, Monodelphis, Marmosa,
                    Micoureus, Thylamys, Lestodelphys, Marmosops, Gracilinanus)
                Family **Caluromyidae** (5 spp., 3 genera: *Caluromys, Glironia* [*Caluromysiops*])
            Order **Paucituberculata** (South America)
                Family **Caenolestidae** (5 spp., 3 genera: *Caenolestes, Rhyncholestes* [*Lestoros*])
            Order **Microbiotheria** (South America)
                Family **Microbiotheriidae** (1 sp., 1 genus: *Dromiciops*)
            Order **Dasyuromorphia** (Australasia)
                Family **Dasyuridae** (61 spp., 15 genera: *Dasyurus, Sarcophilus,*
                    Antechinus, Murexia, Phascogale, Planigale, Sminthopsis [*Dasycercus, Dasykaluta,*
                    Myoictis, Neophascogale, Ningaui, Parantechinus, Phascolosorex, Pseudantechinus])
                Family **Thylacinidae** (1 sp., 1 genus: *Thylacinus**)
                Family **Myrmecobiidae** (1 sp., 1 genus: *Myrmecobius*)
            Order **Notoryctemorphia** (Australia)
                Family **Notoryctidae** (2 spp., 1 genus: *Notoryctes*)
            Order **Peramelina** (Australasia)
                Family **Peramelidae** (19 spp., 7 genera: *Perameles, Isoodon,*
                    Peroryctes, Microperoryctes, Echymipera [*Chaeropus**, *Rhynchomeles*])
                Family **Thylacomyidae** (2 spp., 1 genus: *Macrotis*)
            Order **Diprotodontia** (Australasia)
              Suborder **Vombatiformes**
                Family **Vombatidae** (3 spp., 2 genera: *Vombatus, Lasiorhinus*)
                Family **Phascolarctidae** (1 sp., 1 genus: *Phascolarctos*)
              Suborder **Phalangeriformes**
                Family **Phalangeridae** (18 spp., 6 genera: *Phalanger, Spilocuscus,*
                    Trichosurus [*Ailurops, Strigocuscus, Wyulda*])
                Family **Burramyidae** (5 spp., 2 genera: *Burramys, Cercartetus*)
                Family **Pseudocheiridae** (14 spp., 5 genera: *Pseudocheirus,*
                    Pseudochirops, Petauroides, Hemibelideus [*Petropseudes*])
                Family **Petauridae** (10 spp., 3 genera: *Petaurus, Gymnobelideus, Dactylopsila*)
                Family **Tarsipedidae** (1 sp., 1 genus: *Tarsipes*)
                Family **Acrobatidae** (2 spp., 2 genera: *Distoechurus, Acrobates*)
                Family **Macropodidae**[†] (63 spp., 16 genera: *Macropus, Wallabia,*
                    Dorcopsulus, Dendrolagus, Petrogale, Thylogale, Setonix
                    Aepyprymnus, Bettongia, Hypsiprymnodon [*Caloprymnus**, *Potorous, Dorcopsis,*
                    Lagorchestes, Lagostrophus, Onychogalea])

* Recently or probably extinct.
† All but *Aepyprymnus, Bettongia, Caloprymnus, Potorous* and *Hypsiprymnodon* are here considered members of Subfamily Macropodinae.

nal groupings has greatly influenced the higher classification of marsupials and the interpretation of their morphological and anatomical characteristics.

Key anatomical characters and higher-level marsupial systematics

Prior to Ride (1962, 1964) and Kirsch (1968), traditional views of higher-level marsupial phylogeny were couched in terms of two character-pairs: the diprotodont vs. polyprotodont condition of the lower incisors, and the syndactyl vs. didactyl hindfoot (Bensley 1903; Gregory 1910; Osgood 1921; Simpson 1930, 1945). Diprotodonty is a condition in which the anteriormost pair of lower incisors is enlarged and procumbent, as compared with the polyprotodont state of multiple nibbling or grasping front teeth (Figure 4.2A,B); diprotodonty occurs in Australasian diprotodontians and South American caenolestids, as well as in some extinct South American taxa. The function of diprotodonty varies in different groups. For example, Kirsch (1977a) demonstrated that caenolestids use their procumbent lower incisors to stab prey, while kangaroos and possums employ theirs to browse and graze. As to the pedal feature, syndactyly is a state in which the second and third digits are often reduced and always joined together by a common integument, as opposed to remaining free and more or less equal to the fourth and fifth digits in didactyly (Figure 4.2C, D). Even so, both syndactyl digits retain strong claws and there is no degeneration of bone or muscle tissue. Syndactyly occurs full-blown only in two Australasian orders, Diprotodontia and Peramelina. The syndactyl foot functions as a grooming organ in at least some syndactylous species where digits two and three are greatly diminished (bandicoots, kangaroos), although the selective pressures associated with its origin remain unclear (Hall 1987). Occasional syndactyly in humans (and presumably in other mammals) is caused by a single HOX-gene mutation (Muragaki et al. 1996).

Diprotodonty and syndactyly are both derived character-states but each suggests a different phylogenetic affinity for the Australasian diprotodontians: either with the diprotodont but didactyl South American caenolestids, or with the syndactyl but polyprotodont Australasian bandicoots. Given the rarity of syndactyly among mammals (at least as an invariant taxonomic marker), as well as the common geographic provenance of bandicoots and diprotodontians, syndactyly in these two orders is most often viewed as an homologous character, with the shared diprotodonty of Australasian diprotodontians and American caenolestids attributed to convergence (e. g., Bensley 1903; Marshall 1980; Ride 1962, 1964; Woodburne 1984). Indeed, the geographic isolation of Australia from other continents has prompted a widespread view of a fundamental dichotomy between Australasian and American marsupials, even though numerous attempts have been made to document close relationships between particular Australian and American taxa (e.g., thylacines and borhyaenids).

More recently, Szalay (1982) proposed that South American microbiotheres (*Dromiciops*) are more closely related to Australasian taxa than to other South American forms, based on putative synapomorphies of the ankle region. Australasian species and *Dromiciops* share a continuous lower ankle-joint pattern (CLAJP); this contrasts with the separate lower ankle-joint pattern (SLAJP) seen in American didelphimorphs and caenolestids, in which the sustentacular and calcanoastragular facets are not confluent with each other (Figure 4.2E,F). Szalay (1982) argued that the confluent ankle-joint and foot are highly adapted for arboreal locomotion because they produce a strongly grasping foot, and concluded that the CLAJP must be of utmost phylogenetic significance. We note, however, that Hershkovitz (1992a) regards CLAJP as occurring in some didelphimorphs and possibly in caenolestids, and that bandicoots and kangaroos have (probably secondarily derived) separate joints; the taxonomic distribution of character-states may thus not be as clear-cut as portrayed by Szalay (1982). Disregarding these details, Szalay placed CLAJP marsupials in one cohort comprised of Australasian forms and *Dromiciops*, while the other American

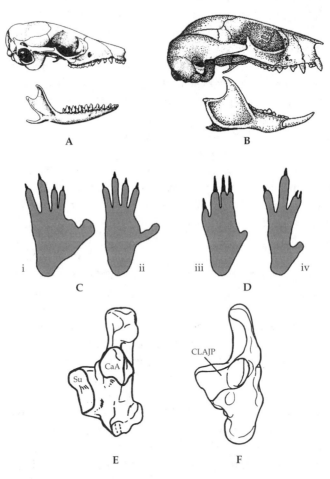

Figure 4.2. Alternative states of three key marsupial characters; the presumed primitive state is shown at left in each case. (**A,B**) Lateral views of the skulls and lower jaws of *Sminthopsis laniger* and *Petaurus breviceps* to illustrate the polyprotodont and diprotodont conditions of the lower incisors, respectively (modified after Archer 1984). (**C,D**) Hindfeet of several marsupials: i and ii are the didactyl feet of *Gracilinanus microtarsus* and *Thylamys elegans*, while iii and iv show the syndactyl feet of *Vombatus ursinus* and *Hypsiprymnodon moschatus* (modified after Hall 1987). (**E,F**) Left calcaneae to illustrate the separate lower ankle-joint pattern and the continuous lower ankle-joint pattern. In E, the sustentacular (Su) and calcanoastragular (CaA) facets are separate (SLAJP). In F, these facets are confluent to produce the continuous lower ankle-joint pattern (CLAJP). Modified after Szalay (1994).

marsupials (didelphimorphs and caenolestids) were combined in a second cohort based on shared possession of the SLAJP condition. This arrangement has been incorporated in other recent classifications (Aplin and Archer 1987; Marshall et al. 1990). However, given that SLAJP is primitive and CLAJP derived, ankle morphology does not provide support for a monophyletic American group, even if convergent or intermediate CLAJP conditions in American marsupials are ignored. In fact, only one character – epididymal sperm-pairing, unique to didelphimorphs and caenolestids across vertebrates – appears to support the unity of Szalay's American cohort (Biggers and DeLamater 1965). Yet caenolestids and didelphimorphs exhibit quite different patterns of expression of even this character: pairing is head-to-head in caenolestids vs. side-by-side in most didelphimorphs (Temple-Smith 1987), suggesting that sperm-pairing may be independently derived in these American groups.

Diprotodontia is the most diverse of the extant marsupial orders, and the only one to contain more than two extant families. Many anatomical features provide strong support for diprotodontian monophyly (see reviews by Kirsch 1977b, Archer 1984, and Aplin and Archer 1987). There are also several morphological synapomorphies that link wombats and the koala (Hughes 1965; Bryant 1977; Kirsch 1977b; Archer 1984) in suborder Vombatiformes. However, there remain contentious issues regarding diprotodontian phylogeny: the supposed monophyly of the suborder Phalangeriformes (traditionally including kangaroos and several possum families), the relationship of the honey-possum to other diprotodontians, and the resolution of an apparent possum/kangaroo/wombat-koala trichotomy.

Even though marsupial anatomists have examined the phylogenetic significance of features from a wide variety of character systems (dental, basicranial, pedal, and sperm ultrastructural), there has been little effort to compile this information into a character matrix that includes all extant orders and families of marsupials. A notable exception is Luckett's (1994) attempt to integrate different types of anatomical as well as molecular characters. However, Luckett (1994) stopped short of employing formal parsimony algorithms and instead relied upon subjective analysis of characters. Consistent with the now-conventional view, Luckett's (1994) phylogeny supports Szalay's (1982) cohorts and a sister-group relationship between the bandicoots and also-syndactyl diprotodontians. We re-analyzed the unordered morphological characters from Luckett's data-set using PAUP 3.0s (Swofford 1991), and found support for the monophyly of the Australasian marsupials plus *Dromiciops* (87% bootstrap value), but not for the monophyly of the remaining two American orders. Within the Australasian marsupials, Diprotodontia and Peramelina were sister-taxa in our analysis, but with only 68% bootstrap support.

We compiled a larger data set of 102 anatomical characters (Appendix 4.1) and subjected it to maximum parsimony analysis as well. Our character coding differs from that of Luckett (1994), who scored all diprotodontian families as having dilambdodont dentition. While we agree that this may be the primitive condition for diprotodontians, it is not the state seen in members of certain living families (e. g., Petauridae). We also note that all comprehensive anatomical analyses, whether informal or formal (our own included), assume monophyly at the family level.

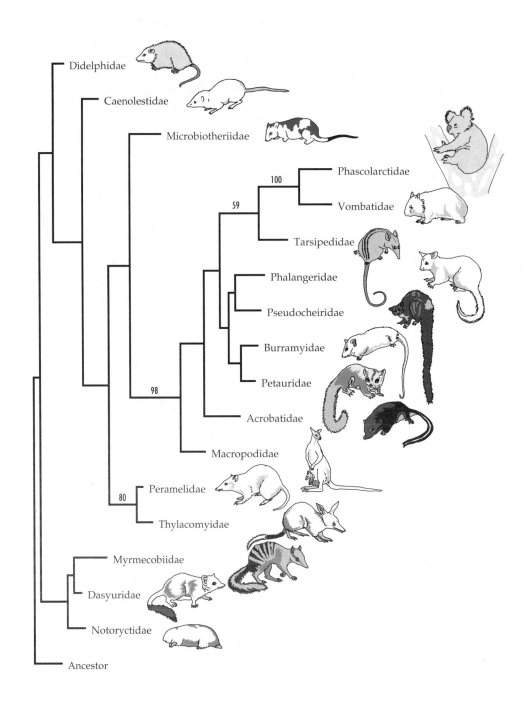

Figure 4.3. Single most parsimonious tree (365 steps) based on the matrix of 102 morphological and anatomical characters shown in Appendix 4.1 (all characters are informative). Numbers refer to bootstrap support for clades at or above 50% using 500 replications. Consistency and retention indices were 0.419 and 0.575, respectively. Animal sketches redrawn from several sources, but mainly Archer (1982), Kirsch and Waller (1979), and Marlow (1965).

Our single most parsimonious tree (Figure 4.3) does not support Szalay's cohorts, or the association of the two syndactyl orders; Microbiotheria appears to be sister to the diprotodontians. However, the interordinal associations in this morphology-based tree all have less than 50% bootstrap support. In fact, only four relationships between families have more than 50% bootstrap support; these include the monophyly of diprotodontians (98%), peramelinans (80%), and the wombat-koala clade (100%).

Molecular approaches to marsupial phylogeny

Phylogenetic analyses

Over the last decade, studies based on single-copy DNA/DNA hybridization and DNA sequencing have provided new data for the analysis of evolutionary relationships among marsupials. DNA hybridization studies have now been been conducted on more than 100 marsupial species, including representatives of the seven extant orders and all but one living family, Myrmecobiidae (Springer and Kirsch 1989, 1991; Kirsch et al. 1990a,b, 1991, 1995a,b; Springer et al. 1990, 1992; Westerman et al. 1990; Westerman 1991; Westerman and Edwards 1991; Edwards and Westerman 1992, 1995; Kirsch and Palma 1995; Lapointe and Kirsch 1995). Hybridization data confirmed many previously established (or assumed) nodes in the marsupial family tree, including several interspecific and intergeneric relationships, a close association of wombats and koalas (Springer and Kirsch 1989, 1991), and the polyphyly of volant taxa (Edwards and Westerman 1992, 1995; Springer et al. 1992). Inter- and intrafamilial relationships within Diprotodontia are especially well resolved by the DNA-hybridization data. Springer and Kirsch (1989, 1991) provided information about most diprotodontian families, many of which were separately examined in some detail using the same technique (Springer et al. 1990, 1992; Edwards and Westerman 1992, 1995; Kirsch et al. 1995b), confirming the monophyly of each.

We reanalyzed an expanded version of the data of Springer and Kirsch (1991); the distance-based tree for diprotodontians and levels of bootstrap support are shown in Figure 4.4. This tree does not resolve the pairwise relationships among the three major clades (possums, kangaroos, koala-wombats), but is otherwise well resolved except within the macropodine kangaroos (*Macropus rufus* and species of *Dorcopsulus* and *Dendrolagus*). Missing from this analysis is the nectarivorous honey-possum *Tarsipes* (Tarsipedidae); however, Edwards and Westerman (1995) used DNA-hybridization data to estimate its position and concluded that it was closest (72% bootstrap support) to one of the glider families, Petauridae (*Petaurus* and *Dactylopsila* in Figure 4.4, although Edwards and Westerman [1995] included only the latter genus).

At the ordinal level, complete resolution has proved as difficult as it has for placental mammals. Still, hybridization data do provide strong support (92% bootstrap value) for a clade containing the orders Dasyuromorphia, Diprotodontia, and Microbiotheria (Kirsch et al. 1991; Lapointe and Kirsch 1995); Notoryctemorphia appears to be a fourth member of this clade (Westerman 1991). Within this group, the South American *Dromiciops* is sister to the Australasian diprotodontians (88% bootstrap support averaged over several analyses, although only one included more than one

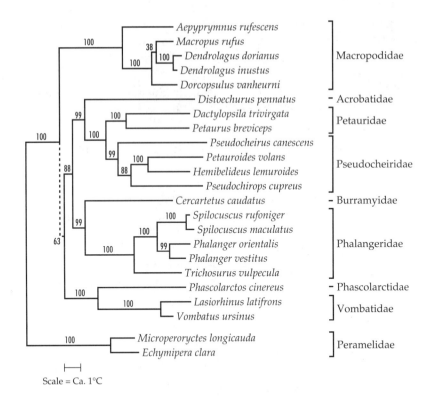

Figure 4.4. DNA-hybridization tree of relationships among 21 species of diprotodontian marsupials, rooted with two peramelid bandicoots (data from Springer 1988 and Springer and Kirsch 1991). All but two species (*Aepyprymnus rufescens* and *Distoechurus pennatus*) are labeled. Missing data were reconstructed by the additive method of Landry et al. (1996), after symmetrization and reflection of known reciprocals. Other variants of the reconstruction algorithm [e.g., ultrametric estimation (Lapointe and Kirsch 1995)] gave qualitatively similar results, differing chiefly in placing *Pseudochirops cupreus* with *Pseudocheirus canescens*. Numbers at nodes are bootstrap percentages based on 1,000 pseudoreplicates, using the adaptation of bootstrapping for distances of Krajewski and Dickerman (1990). The node with 63% bootstrap support was unsupported in the average-consensus trees of single- and multiple-deletion jackknife analyses using the procedure for weighted trees (Lapointe et al. 1994), so that the preceding internode is shown as a dashed line – relationships among possums, kangaroos, and vombatiforms collapse to a trichotomy. Branch lengths are drawn approximately to scale based on the averages of path lengths observed over the 1,000 bootstrap trees.

diprotodontian). Thus, DNA hybridization provides evidence for a close relationship between South American microbiotheres and Australasian marsupials. However, as with our anatomy-based analysis (Figure 4.3), this relationship is between *Dromiciops* and a specific Australasian order, not the Australasian orders as a whole. Contrary to the classical view, the syndactyl bandicoots do not appear to be closely related to the syndactyl diprotodontians. In fact, bandicoots terminate a long branch and their affinities with other marsupials are uncertain (Lapointe and Kirsch 1995).

Parsimony analyses of DNA sequence-variation in the protamine P1 gene (Retief et al. 1995a,b) and mitochondrial 12S rRNA gene (Gemmell and Westerman 1994; Springer et al. 1994) also provide insights into the relationships among marsupial orders and families. Bootstrap values for these studies are not as high as for those involving DNA hybridization, but nevertheless they corroborate several aspects of the hybridization trees and offer further evidence on the phylogenetic position of the marsupial mole *Notoryctes*. The protamine P1 phylogenies of Retief et al. (1995a) confirm the close relationship of diprotodontians, dasyurids, and *Dromiciops* to the exclusion of bandicoots, caenolestids, and didelphids. Furthermore, the marsupial mole groups with Diprotodontia-Dasyuromophia-Microbiotheria and specifically with dasyurids, largely on the basis of a partially repeated intron shared by dasyurids and *Notoryctes*. Partial 12S rRNA sequences also provide limited support for a clade containing Diprotodontia, Dasyuromorphia, Notoryctemorphia, and Microbiotheria (Gemmell and Westerman 1994; Springer et al. 1994). Placental mammals are well separated from marsupials in all gene-sequencing and hybridization trees that include placentals, but the placental orders appear to have diverged earlier than marsupial ones (Springer et al. 1994; Kirsch et al. 1997).

In Figure 4.5 we present a consensus of the trees emerging from these molecular data sets, derived by carefully comparing the DNA-hybridization, protamine P1, and 12S phylogenies for nodes that are moderately well supported (bootstrap support ≥ 70%) in each tree; cytochrome *b* data bearing on dasyuromorph relationships (Krajewski et al. 1992) were also included. In so doing, we adopted the position espoused by Lanyon (1993) that, in assessing trees for congruence, it is essential to take into account the relative support for nodes which superficially may appear to conflict. Thus, an affinity supported at 70% or more in one tree is not contradicted by an alternative arrangement in a second, if that alternative has less than 70% bootstrap support. Taken together, the available DNA trees suggest four major clades of extant marsupials: Didelphimorphia, Paucituberculata, Peramelina, and Microbiotheria-Diprotodontia-Dasyuromorphia-Notoryctemorphia (Figure 4.5). Relationships among these four lineages (and within the last) are mostly unclear, except for possible ties between *Dromiciops* and Diprotodontia and between dasyuromorphs and the marsupial mole.

One implication of the lack of resolution at the base of the marsupial tree is that the orders of marsupials diverged from each other over a short period. Similarly, the three major subgroups of Diprotodontia diverged rapidly from each other. Nonetheless, the much longer branches leading to each clade of therians (marsupials and placentals) are well supported (Springer et al. 1994). The lack of resolution of relationships among marsupial orders thus does not appear to be a result of saturation or the inappropriateness of the molecular techniques employed. In fact, the correlation of the percentage of 12S transversions with time shows a remarkable linearity to at least 120 Mya, when calibrated against divergence points documented in the fossil record (see below).

Character evolution

ANATOMY – Much of the interest and usefulness of DNA-based phylogenies arises when anatomical characters are mapped onto the largely independent molecular trees: further support for the monophyly of some groups or instances of convergence are then highlighted, and plausible scenarios for character evolution are made possible.

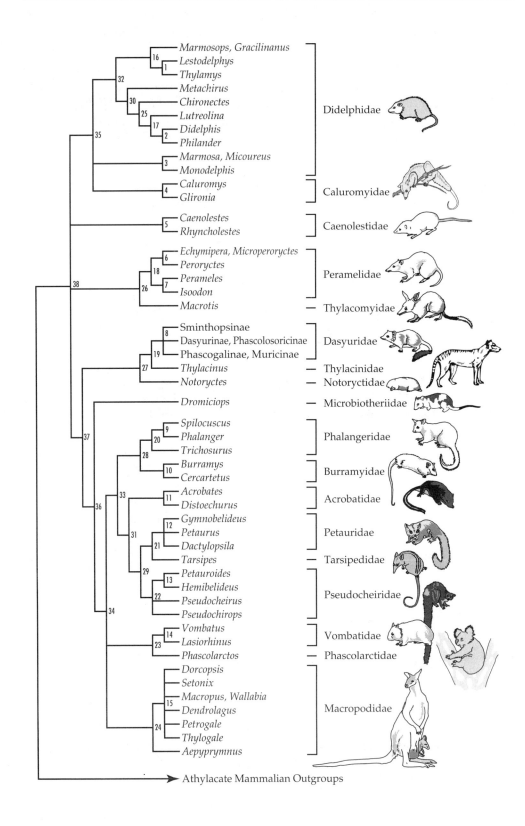

To facilitate such inferences, we used parsimony to overlay six characters on our composite molecular phylogeny using MacClade (Maddison and Maddison 1993).

Even the minimally resolved set of relationships among marsupial orders shown in Figure 4.5 has implications for the evolution of the three classical anatomical characters. First, diprotodonty is not a synapomorphy for South American caenolestids and Australasian diprotodontians (Figure 4.6A); instead, it is convergent in these two groups as has previously been suggested (Bensley 1903; Ride 1962, 1964; Kirsch 1977b; Marshall 1980; Archer 1984; Woodburne 1984). Indeed, the procumbent pair of teeth very likely represent different incisor-positions in Diprotodontia and Paucituberculata (Ride 1962; Archer 1984). However, diprotodonty apparently is a synapomorphy for diprotodontians.

Second, syndactyly appears not to be synapomorphous in Australasian peramelinans and diprotodontians (Figure 4.6b). Rather, this character evolved independently in both groups, or appeared in an earlier common ancestor and was subsequently lost in some of its descendants. Marshall (1972), Hall (1987), and Szalay (1994) all argued that both of these scenarios are unlikely, based on the complexity and detailed similarities of syndactyly in bandicoots and diprotodontians. However, we note that the second and third digits of the hindfoot are reduced and subequal in some didelphid species (Bensley 1903; Goodrich 1935), and that incipient syndactyly has been reported in caluromyids as well (Kirsch 1977b). Thus, there may be a predisposition for the syndactyl foot among marsupials. Furthermore, we repeat that syndactyly in mammals is under simple mutagenic or genetic control (see Hall 1987 and

Figure 4.5. Consensus of DNA-based trees on marsupial taxa, including representatives of all extant families except Myrmecobiidae, showing areas of moderate to strong agreement (bootstrap values ≥70%) among studies (Lanyon 1993). Investigations present data on DNA/DNA hybridizations (Kirsch et al. 1990b, 1991, 1995a,b; Edwards and Westerman 1992, 1995; Springer et al. 1992; Kirsch and Palma 1995; Lapointe and Kirsch 1995; and this paper) and gene sequences of protamine P1 (Retief et al. 1995a), 12S rRNA (Springer et al. 1994), and cytochrome *b* (Krajewski et al. 1992). Unresolved nodes were either resolved in no study or had a different resolution in different studies. Support for each numbered node is as follows: (1) 100% (Kirsch and Palma 1995); (2) 100% (Kirsch and Palma 1995); (3) 99% (Kirsch and Palma 1995; *Marmosa* may be paraphyletic); (4) 93% (Kirsch and Palma 1995); (5) 93% (Springer et al. 1994) and 100% (Lapointe and Kirsch 1995); (6) 95% (Kirsch et al. 1990b; *Echymipera* may be paraphyletic); (7) 95% (Kirsch et al. 1990b); (8) 73% (Krajewski et al. 1992); (9) 100% (this paper); (10) 90% (Edwards and Westerman 1995); (11) 99% (Edwards and Westerman 1995); (12) 100% [this paper, *Gymnobelideus* not included in Figure 4.4 and its position unresolved with respect to *Petaurus* and *Dactylopsila* in Edwards and Westerman (1992)]; (13) 100% (Springer et al. 1992, this paper); (14) 100% (this paper) and 96% (Springer et al. 1994); (15) 100% (this paper, Kirsch et al. 1995); (16) 100% (Kirsch and Palma 1995; *Marmosops* is paraphyletic); (17) 100% (Kirsch and Palma 1995); (18) 95% (Kirsch et al. 1990b); (19) no relevant bootstraps but shown to be distinct from other dasyuromorphs by Krajewski et al. (1992); (20) 100% (this paper); (21) 72% (Edwards and Westerman 1995); (22) 99% (this paper); (23) 100% (this paper); (24) 100% (this paper); (25) 100% (Kirsch and Palma 1995); (26) 100% (unpublished re-analysis of data in Kirsch et al. [1990b]); (27) 73% (estimate based on two bootstrap trees given in Retief et al. [1995a]; although a third bootstrap tree drops this value below 70%, we tentatively retain the clade because of a putatively homologous intron-duplication in the protamine P1 gene); (28) 99% (this paper); (29) 100% (this paper); (30) 100% (Kirsch and Palma 1995); (31) 99% (this paper); (32) 75% (Kirsch and Palma 1995) and 97% in Kirsch et al. (1995a); (33) 88% (this paper); (34) 54% (Kirsch and Lapointe 1995) – we retain this clade because diprotodontian monophyly is broadly consistent with other lines of evidence; (35) 100% (Kirsch and Palma 1995); (36) 88% (Kirsch et al. 1991; Lapointe and Kirsch 1995); (37) 92% (average of values in Kirsch et al. 1991, Lapointe and Kirsch 1995, and Kirsch and Palma 1995); (38) 95% (Springer et al. 1994).

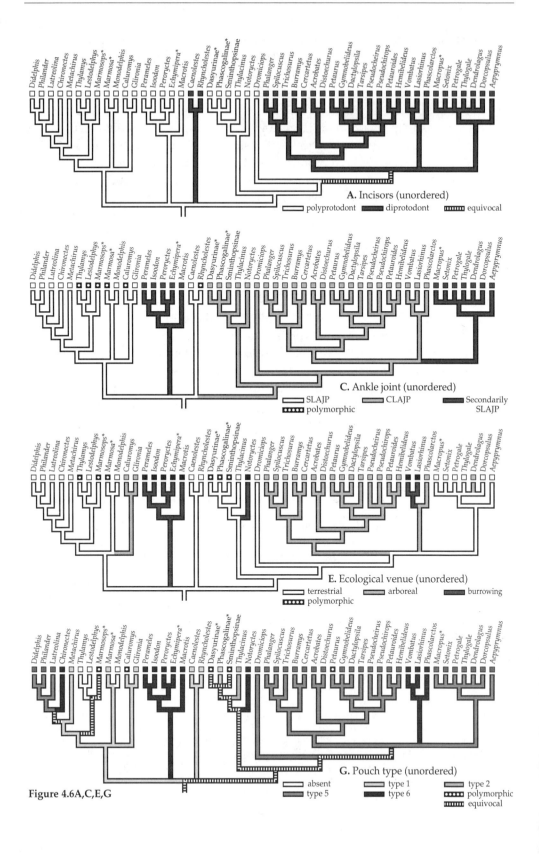

Figure 4.6A,C,E,G

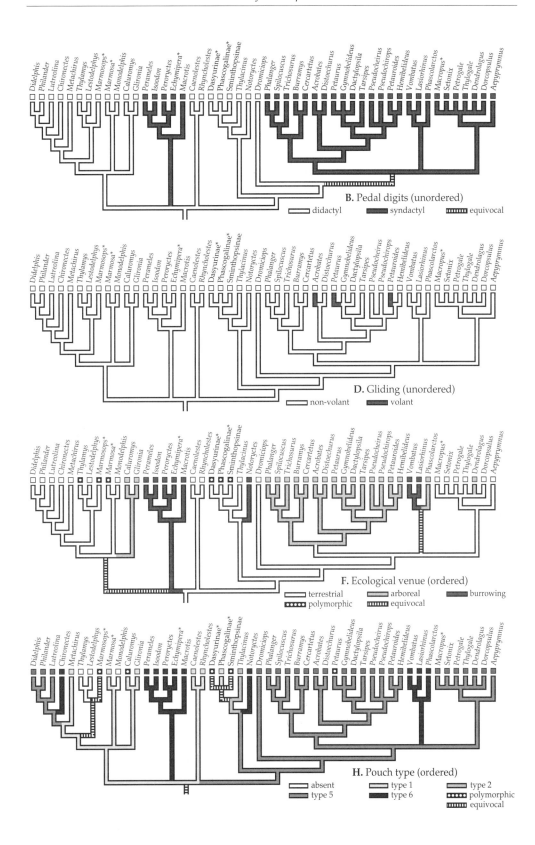

Figure 4.6. (Pages 144 and 145) Mapping of six anatomical characters onto the consensus tree of marsupial relationships (Figure 4.5) using MacClade (Maddison and Maddison 1993). Mappings and number of steps depended on whether character-states were considered ordered or unordered only for ecological venue and pouch type; separate overlays are provided for states of these two characters under each assumption. Putatively primitive states are listed first in each case: (**A**) polyprotodont vs. diprotodont incisors; (**B**) didactyly vs. syndactyly; (**C**) SLAJP vs. CLAJP ankle; (**D**) non-volant vs. volant (gliding); (**E**) terrestrial, burrowing, and arboreal venues, considered as unordered states; (**F**) ecological venues, considered as ordered states; (**G**) pouch absent vs. pouch present, with pouch types (Woolley 1974; Tyndale-Biscoe and Renfree 1987; additional data from Heshkovitz 1992b and M. Renfree and P. Temple-Smith [pers. comm.]) considered as unordered states; (**H**) pouch absent vs. pouch present, with pouch types considered as ordered states. Not all pouch types are shown on (**G**) and (**H**) because types 3 and 4 are present only in polymorphic terminal taxa. For reasons of space, some terminal taxa are omitted (*). Specifically, *Marmosops* includes *Gracilinanus*; *Marmosa* includes *Micoureus*; *Echymipera* includes *Microperoryctes*; Dasyurinae includes Phascolosoricinae; Phascogalinae includes Muricinae; and *Macropus* includes *Wallabia*.

references therein, as well as Muragaki et al. 1996); this fact gives additional credence to the perspective that syndactyly is convergent in peramelinans and diprotodontians. Finally, given the uncertain position of bandicoots in Figure 4.5, the continuous lower ankle-joint pattern might be an ancestral feature of Australian marsupials and microbiotheres (Figure 4.6C), with convergent (and probably more recent) evolution of this feature in select didelphimorphs and caenolestids (Hershkovitz 1992a) and secondary reversion to the SLAJP condition in peramelinans and kangaroos.

ECOLOGY – More illuminating than the preceding patterns of anatomical discordance, perhaps, are evolutionary trends in ecological characteristics related to specific life-styles. For example, molecular data affirm the independent evolution of gliding in three separate groups of marsupials (Figure 4.6D). More surprising is the result of mapping simple characterizations of "venue" (whether taxa are largely arboreal, terrestrial, or burrowing) onto the molecular tree. Marsupials – or at least those now found in Australasia – are frequently said to be derived from an arboreal ancestor (e. g., Huxley 1880, Szalay 1982), but this is not likely given the apparent evolution of venue assuming unordered (Figure 4.6E) or ordered (Figure 4.6F) character-states; the reconstruction involving ordered character-states requires 7 more steps than that involving unordered character-states. Didelphimorphs and dasyurids – both candidates for contemporary approximations to the ancestral marsupials – are polymorphic for venue, and the burrowing habit has evolved at least five times, in wombats, bandicoots, the marsupial mole, some dasyurids, and one kangaroo. Kangaroos are the most interesting case: most are terrestrial (*Bettongia lesueur* constructs burrows); but *Dendrolagus* is arboreal, and one tree-kangaroo species appears, from its ecological habit and cladistic position within the genus, to be in the process of secondarily (or tertiarily!) returning to the ground (Flannery et al. 1995). Many small animals freely alternate between terrestrial and arboreal activity (*Dromiciops gliroides* is one such species [Pearson 1983]), so our scoring may be somewhat arbitrary, and the ordering of states is by no means obvious.

Venue is related to that quintessential marsupial character, the pouch. In fact, many species lack the well-formed structure popularly associated with marsupials. The taxonomic distribution of the six types of pouches recognized by Woolley (1974) and Tyndale-Biscoe and Renfree (1987) – which range from simple, seasonally enlarged folds of skin (Type 1) to the capacious bags of kangaroos and bandicoots (Types 5 and 6) – begs the question of what it means to be marsupial. As with venue, dasyurids and didelphimorphs are polymorphic, collectively including representatives of almost all conditions; species of *Petaurus*, among the otherwise uniform possum families, display two types. Based on the argument of Reig et al. (1987) that possession of pectoral mammae may signal the complete and primitive absence of a pouch, we scored some didelphids as zero (*Thylamys, Lestodelphys, Gracilinanus,* and *Monodelphis*).

Mapping the six kinds of pouches onto the molecular tree requires 16 steps assuming unordered character-states (Figure 4.6G), or 28 steps assuming ordered character-states (Figure 4.6H). Both reconstructions indicate that well-formed marsupia must have evolved several times, in response to distinct – and most likely mechanical – needs; developed pouches are not ineluctably correlated with metatherian reproduction (Reig et al. 1987; Tyndale-Biscoe and Renfree 1987). Thus, typical Type 5 pouches (deep, enclosing the nipples, and with an anterior opening) tend to be associated with violent, ricochetal locomotion (kangaroos) or arboreality (possums). Type 6 pouches (a bag with an opening oriented to the rear) are understandably linked with burrowing in wombats, bandicoots, and the marsupial mole, but are also found in the aquatic didelphid *Chironectes* (Nowak 1991). However, there is a ringer: the koala possesses a backwards-opening pouch, despite the fact that it is almost strictly arboreal and certainly never burrows! We are tempted to regard this instance of an inappropriate container for the neonate as an instance of maladaptive radiation, but perhaps it is just a palimpsest of the koala's shared phylogenetic history with wombats. However, M. Renfree (pers. comm.) makes the interesting suggestion that the koala's Type 6 pouch may facilitate coprophagy by the young, which is necessary to develop the gut flora associated with digestion of *Eucalyptus* leaves. The case of the koala shows that the simple ornamentation of molecular trees with anatomical characters must give way to a more sophisticated comparative analysis (correcting for phylogenetic inertia) if detection of any correlation among traits is the desired goal (Felsenstein 1985).

Timing of the marsupial radiation

Molecular data can provide a means of estimating the time since the divergence of sister clades; the combination of such dates with the paleogeographic record then allows the reconstruction of a reasonable scenario for events of dispersal and vicariance in the history of a particular radiation.

Approximate divergence times based on DNA-hybridization data (tree-fitted, saturation-corrected distances) and complete 12S rRNA sequences (based on corrected transversions [Kimura 1980]) are given in Table 4.2. Hybridization estimates were taken from Kirsch et al. (1997); those for 12S are based on the regression shown in Figure 4.7. Several calibration points (based on fossils and vicariance) were used in determining the initial substitutional rate upon which the DNA-hybridization datings are based (Springer and Kirsch 1991; Kirsch et al. 1997). The 12S sequence regression results from 21 eutherian as well as marsupial divergence dates. Inspection of Table 4.2 shows that estimated dates based on DNA-hybridization data and 12S sequence data are in good agreement not only for mean interordinal divergences, but also for several interfamilial and intergeneric divergences where complementary data are available. For example, DNA-hybridization data suggest that the extant orders of marsupials diverged from each other between 60 and 80 Mya, with the mean time of interordinal divergence roughly 72 Mya; based on 12S transversions, the mean time of interordinal divergence is 66 Mya. With didelphimorphs (sister to all other marsupials on 12S trees [Springer et al. 1994]) excluded, the mean time of divergence for the remaining orders inferred from 12S data is 63 Mya; comparisons between didelphimorphs and other orders give a mean estimate of 75 Mya. Within Diprotodontia, the mean divergence time among families of possums is approximately 51 Mya based on either 12S sequences or DNA-hybridization data. Didelphimorphs also show an early split among living taxa: caluromyids and didelphids separated approximately 50 Mya based on DNA hybridization, or 64 Mya based on 12S transversions. Thus, the radiation that produced extant didelphimorphs apparently commenced shortly after the marsupial orders diverged from each other.

Significantly, both hybridization and sequence data suggest that the oldest extant marsupial lineages trace back only to a Santonian or Campanian common ancestor,

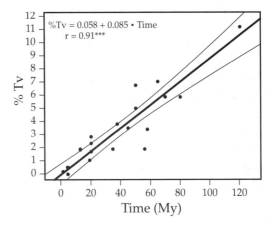

Figure 4.7. Regression of percent transversions (corrected following Kimura 1980) against divergence times estimated from the fossil record for 21 pairs of taxa. Nucleotide positions with ambiguous alignments (Springer and Douzery 1996) were omitted from this analysis. Estimates of divergence times are as follows: *Phoca* to *Halichoerus* = 2 My (Arnason et al. 1993); Pinnipedia to Felidae = 58 My (Garland et al. 1993; also assumes that pinnipeds are closer to ursids than to felids – see Lento et al. 1995); *Balaenoptera physalus* to *Balaenoptera musculus* = 5 My (Arnason and Gullberg 1993); Balaenopteridae to Delphinidae = 40 My (Barnes et al. 1985); *Mus* to *Rattus* = 12 My (Catzeflis et al. 1992); *Cavia* to *Hydrochaeris* = 20 My (Carroll 1988); *Antilocapra* to Cervidae = 19.5 My (Garland et al. 1993); *Bos* to *Capra* = 19.5 My (Garland *et al.* 1993); {Antilocapridae, Cervidae} to Bovidae = 20 My (Garland et al. 1993); Bovidae to Tragulidae = 38 My (Garland et al. 1993); Suidae to Tayasuidae = 45 My (J. Sudre [pers. comm.] to E. Douzery in Springer and Douzery 1996); Ruminantia to Suiformes = 50 My (Garland et al. 1993); *Equus grevyi* to *Equus caballus* = 4 My (MacFadden 1992); Equidae to Rhinocerotidae = 56 My (Garland et al. 1993); Artiodactyla to Perissodactyla = 66 My (Garland et al. 1993); *Elephas* to *Loxodonta* = 5 My (Coppens et al. 1978); Sirenia to Proboscidea = 70 My (Novacek 1993); Tethytheria to Hyracoidea = 80 My (Novacek 1993); Edentata to Pholidota = 65 My (Novacek 1993); Microchiroptera to Megachiroptera = 50 My (Simmons 1994); *Trichosurus* to *Phalanger* = 20 My (Flannery and Archer 1987); Eutheria to Metatheria = 120 My (midpoint of estimates by Rowe 1993, Novacek 1993, and Szalay 1994).

Table 4.2. Estimated time of divergence between pairs of marsupial taxa, based on 12S transversions and single-copy DNA-hybridization measurements.

	Taxa:	Divergence times (My) based on:	
		12S	scDNA
1.	Vombatidae to Phascolarctidae	29.4	41.5
2.	Petauridae to Pseudocheiridae	46.6	41.8
3.	Mean divergence-time among diprotodontian possum families	50.9	50.6
4.	Possums to Macropodidae	50.7	55.1
5.	*Phascogale* to *Murexia*	21.1	25.5
6.	{*Phascogale, Murexia*} to *Planigale*	29.5	35.9
7.	*Isoodon* to *Microperoryctes*	11.5	25.1
8.	*Caenolestes* to *Rhyncholestes*	21.0	12.9
9.	*Lutreolina* to *Didelphis*	28.0	10.8
10.	{*Lutreolina, Didelphis*} to *Caluromys*	64.3	49.8
11.	Mean divergence time among orders of marsupials	66.4	71.9

with divergence of persisting orders occurring in the Late Cretaceous and early Paleocene. Our dates contrast markedly with previous estimates based on globin amino-acid sequences that place the split between American didelphids and Australian taxa as far back as 128 Mya (Richardson 1988; Hope et al. 1990); they also contradict Hershkovitz's (1992a) assertion that microbiotherians diverged from other marsupials in the Jurassic. The branch leading to the extant marsupials is longer than that leading to the extant placentals (Springer et al. 1994; Kirsch et al. 1997), suggesting a more recent common ancestry for marsupials.

Historical biogeography and marsupial adaptive radiation

In conjunction with the molecular divergence times given here, recent fossil discoveries (which show that marsupials were not present in Gondwana prior to the end of the Cretaceous) suggest a new view of early marsupial radiation. Specifically, marsupials appear to have originated and then undergone an initial phase of diversification in Laurasia rather than in the Americas. However, molecular divergence times indicate that the extant marsupial taxa – which are almost exclusively Gondwanan – do not trace back to this initial radiation. Rather, a second wave of diversification in the late Cretaceous and Paleocene resulted in the extant orders that now occupy their respective adaptive zones in Australasia and the Americas. An unresolved polytomy at the base of the marsupial tree is consistent with a rapid entry into these adaptive zones, analogous to the precipitate (but probably earlier) emergence of the eutherian orders. This second phase of adaptive radiation in the marsupials may have occurred entirely in Gondwana before the final separation of South America, Antarctica, and Australia (Woodburne and Zinsmeister 1984). Possibly, a single ancestor reached South America in the late Cretaceous and then radiated into the persistent Gondwanan ordinal assemblage. According to this view, the diversification in South America is tied much more closely, both temporally and phylogenetically, to radiation in Australasia. Furthermore, the Gondwanan phase may have been triggered by empty adaptive zones that were created in the wake of the mass extinctions at the end of the Cretaceous.

Divergence times suggested by our molecular data are consistent with the view of Marshall et al. (1990) that marsupials entered Australia by about 56 Mya. However, if our construction of interordinal relationships is correct, then marsupial dispersal to Australia is more complicated than previously believed (Kirsch et al. 1991; Springer et al. 1994). The initial divergence of taxa that are now Australasian (i. e., bandicoots vs. other orders) may have occurred outside of Australia, with independent dispersals to the Australian craton. Our divergence estimates also allow for the possibility that other interordinal splits among orders now exclusively Australian took place elsewhere.

Intraordinal episodes of adaptive radiation characterized the subsequent evolution of marsupials. For example, in Australia the radiation of diprotodontians apparently commenced in the Eocene and may have been promoted by a key adaptation for herbivory: a complex type of mandibular fossa, structurally analogous to that seen in various ungulates, that permits precise cheektooth occlusion as well as incisive function (Aplin 1987). First, possum lineages originated from a common ancestor over a short period in the middle Eocene. This radiation of arboreal herbivores is apparently correlated with the expansion of *Nothofagus*-dominated closed forests under moist conditions (Case 1989). In turn, terrestrial vombatiforms, most of which are now extinct (e. g., diprotodontids, palorchestids, ilariids), apparently radiated into an adaptive zone that developed in the Oligocene and continued into the Miocene with increased aridity and the augmentation of open-forest habitat (Case 1989; Springer and Kirsch 1991). The diversification of macropodine kangaroos from forest to open-country species within the last 10 My (Raven and Gregory 1946; Kirsch et al. 1997) probably was in response to further drying and the development of extensive grasslands (Case 1989), analogous to but later than the evolution of grazing in horses (MacFadden 1992).

Interspersed among and sometimes at the base of these episodes of inter- and intraordinal adaptive radiation were probably periods of progressive occupation, wherein one or a few unrelated taxa made the transition to a different adaptive zone, often well after that zone was established. Examples of this pattern are, however, less obvious than explosive radiation in marsupial phylogeny, largely because the fossil record is insufficiently complete to allow the critical distinction between age-of-origin and age-of-diversification. Early diprotodontian lineages, representing the ancestors of vombatiformes, possums, and kangaroos, may well have been arboreal (as argued by Szalay 1994), although this interpretation is not supported by our reconstructions of venue evolution (Figure 4.6E,F). If the early diprotodontians were arboreal, then the rather different life-styles adopted by some of their respective descendants would represent progressive adaptation to open forests and grasslands. In South America, the successive but temporally overlapping infiltrations of the medium- and large-carnivore zones by borhyaenids and didelphimorphs (Marshall 1978) – together with the sporadic appearances of a number of unrelated but presumably gliriform diprotodont marsupials – provide further examples of progressive occupation.

Finally, given that marsupials continued to radiate extensively in Australia and South America subsequent to the final sundering of these continents from the once-contiguous Gondwanan landmass, the marsupials also provide examples of the "duplication of adaptive types" (Simpson 1953, p. 224) in Australasia and South America. Examples include the carnivorous thylacines and borhyaenids, the didelphid opossums and dasyurids, and the caluromyid opossums and ringtail possums. The certain convergence of diprotodonty in caenolestids and diprotodontians is emblematic of such repetitive evolution at the level of single, key characters.

Conclusions

The chronicle of marsupial evolution includes elements of adaptive radiation, progressive occupation of habitat, and duplication of adaptive types by unrelated metatherians in different regions. Adaptive radiation occurred at several taxonomic levels, at various times in different geographic areas. The relatively short time over which several major lineages diverged from each other into distinct adaptive zones is underscored by the lack of resolution and short branch lengths on both molecular and morphological trees. Our analysis of marsupial phylogeny based on 102 anatomical features confirms that morphology does not resolve questions regarding higher-order relationships within this group.

We constructed a consensus phylogeny for all orders and most families of extant marsupials, based on the level of bootstrap support in trees based on DNA/DNA hybridization or DNA sequences. This molecular phylogeny supports the inclusion of the South American *Dromiciops* within a clade of Australian diprotodontians, dasyurids, and marsupial moles, and fails to provide evidence for a close relationship between the two other South American orders.

The molecular phylogeny confirms that diprotodonty evolved independently in diprotodontians and caenolestids, and suggests that syndactyly evolved twice, or instead arose once and then was secondarily lost in several groups. This phylogeny is consistent with but provides no positive support for a monophyletic group in which the CLAJP evolved in the common ancestor of *Dromiciops* and certain Australasian marsupials. Deep pouches evolved in arboreal or saltatory taxa; backward-oriented pouches arose in burrowing forms and an aquatic species, and were retained (perhaps to encourage coprophagy and transfer of intestinal flora) in the arboreal folivorous koala.

Molecular data, together with recent paleontological discoveries, suggest that the radiation that produced the extant marsupial orders occurred in the late Cretaceous or early Paleocene, possibly as a result of a single immigration event to Gondwana from Laurasia. The South American radiation of marsupials thus appears to have had much closer ties to the Australasian radiation than to the earlier radiation in North America. The subsequent radiation and diversification of the diprotodontians seems to have closely tracked changes in the Australian environment after the breakup of the southern continents, with the initial evolution of arboreal herbivores under moist conditions, and the subsequent rise of grazers and burrowers under semi-arid conditions. Convergent evolution resulted in different marsupial groups occupying similar adaptive zones in Australia and South America; examples include ringtail possums and caluromyid opossums, thylacines and borhyaenids, and dasyurids and didelphid opossums, respectively.

Acknowledgments

We thank Michael Woodburne, Thomas Givnish, James Hutcheon, and anonymous reviewers for commenting on earlier drafts of this manuscript. MSS acknowledges a grant from the National Science Foundation (DEB-9419617) and an Alfred P. Sloan Young Investigator's Award in Molecular Evolution. The DNA-hybridization studies were supported by awards from the Systematic Biology and International Cooperative Science programs of the US National Science Foundation, including a Dissertation Research Improvement Grant to the senior author; the Graduate Research Committee, Natural History Museums Council, and Nave Fund of the University of Wisconsin-Madison; and private donors. We are also grateful to Marilyn Renfree and Peter Temple-Smith for sharing data on marsupial pouches, and to William Feeny for constructing several of the figures.

References

Aplin, K. 1987. Basicranial anatomy of the Early Miocene diprotodontian *Wynyardia bassiana* Marsupialia: Wynyardiidae and its implications for wynyardiid phylogeny and classification. Pp. 369–391 in *Possums and opossums: studies in evolution*, M. Archer, ed. Chipping Norton, Australia: Surrey Beatty and Sons.

Aplin, K., and Archer, M. 1987. Recent advances in marsupial systematics with a new syncretic classification. Pp. xv–lxxii in *Possums and opossums: studies in evolution*, M. Archer, ed. Chipping Norton, Australia: Surrey Beatty and Sons.

Archer, M. 1982. *Carnivorous marsupials*. Sydney, Australia: Royal Zoological Society of New South Wales.

Archer, M. 1984. The Australian marsupial radiation. Pp. 633–808 in *Vertebrate zoogeography and evolution in Australia*, M. Archer and G. Clayton, ed. Perth, Australia: Hesperian Press.

Archer, M., Godthelp, H., and Hand, S. J. 1993. Early Eocene marsupial from Australia. Kaupia 3:193–200.

Archer, M., Hand, S., and Godthelp, H. 1988. A new order of Tertiary zalambdodont marsupials. Science 239:1528–1531.

Arnason, U., and Gullberg, A. 1993. Comparison between the complete mtDNA sequences of the blue and the fin whale, two species that can hybridize in nature. Journal of Molecular Evolution 37:312–322.

Arnason, U., Gullberg, A., Johnsson, E., and Ledje, C. 1993. The nucleotide sequence of the mitochondrial DNA molecule of the grey seal, *Halichoerus grypus*, and a comparison with mitochondrial sequences of other true seals. Journal of Molecular Evolution 37:323–330.

Avers, C. J. 1989. *Process and pattern in evolution.* New York, NY: Oxford University Press.
Barnes, L. G., Domning, D. P., and Ray, C. E. 1985. Status of studies in fossil marine mammals. Marine Mammal Science 1:15–53.
Bensley, B. A. 1903. On the evolution of the Australian Marsupialia; with remarks on the relationships of the marsupials in general. Transactions of the Linnean Society, London, 2nd Series (Zoology) 9:83–217.
Biggers, J. D., and DeLamater, E. D. 1965. Marsupial spermatozoa pairing in the epididymis of American forms. Nature 208:1602–1603.
Bisacre, M., Carlisle, R., Robertson, D., and Ruck, J. 1989. *Illustrated encyclopedia of plants and animals.* London, England: Marshall Cavendish Limited.
Bonaparte, J. F. 1990. New Late Cretaceous mammals from the Los Alamitos Formation, northern Patagonia. National Geographic Research 6:63–93.
Bryant, B. J. 1977. The development of the lymphatic and immunohematopoietic systems. Pp. 349–385 in The biology of marsupials, D. Hunsaker, ed. New York, NY: Academic Press.
Carroll, R. L. 1988. *Vertebrate paleontology and evolution.* New York, NY: W. H. Freeman and Company.
Case, J. A. 1989. Antarctica: the effect of high latitude heterochroneity on the origin of the Australian marsupials. Geological Society Special Publication 47:217–226.
Catzeflis, F., Aguilar, J.-P., and Jaeger, J.-J. 1992. Muroid rodents: phylogeny and evolution. Trends in Ecology and Evolution 7:122–126.
Chase, J., and Graydon, M. L. 1990. The eye of the northern brown bandicoot, *Isoodon macrourus*. Pp. 117–122 in *Bandicoots and bilbies,* J. H. Seebeck, P. R. Brown, R. L. Wallis, and C. M. Kemper, eds. Chipping Norton, Australia: Surrey Beatty and Sons.
Cifelli, R. L. 1990a. Cretaceous mammals of southern Utah. I. Marsupials from the Kaiparowits Formation Judithian. Journal of Vertebrate Paleontology 10:295–319.
Cifelli, R. L. 1990b. Cretaceous mammals of southern Utah. II. Marsupials and marsupial-like mammals from the Wahweap Formation early Campanian. Journal of Vertebrate Paleontology 10:320–331.
Cifelli, R. L. 1993a. Early Cretaceous mammals from North America and the evolution of marsupial dental characters. Proceedings of the National Academy of Sciences USA 90:9413–9416.
Cifelli, R. L. 1993b. Theria of metatherian-eutherian grade and the origin of marsupial orders. Pp. 205–215 in *Mammal phylogeny,* F. Szalay, M. J. Novacek, and M. C. McKenna, eds. New York, NY: Springer-Verlag.
Cifelli, R. L., and Eaton, J. G. 1987. Marsupial from the earliest Late Cretaceous of western U.S. Nature 325:520–522.
Cockburn, A. 1990. Life history of the bandicoots: developmental rigidity and phenotypic plasticity. Pp. 285–292 in *Bandicoots and bilbies,* J. H. Seebeck, P. R. Brown, R. L. Wallis, and C. M. Kemper, eds. Chipping Norton, Australia: Surrey Beatty and Sons.
Coppens, Y., Maglio, V. J., Madden, C. T., and Beden, M. 1978. Proboscidea. Pp. 336–367 in *Evolution of East African mammals,* V. J. Maglio and H. B. S. Cooke, eds. Cambridge, MA: Harvard University Press.
Eaton, J. G. 1993a. Therian mammals from the Cenomanian Upper Cretaceous Dakota Formation, southwestern Utah. Journal of Vertebrate Paleontology 13:105–124.
Eaton, J. G. 1993b. Marsupial dispersal. National Geographic Research and Exploration 9:436–443.
Edwards, D., and Westerman, M. 1992. DNA-DNA hybridisation and the position of Leadbeater's possum *Gymnobelideus leadbeateri* McCoy in the family Petauridae Marsupialia: Diprotodontia. Australian Journal of Zoology 40:563–571.
Edwards, D., and Westerman, M. 1995. The molecular relationships of possum and glider families as revealed by DNA-DNA hybridisations. Australian Journal of Zoology 43:231–240.
Felsenstein, J. 1985. Phylogenies and the comparative method. American Naturalist 125:1–15.
Flannery, T. F. 1989. Phylogeny of the Macropodoidea: a study in convergence. Pp. 1–46 in *Kangaroos, wallabies and rat-kangaroos,* G. Grigg, P. Jarman, and I. Hume, eds. Chipping Norton, Australia: Surrey Beatty and Sons.
Flannery, T. F., and Archer, M. 1987. *Strigocuscus reidi* and *Trichosurus dicksoni,* two new fossil phalangerids (Marsupialia: Phalangeridae) from the Miocene of Northwestern Queensland. Pp.527–536 in *Possums and opossums: studies in evolution,* M. Archer, ed. Chipping Norton, Australia: Surrey Beatty and Sons.
Flannery, T. F., Boeadi, and A. L. Szalay 1995. A new tree-kangaroo *Dendrolagus* : Marsupialia from Irian Jaya, Indonesia, with notes on ethnography and the evolution of tree-kangaroos. Mammalia 59:65–84.
Garland, T., Jr., Dickerman, A. W., Janis, C. M., and Jones, J. A. 1993. Phylogenetic analysis of covariance by computer simulation. Systematic Biology 42:265–292.

Gayet, M., Marshall, L. G., and Sempere, T. 1991. The Mesozoic and Paleocene vertebrates of Bolivia and their stratigraphic context: a review. Revista Tecnica de YPFB 12:393–433.

Gemmell, N. J., and Westerman, M. 1994. Phylogenetic relationships within the class Mammalia: a study using mitochondrial 12S RNA sequences. Journal of Mammalian Evolution 2:3–23.

Godthelp, H., Archer, M., Cifelli, R., Hand, S. J., and Gilkeson, C. F. 1992. Earliest known Australian Tertiary mammal fauna. Nature 256:514–516.

Goodrich, E. S. 1935. Syndactyly in marsupials. Proceedings of the Zoological Society, London 1935:175–178.

Gregory, W. K. 1910. The orders of mammals. Bulletin of the American Museum of Natural History 27:1–524.

Haight, J. R., and Sanderson, K. J. 1990. An autoradiographic analysis of the organization of the retinal projections of the dorsal lateral geniculate nucleus in two bandicoots, *Perameles gunnii* and *Isoodon obesulus*; do bandicoots see like polyprotodonts or diprotodonts? Pp. 159–173 in *Bandicoots and bilbies*, J. H. Seebeck, P. R. Brown, R. L. Wallis, and C. M. Kemper, eds. Chipping Norton, Australia: Surrey Beatty and Sons.

Hall, L. S. 1987. Syndactyly in marsupials – problems and prophecies. Pp. 245–255 in *Possums and opossums: studies in evolution*, M. Archer, ed. Chipping Norton, Australia: Surrey Beatty and Sons.

Hand, S. J., Novacek, M., Godthelp, H., and Archer, M. 1994. First Eocene bat from Australia. Journal of Vertebrate Paleontology 14:375–381.

Harding, H. R., and Aplin, K. 1990. Phylogenetic affinities of the koala (Phascolarctidae: Marsupialia): a reassessment of the spermatozoal evidence. Pp. 1–31 in *Biology of the koala*, A. K. Lee, K. A. Handasyde, and G. D. Sanson, eds. Chipping Norton, Australia: Surrey Beatty and Sons.

Harland, W.B., Armstrong, R. L., Cox, A. V., Craig, L. E., Smith, A. G., and Smith, D. G. 1989. *A geologic time scale*. New York, NY: Cambridge University Press.

Hershkovitz, P. 1992a. Ankle bones: The Chilean opossum (*Dromiciops glíroides* Thomas), and marsupial phylogeny. Bonner Zoologische Beiträge 43:181–213.

Hershkovitz, P. 1992b. The South American gracile mouse opossums, genus *Gracilinanus* Gardner and Creighton, 1989 (Marmosidae, Marsupialia): a taxonomic review with notes on general morphology and relationships. Fieldiana: Zoology, New Series 70:1–56.

Hope, R., Cooper, S., and Wainwright, B. 1990. Globin macromolecular sequences in marsupials and monotremes. Pp. 147–171 in *Mammals from pouches and eggs: genetics, breeding and evolution of marsupials and monotremes*, J. A. Marshall Graves, R. M. Hope, and D. W. Cooper, eds. Melbourne, Australia: CSIRO.

Hughes, R. L. 1965. Comparative morphology of spermatozoa from five families. Australian Journal of Zoology 13:533–543.

Huxley, T. H. 1880. On the application of the laws of evolution to the arrangement of the Vertebrata and more particularly of the Mammalia. Proceedings of the Zoological Society of London 1880:649–662.

Kimura, M. 1980. A simple method for estimating evolutionary rate of base substitutions through comparative studies of nucleotide sequences. Journal of Molecular Evolution 16:111–120.

Kirsch, J. A. W. 1968. Prodromus of the comparative serology of Marsupialia. Nature 217:418–420.

Kirsch, J. A. W. 1977a. The six-percent solution: second thoughts on the adaptedness of marsupials. American Scientist 65:276–288.

Kirsch, J. A. W. 1977b. The comparative serology of Marsupialia, and a classification of marsupials. Australian Journal of Zoology (Supplementary Series) 38:1–152.

Kirsch, J. A. W., Dickerman, A. W., and Reig, O. A. 1995a. DNA/DNA hybridization studies of carnivorous marsupials. IV. Intergeneric relations among opossums (Didelphidae). Marmosiana 1:57–78.

Kirsch, J. A. W., Dickerman, A. W., Reig, O. A., and Springer, M. S. 1991. DNA hybridization evidence for the Australasian affinity of the American marsupial *Dromiciops australis*. Proceedings of the National Academy of Sciences, USA 88:10465–10469.

Kirsch, J. A. W., Krajewski, C., Springer, M. S., and Archer, M. 1990b. DNA-DNA hybridisation studies of carnivorous marsupials. II. relationships among dasyurids (Marsupialia: Dasyuridae). Australian Journal of Zoology 38:676–696.

Kirsch, J. A. W., Lapointe, F.-J., and Foeste, A. 1995b. Resolution of portions of the kangaroo phylogeny (Marsupialia: Macropodidae) using DNA hybridization. Biological Journal of the Linnean Society 55:309–328.

Kirsch, J. A. W., Lapointe, F.-J., and Springer, M. S. 1997. DNA-hybridisation studies of marsupials and their implications for metatherian classification. Australian Journal of Zoology, submitted.

Kirsch, J. A. W., and Palma, R. E. 1995. DNA/DNA hybridization studies of carnivorous marsupials. V. A further estimate of relationships among opossums (Marsupialia: Didelphidae). Mammalia 59:403–425.

Kirsch, J. A. W., Springer, M. S., Krajewski, C., Archer, M., Aplin, K., and Dickerman, A. W. 1990a. DNA/DNA hybridization studies of the carnivorous marsupials. I. The intergeneric relationships of bandicoots (Marsupialia: Perameloidea). Journal of Molecular Evolution 30:434–448.

Kirsch, J. A. W., and Waller, P. F. 1979. Notes on the trapping and behavior of the Caenolestidae (Marsupialia). Journal of Mammalogy 60:390–395.

Krajewski, C. A., and Dickerman, A. W. 1990. Bootstrap analysis of phylogenetic trees derived from DNA hybridization distances. Systematic Zoology 39:383–390.

Krajewski, C., Driskell, A. C., Baverstock, P. R., and Braun, M. J. 1992. Phylogenetic relationships of the thylacine (Marsupialia: Thylacinidae) among dasyuroid marsupials: evidence from cytochrome-b DNA sequences. Proceedings of the Royal Society of London, Series B 250:19–27.

Krajewski, C., Painter, J., Buckley, L., and Westerman, M. 1994. Phylogenetic structure of the marsupial family Dasyuridae based on cytochrome-b DNA sequences. Journal of Mammalian Evolution 2:25–35.

Landry, P.-A., Lapointe, F.-J., and Kirsch, J. A. W. 1996. Estimating phylogenies from lacunose distance matrices: additive is superior to ultrametric estimation. Molecular Biology and Evolution 13:818–823.

Lanyon, S. M. 1993. Phylogenetic frameworks: towards a firmer foundation for the comparative approach. Biological Journal of the Linnean Society 49:45–56.

Lapointe, F.-J., and Kirsch, J. A. W. 1995. Estimating phylogenies from lacunose distance matrices, with special reference to DNA hybridization data. Molecular Biology and Evolution 12:266–284.

Lapointe, F.-J., Kirsch, J. A. W., and Bleiweiss, R. 1994. Jackknifing of weighted trees: validation of phylogenies constructed from distance matrices. Molecular Phylogenetics and Evolution 3:256–267.

Lento, G. M., Hickson, R. E., Chambers, G. K., and Penny, D. 1995. Use of spectral analysis to test hypotheses on the origin of pinnipeds. Molecular Biology and Evolution 12:28–52.

Luckett, W. P. 1994. Suprafamilial relationships within Marsupialia: resolution and discordance from multidisciplinary data. Journal of Mammalian Evolution 2:255–283.

MacFadden, B. J. 1992. *Fossil horses: systematics, paleobiology, and evolution of the family Equidae.* New York, NY: Cambridge University Press.

Maddison, W. P., and D. R. Maddison 1993. *MacClade: analysis of phylogeny and character evolution, vers. 3.* Sunderland, MA: Sinauer Associates.

Marlow, B. J. 1965. *Marsupials of Australia.* Brisbane, Australia: Jacaranda Press.

Marshall, L. G. 1972. Evolution of the peramelid tarsus. Proceedings of the Royal Society of Victoria 85:51–60.

Marshall, L. G. 1978. Evolution of the borhyaenids, extinct South American predaceous marsupials. University of California Publications in Geological Science 117:1–89.

Marshall, L. G. 1980. Systematics of the South American marsupial family Caenolestidae. Fieldiana Geology, New Series 5:145.

Marshall, L. G., Webb, S. D., Sepkoski, J. J., Jr., and Raup, D. M. 1982. Mammalian evolution and the great American interchange. Science 215:1351–1357.

Marshall, L. G., Case, J. A., and Woodburne, M. O. 1990. Phylogenetic relationships of the families of marsupials. Pp. 433–505 in *Current mammalogy, vol. 2*, H. H. Genoways, ed. New York, NY: Plenum Press.

Muragaki, Y., Mundios, S., Upton, J., and Olsen, B. R. 1996. Altered growth and branching patterns in synpolydactyly caused by mutations in HOXD13. Science 272:548–551.

Novacek, M. J. 1993. Reflections on higher mammalian phylogenetics. Journal of Mammalian Evolution 1:3–30.

Nowak, R. M. 1991. *Walker's mammals of the world, vol. 1 (5th ed.).* Baltimore, MD: Johns Hopkins University Press.

Osgood, W. H. 1921. A monographic study of the American marsupial *Caenolestes*. Field Museum of Natural History Zoological Series 14: 175–232.

Patton, J. L., dos Reis, S. F., and da Silva, M. N. F. 1995. Relationships among didelphid marsupials based on sequence variation in the mitochondrial cytochrome b gene. Journal of Mammalian Evolution 3:3–25.

Pearson, O. P. 1983. Characteristics of a mammalian fauna from forests in Patagonia, southern Argentina. Journal of Mammalogy 64:476–492.

Price, P. W. 1996. *Biological evolution.* New York, NY: Saunders.

Raven, H. C., and Gregory, W. K. 1946. Adaptive branching of the kangaroo family in relation to habitat. American Museum Novitates 1309:1–15.

Reig, O. A., Kirsch, J. A. W., and Marshall, L. G. 1987. Systematic relationships of the living and neocenozoic American "opossum-like" marsupials (suborder Didelphimorphia), with comments on the classification of these and of the Cretaceous and Paleogene New World and European metatherians. Pp. 1–89 in *Possums and opossums: studies in evolution*, M. Archer, ed. Chipping Norton, Australia: Surrey Beatty and Sons.

Retief, J. D., Krajewski, C., Westerman, M., Winkfein, R. J., and Dixon, G. H. 1995a. Molecular phylogeny and evolution of marsupial protamine P1 genes. Proceedings of the Royal Society of London, Series B 259:7–14.

Retief, J. D., Rees, J. S., Westerman, M., and Dixon, G. H. 1995b. Convergent evolution of cysteine residues in sperm protamines of one genus of marsupials, the *Planigales*. Journal of Molecular Evolution 12:708–712.

Richardson, B. J. 1988. A new view of the relationships of Australian and American marsupials. Australian Mammalogy 11:71–73.

Ride, W. D. L. 1962. On the evolution of Australian marsupials. Pp. 281–306 in *The evolution of living organisms*, G. W. Leeper, ed. Melbourne, Australia: Melbourne University Press.

Ride, W. D. L. 1964. A review of Australian fossil marsupials. Journal and Proceedings of the Royal Society of Western Australia 47:97–131.

Ride, W. D. L. 1970. *A guide to the native mammals of Australia*. Melbourne, Australia: Oxford University Press.

Ridley, M. 1993. *Evolution*. Oxford, England: Blackwell Scientific Publications.

Rowe, T. 1993. Phylogenetic systematics and the early history of mammals. Pp. 129–145 in *Mammal phylogeny*, F. S. Szalay, M. J. Novacek, and M. C. McKenna, eds. New York, NY: Springer-Verlag.

Segall, W. 1969a. The middle ear region of *Dromiciops*. Acta Anatomica 72:489–501.

Segall, W. 1969b. The auditory ossicles malleus, incus and their relationships to the tympanic in marsupials. Acta Anatomica 73:176–191.

Segall, W. 1970. Morphological parallelisms of the bulla and auditory ossicles in some insectivores and marsupials. Fieldiana Zoology 51:169–205.

Simmons, N. B. 1994. The case for chiropteran monophyly. American Museum Novitates 3103:1–54.

Simpson, G. G. 1930. Post-Mesozoic Marsupialia. Pp. 1–87 in *Fossilium Catalogus. 1: Animalia. Pars. 47*. Berlin, Germany: W. Junk.

Simpson, G. G. 1945. The principles of classification and a classification of mammals. Bulletin of the American Museum of Natural History 85:1–350.

Simpson, G. G. 1953. *The major features of evolution*. New York, NY: Columbia University Press.

Springer, M. S. 1988. *The phylogeny of diprotodontian marsupials based on single-copy nuclear DNA-DNA hybridization and craniodental anatomy*. Ph.D. dissertation, University of California at Riverside.

Springer, M. S., and Douzery, E. 1996. Secondary structure and patterns of evolution among mammalian mitochondrial 12S rRNA molecules. Journal of Molecular Evolution 43:357–373.

Springer, M. S., and Kirsch, J. A. W. 1989. Rates of single-copy DNA evolution in phalangeriform marsupials. Molecular Biology and Evolution 6:331–341.

Springer, M. S., and Kirsch, J. A. W. 1991. DNA hybridization, the compression effect, and the radiation of diprotodontian marsupials. Systematic Zoology 40:131–151.

Springer, M. S., Kirsch, J. A. W., Aplin, K., and Flannery, T. 1990. DNA hybridization, cladistics, and the phylogeny of phalangerid marsupials. Journal of Molecular Evolution 30:298–311.

Springer, M. S., McKay, G., Aplin, K., and Kirsch, J. A. W. 1992. Relations among the ringtail possums (Marsupialia: Pseudocheiridae) based on DNA-DNA hybridisation. Australian Journal of Zoology 40:423–435.

Springer, M. S., Westerman, M., and Kirsch, J. A. W. 1994. Relationships among orders and families of marsupials based on 12S ribosomal DNA sequences and the timing of the marsupial radiation. Journal of Mammalian Evolution 2:85–115.

Springer, M. S., and Woodburne, M. O. 1989. The distribution of some basicranial characters within the Marsupialia and a phylogeny of the Phalangeriformes. Journal of Vertebrate Paleontology 9:210–222.

Swofford, D. L. 1991. *PAUP: phylogenetic analysis using parsimony, vers. 3.0s*, Champaign, IL: Illinois Natural History Survey.

Szalay, F. S. 1982. A new appraisal of marsupial phylogeny and classification. Pp. 621–640 in *Carnivorous marsupials*, M. Archer, ed. Sydney, Australia: Royal Zoological Society of New South Wales.

Szalay, F. S. 1994. *Evolutionary history of the marsupials and an analysis of osteological characters*. New York, NY: Cambridge University Press.

Temple-Smith, P. 1987. Sperm structure and marsupial phylogeny. Pp. 171–193 in *Possums and opossums: studies in evolution*, M. Archer, ed. Chipping Norton, Australia: Surrey Beatty and Sons.

Trofimov, B. A., and Szalay, F. J. 1994. New Cretaceous marsupial from Mongolia and the early radiation of Metatheria. Proceedings of the National Academy of Sciences, USA 91:12569–12573.

Tyndale-Biscoe, C. H., and Renfree, M. 1987. *Reproductive physiology of marsupials*. Cambridge, England: Cambridge University Press.

Westerman, M. 1991. Phylogenetic relations of the marsupial mole, *Notoryctes typhlops* (Marsupialia: Notoryctidae). Australian Journal of Zoology 39:529–537.

Westerman, M., and Edwards, D. 1991. The relationship of *Dromiciops australis* to other marsupials: data from DNA-DNA hybridisation studies. Australian Journal of Zoology 39:123–130.

Westerman, M., Janczewski, D. N., and O'Brien, S. J. 1990. DNA-DNA hybridisation studies and marsupial phylogeny. Pp. 173–181 in *Mammals from pouches and eggs: genetics, breeding and evolution of marsupials and monotremes*, J. A. Marshall Graves, R. M. Hope, and D. W. Cooper, eds. Melbourne, Australia: CSIRO.

Wilson, D. E., and Reeder, A. M. 1993. *Mammal species of the world: a taxonomic and geographic reference (2nd ed.)*. Washington, DC: Smithsonian Institution Press.

Woodburne, M. O. 1984. Families of marsupials: interrelationships, evolution, and biogeography. Pp. 48–71 in *Mammals; notes for a short course*, P. D. Gingerich and C. E. Badgley, eds. University of Tennessee: Department of Geological Sciences Studies in Geology, Vol. 8.

Woodburne, M. O., MacFadden, B. J., Case, J. A., Springer, M. S., Pledge, N. S., Power, J. D., J. Woodburne, M. O., and Springer, K. B. 1993. Land mammal biostratigraphy and magnetostratigraphy of the Etadunna Formation late Oligocene of South Australia. Journal of Vertebrate Paleontology 13:483–515

Woodburne, M. O., Tedford, R. H., Archer, M., Turnbull, W. D., Plane, M. D., and Lundelius, E. L., Jr. 1985. Biochronology of the continental mammal record of Australia and New Guinea. Special Publications of the South Australia Department of Mines and Energy 5:347–363.

Woodburne, M. O., and Zinsmeister, W. J. 1984. The first land mammal from Antarctica and its biogeographic implications. Journal of Paleontology 58:913–948.

Woolley, P. 1974. The pouch of *Planigale subtilissima* and other dasyurid marsupials. Journal of the Royal Society of Western Australia 57:11–15.

Appendix 4.1. States of 102 anatomical characters among extant marsupial families, excluding Caluromyidae. Description of character-states and transformation types are given below the data matrix; ancestral states are given in the last line of the matrix.

```
Chararcter*:
                              1111111111222222222233333333334444444444
                    1234567890123456789012345678901234567890123456789

     Didelphidae    0001002100011000001100311000010000000000000000000
    Caenolestidae   10?1121011022210001111312000120000001?000?00000000
 Microbiotheriidae  0001011000022100001011310000110000010000?00100000
      Peramelidae   0101002121022100001000211000110000001000000100000
    Thylacomyidae   0101102121022120001000211000110000001000001010000
    Myrmecobiidae   1112022??00112000?000230100001000000?0000100000100
     Notoryctidae   21?1000?00000000002?2???0?00010000101000?00000100
       Dasyuridae   1101002100000000001000210000010000000000100000000
   Phascolarctidae  2321111111111221100100322220101110022100010101111
       Vombatidae   3321121001011220001?2?0012220?00101022200010100011
      Macropodidae  23011210?101122000101?21120010101010021001000001
      Phalangeridae 2211200110112210201?03310112111110010001000001
        (Phalangeridae) 22112001101122102010?33101112111100100010000001
       Burramyidae  2221200210221210?00023111111101100?1000001000
      Pseudocheiridae 2221111111122221201001111012011100111000010000000
       Petauridae   2221210210212210101?031201110011010110100010000
       Acrobatidae  22?012102102122000?011?10000010001011110100000
      Tarsipedidae  43??2????????????????????1222020010???2010000010000
         Ancestor   0001001000000000000021100001000000000000000000000
```
(data values reproduced as visible)

Key to characters and character-states (**O** = ordered, **U** = unordered). Osteological data compiled from Aplin (1987), Aplin and Archer (1987), Archer (1984), Luckett (1994), Marshall et al. (1990), Segall (1969a,b, 1970), and Springer and Woodburne (1989), as well as from original observations. Ankle and pes data derive from Szalay (1982, 1994) as summarized in Luckett (1994); soft anatomical and histological data were taken from Chase and Graydon (1990), Haight and Sanderson (1990), and from earlier studies summarized in Luckett (1994). Sperm-ultrastructure data were taken from Biggers and DeLamater (1965) and Harding and Aplin (1990). A more detailed discussion of this data set will be presented elsewhere (Case and Springer, in prep.).

Dental characters:

1. Number of upper incisors (**O**): 0 = five; 1 = four; 2 = three; 3 = two; 4 = one
2. Number of lower incisors (**O**): 0 = four; 1 = three; 2 = two; 3 = one
3. Eruption of P3 (**O**): 0 = erupts after posterior molar or concomitant with it; 1 = erupts concomitant with M3; 2 = erupts concomitant with M2
4. Number of molars (**O**): 0 = three; 1 = four; 2 = greater than four
5. Upper molar shape (**U**): 0 = triangular; 1 = rectangular; 2 = conical
6. Paracone (pa) and metacone (me) placement (**O**): 0 = medial; 1 = buccal; 2 = buccal margin
7. Paracone versus metacone size (**O**): 0 = pa > me; 1 = pa equal to me; 2 = pa < me
8. Centrocrista shape (**O**): 0 = linear; 1 = V-shaped; 2 = absent
9. Angle between protoconal cristae (if present) (**O**): 0 = acute; 1 = approximately 90 degrees; 2 = obtuse
10. Metaconule (**O**): 0 = absent or not well-developed; 1 = well-developed or enlarged
11. Neometaconule (**O**): 0 = absent; 1 = present
12. Trigonid *versus* talonid length (**O**): 0 = trigonid > talonid; 1 = trigonid equal to talonid; 2 = trigonid < talonid
13. Trigonid *versus* talonid width (**O**): 0 = trigonid > talonid; 1 = trigonid equal to talonid; 2 = trigonid < talonid
14. Trigonid *versus* talonid height (**O**): 0 = trigonid >> talonid; 1 = trigonid > talonid; 2 = trigonid equal to talonid

```
                                                        111
        5555555555666666666677777777778888888888999999999000
        0123456789012345678901234567890123456789012345678012

        00201222220000000000000000000001000000111007100010l
        00202012221010000000000000007000007007?770??????7?1
        000l02111010000000000011210110120?0?????00?11??00
        00211022222010000000000011000112000011001001070l010
        00233022222010100000010007???0??11200007????0??????0
        00002212122010100000010000100000121?00????07????????0
        ?000001220?01000000001000l??0??0120?00????0????????0
        00201012121010100000010000100001210000111001110l010
        11201210l0111l0000010l0011?121?1100111??0??1100110000
        01202220?2110220000011001111210ll001111100?11?0110000
        1010211112212020000201101121211112001ll101?00l1100100
        l0111211120120211112111111212111120011101000l1100100
        00201022211202111121l0111??2??11207l????00l1100100
        00201221221120211112111111??21?1120011110??0011100100
        002012122211202111121111ll??21?1120011??01?0011100100
        002010111001002111121110ll??l??11200?1??1??00l1100l00
        00021222122110121112110l????1??11000?1?????0011101010
        000010010110000000000000000000000000000000000?00000
```

15. Paraconid on M/3-5 (**O**): 0 = distinct; 1 = present; 2 = absent
16. Location of protoconid on 1st molar (**O**): 0 = buccal; 1 = midline of tooth; 2 = lingual
17. Protostylid on 1st molar (**O**): 0 = absent; 1 = present and distinct
18. Angle between paracristid and protocristid (**O**): 0 = acute; 1 = right angle; 2 = obtuse
19. Hypoconid presence and location (**U**): 0 = between protoconid and posterior tooth margin; 1 = at posterobuccal corner of tooth; 2 = absent
20. Entoconid shape (if present) (**O**): 0 = conical; 1 = bladed
21. Entoconid presence and location (**U**): 0 = between metaconid and posterior tooth margin; 1 = at posterolingual corner of tooth; 2 = absent
22. Hypoconulid presence and location (**O**): 0 = at posterolingual corner of tooth; 1 = near posterolingual corner of tooth; 2 = set distinctly buccally from the posterolingual corner of tooth
23. Intersection of cristid obliqua with trigonid (**O**): 0 = lingual to the metaconid; 1 = at the metaconid; 2 = between metaconid and protoconid; 3 = at the protoconid; 4 = at the protostylid
24. Shape and presence of cristid obliqua (**O**): 0 = linear; 1 = simple arc; 2 = C-shaped; 3 = curved and kinked
25. Upper incisor arcade shape (**O**): 0 = U-shape; 1 = broad V-shape; 2 = long narrow V-shape
26. Upper dP1 (**O**): 0 = present; 1 = greatly reduced; 2 = absent
27. Lower dP2 (**O**): 0 = present; 1 = greatly reduced; 2 = absent
28. Upper dP2 (**O**): 0 = present; 1 = greatly reduced; 2 = absent
29. Upper incisors spatulate (**O**): 0 = no; 1 = yes
30. Size I3/ vs. I2/ (**O**): 0 = I3/ > I2/; 1 = I3/ equal to I2/; 2 = I3/ < I2/
31. P/3 bladed and serrated (**O**): 0 = no; 1 = yes
32. M/2 with paralophid shear (**O**): 0 = no; 1 = yes
33. Procumbent lower I2 (**O**): 0 = no; 1 = yes
34. Molar crenulations moderately to extensively developed (**O**): 0 = no; 1 = yes
35. Hypoconulid present or absent (**O**): 0 = present; 1 = absent
36. M5 much smaller than M4 (**O**): 0 = no; 1 = yes

37. dP3 (**O**): molariform or semi-molariform; 1 = reduced, premolariform; 2 = vestigial
38. Lower C, dP1 (**O**): 0 = retained; 1 = greatly reduced; 2 = absent
39. Upper C (**O**): 0 = caniniform; 1 = reduced; 2 = absent
40. Lower P3 (**O**): 0 = not reduced; 1 = reduced
41. dP2 development and eruption (**O**): 0 = earlier than dP1; 1 = later than dP1
42. Bunolophodonty or lophodonty developed (**O**): 0 = no; 1 = yes
43. Selenodonty developed (**O**): 0 = no; 1 = yes

Cranial characters:
44. Parietal-alisphenoid or squamosal-frontact on braincase (**O**): 0 = parietal-alisphenoid; 1 = squamosal-frontal
45. Width of frontals *versus* width of parietals (**O**): 0 = parietal $\geq$ frontal; 1 = frontal > parietal
46. Temporal wing on squamosal reduced (**O**): 0 = no; 1 = yes
47. Angular process inflected (**O**): 0 = yes; 1 = no
48. Mandibular symphysis ankylosed (**O**): 0 = no; 1 = yes
49. Bones surrounding infraorbital canal in the orbit (**O**): 0 = maxilla + lacrimal; 1 = maxilla only
50. Palatine-lacrimal or frontal-maxillary contact in the orbit (**O**): 0 = palatine-lacrimal; 1 = frontal-maxillary
51. Distinct preglenoid process (**O**): 0 = no; 1 = yes
52. Posterior-most point of premaxillonasal contact (**O**): 0 = at the canine; 1 = posterior to the canine; 2 = at or beyond the first premolar
53. Maxillopremaxilla suture on the palate (**O**): 0 = horizontal; 1 = moderately inclined laterlly; 2 = V-shaped; 3 = steeply inclined laterally
54. Maxilla (palatal portion) length/width ratio (**O**): 0 = 1:1; 1 = 2:1; 2 = 3:1; 3 = 4:1
55. Maxillofrontal contact (**U**): 0 = anterior to nasofrontal contact; 1 = even with nasofrontal contact; 2 = little or no contact
56. Angle of maxillojugal contact (**O**): 0 = vertical; 1 = angled posteriorly; 2 = steeply angled posteriorly
57. Anterior extent of lacrimal (**O**): 0 = within the orbit; 1 = just beyond the orbit; 2 = onto the rostrum
58. Lacrimal canal location (**O**): 0 = inside the orbit; 1 = on the orbit border; 2 = on the rostrum
59. Infraorbital canal position (**O**): 0 = at the jugal; 1 = mid-molar row; 2 = anterior to molar row
60. Palatine length (**O**): 0 = short, anterior orbit to M5/; 1 = moderate, anterior orbit to M4/; 2 = long, anterior orbit to M3
61. Medial wall of mandibular fossa (**O**): 0 = formed by alisphenoid; 1 = formed by squamosal
62. Alisphenoid tympanic wing (**O**): 0 = poorly developed; 1 = moderately developed; 2 = well-developed, extending to or near posterior lacerate foramen and paroccipital process
63. Squamosal contributes to bulla (**O**): 0 = no contribution; 1 = minor contribution; 2 = major contribution
64. Ectotympanic broadened or tubelike (**O**): 0 = ring-shaped; 1 = moderately broadened; 2 = tubelike
65. Postglenoid process (**U**): 0 = nonvacuous; 1 = vacuous; 2 = absent
66. Bony external auditory meatus separates ear canal from epitympanic recess (**O**): 0 = no; 1 = yes
67. Lateral expansion of squamosal covers external entrance to epitympanic recess (**O**): 0 = no; 1 = yes
68. Fusion of ectotympanic with other bones of the skull (**O**): 0 = no; 1 = yes
69. Postglenoid foramen (**O**): 0 = located posterior to postglenoid process; 1 = even with postglenoid process; 2 = anterior to postglenoid process and frequently encircled by squamosal
70. Absence, reduction, and/or dorsomedial shift of subsquamosal foramen (**O**): 0 = no; 1 = yes
71. Squamosal epitympanic sinus (**O**): 0 = no; 1 = yes
72. Mastoid epitympanic sinus (**O**) 0 = no; 1 = yes
73. Contact between alisphenoid tympanic wing and mastoid tympanic wing (**O**): 0 = minimal or no contact; 1 = broader contact
74. Position of incisura tympanica (**O**): 0 = caudal or caudodorsal; 1 = located dorsally or anterior crus and posterior crus unite dorsally
75. Size of incisura tympanica (**O**): 0 = wide; 1 = narrow or absent
76. Ossicular axis (**O**): 0 = $\geq$ 20 degrees; 1 = 10 to 20 degrees; 2 = < 10 degrees
77. Malleolar neck (**O**): 0 = long and sharply bent; 1 = short with slight curve; 2 = short and straight
78. Squamosal epitympanic wing (**O**): 0 = no; 1 = small; 2 = well developed
79. Stapedial ratio (**O**): 0 = < 1.5; 1 = $\geq$ 1.5
80. Manubrial-incudal lever-arm ratio (**O**); 0 = $\leq$ 1.6; 1 = > 1.6

Ankle and Pes:
81. Pes (**O**): 0 = didactylous; 1 = syndactylous
82. Lower ankle-joint pattern (**O**): 0 = separate lower ankle-joint pattern; 1 = continuous lower ankle-joint pattern, with loss of SUD
83. Calcaneo-cuboid joint pattern (**O**): 0 = ovoid; 1 = triple-faceted, with CaCum facet
84. Secondary CaAd facet (**O**): 0 = absent; 1 = present and continuous with AN facet

Soft Anatomy:
85. Presence of thinned "respiratory" chorio-allantoic membrane (**O**): 0 = no; 1 = yes
86. Fasciculus aberrans present in brain (**O**): 0 = absent; 1 = present
87. Superficial cervical thymus (**O**): 0 = absent; 1 = present
88. Retinal vascularization (**O**): 0 = vascular; 1 = avascular
89. Retinal conus (**O**): 0 = absent; 1 = present
90. Binocular overlap of retinal terminals in lateral geniculate nucleus (**O**): 0 = no; 1 = yes
91. Percentage of lateral geniculate nucleus occupied by uncrossed input (**O**): 0 = less than 15%; 1 = greater than 15%
92. Distribution of crossed and uncrossed retinal terminals (**O**): 0 = symmetrical; 1 = asymmetrical
93. Thoracic thymus (**O**): 0 = present; 1 = greatly reduced or absent

Sperm:
94. Spermatozoa with recurved head (**O**): 0 = no; 1 = yes
95. Proacrosomal granule (**O**): 0 = present; 1 = absent
96. Midpiece fibre-network (**O**): 0 = absent or present and poorly developed; 1 = present
97. Orientation of the midpiece fibre-network relative to the long axis of the flagellum (**O**): 0 = longitudinal; 1 = oblique (i.e., approximately 45°)
98. Longitudinal columns of the principal-piece fibrous sheath (**O**): 0 = present; 1 = poorly developed or absent
99. Transverse section dimensional ratio of principal-piece fibrous sheath (**O**): 0 = less than or equal to 1; 1 = greater than 1
100. Principal-piece fibre network (**O**): 0 = absent or poorly developed; 1 = present
101. Dense outer fibre-axoneme relationship (**O**): 0 = nil to slight radial displacement; 1 = marked to extreme radial displacement
102. Sperm pairing in epididymis (**O**): 0 = no; 1 = yes

5 Evolutionary Origins of Phenotypic Diversity in *Daphnia*

John K. Colbourne, Paul D. N. Hebert, and Derek J. Taylor

Freshwater zooplankton are extraordinarily favorable targets for evolutionary studies. Their population sizes are so enormous that sample sizes rarely constrain investigations, and their development is so rapid that multigenerational studies are frequently feasible. Many groups also employ unusual breeding systems that are thought to accelerate evolution (Lynch and Gabriel 1983). Their "insular" environments crystallize population boundaries and provide exceptional venues for disruptive selection.

Terrestrial archipelagoes – where the most celebrated examples of rapid adaptive evolution have been documented – provide no analog to the covariation in biological and physical factors that zooplankton encounter in moving from lakes to ponds. Lake zooplankton feed on phytoplankton, are consumed by visual vertebrate predators and occupy a medium favorable to their persistence for millennia. By contrast, pond species often feed on detritus, are exposed to tactile invertebrate predators and require resistant stages to survive recurrent but unpredictable bouts of desiccation. These ecological dichotomies between ponds and lakes would likely create divergent selection pressures similar to those that have provoked explosive radiations seen in many insular fauna (Freed et al. 1987; Kambysellis et al. 1995; Shaw 1996).

Yet, despite the confluence of factors favoring rapid diversification, freshwater zooplankton are notorious for their morphological stasis (Frey 1987). Evolutionary theorists have attributed this lack of divergence to the homogenizing effects of gene flow, arising through the dispersal of their resistant stages (Mayr 1963). However, more recent allozyme and DNA studies have shown a surprising amount of local population differentiation (DeMelo and Hebert 1994; Gómez et al. 1995; Hebert 1995), challenging the notion that stasis is simply a consequence of gene flow, and highlighting the need for further research.

Among more than a hundred genera of freshwater zooplankton, none has attracted more attention than the genus *Daphnia* (Crustacea: Cladocera). Named after the virgin nymph Daphne of Greek mythology, these organisms first attracted interest because of their unusual mode of reproduction, involving both sexual and asexual propagation. Although most of the subtleties of their breeding system were resolved by the mid-19th century, *Daphnia* has maintained its scientific appeal because of its central importance in freshwater foodwebs and its amenability to laboratory culture. As a result of extensive research, we now have a deep understanding of morphological and ecological diversity in the genus. Its taxonomy has been less satisfactorily resolved, because hybridization and phenotypic plasticity have combined to defeat the best efforts of morphologists (Brooks 1957; Hebert 1978). However, the use of molecular markers to detect the genetic discontinuities associated with reproductive isolation is now providing the stable taxonomic platform needed for phylogenetic studies (Hebert and Finston 1993, 1996; Taylor and Hebert 1994). *Daphnia* appears to entail nearly 200 species, most of which are restricted to single continents. Nonetheless, species on different continents are often closely allied, suggesting that sweepstakes dispersal has often extended the geographic ranges of individual taxa which then led to speciation. While it seems likely that most speciation occurs allopatrically (Hebert and Wilson 1994), there is evidence of more exotic processes. Some taxa have apparently arisen through reticulate speciation, fostered by the parthenogenetic amplification of F_1 hybrids and subsequent introgression (Taylor and Hebert 1993). A few polyploid lineages are known, but they are closely related to diploid taxa, and appear to have played a minor role in diversification (Dufresne and Hebert 1994). A number of closely related pairs of species seem to have arisen through sympatric speciation, linked to disruptive selection between habitat types (Taylor et al. 1996).

Despite 200 years of detailed morphological studies, daphniids are seen to possess relatively few morphological characters that are phylogenetically informative, suggesting that their morphological evolution has been constrained. In general, phenotypic stasis has been attributed to a number of genetic factors (e.g., pleiotropy, epistasis, or lack of variation), including ecological restriction (Morris et al. 1995) with an emphasis on the role of interactions with existing competitors in slowing diversification (Paul 1977; Valentine 1980; Conway Morris 1989). Information on the timing of both speciation and phenotypic change in *Daphnia* is of particular value, given the potential for their strong association with habitat shifts or competitive release.

Results from our most recent investigations into *Daphnia* phylogenesis were presented in earlier papers (Colbourne and Hebert 1996; Taylor et al. 1996). This paper examines the origins of morphological diversity in the North American members of the genus. We restrict our analysis to this region because ecological and taxonomic knowledge of *Daphnia* in other biogeographic areas is poor. Fortunately, the North American fauna is thought to include representatives of most major species complexes in the genus. Our work aims to describe the nature and patterning of morphological diversification within *Daphnia*, and to ascertain if this diversity has arisen at a relatively uniform tempo. The ecological conditions associated with character-state transformations are examined to determine the possible adaptive value of variation in several key traits (see Wiley et al. 1991). We begin by summarizing some salient biological features of the genus.

Daphnia in North America – a model system

Species of *Daphnia* are common in all areas of the globe except the tropics. The intensive application of genetic methods has identified 34 species inhabiting North America (Hebert 1995). They have been assigned to one of three subgenera: *Ctenodaphnia*, *Daphnia* and *Hyalodaphnia*. Given the lack of morphological resolution among members of a subgenus, it is not surprising that 19 of these 34 species were unrecognized by earlier taxonomists (see Brooks 1957). Nevertheless, species do differ in body size, coloration, possession of predator-induced headshields and neckteeth, morphology of the carapace and ephippia, and chromosome number (see below). Table 5.1 summarizes some of the key biological attributes of the North American fauna.

Members of the genus typically reproduce by cyclic parthenogenesis, with populations being founded from females that hatch from diapausing eggs. These females ordinarily produce broods of diploid eggs that develop solely into daughters, fostering high rates of reproduction and rapid population growth. However, under conditions that depress their metabolic rate, individuals switch to the production of both male and female offspring (see Hebert 1987). Although broods from individual females are usually single-sexed, the populational shift to a 1:1 sex ratio is remarkably precise. Following this shift in sex ratio, females begin to produce haploid sexual eggs that require fertilization in order to develop. These eggs – packaged in a protective structure termed an ephippium – resist desiccation and freezing, and are the means of long-distance dispersal. The unusual reproductive system of *Daphnia* is important, in that cyclic parthenogenesis may significantly accelerate phenotypic evolution (Lynch and Gabriel 1983; Lynch 1985).

There is only limited information on the evolutionary origins of the genus. Fossil ephippia of *Daphnia* and the closely related *Simocephalus* are known from the early Cretaceous (Fryer 1991a; Smirnov 1992). Biogeographic patterns provide further support for the ancient origin of the genus, as the southern continents are dominated by the subgenus *Ctenodaphnia*, while the faunas in the northern hemisphere consist largely of the subgenera *Daphnia* and *Hyalodaphnia* (Hebert 1978; Benzie 1987). The age of species lineages is ill defined, although Brooks (1957) suggested that many North American species had originated since the Pleistocene. Speciation mechanisms in the group are unclear, although Lynch (1985) proposed that speciation via founder effects was important and reflected the establishment of populations from single ephippia. Other research has suggested the importance of speciation linked to introgressive hybridization (Taylor et al. 1996) as well as geographic isolation (Hebert and Wilson 1994).

The origins of North American *Daphnia*

The uncertainties in the timing of speciation events and the origins of major sublineages within *Daphnia* motivated us to analyze DNA sequence variation within the entire North American fauna (Colbourne and Hebert 1996; Taylor et al. 1996). Sequence comparisons were made on a 503 bp region of the relatively slowly evolving mitochondrial 12S rRNA gene, which was selected for analysis because of the ancient origin of the genus. The arthropod mitochondrial DNA clock (Brower 1994) of 2.3% sequence divergence per million years was used for dating speciation events among closely related taxa. For divergence values in excess of 10%, a 12S rDNA clock was calculated according to Lynch and Jarrell (1993). The estimated asymptotic identity (I_∞) for our data was 0.281, while other parameters, including the nucleotide substitution rate of 0.489% per million years, were obtained from Lynch and Jarrell's (1993) investigation of mitochondrial DNA evolution. The resultant molecular phylogeny (Figure 5.1) shows that the North American fauna consists of three very distinct species assemblages, corresponding to the subgenera proposed by earlier workers. These subgenera show sufficient 12S divergence (>20%) to suggest that they

Table 5.1. Biological and ecological attributes of the 34 *Daphnia* species found in North America. Key to table: **Size** (range of core body length in mm); **breeding system** (C = cyclic parthenogenesis, O = obligate parthenogenesis); **range** (N = Nearctic, H = Holarctic, B = several biogeographic regions); **habitat** (E = ephemeral ponds, P = permanent ponds, L = lakes); **biome** (S = subtropical, W = warm temperate, C = cool temperate, B = boreal, A = arctic, where more than one biome is listed, they represent the distributional boundaries); **tolerance of water turbidity** (Y = yes, N = no); **presence of melanin in carapace** (Y = yes, N = no).

Taxon	Size	Breeding system	Range	Habitat	Biome	Water turbidity	Melanic carapace
Subgenus *Daphnia*							
ambigua	< 1.3	C	N	L	S,C	N	N
latispina	1.8–3.8	C	N	E	C	Y	N
villosa	1.4–2.3	C	N	E	C	N	N
oregonensis	1.6–3.7	C	N	E	C	Y	N
catawba	1.3–2.1	C	N	L	C,B	N	N
minnehaha	1.1–3.2	C	N	E	C,B	N	N
parvula	1.1–1.4	C	N	L	W,C	N	N
retrocurva	1.0–1.8	C	N	L	W,B	N	N
cheraphila	1.3–1.7	C	N	P,L	S,C	Y	N
prolata	1.5–2.0	C	N	E,P	S,C	Y	N
obtusa	1.1–2.2	C	H	E	W,C	N,Y	N
neo-obtusa	1.4–2.8	C	N	E	C	Y	N
pileata	1.6–2.8	C	N	E	S,W	Y	N
middendorffiana	1.9–2.7	O	H	E	A	N	Y
pulicaria	1.4–3.2	C,O	N	P,L	S,A	N	N
pulex	1.1–3.5	C,O	H	E	C,B	N	N
arenata	1.5–3.5	C	N	E	C	N	N
melanica	1.6–2.5	C	N	E	C	N	Y
tenebrosa	1.6–2.5	C,O	H	P,L	A	N	Y,N
Subgenus *Hyalodaphnia*							
laevis	1.1–1.8	C	B	P,L	S,C	N	N
dubia	1.1–1.9	C	N	L	C	N	N
curvirostris	1.3–1.8	C	H	E,P	A	N	N
umbra	1.1–2.7	C	N	P,L	A	N	Y
thorata	1.3–1.6	C	N	L	C	N	N
dentifera	0.9–2.2	C	N	P,L	C,B	N	N
mendotae	1.2–2.8	C	N	L	C,B	N	N
longiremis	0.6–2.4	C	H	L	C,A	N	N
Subgenus *Ctenodaphnia*							
exilis	1.8–4.5	C	N	E	W,C	Y	N
similis	2.0–4.5	C	H	E	C	Y	N
salina	2.0–3.5	C	N	E	C	N	N
lumholtzi	1.6–2.6	C	B	L	S,W	N	N
magna	2.0–5.0	C	B	E	C,B	N	N
ephemeralis	1.3–3.0	C	N	E	C	N	N
brooksi [a]	< 1.6	C	N	E	C	Y	N

[a] The sole species excluded from our study; it is known from only nine individuals and may be an introduced species.

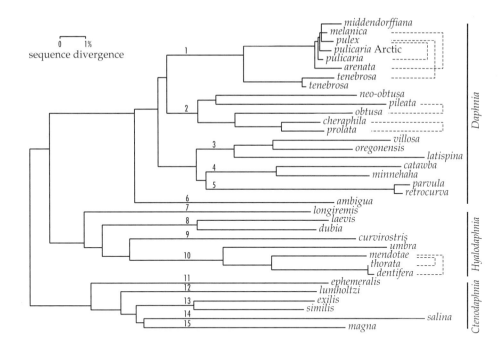

Figure 5.1. A neighbor-joining tree based on sequence variation in the mitochondrial 12S rRNA gene in *Daphnia*. Estimates of sequence divergence were corrected using the Kimura two-parameter model (Kimura 1980) and analyzed using MEGA 1.02 (Kumar et al. 1993). Dashed lines connect taxa known to hybridize. Numbers indicate species complexes within each of the three subgenera. The six complexes within subgenus *Daphnia* are: *pulex* (1), *obtusa* (2), *villosa* (3), *catawba* (4), *retrocurva* (5), *ambigua* (6). Complexes within subgenus *Hyalodaphnia* are: *longiremis* (7), *laevis* (8), *curvirostris* (9), *longispina* (10). *Ctenodaphnia* complexes are: *ephemeralis* (11), *lumholtzi* (12), *similis* (13), *atkinsoni* (14), *magna* (15).

originated during the Mesozoic. Although the subgenus *Ctenodaphnia* appears to have diverged first, all three subgenera apparently differentiated during a brief interval, complicating the delineation of ancestry. For example, the majority rule consensus of the seven shortest cladograms (which is also one of the seven most parsimonious trees; Figure 5.2) suggests that *Hyalodaphnia* could be polyphyletic. However, because only one extra step to the tree length is needed to place the (*laevis, dubia*) clade within *Hyalodaphnia* (dashed line in Figure 5.2), we used this latter tree for character-state optimizations using MacClade vers. 3.04 (Maddison and Maddison 1992). Additional details on the phylogenetic and evolutionary analyses are given in the figure legends and in Colbourne and Hebert (1996).

Each of the three subgenera includes a number of species complexes, with each complex defined as a set of species showing less than 14% sequence divergence in the 12S rRNA gene. This boundary condition was selected because it coincides with the maximum sequence divergence between species of *Daphnia* known to produce viable hybrids. The North American fauna includes five species complexes of *Ctenodaphnia*, four of *Hyalodaphnia*, and six within the subgenus *Daphnia* (Figure 5.1). Based on the expected rates of 12S rDNA evolution (see Lynch and Jarrell 1993), each of these

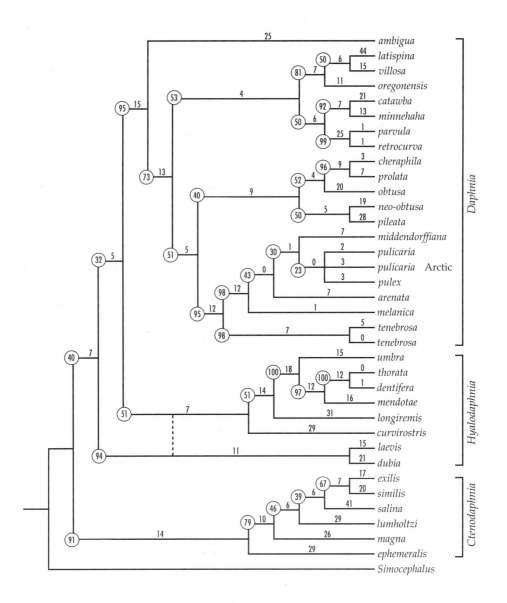

Figure 5.2. Fifty percent majority rule consensus and one of seven most parsimonious trees (length 1,186 steps, CI = 0.41, RI = 0.60) for *Daphnia*, based on 270 variable characters of the 12S rRNA gene. Transversions and transitions were weighted equally. Trees were obtained using a heuristic search in PAUP vers. 3.1.1 (Swofford 1993); taxa were added using the SIMPLE option with MULPARS and steepest descent options invoked; branch swapping used the tree bisection-reconstruction algorithm. No other equally short (or shorter) trees were found when taxa were added randomly in 25 replicate trials, with 10 trees being held at each step. Variation among the seven most parsimonious trees involved only the branches within the *pulex* complex. Four of the seven trees placed *D. melanica* at the basal node and placed *D. arenata* at the next higher branch. *D. middendorffiana* was the sister species to the unresolved (*pulex, pulicaria*) clade in five of the seven trees. The number of characters changing unambiguously (including all variable sites) is shown on each tree branch. Bootstrap percentages from 200 pseudo-replicates are shown in circles (calculated using Random Cladistics, vers. 2.1.0 [Siddall 1994]). Tree length using informative characters only was 1135 steps. A tree with one extra step suggesting the monophyly of *Hyalodaphnia* (dashed line) was chosen for character-state optimization analysis. Sequence alignments are available upon request from the authors.

complexes has persisted for at least 50 million years. It should be noted, however, that 11 of these 16 complexes contain only one or two species in North America. Four of the five remaining complexes (*pulex, obtusa, retrocurva, longispina*) show evidence of more active speciation in the last three million years. But even in these cases, diversification has not exceeded four species. Clearly, speciation rates have been constrained in North American *Daphnia*.

Phenotypic evolution

Despite their low rate of speciation, daphniids undoubtedly encountered diversifying selective pressures among different freshwater habitat types, many of which are sufficiently old for phylogenesis. For example, ca. 2,000 low-pH lakes have existed in north central Florida for at least several hundred thousand years (Hendry and Brezonik 1984). Similarly, some 30,000 lakes of playa origins have been present in the Texas/Oklahoma region since the Pleistocene (Osterkamp and Wood 1987). Water temperature, pH, and salinity, habitat permanency, and the intensity of exposure to ultraviolet radiation all vary on both local and regional scales. Aside from these physical variables, *Daphnia* populations also encounter local variation in predation regime (often linked most clearly to differences in habitat permanency). Lake populations are usually exposed to predation by fish, while those in ponds – where anoxia or complete freezing can kill fish, or where a lack of inlets or outlets may prevent fish dispersal – generally are exposed to predation by invertebrates only. To investigate the extent to which *Daphnia* have adapted to these physical and biological regimes, we analyze the ecological conditions associated with character-state transformations for six key phenotypic traits: cuticular melanization, headshield morphology, carapace setation, ephippial morphology, breeding system, and chromosome number.

CUTICULAR MELANIZATION AND ULTRAVIOLET RADIATION – Four North American species of *Daphnia* deposit melanin in their carapace (Figure 5.3). These pigmented daphniids predominate in the Arctic, but also occur in alpine and coastal sand-dune habitats. At these sites, *Daphnia* are found in shallow, clear-water ponds that are unshaded and lack dissolved humics, which absorb ultraviolet (UV) light in other aquatic habitats. Experimental studies have shown that the melanized carapaces of these daphniids intercept more than 90% of incident UV radiation, and transplantation experiments have shown that unpigmented *Daphnia* cannot survive the high UV exposure in clear-water ponds (Hebert and Emery 1990). Cuticular melanization clearly provides protection from UV radiation in these habitats (see Hobæk and Wolf 1991; Hebert 1995); in other contexts it is likely to be a net disadvantage because melanic carapaces render their bearers more conspicuous to potential predators.

Brooks (1957) hypothesized that melanization evolved only once in North American *Daphnia*, suggesting that the trait was shared among closely related species because of introgressive hybridization. However, character-state optimization of this trait onto the molecular phylogeny reveals that pigmentation in the genus has convergently evolved from a non-melanic state in members of two subgenera (Figure 5.4). Within *Hyalodaphnia*, one species (*D. umbra*) acquired the trait, while within *Daphnia*, three species of the *pulex* complex are melanic. These two subgenera show

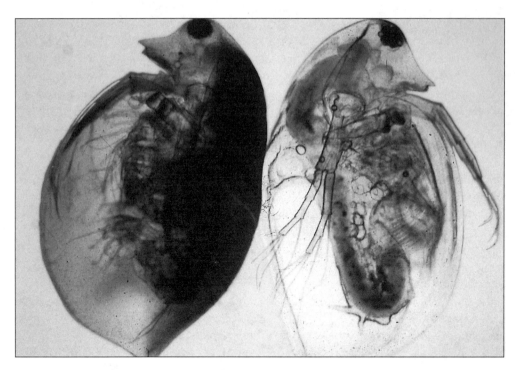

Figure 5.3. Melanic *Daphnia middendorffiana* (left) and non-melanic *D. pulex* from low-Arctic tundra ponds near Churchill, Manitoba, Canada.

approximately 24% sequence divergence at the 12S rRNA gene, representing over 180 million years of independent evolution. Hybridization is unknown between members of these two subgenera, making it apparent that melanization evolved on at least two occasions.

It is, however, much more difficult to ascertain the number of occasions on which melanization evolved in the *pulex* complex. Figure 5.4 suggests three origins of melanization, but this cladogram depicts only one of seven equally parsimonious patterns for the closely related members of the *pulex* complex, based on our 12S rDNA sequence data. In principle, the accumulation of additional sequence information at faster evolving genes could establish a single phylogeny and resolve this uncertainty. However, such efforts to reconstruct the evolutionary origins of melanization are complicated by the occurrence of hybridization among members of the *pulex* complex. It is possible that these three species share the character due to an introgressed gene that diffused rapidly among taxa, possibly because of the adaptive superiority of melanized individuals in clear-water habitats.

Nevertheless, biogeographical and historical evidence suggest that melanization has evolved recently (< 1 Mya) in the subgenus *Daphnia*. *Daphnia middendorffiana* and *D. tenebrosa*, although now endemic to the Arctic, are closely related to more southern pond species (Dufresne and Hebert 1994, 1995), suggesting that the *pulex* complex radiated by adapting to the huge number of northern ponds which were created following deglaciation. Indeed, these taxa appear to have originated during the Pleistocene. Melanization in *Daphnia* may represent a key adaptation which arose following Pleistocene glaciations, and promoted access to vacant, shallow-water, high-radiation habitats.

HEADSHIELD MORPHOLOGY AND PREDATION – While adults of some species of *Daphnia* show little variation in head shape, others exhibit striking seasonal changes in morphology, termed cyclomorphosis (Figure 5.5; see Jacobs 1987). These changes can involve the production of either spikelike or bladelike extensions to the head (helmets) or thornlike spines in the neck region (neckteeth; Figure 5.5D). Although abiotic factors

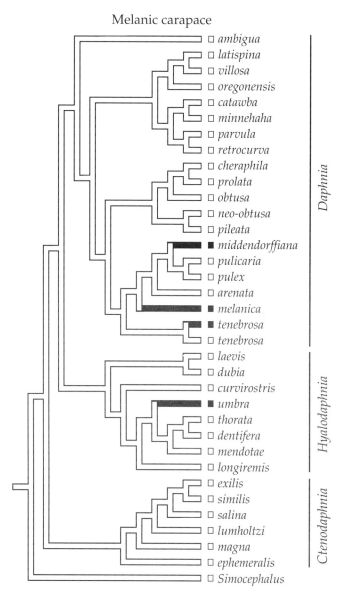

Figure 5.4. Character-state optimization of the presence/absence of cuticular melanization in *Daphnia*. Black shading denotes the presence of melanization.

modulate helmet development (Jacobs 1962; Havel and Dodson 1985), chemical signals from predators are largely responsible for prompting reconfigurations in head shape (Krueger and Dodson 1981; Hebert and Grewe 1985; Parejko and Dodson 1990). The predation regime encountered by *Daphnia* is very dependent on habitat occupancy; pond populations typically encounter only small invertebrate predators, while lake populations are exposed to predaceous fish as well. Studies have shown that even small shape changes provide substantial protection against invertebrate predators (Krueger and Dodson 1981; Havel and Dodson 1984; Tollrian 1995). Larger helmets provide defense against both invertebrate and vertebrate predators (Green 1967; O'Brien et al. 1979; Hanazato 1991), but in North America, they are generally only induced by invertebrate chemicals when both kinds of predators are present (S. Dodson, pers. comm.). This association likely reflects, in part, the effect of visual predation by fish in reducing

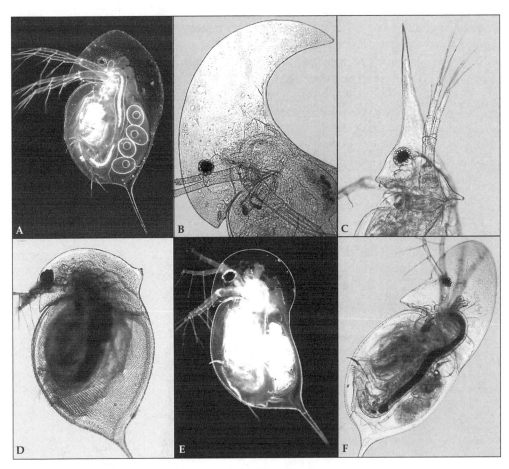

Figure 5.5. Predator-induced headshield morphology of selected adult female daphniids. (**A**) *Daphnia longiremis* from Eskimo Lakes 7, NWT. August 12, 1993. (**B**) *D. retrocurva* from Eagle Lake, Michigan. August 2, 1994. (**C**) *D. lumholtzi* from Grand Lake, Oklahoma. July 20, 1994. (**D**) *D. minnehaha* from a pond near Sault Ste. Marie, Ontario. May 29, 1991. (**E**) *D. pileata* from Santo Domingo, Mexico. February 28, 1992. (**F**) *D. mendotae* from Lake Mendota, Wisconsin. July 4, 1994.

the body size of lake daphniids, and indirectly enhancing their exposure to invertebrate predation (see Dodson 1988).

Figure 5.6 depicts the evolution of helmet and neckteeth formation on the phylogenetic tree for North American *Daphnia*. The plesiomorphic state exhibited by *Ctenodaphnia* appears to be a lack of both helmets and neckteeth. This observation coincides with their habitat, for all species excepting *D. lumholtzi* (Havel and Hebert 1993) are large-bodied animals restricted to ponds lacking vertebrate predators. The most parsimonious evolutionary pattern from accelerated transformation of the character-states (Figure 5.6A) indicates that on the two occasions where *Ctenodaphnia* dispersed into lakes, they evolved defensive helmets. *D. lumholtzi* produces huge spiked helmets when exposed to fish predation (Green 1967). Similarly, the *Ctenodaphnia* species ancestral to *Daphnia* and *Hyalodaphnia* (labeled I in Figure 5.6A), presumably a lacustrine species, also evolved a helmet, as indicated by the accelerated character-state optimization of both the trait and habitat (Figure 5.6B) onto the cladogram. Furthermore, helmet production persisted (with modifications) throughout the evolution of five lacustrine species of *Hyalodaphnia* and the oldest member of the subgenus *Daphnia*.

This association between habitat occupancy and helmet production is reinforced by other character-state changes. For example, one ancestral species of *Daphnia* (labeled II in Figure 5.6A) and three extant members of *Hyalodaphnia* lost the ability to produce helmets following their habitat shift from lakes back to ponds. On two other separate occasions, helmets originated in species of the subgenus *Daphnia*. *D. retrocurva* is one of these species which gained a helmet and fits the general pattern, as it is found only in lakes. However, *D. pileata*, the sole helmeted member of the *obtusa* complex, is restricted to ponds. This exceptional taxon produces a helmet very different in appearance from those of helmeted species in lakes (see Figure 5.5E), but similar to those developed by several Australian *Ctenodaphnia*. The latter species are also pond dwellers and in this case, helmet formation is known both to be induced by notonectids and to provide protection from these predators (Grant and Bayly 1981). In contrast to most other invertebrate predators encountered by *Daphnia*, which preferentially capture prey less than 1 mm in length (Krueger and Dodson 1981; Tollrian 1995), notonectids capture prey as large as 3 mm in length (Barry and Bayly 1985). Hence, helmet development is advantageous even in large-bodied daphniids exposed to notonectid predation (Grant and Bayly 1981; Barry and Bayly 1985). The fact that notonectid populations only reach high densities in warm regions explains the prevalence of helmeted pond daphniids in Australia, and the occurrence of this trait in *D. pileata*, which has one of the most southerly distributions of any North American species (Hebert 1995).

The association between the evolution of helmets and pond to lake habitat shifts is not as evident in optimizations generated by delayed transformation of the character-states onto the cladogram (Figure 5.7). Although Figure 5.7A indicates that helmet formation may have been derived by both *D. lumholtzi* and *D. ambigua* – coinciding with proposed habitat shifts by delayed transformation as well (Figure 5.7B) – the pattern is broken when inferring which (ancestral) taxa invaded lakes among the *Hyalodaphnia*. Nonetheless, this conflicting pattern does not diminish the strong correlation between lake daphniids and defensive headshields. It does, however, suggest another equally parsimonious sequence for helmet evolution.

The ability to produce neckteeth has also arisen independently in both *Daphnia* and *Hyalodaphnia* (Figure 5.6), and occurs in both pond and lake taxa. In species able to produce this structure, neckteeth are induced by a chemical(s) released by larvae of the predatory dipteran *Chaoborus* (Krueger and Dodson 1981; Spitze 1992). The production of neckteeth by both lacustrine and pond species is not unexpected, as this predator occurs in both habitats. However, *Chaoborus* is absent from turbid water habitats and none of the *Daphnia* species from these settings (species belonging to the

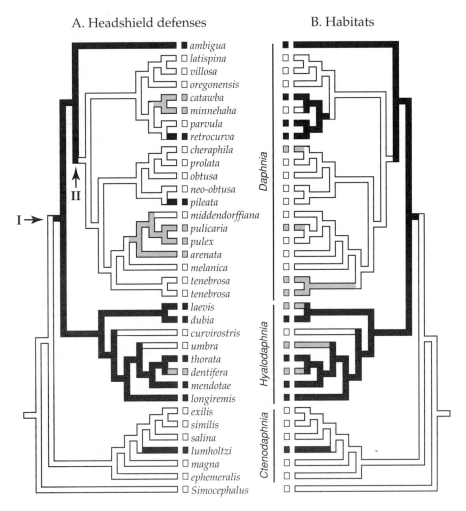

Figure 5.6. (**A**) Character-state optimization of helmet and neckteeth formation in the genus *Daphnia* assuming accelerated transformation. Black shading denotes the ability to produce helmets by adult daphniids, while gray shading denotes the inducible production of neckteeth. (**B**) Proposed habitat shifts associated with *Daphnia* speciation assuming accelerated transformation: ephemeral pond (white), permanent pool/small lake (gray), and larger lake habitats (black). Characters were not polarized prior to the transformations.

obtusa and *villosa* complexes) produces neckteeth, even when exposed to *Chaoborus* in the lab (Beaton 1995). In most taxa which produce them, neckteeth are evanescent, being retained only in the early juvenile instars when *Daphnia* is sensitive to this predator. However, the largest species of *Chaoborus* are restricted to ponds, and two species of *Daphnia* (*arenata*, *minnehaha*) which occur in these habitats have independently evolved the ability to sustain necktooth production even as adults.

This analysis demonstrates that the evolution of helmets is typically correlated with evolutionary shifts from pond to lake habitats, and in North America, apparently reflects an indirect response to vertebrate predation. Conversely, taxa invading pond

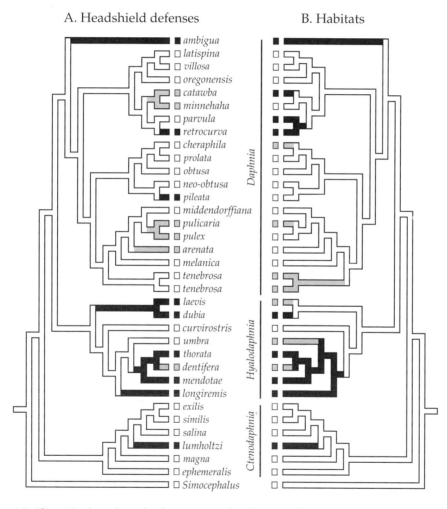

Figure 5.7. Alternative hypothesis for the sequence of evolutionary changes in inducible head morphologies of daphniids. (**A**) Character-state optimization of helmet and neckteeth formation in *Daphnia* assuming delayed transformation. Black shading denotes the ability to produce helmets by adult daphniids, while gray shading denotes the inducible production of neckteeth. (**B**) Proposed habitat shifts associated with *Daphnia* speciation assuming delayed transformation: ephemeral ponds (white), permanent pools/small lakes (gray), and larger lakes (black). Characters were not polarized prior to the transformations.

habitats with lower predation pressure lose helmets, presumably reflecting the cost of producing such structures (Black and Dodson 1990; Walls et al. 1991). In the case of neckteeth, species similarly gain the ability to produce and modulate neckteeth as they enter habitats with *Chaoborus*. The regularity with which daphniids modulate head shape, and the repetitious nature of these modifications suggest the maintenance of a simple proximate mechanism for such change (Beaton 1995). This conclusion differs from the past assumption of shared ancestry between helmeted forms, as shown for example by the subdivision of the *longispina* complex into helmeted and unhelmeted species assemblages before Taylor and Hebert (1994 and references herein).

CARAPACE SETATION AND SUSPENDED PARTICULATES – All *Daphnia* possess spines on the external margin of their ventral carapace, but only 13 species in North America have elongate plumose setae on their internal margin (Figure 5.8; Scourfield 1942; Hebert 1995; Hebert and Finston 1996). The carapace houses the filtering appendages which daphniids use to collect particulate food such as algae, and these plumose setae are believed to prevent debris from entering the carapace chamber (Fryer 1991b). The size of suspended debris encountered by a daphniid varies according to its habitat occupancy and feeding behavior. Species of *Ctenodaphnia*, although they are capable of foraging in the water column (Burns 1968), spend significant time recovering large particulate food from bottom sediments (except *D. lumholtzi*; [Fryer 1991b]). Some members of the subgenus *Daphnia*, such as *D. obtusa*, employ a similar mode of foraging. The latter species uses, like *D. magna*, a specialized scraperlike medial spine on the second thoracic limb to resuspend ingestible materials (Fryer 1991b). By contrast, many other species of subgenus *Daphnia* and all *Hyalodaphnia* filter-feed in open water, which normally contains only smaller particles. It is interesting that daphniids that inhabit turbid habitats with high concentrations of colloidal clay invariably possess elongate setae (Hebert 1995).

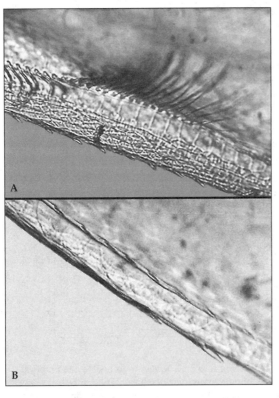

Figure 5.8. (A) Ventral margins of the carapace of *Daphnia* with internal plumose setae. (B) Same, but without plumose setae.

Character-state optimization onto the phylogeny reveals that the presence of setae is the plesiomorphic state (Figure 5.9), coinciding with the benthic feeding behavior of most *Ctenodaphnia*. Although the ancestral *Ctenodaphnia* possess this trait, recent molecular phylogenetic

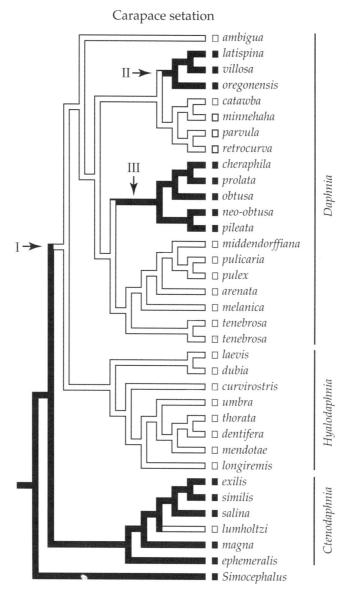

Figure 5.9. Character-state optimization of the presence/absence of plumose setae on the internal margin of the carapace in *Daphnia*. Black shading denotes the presence of plumose setae.

studies of branchiopod crustaceans (D. Taylor, in prep.) suggest that their benthic feeding behavior is derived. Moreover, the character was lost twice by *Daphnia* following habitat transitions from ponds to lakes. *D. lumholtzi* does not have these elongate setae (Fryer 1991b), nor did the species of *Ctenodaphnia* ancestral to the *Hyalodaphnia* and *Daphnia* clades (labeled I on the tree). Despite its absence in ancestral species, the character was regained twice in the subgenus *Daphnia*. The re-acquisition of this trait occurred in pond-dwelling lineages and was likely a consequence of a shift to a detritus-based mode of feeding by members of the *villosa* and *obtusa* lineages (labeled II and III, respectively). However, the presence of this trait subsequently made it possible for these lineages to colonize a new environment – ponds with highly turbid waters – resulting in the origin of new species adapted to these environments.

EPHIPPIAL MORPHOLOGY – Ephippia display conspicuous morphological variation in the number/position of egg chambers, intensity of melanization, shape, and spinescence (Hebert 1995). All North American species have a two-egg ephippium, but at least one Australian species has a one-egg ephippium (Benzie 1988). The orientation of the egg chambers relative to the dorsal surface of the ephippium varies from horizontal to perpendicular, and is sufficiently invariant to discriminate the ephippial eggs of *Ctenodaphnia* from those of the other subgenera. The ephippial matrix varies in color from white to black, with the most melanized ephippia being produced by species showing melanization of the carapace. Ephippial shape varies from an elongate ellipse to subtrapezoidal, but quantification of shape variation is complex. As a result, we focus our analysis on the most easily scored trait – the presence/absence of spinescence on the dorsal margin of the ephippium (Figure 5.10).

Adult females of all species of *Daphnia* have dorsal spinescence on their carapace, but prior to ephippial production, 14 species suppress this trait and produce ephippia lacking spinescence. Our analysis indicates that absence of spines is the plesiomorphic state, as ephippia from both the outgroup (including other Cladocera such as *Daphniopsis* and *Ceriodaphnia*) and the oldest member of *Ctenodaphnia* are without spines. Figure 5.11 shows that ephippial spinescence originated twice in the genus, once in an ancestral *Ctenodaphnia*, and again in the ancestor to the subgenus *Daphnia* (labeled I on the tree). Subsequently, there have also been three independent losses of the trait, twice by ancestors to clades within the subgenus *Daphnia*, and once by a species in the *pulex* complex. As these lineages include both lake and pond species, there is no obvious association between the gain/loss of spines and habitat occupancy.

The adaptive significance of shifts in the spinescence patterns on ephippia is not clear. It has been argued that dorsal spines foster long-distance dispersal by fostering ephippial adhesion to birds and mammals or, contrarily, that they impede movement by serving as anchors (Frey 1982; Sergeev 1990; Fryer 1991b; Hebert and Wilson 1994). The information now available on the extent of gene frequency divergence among local populations of a number of *Daphnia* species may provide some resolution to this controversy. These findings (Colbourne et al., unpubl. data) suggest that ephippial spinescence is associated with pronounced gene frequency divergence among populations, supporting their potential role in anchoring eggs. Species of *Ctenodaphnia*, which possess the most spinescent ephippia, show much greater local differentiation in gene frequencies than species of *Hyalodaphnia*, which produce ephippia lacking spines.

BREEDING SYSTEMS – As noted earlier, most species of *Daphnia* reproduce by cyclical parthenogenesis, but four species have abandoned the sexual phase of the life cycle. These taxa produce apomictic ephippial eggs, in contrast to the sexual ephippial eggs produced by other species (Hebert 1981). It is interesting that three of these species (*pulex, pulicaria, tenebrosa*) include some populations which retain the ancestral breeding system of cyclic parthenogenesis. The obligately parthenogenetic populations of each of these taxa show extraordinary clonal diversity, which seems to be the result of multiple transitions to obligate apomixis (Crease et al. 1989). Experimental studies have linked this shift in breeding system to the transmission of a sex-limited meiosis suppressor (Innes and Hebert 1988).

The *Daphnia* phylogeny indicates that obligate parthenogenesis has evolved in North America only in the *pulex* complex (Figure 5.12). Even though relationships are poorly resolved among members of this complex, it is clear from 12S rDNA sequence divergence data that this breeding system must have originated within the last million years or so. It is important to note that the derived mating system is only shared among species known to hybridize, or species which are themselves of hybrid origin (Hebert et al. 1989; Crease and Lynch 1991; Dufresne and Hebert 1994; Hebert and Finston 1996; P. Hebert, unpubl. data). This fact suggests that the trait may have originated only once in the North American fauna, and then diffused to other taxa by

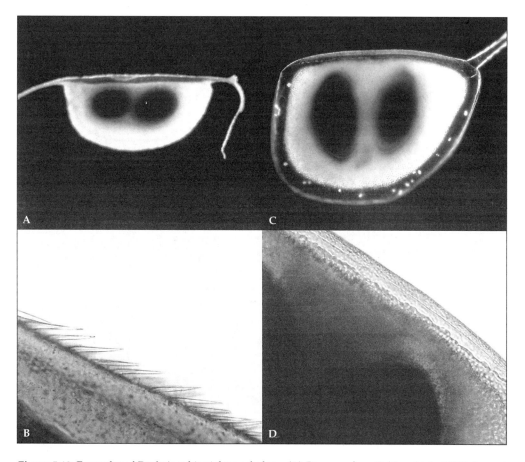

Figure 5.10. Examples of *Daphnia* ephippial morphology. (**A**) *D. magna* from Tuktoyaktuk, NWT. August 16, 1993. (**B**) Dorsal surface of the same ephippium possessing spinescence. (**C**) *D. laevis* from Rondeau Park, Ontario. June 9, 1992. (**D**) Dorsal surface of the same latter ephippium without spinescence.

introgressive hybridization. For this reason, it would be misleading to infer that the trait evolved in the common ancestor of the (*pulex*, (*pulicaria*, *middendorffiana*)) clade. Obligate apomixis may thus have had a very recent origin. The geographic distributions of breeding systems suggest that there may be selection for asexuality in a specific habitat, because all members of the *pulex* complex reproduce by obligate parthenogenesis in the Arctic. However, because these taxa are also polyploids, and

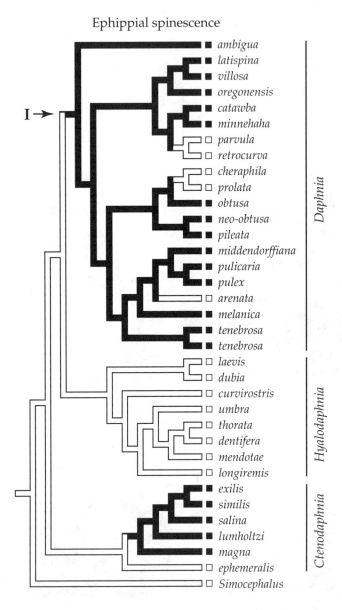

Figure 5.11. Character-state optimization of the presence/absence of ephippial spinescence in the genus *Daphnia* onto the cladogram of North American species. Black shading denotes the presence of ephippial spinescence.

because only diploids are present in temperate localities, the dominance of obligate apomicts at higher latitudes may simply be an indirect consequence of selection for greater ploidy levels (Beaton and Hebert 1988).

CHROMOSOMAL EVOLUTION – Species of *Daphnia* show relatively little variation in chromosome number (Zaffagnini and Trentini 1975; Trentini 1980). Current data indicate that all species of *Ctenodaphnia* and *Hyalodaphnia* possess 20 chromosomes, while all members of subgenus *Daphnia* have 24 (Beaton and Hebert 1994). There are poly-

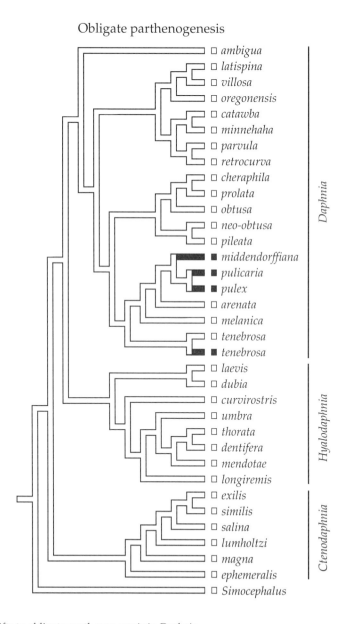

Figure 5.12. Shifts to obligate parthenogenesis in *Daphnia*.

ploid derivatives of several species in the latter group with higher chromosome numbers, but these polyploids are of recent origin and are incapable of sexual reproduction. Consequently, the sole shift in chromosome numbers within cyclically parthenogenetic lineages appears to have occurred early in the evolution of the genus (Figure 5.13). There is some information to suggest that even linkage relationships have been stable, because at least one pair of enzyme loci show similar map distances in species from two different subgenera (Hebert and Moran 1980; Innes 1989)!

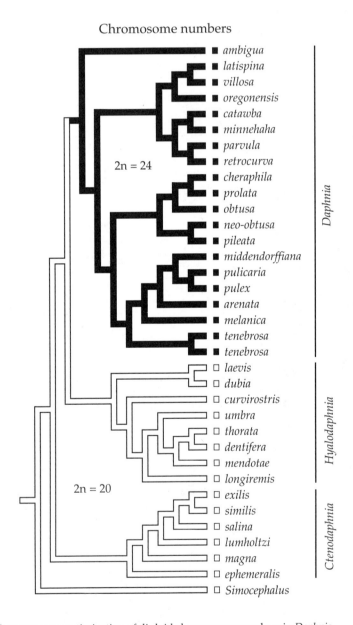

Figure 5.13. Character-state optimization of diploid chromosome numbers in *Daphnia*.

In addition to the stability in chromosome numbers and gene arrangements, genome sizes are relatively invariant within the genus (Beaton 1995). Variation in genome sizes among species of *Ctenodaphnia* and *Hyalodaphnia* is limited, with less than a 50% difference in mean genome size between the subgenera. More variation among species has been detected within subgenus *Daphnia*, with up to a four-fold variation in genome size. The largest genomes occur in obligately apomictic clones of the *pulex* complex that are recently derived polyploids (*middendorffiana, tenebrosa*).

There is only one exception to the small differences in genome size among taxa capable of sexual reproduction. Diploid clones of *D. tenebrosa* have a genome size which is nearly twice as large as that of other members of the *pulex* complex. Our 12S rDNA data suggest that this increase in genome size has occurred within the last million years. *Daphnia tenebrosa* is morphologically exceptional – it is not only the largest member of its subgenus, but it also produces the largest eggs of any species in the genus. Its increased body and egg size is likely a simple consequence of the usual genome size/cell size relationship (Cavalier-Smith 1985a,b; Beaton and Hebert, in prep.). These life history shifts are ecologically important, as they allow *D. tenebrosa* to co-occur with the predatory copepod *Heterocope septentrionalis*, while other small-bodied *Daphnia* are excluded (O'Brien et al. 1979; Hebert and Loaring 1980; Dodson 1984). Unfortunately, there is no understanding of the process which led to the rapid increase in its genome size, but it is clearly not a result of conventional polyploidy, as the species has the same chromosome number as other members of the subgenus *Daphnia*.

Conclusions

This study represents a preliminary effort to reconstruct the tempo and phenotypic pattern of evolution in North American *Daphnia*. Although the phylogeny employed to reconstruct character-state evolution is based on the analysis of sequence divergence in only one gene, our principal conclusions are likely to be robust, given the patterns of character-state variation within vs. between subgenera. For example, the multiple independent origins of cuticular melanization and helmets in two different subgenera – and the monophyletic nature of shifts in chromosome number and breeding systems corresponding to subgeneric or complex limits – are unlikely to be altered by more intensive analyses of sequence divergence at other loci. However, in other cases, such as the presence of melanization in several species of the *pulex* complex, our conclusions are less secure, reflecting in part the general problem of determining evolutionary trajectories where several closely related species share a distinctive trait. Such trait concordance has three possible explanations – its maintenance from a shared ancestor, its repeated independent acquisition, or its secondary acquisition as a consequence of introgressive hybridization. Unambiguous discrimination among these alternatives is only likely to result from the targeted analysis of sequence diversity in the gene(s) controlling the trait. At present, such analysis represents a substantial challenge, because there is usually no simple way to identify the gene(s) in question. Consensus phylogenies, based on the analysis of sequence diversity in a number of genes, are certainly not a panacea for resolving such uncertainty in character-state transitions (also see Leroi et al. 1994), but such evidence is the only tool we now have to confront the vagaries of molecular evolution.

This study provides evidence that *Daphnia* has shown limited phenotypic innovation since the mid-Mesozoic. However, shortly after its origin, the genus showed a brief interval of intense diversification and phenotypic evolution, which led to the establishment of morphologically distinct forms that underlie the three modern subgenera. Following this initial radiation, rates of evolutionary diversification then slowed; pools trod by dinosaurs undoubtedly contained a daphniid fauna similar to extant forms. This result suggests that daphniids have not experienced global competitive release since the Mesozoic, implying that the zooplankton communities of inland waters were little impacted by the general faunal collapse at the Cretaceous boundary. Despite the lack of major evolutionary innovations, morphological diversification within the genus has not ceased. However, the pattern of diversification is one characterized by frequent character-state convergence. Placed in a lake setting, daphniids typically evolve helmets, and when placed with *Chaoborus*, they evolve neckteeth. A similar convergence involving cuticular melanization has occurred in response to intense ultraviolet exposure in two lineages of North American *Daphnia* as well as several Australian daphniids. As many other zooplankters sequester carotenoids gained from their food for apparently the same purpose, the repeated evolution of this adaptation by *Daphnia* suggests that underlying intrinsic factors (as well as the obvious extrinsic factor of UV radiation) favor its recurrent evolution. In this case, it is known that the genes coding for melanin formation antedate the evolution of a melanized carapace, for they are regularly expressed in both the eye and in the epidermal tissue surrounding egg chambers of the ephippium. The development of a melanic *Daphnia* may simply require the activation of these loci in other epidermal cells. Similarly, the loss of ephippial spinescence may only require the suppression of a trait normally present on the adult carapace. The gain of plumose setae may also not be revolutionary, given that all *Daphnia* males possess similar structures which they use to attach themselves to females when copulating. Some character-state transitions have occurred more rarely – for example, obligate apomixis has only evolved in four members of the *pulex* complex. This case may reflect character-state convergence due to introgressive hybridization rather than independent origin. Introgression may also account for the sharing of melanic carapaces by members of the same species complex.

The pattern of evolutionary diversification in the genus *Daphnia* appears similar to that of other groups of planktonic cladocerans, which also show limited species diversity and morphological variation (Frey 1987). However, the striking morphological diversity *among* different genera and families of cladocerans makes it clear that a varied body plan is compatible with success as a zooplankter (see Fryer 1991b). It seems likely that morphological divergence *within* each of these groups has been restrained by the long-term persistence of a stable community assemblage of different zooplankton groups. Further molecular phylogenetic work is now required to verify that the modern zooplankton fauna of inland waters derives from a single burst of diversification in the early Mesozoic. It is interesting that no family of planktonic cladocerans shows as much taxonomic diversity as the Chydoridae (Frey 1987), specialists that occupy benthic and littoral habitats that are much more complex structurally than those occupied by pelagic forms. Molecular evolutionary studies are needed to determine whether chydorids show rates of morphological and genetic divergence that greatly exceed those of their planktonic counterparts. The verification of grossly different rates of differentiation linked to different kinds of habitat (as opposed to differences in competitive regime) would reinforce the situation-dependence of the evolutionary process.

Acknowledgments

We thank L. Boulding, T. Crease, D. McLennan, and C. Wilson for their valuable comments during the review of this manuscript. T. Little provided stimulating discussions about cladoceran ecology and evolution. R. Geddes enhanced the digital images presented in this paper, which were obtained from Hebert (1995). This research was supported by a NSERC Research Grant to PDNH, a NSERC Post-Doctoral Fellowship to DJT and an OGS scholarship to JKC.

References

Barry, M. J., and Bayly, I. A. E. 1985. Further studies on predator induction of crests in Australian *Daphnia*, and the effects of crests on predation. Australian Journal of Marine and Freshwater Research 36:519–535.

Beaton, M. J. 1995. Patterns of endopolyploidy and genome size variation in *Daphnia*. Ph.D. thesis, University of Guelph.

Beaton, M. J., and Hebert, P. D. N. 1988. Geographical parthenogenesis and polyploidy in *Daphnia pulex*. American Naturalist 132:837–845.

Beaton, M. J., and Hebert, P. D. N. 1994. Variation in chromosome numbers of *Daphnia* (Crustacea, Cladocera). Hereditas 120:275–279.

Benzie, J. A. H. 1987. The biogeography of Australian *Daphnia*: clues of an ancient >70 m.y. origin for the genus. Hydrobiologia 145:51–65.

Benzie, J. A. H. 1988. The systematics of Australian *Daphnia* (Crustacea: Daphniidae). Species descriptions and keys. Hydrobiologia 166:95–161.

Black, A. R., and Dodson, S. I. 1990. Demographic costs of *Chaoborus*-induced phenotypic plasticity in *Daphnia pulex*. Oecologia 83:117–122.

Brooks, J. L. 1957. The systematics of North American *Daphnia*. Memoirs of the Connecticut Academy of Arts and Science 13:1–180.

Brower, A. V. Z. 1994. Rapid morphological radiation and convergence among races of the butterfly *Heliconius erato* inferred from patterns of mitochondrial DNA evolution. Proceedings of the National Academy of Sciences, USA 91:6491–6495.

Burns, C. W. 1968. Direct observations of mechanisms regulating feeding behaviour of *Daphnia* in lakewater. Internationale Revue der Gesamten Hydrobiologie. 53:83–100.

Cavalier-Smith, T. 1985a. Introduction: the evolutionary significance of genome size. Pp. 1–36 in *The evolution of genome size*, T. Cavalier-Smith, ed. New York, NY: John Wiley.

Cavalier-Smith, T. 1985b. Cell volume and the evolution of eukaryote genome size. Pp. 105–184 in *The evolution of genome size*, T. Cavalier-Smith, ed. New York, NY: John Wiley.

Colbourne, J. K., and Hebert, P. D. N. 1996. The systematics of North American *Daphnia* (Crustacea: Anomopoda): a molecular phylogenetic approach. Philosophical Transactions of the Royal Society of London, Series B 351:349–360.

Conway Morris, S. 1989. Burgess Shale faunas and the Cambrian explosion. Science 246:339–346.

Crease, T. J., and Lynch, M. 1991. Ribosomal DNA variation in *Daphnia pulex*. Molecular Biology and Evolution 8:620–640.
Crease, T. J., Stanton, D. J., and Hebert, P. D. N. 1989. Polyphyletic origins of asexuality in *Daphnia pulex*. II. Mitochondrial-DNA variation. Evolution 43:1016–1026.
DeMelo, R., and Hebert, P. D. N. 1994. Allozyme variation and species diversity in North American Bosminidae. Canadian Journal of Fisheries and Aquatic Science 51:873–880.
Dodson, S. I. 1984. Predation of *Heterocope septentrionalis* on two species of *Daphnia*: morphological defenses and their cost. Ecology 65:1249–1257.
Dodson, S. I. 1988. Cyclomorphosis in *Daphnia glaeata mendotae* Birge and *D. retrocurva* Forbes as a predator-induced response. Freshwater Biology 19:109–114.
Dufresne, F., and Hebert, P. D. N. 1994. Hybridization and origins of polyploidy. Proceedings of the Royal Society of London, Series B 258:141–146.
Dufresne, F., and Hebert, P. D. N. 1995. Polyploidy and clonal diversity in an arctic cladoceran. Heredity 75:45–53.
Freed, L. A., Conant, S., and Fleischer, R. C. 1987. Evolutionary ecology and radiation of Hawaiian passerine birds. Trends in Ecology and Evolution 2:196–203.
Frey, D. G. 1982. Questions concerning cosmopolitanism in Cladocera. Archiv für Hydrobiologie 93:484–502.
Frey, D. G. 1987. The taxonomy and biogeography of the Cladocera. Hydrobiologia 145:5–17.
Fryer, G. 1991a. A daphniid ephippium (Branchiopoda: Anomopoda) of Cretaceous age. Zoological Journal of the Linnean Society 102:163–167.
Fryer, G. 1991b. Functional morphology and the adaptive radiation of the Daphniidae (Branchiopoda, Anomopoda). Philosophical Transactions of the Royal Society of London, Series B 331:1–99.
Gómez, Á., Temprano, M., and Serra, M. 1995. Ecological genetics of a cyclic parthenogen in temporary habitats. Journal of Evolutionary Biology 8:601–622.
Grant, J. W. G., and Bayly, I. A. C. 1981. Predator induction of crests in morphs of the *Daphnia carinata* King complex. Limnology and Oceanography 26:201–218.
Green, J. 1967. The distribution and variation of *Daphnia lumholtzi* (Crustacea: Cladocera) in relation to fish predation in Lake Albert, East Africa. Journal of Zoology 151:181–197.
Hanazato, T. 1991. Induction of development of high helmets by a *Chaoborus*-released chemical in *Daphnia galeata*. Archiv für Hydrobiologie 122:167–175.
Havel, J. E., and Dodson, S. I. 1984. *Chaoborus* predation on typical and spined morphs of *Daphnia pulex*: behavioral observations. Limnology and Oceanography 29:487–494.
Havel, J. E., and Dodson, S. I. 1985. Environmental cues for cyclomorphosis in *Daphnia retrocurva* Forbes. Freshwater Biology 15:469–478.
Havel, J. E., and Hebert, P. D. N. 1993. *Daphnia lumholtzi* in North America: another exotic zooplankter. Limnology and Oceanography 38:1823–1827.
Hebert, P. D. N. 1978. The population biology of *Daphnia* (Crustacea, Daphnidae). Biological Reviews 53:387–426.
Hebert, P. D. N. 1981. Obligate asexuality in *Daphnia*. American Naturalist 117:784–789.
Hebert, P. D. N. 1987. Genetics of *Daphnia*. In *Daphnia*. eds. R. H. Peters, and R. de Bernardi. Memorie dell'Instituto Italiano di Idrobiologia. 45:439–460.
Hebert, P. D. N. 1995. The *Daphnia* of North America: an illustrated fauna. CD-ROM. Distributed by the author. Department of Zoology, University of Guelph.
Hebert, P. D. N., Beaton, M. J., Schwartz, S. S., and Stanton, D. J. 1989. Polyphyletic origins of asexuality in *Daphnia pulex*. I. Breeding-system variation and levels of clonal diversity. Evolution 43:1004–1015.
Hebert, P. D. N., and Emery, C. J. 1990. The adaptive significance of cuticular pigmentation in *Daphnia*. Functional Ecology 4:703–710.
Hebert, P. D. N., and Finston, T. L. 1993. A taxonomic reevaluation of North American *Daphnia* (Crustacea: Cladocera). I. The *D. similis* complex. Canadian Journal of Zoology 71:908–925.
Hebert, P. D. N., and Finston, T. L. 1996. A taxonomic reevaluation of North American *Daphnia* (Crustacea: Cladocera). II. New species of the *D. pulex* group from southcentral United States and Mexico. Canadian Journal of Zoology 74:632–653.
Hebert, P. D. N., and Grewe, P. M. 1985. *Chaoborus*-induced shifts in the morphology of *Daphnia ambigua*. Limnology and Oceanography 30:1291–1297.
Hebert, P. D. N., and Loaring, J. M. 1980. Selective predation and the species composition of arctic ponds. Canadian Journal of Zoology 58:422–426.

Hebert, P. D. N., and Moran, C. 1980. Enzyme variability in natural populations of *Daphnia carinata* King. Heredity 45:313–321.

Hebert, P. D. N., and Wilson, C. C. 1994. Provincialism in plankton: endemism and allopatric speciation in Australian *Daphnia*. Evolution 48:1333–1349.

Hendry, C. D., and Brezonik, P. L. 1984. Chemical composition of softwater Florida lakes and their sensitivity to acid precipitation. Water Research Bulletin 20:75–86.

Hobæk, A., and Wolf, H. G. 1991. Ecological genetics of Norwegian *Daphnia*. II. Distribution of *Daphnia longispina* genotypes in relation to short-wave radiation and water colour. Hydrobiologia 225:229–243.

Innes, D. J. 1989. Genetics of *Daphnia obtusa*: genetic load and linkage analysis in a cyclic parthenogen. Journal of Heredity 80:6–10.

Innes, D. J., and Hebert, P. D. N. 1988. The origin and genetic basis of obligate parthenogenesis in *Daphnia pulex*. Evolution 42:1024–1035.

Jacobs, J. 1962. Light and turbulence as co-determinants of relative growth rates in cyclomorphic *Daphnia*. Internationale Revue der Gesamten Hydrobiologie. 47:146–156.

Jacobs, J. 1987. Cyclomorphosis in *Daphnia*. In *Daphnia*. eds. R. H. Peters, and R. de Bernardi. Memorie dell'Instituto Italiano di Idrobiologia. 45:325–352.

Kambysellis, M. P., Ho, K.-F., Craddock, E. M., Piano, F., Parisi, M., and Cohen, J. 1995. Patterns of ecological shifts in the diversification of Hawaiian *Drosophila* inferred from a molecular phylogeny. Current Biology 5:1129–1139.

Kimura, M. 1980. A simple method for estimating evolutionary rate of base substitutions through comparative studies of nucleotide sequences. Journal of Molecular Evolution 16:111–120.

Krueger, D. A., and Dodson, S. I. 1981. Embryological induction and predation ecology in *Daphnia pulex*. Limnology and Oceanography 26:219–223.

Kumar, S., Tamura, K., and Nei, M. 1993. *MEGA: Molecular evolutionary genetics analysis, vers. 1.02.* Distributed by the authors. Pennsylvania State University.

Leroi, A. M., Rose, M. R., and Lauder, G. V. 1994. What does the comparative method reveal about adaptation? American Naturalist 143:381–402.

Lynch, M. 1985. Speciation in the Cladocera. Verhandlungen der Internationalen Vereinigung für Theoretische und Angewandte Limnologie. 22:3116–3123.

Lynch, M., and Gabriel, W. 1983. Phenotypic evolution and parthenogenesis. American Naturalist 122:745–764.

Lynch, M., and Jarrell, P. E. 1993. A method for calibrating molecular clocks and its application to animal mitochondrial DNA. Genetics 135:1197–1208.

Maddison, W. P., and Maddison, D. R. 1992. *MacClade: analysis of phylogeny and character evolution, vers. 3.04.* Sunderland, MA: Sinauer and Associates.

Mayr, E. 1963. *Animal species and evolution*. Cambridge, MA: Belknap Press of Harvard University Press.

Morris, P. J., Ivany, L. C., Schopf, K. M., and Brett, C. E. 1995. The challenge of paleoecological stasis: reassessing sources of evolutionary stability. Proceedings of the National Academy of Sciences, USA 92:11269–11273.

O'Brien, W. J., Kettle, D., and Riessen, H. 1979. Helmets and invisible armor: structures reducing predation from tactile and visual planktivores. Ecology 60:287–294.

Osterkamp, W. R., and Wood, W. W. 1987. Playa-lake basins on the Southern High Plains of Texas and New Mexico: part 1. Hydrologic, geomorphic and geological evidence for their development. Geological Society of America Bulletin 99:215–223.

Parejko, K., and Dodson, S. I. 1990. Progress towards characterization of a predator/prey kairomone: *Daphnia pulex* and *Chaoborus americanus*. Hydrobiologia 198:51–59.

Paul, C. R. C. 1977. Evolution of primitive echinoderms. Pp. 123–158 in *Patterns of evolution as illustrated by the fossil record*, A. Hallam, ed. Amsterdam, The Netherlands: Elsevier.

Scourfield, D. J. 1942. The "*pulex*" forms of *Daphnia* and their separation into two distinct series represented by *D. pulex* de Geer and *D. obtusa* Kurz. Ann. Mag. Nat. Hist. 9:202–219.

Sergeev, V. 1990. A new species of *Daphniopsis* (Crustacea: Anomopoda): Daphniidae from Australian salt lakes. Hydrobiology 190:1–7.

Shaw, K. L. 1996. Sequencial radiations and patterns of speciation in the Hawaiian cricket genus *Laupala* inferred from DNA sequences. Evolution 50:237–255.

Siddall, M. E. 1994. *Random cladistics. vers. 2.1.0.* Published electronically on the Internet by ftp at zoo.toronto.edu.

Smirnov, N. N. 1992. Mesozoic Anomopoda (Crustacea) from Mongolia. Zoological Journal of the Linnean Society 104:97–116.

Spitze, K. 1992. Predator-mediated plasticity of prey life-history and morphology: *Chaoborus americanus* predation on *Daphnia pulex*. American Naturalist 139:229–247.

Swofford, D. L. 1993. *PAUP: phylogenetic analysis using parsimony, vers. 3.1.1.* Champaign, IL: Illinois Natural History Survey.

Taylor, D. J., and Hebert, P. D. N. 1993. Habitat dependent hybrid parentage and differential introgression between neighboringly sympatric *Daphnia* species. Proceedings of the National Academy of Sciences, USA 90:7079–7083.

Taylor, D. J., and Hebert, P. D. N. 1994. Genetic assessment of species boundaries in the North American *Daphnia longispina* complex (Crustacea: Daphniidae). Zoological Journal of the Linnean Society 110:27–40.

Taylor, D. J., Hebert, P. D. N., and Colbourne, J. K. 1996. Phylogenetics and evolution of the *Daphnia longispina* group (Crustacea) based on 12S rDNA sequence and allozyme variation. Molecular Phylogenetics and Evolution 5:495–510.

Tollrian, R. 1995. *Chaoborus crystallinus* predation on *Daphnia pulex*: can induced morphological changes balance effects of body size on vulnerability? Oecologia 101:151–155.

Trentini, M. 1980. Chromosome numbers of nine species of Daphniidae (Crustacea, Cladocera). Genetica 54:221–223.

Valentine, J. W. 1980. Determinants of diversity in higher taxonomic categories. Paleobiology 6:444–450.

Walls, M., Caswell, H., and Ketola, M. 1991. Demographic costs of *Chaoborus*-induced defenses in *Daphnia pulex*: a sensitivity analysis. Oecologia 87:43–50.

Wiley, E. O., Siegel-Causey, D., Brooks, D. R., and Funk, V. A. 1991. *The compleat cladist: a primer of phylogenetic procedures.* University of Kansas, Special Publications 19:1–158.

Zaffagnini, F., and Trentini, M. 1975. Osservazioni sul corredo chromosomico di alcuni Dafnidi Crostacei, Cladoceri. Atti Acad. Sci. Ist. Bologna Rc. Cl. Sci. Fis. 13:63–70.

6 EVOLUTIONARY TRENDS IN THE ECOLOGY OF NEW WORLD MONKEYS INFERRED FROM A COMBINED PHYLOGENETIC ANALYSIS OF NUCLEAR, MITOCHONDRIAL, AND MORPHOLOGICAL DATA

Inés Horovitz and Axel Meyer

The New World monkeys (Order Primates, Infraorder Platyrrhini) arose soon after African primates invaded South America some 25 million years ago (Figure 6.1; see Hoffstetter 1972, 1980; Martin 1990; Kay et al. 1997). They subsequently underwent extensive taxonomic diversification coupled with a spectacular adaptive radiation in diet, body size, feeding strategy, and mode of locomotion (Table 6.1). They now include roughly 80 of the 250 primate species worldwide (Mittermeier 1996). New World monkeys range throughout tropical America from Southern Mexico to northern Argentina, inhabiting steamy lowland rain forests, cool cloud forests, seasonally arid dry forests, and sun-baked savannas. They feed on fruits, leaves, nectar, plant exudates, insects, and vertebrates; some are specialized to exploit one or a few of these resources, while others are more generalized (Table 6.1). Most species are active diurnally, except the owl monkey, which is nocturnal over most of its geographical range. Within genera, most species and subspecies are allopatric and show relatively little ecological and morphological divergence from each other, in contrast to the striking divergence seen between genera in these respects. Body size ranges from 120 g to 12 kg; genera characterized by smaller body sizes tend to have more species and subspecies than larger-bodied forms.

Inferences about the adaptive radiation of New World monkeys from ecological and morphological points of view have been derived, so far, solely from phylogenies based on morphological data. Recently, however, extensive data on DNA sequences have become available (e.g., see Schneider et al. 1993, 1995; Harada et al. 1995; Horovitz and Meyer 1995; Meireles et al. 1995) that – when added to existing morphological data – change the inferred phylogeny of New World monkeys and provide a basis for new inferences about their pattern of morphological and ecological evolution. In recent years, several authors have published phylogenetic studies on subsets of New World monkeys, including Goeldi's monkey, marmosets and tamarins (callitrichins, including 5 genera and 32 spp.), howler, spider, woolly, and woolly spider monkeys (atelines, including 4 genera and 14 spp.), sakis and uakaris (pithecins, 3 genera and 9 spp.), owl and titi monkeys (2 genera, 19 spp.), and capuchin and squirrel monkeys (2 genera, 7 spp.), as well as some studies that encompass the whole radiation (e.g., see Rosenberger 1992; Ford and Davis 1992). This paper presents an evolutionary synthesis based on an analysis of combined DNA sequence and

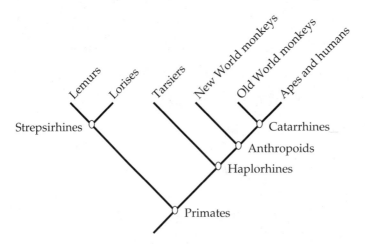

Figure 6.1. Phylogenetic tree of the order Primates, based on several lines of evidence and showing the common names associated with widely recognized clades mentioned in the text.

morphological data, and tests hypotheses about the adaptive radiation of the entire group and each nested subgroup of New World monkeys.

An important question in this kind of study is what data should be included in the phylogenetic analysis. For a character to evolve, it necessarily has to be heritable and therefore have the possibility of containing phylogenetic information. This, in turn, is a good reason to include every heritable character in a phylogenetic analysis. It has been argued that a character's evolution should be analyzed only on phylogenies that are independent of that character, and therefore that the character to be analyzed should be excluded from the phylogenetic analysis (see Avise 1994; Meyer et al. 1994). Here we argue that if the phylogeny changes its topology when the character in question is excluded, more data are necessary before the evolution of this character can be studied, until the topology is stable including or excluding the character in question. A character is evidence; if we exclude evidence from a phylogenetic analysis and obtain a different answer, then the answer with less evidence is likely not to be as reliable as the one based on more evidence.

After providing an overview of phylogenetic methods, we introduce the study organisms and summarize relevant ecological data collected from the literature, including habitat, diet, and locomotory and positional behavior (not all data are available for all genera, however), and then turn to an analysis of their adaptive radiation. The importance of different morphological characters and locomotory or positional behaviors in enabling different monkeys to exploit environments different from those exploited by platyrrhines (or their closest relatives) is an important point in our analysis. We will, also, discuss the plausibility and parsimony of character functions in the ancestors given the current roles of those structures.

Diet, locomotory and positional behavior, diurnality and nocturnality are all aspects in which New World monkeys have diversified extensively in comparison with catarrhines and suggest that they underwent an adaptive radiation following

Table 6.1. Species diversity, body size, diet, ecological distribution, and range of the genera of New World monkeys.[1]

Subfamily/tribe	Genus	Common name	Numbers of spp. (subspp.)	Body mass (g)	Diet	Habitats	Geographical distribution
Atelinae	*Ateles*	Spider monkey	4(16)	7,456 – 9,000	Fruits and leaves	Tall, mature forest with continuous canopy	Southern Mexico to central Bolivia and Brazil
	Brachyteles	Woolly spider monkey	1(1)	9,450 – 12,125	Fruits and leaves	Coastal forest	Restricted area of Atlantic coastal forest of southeastern Brazil
	Lagothrix	Woolly monkey	2(5)	5,750 – 10,000	Fruits and leaves; in exceptional populations, insects	Tall, mature forest with continuous canopy	Central Colombia and a small portion of Venezuela, throughout the upper Amazonian basin, as far south as northern Bolivia
	Alouatta	Howler monkey	7(14)	4,550 – 11,352	Leaves and fruits	Wide spectrum of environments, including savannas	Southern Mexico to northern Argentina
Pithecinae	*Callicebus*	Titi monkey	10(16)	800 – 1,325	Fruits and either insects or leaves as secondary food	Primary forest, varilla and palm forest	Central Colombia and Venezuela to Paraguay
	Pithecia	Saki	5(8)	1,515 – 2,795	Fruits	High primary forest of terra firme	Western Amazonian region and Guianas
	Cacajao	Uakari	2(6)	2,740 – 3,450	Fruits	Floodplains	Western portion of Amazonian basin
	Chiropotes	Bearded saki	2(4)	2,510 – 3,100	Fruits	High primary forest of terra firme	Eastern portion of Amazonian basin
Aotinae	*Aotus*	Owl monkey	9(11)	690 – 1,232	Fruits; insects and leaves as a complement	Most forested areas and marginal areas	Panama to northern Argentina, absent in Guyana Shield and Atlantic and Paranense forests
Cebini	*Cebus*	Capuchin monkey	5(21)	2,220 – 3,868	Fruits and animals	Primary and secondary forest	Belize to northern Argentina
	Saimiri	Squirrel monkey	2(12)	695 – 932	Fruits and animals	Primary and secondary forest	Southern Costa Rica and Panama, central Colombia to Bolivia and northeastern Brazil
Callitrichini	*Callimico*	Goeldi's monkey	1(1)	483 – 502	Diet virtually unstudied, includes insects	Shrub and bamboo forest	Patchy distribution in the upper Amazonian basin; southern Colombia to northern Bolivia
	Callithrix	Marmoset	15(19)	182 – 429	Fruits, insects and exudates	Terra firme, disturbed areas, secondary forest	Eastern Bolivia and Brazil south of the Amazon and east of the Rio Madiera
	Cebuella	Pygmy marmoset	1(1)	126 – 130	Insects and exudates. Fruits are a minor component of their diet	Edges and interiors of seasonally inundated mature floodplain forests and mature nonflooded forests	Upper Amazonian basin, southern Colombia to northwestern Bolivia
	Leontopithecus	Lion tamarin	3(4)	535 – 615	Fruits, insects and plant exudates	Coastal lowlands, inland or low inundated forests	Four restricted areas in southeastern Brazil in the Atlantic coastal region and the Rio Parana basin
	Saguinus	Tamarin	12(33)	403 – 740	Insects, fruits, nectar and plant exudates	Terra firme, disturbed areas, secondary forest; some species in primary forest	Amazonian basin, Panama, and northwestern Colombia

[1] Sources: systematics from Mittermeier et al. (1988); body mass from Fleagle (1988; in prep.) and Ford and Davis (1992); geographic distribution from Wolfheim (1981) and Rylands et al. (1993); sources for habitats and diets in text.

their colonization of the New World (Fleagle 1988). We will use the concept of a "key innovation" or "evolutionary novelty" (Mayr 1963) when there is evidence that the function of the character in question is related with the use of a new range of resources or substrates.

Phylogenetic analysis

Analyses were conducted at the generic level, including all 16 genera of living New World monkeys plus one fossil taxon (*Cebupithecia sarmientoi*) from the late Miocene of La Venta, Colombia (Stirton 1951; Stirton and Savage 1951). Genera of New World monkeys are well-defined clades, whereas the limits between species and/or subspecies are frequently debated (see Napier 1976; Groves and Ramírez-Pulido 1982; Hershkovitz 1983, 1984; Ayres 1985; Thorington 1985; Ford 1994). There are at least 16 fossil genera of New World monkeys, most of which are very poorly known. The phylogenetic position of most of these is currently under debate, and when included in a cladistic analysis the large number of missing characters increases dramatically the number of most parsimonious trees (Novacek 1992, 1994; Forey, pers. comm.). We chose to include only one fossil taxon in our analysis, because its morphology is fairly well known and its phylogenetic position relatively stable.

Outgroups used included representatives of each of the major lines of haplorhine primates: tarsiers (*Tarsius*), macaques (*Macaca*)/proboscis monkeys (*Nasalis*), gibbons (*Hylobates*), humans (*Homo sapiens*), and the fossil anthropoids *Aegyptopithecus, Apidium,* and *Parapithecus* from the Oligocene deposits of Fayum, Egypt (Simons 1962, 1965, 1987; Kay et al. 1981; Fleagle and Kay 1987). Molecular and morphological characters about which we had no information were scored as missing data.

Data used to estimate phylogeny included (i) nuclear DNA sequences of the ε-globin genes (Schneider et al. 1993) (261 informative characters) and interphotoreceptor retinol-binding protein (IRBP) gene, intron 1 (Harada et al. 1995) (332 informative characters); (ii) a fragment of the mitochondrial DNA sequence for the 16S ribosomal gene (Horovitz and Meyer 1995) (142 informative characters); and (iii) 66 morphological characters (see Appendix 6.1).

DNA sequences were aligned using Malign 1.89 (Wheeler and Gladstein 1993). Phylogenetic analyses were conducted using the heuristic algorithm in PAUP 3.1.1 (Swofford 1993), with 50 replicate searches based on randomly assembled starting trees. Bootstrap values for the cladogram obtained from combining all three data sets (the "total evidence" tree [Kluge 1989; Kluge and Wolf 1993]) were also obtained using PAUP, with 1,000 replications. Aligned sequences are available from the authors upon request. Entire gaps were considered characters, not each position separately, and gaps with different lengths were coded in sections. For example, given the alignment in Table 6.2, we distinguish three different gaps at (a) positions 5–7; (b) positions 8–11; and (c) positions 12–15. We cannot know how many events actually happened to create these gaps; there is a large number of possibilities. For example, each position could have undergone a single deletion event, gaps (a) and (b) could have been a single deletion event in taxa B and C, and so forth. According to the auxiliary

Table 6.2. Hypothetical DNA sequences used to illustrate the scoring of gaps in this study (see text).

Taxon	Position	Gap(s)
	1111111111222222 1234567890123456789012345 67	
A	CTTAAACCGTGTGTACTGGGAGAACCA	
B	CTTA————————————CTGGGAGAACCA	abc
C	CTTA————————————CTGGGAGAACCA	abc
D	CTTAAAC—————————CTGGGAGAACCA	bc
E	CTTAAAC—————————CTGGGAGAACCA	bc
F	CTTAAAC—————————CTGGGAGAACCA	bc
G	CTTAAACCGTG—————CTGGGAGAACCA	c
H	CTTAAACCGTG—————CTGGGAGAACCA	c
I	CTTAAACCGTGTGTACTGGGAGAACCA	

principle of Hennig (1966), we will consider gaps in the same positions across taxa as homologous, and therefore we consider gap (a) homologous in B and C, gap (b) homologous in B through F, and gap (c) homologous in B through H. Distinguishing gaps (a), (b), and (c) allows us to capture all the information contained in these alignments and to postulate the smallest number of insertion-deletion events possible, which is the most parsimonious hypothesis.

Nuclear, mitochondrial, and morphological data sets were analyzed separately first, and yielded different topologies (Figure 6.2). The consistency index (CI) and tree length (L) excluding uninformative characters for each tree were as follows: CI = 0.63, L = 1,208 for nuclear DNA; CI = 0.47, L = 506 for mitochondrial DNA; and CI = 0.56, L = 214 for morphological data. The nuclear and morphological trees showed the most congruent topologies. Analysis of the combined data yielded a single tree with CI = 0.57 and L = 1,953 (Figure 6.3). The topology of this tree is not perfectly congruent with that resulting from any of the individual data sets, but most branches of the total evidence tree are supported by each data set. There are three notable exceptions: nodes 1, 2, and 3 are not supported by morphology. When morphological characters are excluded, the topology of the total evidence tree does not change.

Support for the position of the owl monkey (*Aotus*) relative to the capuchin and squirrel monkeys (*Cebus, Saimiri*) and the callitrichins is weak; the completely resolved topology presented (Figure 6.3) is only two steps shorter than the grouping of the owl monkey with either callitrichins or with the capuchin-squirrel monkey dyad; the bootstrap tree does not resolve these relationships. The degree of incongruence between data sets (Mickevich and Farris 1981; Kluge 1989) was very low. The total number of extra steps in the total evidence tree was 833, accounted for by incongruence within and between data sets. Only 25 of them were generated by incongruence between data sets, which represents 3% of the total incongruence. All three data sets required extra numbers of steps when overlaid on the total evidence tree,

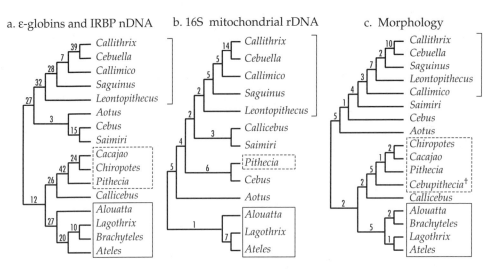

Figure 6.2. Cladograms obtained with (**A**) nuclear sequences of ε-globins (Schneider et al. 1993) and IRBP (Harada et al. 1995) and realigned with Malign, CI = 0.63; (**B**) 16S mitochondrial rDNA sequences (Horovitz and Meyer 1995), CI = 0.47; and (**C**) morphological characters, CI = 0.56 (see Appendix). All consistency indices shown exclude uninformative characters. The number of unambiguous character-state changes is indicated above each branch. Callitrichins are demarcated with a bracket, and pithecins and atelines are enclosed in a dashed and a solid rectangle respectively. †Indicates fossil taxa.

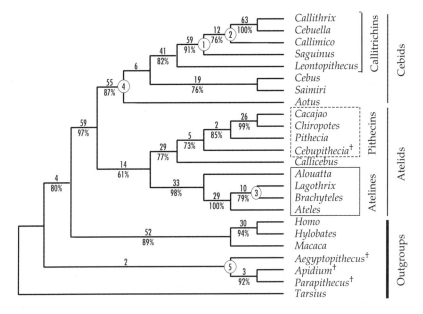

Figure 6.3. Total evidence cladogram (CI = 0.57) obtained by analyzing the combined data from ε-globins (Schneider et al. 1993), IRBP (Harada et al. 1995), 16S mitochondrial rDNA (Horovitz and Meyer 1995), and morphology (see Appendix). Nodes 1, 2, and 3 have no support from morphological characters. Numbers below the branches indicate level of bootstrap support; numbers above branches indicate branch lengths. The bootstrap tree does not support nodes 4 and 5. The two basal platyrrhine clades are Atelidae and Cebidae. The consistency indices, excluding uninformative characters, for the various data sets overlaid on the total evidence tree are 0.63 for nuclear DNA sequences, 0.46 for mitochondrial DNA sequences, and 0.53 for morphology. †Indicates fossil taxa.

relative to the most parsimonious tree based on each data set separately: the nuclear data set required 1 extra step (relative to 593 informative characters), the mitochondrial 11 extra steps (relative to 142 informative characters), and the morphological 11 extra steps (relative to 66 informative characters).

Different data sets reflect a shared history, so the phylogenetic signal they contain should be the same, even if it is obscured by homoplasy. On the other hand, the distribution of homoplasy is likely to be different for each data set, given that each is subject to different constraints (e.g., those pertaining to function). If the data sets are combined, the signal common to all of them is more likely to overwhelm the homoplasy than if each is analyzed separately. Whether this approach is always appropriate is still being debated (Kluge 1989; Kluge and Wolf 1993; Donoghue and Sanderson 1992; Bull et al. 1993; Chippindale and Wiens 1994; Huelsenbeck et al. 1994; Funk et al. 1995; Lockhart et al. 1995; see Chapters 1 and 2 in this volume).

Adaptive radiation in the New World monkeys

The implications of our phylogeny for the interpretation of the adaptive radiation in the New World monkeys are summarized below. The taxonomic categories used are based on Rosenberger (1979) but adjusted to the topology of our preferred tree: the family Atelidae includes subfamilies Atelinae and Pithecinae; its sister-group, the family Cebidae, includes subfamilies Aotinae and Cebinae. The last group contains the tribes Callitrichini and Cebini. For each group, we discuss the implications of our phylogeny for systematic relationships and for shifts in diet, habitat, mode of locomotion, and positional behavior.

Atelidae (Atelinae, Pithecinae)

SYSTEMATICS – This clade is supported by two unambiguous morphological characters: reduction of the pterygoid fossa, and a deciduous lower second premolar with a rounded outline, derived from a mesiodistally elongated outline. The basal dichotomy of this clade implies that the Atelinae and Pithecinae are sister clades.

Atelinae (*Alouatta*, [*Ateles*, {*Lagothrix*, *Brachyteles*}])

SYSTEMATICS – The Atelinae includes the howler (*Alouatta*), spider (*Ateles*), woolly (*Lagothrix*), and woolly spider (*Brachyteles*) monkeys. Three morphological characteristics are unique to atelines among New World monkeys: a prehensile tail covered ventrally by bare skin with friction ridges; a large body; and very long forelimbs relative to hindlimbs (Erikson 1963). They show some convergences with extant apes in their limb and trunk morphology, which are probably related to suspensory habits (Erikson 1963). An ateline can support the weight of its suspended body by its tail.

Relationships within atelines in the nuclear DNA sequence data trees (ε-globin genes, Schneider et al. 1993; γ-globin genes, Meireles et al. 1995; IRBP gene, Schneider et al. 1995 and Harada et al. 1995) and the total evidence tree (Figure 6.4) differ from previous hypotheses: (*Alouatta*, [*Ateles*, {*Lagothrix*, *Brachyteles*}]). We designate the clade composed of the latter three taxa as the Atelini. Members of this group have only four

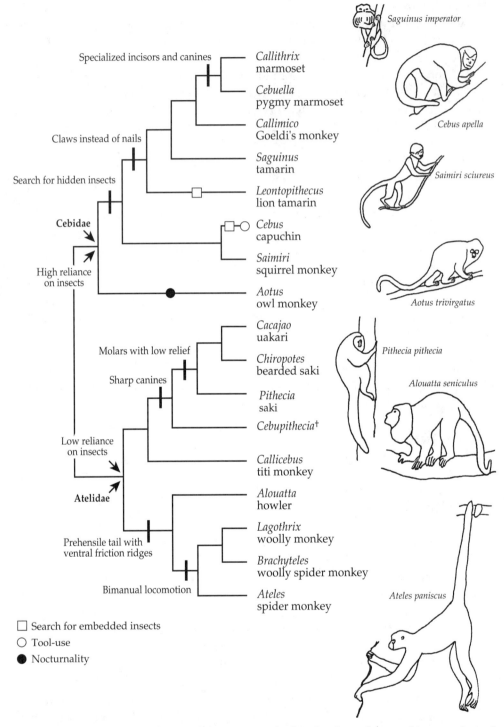

Figure 6.4. Overlay of key morphological innovations and inferred ecology of ancestral forms on the total-evidence phylogeny. Synapomorphies are indicated by vertical bars and associated labels; special symbols (see legend) mark the autopomorphies.

lumbar vertebrae and large ratios of hindlimb relative to trunk length. In contrast, howler monkeys have a mode of five lumbar vertebrae and a smaller hindlimb ratio, falling roughly between that of the owl and the capuchin monkey (Erikson 1963), plus a number of other postcranial synapomorphies listed by Ford (1986b).

Erikson (1963) found several derived characters shared by spider monkeys and woolly spider monkeys, including a reduced thumb (nearly or completely absent as an external character) and a very short lumbar region relative to thoracic length (in terms of both the number and size of individual vertebrae). A few dental characters are shared by the howler monkeys and woolly spider monkeys, such as the presence of a mesoloph on the first upper molar (Zingeser 1973; MacPhee et al. 1995). A mesoloph is a large crest, usually presented as an adaptation for masticating leaves. Under the total evidence tree topology, these characters are either homoplasies, as in the case of characters shared by the howler and woolly spider monkeys, or of ambiguous optimization, as in the case of those shared by the spider and woolly spider monkeys. In other words, the latter are either convergences between the spider and woolly spider monkeys, or have been acquired by the common ancestor of spider, woolly, and woolly spider monkeys and secondarily lost by the woolly monkey.

HABITAT, LOCOMOTION, AND POSITIONAL BEHAVIOR – Howling monkeys show the broadest geographic distribution, from southern Mexico to northern Argentina, and throughout the Amazon basin. They are everywhere sympatric with at least one other genus of atelines, except in the most extreme parts of their range. The woolly spider monkey has the most limited distribution; it occurs in a restricted area of the Atlantic coastal forest of southeastern Brazil. The spider and woolly monkeys occur primarily in tall mature forest with continuous canopy, while howler monkeys are found in a wide range of environments, including savannas (Fooden 1963; Fleagle and Mittermeier 1980; Peres 1990, 1994; Soini 1990; Stevenson et al. 1994). All atelines occur in the middle and high levels of the canopy (Mendel 1976; Fleagle and Mittermeier 1980; Gebo 1992; Defler 1995).

The spider, woolly, and woolly spider monkeys have the ability to travel with bimanual locomotion (brachiating and arm-swinging) and use relatively small supports. In contrast, howlers do not brachiate and travel mostly quadrupedally along bigger supports. Spider and howler monkeys climb frequently; when feeding, all genera use suspensory positions aided by their prehensile tail (Mendel 1976; Fleagle and Mittermeier 1980; Cant 1986; Gebo 1992).

DIET – Atelines are highly folivorous and frugivorous. Howlers are the most folivorous (Milton 1980) regardless of whether they are sympatric with other atelines; the woolly spider monkey is intermediate; and spider and woolly monkeys are mostly frugivorous (Strier 1992). Reports from different study sites are consistent for most species, except in the case of the woolly monkey, for which contrasting degrees of insectivory have been reported from different sites. *Lagothrix lagothricha lugens* was studied in La Macarena, Colombia and reported to eat mostly fruits (60% of its diet throughout the year) (Stevenson et al. 1994). Insects were the second most commonly consumed resource throughout the year (23%) at this single location. This degree of insectivory is unusually high for atelines, and contrasts sharply with reports on a population of *L. l. lagothricha* in Vaupes, Colombia. According to Defler (1995), the diet of this population included 94.6% leaves and fruits (with no report on the remainder). Muñoz Durán (1991) stated this same population included insects in its diet, but provided no percentages. In the headwaters of the Urucu river in Brazil, *L. l. cana* fed on insects during only 0.1% of its feeding time (Peres 1994).

Pithecinae (*Callicebus, Pithecini*)

SYSTEMATICS – This clade is supported by the following unambiguous morphological synapomorphies: trigonid and talonid of subequal height in the lower second molar, and presence of prehypocrista on the first upper molar derived from a primitive condition of absence (subsequently reversed in living pithecins). Pithecines include the titi monkey *(Callicebus)* and the pithecins. The latter include four genera: *Cebupithecia* (fossil pithecin), sakis *(Pithecia)*, uakaris *(Cacajao)*, and bearded sakis *(Chiropotes)*. The sister-group relationship between the titi monkey and pithecins is supported by the total evidence, morphological and nuclear DNA trees. Rosenberger (1979, 1981, 1984) and Horovitz (1995, in prep.) presented hypotheses based on morphology that are closest to our total evidence tree; however the owl monkey *(Aotus)* does not appear to be sister to the titi monkey as Rosenberger (1984) suggested, but instead appears to be sister to the callitrichins, *Cebus*, and *Saimiri* (Figure 6.4).

Callicebus

HABITAT AND POSITIONAL BEHAVIOR – Titi monkeys are found from central Colombia and Venezuela to Paraguay. The yellow-handed titi (*C. torquatus*) forages in different varieties of varillal and palm forest (Kinzey et al. 1977; Kinzey 1977a). Varillal is a non-flooded forest with a relatively closed canopy, abundant vertical tree trunks (a characteristic from which its name is derived), and reduced undergrowth; it comprised 60% of the territory of the troop studied. While the yellow-handed titi forages mostly in the second and emergent stories of the forest canopy, *C. cupreus discolor* (sensu Mittermeier et al. 1988) is generally found lower in the forest (Kinzey 1978).

The most common feeding posture of the yellow-handed titi is sitting except if feeding on berries, when it adopts an erect posture, with the torso parallel to the vertical trunk and the feet inverted and powerfully adducted, with the pollex grasping the trunk (Kinzey 1977a,b).

DIET – Titi monkeys are primarily frugivorous but different species complement their diet in different ways. The second leading food item for yellow-handed titis (*C. torquatus*) are insects (14%) (Kinzey et al. 1977; Kinzey 1977a; Kinzey 1978), whereas that for *C. brunneus* and *C. personatus* are leaves (Kinzey 1978; Kinzey and Becker 1983; Wright 1989). *C. brunneus* spent 10 to 15% of its time sitting and scanning for insects and, although data on actual insect feeding bouts are unavailable, the success rate was apparently low (Wright 1989). The masked titi (*C. personatus*) has not been observed to eat insects (Kinzey and Becker 1983).

The yellow-handed titi opens hard husked fruits by placing them in the corner of its mouth, and cracking them with the canines or the premolars (Kinzey 1977a). Titis catch insects in the air, leaves, or at ant nests; only on rare occasions have they been observed to go down to the ground to obtain insects.

Pithecini (*Cebupithecia*, [*Pithecia*, {*Cacajao*, *Chiropotes*}])

SYSTEMATICS – Pithecini includes sakis (*Pithecia*), uakaris (*Cacajao*), bearded sakis (*Chiropotes*), and the fossil *Cebupithecia*. The monophyly of this group is supported by the presence of a diastema between the lower canine and second incisor, a sharp lingual vertical edge on the lower canines, a reduction of the lower third molar relative to the length of the fourth premolar, proclivious upper incisors, and a high preparacrista on the upper third premolar. Monophyly of living pithecins is further supported by crenulated molar enamel and loss of a lingual cingulum on the fourth upper premolar. A sister-group relationship between uakaris and bearded sakis is supported by a buccolingual enlargement of the fourth upper premolar relative to the first molar, and the loss of the prehypocrista on the first upper molar.

HABITAT, LOCOMOTION, AND POSITIONAL BEHAVIOR – Pithecines live in the Amazon basin and the southern margin of the Orinoco river. Uakaris are always associated with floodplains, whereas bearded sakis live in areas of high primary forests of terra firme (relatively high non-flooded ground) (Fontaine 1981; Ayres 1989). Sakis seem to be the most flexible of the pithecins: they are found primarily in unflooded forest and they are sympatric over a large part of their range with bearded sakis; they overlap with uakaris in narrow bands of flooded forest (Peres 1993).

The white-faced saki (*Pithecia pithecia*) is predominantly a leaper, while bearded sakis and uakaris are primarily quadrupedal. Data on posture while feeding is more limited, but sakis frequently cling to trunks of trees or lianas, and uakaris and bearded sakis apparently feed more commonly in pronograde quadrupedal postures, and less commonly adopt hindlimb suspensory postures (Ayres 1986; Fleagle and Meldrum 1988).

Bearded sakis are upper- and middle-canopy frugivores (Norconk and Kinzey 1994); sakis have more varied habits. Where the white-faced saki (*P. pithecia*) co-occurs with bearded sakis (e.g., in the Guianas), it frequently feeds in the understory and the lower part of the canopy (van Roosmalen et al. 1988; Kinzey and Norconk 1993). On the other hand, where the white saki (*P. albicans*) occurs in the absence of other pithecines (e.g., in Amazonia), it is found mostly in the higher levels of the canopy (Peres 1993). Uakaris live in floodplains and surrounding terra firme, and descend to the ground when water levels drop in order to eat seeds and seedlings (Ayres 1989).

DIET – Pithecins are frugivores (at least 85% of feeding time), but differ from other frugivorous New World monkeys in exploiting unripe fruits, with a harder pericarp and a pulp with lower sugar content and more defensive compounds than ripe fruits. Biochemical analysis of the fruits eaten by sakis indicate that the preferred species have a high lipid content (47 to 50%) and therefore a high nutrient value (Kinzey and Norconk 1993). Pithecins also consume leaves and arthropods (Ayres 1986, 1989; Kinzey 1992).

Pithecins are seed predators, digesting the seeds they ingest. Their strategy contrasts with that of seed dispersers (e.g., spider monkeys) who feed on and digest the pericarp, in most cases the mesocarp, and/or the aril, and either drop the seed before ingestion or allow the seed to pass through their tract undigested. Seeds of consumed fruit are frequently protected by a hard covering (pericarp, usually a hard mesocarp, and/or sometimes the seed coat). Pithecins can break the hard husk with their specialized canines, discard the pericarp, and then masticate the seeds, which are actually the protein-rich part of the fruits. Bearded sakis feed on the immature seeds of a large number of species that are consumed when mature by sympatric spider monkeys. The black-bearded saki *(Chiropotes satanas)* can open fruit with pericarp as much as 15 times harder than those opened by the black spider monkey *(Ateles paniscus)*; and the average crushing pressure (2.77 kg mm^{-2}) exerted by the black-bearded saki is significantly greater than that (0.03 kg mm^{-2}) exerted by the black spider monkey (Kinzey and Norconk 1990). The average crushing resistance of seeds consumed by the black-bearded saki (7.2 ± 0.7 kg) is significantly smaller than that of seeds swallowed by the black spider monkey (17.1 ± 2.6 kg) (Kinzey and Norconk 1990). The hardness of fruits that sakis open is intermediate between those opened by spider monkeys and bearded sakis (Kinzey and Norconk 1990, 1993).

Cebidae (*Aotus*, [{*Cebus, Saimiri*}, Callitrichini])

SYSTEMATICS – This clade is composed of owl monkeys (*Aotus*), capuchin monkeys (*Cebus*), squirrel monkeys (*Saimiri*), and the callitrichins, which include Goeldi's monkey (*Callimico*), marmosets (*Callithrix* and *Cebulla*), and tamarins (*Saguinus* and *Leontopithecus*). The monophyly of callitochins has rarely been questioned, but its relationships to other New World primates has been widely debated. Rosenberger (1979, 1981, 1984) suggested that capuchins (*Cebus*) and squirrel monkeys (*Saimiri*) were sister-groups based on morphology; the same conclusion was reached by Schneider et al. (1993) based on nuclear DNA sequences, and by us (Figure 6.4) based on an analysis of combined molecular and morphological data. The only previous studies that have placed owl monkeys (*Aotus*) in Cebidae (as defined here) are Schneider et al. (1993), Harada et al. 1995, and Horovitz (1995), based on nuclear DNA sequences and morphological data. Our total evidence analysis (Figure 6.4) indicates that *Aotus* is sister to the cebids and callitrichins.

Aotinae (*Aotus*)

HABITAT – Owl monkeys are widespread from Panama to northern Argentina, and live in most forested areas except in the Guyana shield and the Atlantic and Paranense forests. They are also successful in certain seasonally arid environments, such as the Chaco in southern Paraguay and north-central Argentina. Owl monkeys are the only nocturnal members of Anthropoidea, including both Platyrrhini and Catarrhini (see Figure 6.1).

DIET – *Aotus nigriceps* in Cosha-Cashu National Park, Peru, eats mostly small ripe fruits, which it complements with insects and leaves; flowers and nectar are also consumed (Wright 1989). Owl monkeys forage for insects at dawn and dusk and during moonlit nights. They grab insects out of the air with one hand while walking along the branches of tall trees. Owl monkeys ingest more insects on a daily basis than sympatric titi monkeys, based on data from fecal samples (Wright 1989).

Owl monkeys are successful in certain habitats that are marginal for other primates, such as the Chaco. There it is sympatric with the black howler (*Alouatta caraya*) and, at least in some areas, the brown capuchin (*C. apella*) (M. Di Bitetti pers. comm.). Some aerial predators present in Amazonian Peru (e.g., harpy eagles and crested eagles) are rare in the semi-arid Chaco, where the great horned owl is common (Wright 1989). Owl monkeys are not strictly nocturnal in Paraguay (Wright 1989) or Argentina (Arditi pers. comm.). They can commonly be seen foraging for 1 to 3 hours during the day; traveling and feeding can occur at any time of the day or night. In Paraguay, leaf-eating is higher (46%) than in Peru in winter, when fruits are scarce. During the spring, insect and flower consumption is very high (Wright 1989).

Cebini (*Cebus, Saimiri*)

SYSTEMATICS – Capuchin monkeys (*Cebus*) and squirrel monkeys (*Saimiri*) are sister taxa. This is supported by at least two morphological characters: the fourth upper premolar is wider than the first molar and the vomer is exposed in the orbit. Capuchin monkeys have a prehensile tail which can support the entire body weight for short periods of time in adults (10 to 15 s) and for longer periods (> 30 s) in juveniles (M. Di Bitetti and C. Janson, pers. comm.).

HABITAT, LOCOMOTION, AND POSITIONAL BEHAVIOR – *Cebus* is one of the most widely distributed genera of New World monkeys, ranging from Belize to northern Argentina. *Saimiri* occurs in Costa Rica and Panama, and ranges from central Colombia to Bolivia and northeastern Brazil. *Cebus* provide two of the few cases of congeneric sympatry: the tufted capuchin (*C. apella*) co-occurs over part of its range with the white-fronted capuchin (*C. albifrons*) or the wedge-capped capuchin (*C. nigrivitattus*). Capuchin and squirrel monkeys are both found in primary and secondary rain forests (Terborgh 1983; Boinski 1987, 1989b).

Capuchins show great variability in feeding heights in the forest (Terborgh 1983). Data on locomotor behavior of the white-throated capuchin (*C. capucinus*) indicate that it is highly quadrupedal (54%), and secondarily climbs (26%) and leaps (15%). Its positional behavior includes sitting (44%), standing (31%), and reclining (13%) (Gebo 1992). C. Janson (pers. comm.) has occasionally observed the tufted capuchin (*C. apella*) suspended from its tail when feeding on spiny palms.

Saimiri oerstedii typically forages and travels at about 5 to 10 m above the ground, on thin branches (< 5 cm diameter). Squirrel monkeys are basically quadrupedal, with a lower incidence of climbing or leaping and clinging on vertical thin substrates (Boinski 1987, 1989b). Its most common feeding postures are sitting, hanging by the hindlimbs, and sitting in tripod stance (i.e., on its hindlimbs while maintaining its tail as a third point of support on the substrate) (Boinski 1989b).

DIET – Capuchins and squirrel monkeys are mainly frugivorous but also rely heavily on other animals as source of protein (Janson and Boinski 1992). Capuchins also rely on other resources toward the margins of their distribution. For example, in some areas in northern Argentina, their primary resource is bromeliads; in other areas, they feed heavily on fruits during one season, but switch to insects at other times (Brown and Zunino 1992).

Capuchins can open hard husked fruits by holding them in their hands and biting them open, using incisors for smaller fruits (1–3 cm) and premolars or molars for larger fruits (Janson and Boinski 1992). Capuchins are a pre-dispersal seed predator of *Cariniana micrantha*, an emergent member of the Brazil-nut family Lecythidaceae (Peres 1991).

Callitrichini (*Leontopithecus*, [*Saguinus*, {*Callimico*, (*Callithrix*, *Cebuella*)}])

SYSTEMATICS – The Callitrichini is composed of five genera: Goeldi's monkey (*Callimico*), the marmosets (*Callithrix*, *Cebuella*), and the tamarins (*Leontopithecus*, *Saguinus*). Characteristics of this group are that they are among the smallest of the anthropoids, and in absolute terms have the smallest brain volumes; they bear claws on all manual and pedal digits except the hallux. Within this group, the only point on which both morphological and molecular analyses agree is the sister-group relationship between the two genera of marmosets. The most common hypothesis based on morphology alone is that Goeldi's monkey is the earliest diverging taxon of callitrichins (Rosenberger 1981, 1984; Ford 1986b; Kay 1990; this paper). On the other hand, analyses based on DNA and amino-acid sequences (Schneider et al. 1993; Horovitz and Meyer 1995), immunological data (Sarich and Cronin 1980), and cytogenetic data (Seuanez et al. 1989) suggest that Goeldi's monkey is closely related to the marmosets. Some studies suggest the tamarins are monophyletic, but others (including our total evidence analysis, see Figure 6.4) do not.

In addition to the unreversed characters mentioned above, Callitrichini is supported by (i) reduction of the size of the pterygoid fossa from reaching the base of the skull to a shallow space between the lateral pterygoid process and the splinterlike medial process; (ii) loss of the third molar; (iii) loss of hypocone on the first upper molar; and (iv) two offspring at a time (from a primitive condition of one). All four characters are reversed in Goeldi's monkey. No unambiguous morphological characters support nodes 1 and 2 (Figure 6.3). Morphology strongly indicates a basal position for Goeldi's monkey within callitrichins, but this signal is overwhelmed by molecular characters indicating its position as sister to the marmosets. Marmosets share several specializations such as staggered lower incisors of equal height which display meso- and distostyles, mesiodistally compressed canines, and buccolingually compressed deciduous lower incisors.

Lion tamarins have acquired certain specializations, unique among anthropoids, that have been associated with their outstanding manipulative abilities; they have long and slender arms and hands, and partially webbed middle fingers that they use to probe for and extract prey (Coimbra-Filho 1970b; Hershkovitz 1977).

HABITAT, LOCOMOTION, AND POSITIONAL BEHAVIOR – Callitrichins range from southeastern Costa Rica (tamarins) to Bolivia (Goeldi's monkey and marmosets) and southeastern Brazil (lion tamarins). Traits all callitrichins share are the exploitation of low levels of the canopy and understory, and the ability to cling onto big trunks and large branches for feeding and/or traveling purposes. The marmoset *Callithrix* and the tamarin *Saguinus* occur in "terra firme" and are generally absent from floodplains. They use disturbed, edge, or secondary growth forest, except some species of *Saguinus* that also live in primary forests (Rylands 1986; Garber 1993). The habitats of the pygmy marmoset (*Cebuella*) are the edges and interiors of seasonally inundated mature floodplain forests, although it also occurs in mature non-flooded forest (Moynihan 1976b; Soini 1993; Hernández-Camacho and Cooper 1976). Lion tamarins live in coastal lowlands, inland Atlantic forests, and low inundated forests (Coimbra-Filho 1970a,b; 1976). The forests occupied by all species except *L. chrysopy-*

gus chrysopygus have abundant epiphytic bromeliads, a common foraging substrate (Rylands 1993). Goeldi's monkey lives in scrub forest, mostly low and young second growth and in bamboo forests (Moynihan 1976a; Izawa 1979b; Pook and Pook 1981; Buchanan-Smith 1991), although it is also found in primary forest (Christen and Geissmann 1994).

Saguinus fuscicollis, S. geoffroyi, S. mystax, and *S. midas midas* travel mostly quadrupedally, climbing, and leaping (Fleagle and Mittermeier 1980; Garber 1991; Garber and Pruetz 1995). In *S. fuscicollis,* 20% of leaps involve moderate- to large-sized vertical trunks, which are rare in *S. geoffroyi* and *S. mystax* (Garber 1991). Lion tamarins seem to have a pattern of locomotion like that of *Saguinus* excluding *S. fuscicollis* (Coimbra-Filho and Mittermeier 1973). *Cebuella* and *Callimico* leap frequently to and from vertical trunks (Kinzey et al. 1975; Moynihan 1976a,b; Pook and Pook 1981). *Cebuella* spends 77% of its feeding time clinging onto trunks (Kinzey et al. 1975). *Callithrix* probably has very similar locomotor and positional habits, although no quantitative data are available for this genus.

DIET – Marmosets and tamarins feed on plant exudates (sap, gum, resin) as a complement to fruits and insects (Sussman and Kinzey 1984; Ferrari and Lopes Ferrari 1989; Soini 1993). Marmosets possess specialized incisors which they use for gouging holes in tree bark and directly stimulating the flow of gum, and spend a considerably higher percentage of their time feeding on this resource than tamarins. *Leontopithecus rosalia* has also been observed to chew on bark to stimulate gum flow (Peres 1989), despite the fact that it does not possess specialized incisors. *Saguinus* relies on natural damage to the bark or the activity of wood-boring insects to obtain gums.

Lion tamarins forage for insects in a manipulative fashion and catch mainly non-mobile prey concealed in palm crowns, bromeliad axils, wooden crevices, and under bark (Garber 1992; Rylands 1989). *Saguinus* forages for insects on the surface of branches and in vine tangles and foliage under the canopy and *Saguinus fuscicollis* in addition explores tree-trunk bark (Garber 1992). Information on the diet of Goeldi's monkey is limited, and there are no year-round field studies. It feeds on insects in a strategy similar to that of *S. fuscicollis* and *nigricollis* (Garber 1992).

Evolutionary patterns of ecological specialization

The first cladogenetic event in the platyrrhine ancestral lineage gave rise to two successful and diverse groups, the Atelidae and Cebidae (Figure 6.4). No obvious basal morphological innovations appeared in these two clades that allowed them to exploit new resources. Atelidae is comprised of atelines (howler, spider, woolly, and woolly spider monkeys), pithecins (sakis and uakaris), and the titi monkey; Cebidae is comprised of the owl, capuchin, and squirrel monkeys, as well as the callitrichins. The most noticeable ecological difference between the Atelidae and Cebidae is the highly herbivorous diet of virtually all species in the former, and a higher reliance on insects in the latter (Rosenberger 1981, 1992). In the Atelidae, pithecines and atelines generally feed at least 94% of the time on plants (Kinzey 1978, 1992; Rosenberger and Strier 1989). The known exceptions are one population of *Lagothrix lagothricha* and one of *Callicebus torquatus* that feed on insects a high proportion of the time (23 and 14%, respectively).

Cebidae is characterized by a higher consumption of insects than most Atelidae. The heavy consumption of insects by owl monkeys has been attributed to their activity at night when insects are most active (Kinzey 1992). However, insect consumption also seems to be high in areas where owl monkeys are more active during the daylight hours (Wright 1989). The primary food for squirrel monkeys is sometimes insects (Boinski 1989a; Janson and Boinski 1992). Capuchins rely less on insects than squirrel monkeys do; for example, the tufted capuchin (*C. apella*) spends about half of its day manipulating substrates and ingesting prey (Janson 1990) and obtains approximately 16% of its energy from insects (Janson 1985; Janson and Boinski 1992). Callitrichins also rely heavily on insects; according to most reports, they spend more than 13% of their time foraging for insects.

Besides diet, Atelidae and Cebidae generally differ in body weight. Body weight varies from 0.13 to 3.8 kg in Cebidae, and from 1.3 to 12 kg in Atelidae. Curiously, the species with the extreme body weights in either group belong to genera that are deeply nested in the phylogeny: there has been a certain tendency towards reduction in body size in Atelidae, and towards enlargement in body size in Cebidae.

Across primates generally and in New World monkeys in particular, the reliance on animal prey (or other food items of high energy content) decreases with increasing body size (see Ford and Davis 1992). However, within smaller groups the pattern is not always so clear. Within callitrichins, for example, there is no direct relationship between body size and degree of insectivory or exudativory (Garber 1992). The relationship between body weight and diet has traditionally been explained in terms of metabolic rates. Smaller species have higher metabolic rates than larger ones, and therefore the expectation is that they need to consume a higher proportion of energy-rich resources to support their higher needs (Kleiber 1947, 1961; Clutton-Brock and Harvey 1983; Eisenberg 1981, 1990; Martin 1990; Schmidt-Nielsen 1984; Kay 1984; Ford and Davis 1992).

In spite of the apparent absence of key morphological innovations for the Atelidae and Cebidae, subclades of these broad groups seem to have evolved traits that enable them to exploit resources that may not be accessible to other groups, based on the nature of those resources or their arboreal location. This functional radiation may have increased the number of species that can coexist locally, and is discussed below.

Atelinae

ATELINES – the howler, spider, woolly, and woolly spider monkeys – possess an apparent key innovation, a prehensile tail, that opens the possibility of exploiting a range of resources inaccessible to other quadrupeds. Tail prehensility has evolved independently in six orders of mammals with very different ecological roles (e.g., frugivory, folivory, and omnivory), in such groups as the opossums, kinkajous, porcupines, and primates. All use their prehensile tails both for support while feeding on branch-tips or locations of difficult access and as an aid in locomotion, especially on unstable supports or while descending (Grand 1978; Charles-Dominique et al. 1981; Emmons and Gentry 1983). Emmons and Gentry (1983) noted that prehensile-tailed animals occur with a higher frequency in neotropical forests than in Africa or Asia.

Moreover, use of the prehensile tail to move through the canopy seems mostly restricted to the Neotropics (Emmons and Gentry 1983). Tropical forests are thought to be structurally similar across continents (Richards 1952; Leigh 1975), based mostly on measures of biomass and productivity, leaf size and shape, canopy height and degree of stratification, and tree density (Dawkins 1959; Leigh 1975). However, Emmons and Gentry (1983) reasoned that there might be some differences that make the possession of a prehensile tail advantageous in the Neotropics and not the Paleotropics. With the purpose of investigating this question, they quantified several factors, and found two interesting differences: (1) liana density is higher in Africa than in the Neotropics and Borneo; and (2) there are more palm trees in South American forests than on the other continents.

Lianas have two important functions for monkeys, providing food and travel corridors. Prehensile tails are probably not of much use when traveling on lianas (Emmons and Gentry, 1983). But they may be particularly useful in coping with the downward bending of branch tips under the weight of suspended animals. When lianas are present, animals tend to use them to pass from one tree to another, which allows them to bypass droop-prone terminal branches. Therefore, it seems that the extensive presence of lianas in Africa may have reduced the usefulness of prehensile tails.

The second variable considered by Emmons and Gentry (1983) is the frequency of palms, which is far greater in neotropical forests. Palms are rarely invaded by lianas (Putz 1980) and are often surrounded by gaps in the vegetation. Arboreal animals frequently use palms as a pathway through the forest, and some eat palm fruits. Those animals who have prehensile tails use them to gain access to the palm trees. Emmons and Gentry (1983) postulated that frequent climbing on palm trees may have contributed to the selective advantage of prehensile tails in the Neotropics.

Atelines use suspensory feeding postures with the aid of their tail, spreading their weight over several widely dispersed supports, hanging by their long limbs and tail. This allows them access to resources on branches that would be too thin to support their large body weight if they were to stand on them. Spider monkeys (and probably woolly and woolly spider monkeys) travel along smaller supports than one would expect based on body size alone (Fleagle and Mittermeier 1980), by using bimanual locomotion and with the aid of their tails. In contrast, howlers (which travel mostly quadrupedally and are roughly of the same size as spider monkeys) use one large support at a time (Fleagle and Mittermeier 1980). Brachiation and climbing are likely to allow spider monkeys to shorten pathways between or within feeding patches, and suspension would enhance their maneuverability on thin supports. A quadruped would have to follow branches that zig and zag (Grand 1984).

It is likely that the ancestor of atelines was quadrupedal and the ancestor of atelins a brachiator (Figure 6.4). We base this conclusion on the fact that howlers and ateline sister-groups are quadrupedal, and that all atelins are brachiators (see Rosenberger and Strier 1989 for a similar conclusion).

There is a general tendency for larger-bodied mammals to be more folivorous than smaller ones (see above). The atelines are the largest New World monkeys and some of them are the most folivorous. Larger monkeys have absolutely greater nutritional requirements and therefore need to feed on highly abundant resources, such as leaves. On the other hand, leaves are low-quality foods that require longer digestion than other kinds of food. Gaulin (1979) suggested that the lower basal metabolic rates of larger animals permit low rates of digestion. Within atelines, however, the pattern is not so simple. For example, the woolly spider monkey and the mantled howler (*Alouatta palliata*) eat comparable amounts of fruit in two different geographical locations, but the woolly spider monkey is twice as frugivorous as the sympatric brown howler (*Alouatta fusca*) studied during the same period (Strier 1992).

In any habitat where they co-occur, howlers are always more folivorous than atelins. Howlers occur in sympatry with other atelines over most of their range, but even where they occur alone (e.g., northern Argentina), howlers are the most folivorous of all Atelinae, which suggests that their diet became established in evolutionary time, and is not a mere recent ecological condition.

The geographical distributions of spider and woolly spider monkeys are, for the most part, mutually exclusive (Hernández-Camacho and Cooper 1976). Terborgh (1983) noted that they both are present in some regions of Peru, but that they never occur in the exactly the same places; they are always at least a few kilometers apart. This was considered as a good example of competitive exclusion by Waser (1987). But spider and woolly spider monkeys are sympatric in some places in Colombia (Izawa 1975; Stevenson et al. 1991, 1994) and Peru (Herrera, unpubl. data, cited in Peres 1994).

Woolly monkeys (*L. lagothricha lugens*) and long-haired spider monkeys (*Ateles belzebuth*) occur in sympatry at La Macarena, Colombia (Stevenson et al. 1991). Few differences were found in the way these species exploited resources. However, spider monkeys fed heavily on fruits of *Jessenia* while this item was absent from the diet of woolly spider monkeys. In addition, woolly spider monkeys fed heavily on insects, while spider monkeys rarely consumed them; this is the only reported case of heavy insect consumption by *Lagothrix*. These two divergences could be interpreted as a mechanism by which woolly spider monkeys can survive in sympatry with spider monkeys, given that both otherwise have very similar diets and modes of exploiting resources. It remains to be seen whether this characteristic of *L. lagothricha lugens* is an opportunistic strategy to overcome a presumed shortage of fruits, or is a fixed feature of this subspecies that occurs regardless of the availability of its preferred food.

Chapman (1987, 1988) studied the black-handed spider monkey, the mantled howler, and the white-throated capuchin (*Ateles geoffroyi*, *Alouatta palliata*, and *Cebus capucinus* respectively) in sympatry in Costa Rica over two years. Their diets showed high variability and overlap, leading him to conclude that it was unlikely that these species' diets could be influenced by interspecific competition (Chapman 1987). Tomblin and Cranford (1994) studied the white-throated capuchin (*Cebus capucinus*) and the mantled howler (*A. palliata*) elsewhere in Costa Rica, and reported that the two species used the same macrohabitats. However, based on a detailed study of branch use, feeding mode, positional behavior, and diet, they found significant differences between the species, at least during the rainy season. Even when both monkeys used exactly the same tree species at different times, they did not use them in the same way. The relative diameter of the branches used differed significantly, as did the distance from the trunk at which they foraged. Mantled howlers were concentrated on the periphery of the crowns, where they could reach the leaves, whereas capuchins showed a greater variability of branch use, spending part of the time on the periphery and also near the trunk. Capuchins ate a greater variety of food; the major components were fruit (44%) and invertebrates (37%), whereas the howlers were never observed eating fruit but instead ate leaves (94%) and buds (5.8%). Only howlers used their prehensile tail while feeding to suspend themselves near the tips of the branches.

In summary, atelines seem to have acquired key innovations in a stepwise fashion: all of them share a prehensile tail and suspensory positional behavior. A nested subset of atelines – the spider, woolly, and woolly spider monkeys – have developed the ability to travel bimanually. Field observations suggest that tail prehensility opens access to many resources that would be otherwise inaccessible.

Pithecinae

Pithecins possess unique dental characteristics that allow exploitation of resources that other species do not exploit. They have sharp canines which they use to open hard-husked, immature fruits. Bearded sakis possess the most remarkable canines with which they can open the hardest fruits (Kinzey and Norconk 1993). Living pithecins have very procumbent upper incisors (an unknown character in *Cebupithecia*), which they use to open some kinds of fruits (van Roosmalen et al. 1988).

In addition to sharp canines and procumbent incisors, living pithecins share a thin crenulated molar enamel and reduction in molar relief. Pithecins break the husk of some fruits and masticate the seeds inside. These seeds are usually softer than seeds of fruits that do not possess a hard husk and are usually swallowed and dispersed undigested. Low occlusal relief may resist wear well (Rosenberger and Kinzey 1976). The absence of a thick enamel could be correlated with the consistency of seeds: they may be hard to masticate, but they are not brittle (Kinzey 1992). The fossil pithecin *Cebupithecia sarmientoi* seems to be the sister-group to recent pithecins. It displays long sharp canines, similar in shape to those of living pithecins, but the molars have a higher relief and the enamel is not crenulated. This suggests a two-stage evolution of sclerocarpic foraging in pithecins: the characters that enabled them to open fruits arose first, and then molar modifications evolved which were (presumably) advantageous for seed processing; the latter modifications are seen in recent forms only (Setoguchi et al. 1988; Kinzey 1992). As in the atelines, key characters in the pithecines appear to have evolved in steps, not as an integrated complex.

Uakaris and bearded sakis seem unlikely to be sympatric because both are frugivores specialized in seed consumption (Ayres 1989). Sakis and bearded sakis also have similar diets, yet they co-occur throughout most of their geographical range (Kinzey and Norconk 1993). Some differences have been noticed in behavior that might partly explain their capacity to share the same habitats. Guianan sakis tend to forage in lower levels of the canopy (van Roosmalen et al. 1988; Mittermeier and van Roosmalen 1981) and eat softer pericarps (Kinzey and Norconk 1993) than sympatric bearded sakis. Sakis also eat more flowers during the dry season, when fruit availability is depressed, while bearded sakis continue to specialize on fruits (Kinzey and Norconk 1993).

The geographic distribution of spider monkeys also overlaps extensively with those of sakis and bearded sakis. Like bearded sakis, spider monkeys are upper-canopy frugivores (Norconk and Kinzey 1994). Sakis that feed in the understory and lower canopy use different plant species (van Roosmalen et al. 1988). Fruits represent over 90% of the diet of both the black spider monkey (*Ateles paniscus*) and the black bearded saki (*Chiropotes satanas*) and their diets overlap in a number of species. The bearded saki exploits many fruits at an earlier stage than the spider monkey, when they are still unripe and much harder to open. The bearded saki opens fruits significantly harder than those the spider monkey does, whereas the average hardness of the fruits sakis open is intermediate and not significantly different from the other two taxa (Kinzey and Norconk 1990; 1993). Species of sakis and bearded sakis that do not overlap in distribution have more similar ecological characteristics (Peres 1993). Given all factors described above, there is no strong evidence of an evolutionary divergence in the ecology of sakis and bearded sakis, at least at a broad generic level.

In Venezuela, the diets of the black bearded saki (*Chiropotes satanas*), black spider monkey (*Ateles paniscus*), and long-haired spider monkey (*A. belzebuth*) have been studied in sympatry (Kinzey and Norconk 1990; Norconk and Kinzey 1994) and allopatry (Kinzey and Norconk 1993). Although the data available are limited, they give no indication of competitive release (Norconk and Kinzey 1994). This suggests that while sakis and spider monkeys overlap in food sources, they may not compete with each other intensively.

At least one species of titi monkey (*Callicebus brunneus*) has been reported to include immature fruits in its diet, as do pithecins (Wright 1989). The yellow-handed titi monkey (*C. torquatus*) opens most hard fruits with its canines (as do pithecins) or premolars (Kinzey 1977a), in contrast with capuchins that open hard fruits with premolars or molars (Janson and Boinski 1992). This habit of titi monkeys of using their canines to open fruits could be the beginning of a tendency in the clade, despite the fact that titi monkeys have the relatively smallest canines among living platyrrhines. They also share lower molars (especially the second) that show a subequal height of trigonid and talonid with the other pithecines. This character could have some relation with reduction of occlusal relief and mastication of seed and as pointed out for pithecins (Rosenberger and Kinzey 1976) but this is not possible to test at the moment. There are no reports of seed-predation in titi monkeys.

A few subspecies of titi monkeys overlap in their geographical distributions: *C. torquatus torquatus* overlaps with both *C. cupreus discolor* and *C. c. cupreus* in Peru and western Brazil (Soini 1972; Kinzey 1978; nomenclature follows Mittermeier et al. 1988); *C. cupreus* and *C. torquatus* also have been reported to overlap in southern Colombia (Hernández-Camacho and Cooper 1976; Klein and Klein 1976). But geographical overlap does not imply spatial co-occurrence: different habitat preferences were detected, at least for *C. t. torquatus* and *C. c. discolor* (Soini 1972; Moynihan 1976a; Kinzey 1978, 1981; Kinzey and Gentry 1979). These studies also found interspecific differences in dental morphology, diet, and preferences in canopy levels, among other factors (Kinzey 1978). These data suggest an adaptive radiation at the species level.

To summarize, Pithecinae are characterized by a few morphological characters, of which none can be identified as a key innovation at present. In contrast, Pithecini is supported by a suite of dental characters (e.g., sharp canines) that can be considered key innovations. In addition, living pithecins show low cusp relief. The evolution of these characters in two steps suggests that ability to open hard-husked fruits and masticate the enclosed seeds evolved in separate phases. This is another example of evolution in steps through key innovations. Within living pithecins, uakaris show the most remarkable divergence in that they specialize in living in flooded forests.

Cebidae

AOTUS – Owl monkeys bear several traits that may represent key innovations for a nocturnal habit. They have the lowest metabolic rate among the few known for platyrrhines (Le Maho et al. 1981). All nocturnal primates have low metabolic rates, although this is not true for all nocturnal mammals (McNab 1983). Low metabolic rates presumably allow nocturnal animals to live with much lower levels of energy consumption (Crompton et al. 1978).

Owl monkeys have relatively larger eyes than other platyrrhines, which enhance vision at low light levels. Its lens is more spherical than in diurnal forms, a shape that refracts more light onto the retina (Wright 1989, 1994). The iris is located more posteriorly, toward the center of the eye, allowing the pupil to reach a larger diameter and more light to reach the retina (Noback 1975). In addition, according to Ogden (1975), the rod density throughout the retina is several times higher than that of humans. The size of the olfactory lobes of the brain relative to the size of the visual cortex suggests that olfaction is more important in owl monkeys than in other platyrrhines (Wright 1989).

SAIMIRI AND *CEBUS* – Differences in behavior seem to be the key to differential resource exploitation in capuchins (*Cebus*) and squirrel monkeys (*Saimiri*) (Janson and Boinski 1992). Certain behaviors are shared by these genera; an example is the manipulative foraging through foliage and small twigs when searching for insects. But this behavior is also displayed by callitrichins, their sister-group, and so appears to be a primitive condition for *Cebus-Saimiri*.

Body size and biting force may permit capuchins to forage in hidden, mechanically tough substrates, such as palm frond bases, cane, bamboo, dead branches, and termite nests (Janson and Boinski 1992); the most robust species (*Cebus apella*) spends up to 44.3% of its time associated with such substrates (Terborgh 1983). By comparison, squirrel monkeys spend only 0.7% of their time searching in such difficult substrates. Capuchins (especially *C. apella*) also possess very thick dental enamel, relatively more substantial than that in any other living primate (Kay 1981). This may be related to the fact that they feed on very hard plant tissues, such as palm nuts (Izawa and Mizano 1977).

As mentioned earlier, capuchins provide one of the few cases of congeneric sympatry in New World monkeys. This raises the question of whether differences in the way the sympatric capuchins exploit the environment suggest an adaptive radiation within the genus. The tufted capuchin (*C. apella*) and white-fronted capuchin (*C. albifrons*) have been studied in sympatry in Manú National Park in Peru (Terborgh 1983; Janson 1985; Janson and Boinski 1992). These species showed some differences in their dietary preferences: the tufted capuchin was seen more frequently foraging on figs; the white-fronted capuchin, on palm trees. The tufted capuchin showed an ability to break palm nuts not seen in the white-fronted capuchin; the latter always took much longer to break such nuts. The tufted capuchin forages heavily on palm nuts in other parts of its geographical distribution where it is not sympatric with the white-fronted capuchin, such as northeastern Argentina (I. Horovitz, pers. obs.); therefore, this dietary trait does not seem to be a localized specialization of the Manú population, but more likely is a characteristic of the species.

There are also some morphological differences between these two capuchin species. The tufted capuchin has a larger body weight and is therefore stronger; it also has a deeper and more buttressed mandible, larger zygomatic arches, and some individuals possess a sagittal crest, all suggesting a larger biting force. These characters are consistent with the tufted capuchin's frequent habit of breaking dead branches in search of insects and the use of palm nuts as a common food source (Terborgh 1983; Janson 1985; Janson and Boinski 1992).

All capuchin species have an ability to exploit a wide range of food items. This may be derived from their ability to manipulate substrates and employ tools, abilities not possessed by other New World primates (Costello and Fragaszy 1988; Chevalier-Skolnikoff 1989a; Fragaszy et al. 1990; Visalberghi 1990). An animal uses a tool when it employs an unattached environmental object as a functional extension of its own body in attaining an immediate goal (van Lawick-Goodall 1970). Sensorimotor ability, tool use, and omnivorous extractive foraging have a morphological correlate: brain size (Gibson 1986; Janson and Boinski 1992). When seasonal sources are scarce in the low season, capuchins can extract embedded food which is available year-round and has high concentrations of energy and protein (Parker and Gibson 1977). This might be the reason why capuchins can inhabit areas not inhabited by other monkeys that do not have the ability to engage in extractive tasks involving complex, cortically mediated, sensorimotor coordinations for tapping, probing, looking, and listening to locate and recognize bark-embedded insects, ripe palm nuts, frogs, and grasshoppers hidden within tree cavities (Izawa and Mizano 1977; Izawa 1978, 1979a; Terborgh 1983; Gibson 1986). Most studies on tool-use in capuchins have been conducted with captive individuals. In the wild, capuchins have been observed to use sticks as probes and clubs (Boinski 1988; Chevalier-Skolnikoff 1989b), and to employ oyster shells as hammers (Fernandes 1991) although these events are quite rare. They frequently open nuts and other hard fruits by pounding them against tree trunks or by hitting them together (Izawa and Mizano 1977). Tool-use is typical of animals who lack specialized anatomical characteristics and need to extract embedded food (Alcock 1972; Gibson 1986). Extraction *per se* is not correlated with brain size. Extractors who possess a rather specialized anatomy to concentrate on one extractive food (such as the marmosets), tend to have small brain sizes relative to body size (Gibson 1986).

CALLITRICHINI – Claws in this group serve a vital function, allowing individuals to cling to trunks and other vertical supports while exploiting plant exudates and insects, and/or to leap from trunk to trunk as a traveling behavior (Cartmill 1974; Garber 1980, 1992; Rylands and de Faria 1993). Such supports are too large for small primates to span with tiny hands and feet; therefore, the possession of claws seems a character required for clinging onto trunks. Two hypotheses have been suggested for the original function of claws in the Callitrichini, involving either feeding on tree exudates (Sussman and Kinzey 1984) or traveling on large supports (Ford (1986a). Two hypotheses have been suggested for the original function of claws in the Callitrichini, involving either feeding on tree exudates (Sussman and Kinzey 1984) or traveling on large supports (Ford 1986a). To test these two hypotheses we need to assume that feeding and traveling behaviors shown by each species are genetically fixed.

All callitrichins studied in the wild feed on exudates, with the apparent exception of Goeldi's monkey (Sussman and Kinzey 1984; Garber 1992). Ford (1986a) based the second hypothesis on the observation that Goeldi's monkey (which she considered basal to callitrichins) uses its claws to travel by vertical clinging and leaping. Other species that can be characterized as using a clinging-and-leaping mode of progression are *Saguinus fuscicollis,* and *Cebuella pygmaea* (Kinzey et al. 1975; Moynihan 1976b; Castro and Soini 1977; Sussman and Kinzey 1984). Even so, *Saguinus* and *Cebuella* use their claws primarily for clinging to vertical supports while feeding on exudates rather than for locomotor activities (Kinzey et al. 1975; Sussman and Kinzey 1984). Some species rarely cling onto trunks to forage for insects (*Saguinus mystax* and *S. geoffroyi* [Garber 1992]). The answer to whether Goeldi's monkey lost the habit of feeding on exudates, or most other species virtually abandoned the habit of clinging and leaping and hence the use of their claws while traveling, would be not much more than a guess at this point. If Goeldi's monkey were sister to the marmosets (*Callithrix-Cebuella*), then the travel hypothesis for origin of claws would be even less likely. On this basis, we infer that Goeldi's monkey lost its habit of feeding on exudates secondarily.

Callitrichins share an ability to search for hidden insects with capuchins and squirrel monkeys; therefore, it seems likely that this ability evolved in their common ancestor. All species of lion tamarins (*Leontopithecus*) also have the ability to search for embedded insects.

The variety of environments and resources callitrichins exploit are not restricted to individual clades. Species belonging to different genera have converged in their ecological characteristics. There is only one strong tendency that marmosets exhibit which is not found in other groups: the most specialized form of exudativory, evidenced by their habits and morphology.

The geographic distributions of congeneric callitrichine species and subspecies are generally non-overlapping. Where sympatry does occur, there is a relatively sharp differentiation in the way the different types exploit the environment (Ferrari 1993). The saddle-back tamarin (*Saguinus fuscicollis*) is sympatric over part of its distribution with marmosets and some of its congeners. It appears to capture larger prey than its sympatric congeners by foraging in specific sites such as holes and fissures in bark and leaf litter accumulations (Yoneda 1981, 1984; Terborgh 1983), in contrast with the less manipulative techniques of "scan-and-pounce" or leaf-gleaning used by other species. Another important difference is that saddle-back tamarins typically forage at lower levels in the forest than its congeners; this difference persists even in the absence of other callitrichins (Pook and Pook 1981; Terborgh 1983; Yoneda 1984; Soini 1987; Buchanan-Smith 1990; Fang 1990; Heymann 1990; Ferrari 1993). The two smallest tamarin species – the saddle-back tamarin (*S. fuscicollis*) and the black-and-red tamarins (*S. nigricollis*) – forage in the lowest forest strata (< 11 m height), while the larger species occupy mostly the middle strata and lower parts of the main canopy (> 10 m height) (see Soini 1987). We regard this as a genetically fixed preference; it is possible that it facilitates the coexistence of *S. fuscicollis* and other callitrichins.

Possession of claws and a small body enable callitrichins to feed and travel in the lower forest, on substrates that may be inaccessible to other monkeys. Claws seem to have been a key innovation that paved the way for further specializations in the marmosets, including modified canines and incisors. This seems to be a case of progressive specialization for a new niche: exploitation of plant exudates. Goeldi's monkey, the putative sister-group of the marmosets, has lost the habit of exudate-feeding and many of the morphological characteristics inferred to occur in their common ancestor.

All cebids are strongly insectivorous, but their searching strategies vary. Owl monkeys look only for insects exposed on the surface of the branches or in the air. In contrast, capuchins, squirrel monkeys, and callitrichins have evolved manipulative abilities and search for insects hidden under leaves, or (in *Cebus* and *Leontopithecus*) for insects under bark. Did such manipulative abilities evolve more than once in the New World monkeys? According to the currently most parsimonious scenario – in which capuchins, squirrel monkeys, and callithrichins form a monophyletic group (Figures 6.3 and 6.4) – manipulative abilities appear to have evolved only once. However, given that the monophyly of this group is weakly supported, this conclusion should be considered provisional.

Conclusions

We conducted a "total evidence" analysis for New World monkeys at the generic level, combining nuclear and mitochondrial DNA sequences and morphological characters. The tree obtained is congruent with that derived excluding morphology. The New World monkeys appear to have undergone a basal split into two clades: Atelidae = (Atelinae, [*Callicebus*, Pithecini]) and Cebidae = (*Aotus*, [{*Cebus*, *Saimiri*}, Callitrichini]). Neither of these clades seems to display marked morphological "key innovations" – that is, synapomorphies that allow them to exploit resources in a specialized fashion. However, both show an evident difference in ecology: Atelidae are mainly herbivorous, whereas Cebidae have a heavy component of insectivory. The only clades unsupported by morphological characters are (*Brachyteles, Lagothrix*) and the two most basal nodes of (*Saguinus*, [*Callimico*, {*Callithrix, Cebuella*}]).

Owl monkeys (*Aotus*) appear to be sister to the remaining cebids, all of which have a strong component of insectivory. While owl monkeys only search for insects exposed on the surface of the branches or in the air, their sister taxon has evolved manipulative abilities and search for hidden insects and, in some cases, embedded ones. This basal position of *Aotus* is, however, only two steps more parsimonious than alternative topologies (see above), so we consider the conclusions based on this apparent position to be tentative.

Atelines display a prehensile tail and suspensory positional behavior, and the atelin subclade has developed the ability to travel bimanually. These seem to be key innovations for access to and exploitation of certain food resources, particularly fruits and leaves near branch tips. Pithecins possess sharp canines; a nested subset of this group shows low cusp relief, which may be important in the exploitation of hard-husked fruits and mastication of seeds. Possession of claws and small body size may enable callitrichins to feed and travel in the lower strata of rain forests. Claws may also have been a prerequisite for dental adaptations for exploiting plant exudates in marmosets.

In each of the three clades just mentioned, batteries of morphological characters appear to perform specific functions in an integrated fashion. Our phylogenetic analysis shows these batteries appear to have evolved in a stepwise fashion – that is, early diverging taxa possess only one or some of these derived characters, while more derived groups show more of these characters. It appears that these characters often serve the same functions in both basal and derived groups, although in some cases additional functions are observed in the latter. Major morphological characters seem important in several nodes, because they appear associated with the exploitation of new resources. Most ecological studies of New World primates focus on differences between co-occurring species in their use of resources. Given that, at least in some cases, such differences persist in allopatry as well and are somewhat characteristic of the species involved, it appears that most of the ecological variations in this group do in fact represent evolutionary trends. Differentiation in behavior also seems to be important at or below the generic level, and does not always have obvious morphological correlates; inclusion of such behavioral differences promises to be an important new direction for research on the adaptive radiation of the New World monkeys.

Acknowledgments

We are grateful to Drs. Thomas J. Givnish and Kenneth J. Sytsma for inviting us to participate in their symposium. T. J. Givnish contributed significantly to the improvement of this paper. Patricia Escobar-Páramo, John Fleagle, Charles Janson, Mario Di Bitetti, Rob Asher, Dennis Slice, and Luis Chiappe provided valuable comments and discussion. IH thanks Ross MacPhee and the American Museum of Natural History for help and financial support. AM thanks the National Science Foundation for financial support (BSR-9119867, BSR-9107838).

References

Alcock, J. 1972. The evolution of tools by feeding animals. Evolution 26:464–473.
Avise, J. C. 1994. *Molecular markers, natural history and evolution.* New York, NY: Chapman and Hall.
Ayres, J. M. 1985. On a new species of squirrel monkey, genus *Saimiri*, from Brazilian Amazonia (Primates, Cebidae). Papéis Avulsos de Zoologia, Sao Paulo 36:147–164.
Ayres, J. M. 1986. *The white uakaris and the Amazonian flooded forests.* Ph.D. dissertation, Cambridge University.
Ayres, J. M. 1989. Comparative feeding ecology of the uakari and bearded saki, *Cacajao* and *Chiropotes*. Journal of Human Evolution 18:697–716.
Bailey, W. J., Slighton, J. L., and Goodman, M. 1992. Rejection of the "flying primate" hypothesis by phylogenetic evidence from the ε-globin gene. Science 256:86–89.
Boinski, S. 1987. Habitat use by squirrel monkeys *(Saimiri oerstedi)* in Costa Rica. Folia Primatologica 49:151–167.
Boinski, S. 1988. Use of a club by a white-faced capuchin *(Cebus capucinus)* to attack a venomous snake *(Bothrops asper)*. American Journal of Primatology 14:177–179.
Boinski, S. 1989a. The ontogeny of foraging in squirrel monkeys, *Saimiri oerstedi*. Animal Behaviour 37:415–428.
Boinski, S. 1989b. The positional behavior and substrate use of squirrel monkeys: ecological implications. Journal of Human Evolution 18:659–677.
Brown, A. D., and Zunino, G. E. 1990. Dietary variability in *Cebus apella* in extreme habitats: evidence for adaptability. Folia Primatologica 54:187–195.
Buchanan-Smith, H. 1991. Field observations of Goeldi's monkey, *Callimico goeldii*, in northern Bolivia. Folia Primatologia 57:102–105.
Buffon, G. L. L. 1767. Histoire naturelle générale et particulière avec description du cabinet du roi [with supplement by M. Daubenton]. 15:327.
Bull, J. J., Huelsenbeck, J. P., Cunningham, C. W., Swofford, D. L., and Waddell, P. J. 1993. Partitioning and combining data in phylogenetic analysis. Systematic Biology 42:384–397.
Cant, J. G. H. 1986. Locomotion and feeding postures of spider and howler monkeys: field study and evolutionary interpretation. Folia Primatologia 46:1–14.
Cartmill, M. 1974. Pads and claws in locomotion. Pp. 45–83 in *Primate locomotion,* F. A. Jenkins, ed. New York, NY: Academic Press.
Cartmill, M., MacPhee, R. D. E., and Simons, E. L. 1981. Anatomy of the temporal bone in early anthropoids, with remarks on the problem of anthropoid origins. American Journal of Physical Anthropology 56:3–21.
Castro, R., and Soini, P. 1977. Field studies on *Saguinus mystax* and other callitrichids in Amazonian Peru. Pp. 73–78 in *The biology and conservation of the Callitrichidae,* D. G. Kleiman, ed. Washington, DC: Smithsonian Institution Press.
Chapman, C. 1987. Flexibility in diets of three species of Costa Rican primates. Folia Primatologia 49:90–105.
Chapman, C. 1988. Patterns of foraging and range use by three species of Neotropical primates. Primates 29:177–194.
Charles-Dominique, P., Atramentowicz, M., Charles-Dominique, M., Gérard, H., Hladik, A., Hladik, C. M., and Prévost, M. F. 1981. Les mammifères frugivores arboricoles nocturnes d'une forêt guyanaise: interrelations plantes-animaux. Revue d`Ecologie la Terre et la Vie 35:341–435.
Chevalier-Skolnikoff, S. 1989a. Spontaneous tool use and sensorimotor intelligence in *Cebus* compared with other monkeys and apes. Behavioural Brain Science 12:561–588.
Chevalier-Skolnikoff, S. 1989b. Tool use by wild *Cebus* monkeys at Santa Rosa National Park, Costa Rica. Primates 31:375–383.
Chippindale, P. T., and Wiens, J. J. 1994. Weighting, partitioning, and combining characters in phylogenetic analysis. Systematic Biology 43:278–287.
Christen, A., and Geissmann, T. 1994. A primate survey in northern Bolivia, with special reference to Goeldi's monkey, *Callimico goeldii*. International Journal of Primatology 15:239–273.
Clutton-Brock, T. H., and Harvey, P. H. 1983. The functional significance of variation in body size among mammals. Pp. 633–663 in *Advances in the study of mammalian behavior,* J. F. Eisenberg and D. G.. Kleiman, eds. Spec. Publ. #7 of the American Society of Mammalogists.
Coimbra-Filho, A. F. 1970a. Acerca da redescoberta de *Leontideus chysopygus* (Mikan, 1823) e apontamentos sobre sua ecologia (Callitrichidae, Primates). Revista Brasileira de Biologia 30:609–615.

Coimbra-Filho, A. F. 1970b. Considerações gerais e situação atual dos mocosleões escuros, *Leontideus chrysomelas* (Kuhl 1820) e *Leontideus chrysopygus* (Mikan 1823) (Callitrichidae, Primates). Rev. Brasil. Biol. 30:249–268.

Coimbra-Filho, A. F. 1976. *Leontopithecus rosalia chrysopygus* (Mikan 1823), o mico-leao do Estado de Sao Paulo (Callitrichidae, Primates). Silvic. S. Paulo 10:1–36.

Coimbra-Filho, A. F., and Mittermeier, R. A. 1973. Distribution and ecology of the genus *Leontopithecus* Lesson, 1840 in Brazil. Primates 14:47–66.

Costello, M. B., Dickinson, C., Rosenberger, A. L., Boinski, S., and Szalay, F. S. 1993. Squirrel monkey (genus *Saimiri*) taxonomy: a multidisciplinary study of the biology of species. Pp. 177–210 in *Species, species concepts, and primate evolution*, W. H. Kimbel and L. B. Martin, eds. New York, NY: Plenum.

Costello, M. B., and Fragaszy, D. M. 1988. Prehension in *Cebus* and *Saimiri*. American Journal of Primatology 15:235–245.

Crompton, A. W., Taylor, C. R., and Jagger, J. J. 1978. Evolution of homeothermy in mammals. Nature 272:333–336.

Czelusniak, J., Goodman, M., Koop, B. F., Tagle, D. A., Shoshani, J., Braunitzer, G., Kleinschmidt, T. K., de Jong, W. W., and Matsuda, G. 1990. Perspectives from amino acid and nucleotide sequences on cladistic relationships among higher taxa of Eutheria. Pp. 545–572 in *Current mammalogy*, H. H. Genoways, ed. New York, NY: Plenum.

Dawkins, H. C. 1959. The volume increment of natural tropical forest highforest and limitations on its improvements. Empire Forestry Review 38:175–180.

Defler, T. R. 1995. The time budget of a group of wild woolly monkeys *(Lagothrix lagothricha)*. International Journal of Primatology 16:107–120.

Donoghue, M. J., and Sanderson, M. J. 1992. The suitability of molecular and morphological evidence in reconstructing plant phylogeny. Pp. 340–367 in *Molecular systematics of plants*, P. S. Soltis, D. E. Soltis, and J. J. Doyle, eds. New York, NY: Chapman and Hall.

Eisenberg, J. F. 1981. *The mammalian radiations: an analysis of trends in evolution, adaptation, and behavior.* Cambridge, England: Cambridge University Press:

Eisenberg, J. F. 1990. The behavioral/ecological significance of body-size in the mammalia. Pp. 25–37 in *Body size in mammalian paleobiology: estimation and biological implications*, J. Damuth and B. J. MacFadden, eds. Cambridge, England: Cambridge University Press:

Emmons, L. H., and Gentry, A. H. 1983. Tropical forest structure and the dispersion of gliding and prehensile-tailed vertebrates. American Naturalist 121:513–524.

Erikson, G. E. 1963. Brachiation in New World monkeys and in anthropoid apes. Symposia of the Zoological Society of London 10:135–163.

Fang, T. G. 1990. La importancia de los frutos en la dieta de *Saguinus mystax* y *S. fuscicollis* (Primates, Callitrichidae), en el Río Tahuayo, Departamento de Loreto. Pp. 342–358 in *La Primatología en el Perú: investigaciones primatológicas (1973–1985)*, Proyecto Peruano de Primatología "Manuel Moro Sommo", Lima, Peru.

Fernandes, M. E. B. 1991. Tool use and predation of oysters (*Crassostrea rhizophorae*) by the tufted capuchin, cebus apella, in brackish water mangrove swamp. Primates 32:529–531.

Ferrari, S. F. 1993. Ecological differentiation in the Callitrichidae. Pp. 314–328 in *Marmosets and tamarins: systematics, behaviour, and ecology*, A. B. Rylands, ed. Oxford, England: Oxford University Press.

Ferrari, S. F., and Lopes Ferrari, M. A. 1989. A re-evaluation of the social organisation of the Callitrichidae, with special reference to the ecological differences between genera. Folia Primatolologia 52:132–147.

Fleagle, J. G. 1988. *Primate adaptation and evolution.* New York, NY: Academic Press.

Fleagle, J. G., and Kay, R. F. 1987. The phyletic position of the Parapithecidae. Journal of Human Evolution 16:483–531.

Fleagle, J. G., and Meldrum, D. J. 1988. Locomotor behavior and skeletal morphology of two sympatric pitheciine monkeys, *Pithecia pithecia* and *Chiropotes satanas*. American Journal of Primatology 16:227–249.

Fleagle, J. G., and Mittermeier, R. A. 1980. Locomotor behavior, body size and comparative ecology of seven Surinam monkeys. American Journal of Physical Anthropology 22:301–314.

Fontaine, R. 1981. The uakaris, genus *Cacajao*. Pp. 443–493 in *Ecology and behavior of neotropical primates, Vol. 1*, A. F. Coimbra-Filho and R. A. Mittermeier, eds. Rio de Janeiro, Brazil: Brazilian Academy of Sciences.

Fooden, J. 1963. A revision of the woolly monkeys (genus *Lagothrix*). Journal of Mammalogy 44:213–217.
Ford, S. M. 1986a. Comment on the evolution of claw-like nails in callitrichids (marmosets/tamarins). American Journal of Physical Anthropology 71:1–11.
Ford, S. M. 1986b. Systematics of the New World monkeys. Pp. 73–135, In *Comparative primate biology, Vol. 1*, D. R. Swindler and J. Erwin, eds. New York, NY: Alan R. Liss.
Ford, S. M. 1994. Taxonomy and distribution of the owl monkey. Pp. 1–56 in *Aotus: the owl monkey*, J. F. Baer, R. E. Weller, and I. Kakoma, eds. San Diego, CA: Academic Press.
Ford, S. M., and Davis, L. C. 1992. Systematics and body size: implications for feeding adaptations in New World monkeys. American Journal of Physical Anthropology 88:415–468.
Fragaszy, D. M., Visalberghi, E., and Robinson, J. G. 1990. Variability and adaptability in the genus *Cebus*. Folia Primatologia 54:114–118.
Funk, D. J., Futuyma, D. J., Ortí, G., and Meyer, A. 1995. Mitochondrial DNA sequences and multiple data sets: a phylogenetic study of phytophagous beetles (Chrysomelidae: *Ophraella*). Molecular Biology and Evolution 12:627–640.
Garber, P. A. 1980. Locomotor behavior and feeding ecology of the Panamanian tamarin (*Saguinus oedipus geoffroyi*, Callitrichidae, Primates). International Journal of Primatology 1:185–201.
Garber, P. A. 1984. Use of habitat and positional behavior in a neotropical primate, *Saguinus oedipus*. Pp. 112–133 in *Adaptations for foraging in non-human primates*, P. S. Rodman and J. G. H. Cant, eds. New York, NY: Columbia University Press.
Garber, P. A. 1991. A comparative study of positional behavior in three species of tamarin monkeys. Primates 32:219–230.
Garber, P. A. 1992. Vertical clinging, small body size, and the evolution of feeding adaptations in the callitrichinae. American Journal of Physical Anthropology 88:469–482.
Garber, P. A. 1993. Feeding ecology and behaviour of the genus *Saguinus*. Pp. 273–295, In *Marmosets and tamarins: systematics, behavior, and ecology*, A. B. Rylands, ed. New York, NY: Oxford University Press.
Garber, P. A., and Pruetz, J. D. 1995. Positional behavior in moustached tamarin monkeys: effects of habitat on locomotor variability and locomotor stability. Journal of Human Evolution 28:411–426.
Gaulin, S. J. C. 1979. A Jarman/Bell model of primate feeding niches. Human Ecology 7:1–20.
Gebo, D. L. 1992. Locomotor and postural behavior in *Alouatta palliata* and *Cebus capucinus*. American Journal of Physical Anthropology 26:277–290.
Gibson, K. R. 1986. Cognition, brain size and the extraction of embedded food resources. Pp. 93–103 in *Primate ontogeny, cognition and social behavior*, J. G. Else and P. C. Lee, eds. Cambridge, England: Cambridge University Press.
Grand, T. I. 1978. Adaptations of tissue and limb segments to facilitate moving and feeding in arboreal folivores. Pp. 231–241 in *The ecology of arboreal folivores*, G. Montgomery, ed. Washington, DC: Smithsonian Institution Press.
Grand, T. I. 1984. Motion economy within the canopy: four strategies for mobility. Pp. 54–72 in *Adaptations for foraging in non-human primates*, P. S. Rodman and J. G. H. Cant, eds. New York, NY: Columbia University Press.
Groves, C. P., and Ramírez-Pulido, J. 1982. Family Cebidae. In *Mammal species of the world*, J. H. Honacki, K. E. Kinman, and J. W. Koeppl, eds. Lawrence, KS: Allen Press.
Harada, M., Schneider, H., Schneider, M., Sampaio, I., Czelusniak, J., and Goodman, M. 1995. DNA evidence on the phylogenetic systematics of New World monkeys: support for the sister-grouping of *Cebus* and *Saimiri* from two unlinked nuclear genes. Molecular Phylogenetics and Evolution 4:331–349.
Hennig, W. 1966. *Phylogenetic systematics*. Urbana, IL: University of Illinois Press.
Hernández-Camacho, J., and Cooper, R. W. 1976. The nonhuman primates of Colombia. Pp. 35–69 in *Neotropical primates, field studies and conservation*, R. W. Thorington,. Jr. and P. G. Heltne, eds. Washington, DC: National Academy of Sciences.
Hershkovitz, P. 1977. *Living New World monkeys (Platyrrhini) with an introduction to primates*. Chicago, IL: Chicago University Press.
Hershkovitz, P. 1983. Two new species of night monkeys, genus *Aotus* (Cebidae, Platyrrhini): a preliminary report on *Aotus* taxonomy. American Journal of Primatology 4:209–243.
Hershkovitz, P. 1984. Taxonomy of squirrel monkeys, genus *Saimiri* (Cebidae, Platyrrhini): a preliminary report with the description of a hitherto unnamed form. American Journal of Primatology 7:155–210.
Heymann, E. W. 1990. Interspecific relations in a mixed-species troop of moustached tamarins, *Saguinus mystax*, and saddle-back tamarins, *Saguinus fuscicollis* (Platyrrhini: Callitrichidae), at the Río Blanco, Peruvian Amazon. American Journal of Primatology 21:115–127.

Hill, J. P. 1926. Demonstration of the embryologia varia (Development of *Hapale jacchus*). Journal of Anatomy 60:486–487.
Hoffstetter, R. 1972. Relationships, origins, and history of the ceboid monkeys and caviomorph rodents: a modern reinterpretation. Evolutionary Biology 6:323–347.
Hoffstetter, R. 1980. Origin and development of New World monkeys emphasizing the southern continents route. Pp. 103–122 in *Evolutionary biology of New World monkeys and continental drift*, R. Ciochon and A. B. Chiarelli, eds. New York, NY: Plenum Press.
Horovitz, I. 1995. A phylogenetic analysis of the basicranial morphology of New World monkeys. American Journal of Physical Anthropology 20(Suppl.):113.
Horovitz, I., and Meyer, A. 1995. Systematics of the New World monkeys (Platyrrhini, Primates) based on 16S mitochondrial DNA sequences: a comparative analysis of different weighting methods in cladistic analysis. Molecular Phylogenetics and Evolution 4:448–456.
Huelsenbeck, J. P., Swofford, D. L., Cunningham, C. W., Bull, J. J., and Waddell, P. J. 1994. Is character weighting a panacea for the problem of data heterogeneity in phylogenetic analysis? Systematic Biology 43:288–291.
Izawa, K. 1975. Foods and feeding behavior of monkeys in the upper Amazon basin. Primates 16:295–316.
Izawa, K. 1978. Frog-eating behavior of wild black-capped capuchin. Primates 19:633–642.
Izawa, K. 1979a. Foods and feeding behavior of wild black-capped capuchin. Primates 20:57–76.
Izawa, K. 1979b. *Studies on peculiar distribution pattern of* Callimico. Kyoto, Japan: Kyoto University Overseas Research 1:1–19.
Izawa, K., and Mizano, A. 1977. Palm fruit cracking behavior of wild black-capped capuchin (*Cebus apella*). Primates 18:773–792.
Janson, C. H. 1985. Aggressive competition and individual food intake in wild brown capuchin monkeys. Behavioral Ecology and Sociobiology 18:125–138.
Janson, C. H. 1990. Ecological consequences of individual spatial choice in foraging groups of brown capuchin monkeys *Cebus apella*. Animal Behaviour 40:922–934.
Janson, C. H., and Boinski, S. 1992. Morphological and behavioral adaptations for foraging in generalist primates: the case of the cebines. American Journal of Physical Anthropology 88:483–498.
Kay, R. F. 1980. Platyrrhine origins: a reappraisal of the dental evidence. Pp. 159–188 in *New World monkeys and continental drift*, R. L. Ciochon and A. B. Chiarelli, eds. New York, NY: Plenum Press.
Kay, R. F. 1981. The nut-crackers: a new theory of the adaptations of the Ramapithecinae. American Journal of Physical Anthropology 55:141–151.
Kay, R. F. 1984. On the use of anatomical features to infer foraging behavior in extinct primates. Pp. 21–53 in *Adaptations for foraging in nonhuman primates*, P. S. Rodman and J. G. H. Cant, eds. New York, NY: Columbia University Press.
Kay, R. F. 1990. The phyletic relationships of extant and fossil Pitheciinae. Journal of Human Evolution 19:175–208.
Kay, R. F., Fleagle, J. G., and Simons, E. L. 1981. A revision of the Oligocene apes from the Fayum Province, Egypt. American Journal of Physical Anthropology 55:293–322.
Kay, R. F., Ross, C., and Williams, B. A. 1977. Anthropoid origins. Science 275:797–804.
Kay, R. F., and Williams, B. A. 1994. Dental evidence for anthropoid origins. Pp. 361–445 in *Anthropoid origins*, J. G. Fleagle and R. F. Kay, eds. New York, NY: Plenum Press.
Kinzey, W. G. 1973. Reduction of the cingulum in the Ceboidea. Pp 101–127 in *Symposium of the Fourth International Congress of Primatology, Vol. 3*, W. Montagna, ed. Basel, Switzerland: Karger.
Kinzey, W. G. 1977a. Diet and feeding behavior in *Callicebus torquatus*. Pp. 127–151 in *Primate ecology*, T. H. Clutton-Brock, ed. New York, NY: Academic Press.
Kinzey, W. G. 1977b. Positional behavior and ecology in *Callicebus torquatus*. Yearbook of Physical Anthropology 20:468–480.
Kinzey, W. G. 1978. Feeding behaviour and molar features in two species of titi monkey. Pp. 373–385 in *Recent advances in primatology, Vol. 1*, D. J. Chivers and J. Herbert, eds. London, England: Academic Press.
Kinzey, W. G. 1981. The titi monkey, genus *Callicebus*. Pp. 241–276 in *Ecology and behavior of neotropical primates*, A. F. Coimbra-Filho and R. A. Mittermeier, eds. Rio de Janeiro, Brazil: Brazilian Academy of Sciences.
Kinzey, W. G. 1992. Dietary and dental adaptations in the Pitheciinae. American Journal of Physical Anthropology 88:499–514.
Kinzey, W. G., and Becker, M. 1983. Activity pattern of the masked titi monkey, *Callicebus personatus*. Primates 24:337–343.

Kinzey, W. G., and Gentry, A. H. 1979. Habitat utilization in two species of *Callicebus*. Pp. 89–100 in *Primate ecology: problem-oriented field studies*, R. W. Sussman, ed. New York, NY: Wiley.

Kinzey, W. G., and Norconk, M. A. 1990. Hardness as a basis of fruit choice in two sympatric primates. American Journal of Physical Anthropology 81:5–15.

Kinzey, W. G., and Norconk, M. A. 1993. Physical and chemical properties of fruits and seed eaten by *Pithecia* and *Chiropotes* in Surinam and Venezuela. International Journal of Primatology 20:204–205.

Kinzey, W. G., Rosenberger, A. L., and Ramírez, M. 1975. Vertical clinging and leaping in a neotropical anthropoid. Nature 255:327–328.

Kinzey, W. G., Rosenberger, P. S., Heisler, D. L., Prowse, D. L., and Trilling, J. S. 1977. A preliminary field investigation of the yellow-handed titi monkey, *Callicebus torquatus torquatus*, in northern Perú. Primates 18:159–181.

Kleiber, M. 1947. Body size and metabolic rate. Physiological Review 27:511–541.

Kleiber, M. 1961. *The fire of life*. New York, NY: Wiley.

Klein, L. L., and Klein, D. J. 1976. Neotropical primates: aspects of habitat usage, population density, and regional distribution in La Macarena, Colombia. Pp. 70–78 in *Neotropical primates, field studies and conservation*, R. W. Thorington, Jr. and P. G. Heltne, eds. Washington, DC: National Academy of Sciences.

Kluge, A. G. 1989. A concern for evidence and a phylogenetic hypothesis of relationships among *Epicrates* (Boidae, Serpentes). Systematic Zoology 38:7–25.

Kluge, A. G., and Wolf, A. J. 1993. Cladistics: what's in a word? Cladistics 9:183–199.

Leigh, E. G. 1975. Structure and climate in tropical rain forest. Annual Review of Ecology and Systematics 6:67–86.

LeMaho, G., Rochas, M., Felbabel, H., and Chatonnet, J. 1981. Thermoregulation in the nocturnal simian: the night monkey *Aotus trivirgatus*. Journal of Physiology 240:R156–R165.

Lewis, O. J. 1974. The wrist articulation of the Anthropoidea. Pp 143–169 in *Primate locomotion*, F. A. Jenkins, ed. New York, NY: Academic Press.

Lewis, O. J. 1977. Joint remodelling and the evolution of the human hand. Journal of Anatomy 123:157–201.

Lockhart, P. D., Penny, D., and Meyer, A. 1995. Testing the phylogeny of swordtail fishes using split decomposition of spectral analysis. Journal of Molecular Evolution 41:666–674.

MacNab, B. 1983. Ecological and behavioral consequences of adaptation to various food sources. Pp. 664–697 in *Advances in the study of mammalian behavior*, J. Eisenberg and D. Kleiman, eds. Special Publication No. 7 of the American Society of Mammalogists.

MacPhee, R. D. E., Horovitz, I., Arredondo, O., and Jiménez Vázquez, O. 1995. A new genus of the extinct Hispaniolan monkey *Saimiri bernensis* Rímoli, 1977, with notes on its systematic position. American Museum Novitates 3134:1–21.

Martin, R. D. 1990. *Primate origins and evolution: a phylogenetic reconstruction*. Princeton, NJ: Princeton University Press.

Martin, R. D. 1992. Goeldi and the dwarfs: the evolutionary biology of the small New World monkeys. Journal of Human Evolution 22:367–393.

Mayr, E. 1963. *Animal species and evolution*. Cambridge, MA: Harvard University Press.

Meireles, C. M. M., Sampaio, I., Schneider, H., Chiu, C., Slightom, J. L., and Goodman, M. 1995. Use of g-globin loci nucleotide sequences to investigate ateline phylogeny. American Journal of Physical Anthropology 20(Suppl.):151.

Mendel, F. 1976. Postural and locomotor behavior of *Alouatta palliata* on various substrates. Folia Primatologia 26:36–53.

Meyer, A., Morrisey, J. M., and Schartl, M. 1994. Recurrent origin of a sexually selected trait in *Xiphophorus* fishes inferred from a molecular phylogeny. Nature 368:539–542.

Mickevich, M. F., and Farris, M. F. 1981. The implications of congruence in *Menidia*. Systematic Zoology 30:351–370.

Milton, K. 1980. *The foraging strategy of the howler monkey: a study in primate economics*. New York, NY: Columbia University Press.

Mittermeier, R. A. 1996. Introduction. In *The pictorial guide to the living primates*, N. Rowe, ed. East Hampton, NY: Pogonias Press.

Mittermeier, R. A., Rylands, A. B., and Coimbra-Filho, A. F. 1988. Systematics: species and subspecies – an update. Pp. 13–75 in *Ecology and behavior of neotropical primates*, Vol. 2, R. A. Mittermeier, F. Coimbra-Filho, and G. A. B. da Fonseca, ed. Washington DC: World Wildlife Fund.

Mittermeier, R. A., and van Roosmalen, M. G. M. 1981. Preliminary observations on habitat utilization and diet of eight Surinam monkeys. Folia Primatologia 36:1–39.
Miyamoto, M. M., and Goodman, M. 1990. DNA systematics and evolution of primates. Annual Review of Ecology and Systematics 21:197–220.
Moynihan, M. 1976a. *The New World primates, adaptive radiation and the evolution of social behavior, languages, and intelligence.* Princeton, NJ: Princeton University Press.
Moynihan, M. 1976b. Notes on the ecology and behaviour of the pygmy marmoset (*Cebuella pygmaea*) in the Amazonian Colombia. Pp. 79–84 in *Neotropical primates, field studies and conservation*, R. W. Thorington, Jr. and P. G. Heltne, eds. Washington, DC: National Academy of Sciences.
Muñoz Durán, J. V. 1991. *Algunos aspectos de la dispersión, estructura social y uso del espacio habitado, en un grupo de* Lagothrix lagothricha *(Humboldt, 1812) –Primates Cebidae –en la Amazonia Colombiana.* Tesis de Licenciatura, Departamento de Biología, Universidad Nacional de Colombia.
Napier, P. H. 1976. *Catalogue of primates in the British Museum. Part 1: families Callitrichidae and Cebidae.* London, England: British Museum (Natural History).
Noback, C. R. 1975. The visual system of primates in phylogenetic studies. Pp. 199–218 in *Phylogeny of the primates*, P. Lickett and F. S. Szalay, eds. New York, NY: Plenum Press.
Norconk, M. A., and Kinzey, W. G. 1994. Challenge of neotropical frugivory: travel patterns of spider monkeys and bearded sakis. American Journal of Primatology 34:171–183.
Novacek, M. J. 1992. Fossils as critical data for phylogeny. Pp. 46–88 in *Extinction and phylogeny*, M. J. Novacek and Q. D. Wheeler, eds. New York, NY: Columbia University Press.
Novacek, M. J. 1994. Morphological and molecular inroads to phylogeny. Pp. 85–131 in *Interpreting the hierarchy of nature: from systematic patterns to evolutionary process*, E. Grande and O. Rieppel, eds. San Diego, CA: Academic Press.
Ogden, T. E. 1975. The receptor mosaic of *Aotus trivirgatus*: distribution of rods and cones. Journal of Comparative Neurol. 163:193–202.
Parker, S. T., and Gibson, K. R. 1977. Object manipulating, tool use and sensorimotor intelligence as feeding adaptations in *Cebus* monkeys and great apes. Journal of Human Evolution 6:623–641.
Peres, C. A. 1989. Exudate-feeding by wild golden lion tamarins, *Leontopithecus rosalia.* Biotropica 21:287–288.
Peres, C. A. 1990. Effects of hunting on western Amazonian primate communities. Biological Conservation 54:47–59.
Peres, C. A. 1991. Seed predation of *Cariniana micrantha* (Lecythidaceae) by brown capuchin monkeys in Central Amazonia. Biotropica 23:262–270.
Peres, C. A. 1993. Notes on the ecology of buffy saki monkeys (*Pithecia albicans*, Gray 1860): a canopy seed-predator. American Journal of Primatology 31:129–140.
Peres, C. A. 1994. Diet and feeding ecology of gray woolly monkeys (*Lagothrix lagothricha cana*) in Central Amazonia: comparisons with other atelines. International Journal of Primatology 15:333–372.
Pocock, R. I. 1925. Additional notes on the external characters of some platyrrhine monkeys. Proceedings of the Zoological Society of London 1925:27–42.
Pook, A., and Pook, G. 1981. A field study of the socio-ecology of the goeldi's monkey (*Callimico goeldii*) in northern Bolivia. Folia Primatologia 35:288–312.
Putz, F. E. 1980. Lianas vs. trees. Biotropica 12:224–225.
Richards, P. W. 1952. *The tropical rain forest.* Cambridge, England: Cambridge University Press.
Rosenberger, A. L. 1979. *Phylogeny, evolution and classification of New World monkeys (Platyrrhini, Primates).* Ph. D. dissertation, City University of New York.
Rosenberger, A. L. 1981. Systematics: the higher taxa. Pp. 111–168 in *Ecology and behavior of neotropical primates, Vol. 1*, A. F. Coimbra-Filho and R. A. Mittermeier, eds. Rio de Janeiro, Brazil: Brazilian Academy of Sciences.
Rosenberger, A. L. 1984. Fossil New World monkeys dispute the molecular clock. Journal of Molecular Evolution 13:737–742.
Rosenberger, A. L. 1992. Evolution of feeding niches in New World monkeys. American Journal of Physical Anthropology 88:525–562.
Rosenberger, A. L., and Kinzey, W. G. 1976. Functional patterns of molar occlusion in platyrrhine primates. American Journal of Physical Anthropology 45:281–298.
Rosenberger, A. L., and Strier, K. B. 1989. Adaptive radiation of the ateline primates. Journal of Human Evolution 18:717–750.
Rylands, A. B. 1986. Ranging behaviour and habitat preference of a wild marmoset group, *Callithrix humeralifer* (Callitrichidae, Primates). Journal of the Zoological Society of London 210:489–514.

Rylands, A. B. 1989. Sympatric Brazilian callitrichids: the black tufted-ear marmoset, *Callithrix kuhli*, and the golden-headed lion tamarin, *Leontopithecus chrysomelas*. Journal of Human Evolution 18:679–695.

Rylands, A. B. 1993. The ecology of the lion tamarins, *Leontopithecus*: some intergeneric differences and comparisons with other callitrichids. Pp. 296–313 in *Marmosets and tamarins: systematics, behaviour, and ecology*, A. B. Rylands, ed. New York, NY: Oxford University Press.

Rylands, A. B., Coimbra-Filho, A. F., and Mittermeier, R. A. 1993. Systematics, geographic distribution, and some notes on the conservation status of the Callitrichidae. Pp. 11–77 in *Marmosets and tamarins: systematics, behaviour, and ecology*, A. B. Rylands, ed. New York, NY: Oxford University Press.

Rylands, A. B., and de Faria, D. S. 1993. Habitats, feeding ecology, and home range size in the genus *Callithrix*. Pp. 262–272 in *Marmosets and tamarins: systematics, behaviour, and ecology*, A. B. Rylands, ed. New York, NY: Oxford University Press.

Sarich, V. M., and Cronin, J. E. 1980. South American mammal molecular systems, evolutionary clocks, and continental drift. Pp. 399–421 in *Evolutionary biology of the New World monkeys and continental drift*, R. L. Ciochon and A. B. Chiarelli, eds. New York, NY: Plenum Press.

Schmidt-Nielsen, K. 1984. *Scaling: why is animal size so important?* Cambridge, England: Cambridge University Press.

Schneider, H., Schneider, M. P. C., Sampaio, I., Harada, M. L., Stanhope, M., Czelusniak, J., and Goodman, M. 1993. Molecular phylogeny of the New World monkeys (Platyrrhini, Primates). Molecular Phylogenetics and Evolution 2:225–242.

Schneider, H., Schneider, M. P. C., Sampaio, I., Harada, M. L., Barroso, C. M. L., Czelusniak, J., and Goodman, M. 1995. DNA evidence on platyrrhine phylogeny from two unlinked nuclear genes. American Journal of Physical Anthropology 20 (Suppl.):191.

Schultz, A. H. 1930. The skeleton of the trunk and limbs of higher primates. Human Biology 2:303–438.

Schultz, A. H. 1961. Vertebral column and thorax. Primatologia 4:1–66.

Setoguchi, T., Takai, M., Villarroel, C., Shigehara, N., and Rosenberger, A. l. 1988. New specimen of *Cebupithecia* from La Venta, Miocene of Colombia, South America. Kyoto Overseas Reports on New World Monkeys 6:7–9.

Seuanez, H. N., Forman, L., Matayoshi, T., and Fanning, T. G. 1989. The *Callimico goeldii* (Primates, Platyrrhini) genome: karyology and middle repetitive (LINE1) DNA sequences. Chromosoma 98:389–395.

Shoshani, J., Groves, C. P., Simons, E. L., and Gunnell, G. F. 1996. Primate phylogeny: morphological vs. molecular results. Molecular Phylogenetics and Evolution 5:102–154.

Simons, E. L. 1962. Two new primate species from the African Oligocene. Postilla 166:1–12.

Simons, E. L. 1965. New fossil apes from Egypt and the initial differentiation of Hominoidea. Nature 205:135–139.

Simons, E. L. 1987. New faces of *Aegyptopithecus* from the Oligocene of Egypt. Journal of Human Evolution 16:273–289.

Soini, P. 1972. The capture and commerce of live monkeys in the Amazonian region of Perú. International Zoo Yearbook 12:26–35.

Soini, P. 1987. Ecology of the saddle-back tamarin *Saguinus fuscicollis illigeri* on the Río Pacaya, northeastern Perú. Folia Primatologia 49:11–32.

Soini, P. 1990. Ecologia y dinamica poblacional del "Choro" (*Lagothrix lagothricha*, Primates) en Rio Pacaya, Peru. Pp. 382–396 in *La Primatología en el Perú*, Lima: Proyecto Peruano de Primatología.

Soini, P. 1993. The ecology of the pygmy marmoset, *Cebuella pygmaea*: some comparisons with two sympatric tamarins. Pp. 257–261 in *Marmosets and tamarins: systematics, behaviour, and ecology*, A. B. Rylands, ed. New York, NY: Oxford University Press.

Stevenson, P. R., Quiñones, M. J., and Ahumada, J. A. 1991. *Relación entre la abundancia de frutos y las estrategias alimenticias de cuatro especies de primates en La Macarena*. Informe final presentado al Fondo para la Promoción de la Investigación y la Tecnología del Banco de la República. Santa Fé de Bogotá, Colombia.

Stevenson, P. R., Quiñones, M. J., and Ahumada, J. A. 1994. Ecological strategies of woolly monkeys (*Lagothrix lagothricha*) at Tinigua National park, Colombia. American Journal of Primatology 32:123–140.

Stirton, R. A. 1951. Ceboid monkeys from the Miocene of Colombia. University of California Publications, Bulletin of the Department of Geological Sciences 28:315–356.

Stirton, R. A., and Savage, D. E. 1951. *A new monkey from the La Venta late Miocene of Colombia*. Bogotá, Colombia: Servicio Geológico Nacional.

Strier, K. B. 1992. Atelinae adaptations: behavioral strategies and ecological constraints. American Journal of Physical Anthropology 88:515–524.

Sussman, R. W., and Kinzey, W. G. 1984. The ecological role of the Callitrichidae. American Journal of Physical Anthropology 64:419–449.

Swofford, D. L. 1993. *PAUP: phylogenetic analysis using parsimony, vers. 3.1.1.* Champaign, IL: Illinois Natural History Survey.

Szalay, F. S., and Delson, E. 1979. *Evolutionary history of the primates.* New York, NY: Academic Press.

Terborgh, J. 1983. *Five New World monkeys: a study in comparative ecology.* Princeton, NJ: Princeton University Press.

Thorington, R. W., Jr. 1968. Observations of the tamarin, *Saguinus midas*. Folia Primatologia 9:85–98.

Thorington, R. W., Jr. 1985. The taxonomy and distribution of the squirrel monkeys. Pp. 1–33 in *Handbook of squirrel monkey research*, L. A. Rosenblum and C. L. Coe, eds. New York, NY: Plenum Press.

Tomblin, D. C., and Cranford, J. A. 1994. Ecological niche differences between *Alouatta palliata* and *Cebus capucinus* comparing feeding modes, branch use, and diet. Primates 35:265–274.

van Lawick-Goodall, J. 1970. Tool-using in primates and other vertebrates. Pp. 195–249 in *Advances in the study of behavior, Vol. 3*, D. Lehrman, R. A. Hinde, and E. Shaw, eds. New York, NY: Academic Press.

van Roosmalen, M. G. M., Mittermeier, R. A., and Fleagle, J. G. 1988. Diet of the northern bearded saki (*Chiropotes satanas chiropotes*): A neotropical seed predator. American Journal of Primatology 14:11–35.

Visalberghi, E. 1990. Tool use in *Cebus*. Folia Primatologia 54:146–154.

Wheeler, W. C., and Gladstein, D. 1993. *Malign, vers. 1.95.* (Computer program distributed by authors.)

Wright, P. C. 1989. The nocturnal niche in the New World. Journal of Human Evolution 18:635–658.

Wright, P. C. 1994. Behavior and ecology of the owl monkey. Pp. 97–112 in *Aotus: the owl monkey*, R. E. Weller and I. Kakoma, eds. San Diego, CA: Academic Press.

Yoneda, M. 1981. Ecological studies of *Saguinus fuscicollis* and *S. labiatus* with reference to habitat segregation and height preference. Kyoto University Overseas Research Reports 2:43–50.

Yoneda, M. 1984. Comparative studies on vertical separation, foraging behavior and traveling mode of saddle-backed tamarins (*Saguinus fuscicollis*) and red-chested moustached tamarins (*Saguinus labiatus*) in northern Bolivia. Primates 25:414–442.

Zingeser, M. R. 1973. Dentition of *Brachyteles arachnoides* with reference to Alouattine and Atelinine affinities. Folia Primatologia 20:351–390.

Appendix 6.1. Morphological and molecular characters used to analyze the phylogeny of New World monkeys (see text).

Morphological characters

Taxa	Character number
| | ```
          1111111111222222222233333333334444444444555555555566666666
 123456789012345678901234567890123456789012345678901234567890123456
``` |
Tarsius	`000000020000020001101071000101001010000001000030007100120003100100`
Leontopithecus	`001000021111111002012000001011110111000100070000110001210001101010`
Saguinus	`00100002101111100201200000101111011100011000?00001000120000110102 1`
Callimico	`00100000101011100211200000101111021110110000010000100110101110010 1`
Callithrix	`0010000211111110020011101001001010100001000?00001000120000110101`
Cebuella	`0010000211111110020011101001001010100010000000001001122000110101`
Aotus	`00000002101011110201200000100011121110000000020000000000101100010 11 2 1 1`
Cebus	`00000002101011010201200000010111321100001001001000010011111001110`
Cacajao	`0000000010011110102022010101100101111100001102111000101111120001 1 2`
Pithecia	`00000011011111010202201010110010111110000110111100000101113000101`
Chiropotes	`0000000010011101020220101011001012111100001102111000101111120001 1 1 2`
Saimiri	`00000002001011001020120000010011102111000100001000010101100111011 1 1 1 1`
Alouatta	`10000011001110010011100000101010110110110000031001000201011300011 2 1`
Lagothrix	`11000000100111001011120000010001012101000000003100100000110130001 1 1 1 1`
Brachyteles	`110000011001110010117?00000010100?0?110101000000370010002011?12?0001 1`
Callicebus	`000000001011111110211200001101010111100001003100010001011120001 2`
Ateles	`110000001001010010111200000101010120000000031001000001101200011 1 2`
Homo	`20011100100001011001200001007???100100000100300000000011101300101 1 1 1`
Hylobates	`21011100000001011101200000010107?70111000000003100000001110130001 1`
Cercopithecoids	`0001000000000101100120000010117?70111010000003110000001121310010`
Aegyptopithecus	`???0????2?00001?01100?20000110107??000000010001310001001100131???? 1 1`
Apidium	`?????????01?0?????1??2000011010???100000110001?3??00000011123171???? 1`
Parapithecus	`??????????????????????????0110107??100000110001?3??00000011112131????`
Cebupithecia	`???????0100?11????20??0??1011???????100000??72?1101000101?111????`
```

Key to morphological characters. Symbols used: i, I lower and upper incisors; c, C canines; m, M molars; and p, P premolars. Citations refer to the describer(s) of the characters, not necessarily to the scoring of all taxa, which was done by the current authors.

(1) Tail with ventral glabrous surface (Pocock 1925): 0 = tail present without glabrous surface, 1 = tail present with glabrous surface, 2 = tail absent.
(2) Relative forelimb length (humerus+ radius)/(femur+ tibia) (Erikson 1963): 0 = short, 1 = long.
(3) Claws in digits (except hallux) (Buffon 1767): 0 = absent (nails in all), 1 = present.
(4) Carpo-metacarpal joint (Lewis 1977): 0 = non-saddle type, 1 = saddle type.
(5) Rib cage (Schultz 1961): 0 = dorsoventrally larger, 1 = laterally larger.
(6) Ulna-carpal articulation (Lewis 1974): 0 = present, 1 = absent.
(7) Sternebrae fusion (Schultz 1930): 0 = no fusion, 1 = fusion.
(8) Postglenoid foramen (ord.) (MacPhee et al. 1995): 0 = absent, 1 = reduced, 2 = large (inside of braincase visible through it).
(9) Ossified tentorium cerebelli (Hershkovitz 1977; Horovitz 1995): 0 = absent, 1 = present.
(10) Pneumatization in the anteroventral region of the middle ear (Cartmill et al. 1981): 0 = absent, 1 = present.
(11) Number of prominences on promontorium (Horovitz in prep.): 0 = one, 1 = two.
(12) Pterygoid fossa depth (Hershkovitz 1977; MacPhee et al. 1995): 0 = deep, 1 = not reaching basicranium.
(13) Canal connecting sigmoid sinus and subarcuate fossa (Horovitz 1995): 0 = absent, 1 = present.
(14) Ectotympanic shape (Buffon 1767): 0 = tube, 1 = ring, 2 = tube II.
(15) Temporal emissary foramen (MacPhee et al. 1995): 0 = large and above plane of infratemporal margin of zygomatic process of squamosal, 1 = small, absent, or below that plane.
(16) Eyeball physically enclosed (Martin 1992): 0 = absent, 1 = present.
(17) Cranial capacity (Horovitz in prep.): 0 = less than 15cc, 1 = more than 15cc.
(18) Pterion region contact (Pocock 1925): 0 = zygomatico-parietal, 1 = fronto-sphenoid.
(19) Infraorbital foramen, vertical position relative to maxillary cheek teeth in Frankfurt plane (ord.) (MacPhee et al. 1995): 0 = above interval between M1 and P4 or caudal to this position, 1 = above interval between P4 and P3, 2 = above anteriormost premolar or rostral to this position.
(20) Zygomatico-facial foramen size, relative to M1 breadth (MacPhee et al. 1995): 0 = small, 1 = large.
(21) Second lower deciduous incisor shape (Horovitz in prep.): 0 = blade-like, 1 = spatulate (with heel), 2 = styliform (without heel).
(22) Relative height of lower i1 and i2 (Rosenberger 1979): 0 = i1 absent, 1 = i1 lower than i2, 2 = subequal.
(23) Alignment of i1 and i2 (Rosenberger 1979): 0 = transversely arcuate, 1 = staggered.
(24) i1,2 shape (Rosenberger 1979): 0 = spatulate, 1 = styliform.
(25) Meso/distostyles on i1,2 (Hershkovitz 1977): 0 = absent, 1 = present.
(26) Diastema between c and i2 (Rosenberger 1979): 0 = absent, 1 = present.
(27) c cross section (MacPhee et al. 1995): 0 = round to oval, 1 = highly compressed.
(28) Lingual cingulum on c (Kinzey 1973): 0 = incomplete or absent, 1 = complete.
(29) c lingual crest shape (Kay 1990): 0 = rounded, 1 = sharp.
(30) c with lingual cingulum mesial elevation (Horovitz in prep.): 0 = not elevated, 1 = elevated.
(31) c lingual cingulum forming a spike on mesial edge of tooth (Horovitz in prep.): 0 = absent, 1 = present.
(32) Deciduous p2 lingual basin (Horovitz in prep.): 0 = absent, 1 = present.
(33) Deciduous p2 cross section shape (Horovitz in prep.): 0 = rounded, 1 = mesio-distally elongated.
(34) Deciduous p2 postprotocrista/longitudinal axis angle (ord.) (Horovitz in prep.): 0 = smaller than 45 deg, 1 = bigger than 45 deg.
(35) p3 metaconid (ord.) (Rosenberger 1979): 0 = metaconid absent, 1 = metaconid smaller than protoconid, 2 = metaconid and protoconid subequal, 3 = metaconid larger than protoconid.
(36) p4 metaconid (ord.) (Rosenberger 1979): 0 = metaconid smaller than protoconid, 1 = metaconid and protoconid subequal, 2 = metaconid larger than protoconid.
(37) p4 cuspiform hypoconid (Kay and Williams 1994): 0 = absent, 1 = present.
(38) p4 cuspiform entoconid (Kay and Williams 1994): 0 = absent, 1 = present.

(39) Lower molar enamel surface (Rosenberger 1979): 0 = smooth, 1 = crenulated.
(40) Lower molar, projection of distobuccal quadrant (MacPhee et al. 1995): 0 = not projecting, 1 = projecting (crown sidewall hidden).
(41) m1 cristid obliqua (Kay and Williams 1994): 0 = fully distal to protoconid, 1 = distolingual to protoconid.
(42) m1,2 buccal cingulum (Kinzey 1973): 0 = absent, 1 = present.
(43) m1,2 sulcus between protoconid and metaconid (Fleagle and Kay 1987): 0 = absent, 1 = present.
(44) m2 trigonid/talonid relative height (Horovitz in prep.): 0 = trigonid taller than talonid, 1 = subequal.
(45) m2 mesoconid (Horovitz in prep.): 0 = absent, 1 = present.
(46) m3 distolongual fovea (Kay 1980): 0 = absent, 1 = present.
(47) m3/ p4 relative length (ord.) (Horovitz in prep.): 0 = m3 absent, 1 = m3 shorter, 2 = subequal, 3 = m3 longer.
(48) I1 lingual heel (Rosenberger 1979): 0 = absent, 1 = present.
(49) I2 orientation (Rosenberger 1979): 0 = vertical, 1 = proclivious.
(50) P3 preparacrista (Horovitz in prep.): 0 = absent or vestigious, 1 = high.
(51) P4 protocone position (MacPhee et al. 1995): 0 = on widest point of trigon, 1 = mesial to widest point.
(52) P4 ligual cingulum (Kinzey 1973): 0 = absent, 1 = present.
(53) P4 hypocone (Kay 1990): 0 = absent, 1 = present on lingual cingulum, 2 = present on postprotocrista.
(54) P4 and M1 relative buccolingual breadth (MacPhee at al. 1995): 0 = P4 smaller, 1 = P4 subequal or bigger than M1.
(55) M1 mesostyle (Kinzey 1973): 0 = absent, 1 = present, 2 = replaced by mesoloph.
(56) M1 cingulum lingual to protocone (Kinzey 1973): 0 = absent, 1 = present.
(57) M1 hypocone/prehypocrista presence (Rosenberger 1979; MacPhee et al. 1995): 0 = hypocone and prehypocrista present, 1 = hypocone present and prehypocone absent, 2 = hypocone present and prehypocrista absent.
(58) M1 postmetacrista slope (MacPhee et al. 1995): 0 = buccal slope, 1 = distolingual slope, 2 = absent.
(59) M2 cristae on distal margin of trigon (MacPhee et al. 1995): 0 = cristae form distinct, continuous wall between protocone and metacone, 1 = cristae interrupted by a fossette or do not form a distinct wall, 2 = cristae absent or differently organized.
(60) M2 hypocone (Rosenberger 1979): 0 = absent, 1 = present.
(61) M3 length (Rosenberger 1979; Horovitz in prep.): 0 = M3 absent, 1 = M3 shorter than P4, 2 = M3 and P4 subequal, 3 = M3 longer than P4.
(62) M's parastyles (Horovitz in prep.): 0 = absent, 1 = present.
(63) Number of offspring at a time (Hill 1926): 0 = one, 1 = two.
(64) Exposure of vomer in orbit (Rosenberger 1979): 0 = absent, 1 = present.
(65) Thumb degree of development (Pocock 1925): 0 = absent or reduced, 1 = present.
(66) Number of lumbar vertebrae (Erikson 1963): 0 = more than five, 1 = five or fewer.

Molecular characters

Molecular characters were obtained by aligning the following sequences, deposited in the GenBank Data Libraries under the accession numbers indicated:

Sequence	Authors	Accession numbers
ε-globin genes	Schneider et al. 1993	L25354–L25371
RIBP genes	Harada et al. 1995	U18601–U18609, U18611–U18619, U19748–U19753
16S rDNA fragment	Horovitz and Meyer 1995	U38997–U39012

7 Adaptive radiation in the aquatic plant family Pontederiaceae: insights from phylogenetic analysis

Spencer C. H. Barrett and Sean W. Graham

The invasion of aquatic environments from land has occurred repeatedly during the evolutionary history of the flowering plants. The precise number of transitions from land to water is not known with certainty, although Cook (1990) recently estimated that it had taken place at least 50 times. Approximately 33 diverse families of monocotyledons and dicotyledons are exclusively aquatic, and numerous aquatic genera are found in predominantly terrestrial plant families. Aquatic plants constitute only 1–2% of angiosperms but they have received considerable attention from botanists and ecologists, primarily because of the high degree of ecological specialization that they exhibit. Adaptation to life in water has demanded the evolution of a distinctive array of morphological, anatomical, physiological and biochemical attributes that have developed on multiple occasions among the lineages that have invaded aquatic habitats.

Depending on the degree to which the life cycle of an aquatic plant is spent in water, individual taxa show increasing divergence from their terrestrial ancestors. At one extreme are species that spend their entire lives submersed below the water surface and are most distinct from land plants, to amphibious taxa that are equally at home on land or in water and that closely resemble their strictly terrestrial relatives. Aquatic groups often display considerable evolutionary diversification resulting from adaptation to the wide range of ecological conditions that occur in wetland habitats. This diversity offers excellent opportunities for relating form to function (reviewed in Arber 1920; Sculthorpe 1967; Crawford 1987; Barrett et al. 1993).

While the concept of adaptive radiation is central to evolutionary theory, there is a wide range of viewpoints as to what it entails. Futuyma (1986) states that adaptive radiation is simply diversification into different ecological niches by species derived from a common ancestor. According to Simpson (1953), however, such diversification is a direct response to a novel ecological or geographic circumstance experienced by the common ancestor of species involved in the radiation. More recently, the idea that increased species richness may (or must) be associated with adaptive radiation has become prevalent among phylogenetic systematists (e.g., Brooks and McLennan 1991). In this paper however, we follow Simpson's perspective that adaptive radiation occurs through character diversification among different lineages in response to a novel set of ecological circumstances or a key innovation. This process may involve an increase in speciation rate, no change in speciation rate, or even a reduction in speciation rate. In our view a radiation in slow motion is still a radiation – the number of lineages arising from an adaptive radiation is of secondary importance to the patterns of character diversification among lineages.

Aquatic plant groups have rarely been investigated from the perspective of adaptive radiation. While in part this is undoubtedly associated with the paucity of phylogenetic data available for most angiosperm families, it may also have been because of a widespread belief that aquatic environments are relatively homogeneous compared with those on land, thus providing less opportunity for evolutionary diversification. Indeed, such arguments have frequently been used to explain the apparently conservative macroevolutionary patterns found in certain aquatic groups (Sculthorpe 1967; Hutchinson 1975; Les 1988; Cook 1990). Of the 33 strictly aquatic families, 30 include fewer than 10 genera, 17 contain only one genus and three consist of a single species (Sculthorpe 1967). Increasing commitment to an exclusively aquatic existence appears to be associated with reduced taxonomic differentiation, as groups containing primarily amphibious and emergent aquatics show little evidence of reduced species diversity. The suggestion that some aquatic radiations are associated with reduced species richness (via decreased speciation rates or increased extinction rates) is intriguing, but requires detailed phylogenetic analysis of the sort suggested by Sanderson and Donoghue (1994) and Nee and Harvey (1994).

The wide spectrum of life-forms and diversity of reproductive strategies found in aquatic plants suggests that extensive character diversification has occurred in response to the novel ecological opportunities afforded by possession of the aquatic habit. The breadth of adaptations implies that the habitats occupied by aquatic plants are far from ecologically uniform, as is often supposed. Because of their many specialized features we believe that aquatic plant groups can provide outstanding opportunities for studies of adaptive radiation and character evolution, as has been undertaken in many animal groups that are restricted to aquatic environments (see Chapters 5 and 12).

Pontederiaceae is a small monocotyledonous family of exclusively freshwater aquatics, composed of approximately six to nine genera and 35 to 40 species, most of which are native to the New World tropics (Barrett 1978a). Members of the family display a remarkable diversity of life history and reproductive strategies ranging from highly clonal, long-lived taxa that inhabit permanent marshes and river systems to exclusively sexual species that are annual and occur in ephemeral pools, ditches and ricefields. Linking these extremes are species with various combinations of sexual and asexual reproduction and a variety of different pollination and mating systems. Evolutionary studies of the family over the past two decades have focused primarily on the floral biology and sexual systems of selected taxa (reviewed in Barrett 1988, 1993; Barrett et al. 1992). More recently, phylogenetic reconstructions using both morphological (Eckenwalder and Barrett 1986) and molecular data (Graham and Barrett 1995; Kohn et al. 1996; Graham et al., unpubl. data) have been employed to investigate character evolution and the systematic relationships of taxa within the family and its close relatives.

The diversity of life history traits in Pontederiaceae suggests that this family might provide a valuable opportunity for investigating the processes of adaptive radiation in an aquatic plant group. Carson (1985) and Johnson (1996) suggested two major modes of adaptive radiation in plants: habitat-driven and pollinator-driven.

Below we review the patterns of character variation and ecological differentiation in vegetative traits (with a particular focus on traits important for growth under aquatic conditions) and reproductive characters in Pontederiaceae. These lines of evidence suggest that selection acting on both reproductive and vegetative characters has contributed to the radiation of taxa in this family.

We begin by providing a brief review of the taxonomy and natural history of Pontederiaceae, emphasizing the diversity of life history and morphological adaptations to life in and out of water that occur in the family. We then perform phylogenetic reconstructions to examine the origins and evolution of a range of life history and reproductive attributes, including aquatic life-form, life-cycle duration, patterns of leaf development, types of clonality, floral form and self-incompatibility system. Throughout our chapter two particular issues form the basis of much of the discussion: (1) What is the ecological evidence that the various morphological characters we consider are adaptations in response to an aquatic life-style? (2) What is the phylogenetic pattern of diversification in these characters and which traits have evolved on multiple occasions within the family?

Taxonomy and natural history

Taxonomy

Pontederiaceae is composed of four main genera: *Eichhornia* (8–9 spp.), *Pontederia* (6 spp.), *Heteranthera* (10–12 spp.), and *Monochoria* (7–8 spp.), and several smaller segregate genera: *Eurystemon* (1 sp.), *Hydrothrix* (1 sp.), *Scholleropsis* (1 sp.), and *Zosterella* (1–2 spp.) allied with *Heteranthera*, and one segregate genus, *Reussia* (2–3 spp.) allied with *Pontederia*. The taxonomic affinity of this family to other monocotyledons is not clear-cut (see Dahlgren and Clifford 1982; Dahlgren et al. 1985; Rosatti 1987; Simpson 1987; Goldberg 1989) but recent treatments suggest a close affinity to Philydraceae and Haemodoraceae (Hamann 1966; Huber 1977; Simpson 1990; Thorne 1992a,b).

Biogeography

Members of Pontederiaceae are largely tropical in distribution with the primary concentration of species occurring in the Neotropics, particularly lowland South America and especially Brazil. Several taxa occur in North America, with some reaching as far north as Canada (e.g., *Pontederia cordata* [Figure 7.1A] and *Heteranthera* [*Zosterella*] *dubia* [Figure 7.1F]). In common with many other aquatic plants (see Ridley 1930; Sculthorpe 1967; Cook 1987), most species of Pontederiaceae have widespread distributions, often involving strikingly disjunct areas (e.g., *Eichhornia paradoxa*, *Eichhornia paniculata*; Barrett 1988). All members of the family occur in freshwater habitats frequented by waterfowl and wading birds that are capable of mediating long-distance dispersal. Aside from *Pontederia* and *Reussia* which are relatively large-seeded, all species have small diaspores that are likely to adhere to mud and be easily transported on the feet of water birds. In some cases long-distance dispersal can be achieved by stem fragments acting as floating vegetative propagules. The occurrence

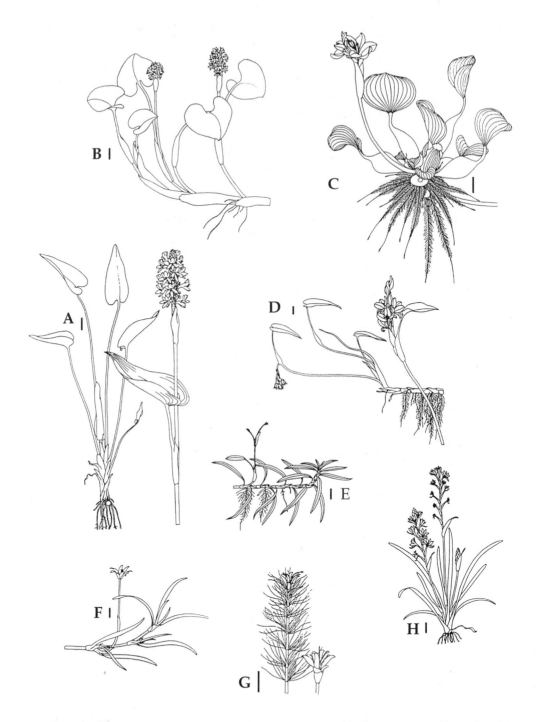

Figure 7.1. Pontederiaceae: diversity of aquatic life-forms and range of leaf types. (**A**) *Pontederia cordata*; (**B**) *Pontederia (Reussia) rotundifolia*; (**C**) *Eichhornia crassipes*; (**D**) *Monochoria vaginalis*; (**E**) *Heteranthera zosterifolia*; (**F**) *Heteranthera (Zosterella) dubia*; (**G**) *Hydrothrix gardneri*; (**H**) *Heteranthera (Eurystemon) mexicana*. All species except *H. mexicana* are represented in the phylogenetic reconstructions. Illustrations are from Cook (1990), with permission. All scale bars are 1 cm (except for **C**, where it is 3 cm).

of *Eichhornia crassipes* (Figure 7.1C) and *Eichhornia azurea* throughout the large river systems of South America and also on some Caribbean islands may have been largely the result of dispersal by vegetative means (Barrett 1978b; Barrett and Forno 1982). The natural distributions of a handful of Pontederiaceae have been extended over the past century due to human influences. Several New World *Heteranthera* species *(Heteranthera limosa, Heteranthera rotundifolia, Heteranthera reniformis)* occur as weeds of rice in Europe and Asia (C. Horn, pers. comm.; S. W. Graham, pers. obs.). *Monochoria vaginalis* (Figure 7.1D), a noxious weed of Asian rice, has also been introduced to Californian rice fields, probably as a seed contaminant (Barrett and Seaman 1980). The most widespread and economically important member of the family is the notorious clonal weed water hyacinth *(E. crassipes)*. Originally native to lowland South America, vast floating mats of this species now infest lakes, rivers, reservoirs, and drainage canals in many parts of the warmer regions of the world (Barrett 1989).

Aquatic habitats and ecological differentiation

Members of Pontederiaceae can be found in a wide variety of natural and man-made habitats provided by lakes, rivers, streams, permanent marshlands, bogs and fens, seasonal pools, drainage ditches, low-lying pastures, and rice fields, indicating a wide range of habitat preferences within the family. Aquatic habitats can be exceptionally diverse and therefore provide considerable opportunities for ecological differentiation by aquatic plants. Extensive field observations of the family over the past two decades by the first author indicate that the most significant features of aquatic environments that determine whether a particular species of Pontederiaceae can persist relate to the permanency of the habitat, depth of water, extent of water-level fluctuations, amount of nutrient loading, and the degree of interspecific competition from other aquatic plants. The overall depth of water, the predictability of the habitat, and the degree of interspecific competition are of particular importance in determining the aquatic life-form, duration of the life cycle and degree of clonality of individual species.

We outline below the variation in growth forms, life histories and reproductive strategies found in members of the Pontederiaceae and discuss their likely ecological and evolutionary bases. In order for phylogenetic reconstructions of character evolution to be conducted, it is necessary to classify the range of observed variation in traits. This exercise can be difficult, particularly where apparently continuous variation occurs or where detailed morphological and developmental information on the homologies of differing structures are not available. Nevertheless, we present such classifications below to begin exploration of the patterns of adaptive radiation in life history traits in the family.

Aquatic life-forms

Life-form classifications of aquatic plants are many and varied (reviewed in Raunkiaer 1934; Den Hartog and Segal 1964; Sculthorpe 1967; Hutchinson 1975). Here we adopt the classification of Sculthorpe (1967) which distinguishes four main classes of aquatic life-forms: emergent, floating-leaved, free-floating, and submergent. All occur in Pontederiaceae (Figure 7.1) and are closely associated with the habitat preferences of individual taxa. The emergent life-form, in which the plant is rooted in soil below water, but grows above the water surface to varying degrees, is the most common aquatic form and occurs in the majority of taxa in the family. Populations with this habit can be found over a broad range of aquatic conditions from temporary pools to more permanent wetland habitats. Leaf-blades of emergent taxa are held by self-supporting petioles. This allows them to overshadow and outcompete floating-leaved and submerged species in shallow waters. In deeper waters such petioles would be too costly to produce (Givnish 1995). The emergent life-form is therefore restricted to shallow locations at the edges of ponds, lakes, or rivers. In this study we distinguish two subclasses of emergents depending on whether the growth form is largely erect or procumbent. This dichotomy is somewhat artificial since a few species occur that link the two extremes (e.g., *Monochoria vaginalis, Pontederia (Reussia) subovata*) and considerable plasticity in the degree of erectness is evident depending on water depth (e.g., *Heteranthera seubertiana;* Horn 1988) and stand density. Nevertheless, we believe that this is a useful distinction because most taxa are distinguished by whether or not internodal elongation is extensive, producing plants that have either a creeping stem or a compact, erect rosette (Figure 7.1).

The only members of the family exhibiting the floating-leaved life-form in mature plants are *Eichhornia diversifolia, Eichhornia natans,* and *Scholleropsis lutea.* Individuals of these species are rooted to the substrate, with the stems and leaves floating on the water surface. We distinguish this growth form from the procumbent class of emergents by the predominance of truly floating leaves possessed by species in this category. While species such as *Pontederia (Reussia) rotundifolia* (Figure 7.1B), *E. azurea,* and *H. reniformis* frequently grow out from land over the surface of water, the majority of leaves that they produce are held erect as a result of upturned petioles and laminas. Many species of Pontederiaceae with emergent life-forms produce a small number of floating leaves as they emerge from below the water surface, following seed germination or perennation (Horn 1988). However, these leaves can be viewed as transitional, since the majority of mature leaves produced by these forms are adapted for terrestrial rather than aquatic conditions.

Eichhornia crassipes is the only species in the family that is truly free-floating (Figure 7.1C). The free-floating life-form is characterized by only a brief dependence on solid substrate to enable seed germination and establishment. Once established, young seedlings sever their connection with the sediments in which they germinated and float to the water surface. Floating is accomplished via swollen, aerenchymous petioles. Subsequent growth, clonal propagation, and dispersal occur entirely independently of land. While many taxa of Pontederiaceae with procumbent or floating-leaved growth forms can form floating mats, these are incapable of extensive growth and regeneration unless rooted to the substrate.

The final aquatic life-form in Pontederiaceae is the submersed life-form, represented by *Heteranthera zosterifolia* (Figure 7.1E), *H. dubia* (Figure 7.1F), and *Hydrothrix gardneri* (Figure 7.1G). In these species the entire plant body is submersed below the water surface, except during flowering when reproductive parts may be elevated just above the water (Wylie 1917; Rutishauser 1983). *Heteranthera dubia* and *H. zosterifolia* can tolerate partial emergence (e.g., mud-flat ecotypes of *H. dubia*; Horn 1983). Apart from occasional flowers above water, *Hydrothrix gardneri* is obligately submersed. The four main aquatic life-forms therefore appear to show different degrees of adaptation to the aquatic environment and could be thought of as involving an evolutionary transition from a terrestrial ancestor through an amphibious existence to a fully aquatic habit. Phylogenetic reconstruction may assist in evaluating this hypothesis by determining the direction and sequence of evolutionary change within Pontederiaceae.

Life-cycle duration

The adaptive basis of life-cycle duration in flowering plants has been the subject of much discussion, with a variety of ecological and demographic factors invoked as important selective agents (reviewed in Harper 1977; Grime 1979). Members of Pontederiaceae display a spectrum of life histories that are frequently associated with the permanency of the aquatic habitat occupied. These range from annual species that occur in ephemeral aquatic habitats such as seasonal pools, ditches, and rice fields (e.g., *E. diversifolia* and *Heteranthera* spp.) to very long-lived taxa that are largely restricted to permanent marshlands (*Pontederia* spp.), or to large river and lake systems (*E. azurea*) such as those found in Amazonia and the Pantenal region of South America.

We distinguish three categories of life-cycle duration: annual, short-lived perennial, and long-lived perennial. Annual species are those in which the majority of populations of a species complete their life cycle within a year. Short-lived perennials may persist for up to five years and long-lived perennials often live for considerably longer time periods. These categories are not rigid, because altered ecological conditions may modify patterns of longevity in any species. For example, several of the species that we classify as annuals (e.g., *E. paniculata*), because they usually cannot persist vegetatively in their native habitats from season to season as a consequence of severe desiccation, can continue growing almost indefinitely in the glasshouse if provided with suitable conditions. In contrast, populations of some annual species (e.g., *Eichhornia meyeri, H. limosa*) display obligate annualness, undergoing programmed senescence regardless of growing conditions. Among several of the species we classify as short-lived perennials are populations that appear to be annual when grown under glasshouse conditions (e.g., *M. vaginalis* from Californian rice fields).

Clonality

A considerable literature has been devoted to addressing questions concerned with the ecology and evolution of clonal versus sexual reproduction (Williams 1975; Maynard Smith 1978; Bell 1982). Valuable perspectives on the adaptive basis of clonality in plants have been provided by Abrahamson (1980), Leakey (1981), and Cook (1985). Aquatic plants are of particular interest in these discussions because of their heavy reliance on asexual methods of propagation (Arber 1920; Hutchinson 1975), and it has often been suggested that cloning may be favored in aquatic environments where regular seed reproduction is difficult in deep or turbulent water (Sculthorpe 1967). However, in a recent review Grace (1993) drew attention to the variety of clonal strategies found in aquatic plants and argued that at least six major selective forces may be involved in the evolution of clonal growth in aquatics: (i) numerical increase, (ii) dispersal, (iii) resource acquisition, (iv) storage, (v) protection, and (vi) anchorage.

Clonality in members of the Pontederiaceae appears to be closely linked with the life-form and longevity of individual taxa. Propagation in annual species is typically entirely sexual but most perennials in this family possess some form of clonal growth. This includes local colony expansion through rhizome growth in erect, emergent taxa such as *Pontederia sagittata* and *Monochoria hastata*, fragmentation of creeping stems in procumbent taxa with extensive internodal elongation [e.g., *P. rotundifolia* (Figure 7.1B) and *E. azurea*], fragmentation of stems in submersed taxa (e.g., *H. dubia*) and the formation of slender stolons with daughter rosettes in the free-floating *E. crassipes*. As in many other perennial aquatics (see Eckert and Barrett 1993) the balance between sexual and asexual reproduction in Pontederiaceae can vary with habitat conditions and the combination of growth form and clonality that occurs (Richards 1982). Seed reproduction is common in most emergent taxa with rhizomatous growth or stem fragmentation because they usually occupy habitats suitable for seed germination and seedling establishment. In contrast, in submersed and free-floating taxa sexual recruitment probably occurs less often, despite seed formation, because of deep water conditions that restrict seedling establishment, and in some taxa there are populations that regenerate exclusively through clonal propagation (e.g., *E. crassipes*; Barrett 1980a,b).

The various clonal strategies displayed by members of the Pontederiaceae serve different functions. One of these is numerical increase (i.e., reproduction via ramet formation), which is most obvious in taxa with regular stem fragmentation or stolon production. Dispersal of these vegetative structures by water currents also enables exploitation of new environments, with the free-floating daughter rosettes in *E. crassipes* representing the most specialized adaptation for vegetative dispersal in the family. For species that experience long periods with unfavorable growing conditions, such as during winters in eastern North America, rhizomes and stem fragments are also used as perennating structures (e.g., *P. cordata* and *H. dubia*; Lowden 1973, Horn 1983). However, these structures are also capable of withstanding considerable desiccation and in tropical habitats prone to drought can contribute to persistence during dry periods. Finally, in taxa of Pontederiaceae with creeping stems or stolons, the structures involved in clonal growth are also photosynthetic and produce roots. Thus they are highly effective in resource acquisition, the exploitation of suitable habitat patches, and competition with coexisting aquatic species.

Patterns of leaf development

Aquatic plants display striking foliar plasticity involving continuous variation in leaf shape and the formation of discrete leaf types with very distinct morphologies on a single individual. The latter condition has been referred to as heterophylly and a considerable literature exists on the proximate ecological, physiological, and developmental mechanisms that control changes in leaf shape in heterophyllous species (Arber 1919; Sinnott 1960; Sculthorpe 1967; Lee and Richards 1991). Less attention has been paid to the genetic and evolutionary basis of such patterns (although see Bradshaw 1965; Cook and Johnston 1968). It is usually assumed that in aquatic plants the formation of flaccid, ribbon-shaped (Figure 7.1E,F), or highly dissected leaves or leaf whorls (Figure 7.1G) represents an adaptive response to submersed conditions. Heterophylly often has been considered to be a manifestation of heteroblastic leaf development – the ontogenetic sequence in which early-formed "juvenile" leaves are markedly different in appearance from later "adult" ones. However, because the distinction between leaf types often is not clear-cut and so-called "juvenile" leaves often can be retained throughout the life cycle by neoteny (Sculthorpe 1967, and see below), it is important to realize that considerable diversity exists in the patterns of leaf development found in aquatic plants and that any attempt at classification is likely to be somewhat artificial.

With the exception of a detailed investigation of leaf ontogeny in *E. crassipes* (Richards 1983) and descriptions of heterophylly in *Heteranthera* (Horn 1988), there has been little work on the developmental basis of leaf-shape variation in Pontederiaceae. For the purpose of our study we tentatively recognize five basic classes (referred to as patterns A–E) that differ primarily in the duration of the "juvenile" phase. In the first and most common type (pattern A) plants first produce a small number (one to four) of juvenile, linear, strap-shaped leaves, the width and size of which vary with species and degree of submersion, before producing "adult" aerial leaves (Figure 7.1A–D) with distinct petioles and laminas. In *Heteranthera* the juvenile leaves can be very narrow, whereas in *Pontederia* they can be up to several centimeters in width. Species with this type of leaf development are usually amphibious with seedlings commonly developing in shallow water or on wet mud. The important feature of this leaf development strategy is a rapid transition to the formation of aerial leaves, the characteristics of which vary according to species.

In the second class (pattern B), this transition is much slower and a greater number (more than twenty) of submersed ribbonlike leaves are produced before the transition to aerial or floating leaves. This pattern of leaf development is quite restricted in the family and occurs only in *Eichhornia azurea*, *E. diversifolia*, *E. heterosperma*, and *E. natans*, all species that commonly germinate in deep water and experience extended periods of seedling development under water. While ribbon-shaped leaves are always the first leaves to be produced by species in class B, damage to the shoot apex through herbivory or disease in adult plants can result in a temporary reversion to the "juvenile" phase, indicating that heteroblastic development is not necessarily developmentally fixed.

The next two classes involve taxa in which all foliar leaves retained throughout the life-cycle are sessile, linear, and ribbon-shaped and are similar in appearance to the "juvenile" leaves initially produced by taxa in the first two categories. We chose to recognize two separate classes of these (presumably) "retained juvenile" or paedomorphic forms, because it seems probable that they have different developmental origins (neotenic versus progenetic; J. H. Richards, pers. comm.) and different ecological significances. Petiolate leaves are never produced in these taxa (the lower spathe in inflorescences of *H. zosterifolia* being the sole exception).

Pattern C occurs in the perennial, submersed aquatics *Heteranthera zosterifolia* (Figure 7.1E) and *H. dubia* (Figure 7.1F), and presumably reflects their predominantly submersed existence. The pattern appears to have arisen through neoteny (i.e., through slower somatic development relative to the onset of reproductive maturity), analogous to but much more accentuated than in pattern B. Such a neotenic shift may be a consequence of direct selection for retention of the juvenile leaf form throughout the life cycle. Ribbon-shaped leaves have been interpreted as a mechanical (antidrag) adaptation to moving water (Sculthorpe 1967), but may play a role in counteracting diffusive limitations on photosynthesis underwater (see below). Pattern D is found in two annual taxa, *Heteranthera seubertiana* and *Heteranthera* (*Eurystemon*) *mexicana* (Figure 7.1H). These two species are primarily emergent, and are found in ephemeral pools. This pattern could have arisen through an earlier onset of reproduction (i.e., progenesis; see Alberch et al. 1979). Such precocious reproduction may be an adaptive response to the ephemeral nature of some aquatic environments (Arber 1920; Van Steenis 1957; Sculthorpe 1967). Emersed leaves from the four species exhibiting these two patterns are not flaccid but are stiff and erect and clearly adapted for terrestrial conditions.

The final class of leaf development (pattern E) is restricted to the submersed *Hydrothrix* (Figure 7.1G). This monotypic genus is characterized by whorls of small, threadlike leaves whose developmental origin is highly unusual (Rutishauser 1983). The leaf whorl, which has analogues in many plant groups (including some algae), is an example of the "*Hippuris* syndrome" (Cook 1978). The occurrence of this syndrome in *Hydrothrix* is virtually unique within monocotyledons (Cook 1978).

The functional significance of the *Hippuris* syndrome may be to reduce self-shading in deep water (Arber 1920). It has also been interpreted as a means of reducing drag in flowing water (Arber 1920, but see Sculthorpe 1967) or for increasing total assimilating area (Sculthorpe 1967, but see Arber 1920). A major function of ribbon- or threadlike leaves in submersed aquatics may be to decrease the impedance of leaf-surrounding boundary layers on CO_2 diffusion, which is far slower in water than in air (Givnish 1987). The more effectively "narrow" a leaf or leaf division is, the smaller its boundary layer will be, and the less it will impede diffusion and limit photosynthesis (Givnish 1987). More slowly flowing waters should therefore favor narrower leaves. This prediction is supported by comparative data on aquatic plants in ponds vs. streams (see Madsen 1986; Givnish 1987). The cordate bases seen in the petiolate leaves of several emergent taxa may represent the minimization of vein construction costs in lamina atop erect petioles (Givnish and Vermeij 1977).

Floral ecology, pollination, and mating systems

Angiosperms display a spectacular array of floral diversity associated with the pollination biology and mating systems of individual species. Indeed, reproductive adaptations associated with different pollen vectors are among the few plant traits that have been considered explicitly in the context of adaptive radiation (Grant and Grant 1965; Stebbins 1970). In addition to descriptive studies of pollination syndromes, recent functional interpretations of floral radiation have emphasized the importance of individual selection for fitness gain through female and male reproductive function (e.g., Bell 1985; Campbell 1989; Devlin and Ellstrand 1990). Using this approach it is important to distinguish aspects of floral display and design that reduce the incidence of self-pollination and inbreeding depression (Charlesworth and Charlesworth 1987) from those that promote outcrossed siring success through more effective pollen dispersal (Harder and Barrett 1996). This is of particular importance in species that possess physiological self-incompatibility systems (Lloyd and Webb 1986; Bertin and Newman 1993; Harder and Barrett 1995).

The flowers of Pontederiaceae are showy and blue, mauve, yellow, or white and can be solitary or displayed in inflorescences. They exhibit a broad range of morphological specializations associated with a variety of pollination mechanisms and mating systems. Despite being an entirely aquatic family, the floral syndromes of Pontederiaceae involve either animal pollination or self-pollination, with no evidence of the kinds of adaptations towards hydrophily that occur in many other exclusively aquatic taxa (Cox 1988). Flowers in the family are largely pollinated by bees, and to a lesser extent by butterflies. In *Pontederia* and *Eichhornia*, the flowers are tubular and moderately zygomorphic and pollination is largely achieved through the services of long-tongued bees that feed primarily on nectar (Wolfe and Barrett 1988; Husband and Barrett 1992). In *Heteranthera* and *Monochoria*, floral visitors are mostly pollen-collecting bees (Iyengar 1923; Wang et al. 1995; S. C. H. Barrett, pers. obs.). In most species of *Heteranthera*, *Hydrothrix gardneri*, *E. diversifolia*, *E. natans*, and *M. vaginalis*, flowers are also produced which develop underwater and hence are completely self-fertilized. This phenomenon is known as pseudo-cleistogamy and is commonly reported in other aquatic groups (Sculthorpe 1967). Amphibious flowers with underwater ovaries but aerial pollination organs are found in *Heteranthera dubia* (Wylie 1917; Horn 1985). Floral tubes in this species are very variable in length (Horn 1985) and can reach over 11 cm long (R. Rutishauser, pers. comm. in Endress 1995). Amphibious flowers and pseudocleistogamy represent the only obvious examples of shifts in reproductive characters in response to the aquatic habit of the family.

Two conspicuous floral polymorphisms (tristyly and enantiostyly) occur in Pontederiaceae (Figure 7.2). The genetic polymorphism tristyly occurs in all species of *Pontederia* s. lat. except *Pontederia parviflora*, and in three species of *Eichhornia* (*E. azurea*, *E. crassipes*, and *E. paniculata*). Tristylous species possess a reciprocal arrangement of style and anther heights (Figure 7.2A) and an associated syndrome of ancillary characters exhibiting polymorphisms of pollen and stigmas (reviewed in Barrett 1992). The remaining taxa in both genera are small-flowered and monomorphic for style and stamen length. A self- and intramorph-incompatibility system accompanies the floral heteromorphism in all tristylous species, except for *E. crassipes* and *E. paniculata*, which are tristylous but highly self-fertile (Barrett 1979; Barrett and Anderson 1985; Barrett 1988). Experimental studies of *Pontederia* and *Eichhornia* support Darwin's (1877) original hypothesis that the tristylous polymorphism functions to promote proficient cross-pollination among plants through the reciprocal arrangement of male and female sex organs (Barrett and Glover 1985; Kohn and Barrett 1992; Lloyd and Webb 1992).

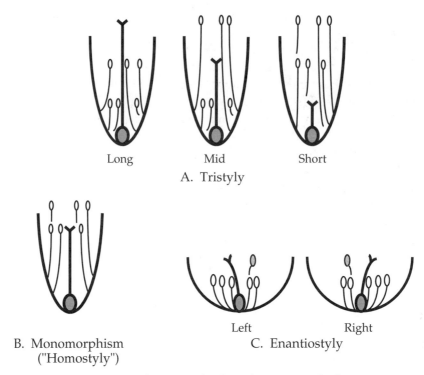

Figure 7.2. Schematic representation of stamen and style configurations in the three most common floral forms in Pontederiaceae (after Graham and Barrett 1995). (**A**) **Tristyly** – individuals produce either long-, mid- or short-styled flowers, depending on their genotype at two diallelic loci controlling this genetic polymorphism. (**B**) **Floral monomorphism** (referred to as "homostyly" when evolutionarily derived from heterostyly) – in *Eichhornia*, monomorphic species have three or six stamens at the same level as the stigma (Barrett 1988). Typically, homostylous variants found *within* tristylous species of *Eichhornia* have only one stamen at the same level as the stigma (referred to as "semi-homostyly" [not shown]). (**C**) **Enantiostyly** – flowers have either left- or right-bending styles, with a single stamen (one of six in *Monochoria*, one of three in *Heteranthera*) bending in the opposite direction. In contrast with heterostyly, individuals can produce both floral forms (after Graham and Barrett 1995).

The second floral polymorphism, enantiostyly (Figure 7.2C), occurs in species of *Heteranthera* s. lat. and *Monochoria* (Iyengar 1923; Eckenwalder and Barrett 1986; Wang et al. 1995). The outward-facing flowers possess either left- or right-bending styles with a single stamen reflexed in a lateral position opposite the stigma. This condition rarely occurs as a true genetic polymorphism with populations composed of plants with either right- or left-handed flowers, but not both (e.g. in *Wachendorfia*; Ornduff and Dulberger 1978). More commonly, however, it exists as a somatic polymorphism with individual flowers possessing both flower types. All enantiostylous members of Pontederiaceae have the somatic form of the polymorphism. They also display a striking anther dimorphism with the reflexed stamen cryptically colored and larger than the remaining stamens. Such dimorphism is termed heteranthery and represents a functional division of labor into attractive "feeding" anthers and a cryptically colored "pollinating" anther (Vogel 1978; Buchmann 1983; Lloyd 1992). Enantiostyly has most often been interpreted as an adaptation to increase the effectiveness of cross-pollination, in a manner analogous to heterostyly. However, there is little empirical evidence to support this hypothesis (e. g., Fenster 1995) and where the polymorphism is somatic other factors must be involved in its origin and maintenance (see Graham and Barrett 1995 for further details).

While mating patterns have not been quantified in the majority of Pontederiaceae, some inferences can be drawn from information on floral biology and experimental studies conducted by the first author over the past two decades. All species of *Monochoria* and *Heteranthera* s. lat. are highly self-compatible. Undisturbed flowers of species in these two groups can usually achieve full seed-set through autonomous self-pollination. Although pollen-collecting bees visit flowers, it seems likely that populations of these taxa experience considerable self-pollination, particularly in taxa with pseudo-cleistogamous flowers. In contrast, tristylous species of *Pontederia* and *Eichhornia* with self-incompatibility must be largely outcrossing because of their physiological barrier to self-fertilization. Even where tristyly is associated with self-compatibility, marker-gene studies indicate that populations can exhibit high outcrossing rates (Barrett et al. 1993). Among non-heterostylous species of *Pontederia* and *Eichhornia*, self-fertilization is likely to predominate, since these taxa are self-compatible and homostylous, with anthers and stigmas close together within each flower (Figure 7.2B; see Barrett 1988).

Phylogenetic systematics of Pontederiaceae

Morphological and molecular evidence of phylogenetic relationships

Molecular evidence from the chloroplast gene *rbc*L strongly supports the monophyly of Pontederiaceae (Graham and Barrett 1995). Three highly congruent data sets derived from the chloroplast genome (based on restriction-site variation and sequence data from *rbc*L and *ndh*F) yield robust and well-resolved estimates of the phylogenetic history of the family (Graham et al., unpubl. data; Figure 7.3). These estimates indicate that *Monochoria*, *Pontederia* s. lat., and *Heteranthera* s. lat. are monophyletic, but that *Eichhornia* is not. Morphological evidence concerning the phylogenetic history of Pontederiaceae also rejects the monophyly of *Eichhornia* (Eckenwalder and Barrett 1986), but is largely insufficient for estimating a robust phylogenetic history of the family (Graham et al., unpubl. data). In combined analyses with the molecular evidence, the morphological evidence has almost no impact on phylogenetic reconstructions (Graham et al., unpubl. data).

Despite being swamped by the molecular evidence, there is statistically detectable incongruence between the molecular and morphological data (Graham et al., unpubl. data). Eckenwalder and Barrett (1986) hypothesized that a "selfing syndrome" in the family (involving multiple parallel shifts in reproductive characters during the origin of predominantly self-fertilizing species) could result in incorrect phylogenetic reconstructions using the morphological data. However, as this hypothesis is not actually supported by the morphological evidence and is contradicted by the molecular data, it cannot be the source of the incongruence between these two

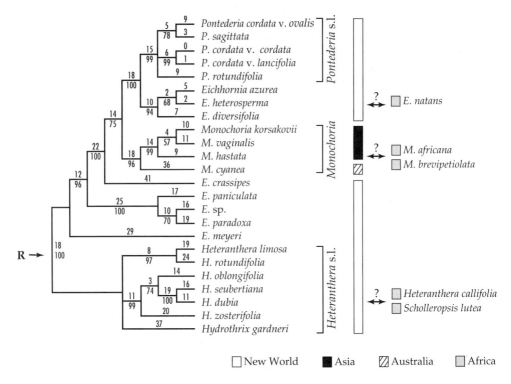

Figure 7.3. Reconstructed phylogenetic history of Pontederiaceae using combined evidence from DNA sequence variation in the chloroplast genes *rbc*L and *ndh*F, and restriction-site variation in the chloroplast genome (Graham et al., unpubl. data). The tree is a strict consensus of four shortest unrooted trees. Branch lengths were determined using ACCTRAN optimization for one of the four shortest trees and are indicated above each branch. Bootstrap values (from 100 bootstrap replicates) are below branches. The root indicated R was found by a variety of closely related taxa using the two chloroplast DNA sequence-based data sets (see text). For the four most parsimonious trees found with the combined chloroplast data, tree length (including autapomorphies) = 609 steps, CI (including autapomorphies) = 0.637, CI (excluding autapomorphies) = 0.525, RI = 0.775 (Graham et al., unpubl. data). The bars indicate whether each taxon is found in the Old World (Asia, Australia, or Africa), or the New World. Question marks indicate possible long-distance dispersal events involving several taxa missing from this analysis (see text).

major lines of evidence (Graham et al., unpubl. data). There is little evidence of hybridization among modern species of Pontederiaceae, but it is not impossible that undetected ancient hybridization events have contributed to the observed incongruence between the morphological and chloroplast data (see Doyle 1992). An improved morphological data base (or new molecular evidence from the nuclear genome) is needed to pinpoint the precise source and extent of the incongruence between these two sources of data. In the meantime, the chloroplast evidence is our only substantial source of evidence concerning the phylogenetic history of Pontederiaceae, and we use it here to investigate the radiation of morphological characters in the family.

Given the trees found with the morphological data set of Eckenwalder and Barrett (1986) as modified by Graham et al. (unpubl. data), the morphological data have levels of homoplasy very close to that expected for this number of taxa. The observed CI (excluding autapomorphies) for the 24 ingroup taxa in Figure 7.3 is 0.474. The expected CI for this number of taxa is 0.495, based on Sanderson and Donoghue's (1989) survey of 60 data sets. However, the trees derived from the molecular data indicate that the amount of homoplasy in the morphological data is substantially higher than this. The CI (excluding autapomorphies) of the morphological data on the four shortest trees found with the combined chloroplast evidence ranges from 0.404 to 0.410. This lower value may partly be a reflection of the incongruence between the morphological and molecular sources of data, and given the low number of informative morphological characters (only 33), it is possible that these estimates of homoplasy in the morphological data are a quite imprecise reflection of the real levels of homoplasy in morphological characters. The patterns of homoplasy among subsets of the morphological characters are nonetheless intriguing. On the molecular trees, the CI for 17 informative floral characters (excluding autapomorphies and post-anthesis characters) ranges from 0.446 to 0.453; for 11 informative vegetative characters, the CI (excluding autapomorphies) ranges from 0.310 to 0.316. The trend toward high levels of homoplasy relative to that expected for this number of taxa (Sanderson and Donoghue 1989) in both classes of data suggests that they have been subject to elevated levels of character divergence in this family, at least compared to unrelated groups.

The location of the family's root was one of the few elements of the phylogenetic structure of Pontederiaceae left unresolved by the molecular studies of Kohn et al. (1996) and Graham et al. (unpubl. data). The local position and membership of Pontederiaceae within a cluster of superorders consisting of Arecanae, Bromelianae, Commelinanae and Zingiberanae (*sensu* Dahlgren et al. 1985) also was not well supported by evidence from *rbc*L (Duvall et al. 1993, Graham and Barrett 1995) or surveys of chloroplast restriction-site variation (Davis 1995). A range of taxa in this complex of superorders was surveyed for variation in the chloroplast genes *ndh*F and evidence from this gene was employed in tandem with the *rbc*L evidence to determine which taxa are most closely related to the family, and to establish the position of the root of the family (Graham and Barrett, unpubl. data). Using equal weighting of all characters and a variety of different unequal weighting schemes to correct for among-site rate variation, combined analyses of these two genes indicated that the sister-group of the family is a clade consisting of Commelinaceae and Haemodoraceae, and that the next most closely related clade is Philydraceae (Graham and Barrett, unpubl. data). These three most closely related families (Commelinaceae, Haemodoraceae, and Philydraceae) also converged upon a single most parsimonious root location of Pontederiaceae (Graham and Barrett, unpubl. data), which is the *a posteriori* rooting employed in character reconstructions here (Figure 7.3).

Implications of the phylogenetic data and fossil evidence for the biogeography of the family

As discussed earlier, the modes of dispersal of taxa in Pontederiaceae provide substantial opportunities for long-range dispersal. With a few isolated exceptions, the basal clades in the family are all currently limited to the New World, and the exclusively Old World genus *Monochoria* is located quite far from the base of the tree, suggesting an ancient inter-continental dispersal (Figure 7.3). The fossil record of the family reaches back into the Eocene (ca. 50 Mya) when Africa, Australia, and South America were no longer in direct contact (Briggs 1987). *Monochoria* is the only genus in Pontederiaceae currently restricted to the Old World (Figure 7.3). Fossilized seeds and leaf material similar to modern *Eichhornia*, and seeds like those of modern *Monochoria*, are known from the upper Eocene onwards in Europe (see Collinson et al. 1993). The presence of *Eichhornia*-like fossils in Europe raises the intriguing possibility that the current limitation of this genus to the New World may be a consequence of ancient extinctions in the Old World. Fossilized root and stem fragments of Pontederiaceae are known from the Eocene in India (Patil and Singh 1978). However, the uncertain generic affinity of this material (Eckenwalder and Barrett 1986) means that it is not clear whether this represents a lineage that arose before or after the divergence of the extant members of the family. Other fossils ascribed to the family (Bureau 1892; Knowlton 1922; Fritel 1928) are from North American and European sites, but as these reports are based solely on leaf material, they must also be treated as being of unclear affinity to modern genera. Philydraceae is currently limited to Australia and Asia (Dahlgren et al. 1985; Adams 1987), but the two families constituting the sister-group (Commelinaceae and Haemodoraceae) are distributed throughout the Old and New World (Dahlgren et al. 1985; Simpson 1990). Aside from the *Monochoria*, the only other species in Pontederiaceae with Old World distributions are *Heteranthera callifolia*, *Scholleropsis lutea*, and *Eichhornia natans*. These three species, together with two species of *Monochoria*, are limited to Africa. All of these taxa are missing from the current study, but their taxonomic affinities and positions in the morphology-based analysis of Eckenwalder and Barrett (1986) suggest that their current distributions are a consequence of several long-range dispersal events (Figure 7.3). However, the possibility that some of these species are relicts from a more cosmopolitan distribution of these genera cannot be ruled out, although this is unlikely for *E. natans* given its probable close relationship to *E. diversifolia* (see below). Multiple long-range dispersal events probably also contributed to the modern distributions of the various New World taxa. North American taxa in general seem to be more morphologically apomorphic than those in tropical South America, suggesting that they may have migrated north after intercontinental contact in the Miocene (Eckenwalder and Barrett 1986).

Character diversification and adaptive radiation in vegetative and reproductive characters

Outgroups and their effect on character reconstruction in Pontederiaceae

The inclusion of outgroup taxa in phylogenetic analysis serves two major purposes: locating the position of the root of ingroup, and polarizing character-state transformations within this group. Although these analytical goals are often addressed simultaneously, they need not be if the characters used to reconstruct a phylogeny (as in this study) are not the ones in whose evolutionary transformation we are interested (e.g., Brooks and McLennan 1991; Maddison and Maddison 1992).

The sister-group plays a major role in polarizing character reconstructions, but other less closely related outgroups also play an important role in this (Maddison et al. 1984; Nixon and Carpenter 1993).

We employed the three most closely related families to Pontederiaceae to provide information concerning the polarity of the character transformations discussed below: Commelinaceae and Haemodoraceae (which together constitute the sister-group) and Philydraceae (the next most closely related taxon). In cases where there was character-state variation among the constituent taxa of individual outgroups, knowledge of the internal phylogenetic structure of each outgroup would be valuable for obtaining "globally parsimonious" reconstructions of the evolution of such characters (Maddison et al. 1984; Maddison and Maddison 1992, p. 47). Simpson (1990) provided a phylogeny of Haemodoraceae based on morphological data, but there are no published phylogenies for Commelinaceae or Philydraceae. It was consequently often necessary in this study to code individual families as polymorphic for character-states for which there was known to be variation among different species within each family. Using polymorphic coding to account for this variation is a less-than-ideal solution to lack of knowledge concerning the phylogenetic structure within individual outgroup families (Nixon and Davis 1991; Maddison and Maddison 1992). However, the reconstructions obtained here, while conditional on increased knowledge of the phylogenetic structure of these groups and improved knowledge of character distributions in them, are nonetheless the most parsimonious ones given our current state of knowledge (Maddison and Maddison 1992, p. 47).

The reconstructions of character diversification were performed using MacClade version 3.0 (Maddison and Maddison 1992) and employed the four most parsimonious unrooted trees of the family found in the combined analysis of the three chloroplast sources of evidence (Graham et al., unpubl. data), with the rooting determined using combined evidence from *rbc*L and *ndh*F. Only minor differences existed among the four trees concerning the placements of *Pontederia rotundifolia* and *Hydrothrix gardneri* in *Pontederia* s. lat. and *Heteranthera* s. lat., respectively. All reconstructions were performed using MacClade version 3.0 (Maddison and Maddison 1992) and used Fitch optimization (Fitch 1971), in which all character-state changes were treated as equally likely events (i.e., unordered or equally weighted), apart from an analysis of reproductive characters in which "relaxed Dollo" schemes of character evolution (Swofford and Olsen 1990) were also assessed (see below). The results are indicated in legends on each figure. With the aid of the "equivocal cycling" tool in MacClade, we obtained counts of the number of gains of each character-state for each character within Pontederiaceae, and determined the primitive state of the family, for all most parsimonious reconstructions of each character on the four trees. The character reconstructions in Figures 7.4–7.8 exemplify much of the diversification observed for each character: the tree used in these figures is one of the four most parsimonious ones, and is the most highly converged-upon tree found in analyses of several different combinations of the available chloroplast evidence (Graham et al., unpubl. data).

Character codings

A total of 24 species were surveyed, representing approximately two-thirds of the family and including all major taxonomic groups (for source and voucher information see Kohn et al. [1996]; Graham 1997). Except for leaf developmental pathway, the character codings for the taxa of Pontederiaceae considered here are derived from Eckenwalder and Barrett (1986), Graham and Barrett (1995), and Graham et al. (unpubl. data). The codings for the three outgroup taxa we included are presented here. Of the 50 genera in Commelinaceae, a few are found in wet places, but only *Murdannia* possesses aquatic species (at least two of 50 species; Cook 1990). Cook (1990) lists only one species of Philydraceae as being helophytic (*Philydrum lanuginosum*), but all six species in the family are found in marshes and wet rain forest habitats (Adams 1987). Species of Haemodoraceae are almost all xeric (M.G. Simpson, pers. comm.), although *Tribonanthes* is found in similar habitats (low, winter-wet flats) to *Philydrella* (Philydraceae) (Simpson 1990). Commelinaceae, Haemodoraceae, and Philydraceae are therefore almost exclusively terrestrial groups, and most closely approach the "emergent" condition in Pontederiaceae. Procumbent and erect life-forms are known in Commelinaceae (Faden 1988), and so this outgroup is coded as polymorphic for these two forms. Haemodoraceae and Philydraceae contain only erect taxa (Adams 1987; Cook 1990; M. G. Simpson, pers. comm.) and these families are coded accordingly for life-form.

For life-cycle duration, Commelinaceae and Philydraceae both are coded as polymorphic for annuality and short- and long-lived perenniality (Dahlgren et al. 1985; Faden 1988; Cook 1990). A "long-lived perennial" coding is appropriate for Haemodoraceae (M. G. Simpson, pers. comm.). Non-clonal species are found in Commelinaceae (Faden 1988), and some species in this family express clonality via rhizomes, stolons, or spreading stem fragmentation (Faden 1988). Commelinaceae is coded as polymorphic for "non-clonality," and for these three kinds of clonality ("via rhizomes," "via stolons," and "via stem fragmentation"). There is no direct evidence of clonality in Haemodoraceae, but extensive underground rhizome/stolon systems are known in this family (M. G. Simpson, pers. comm.). Haemodoraceae is provisionally coded as polymorphic for "non-clonality," "clonality via rhizomes," and "clonality via stolons." Species of Philydraceae are rhizomatous or cormous (Adams 1987), but it is not known if these structures are involved in regeneration. Philydraceae is provisionally coded as polymorphic for "non-clonality" and "clonality via rhizomes."

Information on the timing of the transition to adult leaves, and on the homology of such pathways among the outgroup families and Pontederiaceae is mostly lacking. However, Tillich (1994, 1995) noted that the seedlings of Pontederiaceae and Philydraceae are very alike and that the primary leaves in both families are ribbonlike. The homology of adult leaf-form among these families is uncertain. Members of Commelinaceae possess bifacial leaves, but Haemodoraceae and Philydraceae possess unifacial, ensiform leaves (Dahlgren and Rasmussen 1983). Anatomical data suggest that the bifacial leaves typical of taxa in Pontederiaceae have a unifacial origin (see Arber 1925), and Simpson (1990) hypothesized that an origin of the bifacial leaf was associated with the shift to an aquatic environment in Pontederiaceae. We provisionally coded the three families as "unknown" for leaf developmental pathway.

Showy, insect-pollinated flowers are typical of species in Commelinaceae, Haemodoraceae, and Philydraceae. Species lacking either somatic or genetic polymorphisms in stylar class (referred to here as "monomorphic flowers") predominate in Commelinaceae and Haemodoraceae and in most monocotyledons. Tristyly is only known in Pontederiaceae, but enantiostyly is found in some species of Commelinaceae (Faden 1991) and Haemodoraceae (Simpson 1990) and all species of Philydraceae (Simpson 1990). We did not count as enantiostylous those species that possess flowers with bent styles but that lack a "handedness," such as *Hydrothix gardneri* (Rutishauser 1983) and *H. dubia* in Pontederiaceae. Only two genera in Haemodoraceae (*Schiekia* and *Wachendorfia*) possess flowers with bent styles that have true left- versus right-handedness, i.e., with zygomorphic, outward-facing flowers (Simpson 1990). The non-basal position of these taxa in morphology-based cladograms of Haemodoraceae (Simpson 1990) suggests that monomorphic flowers were ancestral in this family. Haemodoraceae was therefore coded as "monomorphic" for floral form, Philydraceae as "enantiostylous," and Commelinaceae as polymorphic for these two conditions.

Owens (1981) reported self-compatible and self-incompatible species in Commelinaceae. The form of self-incompatibility in Commelinaceae is gametophytic and non-heteromorphic and consequently is highly unlikely to be homologous to that in Pontederiaceae (Graham and Barrett 1995). Commelinaceae was therefore coded as polymorphic for gametophytic SI and self-compatibility. The self-incompatibility status of species in the two other outgroup families is largely unknown. *Philydrum lanuginosum* is fully autogamous (S. C. H. Barrett, pers. obs.). Hamann (1966) also reported autogamy in the family, but no explicit surveys for self-incompatibility have been performed. Philydraceae is therefore coded as "unknown" for this character. There is a single report of a weakly developed incompatibility system in *Wachendorfia paniculata* that appears to be associated with true genetic enantiostyly, in a manner analogous to distylous or tristylous self-incompatibility systems (Ornduff and Dulberger 1978; see also Wilson 1887). However, we are aware of no other data concerning the self-incompatibility status of species in Haemodoraceae, and so code it as "unknown" for this character.

Reconstructions of character evolution

AQUATIC HABIT – Is the capacity to thrive in an aquatic habitat a synapomorphy for the taxa in Pontederiaceae? Of the taxa that are closely related to Pontederiaceae, the two families that constitute its sister-group (Haemodoraceae and Commelinaceae; see above) are almost exclusively terrestrial. For each of these families, it is almost certain that the primitive forms were adapted to a completely terrestrial existence (see above). However, the next most closely related clade (Philydraceae) is semi-aquatic. Whether the aquatic habit is homologous in Philydraceae and Pontederiaceae depends partly on the distribution of aquatic versus terrestrial taxa in clades more distant to Pontederiaceae than its sister-group and Philydraceae. If we assume that these more distant taxa are exclusively terrestrial, then the parsimony criterion indicates that the aquatic habit either arose independently in the two families, or it arose prior to the origin of

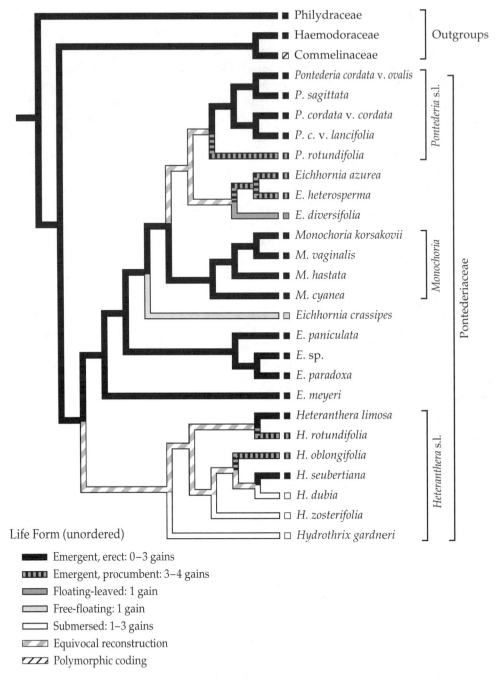

Figure 7.4. Reconstruction of diversification in aquatic life-form in Pontederiaceae and its closest relatives. The ingroup tree is one of four most parsimonious trees found using the combined chloroplast evidence from *rbc*L, *ndh*F, and a survey of restriction-site variation. The root of the ingroup tree is that indicated by a variety of closely related taxa using the two chloroplast DNA sequence-based data sets. See text for a description of the character-states and codings. Commelinaceae was coded as polymorphic. The reconstructed numbers of gains in each character-state are indicated in the legend. They refer to changes within Pontederiaceae (not the whole tree). These values were the same across all four shortest trees.

Commelinaceae, Haemodoraceae, Philydraceae, and Pontederiaceae, with a subsequent loss prior to the split of Commelinaceae and Haemodoraceae. If more distant clades than Philydraceae are aquatic, then homology of the aquatic habit between the two families is more parsimonious than non-homology. In either case, homology of the aquatic habit between Philydraceae and Pontederiaceae is a definite possibility. However, aquatic adaptations in Pontederiaceae are much more complete and diverse than in Philydraceae, indicating that the majority of the character diversification constituting this radiation has taken place in this family alone.

LIFE-FORM – An emergent, erect habit is reconstructed as the primitive condition in Pontederiaceae for all four shortest trees (see the branch that connects the outgroup taxa to Pontederiaceae; Figure 7.4). The free-floating form typical of *Eichhornia crassipes* arose directly from the emergent, erect habit, and the floating-leaved form arose from an emergent, erect or an emergent, procumbent form. In *Eichhornia* and *Pontederia* the emergent, procumbent form arose from an emergent, erect form on one or two occasions, although an origin of the former condition from a floating-leaved form in *Eichhornia* is also possible. Surprisingly, there was no phylogenetic record of any transitional forms between the emergent, erect, and submersed life-forms in *Heteranthera*. There were up to three independent origins of the submersed life-form from the emergent, erect form in this genus. There were no parsimonious reconstructions for any of the trees in which the emergent, procumbent habit was homologous between *Heteranthera* versus *Pontederia* and *Eichhornia* (Figure 7.4). There were two independent origins of this life-form in *Heteranthera* s. lat. under all reconstructions, and up to two independent origins of this form in species of *Pontederia* and *Eichhornia*. Under some of the most parsimonious reconstructions, the emergent, erect life-form within *Heteranthera* and *Pontederia* represented a reversion from a submersed or emergent, procumbent form. The emergent, erect habit in *Heteranthera* is thus potentially not homologous with the occurrence of this life-form outside the genus (Figure 7.4).

A number of taxa that are currently missing from the phylogenetic estimate of Pontederiaceae are likely to make an impact on future reconstructions of this character. These include several emergent, procumbent taxa in *Heteranthera* (*H. reniformis* and allies) and two emergent, erect species (*Heteranthera spicata* and *H. mexicana*). One missing species with a floating-leaved life-form is *Scholleropsis lutea*. This species probably belongs in *Heteranthera* s. lat. (Eckenwalder and Barrett 1986) and consequently may well represent an additional origin of the floating-leaved form in the family. Inclusion of these taxa may also indicate that some or all of the submersed taxa in *Heteranthera* did not arise directly from emergent, erect forms, a finding that in any case seems to us to be biologically implausible. The other missing floating-leaved form is *Eichhornia natans*, but as this species appears to be very closely related to *E. diversifolia* (Verdcourt 1968; Eckenwalder and Barrett 1986), it probably does not represent a novel origin of this life-form.

LIFE-CYCLE DURATION – Despite uncertainty at the base of the tree, with a variety of most parsimonious reconstructions of shifts in life history among the outgroup taxa, an annual life history was reconstructed as the primitive condition of Pontederiaceae for all shortest trees (Figure 7.5). Under all most parsimonious reconstructions on

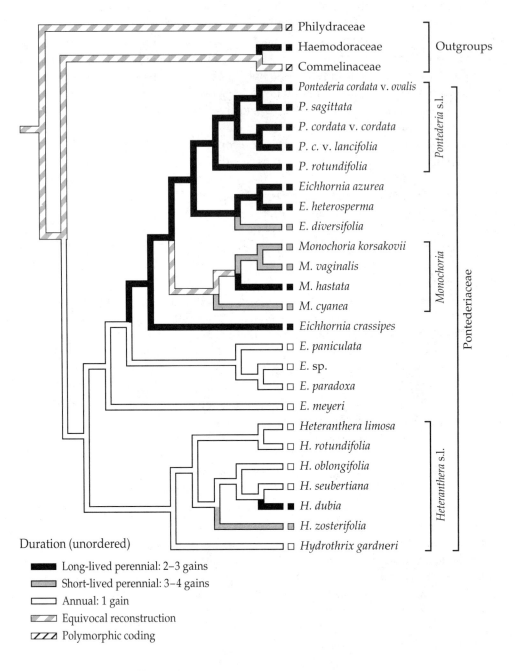

Figure 7.5. Reconstruction of diversification in life-cycle duration in Pontederiaceae and its closest relatives. The ingroup tree is one of four most parsimonious trees found using the combined chloroplast evidence from *rbc*L, *ndh*F, and a survey of restriction-site variation. The root of the ingroup tree is that indicated by a variety of closely related taxa using the two chloroplast DNA sequence-based data sets. See text for a description of the character-states and codings. Commelinaceae and Philydraceae were coded as polymorphic. The reconstructed numbers of gains in each character-state are indicated in the legend, and refer to changes within Pontederiaceae. These values were the same across all four shortest trees.

these trees, the annual life history was homologous for all species exhibiting this life history. It should be borne in mind, however, that many of the species that we coded as annual are capable of growing as short-lived perennials and are therefore "facultative" annuals. In contrast, species coded as short-lived perennials are incapable of an annual existence. Most workers assume that in herbaceous groups annuals are derived from perennials (Stebbins 1974), a shift in life history that has normally been invoked in the context of adaptive radiations from mesic to arid environments. However the reverse shifts may well have occurred in Pontederiaceae during invasion of permanent aquatic habitats. Such environments require specialized aquatic adaptations and may have represented relatively unsaturated niches. Under these circumstances, lack of species competition could have aided an evolutionary transition that for terrestrial groups occurs less frequently, a situation that may be analogous to the evolution of perenniality in island floras (see Carlquist 1974).

Long-lived perennials arose two or three times in the family, and long-lived perenniality in *Heteranthera dubia* was not homologous with other species in the family for any most parsimonious reconstruction. Instances of long- and short-lived perenniality in *Monochoria* were homologous with the occurrences of these forms in *Eichhornia* and *Pontederia* in some reconstructions, but not in others (Figure 7.5). Short-lived perenniality in *E. diversifolia* and *H. zosterifolia* arose uniquely in the terminal lineages leading to these species. Missing perennial taxa from this study include *H. reniformis* and its allies in *Heteranthera*, and several species of *Monochoria* and *Pontederia* s. lat.

CLONALITY – A non-clonal form is reconstructed as the most primitive condition in Pontederiaceae. Vegetative reproduction via stolons (typical only of *Eichhornia crassipes*) arose directly from this form, as did the instance of clonal reproduction via rhizomes in *Monochoria hastata*, which consequently must have arisen independently from the other instances of this clonal form in *Pontederia*. Species of *Pontederia* and *Eichhornia* that express clonality via stem fragmentation can also clone via rhizomes (we scored them using the former coding only), so the question of the number of origins or interconversions between these two clonality modes in the local part of the tree containing these taxa is somewhat moot. Examination of the most parsimonious reconstructions on the four shortest trees indicates that clonality via stem fragmentation arose independently in *Heteranthera* versus *Pontederia* and *Eichhornia*, and may have arisen from one to three times (Figure 7.6) in the former genus. Several missing taxa of Pontederiaceae in this study that are clonal include *H. reniformis* and its allies (which all express clonality via stem fragmentation) and several species in *Pontederia* s. lat. that can reproduce via rhizomes or stem fragmentation. The only taxa to reproduce via stem fragmentation are those with a growth form that could broadly be described as procumbent (emergent, submersed, or floating-leaved).

LEAF DEVELOPMENTAL PATHWAY – Pattern A, with a rapid transition to adult leaves, is the primitive form in the family, although it is possible that it also arose once by reversion from pattern C or D in the terminal lineage leading to *Heteranthera oblongifolia* (Figure 7.7). Pattern B, in which the production of petiolate leaves is more delayed than in pattern A, evolved once from pattern A in *Eichhornia*, supporting the idea that this represents a neotenic shift. The suggestion that the other two patterns (C and D,

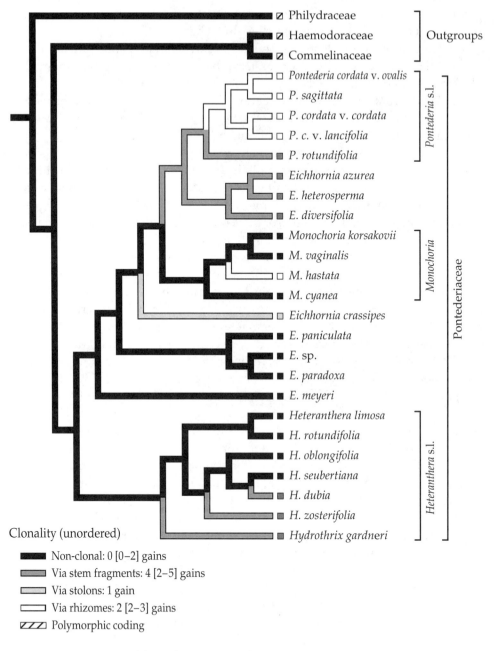

Figure 7.6. Reconstruction of diversification in clonality in Pontederiaceae and its closest relatives. The ingroup tree is one of four most parsimonious trees found using the combined chloroplast evidence from *rbc*L, *ndh*F, and a survey of restriction-site variation. The root of the ingroup tree is that indicated by a variety of closely related taxa using the two chloroplast DNA sequence-based data sets. See text for a description of the character-states and codings. The three outgroup taxa were coded as polymorphic. The reconstructed numbers of gains in each character-state are indicated in the legend. They refer to changes within Pontederiaceae. Where numbers of gains differ across the four shortest trees, this is indicated in square brackets.

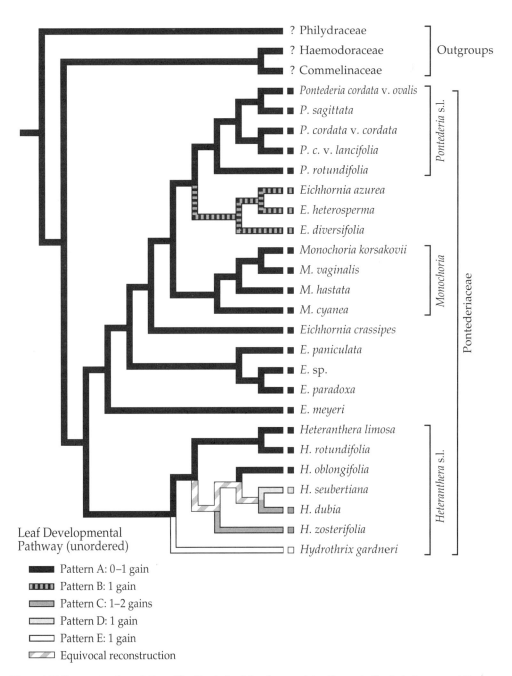

Figure 7.7. Reconstruction of diversification in leaf developmental pathway in Pontederiaceae and its closest relatives. The ingroup tree is one of four most parsimonious trees found using the combined chloroplast evidence from *rbc*L, *ndh*F, and a survey of restriction-site variation. See text for a description of the character-states and codings. The root of the ingroup tree is that indicated by a variety of closely related taxa using the two chloroplast DNA sequence-based data sets. The three outgroup taxa were coded as "unknown" (missing data) for this character (see text). The reconstructed numbers of gains in each character-state are indicated in the legend. They refer to changes within Pontederiaceae (not the whole tree). These values were the same across all four shortest trees.

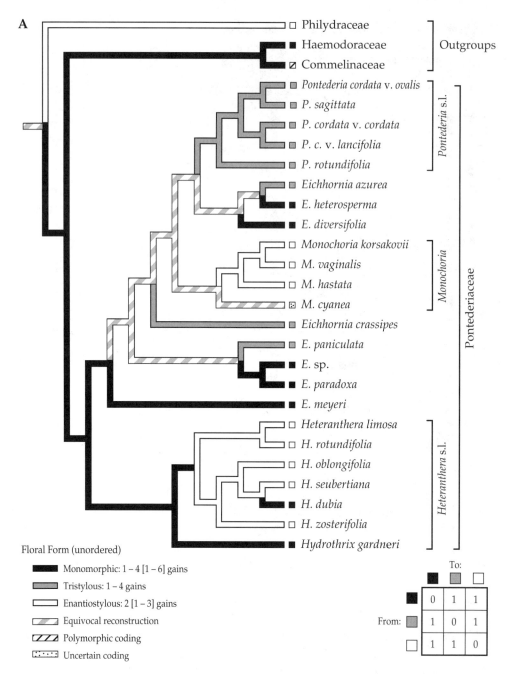

Figure 7.8. Reconstruction of diversification in floral form in Pontederiaceae and its closest relatives (Commelinaceae, Haemodoraceae, and Philydraceae). The ingroup tree is one of four most parsimonious trees found using the combined chloroplast evidence from *rbc*L, *ndh*F, and a survey of restriction-site variation. The root of the ingroup tree is that indicated by a variety of closely related taxa using the two chloroplast DNA sequence-based data sets. *Monochoria cyanea* was coded as "enantiostylous or monomorphic," and Commelinaceae was coded as polymorphic for this character (see text for descriptions of the character-states and codings). The reconstructed numbers of gains in each character-state are indicated in the legend. They refer to changes within Pontederiaceae (not the whole tree). Where numbers of gains differ

Adaptive Radiation in the Pontederiaceae 251

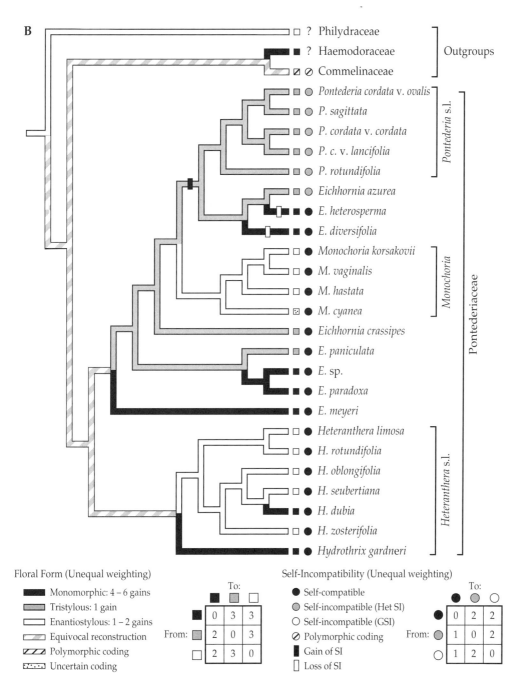

across the four shortest trees, this is indicated in square brackets. (**A**) Reconstructions of shifts in floral form, performed using Fitch optimization. (**B**) Reconstructions of shifts in floral form and self-incompatibility (SI) status performed using a "relaxed Dollo" weighting scheme described in the step matrix below the tree. Gains of either polymorphic form (tristyly or enantiostyly) were coded more heavily (3:2 weighting, see step matrix) than reversions to floral monomorphism. Gains of SI (solid bars) were coded more heavily (2:1, ACCTRAN optimization, see step matrix) than reversions to self-compatibility (hollow bars). GSI = gametophytic self-incompatibility; Het SI = heteromorphic, sporophytic self-incompatibility; see section on character codings.

which possess adult leaves resembling the juvenile leaves of pattern A) arose by various paedomorphic processes was supported by only some of the reconstructions on the shortest trees. A variety of shifts between patterns A, C, D, and E were seen in different most parsimonious reconstructions on the four shortest trees: pattern D arose from patterns A and C in different reconstructions, pattern C arose from patterns A, D, or E (with either one or two origins), and the leaf form unique to *Hydrothrix gardneri* (pattern E, with whorls of highly reduced leaves) could have arisen from patterns A or C on some shortest trees. To our knowledge, all of the taxa missing from the current study exhibit pattern A, apart from *Heteranthera mexicana* (pattern D) and *E. natans* (pattern B; probably homologous with the other three instances of this pattern). An increased sampling of taxa within *Heteranthera* s. lat. would be valuable for obtaining a less equivocal reconstruction of the evolution of patterns C to E, since so much of the variation in leaf developmental pathway is in this genus.

FLORAL FORM AND SELF-INCOMPATIBILITY – We considered two types of evolutionary schemes for the shifts in floral form and self-incompatibility status: (i) Fitch optimization, in which all character-state shifts are equiprobable and there is no implied order of change; and (ii) "relaxed Dollo" schemes that weight against the origin of self-incompatibility and the two polymorphic floral forms in the family, enantiostyly and tristyly. A large body of comparative and microevolutionary evidence indicates that evolutionary gains of tristyly are much more difficult than their loss (reviewed in Graham and Barrett 1995; Kohn et al. 1996), and it seems likely that the same is also true for self-incompatibility systems. Schemes that disfavor the origin of such complex characters are probably more biologically and historically accurate than those that weight all character-state shifts equally.

The choice of weighting scheme can have a profound influence on reconstructions of character evolution (compare Figures 7.8A,B). The scheme that weights all character shifts equally (Fitch optimization) indicates a wide possible range of gains in each floral form, including up to four origins of tristyly (Figure 7.8A). Not surprisingly, the scheme that weights gains of the floral polymorphisms more heavily than their reversion to monomorphism indicates a single origin of tristyly, up to two origins of enantiostyly, and multiple (four to six) origins of floral monomorphism in the family (Figure 7.8B). Enantiostylous flowers in *Monochoria* and *Heteranthera* were not homologous under either scheme, for any of the most parsimonious reconstructions on the shortest trees. Floral monomorphism and enantiostyly were the primitive floral forms in the family in different most parsimonious reconstructions (using the Fitch or relaxed Dollo optimization schemes). Of course, the precise weights that correspond to the actual probabilities of change in these floral forms are unknown, but only very small weighting biases (around 3:2 to 2:1; Graham and Barrett 1995; Kohn et al. 1996) were required to reconstruct a single origin of tristyly in the family.

The reconstructed shifts in self-incompatibility (SI) status indicates that it arose at most twice in the clade consisting of *Pontederia* s. lat. and the several species of *Eichhornia* associated with *E. azurea*. One most parsimonious reconstruction is shown in Figure 7.8B under a small weighting bias. Under equal weighting (Fitch optimization), or unequal weighting (2:1 bias against the origin of SI with DELTRAN optimization), two origins of SI are indicated on the tree (see Kohn et al. 1996). Under unequal weighting (2:1 bias with ACCTRAN optimization), a single origin and two losses of SI are implied (Figure 7.8B, and see Kohn et al. 1996). In either case, SI must have arisen subsequent to the origin of tristyly, given reconstructions of floral form that indicate a single origin of tristyly.

This evolutionary sequence casts doubt on models of the evolutionary origin of heterostyly (Charlesworth and Charlesworth 1979; Charlesworth 1979) in which an origin of self-incompatibility is required prior to the origin of floral heteromorphisms (Graham and Barrett 1995; Kohn et al. 1996). Floral shifts between the three floral types are undoubtedly associated with shifts in pollination mode. Tristylous species of *Eichhornia* and *Pontederia* are predominantly pollinated by nectar-collecting bees, enantiostylous species of *Monochoria* and *Heteranthera* are pollinated by pollen-collecting bees, and monomorphic species throughout the family are predominantly self-pollinating (see above). The reconstructions also indicate that at least some predominantly selfing lineages of *Eichhornia* have existed for substantial evolutionary periods and were even capable of speciation (see also Kohn et al. 1996), a finding at odds with Stebbins' (1957) view that selfing species are evolutionary dead ends. Missing taxa that could influence reconstructions of character evolution include *Eichhornia natans* and *Pontederia parviflora* (both monomorphic and presumably self-compatible).

Conclusions

There is substantial ecological evidence that a number of life-history traits in Pontederiaceae are involved in or affected by an aquatic existence, and that a variety of reproductive characters have undergone diversification in response to shifts in their mode of pollination. We used currently available molecular evidence of phylogenetic relationships within this family to reconstruct patterns of character diversification associated with adaptive radiations in vegetative and floral characters. Shifts in pollination mode, particularly those resulting in predominant self-pollination, occurred on multiple occasions in the family. Vegetative characters in the family are particularly prone to convergence (see also Eckenwalder and Barrett 1986). The extensive homoplasy in vegetative characters suggests that aquatic habitats are far from ecologically uniform, as has often been supposed. As with all phylogenetic analyses, these findings are liable to new interpretations when more taxa are sampled inside and outside the family, and when new sources of phylogenetic evidence become available. More detailed developmental and ecological work is also needed to determine how plastic some of our character classes are in different taxa. Several of the reconstructions of character diversification in our studies of Pontederiaceae challenge widely held views on the course of plant evolution. These include the shift from the annual to the perennial habit, the evolutionary longevity of some predominantly selfing lineages and the sequence in which morphological and physiological traits became associated in the heterostylous syndrome. In challenging these orthodox views, we hope that our analyses may help to provoke future research on these topics in Pontederiaceae and other aquatic plant families.

Acknowledgments

We thank the many colleagues who have helped us in collecting specimens of Pontederiaceae and Nancy Dengler, Jennifer Richards, Michael Simpson, and Connie Soros for helpful advice. The Natural Sciences and Engineering Research Council of Canada and the Province of Ontario supplied financial support in the form of research grants to SCHB and postgraduate fellowships to SWG.

References

Abrahamson, W. G. 1980. Demography and vegetative reproduction. Pp. 90–106 in *Demography and evolution in plant populations*, O. T. Solbrig, ed. Oxford, England: Blackwell Scientific Publications.

Adams, L. G. 1987. Philydraceae. Pp. 40–46 in *Flora of Australia, vol. 45*, A. S. George, ed. Canberra, Australia: Australian Government Publishing Service.

Alberch, P. S., Gould, S. J., Oster, G. F., and Wake, D. B. 1979. Size and shape in ontogeny and phylogeny. Paleobiology 5:296–317.

Arber, A. 1919. On heterophylly in water plants. American Naturalist 53:272–278.

Arber, A. 1920. *Water plants*. Cambridge, England: Cambridge University Press.

Arber, A. 1925. *Monocotyledons, a morphological study*. Cambridge, England: Cambridge University Press.

Barrett, S. C. H. 1978a. Pontederiaceae. Pp. 309–311 in *Flowering plants of the world*, V. H. Heywood, ed. Oxford, England: Oxford University Press.

Barrett, S. C. H. 1978b. Floral biology of *Eichhornia azurea* (Swartz) Kunth (Pontederiaceae). Aquatic Botany 5:217–228.

Barrett, S. C. H. 1979. The evolutionary breakdown of tristyly in *Eichhornia crassipes* (Mart.) Solms (Water Hyacinth). Evolution 33:499–510.

Barrett, S. C. H. 1980a. Sexual reproduction in *Eichhornia crassipes* (Water Hyacinth). I. Fertility of clones from diverse regions. Journal of Applied Ecology 17:101–112.

Barrett, S. C. H. 1980b. Sexual reproduction in *Eichhornia crassipes* (Water Hyacinth). II. Seed production in natural populations. Journal of Applied Ecology 17:113–124.

Barrett, S. C. H. 1988. Evolution of breeding systems in *Eichhornia*: a review. Annals of the Missouri Botanical Garden 75:741–760.

Barrett, S. C. H. 1989. Waterplant invasions. Scientific American 260:90–97.

Barrett, S. C. H. (ed.) 1992. *Evolution and function of heterostyly*. Berlin: Springer-Verlag.

Barrett, S. C. H. 1993. The evolutionary biology of tristyly. Oxford Surveys in Evolutionary Biology 9:283–326.

Barrett, S. C. H., and Anderson, J. M. 1985. Variation in expression of trimorphic incompatibility in *Pontederia cordata* L. (Pontederiaceae). Theoretical and Applied Genetics 70:355–362.

Barrett, S. C. H., and Forno, I. W. 1982. Style morph distribution in New World populations of *Eichhornia crassipes* (Mart.) Solms-Laubach (Water Hyacinth). Aquatic Botany 13:299–306.

Barrett, S. C. H., and Glover, D. E. 1985. On the Darwinian hypothesis of the adaptive significance of tristyly. Evolution 37:745–760.

Barrett, S. C. H., and Seaman, D. E. 1980. The weed flora of Californian rice fields. Aquatic Botany 9:351–376.

Barrett, S. C. H., Eckert, C. G., and Husband, B. C. 1993. Evolutionary processes in aquatic plant populations. Aquatic Botany 44:105–145.

Barrett, S. C. H., Kohn, J. R., and Cruzan, M. B. 1992. Experimental studies of mating-system evolution: the marriage of marker genes and floral biology. Pp. 192–230 in *Ecology and evolution of plant reproduction: new approaches*, R. Wyatt, ed. New York, NY: Chapman and Hall.

Bell, G. 1982. *The masterpiece of nature: the evolution and genetics of sexuality*. Berkeley, CA: University of California Press.

Bell, G. 1985. On the function of flowers. Proceedings Royal Society of London. Series B. Biological Sciences 224:223–265.

Bertin, R. I., and Newman, C. M. 1993. Dichogamy in angiosperms. Botanical Review 59:112–152.

Bradshaw, A. D. 1965. Evolutionary significance of phenotypic plasticity in plants. Advances in Genetics 13:115–155.

Briggs, J. C. 1987. *Biogeography and plate tectonics*. Amsterdam, The Netherlands: Elsevier.

Brooks, D. R., and McLennan, D. A. 1991. *Phylogeny, ecology, and behavior: a research program in comparative biology*. Chicago, IL: University of Chicago Press.

Buchmann, S. L., 1983. Buzz pollination in angiosperms. Pp. 73–113 in *Handbook of experimental pollination biology*, C. E. Jones and R. J. Little, eds. New York, NY: Van Nostrand.

Bureau, E. 1892. Sur la presence d'une Araliacee et d'une Pontederiacee fossile dans le calcaire grossier parisien. Comptes Rendus Hebdomadaires des Seances. Academie des Sciences 115:1335–1337.

Campbell, D. R. 1989. Measurements of selection in a hermaphrodite plant: variation in male and female pollination success. Evolution 45:1965–1968.

Carlquist, S. 1974. *Island biology*. New York, NY: Columbia University Press.

Carson, H. L. 1985. Unification of speciation theory in plants and animals. Systematic Botany 10:380–390.

Charlesworth, D. 1979. The evolution and breakdown of tristyly. Evolution 33:486–498.

Charlesworth, D., and Charlesworth, B. 1979. A model for the evolution of distyly. American Naturalist 114:467–498.

Charlesworth, D., and Charlesworth, B. 1987. Inbreeding depression and its evolutionary significance. Annual Review of Ecology and Systematics 18: 237–268.

Chase, M. W., Soltis, D. E., Olmstead, R. G., Morgan, D., Les, D. H., Mishler, B. D., Duvall, M. R., Price, R. A., Hills, H. G., Qiu, Y., Kron, K. A., Rettig, J. H., Conti, E., Palmer, J. D., Manhart, J. R., Sytsma, K. J., Michaels, H. J., Kress, W. J., Karol, K. G., Clark, W. D., Hedrén, M., Gaut, B. S., Jansen, R. K., Kim, K.- J., Wimpee, C. F., Smith, J. F., Furnier, G. R., Strauss, S. H., Xiang, Q.- Y., Plunkett, G. M., Soltis, P. S., Swensen, S. M., Williams, S. E., Gadek, P. A., Quinn, C. J., Eguiarte, L. E., Golenberg, E., Learn, G. H., Graham, S. W., Barrett, S. C. H., Dayanandan, S., and Albert, V. A. 1993. Phylogenetics of seed plants: an analysis of nucleotide sequences from the plastid gene *rbc*L. Annals of the Missouri Botanical Garden 80:528–580.

Collinson, M. E., Boulter, M. C., and Holmes, P. L. 1993. Magnoliophyta ("Angiospermae"). Pp. 809–841 in *The fossil record 2*, M. J. Benton, ed. London, England: Chapman and Hall.

Cook, C. D. K. 1978. The *Hippuris* syndrome. Pp. 163–176 in *Essays in plant taxonomy*, H. E. Street, ed. London, England: Academic Press.

Cook, C. D. K. 1987. Dispersion in aquatic and amphibious vascular plants. Pp. 179–192 in *Plant life in aquatic and amphibious habitats*, R. M. M. Crawford, ed. Oxford, England: Blackwell Scientific Publications.

Cook, C. D. K. 1990. *Aquatic plant book*. The Hague, The Netherlands: SPB Academic Publishing.

Cook, R. E. 1985. Growth and development in clonal plant populations. Pp. 259–296 in *Population biology and evolution of clonal organisms*, J. B. C. Jackson, L. W. Buss, and R. E. Cook, eds. New Haven, CT: Yale University Press.

Cook, S. A., and Johnston, M. P. 1968. Adaptation to heterogeneous environments I. Variation in heterophylly in *Ranunculus flammula* L. Evolution 22:496–516.

Cox, P. A. 1988. Hydrophilous pollination. Annual Review of Ecology and Systematics 19: 261–280.

Crawford, R. M. M. (ed.) 1987. *Plant life in aquatic and amphibious habitats*. Oxford, England: Blackwell Scientific Publications.

Dahlgren, R. M. T., and Clifford, H. T. 1982. *The monocotyledons: a comparative study*. London, England: Academic Press.

Dahlgren, R. M. T., and F. N. Rasmussen. 1983. Monocotyledon evolution: characters and phylogenetic estimation. Evolutionary Biology 16:255–295.

Dahlgren, R. M. T., Clifford, H. T., and Yeo, P. F. 1985. *The families of the monocotyledons: structure, evolution and taxonomy*. Berlin, Germany: Springer-Verlag.

Darwin, C. 1877. *The different forms of flowers on plants of the same species*. London, England: Murray.

Davis, J. I. 1995. A phylogenetic structure for the monocotyledons, as inferred from chloroplast DNA restriction site variation, and a comparison of measures of clade support. Systematic Botany 20:503–527.

Den Hartog, C., and Segal, S. 1964. A new classification of water-plant communities. Acta Botanica Neerlandica 13:367–393.

Devlin, B., and Ellstrand, N. C. 1990. Male and female fertility variation in wild radish, a hermaphrodite. American Naturalist 136:87–107.

Doyle, J. J. 1992. Gene trees and species trees: molecular systematics as one-character taxonomy. Systematic Botany 11:373–391.

Duvall, M. R., Clegg, M. T., Chase, M. W., Clark, W. D., Kress, W. J., Hills, H. G., Eguiarte, L. E., Smith, J. F., Gaut, B. S., Zimmer, E. A., and Learn, G. H. Jr. 1993. Phylogenetic hypotheses for the monocotyledons constructed from *rbc*L sequence data. Annals of the Missouri Botanical Garden 80:607–619.

Eckenwalder, J. E., and Barrett, S. C. H. 1986. Phylogenetic systematics of Pontederiaceae. Systematic Botany 11:373–391.

Eckert, C. G., and Barrett, S. C. H. 1993. Clonal reproduction and patterns of genotypic diversity in *Decodon verticillatus* (Lythraceae). American Journal of Botany 80:1175–1182.

Endress, P. K. Major evolutionary traits of monocot flowers. Pp. 43–79 in *Monocotyledons: systematics and evolution*, P. J. Rudall, P. J. Cribb, D. F. Cutler, and C. J. Humphries, eds. Kew, England: Royal Botanical Gardens.

Faden, R. B. 1988. Vegetative and reproductive features of forest and nonforest genera of African Commelinaceae. Monographs in Systematic Botany, Missouri Botanical Garden 25:521–526.

Faden, R. B. 1991. The morphology and taxonomy of *Aneilema* R. Brown (Commelincaeae). Smithsonian Contributions in Botany 76:1–166.

Fenster, C. B. 1995. Mirror image flowers and their effect on outcrossing rate in *Chamaecrista fasciculata* (Leguminosae). American Journal of Botany 82:46–50.

Fitch, W. M. 1971. Toward defining the course of evolution: minimal change for a specific tree topology. Systematic Zoology 20:406–416.

Fritel, P. H. 1928. Sur la presence des genres *Salvinia* Mich., *Nymphaea* Tourn. et *Pontederia* Linn. dans les argiles sparnaciennes du Montois. Comptes Rendus Hebdomadaires des Seances, Academie des Sciences 147:724–725.

Futuyma, D. J. 1986. *Evolutionary biology*. Sunderland, MA: Sinauer Associates.

Givnish, T. J. 1987. Comparative studies of leaf form: assessing the relative roles of selective pressures and phylogenetic constraints. New Phytologist 106 (suppl.):131–160.

Givnish, T. J. 1995. Plant stems: biomechanical adaptation for energy capture and influence on species distributions. Pp. 3–49 in Plant stems: physiology and functional morphology, B. L. Gaertner, ed. San Diego, CA: Academic Press.

Givnish, T. J., and Vermeij, G. J. 1977. Sizes and shapes of liane leaves. American Naturalist 110:743–778.

Goldberg, A. 1989. Classification, evolution and phylogeny of the families of monocotyledons. Smithsonian Contributions in Botany 71:1–73.

Grace, J. B. 1993. The adaptive significance of clonal reproduction in angiosperms: an aquatic perspective. Aquatic Botany 44:159–180.

Graham, S. W. 1997. Phylogenetic analysis of breeding-system evolution in heterostylous monocotyledons. Ph.D. dissertation, University of Toronto.

Graham, S. W., and Barrett, S. C. H. 1995. Phylogenetic systematics of Pontederiales: implications for breeding-system evolution. Pp. 415–441 in *Monocotyledons: systematics and evolution*, P. J. Rudall, P. J. Cribb, D. F. Cutler, and C. J. Humphries, eds. Kew, England: Royal Botanical Gardens.

Grant, V., and Grant, K. 1965. *Flower pollination in the phlox family*. New York, NY: Columbia University Press.

Grime, J. P. 1979. *Plant strategies and vegetation processes*. London, England: John Wiley.

Hamann, U. 1966. Embryologische, morphologisch-anatomische und systematische Untersuchungen an Philydraceen. Willdenowia 4:1–178.

Harder, L. D., and Barrett, S. C. H. 1995. Mating cost of large floral displays in hermaphrodite plants. Nature 373:512–515.

Harder, L. D., and Barrett, S. C. H. 1996. Pollen dispersal and mating patterns in animal-pollinated plants. Pp. 140–190 in *Floral biology: studies on floral evolution in animal-pollinated plants*, D. G. Lloyd and S. C. H. Barrett, eds. New York, NY: Chapman and Hall.

Harper, J. L. 1977. *Population biology of plants*. London, England: Academic Press.

Horn, C. N. 1983. The annual growth cycle of *Heteranthera dubia* in Ohio. The Michigan Botanist 23:29–34.

Horn, C. N. 1985. A systematic revision of the genus *Heteranthera* (*sensu lato*). Ph.D. dissertation, University of Alabama.

Horn, C. N. 1988. Developmental heterophylly in the genus *Heteranthera* (Pontederiaceae). Aquatic Botany 31:197–209.

Huber, H. 1977. The treatment of the monocotyledons in an evolutionary system of classification. Plant Systematics and Evolution Supplement 1:285–298.

Husband, B. C., and Barrett, S. C. H. 1992. Pollinator visitation in populations of tristylous *Eichhornia paniculata* in northeastern Brazil. Oecologia 89:365–371.

Hutchinson, G. E. 1975. *Treatise on limnology* III. *limnological botany*. New York, NY: John Wiley.
Iyengar, M. O. T. 1923. On the biology of the flowers of *Monochoria*. Journal of the Indian Botanical Society 3:170–173.
Johnson, S. D. 1996. Pollination, adaptation and speciation models in the Cape flora of South Africa. Taxon 45:59–66.
Knowlton, F. H. 1922. Revision of the flora of the Green River Formation, with descriptions of new species. United States Geological Survey, Professional Paper 131:133–182.
Kohn, J. R., and Barrett, S. C. H. 1992. Experimental studies on the functional significance of heterostyly. Evolution 46:43–55.
Kohn, J. R., Graham, S. W., Morton, B., Doyle, J. J., and Barrett, S. C. H. 1996. Reconstruction of the evolution of reproductive characters in Pontederiaceae using phylogenetic evidence from chloroplast DNA restriction-site variation. Evolution 50:1454–1469.
Leakey, R. R. B. 1981. Adaptive biology of vegetatively regenerating weeds. Advances in Applied Biology 6:57–90.
Lee, D. W., and Richards, J. H. 1991. Heteroblastic development in vines. Pp. 205–243 in *The biology of vines*, F. E. Putz and J. H. Richards, eds. New York, NY: Cambridge University Press.
Les, D. H. 1988. Breeding systems, population structure, and evolution in hydrophilous angiosperms. Annals of the Missouri Botanical Garden 75:819–835.
Lloyd, D. G. 1992. Evolutionary stable strategies of reproduction in plants: who benefits and how? Pp. 137–168 in *Ecology and evolution of plant reproduction*, R. Wyatt, ed. New York, NY: Chapman and Hall.
Lloyd, D. G., and Webb, C. J. 1986. The avoidance of interference between the presentation of pollen and stigmas in angiosperms. I. Dichogamy. New Zealand Journal of Botany 24:135–162.
Lloyd, D. G., and Webb, C. J. 1992. The evolution of heterostyly. Pp. 151–178 in *Evolution and function of heterostyly*, S. C. H. Barrett, ed. Berlin, Germany: Springer-Verlag.
Lowden, R. M. 1973. Revision of the genus *Pontederia* L. Rhodora 75:427–487.
Maddison, W. P., and Maddison, D. R. 1992. *MacClade: analysis of phylogeny and character evolution, ver. 3.0*. Sunderland, MA: Sinauer Associates.
Maddison, W. P., Donoghue, M. J., and Maddison, D. R. 1984. Outgroup analysis and parsimony. Systematic Zoology 33:83–103.
Madsen, J. 1986. The production and physiological ecology of the submerged aquatic macrophyte community in Badfish Creek, Wisconsin. Ph.D. dissertation, University of Wisconsin, Madison.
Maynard Smith, J. 1978. *The evolution of sex*. Cambridge, England: Cambridge University Press.
Nee, S., and Harvey, P. 1994. Getting to the roots of flowering plant diversity. Science 264:1549–1550.
Nixon, K. C., and Carpenter, J. M. 1993. On outgroups. Cladistics 9:314–426.
Nixon, K. C., and Davis J. I. 1991. Polymorphic taxa, missing values and cladistic analysis. Cladistics 7:233–241.
Ornduff, R., and Dulberger, R. 1978. Floral enantiomorphy and the reproductive system of *Wachendorfia paniculata* (Haemodoraceae). New Phytologist 80:427–434.
Owens, S. J. 1981. Self-incompatibility in the Commelinaceae. Annals of Botany 47:567–581.
Patil, G. V., and R. B. Singh. 1978. Fossil *Eichhornia* from the Eocene Deccan Intertrappean beds, India. Palaeontographica. Abteilung B. Palaeophytologie 167:1–7.
Raunkiaer, C. 1934. *The life forms of plants and statistical plant geography*. Oxford, England: Clarendon Press.
Richards, J. H. 1982. Developmental potential of axillary buds of water hyacinth, *Eichhornia crassipes* Solms. (Pontederiaceae). American Journal of Botany 69:615–622.
Richards, J. H. 1983. Heteroblastic development in the water hyacinth *Eichhornia crassipes* Solms. Botanical Gazette 144:247–259.
Ridley, H. N. 1930. *The dispersal of plants throughout the world*. Ashford, U.K.: Reeve.
Rosatti, T. J. 1987. The genera of Pontederiaceae in the southeastern United States. Journal of the Arnold Arboretum 68:35–71.
Rutishauser, R. 1983. *Hydrothrix gardneri* Bau und Entwicklung einer einartigen Pontederiacee. Botanische Jahrbücher für Systematik Pflanzengeschichte und Pflanzengeographie 104:115–141.
Sanderson, M. J., and Donoghue M. J. 1989. Patterns of variation in levels of homoplasy. Evolution 43:1781–1795.
Sanderson, M. J., and Donoghue, M. J. 1994. Shifts in diversification rate with the origin of the angiosperms. Science 264:1590–1593.
Sculthorpe, C. D. 1967. *The biology of aquatic vascular plants*. London, England: Edward Arnold.
Simpson, G. G. 1953. *The major features of evolution*. New York, NY: Columbia University Press.

Simpson, M. G. 1987. Pollen ultrastructure of the Pontederiaceae: evidence from exine homology with the Haemodoraceae. Grana 26:113–126.

Simpson, M. G. 1990. Phylogeny and classification of the Haemodoraceae. Annals of the Missouri Botanical Garden 77:722–784.

Sinnot, E. W. 1960. *Plant morphogenesis*. New York, NY: McGraw-Hill.

Stebbins, G. L. 1957. Self-fertilization and population variability in the higher plants. American Naturalist 41:337–354.

Stebbins, G. L. 1970. Adaptive radiation of reproductive characteristics in angiosperms. 1. Pollination mechanisms. Annual Review of Ecology and Systematics 1:307–326.

Stebbins, G. L. 1974. *Flowering plants: evolution above the species level*. Cambridge, MA: Belknap Press.

Swofford, D. L., and Olsen, G. J. 1990. Phylogeny reconstruction. Pp. 411–500 in *Molecular systematics*, D. M. Hillis and C. Moritz, eds. Sunderland, MA: Sinauer Associates.

Thorne, R. F. 1992a. An updated phylogenetic classification of the flowering plants. Aliso 13:365–389.

Thorne, R. F. 1992b. Classification and geography of the flowering plants. Botanical Review 58:225–348.

Tillich, H.-J. 1994. Untersuchungen zum Bau der Keimpflanzen der Philydraceae und Pontederiaceae (Monocotyledoneae). Sendtnera 2:171–186.

Tillich, H.-J. 1995. Seedlings and systematics in monocotyledons. Pp. 303–352 in *Monocotyledons: systematics and evolution*, P. J. Rudall, P. J. Cribbs, D. F. Cutler, and C. J. Humphries, eds. Kew, England: Royal Botanic Gardens.

Van Steenis, C. G. J. 1957. Specific and infraspecific delimitation. Flora Malesiana, Series 1, Spermatophyta 5:167–234.

Verdcourt, B. 1968. Pontederiaceae. Pp. 1–8 in *Flora of tropical east Africa*, E. Milne-Redhead and R. M. Polhill, eds. London, England: Crown Agents for Overseas Governments and Administrations.

Vogel, S. 1978. Evolutionary shifts from reward to deception in pollen flowers. Pp. 89–96 in *The pollination of flowers by insects*, A. J. Richards, ed. London, England: Academic Press.

Wang, G., Miura, R., and Kusanagi, T. 1995. The enantiostyly and the pollination biology of *Monochoria korsakowii* (Pontederiaceae). Acta Phytotaxonomica Geobotanica 46: 55–65.

Williams, G. C. 1975. *Sex and evolution*. Monographs in Population Biology. Princeton, NJ: Princeton University Press.

Wilson, J. 1887. On the dimorphism of flowers of *Wachendorfia paniculata*. Transactions and Proceedings of the Botanical Society of Edinburgh 17:73–77.

Wolfe, L. M., and Barrett, S. C. H. 1988. Temporal changes in the pollinator fauna of tristylous *Pontederia cordata*, an aquatic plant. Canadian Journal of Zoology 66:1421–1424.

Wylie, R. B. 1917. Cleistogamy in *Heteranthera dubia*. Bulletin from the Laboratories of Natural History, University of Iowa 7:48–58.

8 Molecular evolution and adaptive radiation in *Brocchinia* (Bromeliaceae: Pitcairnioideae) atop tepuis of the Guayana Shield

Thomas J. Givnish, Kenneth J. Sytsma,
James F. Smith, William J. Hahn, David H. Benzing,
and Elizabeth M. Burkhardt

Adaptive radiation – the rise of a diversity of ecological roles and attendant adaptations within a lineage (Osborn 1902; Simpson 1953) – is one of the most important processes bridging ecology and evolution (Chapter 1). As exemplified by Darwin's finches (Lack 1947; Grant 1986), it has profound implications for the origin of adaptations, the genesis of patterns in species richness and community structure, and the study of phylogeny and its relationship to morphology and geography. Several well-known studies of adaptive radiation – in such organisms as Darwin's finches, African rift lake cichlids, Australian marsupials, Hawaiian honeycreepers and tarweeds – have helped shape the conceptual basis of modern ecology, evolutionary biology, systematics, and biogeography (see Amadon 1950; Grant 1963; Mayr 1963, 1970; Carlquist 1965, 1970; Carson et al. 1970; Fryer and Iles 1972; Greenwood 1974, 1978; Grant 1981, 1986; Schluter et al. 1985; Schaefer and Lauder 1986; Carr et al. 1987; Liem 1990; Wilson 1992).

Recently, however, it has become clear that few of these classic cases were studied rigorously, at least from a phylogenetic viewpoint. The fundamental problem is that, in almost every case, the very characters whose radiation was being studied (e.g., beak size and shape) were also used to help classify the organisms in question (e.g., Lack 1947). This exercise can easily become circular, leading to traits being traced down evolutionary pathways determined, at least in part, by the traits themselves (Givnish 1987; Skinner 1988; Sytsma et al. 1991; Albert et al. 1992b; Baldwin 1992; Brooks and McLennan 1994; Crisp 1994). Difficulties can arise even if the characters under study are excluded from phylogenetic analysis, if the remaining traits are themselves subject to high levels of convergence or chance recurrence (Givnish et al. 1995; Givnish and Sytsma 1997: Chapter 1).

Given these fundamental problems with past approaches to adaptive radiation, many important findings in ecology and evolutionary biology may need re-examination. The primary challenge is systematic: **any rigorous, non-circular study of adaptive radiation must be based on a phylogeny that has been derived independently of the traits involved in that radiation.** The emergence of molecular systematics over the last decade has provided the basis for deriving such phylogenies, and has made the time ripe for a revolution in the analysis of adaptive radiation.

Several studies have shown that sequence and restriction-site variation in nuclear and organellar DNA provide a powerful tool for inferring relationships among species, genera, families, and higher taxonomic categories of plants (see Sytsma 1990; Clegg and Zurawski 1992; Hamby and Zimmer 1992; Soltis et al. 1992; Chase et al. 1993; Clegg 1993; Doyle 1993; Hillis et al. 1993, 1994; Sytsma and Hahn 1994, 1996). Phylogenetic analyses based on chloroplast DNA (cpDNA) restriction-site variation show significantly less homoplasy than those based on morphology; cpDNA sequence data show the same low level of homoplasy when the same species are compared (Chapter 2). The large number of independent characters sampled in cpDNA analyses and their low level of homoplasy provide detailed, precise data on phylogenetic relationships. Comparisons of phylogenies based on plastid DNA (usually inherited uniparentally) with those based on nuclear DNA (inherited biparentally) can provide a direct means of screening for hybridization and/or introgression, and for identifying the parental taxa involved (Rieseberg and Morefield 1995).

Over the past seven years, several researchers have begun to use molecular systematics as a basis for studies of adaptive radiation (Baldwin et al. 1990; Meyer et al. 1990, 1991; Givnish et al. 1990a,b, 1994, 1995; Springer and Kirsch 1991; Sytsma et al. 1991; Albert et al. 1992b; Baldwin 1992; Sturmbauer and Meyer 1992; Kocher et al. 1993; Knox et al. 1993; Moran et al. 1993; Baldwin and Robichaux 1995; Hodges and Arnold 1994a,b; Kress et al. 1994). We believe that the study of adaptive radiation based on molecular systematics can provide a powerful approach to several key questions at the interface of ecology, evolutionary biology, systematics, and biogeography, including:

- Do phenotypic similarities among species in a lineage reflect ecological similarities, close relationships, or both? How reliable a guide to phylogeny are traditional morphological characters?
- What has been the evolutionary pathway to various adaptations? Did key innovations arise once or repeatedly within a lineage? Did they accelerate rates of speciation and/or genetic evolution? To what extent do ecology and morphology reflect phylogenetic history?
- What is the relationship between phylogeny and geographic distribution? Are distributional patterns more reflective of the history of speciation and dispersal/vicariance (and thus, of geological history), or of the adaptations of individual species to specific environments?

To show how molecular systematics can help address these questions, here we present a molecular phylogeny for *Brocchinia* (Bromeliaceae: Pitcairnioideae), and then use it to analyze patterns of adaptive radiation and geographic diversification in response to extreme nutrient poverty in this remarkable but heretofore little studied group (Plate 1). *Brocchinia*, with roughly 20 species (Smith 1984, 1986; Holst 1997), is of extraordinary evolutionary interest for four reasons:

First, it is endemic to the Guayana Shield (Figure 8.1), one of the two ancient cores of South America and a center of speciation and endemism for many plant and animal groups (Mayr and Phelps 1967; Maguire 1970; Smith and Downs 1974; Steyermark 1986; Huber 1988; McDiarmid and Gorzula 1989; Funk and Brooks 1990; Berry et al. 1995). Several species of *Brocchinia* are restricted to the remote, often cliff-lined summits (> 1,500 m) of the tepuis – or sandstone table mountains – of this region, and many occur on only one or a few tepuis (Table 8.1). Several are ecological dominants in bogs, moist sandy savannas, and scrub, and on sandstone outcrops (Steyermark 1961, 1967; Givnish et al. 1984; Varadarajan 1986a; Givnish 1989). The tepuis are composed of Precambrian sandstones of the Roraima Supergroup and overlie the older granitic basement of the Guayana Shield. They were uplifted during the late Cretaceous (ca. 100 Mya) as a consequence of the rifting of the Atlantic; the Roraima sediments were subsequently dissected into separate tablelands by erosion and chemical dissolution (Ghosh 1985; Pouyllau and Seurin 1985; Briceño and Schubert 1990; Gibbs and Barron 1993). Today, the tepuis are "islands in the sky," with a cool, extremely wet climate and exceedingly nutrient-poor soils; they tower 500 to 2,500 m above the surrounding tropical lowlands, and are separated from each other by up to 200 km (Mayr and Phelps 1967; Maguire 1970; Brewer-Carías 1978; Steyermark 1986; George 1988; Huber 1995a).

Second, *Brocchinia* may have diverged from other bromeliad lineages at or near the base of the Pitcairnioideae, based on a cladistic analysis of morphological variation (Varadarajan and Gilmartin 1988). Six of the other 16 genera in this subfamily are also endemic to the Guayana Shield (Smith and Downs 1974; Smith 1984, 1986). The Pitcairnioideae is considered the ecologically least specialized of the three bromeliad subfamilies (Pittendrigh 1948; Benzing 1980); some believe it diverged from the more epiphytic Bromelioideae and Tillandsioideae before they diverged from each other, early in the evolution of an almost exclusively neotropical family with over 2,400 species (Smith and Downs 1974; Gilmartin and Brown 1987; Ranker et al. 1990).

Third, and most remarkably, *Brocchinia* has undergone an adaptive radiation in mechanisms of nutrient capture unparalleled at the generic level in angiosperms (Table 8.1). *Brocchinia* includes carnivores, ant-fed myrmecophytes, species with N_2-fixing symbionts, tank epiphytes, and non-impounding terrestrial forms (Givnish et al. 1984; Benzing et al. 1985; Givnish 1989; Gonzales et al. 1991). It is one of only two genera of flowering plants in which carnivory is present but not universal (Givnish 1989). *Brocchinia* contains some of the shortest (< 5 cm) and tallest (> 8 m) members of the Bromeliaceae, and includes herbs, shrubs, rosette trees, and one vinelike taxon (see Plate 1).

Finally, *Brocchinia*'s diversity of nutritional strategies is accompanied by extensive variation in the form and nutrient uptake capacity of its foliar scales, or trichomes (Givnish et al. 1984; Benzing et al. 1985; Owen et al. 1988). Trichomes are an organ of considerable systematic and ecological significance in bromeliads (Robinson 1969; Tomlinson 1969; McWilliams 1974; Smith and Downs 1974; Benzing 1980; Varadarajan and Gilmartin 1987, 1988). Benzing (1980, 1990a) has argued that absorptive leaf trichomes were a "key innovation" for bromeliad evolution, permitting several bromelioid and tillandsioid lineages to invade the epiphytic adaptive zone in neotropical rain forests and cloud forests, speciate explosively, and thereby generate one of the larger families of flowering plants. Absorptive trichomes are present in *Brocchinia* but otherwise absent from the pitcairnoids, most of which are spiny, drought-adapted, terrestrial species (McWilliams 1974; Benzing 1980, 1990a; Benzing et al. 1985; Varadarajan and Gilmartin 1987). *Brocchinia* is also the only pitcairnioid genus to have evolved the tank-forming habit, in which strongly overlapping leaf bases impound rain water and falling vegetable debris; the tank habit is

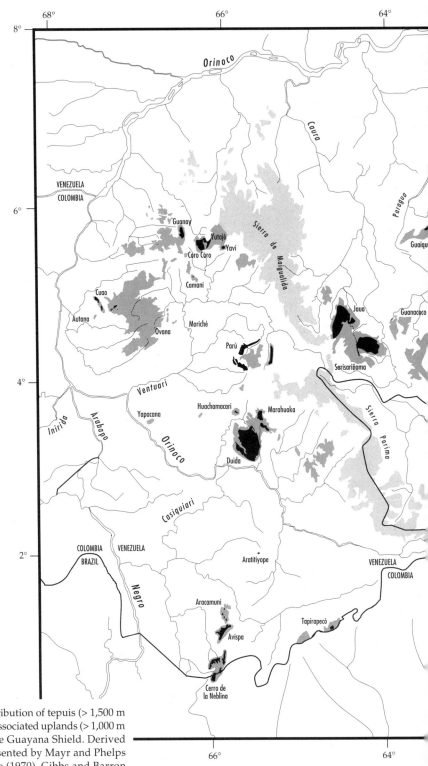

Figure 8.1. Distribution of tepuis (> 1,500 m elevation) and associated uplands (> 1,000 m elevation) in the Guayana Shield. Derived from maps presented by Mayr and Phelps (1967), Maguire (1970), Gibbs and Barron (1993), and Huber and Berry (1995).

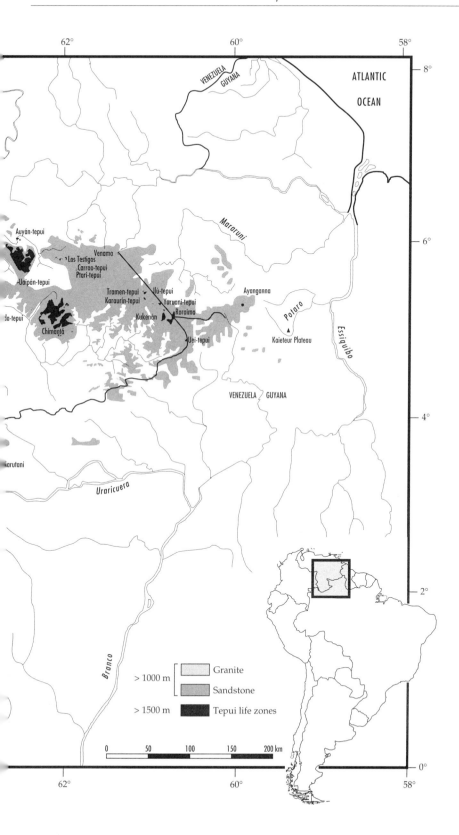

Table 8.1. Ecology, elevational range, habitat, and geographic distribution of *Brocchinia* species. Data compiled from Steyermark (1961), Smith and Downs (1974), Smith (1984, 1986), Varadarajan (1986a), Givnish et al. (1984, 1986), Holst (1997), and personal observations. The first five growth forms are all tank-forming species, with tightly overlapping leaf bases that impound rainwater; the remaining two growth forms are non-impounding.

Species	Elevation (m)	Habitat	Geographical range
Carnivores:			
B. reducta	900–2,200	Bogs, wet sandy savannas, sandstone outcrops	Eastern tepuis, Gran Sabana
B. hechtioides	1,400–2,125	Bogs, wet sandy savannas, sandstone outcrops	Eastern and western tepuis
Tank epiphytes:			
B. hitchcockii	2,000	Cloud forest epiphyte	Cerro Parú
B. tatei	500–2,400	Cloud forest epiphyte, terrestrial on moist sand/sandstone	Eastern and western tepuis
Tank terrestrial:			
B. gilmartinii	1,300	Terrestrial on moist sand/sandstone	La Escalera (N escarpment of Gran Sabana)
Ant-fed myrmecophyte:			
B. acuminata	600–2,100	Tepui scrub, edges of bogs and cloud forests	Eastern and western tepuis, Sierra de Maiguilida, mesetas of SE Colombia
Arborescent species:			
B. micrantha	500–1,200	Gaps in cloud forest	Slopes of eastern tepuis, Gran Sabana
B. paniculata	400–1,500	Gaps in cloud forest	Slopes of western tepuis, adjacent areas in SE Colombia
Paludophytes:			
B. melanacra*	1,300–2,000	Wet meadows and bogs	Western tepuis, Sierra de Maiguilida, Sierra Parima
B. prismatica	100–200	Amazonian savannas	Lowlands on wet white sand along Rios Orinoco and Ventuari near their confluence
B. steyermarkii	460–1,220	Wet meadows and bogs	Gran Sabana region
Saxicoles:			
B. amazonica	1,000	Exposed sandstone	Cerro Aracá
B. cataractarum	60	Exposed sandstone	Potaro River gorge
B. cowanii	1,200–1,300	Exposed sandstone	Cerro Moriche
B. delicatula	400–1,100	Exposed sandstone	Cañon Grande of Cerro de la Neblina
B. maguirei†	1,400–1,800	Exposed sandstone, wet tepui scrub	Autana, Sipapo, Duida
B. pygmaea	150	Exposed sandstone	Rio Siapa
B. rupestris	1,000	Exposed sandstone	Korupung River
B. serrata†	300–450	Exposed sandstone	Mesetas of SE Colombia
B. vestita	1,200–2,000	Exposed sandstone, often near streams	Duida, Cerro de la Neblina

* Pyrophyte with highly sclerotized leaf tips that protect the terminal bud from occasional fires (Givnish et al. 1986).
† Caulescent saxicoles, 1 m or greater in height when flowering

thought to have been another key innovation for the evolution of epiphytism in bromeliads (Benzing 1980, 1990a,b).

Brocchinia thus presents several exciting opportunities for research on the systematics, ecology, evolution, and biogeography of bromeliads, in a group that may constitute a kind of "Darwin's finches" near the base of the family, involving adaptive radiations in nutritional ecology, habit, and the form and function of foliar scales. It affords unique opportunities for (1) tracing the evolutionary pathways to carnivory, the tank-forming habit, and other unusual modes of nutrient capture; (2) evaluating the relative roles of phylogeny and ecology in shaping the form and physiology of foliar trichomes; and (3) examining the relationships among phylogeny, geographic distribution, and ecology in a group endemic to the Guayana Shield.

We begin with an overview of the remarkable adaptive radiation in *Brocchinia*. A molecular phylogeny for the genus is presented, based on restriction site variation in chloroplast and nuclear ribosomal DNA from 27 populations of 15 species. To analyze patterns of morphological evolution and adaptive radiation, we overlay various morphological and ecological character-states on this molecular phylogeny, and determine the most parsimonious scenario for the origin(s) of those states (Kocher et al. 1993; Givnish et al. 1994, 1995; Brooks and McLennan 1994; Crisp 1994). Patterns of geographic diversification are analyzed in similar fashion. We conclude with a synthesis of proposed relationships among phylogeny, morphology, ecology, and biogeography in *Brocchinia*, the first such analysis for a group of tepui organisms. Broader patterns of systematic relationships and adaptive radiation among the remaining genera of the Pitcairnioideae will be treated elsewhere.

Natural history of *Brocchinia*

Systematic position

Brocchinia differs from other pitcairnioid genera in having a combination of (i) small, white, perfect, actinomorphic flowers; (ii) a partly to wholly inferior ovary; (iii) imbricate sepals; and (iv) an open, racemose inflorescence (Smith and Downs 1974; Smith 1986; Varadarajan and Gilmartin 1988). Monotypic *Ayensua* (Smith 1969) differs in having a capitate inflorescence, an epigynous floral tube, and stiff, spreading leaves. All but one species of *Brocchinia* have relatively soft, mesomorphic leaves with entire margins, unlike other pitcairnoids except some species of *Fosterella*. *Brocchinia serrata* is exceptional in this and other respects, having tough, long, narrow leaves with spinose margins, an ovary that is only 1/3 inferior, and loculicidal as well as septicidal dehiscence. *Brocchinia serrata* grows on exposed sandstone on low mesas (300–400 m) in southeastern Colombia (Smith and Downs 1974).

Distribution

Brocchinia is restricted to the Guayana Shield, occurring mainly in southern Venezuela but with a few populations and species found in adjacent parts of Brazil, Colombia, and Guyana (Table 8.1). All but two species are restricted to infertile substrates derived from sandstones of the Roraima Supergroup or accumulated peat

(Steyermark 1961; Smith and Downs 1974; Givnish et al. 1984; Benzing et al. 1985; Varadarajan 1986a; Holst 1997). *Brocchinia acuminata* also occurs on granitic exposures near the edge of the Guayana Shield and at lower elevations between tepuis; *B. melanacra* also occurs on igneous rocks in the Sierra de Maigulida and Sierra Parima (Figure 8.1).

Species of *Brocchinia* occupy a wide elevational range, from 60 to 2,900 m above sea level, with a peak in diversity between 800 and 2,000 m (Table 8.1). All but four species occur on the slopes and summits of tepuis above 500 m elevation; such sites are often wreathed in clouds and receive heavy rainfall throughout much of the year (Brewer-Carías 1978; Huber 1995a,b). High levels of precipitation, humidity, and cloudiness lead to high rates of leaching and exacerbate substrate infertility, creating a unique combination of extremely wet, infertile, cloudy, and cool conditions atop tepuis to which plants must adapt.

The major tepuis (> 1,500 m elevation) of the Guayana Shield fall into two major geographic clusters, centered in southwestern and southeastern Venezuela. These clusters are separated from each other by over 200 km of lowlands centered along the Rio Caura, and by the largely igneous Sierra de Maiguilida and Sierra Parima (Figure 8.1).

WESTERN/PERIPHERAL CLUSTER – Tepuis of southwestern Venezuela and nearby Brazil are isolated towers, rising 800 to 3,000 m above rain forests and edaphically determined Amazonian savannas (Huber 1982) near sea level on the upper drainages of the Rio Orinoco, Rio Ventuari, and Rio Negro. Ten species of *Brocchinia* are restricted to this region, with three occurring in Amazonian savannas and rocky streamsides below 800 m elevation. The remaining seven species occur at higher elevations, on sandstone outcrops and in peaty savannas, bogs, scrub, and cloud forests (Table 8.1). Several *Brocchinia* species in this region are narrowly endemic to one or two tepuis each. Three tepuis on the upper Rio Caura (Jaua, Guanacoco, Sarisari–ama) are included in the western/peripheral cluster based on the occurrence there of the widespread western species *B. melanacra*; no *Brocchinia* species are endemic to these tepuis.

EASTERN/CENTRAL CLUSTER – An extensive but rather closely packed group of tepuis rises from the Gran Sabana, an extensive plateau in southeastern Venezuela and adjacent Guyana (Figure 8.1). The Gran Sabana climbs abruptly from near sea level along the Rio Cuyuni and Rio Caroni to roughly 1,400 m, and then descends gradually southward to ca. 400 m near the headwaters of the Rio Branco (Schubert and Huber 1989). The bedding of the eastern tepuis is nearly horizontal and unfolded, contrasted with the strong folding and metamorphosis that marks the western group (Tate and Hitchcock 1930; Tate 1931a, 1938; Hitchcock 1947; Maguire 1955; Schaefer and Dalrymple 1995). Intriguingly, many eastern tepuis (e.g., Auyán-tepui, Chimanta, Kukenam, Roraima) also have vast areas of nearly bare rock on their summits, often dissected along networks of faults and solution cracks into sandstone labyrinths (Briceño and Schubert 1990; Huber 1995a). Five species of *Brocchinia* are endemic to this region, of which three are relatively widespread at mid to high elevations; one species is restricted to sandstone outcrops near sea level along the Rio Potaro in Guyana (Table 8.1). Three ecologically specialized species occur in both the eastern and western regions, and occupy wide altitudinal ranges of moderate to high elevations. These species include the tank epiphyte *B. tatei* (500–2,400 m), the ant-fed myrmecophyte *B. acuminata* (600–2,100 m), and the carnivore *B. hechtioides* (1,400–2,000 m).

Adaptive radiation in nutritional ecology, habit, and trichome form and function

CARNIVORY – *Brocchinia reducta* (Givnish et al. 1984; Benzing et al. 1985; Gonzales et al. 1991) and *B. hechtioides* (this report) are carnivorous plants. Carnivory is a very rare mechanism of nutrient capture in flowering plants, occurring in only 16 genera (Givnish 1989). Among bromeliads, carnivory is known only in these two species and (in less pronounced form) the tillandsioid *Catopsis berteroniana* (Fish 1976; Frank and O'Meara 1984; Benzing 1986; Givnish 1989; Juniper et al. 1989).

In *Brocchinia reducta* and the taller *B. hechtioides*, the leaf bases overlap strongly and form a tank 2 to 8 cm in diameter, impounding roughly 10 to 60 ml of rain water, respectively. The leaves are erect and form a conspicuous, bright yellow-green cylindrical rosette (Plate 1A,B); the rosette is slightly more open and spreading in *B. hechtioides*. The tank fluid impounded by the leaves is highly acid (ca. pH 3.0); this level of acidity is lower than that seen in other bromeliads and close to the optima measured for certain proteases in *Nepenthes* pitcher-plants (Heslop-Harrison 1976; Benzing 1986; Givnish 1989). Both *B. hechtioides* and *B. reducta* emit an unidentified, nectarlike scent from their tank fluid in the field; a qualitatively similar scent is emitted from the bruised leaf bases of *B. acuminata*, *B. micrantha*, *B. steyermarkii*, and *B. tatei* (Givnish et al. 1984 and pers. obs.). Gonzales et al. (1991) report that, in *B. reducta*, only small (< ca. 25 cm) plants release a nectarlike scent from their tank fluid.

The inner surfaces of the leaves are coated with a fine, waxy powder that readily exfoliates and acts as a lubricant to prevent the escape of arthropods from the tank (Givnish et al. 1984). Ants are the main prey of *B. reducta* (Givnish et al. 1984; Gonzales et al. 1991), whereas bees and wasps are the main prey of *B. hechtioides* (T. J. Givnish, unpubl. data for populations on Auyán-tepui, Duida, and Perai-tepui). Dead prey are often so abundant that they form plugs 2 to 4 cm deep in the tanks of carnivorous *Brocchinia* (Givnish et al. 1984; Gonzales et al. 1991).

The foliar scales of *B. hechtioides* and *B. reducta* are plug-shaped in cross-section; their shield cells are arranged with a high degree of radial symmetry (Figure 8.2A,B), a trait rare in pitcairnioids but progressively more common in bromelioids and tillandsioids (Benzing et al. 1985). Scales on leaf bases can absorb leucine (and, by inference, other amino-acid breakdown products from dead prey) at high rates, and are the principal sites where such absorption occurs (Givnish et al. 1984; Benzing et al. 1985). Such absorptive capacity in foliar scales is unknown in pitcairnioids outside *Brocchinia*, but occurs often among bromelioids and in all tillandsioids examined thus far (Benzing et al. 1976; Benzing and Pridgeon 1983; Benzing 1987; Martin 1994). The uptake capacity of foliar scales in *B. reducta* is associated with an ultrastructural labyrinth of apoplastic passages in the shield cell walls (Owen et al. 1988).

Compared with other species of *Brocchinia*, the two carnivores have rather large foliar scales (ca. 30 to 45 µm in diameter) borne at relatively high densities (ca. 200 mm^{-2}) on the adaxial surface of their leaf bases; as a result, these species have a very

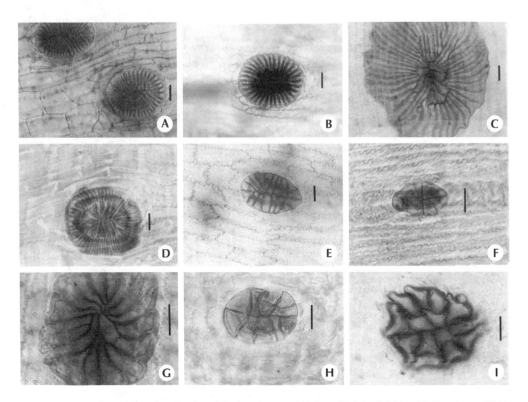

Figure 8.2. Paradermal (overhead) view of leaf trichomes: (**A**) *Brocchinia hechtioides*; (**B**) *B. reducta*; (**C**) *B. acuminata*; (**D**) *B. tatei*; (**E**) *B. micrantha*; (**F**) *B. maguirei*; (**G**) *B.* aff. *maguirei*; (**H**) *B. melanacra*; (**I**) *B. prismatica*. Scale bars represent 10 μm.

high proportion (5.9 to 11.8%) of their leaf-base surfaces covered by absorptive trichome shields (Table 8.2). The foliar scales on the leaf bases of all tank-forming species of *Brocchinia* examined retain their cytoplasmic contents (Givnish et al. 1984; Benzing et al. 1985; Owen et al. 1988); however, those on the upper portions of leaves – above the tank and thus infrequently wetted – appear shriveled, dead, and devoid of cellular contents.

Both carnivores are common in open sites at higher elevations on tepuis, and grow on moist sand or peat in damp savannas and bogs. Both species (and especially *B. reducta*) are virtually the only plants to invade bare sandstone atop some tepuis (Brewer-Carías 1978; Givnish 1989). Both also appear to have smaller root systems than other species of *Brocchinia* of the same size (Givnish 1989).

ANT-FED MYRMECOPHILY – *Brocchinia acuminata* has swollen, achlorophyllous, extremely tough leaf bases that enclose a bulbous, chambered cavity; the leaf bases contract dramatically toward their summit, become chlorophyllous, and flare as elongate leaf tips (Plate 1C). Roughly half of the plants examined in populations on the Gran Sabana (ca. 1,300 m), Auyán-tepui (ca. 1,800 m), and Cerro de la Neblina (ca. 2,000 m) contained active colonies of ants (mainly *Camponotus*) with workers tending live brood. The ants nest in the spaces formed by the nearly vertical walls of adja-

Table 8.2. Trichome area and density in relation to ecological habit in *Brocchinia*.

Ecological habit	Species	Trichome area (μm2)	Density (mm-2)	Trichome area / leaf area (%)
Carnivores	B. reducta	6,107	194	11.8
	B. hechtioides	2,499	236	5.9
Myrmecophyte	B. acuminata	7,605	88	6.7
Tank epiphyte	B. tatei	2,667	141	3.8
Impounding tree	B. micrantha	5,730	32	1.8
Non-impounding terrestrials	B. prismatica	2,667	141	3.8
	B. steyermarkii	1,292	32	0.4
	B. cowanii	4,377	18	0.8
	B. maguirei	2,459	56	1.4
	B. melanacra	1,873	32	0.6

cent leaf bases; the colony overhangs the fluid impounded among the bottoms of the leaf bases. The ant inhabitants of *B. acuminata* are rather timid and do not bite or sting, even when their host plant is torn apart; we have not observed ants to patrol the foliage by day. Dead, disarticulated ants and other arthropods are found in the tank fluid, presumably having fallen there with excreta and occasional food particles.

Foliar scales on the leaf bases of *B. acuminata* are the largest and most elaborate in the genus, with a flaring, nearly peltate shield (Figure 8.2C); they can absorb leucine at high rates (Figure 8.3). The scales occur at a lower density than those of the carnivores, so that a somewhat lower proportion (6.7%) of the leaf base is covered by the absorptive shields (Table 8.2). Adventitious roots often grow into the moist spaces between the overlapping leaf bases and help absorb nutrients.

Brocchinia acuminata is apparently an ant-fed myrmecophyte similar to a number of other tropical plants, many of them epiphytes (e.g., *Tillandsia caput-medusae* and *T. butzii* of the Bromeliaceae [Benzing 1970] and *Hydnophytum* and *Myrmecodia* of the Rubiaceae [Rickson 1978; Huxley 1978]), which obtain nutrients but not defense from ants inhabiting domiciles within the host plant (Huxley 1980, 1986; Thompson 1981; Davidson et al. 1989, 1990; Benzing 1990b; Hölldobler and Wilson 1990; Dejean et al. 1995).

Brocchinia acuminata has the widest geographical range of any species in the genus (Table 8.1), can occur on granitic and igneous substrates as well as on sandstone, and has a broad elevational range as well (600 to 2,100 m). At a few sites, *B. acuminata* invades shaded forest understories, produces a vinelike growth form with long internodes (Plate 1D), and loses both the swollen leaf bases and ant guests. Such a loss is expected on economic grounds, given the lower potential increment to photosynthesis expected to accrue from an increased mineral supply in shaded understories (Givnish 1989).

Tank epiphytism – *Brocchinia* contains eight species capable of impounding rainwater among their closely overlapping leaf bases (Table 8.1), the only such

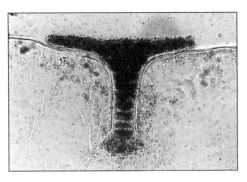

Figure 8.3. Autoradiomicrograph of leaf cross-section of *B. acuminata*, showing rapid uptake of H^3-labeled leucine by leaf trichomes.

tank-forming species in the Pitcairnioideae. Of these, two (*B. tatei* and *B. hitchcockii*) are epiphytic (Steyermark 1961; Smith and Downs 1974). *Brocchinia tatei* is – with the carnivore *B. hechtioides* and ant-fed *B. acuminata* – one of the three most widespread species, occurring on almost all tepuis in the western and eastern clusters. *Brocchinia hitchcockii* is one of the most narrowly distributed species, known only from Cerro Parú at ca. 2,000 m, and has yet to be studied by our group.

Brocchinia tatei is a facultative epiphyte, and sometimes grows as a terrestrial on exposed sand or sandstone, between 500 and 2,500 m elevation. Its leaves are green and covered with a waxy powder, similar to (but less dense than) that covering the foliage of the carnivores. The leaves of *B. tatei* form a spreading rosette and impound moderate amounts of precipitation (ca. 100 to 300 ml) among their bases (Plate 1e). When epiphytic (Plate 1F), *B. tatei* captures mainly falling vegetable debris in its tank (e.g., dead leaves, flowers, fruit, twigs). Under such conditions, it probably obtains mineral nutrients partly from the breakdown of such debris, like many other tank epiphytes (Benzing 1980, 1990a).

Its foliar trichomes are somewhat smaller than those of the carnivorous and ant-fed species, and occur at an intermediate density, resulting in a moderate fraction (3.8%) of adaxial leaf surfaces being covered (Table 8.2). Trichomes share the same strong pattern of subradial symmetry seen in the carnivores (Figure 8.2D) and can absorb leucine at moderate rates. *Brocchinia tatei*, like all species examined thus far, has the C_3 photosynthetic pathway (Martin 1994). It is interesting that the carnivorous epiphyte *Utricularia humboldtii* – endemic to the tepui region and visually conspicuous with large, dazzling blue flowers – is frequently found growing inside the tanks of *B. tatei*; it also sometimes grows in the tanks of *B. hechtioides*, with which it may compete for nutrients obtained from prey in its tanks (Givnish 1989).

NITROGEN FIXATION – In one terrestrial population of *B. tatei* in the northern Gran Sabana, the plants grow in open sites on exposed sandstone; their leaves are a bright yellow-green, and possess an apiculate tip like that of *B. hechtioides* and *B. reducta*. Several of these plants do not have access to falling vegetable debris, and instead possess a highly unusual, gelatinous algal plug in their central tank. This plug contains several genera (*Calothrix, Scytonema, Stigonema*) of heterocystous cyanobacteria (Givnish and Burkhardt, unpubl. data); all heterocystous taxa known fix atmospheric nitrogen. This population of *B. tatei* is the only known instance in which cyanobacteria occur in bromeliad tanks, and may represent a mutualistic association with N_2-fixing prokaryotes. N_2-fixing cyanobacteria often occur on exposed sandstone in the Guayana Shield (Büdel et al. 1994), and Puente and Bashan (1994) have recently reported *Pseudomonas stutzeri*, a weak N_2-fixer, growing inside the leaves of the epiphytic bromeliad *Tillandsia recurvata*. However, for *B. tatei*, we do not yet know if nitrogenous products leak from the plug into the tank, if the trichomes of the N_2-fixing population differ from those of other populations, or if that population is of hybrid origin.

ARBORESCENT SPECIES – *Brocchinia micrantha* is a gigantic, monocarpic species (Steyermark 1961) that can impound several liters of rainwater in its gutterlike leaf axils. It can achieve 5 m in height when vegetative (8 m when in flower), second only to some species of *Puya* in stature among bromeliads (Smith and Downs 1974) (Plate 1G,H). Its leaves are often more than 1.2 m long, and 15 to 35 cm in width. *Brocchinia micrantha* occurs between 500 to 1,200 m in openings in cloud forests in southeastern Venezuela and adjacent Guyana, on shallow or exposed sandstone; the analogous *B. paniculata* grows in similar settings between 200 and 1,500 m in southwestern Venezuela and adjacent Colombia.

Dead leaves and other vegetable debris are the principal tank contents in *B. micrantha*, together with live larvae of *Runchomyia* and *Wyeomyia* (Culicidae). The foliar scales of *B. micrantha* are rather large, have shield cells with radial symmetry (Figure 8.2E), and absorb leucine at moderate rates. However, they occur at low densities on adaxial leaf surfaces, so that only 1.8% of the adaxial leaf bases are covered with absorptive shield surfaces (Table 8.2).

NON-IMPOUNDING SPECIES – The remaining 12 species of *Brocchinia* (Table 8.1) impound essentially no rainwater among their leaf bases, have relatively massive root systems, and apparently absorb most or all of their mineral nutrients from the soil (Plate 1I–M). Their foliar scales are relatively small and/or sparse (Table 8.2), and many have a rather simple arrangement of cap cells, with only a weak pattern of radial symmetry (Figure 8.2F–I). On average, only 1.4% of the adaxial leaf surfaces were covered by trichomes in the five species studied (Table 8.2). Foliar scales with an especially simple arrangement of shield cells occur in *B. maguirei* (Autana), *B.* aff. *maguirei* (Cerro de la Neblina), and *B. vestita* (Duida). Larger trichomes with moderate radial symmetry occur in *B. steyermarkii*. *Brocchinia prismatica* has slightly stellate trichomes, similar to those seen in the pitcairnioid genus *Fosterella* (see illustrations in Benzing 1980 and Varadarajan and Gilmartin 1988).

Several non-impounding species have leaves that are even above, lacking the prominent adaxial nerves seen in such species as *B. acuminata*, *B. hechtioides*, *B. micrantha*, *B. reducta*, and *B. tatei* (Smith and Downs 1974). Four species (*B. cataractarum*, *B. delicatula*, *B. pygmaea*, *B. rupestris*) native to streamside outcrops are among the smallest in the Bromeliaceae, flowering when 5 cm or less in height (Holst 1997). Two other saxicolous species found on exposed sandstone away from streams (*B. amazonica*, *B. cowanii*) are not much larger, forming rosettes less than 10 cm in diameter; *B. serrata* and *B. vestita* also grow on sandstone, but achieve far greater statures. Three of the four remaining species (*B. maguirei*, *B. melanacra*, *B. steyermarkii*) grow on wet peat or muck at moderate to high elevations on tepuis; *B. prismatica* occurs between 100 and 200 m, on permanently wet sand in Amazonian savannas (Huber 1982).

Brocchinia melanacra seems adapted to fire in tepui summit habitats (Givnish et al. 1986). It has extremely tough, sharp leaf apices (Smith and Downs 1974; Holst (1997), a feature that might suggest an adaptation against vertebrate herbivores. However, after the leaves expand, a zone of cells below the pungent apex die, causing it to droop and/or flex in the wind and arguing against an anti-herbivore function. The sclerotized leaf tips may play their primary role in immature leaves: they overlap and completely ensheath the terminal meristem in extremely tough tissue. *Brocchinia melanacra* is one of only four plant species to resprout after fire on Cerro de la Neblina; the others had terminal meristems protected by ensheathing pitchers (Sarraceniaceae: *Heliamphora*), leaf bases (Rapateaceae: *Stegolepis*), or leaf rosettes (Theaceae: *Neblinaria*) (Givnish et al. 1986).

Molecular systematics

Methods

Plant material was gathered for 15 species and 27 populations of *Brocchinia* during expeditions to southern Venezuela and adjacent parts of Brazil, Colombia, and Guyana by the senior author and his colleagues and/or correspondents (Table 8.3). Tepui summits were reached by helicopter or on foot, often after extended travel by boat; the Gran Sabana was reached by car. Herbarium specimens and live plants (or freshly detached leaf tissue) were collected for each population and shipped promptly to Madison.

DNA EXTRACTION – Total DNA's were extracted from fresh and –80° C frozen leaf tissue collected from individual plants, using a 6% CTAB procedure (Smith et al. 1991) modified from the protocols of Saghai-Maroof et al. (1984) and Doyle and Doyle (1987). Use of the higher CTAB concentration and precooling of tissue in liquid nitrogen for 10 minutes greatly enhanced DNA yields from the tough, fibrous leaf tissue of *Brocchinia* and other bromeliads studied.

OUTGROUP ANALYSIS – We conducted a two-step screening process to identify appropriate outgroups with which to polarize cpDNA character-states in *Brocchinia*. First, to establish collinearity between autologous (sample) and heterologous (probe) cpDNA's, and to demonstrate homology of restriction sites between ingroup and outgroup taxa, we constructed detailed restriction-site maps of the entire chloroplast genome for *Brocchinia micrantha* and representative species of the Pitcairnioideae, Tillandsioideae, and Bromelioideae. For each species, DNA was cleaved using each of seven restriction endonucleases (*Sst* I, [*Pst* I, *Bgl* II], [*Mlu* I, *Sal* I], [*Xho* I, *Pvu* II]). For the endonucleases paired in parentheses, all three double-digests involving them and *Sst* I were formed as well. Cleaved DNA's were size-separated by electrophoresis on agarose gels, transferred to nylon filters, and probed with heterologous cpDNA clones to map cpDNA restriction sites through comparisons of fragments visualized autoradiographically using standard procedures (e.g., Sytsma et al. 1990). To increase the likelihood of collinearity between autologous and heterologous cpDNA's, we used mapped *Oncidium* (Orchidaceae) cpDNA clones (Chase and Palmer 1989; clones kindly supplied by M. Chase), using combinations designed to cover nearly the entire chloroplast genome and aid probing efficiency. Use of single- and double-digests permits an accurate map of autologous cpDNA to be constructed in spite of minor to major re-arrangements of a sample's chloroplast genome relative to the heterologous probes, and permits any such re-arrangement to be visualized and incorporated in subsequent analyses (Sytsma and Gottlieb 1986).

Second, a set of three operational outgroups were chosen from among those screened, based on their apparent degree of cpDNA restriction-site homology (i.e., positional identity) with *Brocchinia micrantha*, and their position close to *Brocchinia* in a cpDNA phylogeny of the Pitcairniodeae developed as part of an ongoing study (T. J. Givnish et al., unpubl. data). We used *Brewcaria reflexa*, *Lindmania geniculata*, and *Navia splendens*, members of three of the other six pitcairnioid genera endemic to the Guayana Shield. These genera represent both of the clades closest to the one including *Brocchinia* (T. J. Givnish et al., unpubl. data), and should thus be useful for polarizing character-states within the genus.

Table 8.3. Accessions of *Brocchinia* and outgroups used in the molecular analysis. Unless otherwise indicated, all localities are in southern Venezuela.

Taxon	Location
Brewcaria reflexa	Bana ca. 2 km from San Carlos de Rio Negro, ca. 200 m
Brocchinia acuminata A	Auyán-tepui, near head of Angel Falls, ca. 1,600 m
B. acuminata G	Gran Sabana, near Km. 142 on El Dorado-Santa Elena de Uairen Road, ca. 1,300 m
B. acuminata D	Cerro Duida, Camp 2 near southern rim, ca. 2,000 m
B. acuminata Y	Cerro Yapacana, ca. 900 m
B. amazonica	Serra do Aracá, Brazil, ca. 1,600 m
B. cowanii	Cerro Moriche, ca. 1,100 m
B. gilmartinii	La Escalera via Selby Botanical Garden
B. hechtioides A	Auyán-tepui, near head of Angel Falls, ca. 1,600 m
B. hechtioides D	Cerro Duida, Camp 2 near southern rim, ca. 2,000 m
B. hechtioides AUT	Cerro Autana, summit
B. maguirei AUT	Cerro Autana, summit
B. aff. *maguirei* N	Cerro de la Neblina, northwest end of canyon
B. melanacra D	Cerro Duida, Camp 2 near southern rim, ca. 2,000 m
B. melanacra N	Cerro de la Neblina, Camp 2
B. micrantha G	La Escalera, northern escarpment of Gran Sabana
B. micrantha K	Kaieteur Plateau near falls, Guyana
B. paniculata	Cerro Autana, summit
B. prismatica	Amazonian savanna III, at foot of Cerro Yapacana
B. reducta A	Auyán-tepui, near head of Angel Falls, ca. 1,600 m
B. reducta G	Gran Sabana, near Km. 142 on El Dorado-Santa Elena
B. reducat G2	Gran Sabana, near Km. 138 on El Dorado-Santa Elena de Uairen Road, ca. 1,300 m
B. serrata	Cerro de Circasia, Colombia
B. steyermarkii	Gran Sabana, near Km. 142 on El Dorado-Santa Elena de Uairen Road, ca. 1,300 m
B. tatei A	Auyán-tepui, near head of Angel Falls, ca. 1,600 m
B. tatei D	Cerro Duida, Camp 2 near southern rim, ca. 2,000 m
B. tatei Y	Cerro Yapacana, near summit on west slope
B. vestita	Cerro Duida, Camp 1 (Savanna Hills) near center of plateau, ca. 1,400 m
Lindmania longipes	Forest edge along blackwater stream, southern portion of Auyán-tepui near Whirlpool Camp, ca. 2,000 m
Navia splendens	Selby Botanical Garden

INGROUP ANALYSIS – DNA's of 15 species (27 populations) of *Brocchinia* and 3 outgroup species (Table 8.3) were cleaved using 30 restriction endonucleases (*Afl* II, *Apa* I, *ApaL* I, *Ava* I, *BamH* I, *Bcl* I, *Bgl* I, *Bgl* II, *BstB* I, *BstE* I, *Cla* I, *Dra* I, *EcoN* I, *EcoO* 109, *EcoR* I, *EcoR* V, *Hind* III, *Kpn* I, *Mlu* I, *Pvu* II, *Sal* I, *Sca* I, *Sma* I, *Sst* I, *Stu* I, *Sty* I, *Tth111* I, *Xba* I, *Xho* I, *Xmn* I) that cut at sites corresponding to specific, 6-nucleotide sequences. Cleaved DNAs were size-separated by electrophoresis on agarose gels, transferred to nylon filters, and probed sequentially with the entire suite of *Oncidium* cpDNA clone combinations to recognize cpDNA restriction-site variation using standard techniques (Sytsma and Smith 1988).

To screen for hybridization and introgression events, we analyzed restriction-site variation in nuclear ribosomal DNA (nrDNA), probing the filters used in the cpDNA study with *Glycine* nrDNA (kindly provided by Dr. Elizabeth Zimmer). Because cpDNA is inherited maternally and nrDNA biparentally, a comparison of patterns of variation in the two permits the identification of hybrids and paternal and maternal sources.

Restriction-site variation was analyzed cladistically with PAUP 3.1.1 (Swofford 1993), using global parsimony (Maddison et al. 1984). The branch-and-bound search strategy was used, and both the ACCTRAN and DELTRAN options of PAUP were implemented. To determine whether *Brocchinia* is para- or polyphyletic with respect to the outgroups chosen, we did not constrain taxa of *Brocchinia* to fall within the ingroup. Similarly, to determine whether individual species of *Brocchinia* appear to be para- or polyphyletic based on molecular data, we did not constrain populations of a given species to cluster together.

Bootstrap analysis (Felsenstein 1985, 1988) was conducted on the matrix of potentially informative characters to provide a sense of the strength of support for each clade in the resulting majority-rule tree. We are aware of the limitations and potential misuse of bootstrap analysis (see Sanderson 1989; Wendel and Albert 1992), but believe that it can provide useful insights. Decay analysis was also conducted on the total data set, by forming the strict consensus of all trees that are n steps longer than the shortest tree. A comparison of the latter indicates which clades collapse to an unresolved polytomy with each extra step, providing a measure of the robustness of the monophyly of each clade recognized in the minimal tree (see Bremer 1988; Smith and Sytsma 1990; Kadereit and Sytsma 1992).

Character-state weighted parsimony analyses (Albert et al. 1992a; Kadereit and Sytsma 1992) can yield slightly different (and often, slightly more resolved) trees than unweighted Wagner parsimony. Character-state weighted parsimony permits the use of biases against convergent gains or loss-gains of sites, which are evolutionarily less probable than convergent losses or gain-losses. Consequently, we conducted a weighted parsimony analysis using the step-matrix approach (Albert et al. 1992a) in PAUP 3.1, giving convergent gains and loss-gains a weight of 1.05, 1.10, or 1.50 relative to that for convergent losses and gain-losses (Albert et al. 1992a; Kadereit and Sytsma 1992) and defining ancestral states as unknown (Albert et al. 1992a). Comparisons of the topologies of the rooted trees obtained with weighted and unweighted parsimony were used to determine which relationships indicated by these methods are mutually consistent and/or congruent with each other.

EVOLUTIONARY ANALYSIS – In order to analyze patterns of morphological evolution in *Brocchinia*, we overlaid morphological and ecological character-states on our molecular phylogeny, and then used MacClade 3.0 (Maddison and Maddison 1992) to determine the most parsimonious scenario for the origin(s) of those states (Olmstead and Palmer 1992; Kocher et al. 1993; Givnish et al. 1994, 1995; Brooks and McLennan 1994; Crisp 1994). Geographic patterns of speciation were analyzed in similar fashion. These results were then synthesized to determine relationships among phylogeny, morphology, ecology, and geography, and to evaluate the roles of key innovations (in trichome form and impounding habit) and extreme environments in shaping the evolution of *Brocchinia*.

Results

Phylogenetic reconstruction

HOMOLOGY OF *Brocchinia* cpDNA TO HETEROLOGOUS PROBES AND OUTGROUP cpDNA – The double-digest map of cpDNA restriction sites for *Brocchinia micrantha* indicated that its chloroplast genome was collinear with that of *Oncidium excavatum* (Orchidaceae), source of the heterologous probes used to study restriction-site variation across *Brocchinia* and outgroups. Due to an error in the initial mapping of the probes supplied by M. Chase, our probe combination I (containing Chase's probes 22, 21, 20, 19b) straddled both sides of combination II (containing Chase's probe 19a). We found no evidence of any major re-arrangement (inversion or translocation) of *Brocchinia's* cpDNA relative to that of *Oncidium* once this peculiarity of the heterologous probes was taken into account.

We found that the chloroplast genome of *Brocchinia* (171.4 kb) was substantially larger than that of *Oncidium* (143.3 kb). The length of the large single-copy region in *B. micrantha* was 92.4 kb (82.7 kb in *Oncidium*); the length of the small single-copy region was 21.4 kb (13.4 kb in *Oncidium*); and the length of one copy of the inverted repeat was 28.8 kb (23.6 kb in *Oncidium*). It appears that there are insertions of DNA relative to *Oncidium* throughout much of the chloroplast genome of *Brocchinia*, especially in the small single-copy region, which is unusually small in *Oncidium* (Chase and Palmer 1989, 1992). In addition, there are some deletions in *Brocchinia* cpDNA relative to *Oncidium* in the area homologous to probe combinations VII and VIII. In general, it was easy to homologize restriction sites across potential outgroup taxa; sites were highly conserved, and there were no major re-arrangements of the chloroplast genome evident relative to *Brocchinia*.

CHLOROPLAST DNA RESTRICTION-SITE VARIATION – Using 30 restriction enzymes, we detected an average total of 684 cpDNA restriction sites in the chloroplast genomes of 15 species (27 populations) of *Brocchinia* and the 3 designated pitcairnioid outgroups. Of these, 461 varied across taxa and 263 were phylogenetically informative; 198 mutations were autapomorphies, unique to single taxa. A modest number of small insertions and deletions were also detected, but insofar as it is difficult to establish the homology of such mutations *a priori* (Palmer et al. 1985; Sytsma and Gottlieb 1986; Sytsma et al. 1990), they were omitted from phylogenetic analysis.

The restriction-site survey sampled roughly 2.4% of the entire chloroplast genome. The maximum amount of cpDNA divergence between pairs of species in *Brocchinia*, as measured by the p value of Nei and Li (1979), is 0.035. This is moderately high compared with the levels seen in other angiosperm genera, and is 63% of that detected in an ongoing study that involves 15 of the 17 genera of the entire subfamily Pitcairnioideae (T. J. Givnish et al., unpubl. data).

Mutation rates tended to be much higher in the large and small single-copy regions of the chloroplast genome than in the inverted repeat, paralleling the pattern seen across green plants (Palmer 1987, 1990; Downie and Palmer 1992). If we sum across taxa, the inverted repeat displays an average mutation rate of 1.08 kb^{-1}, roughly 25% of the rates observed in the large single-copy region (3.71 kb^{-1}) and the small single-copy region (4.07 kb^{-1}). Within the large single-copy region, mutations are especially abundant in the areas homologous to probes III and VII, and especially sparse in the areas homologous to probes V and VIII.

PHYLOGENETIC ANALYSIS – Cladistic analysis of the cpDNA restriction-site data using unweighted Wagner parsimony yielded 4 equally most parsimonious trees of length 530 steps; the data are adequate to resolve all but four nodes (of which two are intraspecific) in the strict consensus phylogeny (Figure 8.4). The restriction-site phylogeny has a consistency index (CI) of 0.870 (0.792, excluding autapomorphies), reflecting a low level of homoplasy in the underlying data compared with other molecular studies (Chapter 2). Of the 69 inferred reversals in character-state, 35 are convergent site gains and loss-gains; the remaining 34 are convergent site losses and gain-losses.

Our analysis indicates that *Brocchinia serrata* is clearly divergent from all other species within the genus, and is embedded among the outgroups by several synapomorphies (Figure 8.4). As noted, *B. serrata* is morphologically anomalous within *Brocchinia*, being the only species with a serrate leaf margin, an ovary that is 2/3 superior, and both septicidal and loculicidal dehiscence of the seed capsule (Smith and Downs 1974). Based on these differences and our molecular results, we will soon describe *B. serrata* as a new, monotypic genus in collaboration with Bruce Holst.

The remaining 26 populations of *Brocchinia* constitute a monophyletic clade, and fall into four well-marked sublineages (Figure 8.4):

- The **Prismatica clade** consists solely of *B. prismatica*, a terrestrial species with narrow leaves and non-impounding leaf bases, from low-elevation (< 300 m) Amazonian savannas on wet sand in southwestern Venezuela. This species is the sister-group to the remaining species of *Brocchinia*, having diverged from their common ancestor before they diverged from each other.
- The **Melanacra clade** consists of a group of broad-leaved terrestrial species occurring at middle (500–1,300 m) and high (1,300–2,600 m) elevations on the isolated tepuis of southwestern Venezuela. The group is sister to the two remaining clades, and contains *B. amazonica*, *B. cowanii*, *B. maguirei*, *B. melanacra*, *B. paniculata*, and *B. vestita*. Only *B. paniculata*, a treelike form, impounds rainwater among its leaf bases; it occurs at mid-elevation and is basal to the Melanacra clade. The remaining taxa are all non-impounding, and generally less than 1.5 m in height.

Molecular Evolution and Adaptive Radiation in Brocchinia

cpDNA restriction-site data

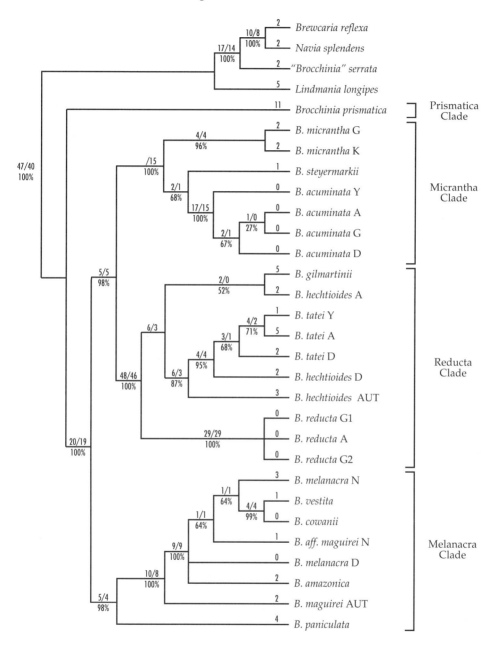

Figure 8.4. Molecular phylogeny of *Brocchinia* based on cpDNA restriction-site variation. Tree is one of four equally most parsimonious trees (length = 530 steps, CI = 0.870 including autapomorphies; length = 307, CI' = 0.792 excluding autapomorphies). Branch length/decay value is shown above each node; the corresponding bootstrap value (%) is shown below the node; terminal branches show length only. Branches that collapse in the strict consensus have zero decay values.

- The **Micrantha clade** consists of three terrestrial species: *Brocchinia acuminata*, *B. micrantha*, and *B. steyermarkii*. *Brocchinia micrantha* is sister to the remaining taxa, and has an arborescent growth form similar to that of *B. paniculata*; it also occurs at mid-elevation, but is restricted to southeastern Venezuela and adjacent Guyana. *Brocchinia steyermarkii,* the only non-impounding species in the group, is also restricted to southeastern Venezuela. *Brocchinia acuminata* is an ant-fed myrmecophyte, with the broadest geographic and elevational range of any species in the genus.
- The **Reducta clade** consists of populations of four impounding species found at high elevations: the carnivores *B. reducta* and *B. hechtioides;* the facultative tank epiphyte *B. tatei,* and the ecologically unstudied *B. gilmartinii.*

Our analysis groups most conspecific populations in monophyletic lineages (Figure 8.4). *Brocchinia hechtioides* is at least paraphyletic (containing *B. tatei* as a monophyletic sublineage) and may be polyphyletic. Populations of *B. maguirei* and *B. melanacra* fail to group together, either because these morphological taxa include cryptic species, or (more likely) because genetic variation within the Melanacra clade is inadequate to reliably identify relationships between certain populations.

Most nodes in the cpDNA restriction-site phylogeny are strongly supported by both bootstrap and decay analyses (Figure 8.4). Each of the four major clades has a bootstrap value of at least 95%, and each persists in the strict consensus of trees 4 steps longer than the most parsimonious trees. Most nodes with lower bootstrap values involve intraspecific variation. The placement of *B. steyermarkii* and a few taxa in the Melanacra clade are also less certain. Low bootstrap values are associated with the occurrence of few synapomorphies at a node; the nodes that disappear in a strict consensus of all trees 1, 2, 3, or 4 steps longer than the minimal trees are those that have a lower degree of bootstrap certainty.

Weighted parsimony analyses of the cpDNA data, giving convergent gains and loss-gains a weight of 1.05, 1.10, or 1.50 relative to convergent losses and gain-losses, yielded a consensus tree identical to that obtained using unweighted Wagner parsimony. Taken together, these results strongly suggest that the phylogeny inferred from cpDNA restriction-site variation is robust, and little affected by varying the assumptions used to infer relationships.

EVOLUTIONARY RATES – Molecular evolution within the chloroplast genome of *Brocchinia* deviates strongly from clocklike behavior. Mutations occurred at roughly twice the average rate in the Reducta clade, and at roughly one-half the average rate in the Melanacra clade (Figure 8.4). Taken as a whole, the two non-impounding clades (Prismatica and Melanacra) show much less divergence from the inferred common ancestor than the two clades (Micrantha and Reducta) in which almost all species impound rainwater among their leaves.

HYBRIDIZATION – To screen for hybridization and introgression, we analyzed restriction-site variation in nrDNA for the 14 endonucleases that cut methylated nrDNA. Our data indicate a much lower maximum absolute amount of site divergence (38 variable characters) in nrDNA than in cpDNA (461 variable characters), and result in a consensus of six shortest trees that, with two exceptions, is roughly consistent with the cpDNA tree though substantially less resolved (Figure 8.5). The exceptions are that the nrDNA data (i) place populations of *B. reducta* and *B. hechtioides* sister to each other, with populations of *B. tatei* sister to *B. reducta-B. hechtioides*; and (ii) fail to place *B. gilmartinii* (endemic to La Escalera at the northern rim of the Gran Sabana) in any of the four cpDNA clades (Figure 8.5).

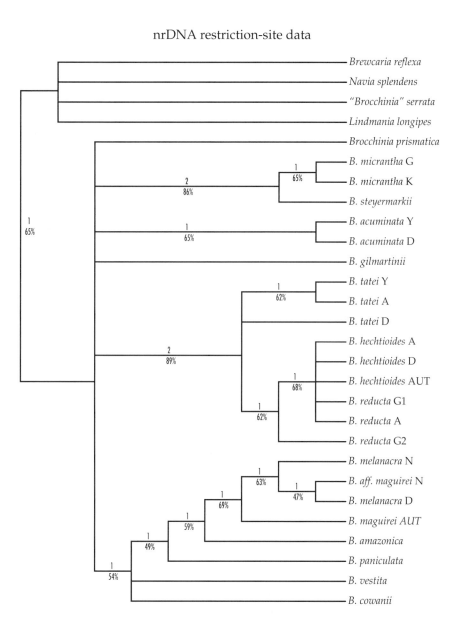

Figure 8.5. Molecular phylogeny of *Brocchinia* based on nrDNA restriction-site variation. Tree is one of six most parsimonious trees (length = 41 steps, CI = 0.927 including autapomorphies; length = 21, CI' = 0.857 excluding autapomorphies). Decay values are shown above each node; bootstrap values (%) are shown below each node.

These data suggest that hybridization and introgression generally have been unimportant in *Brocchinia*, but that *B. gilmartinii* and possibly other members of the Reducta clade may be of hybrid origin. The data presently available are inadequate to support an analysis of possible hybridization involving *B. hechtioides*, *B. reducta*, and *B. tatei*, given that changes in only 2 mutations would yield a topology for this group consistent with that obtained from the cpDNA data. However, the divergence between the cpDNA and nrDNA data sets is greater for *B. gilmartinii*, and suggests that this species may have arisen from a cross between a paternal *B. acuminata* and a maternal *B. hechtioides*. This hypothesis would accord with *B. gilmartinii*'s morphology, which appears intermediate between that of *B. acuminata* and certain other species (Holst 1997); it is also compatible with the narrow geographic distribution of *B. gilmartinii*. However, Holst (1997) suggests that, based on morphology, the most likely parents of *B. gilmartinii* are *B. acuminata* and *B. micrantha*. There are no known populations of *B. hechtioides* near those of *B. gilmartinii* (the former usually occurs atop tepuis, at higher elevations than the latter), although there are populations nearby of *B. acuminata*, *B. micrantha*, and *B. tatei* (Varadarajan 1986b; Holst 1997; T. J. Givnish, pers. obs.).

This seeming conflict between the molecular data and morphology/geography may reflect either an unknown (or extinct) population of *B. hechtioides*, or a two-step hybrid origin of *B. gilmartinii*. We suggest that the "father" of *B. gilmartinii* was indeed *B. acuminata*, but that its "mother" was the product of a cross between a *B. tatei* maternal grandfather and a *B. hechtioides* maternal grandmother. *Brocchinia tatei* and *B. hechtioides* frequently co-occur at higher elevations on nearby tepuis (e.g., Auyán-tepui, Ilú-tepui), and Holst (1997) notes that several populations are not easily assignable to one or the other taxon. *Brocchinia tatei* also shares a spreading rosette form with *B. micrantha*. Seeds from a *B. tatei-B. hechtioides* cross could have blown into the Escalera region, and subsequently hybridized with *B. acuminata* there to form *B. gilmartinii*. Such a scenario seems more likely than others, given the apparent lack of *B. hechtioides* populations (and hence, of maternal grandparents) in the Escalera region. There are indeed populations of *Brocchinia* near those of *B. gilmartinii* which (though sterile) share most of the vegetative characteristics of *B. tatei*, but which also have strongly apiculate leaves and an erect rosette reminiscent of *B. hechtioides* (T. J. Givnish, pers. obs.). Molecular analyses of these populations, which were not sampled as part of this study, should help test our hypothesis for the hybrid origin of *B. gilmartinii*, which is consistent with the morphological, molecular, and geographical data now available.

If we exclude *B. gilmartinii* as a putative hybrid, a cladistic analysis of the combined cpDNA and nrDNA data sets yields a single shortest tree (see Figure 8.6) that is quite similar to the cpDNA phylogeny, but identifies the two populations of *B. melanacra* as possibly monophyletic and places *B. steyermarkii* sister to *B. micrantha* (68% bootstrap support, decay value of 1). For the remainder of this paper, we will base our analyses on this combined phylogeny, recognizing that additional data from the nuclear genome are needed and may shed more light on the relationships among *B. hechtioides*, *B. reducta*, and *B. tatei*.

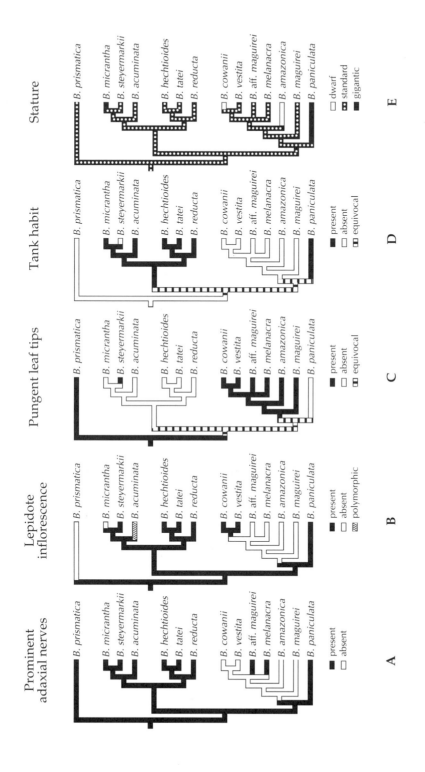

Figure 8.6. Reconstructions of morphological evolution in *Brocchinia* using parsimony. Black indicates inferred presence of trait along branch; white indicates absence; cross-hatchings indicate uncertainty or characteristics shown in legend. (A) Prominent adaxial nerves on leaves; (B) scaly inflorescences; (C) pungent (i.e., piercing) leaf tips; (D) tank habit; (E) stature.

Patterns of morphological evolution

We overlaid various morphological, anatomical, and ecological character-states on the cpDNA restriction-site phylogeny, and then determined the most parsimonious scenario for the origin(s) of those states using delayed character-state transformation (DELTRAN). Based on this analysis, several character-states appear to have arisen or been lost independently two or more times, often in association with particular ecological conditions:

PROMINENT NERVES ON ADAXIAL LEAF SURFACES – This character (used by Smith and Downs [1974] to split *Brocchinia* into two informal groups) appears to have arisen at the base of the genus, been lost in the Melanacra clade subsequent to the divergence of *B. paniculata*, and then regained in *B. melanacra* (Figure 8.6A). Based on our preliminary anatomical studies, the absence of visually prominent adaxial "nerves" is a result of (i) small vascular bundles wholly embedded within the chlorenchyma, in a thick (ca. 500 µm) leaf cross-section; (ii) a thick layer of colorless ground tissue above the chlorenchyma (effectively acting as a "diffusing filter"); and (iii) the absence of prominent aerenchyma between the bundles (Figure 8.7). Prominent nerves in the Micrantha and Reducta clades reflect (i) large bundles extending beyond the chlorenchyma in relatively thin (ca. 300 µm) leaf cross-sections; (ii) a thin adaxial layer of colorless ground tissue; and (iii) prominent aerenchyma, representing an expansion of the substomatal cavities that lie below the longitudinal files of stomata in all species. *Brocchinia melanacra* and *B. prismatica* have small bundles and a thick adaxial layer of colorless ground tissue but also have massive aerenchyma, surrounded by chlorenchyma, which contrast with the intervening bundles and create the impression of nerves (Figure 8.7).

Prominent adaxial nerves are strongly correlated with the tank habit (*B. paniculata*, Micrantha and Reducta clades exclusive of *B. steyermarkii*) or occurrence on wet, sodden substrates (*B. prismatica, B. steyermarkii, B. melanacra*); the little known tank epiphyte *B. hitchcockii* is the sole exception to this rule. Adaxial nerves may reflect selection for aerenchyma in plants exposed to saturated conditions adjacent to leaves and/or roots. In addition, selection for thin, conformable leaves in tank-forming species (see below) would favor the concentration of support tissue in relatively thick vascular bundles (Givnish 1979). The dry, brightly lit conditions to which saxicolous species at sunny mid-elevations are exposed should favor thick leaves (Givnish 1979, 1987), and may favor a thick layer of adaxial ground tissue to screen the chlorenchyma and prevent damage from intense UV radiation (see Robberecht and Caldwell 1978; Day et al. 1993).

LEPIDOTE INFLORESCENCES – Lepidote or scaly inflorescences are primitive in *Brocchinia* and appear to have been lost four times (in *B. prismatica, B. micrantha*, and *B. acuminata*, and in the Melanacra clade subsequent to the divergence of *B. paniculata*), and then regained in the common ancestor of *B. vestita* and *B. cowanii* (Figure 8.6B). The adaptive significance of this trait is unknown.

PUNGENT LEAF TIPS – Sharp, piercing leaf tips appear to have been lost twice (in *B. paniculata* and the common ancestor of the Micrantha and Reducta clades), and regained in *B. steyermarkii* (Figure 8.6C). Such leaf tips may have served to deter her-

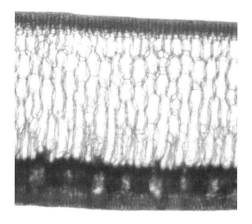
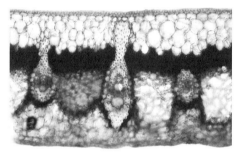

Figure 8.7. Leaf cross-sections from live material of *B. cowanii* (top), *B. micrantha* (middle), and *B. melanacra* (bottom). Adaxial (upper) leaf surfaces at top of photographs. Dark areas in lower parts of leaf cross-sections are chlorenchyma; light areas above are colorless ground tissue which may serve as a UV shield or water-storage tissue. Note prominent vascular bundles in *B. micrantha,* which has very long, thin, soft leaves; note also stellate cells in prominent aerenchyma in *B. melanacra* and *B. micrantha*.

bivory by vertebrates (but not much smaller invertebrates), surrounding the leaf rosette and sole terminal meristem with a phalanx of piercing spearheads. Investment in such defenses should be greatest in the least productive/fertile habitats, where loss of tissue to herbivores would be most costly to replace (Janzen 1974; Coley 1983; Bryant et al. 1985) – provided that such defenses are compatible with other aspects of a plant's morphology.

TANK HABIT – The tank habit, caused by the impounding of rainwater by strongly overlapping leaf bases/sheaths, is inversely related to the presence of pungent leaf tips, and appears to have arisen twice (in the common ancestor of the Micrantha and Reducta clades, and in *B. paniculata*), and to have been lost secondarily in *B. steyermarkii* (Figure 8.6D). Selection for the tank habit probably selected against pungent leaf tips and resulted in the complementary distribution of both traits relative to each other. To form a tank, leaf bases and/or sheaths must be relatively soft in order to conform to each other and be able to impound rainwater. However, for a pungent leaf tip to be effective against vertebrate herbivores, the leaf itself must be stiff to provide an effective mount for its terminal lance (Givnish 1979).

Selection for the tank habit in *Brocchinia* appears related, in part, to elevation and climate. The number of tank-forming species shows a strong peak between 900 and 2,100 m elevation, while the number of non-impounding species shows little variation across the elevational range occupied by the genus (Figure 8.8). The elevational zone from 900 to 2,100 m occupies the upper slopes of most large tepuis, and corresponds to a belt of cloud forests, bogs, and rock outcrops that experience exceedingly high humidity, rainfall, and soil leaching. Within this zone, the appearance in ancestral taxa of the ability to impound even modest amounts of rainwater could lead to the formation of persistent tanks, fed by high

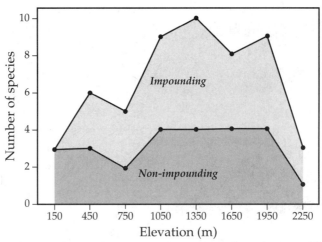

Figure 8.8. Number of impounding and non-impounding species of *Brocchinia* as a function of elevation (see Table 8.1).

precipitation and maintained by high humidity. The persistence of rainwater in tanks appears to have been key to the evolution in *Brocchinia* of four specialized mechanisms for capturing nutrients (see below), which themselves are favored by open, moist, mineral-poor environments (Givnish 1989).

STATURE – Size or stature of the plants has undergone several shifts (Figure 8.6E). If we arbitrarily group species into **dwarf** (< 0.1 m tall in the vegetative state), **standard** (between 0.1 and 1.5 m tall), and **gigantic** (> 1.5 m tall) growth forms, then it would appear that (i) the standard condition is primitive; (ii) the dwarf growth form arose in the Melancra clade (at least among the taxa studied thus far); and (iii) the gigantic growth form arose at least three times, in *B. paniculata*, *B. micrantha*, and certain populations of *B. tatei* (Figure 8.6E). Differences in stature seem closely related to differences in ecological distribution. The dwarf species are all saxicolous, and vice versa. Open rock surfaces have extremely low productivity and coverage; low coverage, in turn, selects for plants of short stature (Givnish 1982; Tilman 1988). Gigantic species all occur in gaps in cloud forests, where dense coverage by woody species favors greater stature (Givnish 1982; Tilman 1988; King 1990) and competition for small, ephemeral openings favors unbranched trees (Givnish 1979, 1995).

Differences in stature are also closely related to the absence or presence of the tank habit: none of the dwarf species is impounding, whereas all the gigantic species are. The miniscule volume of any conceivable tank in a dwarf species – and its high surface area/volume ratio – militates against the evolution of the tank habit in very small bromeliads (Tomlinson 1969; Benzing 1980, 1990a). Conversely, the leaves of gigantic species are both **long** (to sustain a large photosynthetic crown in tall, unbranched woody plants [see Carlquist 1970; Givnish et al. 1995]) and **broad** (to adapt to moist, shady conditions in cloud forest gaps). These adaptations lead almost inevitably – when combined with the rosette leaf arrangement of bromeliads and a constantly humid, rainy environment at moderate elevations – to the impoundment of rainwater in large tanks.

ECOLOGY – **Carnivory** appears to have evolved at the base of the Reducta clade, and then lost in *B. tatei* and the hybrid *B. gilmartinii* (Figure 8.9). **Tank epiphytism** may have arisen twice, in *B. tatei* of the Reducta clade and in (the as yet unstudied) *B. hitchcockii* of the Melanacra clade. **Ant-fed myrmecophily** and **nitrogen fixation** are both known from only one taxon each and thus both appear to have arisen only once; it is not clear whether the loss of myrmecophily in certain shaded populations of *B. acuminata* represents a genetic or developmental change. It is important to note that all four specialized mechanisms for obtaining mineral nutrients evolved in lineages with the impounding habit (Figure 8.9). Given that absorbent trichomes in *Brocchinia* remain alive and subject to desiccation (Givnish et al. 1984; Benzing et al. 1985; Owen et al. 1988), a tank is important not only as a vessel for housing mutualists or capturing prey/vegetable debris, but also as a means of retaining live trichomes and placing them in contact with nutrient-bearing fluids.

TRICHOMES ON ADAXIAL LEAF SURFACES – Size and density of trichomes on adaxial leaf surfaces, lastly, appear closely related to both ecology and phylogeny. Carnivores and ant-fed myrmecophytes have the largest and/or densest trichomes; tank epiphytes, the next largest and/or densest; and non-impounding species the smallest and/or

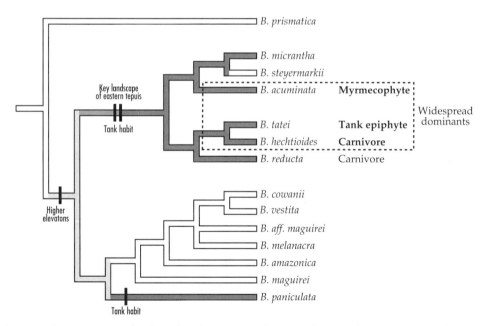

Figure 8.9. Reconstruction of ecological evolution in *Brocchinia*, in relation to the key innovation of the tank habit (gray shading, bars), key landscape of the eastern/central tepuis (bar), and rise to ecological dominance and widespread geographic distribution (dashed rectangle). Uncertainty as to whether the tank habit arose independently in *B. paniculata* and the ancestor of the Micrantha-Reducta clade, or in their common ancestor (see Figure 8.6D), is indicated by a lighter shade of gray. In three of the four species in which highly specialized mechanisms of nutrient capture evolved, ecological dominance and widespread geographic distribution ensued.

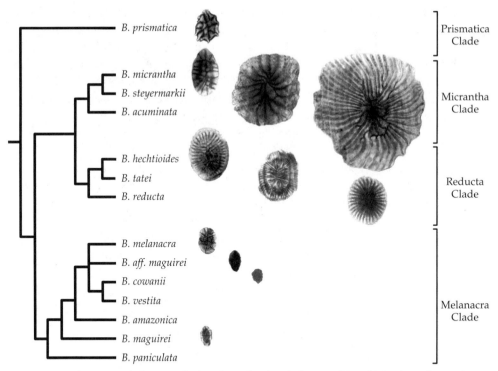

Figure 8.10. Trichome morphology overlaid on the molecular phylogeny of *Brocchinia*. The stellate trichomes of *B. prismatica*, basal to *Brocchinia*, are similar to those of *Fosterella*, which appears to be sister to *Brocchinia* based on a cpDNA-based analysis of relationships in the Pitcairnioideae (T. J. Givnish et al., unpubl. data). The trichomes of the species with the highest documented capacity for nutrient uptake – *B. acuminata, B. hechtioides, B. reducta* – all share a cap-cell arrangement with a high degree of radial symmetry.

least densely packed (Table 8.2). The smallest trichomes with the least regular arrangements of cap cells are concentrated in the Melanacra clade, whereas the densest trichomes with the most regular cell arrangements are concentrated in the Reducta clade (Figure 8.10). The transition to a highly geometric, radial pattern of cap cells appears to have occurred in the common ancestor of the Reducta and Micrantha clades (Figure 8.10); the irregular, slightly stellate cap-cell arrangement seen in the basal species *B. prismatica* is quite similar to that seen in the Andean pitcairnioid genus *Fosterella* (see Smith and Downs 1974; Benzing 1980; Varadarajan and Gilmartin 1988).

Geographic diversification

The tepuis of southern Venezuela and immediately adjacent areas (to which most populations of *Brocchinia* are restricted) divide naturally into two clusters centered on southeastern and southwestern Venezuela, east and west of the Rio Caura, respectively (Figure 8.1). Almost all species of *Brocchinia* are restricted to one cluster or the other, or are widespread in both (Table 8.1), suggesting a categorization of species having "eastern," "western," or "eastern and western" ranges. Several species of *Brocchinia* with an eastern range are known only from tepuis east of the Rio Caroni and its Icabaru affluent, in and near the Gran Sabana. Of the outgroups, *Brewcaria* and *"Brocchinia" serrata* are known only from the western tepuis; most (85%) species of *Navia* are known only from the western tepuis, and *Lindmania* is known from both western and eastern tepuis.

A cladistic analysis of the species ranges of *Brocchinia* and their relationship to topographic features (Figure 8.11) suggests (i) that *Brocchinia* arose west of the present-day Rio Caura; (ii) that the Micrantha-Reducta clade evolved following dispersal or isolation east of the Caroni; and (iii) that subsequently three of the four species highly specialized for life in nutrient-poor environments – the carnivore *B. hechtioides*, the ant-fed myrmecophyte *B. acuminata*, and the more or less saprophytic tank epiphyte *B. tatei* – reinvaded the west and became widespread ecological dominants. The genetic divergence among eastern and western populations of each of the latter species (Figure 8.4) is small relative to that seen across species, consistent with the

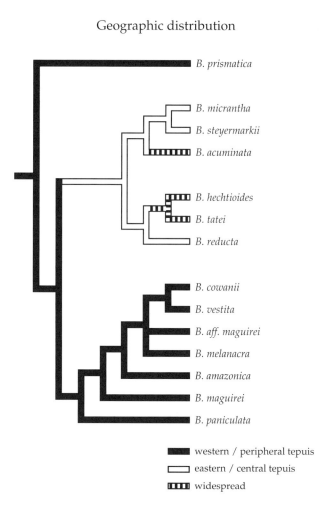

Figure 8.11. Reconstruction of inferred distribution in the eastern/central tepui cluster vs. the western/peripheral cluster using parsimony.

inference that these species expanded their range relatively recently and do not owe their broad range mainly to vicariance and the erosional dissection of the Guayana Highlands atop which they ride.

It is interesting that the four species of *Brocchinia* with the most specialized adaptations for nutrient capture – *B. acuminata, B. hechtioides, B. reducta,* and *B. tatei* – each occur over an exceptionally wide elevational range (mean = 1,344 m) as well, ranking first, second, third, and sixth among twenty species in this respect (Table 8.1). The probability of all four nutritional specialists ranking this high or higher in elevational range is less than 0.07% (Fisher exact test, 19 d.f.).

Other species with wide elevational ranges include both arborescent species, *B. paniculata* (1,100 m) and *B. micrantha* (700 m), geographically wide-ranging in the southwestern and southeastern Guayana Highlands, respectively; *B. vestita* (800 m), the geographically most wide-ranging saxicole, endemic to the southwestern tepuis; *B. melanacra* (700 m), the fire-adapted species of peaty savannas on the southwestern tepuis; and *B. steyermarkii* (760 m), the dominant species of peaty savannas in the Gran Sabana. *Brocchinia delicatula* (700 m), a saxicole endemic to Cerro de la Neblina, completes the list of species with an elevational range at or above the median (Table 8.1). The remaining species are each endemic to narrow geographic areas, one or two tepuis or a single river drainage.

The apparent correlation between the extent of a species' range and its elevational amplitude might reflect (i) the greater chance of successful dispersal (or vicariance) by an elevational generalist; (ii) the greater chance of conspecific populations isolated on different tepuis invading different elevational bands over evolutionary time and increasing the species' total elevational range; (iii) selection for greater dispersal ability in epiphytes (Benzing 1990a) and colonists of exposed, infertile sites; or (iv) a sampling artifact, reflecting the greater chance of a species being collected at a wide range of elevations if it occurs on several different tepuis studied by botanists in search of narrow endemics. Of these possible explanations, we believe that (i) and (ii) are the most plausible, but note that epiphytic *B. tatei* has bicaudate seed appendages divided into capillary strips, perhaps better adapted for wind dispersal than the undivided appendages seen in other *Brocchinia,* and perhaps a crude step toward plumose propagules like those seen in the Tillandsioideae.

Discussion

Our findings provide important insights into the three sets of issues raised at the outset of this study:

1. Do phenotypic similarities among species in a lineage reflect ecological similarities, close relationships, or both? How reliable a guide to phylogeny are traditional morphological characters?

Many of the morphological/ecological character-states that have been studied across the genus – adaxial leaf nerves, pungent leaf tips, tank habit, carnivory, arborescent growth form – have arisen and/or been lost independently at least twice.

In each of these cases, we have provided an argument (and supporting comparative data) that the character-state in question has adaptive value in a particular ecological context or in association with one or more additional character-states. Adaxial leaf nerves may reflect selection for aerenchyma in tank-forming species and those inhabiting wet, sodden soils, as well as selection for thin leaves (with prominent vascular bundles) and a thin adaxial layer of ground tissue in tank-forming species inhabiting cloudy, high-elevation sites. Pungent leaf tips are negatively associated with the tank habit, perhaps because the latter requires relatively soft, broad leaves that conform tightly to each other, while terminal leaf spines must be mounted on tough, stiff leaves to be effective. The tank habit is positively associated with heavy rainfall at middle to high (900–2,100 m) elevations, and with moderate to great stature in the plants possessing it. Saxicolous species have a dwarf growth form, befitting their occurrence on exposed, unproductive rock surfaces with sparse coverage by competitors; several species growing in gaps in dense cloud forest, by contrast, have evolved an arborescent growth form.

Even though each of these morphological character-states shows evidence of convergence – and thus, of a strong relationship to ecology – they display a strong relationship to phylogeny as well (Figures 8.6–8.10). However, while the pattern of variation in these character-states is consistent with the phylogeny for *Brocchinia* inferred from molecular data, the morphological data compiled by other authors (Table 8.4) appear to be too homoplastic to generate that phylogeny (or any other) on reliable grounds by themselves. A cladistic analysis based on these morphological data results in a poorly resolved tree (Figure 8.12) which disagrees at several points with the tree obtained using our cpDNA and nrDNA data sets. This morphological tree is based on a low value of CI (0.581 excluding autapomorphies) and so few variable characters (17) that the likelihood of obtaining the correct tree thereby is quite low ($P < 0.1$) based on the equations of Givnish and Sytsma (1997).

2. What has been the evolutionary pathway to various adaptations? Did key innovations arise once or repeatedly within a lineage? Did they accelerate rates of speciation and/or genetic evolution? To what extent do ecology and morphology reflect phylogenetic history?

The four specialized mechanisms for capturing mineral nutrients in *Brocchinia* – carnivory, ant-fed myrmecophily, tank epiphytism, and nitrogen fixation – all appear to have evolved subsequent to the origin of the tank habit at the base of the Micrantha and Reducta clades (Figure 8.9). The tank habit was essential for the evolution of each of these unusual adaptations, in that it (i) ensured that uncutinized trichomes would remain hydrated, alive, and capable of mineral uptake, at least on the submerged leaf bases; (ii) provided a cavity in which prey, ant wastes, fallen vegetable debris, or algal plugs could collect; and (iii) impounded water into which mineral nutrients could be released from the contents of the tank, and from which the plant could then absorb those nutrients.

The tank habit itself appears to have been favored by the high rainfall and humidity at moderate to high elevations on tepuis, based on the ecological restriction of tank-forming species to such altitudes (Figure 8.8), and on the likelihood of small

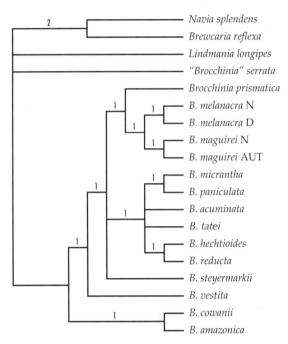

Figure 8.12. Phylogeny for *Brocchinia* based on morphology (see Table 8.5). Tree is strict consensus of six equally shortest trees (length = 40 steps, CI = 0.750 including autapomorphies; length = 36 steps, CI' = 0.722 excluding autapomorphies). Numbers indicate decay values associated with each node; bootstrap values are uniformly low and are not shown. Note relatively poor resolution of relationships within *Brocchinia*, inability to exclude *B. serrata* from the genus, and general lack of concordance of relationships with those seen in molecular phylogenies (see Figures 8.5–8.8).

plant cavities retaining rainwater between storms under such conditions (see above). It is interesting that carnivorous pitcher plants – members of the Cephalotacaceae, Nepenthaceae, and Sarraceniaceae, which retain rainwater in their peltate leaves – also tend to be restricted to areas with high rainfall (usually greater than 1,500 mm/year) relative to evaporation (Givnish 1989).

Based on the phylogenetic relationships among extant species, and their morphology, ecology, and geographical distribution (Tables 8.1–8.3, Figures 8.6–8.11), we envision the following, five-step evolutionary scenario for *Brocchinia*:

A. Short, narrow-leaved ancestors in low-elevation savannas in the western Guayana Shield gave rise to taller treelike forms with large, broad leaves – and, perforce, strongly overlapping leaf bases – growing in gaps in cloud forests at moderate elevations in the eastern and western Guayana Shield. Trichome size and density increased to absorb nutrients from fallen vegetable material.

Both basal lineages – the Prismatica and Melanacra clades – are restricted to the western Guayana Shield, as are *Brewcaria*, "*Brocchinia*" *serrata*, and much of *Navia*

Table 8.4. Morphological character-states used for a cladistic analysis of relationships in *Brocchinia*. Data derived from Smith and Downs (1974) and Smith (1984, 1986). By definition, populations of morphologically defined species would be grouped by a cladistic analysis based on morphology. Hence, ingroup and outgroup species have been represented by a single population except in those instances in which the molecular analysis indicated that conspecific populations were not monophyletic.

Taxon:	Characters:																		
	1	2	3	4	5	6	7	8	9	10	11	12	13	14	15	16	17	18	19
Navia spendens	1	0/1	1	1	0	1	1	0/1	1	0	0	1	0	0/1	0	0	0	0	?
Brewcaria reflexa	1	1	1	0	0	1	1	1	0	0	1	1	1	1/2	0	0	0	0	0
"*Brocchinia*" *serrata*	0	1	0	0	0	2	0	1	0	0	1	1	1	2	0	0	0	0	0
Lindmania	0	1	0	0	1	0	0	1	0	0	0/1	1	0/1	0/1	0	0	0	0	0
B. prismatica	0	0	0	0	0	2	0	0	0	1	0	1	0	1	0	0	0	0	0
B. micrantha	0	0	0	0	0	2	0	0	0	1	0	0	1	2	1	0	0	0	1
B. steyermarkii	0	0	0	0	0	2	0	0	0	1	1	1	0	1	0	0	0	0	1
B. acuminata	0	0	0	0	0	2	0	0	0	1	0/1	0	0	1	1	0	0	0	1
B. tatei	0	0	0	0	0	2	0	0	0	1	1	0	0/1	1	1	0/1	0	0	1
B. reducta	0	0	0	0	0	2	0	0	0	1	1	0	0	1	1	0	1	0	1
B. hechtioides	0	0	0	0	0	2	0	0	0	1	1	0	0	1	1	0	1	0	0
B. melanacra N	0	0	0	0	0	2	0	0	0	1	0	1	1	1	0	0	0	1	1
B. vestita	0	0	0	0	0	2	0	0	0	0	1	1	0	1	0	0	0	0	1
B. cowanii	0	0	0	0	0	2	0	0	0	0	1	1	0	0	0	0	0	0	0
B. maguirei N	0	0	0	0	0	2	0	0	0	0	0	1	1	1	0	0	0	0	?
B. melanacra D	0	0	0	0	0	2	0	0	0	1	0	1	1	1	0	0	0	1	1
B. amazonica	0	0	0	0	0	2	0	0	0	0	1	0	0	0	0	0	0	0	0
B. maguirei AUT	0	0	0	0	0	2	0	0	0	0	0	1	1	1	0	0	0	0	1
B. paniculata	0	0	0	0	0	2	0	0	0	1	1	0	1	2	1	0	0	0	1

Key to characters and character-states: **1** Seed appendages absent (0) vs. present (1); **2** Ovary position ± inferior (0) vs. ± superior (1); **3** Flowers sessile (0) vs. pedicellate (1); **4** Inflorescence racemose (0) vs capitate (1); **5** Sepal aestivation cochlear (0) vs. imbricate (1); **6** Anthers free (0), one series adnate (1), or two series adnate (2); **7** Floral bracts inconspicuous (0) vs. conspicuous (1); **8** Leaf margins entire (0) vs. serrate (1); **9** Leaf bases green (0) vs. brightly pigmented (1); **10** Adaxial leaf nerves unapparent (0) vs. apparent (1); **11** Inflorescence glabrous (0) vs. lepidote (1); **12** Leaf tip texture typical of rest of lamina (0) vs. pungent (sharp, piercing) (1); **13** Plants sessile (0) vs. caulescent (1); **14** Stature dwarf (0), typical (1), or gigantic (2); **15** Tank (impounding leaf bases) absent (0) vs. present; **16** Plants terrestrial (0) vs. epiphytic (1); **17** Plants non-carnivorous (0) vs. carnivorous (1); **18** Leaf tip coloration, margin typical of rest of lamina (0) vs. castaneous to black with tough, inrolled margin (1); **19** Inflorescence bipinnate (0) vs. tri- (or more) pinnate (1)

among the outgroups (see **Geographic diversification**). In addition, the closest relative of *Brocchinia* is *Fosterella* (Givnish et al., unpubl. data), whose range in the northern Andes is close to the western margin of the Guayana Shield. It thus seems likely that the forms ancestral to *Brocchinia* arose in the western Guayana Shield as well. *Brocchinia prismatica* and several saxicolous species known to be part of the Melanacra clade (and all other saxicoles remaining to be studied) are restricted to relatively low elevations (Table 8.1), and have the narrow leaves expected under the hotter and somewhat less humid conditions prevailing there, in scrub on sandstone outcrops and in Amazonian savannas on inundated sand (Varadarajan 1986a; Huber 1982).

The greater rainfall and lower evaporation prevailing at moderate (ca. 500–1,500 m) elevations result in greater canopy coverage in cloud forests, perhaps by reducing drought stress or suppressing fire on well-drained sites. The greater density of coverage would favor the evolution of taller growth forms (Givnish 1982, 1984, 1995; Tilman 1988) in *Brocchinia* invading forest gaps in this zone, perhaps similar to those of *B. micrantha* (at the base of the Micrantha clade) or *B. paniculata* (at the base of the Melanacra clade), with broad, long leaves and a woody stem.

High rainfall and cloudy conditions favor broad, relatively thin leaves (Givnish 1987, 1988). Among unbranched plants, taller species often have longer leaves, perhaps to have a broader canopy with which to balance the increased costs of stem tissue (Givnish et al. 1995). Broad, long leaves with thin cross-sections (and hence, readily conformable bases) would have militated toward the permanent impoundment of rainwater among the leaf axils in species found in the rainy, humid climate at moderate to high elevations. With the retention of rainwater and formation of a series of axillary "tanks" would come the possibility that the trichomes on the leaf bases would remain alive and capable of absorbing nutrients from vegetable material that has fallen into the tanks and begun to decompose. An increase in the tightness of leaf packing along the stem – together with the secretion of mucilage to permit nearby leaves to slide past each other while expanding – would have increased the watertightness and efficacy of the tanks. Species with the growth form postulated – *B. micrantha* and *B. paniculata* – bear live, absorbent trichomes that are somewhat denser and larger than those seen in the non-impounding species (Table 8.2; Figure 8.10).

B. Descendants of arborescent forms in the eastern Shield invaded higher elevations and evolved into shorter, unspecialized tank-forming species.

Cooler temperatures, heavier rainfall, and more intense leaching at higher elevations (1,500–2,500 m) on tepuis should lead to extremely infertile soils and very low rates of plant growth, favoring shorter growth forms in *Brocchinia* species there. Lower evaporation and higher rainfall at these elevations would permit smaller species – with smaller tanks – to retain rainwater more or less permanently. No extant species of *Brocchinia* has the growth form envisioned for this stage of evolution while lacking a specialized mechanism of nutrient capture, although some terrestrial populations of *B. tatei* approach it. Our phylogeny (Figure 8.9) suggests that this growth form evolved near the base of the Micrantha and Reducta clades; the common ancestor of these lineages appears to have arisen in the eastern Shield (Figure 8.13).

C. Modifications of leaf form and further increases in trichome area and density led to high-elevation species adapted for carnivory, myrmecophily, epiphytism, and nitrogen fixation; the advantages of these species on infertile soils led to their ecological dominance atop tepuis in the eastern and western Guayana Shield.

The extremely mineral-poor soils atop tepuis, together with abundant supplies of light and moisture in open habitats, favored the evolution of carnivory and other specialized mechanisms of nutrient capture by increasing the energetic advantage of obtaining a nutrient subsidy, relative to the costs (e.g., traps, lures) of obtaining that subsidy (Givnish et al. 1984; Givnish 1989). Indeed, the Guayana Shield supports

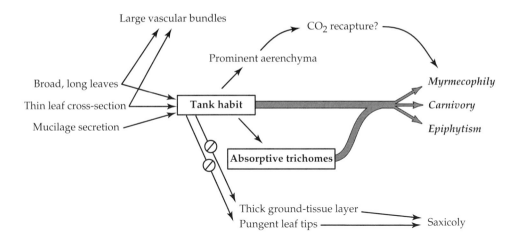

Figure 8.13. Tank habit as a key innovation in *Brocchinia*, showing entrainment of several co-adapted morphological traits and the evolutionary path leading to specialized mechanisms of nutrient capture and to subsequent ecological dominance and widespread distribution.

more genera of carnivorous plants (*Brocchinia, Catopsis, Drosera, Genlisea, Heliamphora, Utricularia*) than any other region of the world (Givnish 1989).

In *Brocchinia*, small changes in leaf shape and orientation from that of the ancestral tank-former could have generated the architecture needed to capture nutrients through carnivory or myrmecophily. A modest increase in radial curvature (and attendant increase in leaf inclination) could shift an open rosette (like that seen in *B. tatei*) to a leafy cylinder adapted for carnivory like that seen in *B. hechtioides*, *B. reducta*, and a wide variety of pitcher plants. A larger increase in the radial and longitudinal curvatures of the leaf sheath could create a bulbous rosette adapted for myrmecophily, like that seen in *B. acuminata* and a wide variety of other bromeliads with ant guests (e.g., *Tillandsia caput-medusae* [Benzing 1980]).

CARNIVORY – Givnish et al. (1984) and Benzing et al. (1985) proposed that carnivorous species of *Brocchinia* evolved from a tank-forming ancestor following invasion of extremely mineral-poor, open terrestrial sites. They argued that intensely lit, infertile conditions would favor the rise of (i) steeply inclined leaves, forming a cylindrical rosette; (ii) a dense waxy cuticle; and (iii) a yellowish, chlorotic coloration. These traits, together with (iv) the release into the tank fluid of an unknown volatile substance with a sweet, nectarlike odor (a similar odor is released when the leaf bases of other *Brocchinia* are bruised [Givnish et al. 1984]), would tend to increase the efficiency of prey attraction and capture. The ancestral tank habit and absorptive trichomes would permit the absorption of mineral nutrients from prey. Increases in the size and density of the trichomes would increase a carnivore's absorptive capacity and its ability to compete, directly and indirectly, for mineral nutrients with the commensal insects and epiphytic *Utricularia* that might grow in its tank.

Our phylogenetic data are consistent with this hypothetical pathway to carnivory, but are inadequate to determine the evolutionary sequence with which traits (i)–(iv) were acquired. The principal deviation of our results from the predictions of Givnish et al. (1984) is that the facultative tank epiphyte *B. tatei* appears to have evolved from within a carnivorous lineage, rather than the reverse (Figure 8.9). That is, Givnish et al. (1984) hypothesized an evolutionary sequence from tank epiphyte tank-forming terrestrial carnivore, whereas our phylogenetic results instead imply a sequence from tank-forming terrestrial carnivore tank epiphyte within the Reducta clade (Figure 8.8). However, this finding needs confirmation, given that our nrDNA data – though meager – *is* consistent with the hypothesis of Givnish et al. (1984).

The loss/lack of carnivory associated with the epiphytic habit in *Brocchinia* parallels the near absence of carnivory among other epiphytes, and the lower expected benefit of carnivory in droughty, partly shaded perches atop other plants (Givnish et al. 1984). The more open, less tightly cylindrical form of *B. hechtioides* may be related to (i) its occurrence at higher elevations than *B. reducta* (so that the latter faces higher rates of evaporation, which can be controlled via a narrow tank opening [see McWilliams 1974]); and/or (ii) its specialization (via greater stature) on flying insects, which require landing surfaces (likely to be greater in species in which the leaves are not perfectly furled into a cylinder along their entire length). The conspicuous yellow leaf color seen in both *B. hechtioides* and *B. reducta* is probably not merely chlorosis, but this issue awaits investigation.

ANT-FED MYRMECOPHILY – This habit evolved within the Micrantha clade, and it is easy to envision *B. acuminata* as a neotenic derivative of an ancestor much like the arborescent *B. micrantha*. During its juvenile stage, *B. micrantha* has spreading, broad leaves whose margins "pinch together" near their base, and then flair as the leaf base and sheath plunge into the heart of the rosette (Plate 1h). This trait disappears in the adult form, and creates small, partially enclosed cavities near the leaf base. These may represent an adaptation to reduce evaporation from the axillary tanks of small plants (with high surface area/volume ratios) growing on exposed, hot sandstone outcrops at mid elevations (500–1,200 m). These partial cavities might, in some instances, have attracted ants which made nests within their shelter. The ants, in turn, by inputting wastes, dead nestmates, and/or fallen food into the axillary tanks may have increased the growth of their hosts, and favored the evolution of more completely enclosed cavities created by more bulbous leaf bases/sheaths, and selected against the shift to the arborescent form in which the bulbous cavities would be lost. Recent ecological research by Dejean et al. (1995) on tillansioid bromeliads with a continuum of growth forms is consistent with this hypothesis. They found that, while ants occur occasionally in tillansioids with impounding tanks, they occur more frequently and with a greater number of species in myrmecophytic species with bulbous leaf bases that enclose a central cavity. An enclosed cavity may protect the ant guests not only from the elements; Davidson et al. (1989) suggest the long hairs characteristic of many myrmecophytes help exclude marauding ants as well.

Continued association with ants probably also favored the evolution of extremely tough leaf bases, both to defend the nests of beneficial guests, and to prevent damage to the plant itself. *Brocchinia acuminata* is similar to tillandsioid ant-plants (e.g., *Tillandsia caput-medusae, T. pruinosa*) in having a highly constricted opening at the summit of the leaf bases (Plate 1c), which has been interpreted as an adaptation to prevent drowning of the ant guests in tillandsioids (Benzing 1970, 1980). However, unlike the tillandsioids, *B. acuminata* lacks a pendent habit and has a less tightly constricted opening; its growth form permits some rainwater to enter the leaf axils. Presumably, this reflects the fact that nutrient absorption in *Brocchinia* occurs mainly through live shield cells (which can die if held above water [Givnish et al. 1984]), but in tillandsioids occurs mainly through dead shield cells (which incorporate vapor gaps to prevent desiccation of the underlying, live stalk cells [Benzing 1980]).

Brocchinia acuminata often has extensive adventitious roots among the leaf bases, a trait absent in the Pitcairnioideae outside *Brocchinia* but often found in epiphytic bromelioids (Benzing 1980); presumably, this trait evolved in *B. acuminata* (and *B. hechtioides*) to supplement nutrient absorption by leaf trichomes. Finally, we would like to propose that *B. acuminata* may have a highly unusual adaptation to enhance photosynthesis, based on capture of the elevated levels of carbon dioxide released by its ant guests. In this species, the stomata on the swollen leaf bases are often borne in clusters evident to the naked eye (Holst 1997). The functional significance of these clusters is unknown, but they connect (as in related species) to channels of aerenchyma; these latter are especially large in leaf sheaths of *B. acuminata,* accounting for roughly half of their cross-sectional area (Figure 8.7). Recent research by Treseder et al. (1995) shows that 39% of photosynthesis by *Dischidia major* (Asclepiadaceae) – an ant-fed myrmecophyte that houses ants inside its hollow leaves – is due to the capture of carbon dioxide released by its guests. In *B. acuminata*, the chlorophyllous parts of the leaf can be 30 cm or more from the non-photosynthetic leaf bases that house the ants. The prominent aerenchyma and stomatal clusters of this species may thus serve to carry enriched levels of carbon dioxide to the photosynthetic leaf tips, further increasing the benefit that the myrmecophyte *B. acuminata* receives from its guests. One question that should be explored is whether internal differentials in the partial pressure of water vapor within the aerenchyma – perhaps created by warming of the dark, broad leaf bases relative to the narrow leaf tips – help drive the movement of CO_2, in much the same way that internal "winds" drive CO_2 movement from roots to leaves in water lily aerenchyma (Dacey 1980).

The remaining member of the Micrantha clade – *B. steyermarkii* – is a non-impounding species of wet peaty savannas in the eastern Guayana Shield. It shows remarkable similarity in gross vegetative form to *B. prismatica*, native to the wet sands of Amazonian savannas in the western Shield. Both species possess tough, thick, terete leaves with abundant aerenchyma and pungent leaf tips (walking through dense stands of either species can be a cruel experience), and both possess massive rhizomes. However, the individual trichomes of *B. steyermarkii* are similar in size and absorptive ability to those of its two closest relatives, *B. acuminata* and *B. micrantha*, and completely unlike those of *B. prismatica* (Figure 8.10). With the evolutionary loss of the impounding habit in *B. steyermarkii*, the density of trichomes – and hence, the total absorptive surface per unit leaf area – has dwindled to a level similar to that seen in other, less closely related species that lack the impounding habit. The convergent evolution of low trichome/leaf surface ratios in non-impounding species (in the Prismatica, Melanacra, and Micrantha clades), and of much higher ratios in impounding species with specialized mechanisms for obtaining mineral nutrients (in the Micrantha and Reducta clades), provides strong evidence for the adaptive value of a high ratio for absorbing mineral nutrients, and of a low ratio in species lacking either the ability (physiological capacity of individual trichomes) or opportunity (presence of tank) to absorb such nutrients.

D. A separate lineage in the western Guayana Shield evolved the saxicolous habit, marked by short stature, thick leaves with pungent tips, small vascular bundles, inconspicuous aerenchyma, and a thick layer of colorless ground tissue over the chlorenchyma. Species lacking the tank habit did not evolve special modes of nutrient capture and have narrow ranges. One exception is *B. melanacra*, a fire-resistant species of wet, peaty savannas and bogs found at higher elevations on tepuis throughout western Venezuela.

Excluding the tank-forming *B. paniculata* at the base of the Melanacra lineage, the remaining species of this group are mostly found growing on exposed sandstone at low to middle elevations (Table 8.1). Their morphological features can be seen as adaptations to the relatively dry, strongly insolated conditions which such species face (see above). Most species are short in stature and grow at low to middle elevations (tentatively including dwarf species for which molecular data are now unavailable, such as *B. cataractarum, B. delicatula, and B. pygmaea*), but *B. maguirei, B.* aff. *maguirei,* and *B. melanacra* can develop woody stems up to 1 m tall while growing in montane bogs or on shallow peat over sandstone at higher elevations. The significance of the sharply defined abscission zone at the base of the leaves of evergreen *B. maguirei* is unknown; it is interesting that such an abscission zone is also known from *Ayensua uaipanensis* (Smith and Downs 1974), a shrub with thick, terete leaves native to moist, windy sites at high elevations on Auyán-tepui and Uaipan-tepui in the eastern Guayana Shield.

Most species of the Melanacra clade are narrowly endemic to one or two tepuis (Table 8.1), but *B. melanacra* is widespread throughout the western Guayana Shield, and indeed has been found on volcanic intrusives in the Sierra Parima (Holst 1997). *Brocchinia melanacra*, with strongly sclerotized leaf tips that enclose the terminal meristem, is known to be fire-adapted (Givnish et al. 1986). Adaptation to fire may seem paradoxical in plants atop tepuis, given that tepui summits are probably among the rainiest, most humid habitats on earth. However, three factors favor an important ecological role for fire on tepuis. First, the extremely infertile conditions atop tepuis – exacerbated by heavy rainfall and leaching – favors the evolution of tough, heavily defended leaves (Janzen 1974; Coley 1983; Bryant et al. 1985). Such leaves, with thick cell walls and high concentrations of chemical defenses, are likely to decompose very slowly, and decomposition rates are likely to be retarded further by the cool temperatures in tepui environments. As a consequence, leaf litter is likely to build up. Second, rainy and cloudy conditions atop tepuis are occasionally punctuated by clear weather, especially during the dry season, during which plant and soil surfaces can rapidly dry. Third, the strong upward convection of air, particularly near escarpment faces, is liable to induce lightning. Fourth, given the highly infertile conditions atop tepuis, even if fires occur rarely, they may have an important ecological effect, given the slow rate of plant regrowth that would follow any fire.

Givnish et al. (1986) observed a pair of large (ca. 1 km^2), recently burned areas atop Cerro de la Neblina in an area unlikely to have been visited by humans. Plants directly observed to resist fire, based on survival or resprouting – *Neblinaria* (thick bark), *Stegolepis* (thick, overlapping leaf bases), *Brocchinia melanacra* (sclerotized leaf tips enclosing the terminal meristem), and *Heliamphora tatei* (water-filled pitchers adjacent to meristems) – dominated large areas lacking topographic barriers to fire, while several other species were largely restricted to areas topographically sheltered from fire. While camping near the base of Cerro Duida, von Humboldt (1819–1829), Spruce (1908), and Tate and Hitchcock (1930) reported seeing a "glow" atop the mountain at night; it seems likely (Tate and Hitchcock 1930; Tate 1931b) that they were seeing a manifestation of a brush fire in progress (though whether ignited by lightning or Amerindians is unknown). The highly fire-resistant *Vellozia tubiflora* (T. J. Givnish, pers. obs.) is often common near exposed ridges and escarpments on Duida (Tate 1931a,b). The senior author has observed a large series of open, boggy habitats along the precipitous edge of Auyán-tepui near Angel Falls, dominated by *B. hechtioides* and *B. reducta*, and noted obvious signs of fire (e.g., charred shrub bases, charcoal) that may have opened these communities. Using data on fossil pollen assemblages, Rull (1991) showed that *Bonnettia* forests were displaced by meadows of *Stegolepis* – native to wet microsites but fire-resistant (Givnish et al. 1986) – during drought cycles atop Chimantá and Auyán-tepui in the last 3,000 years. Thus, while Huber (1995c) downplays the likelihood of fire atop tepuis, there seem to be compelling theoretical reasons and some empirical evidence suggesting that fire is indeed important there. The ecological importance of fire, in turn, may help explain the success of *B. melanacra* in becoming a widespread dominant. We believe that recurrent (albeit infrequent) fire may also help explain part of the ecological success of *B. acuminata, B. hechtioides,* and *B. reducta*, in that fire would tend to create open sites and impoverish the soil over time, favoring invasion by carnivores and ant-fed myrmecophytes there as elsewhere (Givnish 1989). The impounding, pitcher-like habit of these species – with the terminal meristem buried deep inside the rosette and/or below water – might also help ensure their ability to survive fires, much as the sun-pitcher (*Heliamphora duidae*) does (Givnish et al. 1986).

Key innovations

We envision the tank habit and absorptive trichomes as the two key innovations in *Brocchinia* that allowed the evolution of a great diversity of specialized mechanisms of nutrient capture, more than have evolved in any other genus of flowering plants. Of these, we believe that the tank habit had the most pervasive effect because (i) it was essential for trichome survival, and hence, for the evolution of trichome absorptivity and the diversity of nutrient capture mechanisms that led several species to ecological dominance and wide geographical ranges; and (ii) it helped entrain the evolution of several other aspects of plant morphology, some only distantly related to nutrient capture, and others possibly serving as pre-adaptations for particular nutrient capture strategies.

The latter point is important. We have argued that the tank habit was associated with, and helped promote the evolution of (i) broad, thin, long leaves; (ii) prominent vascular bundles; (iii) abundant aerenchyma; (iv) mucilage secretion around emerging leaves and closely packed leaf bases; (v) high densities of large, absorptive trichomes; and (vi) four different specialized mechanisms of nutrient capture – carnivory, ant-fed myrmecophily, tank epiphytism, and nitrogen fixation. We have also suggested that the tank habit is incompatible, either functionally or ecologically, with (vii) pungent leaf tips and (viii) a thick adaxial layer of colorless ground tissue in the leaves. Thus, the tank habit can be seen as a key innovation for *Brocchinia*, not only because it opened important new avenues for ecological diversification, but also because it acted as a central organizing factor that helped constrain and/or shape the evolution of several other characteristics (Figure 8.13). The role of the tank habit as an "evolutionary prime mover" is significant both **functionally** (in terms of the coordination and integration among traits it may have fostered) and **taxonomically** (in terms of the phylogenetic uncertainty it could have created by causing "concerted convergence" in such an integrated suite of traits [Chapters 1 and 2]). Whenever a key innovation arises more than once, it creates a strong possibility for concerted convergence and ensuing confusion regarding evolutionary relationships and phenotypic trends. In the case of *Brocchinia*, our molecular data indicate that the tank habit either arose twice (in the common ancestor of the Micrantha and Reducta clades, and in *B. paniculata*), or arose once (in the ancestor of Melanacra, Micrantha, and Reducta clades) and then was secondarily lost (in the Melanacra clade subsequent to the divergence of *B. paniculata*). In either case, our present data imply that only one significant radiation involving the tank habit occurred within *Brocchinia*, so that possession of this habit defines a group of species that are related both ecologically **and** phylogenetically. However, given that one unstudied species (*B. hitchcockii*) also evolved the tank habit, it will be crucial to study this taxon to determine whether it represents a second origin of tank epiphytism and, if so, whether its various characteristics reflect ecology (i.e., are convergent with members of the other tank-forming clade) or phylogeny (i.e., are similar to those of its relatives).

Our results on the ecology and phylogeny of the tank habit in *Brocchinia* (the only pitcairnioid genus with tank-forming species) may shed some light on the long-standing question of where this key innovation arose in bromelioids and tillandsioids. Pittendrigh (1948) and McWilliams (1974) suggested that tanks might have arisen under (i) rainy, humid conditions, where passive impoundment of abundant rain in a relatively unspecialized rosette could foster tolerance of short droughts; or under (ii) much drier conditions in saxicolous or epiphytic species, where impoundment in a narrow, modified rosette could insure survival through recurrent droughts. In *Brocchinia*, the first of these scenarios seems more plausible – the tank habit appears to have arisen in association with the invasion of rainy, extremely humid environments at higher elevations. There is no guarantee that the tank habit arose in other bromeliads under the same conditions. However, we believe an origin under relatively moist conditions seems more likely, given that smaller morphological changes would be required for plants to impound rainwater under such conditions, and – more important – even small juveniles (potentially subject to the greatest desiccation) could impound rainwater for considerable periods. We see the narrow, tubular tanks of *Billbergia* and other bromelioids on sun-baked granitic domes of the Brazilian Shield as highly specialized, drought-tolerant end-products of a lineage that arose under more humid conditions at the edge of the Amazonian or Atlantic rain forests.

3. What is the relationship between phylogeny and geographic distribution? Are distributional patterns more reflective of the history of speciation and dispersal/vicariance (and hence, of geological history), or of the adaptations of individual species to specific environments?

The Prismatica and Melanacra clades – which diverged earliest from other lineages within *Brocchinia* – are today largely restricted to the western Guayana Shield. In addition, three of the four outgroup genera (as well as the morphologically allied *Steyerbromelia*) are mainly or entirely restricted to the western Shield, suggesting that *Brocchinia* itself arose there.

This inference is consistent with the conclusion, based on morphology and cpDNA restriction-site data, that *Fosterella* is the closest relative of *Brocchinia* (Givnish et al., unpubl. data). *Fosterella* is native to the northern Andes, close to the western edge of the Guayana Shield and the inferred cradle of *Brocchinia*. The rifting of the Atlantic Ocean approximately 100 million years ago caused the simultaneous uplifting of the Guayana Shield close to the mid-Atlantic rift (George 1988), and later led to the Miocene uplift of the Andean chain at the leading edge of South America (Taylor 1995). The intense folding of the Roraima sandstone in the western tepuis (Ghosh 1985) may have been caused partly by buckling as the western edge of Guayana Shield pushed against non-Shield sediments to its west.

Our molecular data imply that the Prismatica and Melanacra clades arose in the western Guayana Shield, while the Micrantha-Reducta clade arose in the east. Several species in the latter, eastern clade evolved specialized adaptations (carnivory, myrmecophily, epiphytism) for gaining mineral nutrients, and subsequently were able to spread widely throughout the Guayana Shield and occupy an unusually wide range of elevation. It is interesting to note that the tall tepuis east of the Caroni and Icabaru (e.g., Auyán-tepui, Chimanta, Ilú-tepui, Ptari-tepui, Kukenam, Roraima) differ from others in having large areas of bare sandstone atop their summits, often weathered along numerous solution cracks into rocky labyrinths and towers (Briceño and Schubert 1990). If the current abundance of such extremely infertile sites at high elevations has been a long-standing characteristic of the eastern tepuis, it might have provided a powerful force favoring the evolution there of the tank habit, of dense arrays of absorptive trichomes, and of specialized mechanisms of obtaining mineral nutrients, such as carnivory, myrmecophily, and tank epiphytism.

The horizontal bedding of the eastern tepuis and their numerous networks of faults (Gibbs and Barron 1993) are two geological characteristics that would have resulted in more infiltration and consequent chemical weathering along solution cracks (see Briceño and Schubert 1990) than would the tectonically folded surfaces of the western tepuis. Such weathering may have contributed to the formation of labyrinths atop the eastern tepuis, leading to the loss of soil via wind or water erosion to adjacent cracks, and interfering with the formation of peat by improving drainage. Catastrophic fire – perhaps favored by the greater rainfall seasonality in the eastern Shield (see data presented by Huber 1995a) – may also have played an important role, removing some or all of the nutrients stored in peaty soils and resetting succession to bare sandstone.

The barren sandstone exposures of the eastern Guayana Shield may thus have acted as "key landscapes" (Chapter 1), favoring the rise of adaptations for nutrient capture that led to diversification within *Brocchinia*. In this connection, it is extremely interesting to note that the available evidence suggests the carnivorous sun-pitchers (Sarraceniaceae: *Heliamphora*) also arose in the eastern Guayana Shield. Jaffe et al. (1995) show that four of the five species of *Heliamphora* (*H. heterodoxa, H. ionasii, H. minor, H. nutans*) do not secrete digestive enzymes and lack the zone of waxy scales that helps precipitate prey into the tank in other pitcher-plants. Each of these species is restricted to the eastern Guayana Shield (Steyermark 1984). By contrast, the single remaining species (*H. tatei*) is endemic to tepuis of the western Shield, secretes digestive enzymes, and has specialized waxy scales to trap prey (Jaffe et al. 1995). We interpret these data as suggesting that carnivory in *Heliamphora* originated on the eastern tepuis, and then became increasingly perfected following dispersal to the western tepuis. This hypothesis could be tested by constructing a molecular phylogeny for *Heliamphora* and determining whether the seemingly rudimentary adaptations of the eastern species are primitive or are instead reversals. Preliminary ITS sequence data for a small sampling of *Heliamphora* taxa (Bayer et al. 1996) are consistent with our hypothesis, but a more extensive study is required for a conclusive test.

The fact that geographically widespread populations of the three ecologically dominant species of *Brocchinia* – *B. acuminata, B. hechtioides,* and *B. tatei* – diverge genetically very little from each other, relative to the amount of divergence seen between species (Figure 8.4), strongly suggests that their distributions result from recent long-distance dispersal, not vicariance, and that they reflect the ecology of the species involved and not the history of topographic connections between tepuis. This contrasts sharply with the pattern seen in tepui-dwelling genera of the Mutisieae (Asteraceae), most of which are narrowly endemic to single tepuis. Funk and Brooks (1990) state that phylogeny and geographic distribution are congruent in this group, with the morphology-based cladogram for the genera being identical to a family tree for the tepuis based on size and stream order of the rivers dividing one tepui (or set of tepuis) from another. In *Brocchinia*, individual species of the Melanacra clade are mainly restricted to one or two tepuis, so further research on this group may show a similar congruence between phylogeny and geography. However, no such pattern is now evident, though this may be an artifact of the small rate of genetic divergence in the Melanacra clade. Missing from the puzzle are four pygmy species (*B. cataractarum, B. delicatula, B. pygmaea, B. rupestris*), which appear morphologically allied to the Melanacra clade and occur in geographically peripheral, low-elevation sites on both sides of the Guayana Shield; if *B. cataractarum* and *B. rupestris* in Guyana are indeed part of the Melanacra clade, it might be necessary to emphasize a peripheral → central evolutionary movement rather than a west → east migration.

Based on current data, one might ask whether geographic distribution is more reflective of (i) the history of speciation, dispersal/vicariance, and geography, or (ii) the ecology of individual species. As a general principle – and in the specific case of *Brocchinia* – this question may be inappropriately dichotomous and impossible to answer correctly. Particular geographic areas may favor particular ecologies in the species originating there, which may make such species more or less likely to spread subsequently into different areas, and thus have present-day distributions that are more or less reflective of phylogenetic relationships vs. geographical history. In *Brocchinia*, the more infertile upland habitats atop the eastern tepuis may have fostered the evolution of specialized mechanisms of nutrient capture that enabled the species possessing them – restricted to the Micrantha-Reducta clade – to spread throughout the Guayana Shield. Clearly, in this case, geographic distribution reflects phylogeny, ecology, **and** geography of origin. Similarly, the narrow endemism of species in the Melanacra clade may reflect the short stature – and presumably low dispersal capacity – of saxicolous species that arose at intermediate elevations on the western tepuis; the greater physical isolation of the western tepuis from each other, and the lack of specialized mechanisms for nutrient capture among saxicolous species. Geographic distribution again appears to reflect phylogeny, ecology, and geography of origin.

Finally, the geographic distribution of individual species within a lineage may reflect the distribution (and, thus, the phylogeny and ecology) of others, based on competitive interactions and the extent of divergence among species (Grant 1981, 1986; Schluter et al. 1985; Moulton and Pimm 1986; Schluter 1988a,b; Schluter and McPhail 1992; Day et al. 1994; Chapter 18). Regular patterns of coexistence coinciding with the extent of ecological divergence between species in a lineage are expected when such species are each other's most important competitor. This seems most likely to occur in lineages that invade islands or islandlike habitats and then become ecologically dominant – as has *Brocchinia* atop tepuis. Indeed, most tepuis have a (i) **carnivore** (*B. reducta* at lower elevations in the east, *B. hechtioides* at higher elevations throughout) in bogs, wet savannas, or on exposed sandstone; (ii) an **ant-fed myrmecophyte** (*B. acuminata*) in scrubby habitats and forest edges; (iii) a **tank epiphyte** (*B. tatei* or *B. hitchcockii*) in cloud-forest canopies; (iv) an **arborescent species** (*B. paniculata* in the west, *B. micrantha* in the east) in cloud-forest gaps; and (v) a **non-impounding paludophyte** (*B. melanacra* in the west, *B. steyermarkii* in the east) at mid to high elevations in bogs and wet meadows (Table 8.1). It is interesting that saxicoles do not show a particularly regular pattern of distribution, occuring only at low elevations in some areas in the eastern and western Shield, and at intermediate and high elevations only on western tepuis. This seemingly irregular pattern within *Brocchinia* might reflect interactions with species of closely related genera of bromeliads endemic to the Guayana Shield (*Brewcaria, Connellia, Lindmania, Navia*) that appear to be ecologically equivalent to saxicolous *Brocchinia* species.

Conclusions

Brocchinia presents a striking example of an adaptive radiation that could not have been studied effectively without molecular systematics. Our cpDNA phylogeny shows that convergent evolution in morphology is rampant within *Brocchinia* and appears related to environmentally determined selection pressures. Even if we were to discount this molecular evidence for extensive convergence in morphology, the small number of morphological characters currently documented and their high degree of homoplasy would make it impossible to resolve relationships with any degree of reliability based on morphology alone. By contrast, our molecular data involve numerous characters with much lower levels of homoplasy, providing strong support for our molecular phylogeny – and hence, for the patterns of morphological evolution and geographic diversification inferred from that phylogeny.

The evolution of the tank habit appears to have been a key innovation in *Brocchinia*, opening the way to the evolution of carnivory, ant-fed myrmecophily, nitrogen fixation, and tank epiphytism and leading toward ecological dominance in many nutrient-poor habitats and microsites atop the tepuis of the Guayana Shield. The tank habit also appears to have entrained the evolution of several other co-adapted traits, including broad, thin, long leaves, prominent vascular bundles, abundant aerenchyma, mucilage secretion around emerging leaves, closely packed leaf bases, a thin adaxial layer of colorless ground tissue, and high densities of large, absorptive trichomes.

Our preliminary data support an evolutionary sequence leading from non-impounding species at low elevations to one lineage of non-impounding, rock-dwelling species at moderate to high elevations, and another lineage involving a progression from impounding arborescent forms in cloud forests, to impounding stemless forms at mid elevations, to carnivores, tank epiphytes, and ant-fed myrmecophytes at mid to high elevations. The occurrence of impounding forms at intermediate to high elevations with heavy precipitation parallels the general restriction of carnivorous pitcher-plants with impounding leaves to areas with heavy precipitation (see Givnish 1989).

There is a peculiar tie between phylogeny and geographic distribution in *Brocchinia*, with the "key landscape" of vast areas of bare and nearly bare sandstone in the eastern tepuis – perhaps associated with their horizontal bedding, extensive systems of faults and resulting solution cracks, and exposure to a more marked dry season – appearing to have favored the evolution of specialized mechanisms of nutrient capture. Once acquired, these adaptations led to ecological dominance and widespread geographic distribution for most of the species bearing them. The extremely infertile, humid, rainy conditions atop the eastern tepuis also appear to have spawned the rise of carnivory in an unrelated group of tepui plants, the South American sun-pitchers.

Geographic distributions in *Brocchinia* reflect both phylogeny and ecology, with a general evolutionary progression from ecologically unspecialized species in the western (peripheral) Guayana Shield, to the rise of ecologically specialized and dominant species in the eastern (central) Shield, followed by dispersal of the latter to tepuis throughout the Shield. The extent of genetic divergence among populations of widespread ecological dominants is consistent with their being a result of recent long-distance dispersal rather than ancient vicariance. This study is the first to use molecular systematics to examine the relationship between phylogeny and geography in any group of organisms native to the ancient tepuis, and focuses on a group with light

PLATE LEGENDS

Plate 1. Adaptive radiation in growth form and mode of nutrient capture in the bromeliad genus *Brocchinia* (see Chapter 8 for map of localities). **(A)** *B. hechtioides*, Auyán-tepui (carnivore). **(B)** *B. reducta*, Gran Sabana (carnivore). **(C)** *B. acuminata*, Gran Sabana (ant-fed myrmecophyte. **(D)** *B. acuminata*, La Escalera, showing loss of swollen leaf bases and adoption of sprawling, vine-like growth form in cloud-forest understories. **(E)** *B. tatei*, La Escalera, fallen from tree with D. H. Benzing (tank epiphyte). **(F)** *B. tatei*, Gran Sabana, growing epiphytically. **(G)** *B. micrantha*, La Escalera, > 2.5 m tall (impounding treelet). **(H)** *B. micrantha* seedling, showing constricted leaf bases. **(I)** *B. vestita*, Cerro Duida (saxicole). **(J)** *B. cowanii* (lower right), Cerro Moriche, < 8 cm across (saxicole). **(K)** *B. melanacra*, Cerro de la Neblina, with E. M. Burkhardt (fire-adapted paludophyte). **(L)** *B. steyermarkii*, Gran Sabana (paludophyte). **(M)** *B. prismatica*, Amazonian savanna at foot of Cerro Yapacana (paludophyte).

Photographs: T. J. Givnish

Plate 2. Cichlid species involved in the highly localized, parallel adaptive radiations within the *Pseudotropheus tropheops* complex of Lake Malawi, East Africa (see Chapter 12 for discussion of species ecologies and map of localities). **(A)** *Pseudotyropheus tropheops* "orange chest," Otter Point. **(B)** *P. t.* "broad-mouth," Otter Point. **(C)** *P. gracilior*, Otter Point. **(D)** *P. microstoma*, Otter Point. **(E)** *P. tropheops* "red cheek," Likoma Island. **(F)** *P. tropheops* "chinyamwezi," Chinyamwezi Rock. **(G)** *P. t.* "orange chest," OB morph, Otter Point. **(H)** *P. t.* unnamed, Chirwa-Chiluuba.

Photographs: P. N. Reinthal and A. Meyer

Plate 3. Floral diversity in *Platanthera* (see chapter 15). **(A)** *Papilio glaucus* pollinating **(B)** *Platanthera ciliaris*, one of the brightly-colored, diurnally pollinated species. Note the pollinium on the eye of the butterfly in **(A)**. **(C)** *Platanthera grandiflora*, another fringe-lipped species. Note the tripartite fringed labellum in this species, as opposed to the unipartite fringed labellum of *P. ciliaris* in **(A)** and **(B)**. **(D)** *Platanthera integra*, a species purported to be pollinated by bumblebees. **(E)** *Platanthera mandarinorum* ssp. *hachijoensis*, pollinated by hawkmoths. **(F)** *Platanthera flava* var. *herbiola*, one of a group of species that have a tubercle on the labellum. **(G)** *Platanthera orbiculata*, one of the many basal-leaved taxa in the genus. It is apparently pollinated by both noctuid moths and hawkmoths. **(H)** *Platanthera hyperborea* var. *hyperborea*. This species is pollinated by a variety of insects, and is also capable of selfing, as seen in **(I)** where the pollinia have fallen from the clinandria and are contacting the stigma. **(J)** *Platanthera leucophaea*, a species of the prairies of the United States east of the Mississippi River. It is pollinated by long-tongued hawkmoths (note the long spurs). This species is on the verge of extinction through loss of habitat and pollinators. **(K)** *Platanthera tipuloides* ssp. *nipponica*, pollinated by noctuid moths which approach the flower from above, the pollinia being placed on the underside of the proboscis. **(L)** and **(M)** close-up and habit shots, respectively, of *Platanthera. dilatata* var. *dilatata*. This species is typical of the spicy-fragrant, noctuid-pollinated Platantheras. **(N)** *Platanthera peramoena* being pollinated by *Hemaris thysbe*, a small diurnal hawkmoth. Like all the diurnally-pollinated species of *Platanthera*, *P. peramoena* is brightly colored. **(O)** a close-up shot of a *Hemaris* individual with numerous pollinia attached to its eyes.

Photos: **A–D, F–J, L, N, O**, J. R. Hapeman; **E, K**, K. Inoue; **M**, T. J. Givnish

Plate 4. Examples of Caribbean *Anolis* ecomorphs placed in their appropriate structural habitats (see Chapter 19). **(A)** *A. luteogularis* (crown-giant, Cuba). **(B)** *A. chlorocyanus* (trunk-crown, Hispaniola). **(C)** *A. loysiana* (trunk, Cuba). **(D)** *A. occultus* (twig, Puerto Rico). **(E)** *A. lineatopus* (trunk-ground, Jamaica). F) *A. alutaceus* (grass-bush, Cuba).

Photographs: T. Jackman

Plate 1. Adaptive radiation in *Brocchinia* (see Chapter 8).

Plate 2. Cichlids of the localized adaptive radiations within the *Pseudotyropheus tropheops* complex in Lake Malawi (see Chapter 12).

Plate 3. Floral diversity in *Platanthera* (see Chapter 15).

Plate 4. *Anolis* ecomorphs and their microhabitats in Caribbean forests (see Chapter 19).

seeds easily dispersed by wind. Investigations of other plant or animal groups with more limited powers of dispersal and strong tendencies toward extremely narrow endemism (e.g., Asteraceae, Bonnetiaceae, Rapateaceae, *Oreophrynella*) may show a stronger congruence between phylogeny and geography based on vicariance.

Different lineages within *Brocchinia* invaded different adaptive zones, and each spread differentially across the tepui landscape. However, the warp of ecology and the woof of geography mesh to form a fairly regular evolutionary fabric of community structure in *Brocchinia*, with one or two carnivores, one myrmecophyte, one tank epiphyte, one arborescent species, and one non-impounding paludophyte found on most tepuis. Such a mesh is expected only in ecologically dominant species that have invaded different adaptive zones largely in the absence of competition from other lineages. As expected, the distribution of saxicolous species of *Brocchinia* is far less regular, perhaps reflecting the fact that several other, closely related genera of pitcairnioids (*Brewcaria, Connellia, Lindmania, Navia*) have adopted this ecological role atop the tepuis.

The evolution of specialized mechanisms of nutrient capture in *Brocchinia* – the essence of **adaptive** radiation in this genus – accounts for only 7 of its 20 species. So, based on the narrow (and what we consider to be biologically unmeaningful) view of adaptive radiation advanced by Guyer and Slowinski (1993) and Sanderson and Donoghue (1994) as the extent of species multiplication within a lineage relative to its sister-group, not much adaptive radiation has occurred in *Brocchinia*. But this is clearly not the case: *Brocchinia* has, in an extremely infertile and humid environment, evolved more different specialized mechanisms of nutrient capture than any other genus of angiosperms, and the specialists have become dominant ecologically, have spread geographically, and can now coexist locally. While many of *Brocchinia* species are rather unspecialized rock-dwellers, their ranges are often restricted to one or two tepuis and rarely overlap. The relatively large number of non-impounding species (13) in *Brocchinia* reflects the narrow distribution of most individual species in this group, and may ultimately be a consequence of high rates of speciation (perhaps fostered by limited seed dispersal, born of extremely short stature) and local competitive exclusion by other non-impounding pitcairnioids. The pitcairnioid genus endemic to the tepuis with the largest number of species is *Navia* (ca. 50 in *Navia sensu stricto*, ca. 95 in *Navia* combined with *Brewcaria* [Holst 1997]). Most of these taxa are micro-endemics, restricted to single tepuis or parts of tepuis. Although *Navia* has not been studied carefully, our impression is that most species are saxicoles and are ecologically equivalent, aside from obvious differences in geographic (and to a lesser extent, elevational) distribution. It is interesting that *Navia* has relatively large, naked seeds with no means of long-distance transport (Smith and Downs 1974). Thus, almost surely, the great species richness of *Navia* is the result not of *adaptive* radiation but of *non-adaptive* radiation, of rapid speciation caused by limited dispersal and unaccompanied by substantial ecological divergence. Differences between lineages in dispersal ability and other lineage- and environment-specific factors that influence their intrinsic tendency to speciate (Mayr 1970; Diamond et al. 1976) must be considered in addition to any differences in ecological differentiation in determining the role of adaptive radiation in generating biological diversity. Contrary to Guyer and Slowinski (1993) and Sanderson and Donoghue (1994), adaptive radiation **is not** speciation; it is just one of several factors that influence the relative balance of speciation and extinction within a lineage, and may prove not to be the predominant driving force in many instances. However, adaptive radiation **is** likely to play a pivotal role in promoting morphological innovation, ecological specialization, and species coexistence in most lineages, especially on islands and islandlike environments.

Acknowledgments

This research was made possible by grants from the National Science Foundation (BSR-8806520 and BSR-9007293), National Geographic Society, and the Nave Fund of the University of Wisconsin. Additional support for travel to remote reaches of southern Venezuela was provided by the Smithsonian Institution, Fundación Terramar, Fundación para el Desarrollo de las Ciencias Fisicas, Matematicas y Naturales, Universidad de los Andes, and Universidad de los Llanos. We wish to express our heartfelt thanks to our colleagues Basil Stergios, Gustavo Romero, Charles Brewer, Fabian and Armando Michelangeli, and Juan Silva for helping arrange permits and logistics in connection with our expeditions to Duida, Marahuaca, Autana, Yapacana, Cerro de la Neblina, Auyán-tepui, Perai-tepui, and the Gran Sabana. We are especially grateful to James Estes of the NSF Systematics Program and Vicki Funk of the Smithsonian Institution for the critical additional financial support they provided after a helicopter crash aborted one of our missions. The Sisters of La Esmeralda kindly provided a base of operations for two expeditions to the Duida region.

In Brazil, Bruce Nelson (Instituto Nacional Pesquisas da Amazonia, Manaus) provided material of *B. amazonica* from Serra da Araca; in Colombia, Julio Betancour (Universidad de Antioquia) collected *B. serrata* on Cerro de Circasia. We thank both of these colleagues for their valuable assistance in obtaining key material from remote (and sometimes dangerous) areas. We also wish to thank Vicki Funk and Carol Kelloff of the Smithsonian Institution, the staff of the Flora of the Guianas project at the University of Guyana, Dean Charles Ramdass, Malcolm and Margaret Chan-a-sue, Harold Ameer, and Mark Wilson for logistical support in Guyana.

We thank Charles Brewer, Vicki Funk, Jorge Gonzales, Otto Huber, Roy McDiarmid, Fabian Michelangeli, Klaus Jaffe, Juan Silva, Basil Stergios, James Zarucchi, and the late Bassett Maguire for useful discussions on scientific and logistical matters. Paul Berry and Bruce Holst of the Missouri Botanical Garden provided unstinting aid on taxonomic and biogeographic questions. Finally, TJG would like to acknowledge the inspiration provided by the late Julian Steyermark, who introduced him to four remarkably divergent species of *Brocchinia* in the Gran Sabana in 1981.

References

Albert, V. A., Mishler, B. A., and Chase, M. W. 1992a. Character-state weighting for restriction site data in phylogenetic reconstruction, with an example from chloroplast DNA. Pp. 369–403 in *Molecular systematics of plants*, P. S. Soltis, D. E. Soltis, and J. J. Doyle, eds. New York, NY: Chapman and Hall.

Albert, V. A., Williams, S. E., and Chase, M. W. 1992b. Carnivorous plants: phylogeny and structural evolution. Science 257:1491–1495.
Amadon, D. 1950. The Hawaiian honeycreepers (Aves, Drepaniidae). Bulletin of the American Museum of Natural History 95:151–262.
Baldwin, B. G. 1992. Phylogenetic utility of the internal transcribed spacers of nuclear ribosomal DNA in plants: an example from the Compositae. Molecular Phylogenetics and Evolution 1:3–16.
Baldwin, B. G., Kyhos, D. W., and Dvorak, J. 1990. Chloroplast DNA evolution and adaptive radiation in the Hawaiian silversword alliance (Madiinae, Asteraceae). Annals of the Missouri Botanical Garden 77:96–109.
Baldwin, B. G., and Robichaux, R. H. 1995. Historical biogeography and ecology of the Hawaiian silversword alliance (Asteraceae): new molecular phylogenetic perspectives. Pp. 259–287 in *Hawaiian biogeography: evolution on a hot spot archipelago*, W. L. Wagner and V. A. Funk, eds. Washington, DC, Smithsonian Institution Press.
Bayer, R. J., Hufford, L., and Soltis, D. E. 1996. Phylogenetic relationships in Sarraceniaceae based on *rbc*L and ITS sequences. Systematic Botany 21:121–134.
Benzing, D. H. 1970. An investigation of two bromeliad myrmecophytes: *Tillandsia butzii* Mez, *T. caput-medusae* E. Morren and their ants. Bulletin of the Torrey Botanical Club 97:109–115.
Benzing, D. H. 1980. *Biology of the bromeliads*. Eureka, CA: Mad River Press.
Benzing, D. H. 1986. Foliar specialization for animal-assisted nutrition in Bromeliaceae. Pp. 235–256 in *Insects and the plant surface*, B. Juniper and R. Southwood, eds. London, England: Edward Arnold.
Benzing, D. H. 1987. The origin and rarity of botanical carnivory. Trends in Ecology and Evolution 2:364–369.
Benzing, D. H. 1990a. *Vascular epiphytes: general biology and associated biota*. New York, NY: Cambridge University Press.
Benzing, D. H. 1990b. Myrmecotrophy: origins, operation, and importance. Pp. 353–373 in *Ant-plant interactions*, C. R. Huxley and D. F. Curley, eds. Oxford, England: Oxford University Press.
Benzing, D. H., Henderson, K., Kessell, B., and Sulak, J. 1976. The absorptive capacities of bromeliad trichomes. American Journal of Botany 63:1009–1014.
Benzing, D. H., Givnish, T. J., and Bermudes, D. 1985. Absorptive trichomes in *Brocchinia reducta* (Bromeliaceae) and their evolutionary and systematic significance. Systematic Botany 10:81–91.
Benzing, D. H., and Pridgeon, A. 1983. Foliar trichomes of Pleurothallidinae (Orchidaceae): functional significance. American Journal of Botany 70:173–180.
Berry, P. E., Huber, O., and Holst, B. K. 1995. Floristic analysis and phytogeography. Pp. 161–192 in *Flora of the Venezuelan Guayana, vol. 1: introduction*, P. E. Berry, B. K. Holst, and K. Yatskievych, eds. Portland OR: Timber Press.
Bremer, K. 1988. The limits of amino acid sequence data in angiosperm phylogenetic reconstructions. Evolution 42:795–803.
Brewer-Carías, C. 1978. *Vegetación del mundo perdido*. Caracas, Venezuela: Fundación Eugenio Mendoza.
Briceño, H. O., and Schubert, C. 1990. Geomorphology of the Gran Sabana, Guayana Shield, southeastern Venezuela. Geomorphology 3:125–141.
Brooks, D. R., and McLennan, D. A. 1994. Historical ecology as a research programme: scope, limitations and the future. Pp. 1–27 in *Phylogenetics and ecology*, P. Eggleton and R. Vane-Wright, eds. London, England: Academic Press.
Bryant, J. P., Chapin III, F. S., Reichardt, P., and Clausen, T. 1985. Adaptation to resource availability as a determinant of chemical defense strategies in woody plants. Pp. 219–237 in *Chemically mediated interactions between plants and other organisms*, G. A. Cooper-Driver, T. Swain, and E. E. Conn, eds. New York, NY: Plenum Press.
Büdel, B., Lüttge, U., Stelzer, R., Huber, O., and Medina, E. 1994. Cyanobacteria of rocks and soils of the Orinoco lowlands and the Guayana uplands, Venezuela. Botanica Acta 107:422–431.
Carlquist, S. 1965. *Island life*. New York, NY: Natural History Press.
Carlquist, S. 1970. *Hawaii: a natural history*. New York, NY: Natural History Press.
Carr, G. D., Robichaux, R. H., Witter, M. S., and Kyhos, D. W. 1987. Adaptive radiation of the Hawaiian silversword alliance (Compositae: Mutisieae): a comparison with the Hawaiian picture-winged *Drosophila*. Pp. 79–97 in *Genetics, speciation, and the founder principle*, L. V. Giddings, K. Y. Kaneshiro, and W. W. Anderson, eds. London, England: Oxford University Press.
Carson, H. L., Hardy, D. E., Spieth, H. T., and Stone, W. S. 1970. The evolutionary biology of the Hawaiian Drosophilidae. Pp. 437–543 in *Essays in evolution and genetics in honor of Theodosius Dobzhansky*, M. K. Hecht and W. C. Steere, eds. Amsterdam, Netherlands: North Holland.

Chase, M. W., and Palmer, J. D. 1989. Chloroplast DNA systematics of Lilioid monocots: resources, feasibility, and an example from the Orchidaceae. American Journal of Botany 76:1720–1730.

Chase, M. W., and Palmer, J. D. 1992. Floral morphology and chromosome number in subtribe Oncidiinae (Orchidaceae): evolutionary insights from a phylogenetic analysis of chloroplast DNA restriction site variation. Pp. 324–339 in *Molecular systematics of plants*, P. S. Soltis, D. E. Soltis, and J. J. Doyle, eds. New York, NY: Chapman and Hall.

Chase, M. W., Soltis, D. E., Olmstead, R. G., Morgan, D., Les, D. H., Mishler, B. D., Duvall, M. R., Price, R. A., Hills, H. G., Qiu, Y., Kron, K. A., Rettig, J. H., Conti, E., Palmer, J. D., Manhart, J. R., Sytsma, K. J., Michaels, H. J., Kress, W. J., Karol, K. G., Clark, W. D., Hedr≥n, M., Gaut, B. S., Jansen, R. K., Kim, K.- J., Wimpee, C. F., Smith, J. F., Furnier, G. R., Strauss, S. H., Xiang, Q.- Y., Plunkett, G. M., Soltis, P. S., Swensen, S. M., Williams, S. E., Gadek, P. A., Quinn, C. J., Eguiarte, L. E., Golenberg, E., Learn, G. H., Graham, S. W., Barrett, S. C. H., Dayanandan, S., and Albert, V. A. 1993. Phylogenetics of seed plants: an analysis of nucleotide sequences from the plastid gene rbcL. Annals of the Missouri Botanical Garden 80:528–580.

Clegg, M. T. 1993. Chloroplast gene sequences and the study of plant evolution. Proceedings of the National Academy of Sciences, USA 90:363–367.

Clegg, M. T., and Zurawski, G. 1992. Chloroplast DNA and the study of plant phylogeny: present status and future prospects. Pp. 1–13 in *Plant molecular systematics*, D. E. Soltis, P. S. Soltis, J. J. Doyle, eds. New York, NY: Chapman and Hall.

Coley, P. D. 1983. Herbivory and defensive characteristics of tree species in a lowland tropical rain forest. Ecological Monographs 53:209–233.

Crisp, M. D. 1994. Evolution of bird-pollination in some Australian legumes (Fabaceae). Pp. 281–309 in *Phylogenetics and ecology*, P. Eggleton and R. Vane-Wright, eds. London, England: Academic Press.

Dacey, J. W. H. 1980. Internal winds in water lilies: an adaptation for life in anaerobic sediments. Science 210:1017–1019.

Davidson, D. W., Foster, R. B., Snelling, R. R., and Lozada, P. W. 1990. Variable composition of some tropical ant-plant associations. Pp. 145–162 in *Herbivory: tropical and temperate perspectives*, P. W. Price, ed. New York, NY: John Wiley and Sons.

Davidson, D. W., Snelling, R. R., and Longino, T. 1989. Competition among ants for myrmecophytes and the significance of plant trichomes. Biotropica 21:64–73.

Day, T. A., Martin, G., and Vogelmann, T. C. 1993. Penetration of UV-B radiation in foliage: evidence that the epidermis behaves as a non-uniform filter. Plant, Cell and Environment 16:735–741.

Day, T., Pritchard, J., and Schluter, D. 1994. A comparison of two sticklebacks. Evolution 48:1723–1734.

Dejean, A., Olmsted, I., and Snelling, R. R. 1995. Tree-epiphyte-ant relationships in the low inundated forest of Sian Ka'an Biosphere Reserve, Quintana Roo, Mexico. Biotropica 27:57–70.

Diamond, J. M., Gilpin, M. E., and Mayr, E. 1976. Species-distance relation for birds of the Solomon archipelago, and the paradox of the great speciators. Proceedings of the National Academy of Sciences, USA 73:2160–2164.

Downie, S. R., and Palmer, J. D. 1992. Use of chloroplast DNA rearrangements in reconstructing plant phylogeny. Pp. 14–35 in *Molecular systematics of plants*, P. S. Soltis, D. E. Soltis, and J. J. Doyle, eds. New York, NY: Chapman and Hall.

Doyle, J. J. 1993. DNA, phylogeny, and the flowering of plant systematics. BioScience 43:380–389.

Doyle, J. J., and Doyle, J. L. 1987. A rapid DNA isolation procedure for small quantities of fresh leaf tissue. Phytochemical Bulletin 19:11–15.

Felsenstein, J. 1985. Confidence limits on phylogenies: an approach using the bootstrap. Evolution 39:783–791.

Felsenstein, J. 1988. Phylogenies from molecular sequences: inference and reliability. Annual Review of Genetics 22:521–565.

Fish, D. 1976. Structure and composition of the aquatic invertebrate community inhabiting bromeliads in southern Florida and the discovery of an insectivorous bromeliad. Ph.D. dissertation, University of Florida, Gainesville.

Frank, J. H., and O'Meara, G. F. 1984. The bromeliad *Catopsis berteroniana* traps terrestrial arthropods but harbors *Wyeomyia* larvae (Diptera: Culicidae). Florida Entomologist 67:418–424.

Fryer, G., and Iles, T. D. 1972. *The cichlid fishes of the great lakes of Africa*. Neptune NJ: Tropical Fish Hobbyist Publications.

Funk, V. A., and Brooks, D. R. 1990. *Phylogenetic systematics as the basis of comparative biology*. Washington, D. C.: Smithsonian Institution Press.

George, U. 1988. *Inseln in der Zeit*. Hamburg, Germany: Geo im Verlag.
Ghosh, S. K. 1985. Geology of the Roraima Group and its implications. Pp. 33–50 in *Memoria I symposium Amazónico*, M. I. Muñoz, ed. Caracas, Venezuela: Ministerio de Energía y Minas.
Gibbs, A. K., and Barron, C. N. 1993. *The geology of the Guiana shield*. Oxford Monographs on Geology and Geophysics, Number 22. Oxford, England: Oxford University Press.
Gilmartin, A. J., and Brown, G. K. 1987. Bromeliales, related monocots, and resolution of relationships among Bromeliaceae subfamilies. Systematic Botany 12:493–500.
Givnish, T. J. 1979. On the adaptive significance of leaf form. Pp. 375–407 in *Topics in plant population biology*, O. T. Solbrig, S. Jain, G. B. Johnson, and P. H. Raven, eds. New York, NY: Columbia University Press.
Givnish, T. J. 1982. On the adaptive significance of leaf height in forest herbs. American Naturalist 120:353–381.
Givnish, T. J. 1984. Leaf and canopy adaptations in tropical forests. Pp. 51–84 in *Physiogical ecology of plants of the wet tropics*, E. Medina, H. A. Mooney, and C. Vásquez-Yánes, eds. The Hague, The Netherlands: Dr. Junk.
Givnish, T. J. 1987. Comparative studies of leaf form: assessing the relative roles of selective pressures and phylogenetic constraints. New Phytologist 106 (Suppl.):131–160.
Givnish, T. J. 1988. Adaptation to sun vs. shade: a whole-plant perspective. Australian Journal of Plant Physiology 15:63–92.
Givnish, T. J. 1989. Ecology and evolution of carnivorous plants. Pp. 243–290 in *Plant-animal interactions*, W. G. Abrahamson, ed. New York, NY: McGraw-Hill.
Givnish, T. J. 1995. Plant stems: biomechanical adaptation for energy capture and influence on species distributions. Pp. 3–49 in *Plant stems: physiology and functional morphology*, B. L. Gartner, ed. New York, NY: Chapman and Hall.
Givnish, T. J., Burkhardt, E. L, Happel, R. E., and Weintraub, J. W. 1984. Carnivory in the bromeliad *Brocchinia reducta*, with a cost/benefit model for the general restriction of carnivorous plants to sunny, moist, nutrient-poor habitats. American Naturalist 124:479–497.
Givnish, T. J., McDiarmid, R. W., and Buck, W. R. 1986. Fire adaptation in *Neblinaria celiae* (Theaceae), a high-elevation rosette shrub endemic to a wet equatorial tepui. Oecologia 70:481–485.
Givnish, T. J., and Sytsma, K. J. 1997. Consistency, characters, and the likelihood of correct phylogenetic inference. Molecular Phylogenetics and Evolution 7:320–333.
Givnish, T. J., Sytsma, K. J., and Smith, J. F. 1990a. Adaptive radiation, plant-animal interactions, and molecular evolution in the bromeliad genus *Brocchinia*. American Journal of Botany 77 (suppl.):174–175.
Givnish, T. J., Sytsma, K. J., and Smith, J. F. 1990b. A re-examination of phylogenetic relationships among bromeliad subfamilies using cpDNA restriction site variation. American Journal of Botany 77 (suppl.):133.
Givnish, T. J., Sytsma, K. J., Smith, J. F., and Hahn, W. J. 1994. Thorn-like prickles and heterophylly in *Cyanea*: adaptations to extinct avian browsers on Hawaii? Proceedings of the National Academy of Sciences, USA 91:2810–2814.
Givnish, T. J., Sytsma, K. J., Smith, J. F., and Hahn, W. J. 1995. Molecular evolution, adaptive radiation, and geographic speciation in *Cyanea* (Campanulaceae), the largest plant genus endemic to Hawaii. Pp. 288–337 in *Hawaiian biogeography: evolution on a hot spot archipelago*, W. L. Wagner and V. A. Funk, eds. Washington, DC: Smithsonian Institution Press.
Gonzales, J. M., Jaffe, K., and Michelangeli, F. 1991. Competition for prey between the carnivorous Bromeliaceae *Brocchinia reducta* and Sarraceniaceae *Heliamphora nutans*. Biotropica 23:602–604.
Grant, P. R. 1981. Speciation and the adaptive radiation of Darwin's finches. American Scientist 69:653–663.
Grant, P. R. 1986. *Ecology and evolution of Darwin's finches*. Princeton, NJ: Princeton University Press.
Grant, V. 1963. *The origin of adaptations*. New York, NY: Columbia University Press.
Greenwood, P. H. 1974. Cichlid fishes of Lake Victoria, East Africa: the biology and evolution of a species flock. Bulletin of the British Museum of Natural History (Zoology), Suppl. 6:1–134.
Greenwood, P. H. 1978. A review of the pharyngeal apophysis and its significance in the classification of African cichlid fishes. Bulletin of the British Museum of Natural History (Zoology) 33:297–323.
Guyer, C., and Slowinski, J. B. 1993. Adaptive radiation and the topology of large phylogenies. Evolution 47:253–263.
Hamby, R. K., and Zimmer, E. A. 1992. Ribosomal RNA as a phylogenetic tool in plant systematics. Pp. 50–91 in *Plant molecular systematics*, D. E. Soltis, P. S. Soltis, and J. J. Doyle, eds. New York, NY: Chapman and Hall.

Heslop-Harrison, Y. 1976. Enzyme secretion and digest uptake in carnivorous plants. Pp. 463–476 in *Perspectives in experimental biology*, N. Sunderland, ed. Oxford, England: Pergamon Press.

Hillis, D. M., Allard, M. W., and Miyamoto, M. M. 1993. Analysis of DNA sequence data: phylogenetic inference. Methods in Enzymology 224:456–487.

Hillis, D. M., Huelsenbeck, J. P., and Cunningham, C. W. 1994. Application and accuracy of molecular phylogenies. Science 264:671–677.

Hitchcock, C. B. 1947. The Orinoco-Ventuari region, Venezuela. Geographic Review 37:525–566.

Hodges, S. A., and Arnold, M. L. 1994a. Floral and ecological isolation between *Aquilegia formosa* and *Aquilegia pubescens*. Proceedings of the National Academy of Sciences, USA 91:2493–2496.

Hodges, S. A., and Arnold, M. L. 1994b. Columbines: a geographically wide-spread species flock. Proceedings National Academy of Sciences, USA 91:5129–5132.

Hölldobler, B., and Wilson, E. O. 1990. *The ants*. Cambridge MA: Belknap Press.

Holst, B. K. 1997. Bromeliaceae. Pp. 548–676 in *Flora of the Venezuelan Guayana, vol. 3*, J. A. Steyermark, P. E. Berry, and B. K. Holst, eds. Portland OR: Timber Press.

Huber, O. 1982. Significance of savanna vegetation in the Amazon Territory of Venezuela. Pp. 221–244 in *Biological diversification in the tropics*, G. T. Prance, ed. New York, NY: Columbia University Press.

Huber, O. 1988. Guayana Highlands vs. Guyana Lowlands: a reappraisal. Taxon 37:595–614.

Huber, O. 1995a. Geographical and physical features. Pp. 1–61 in *Flora of the Venezuelan Guayana, vol. 1: introduction*, P. E. Berry, B. K. Holst, and K. Yatskievych, eds. Portland, OR: Timber Press.

Huber, O. 1995b. Vegetation. Pp. 97–160 in *Flora of the Venezuelan Guayana, vol. 1: introduction*, P. E. Berry, B. K. Holst, and K. Yatskievych, eds. Portland, OR: Timber Press.

Huber, O. 1995c. Conservation of the Venezuelan Guayana. Pp. 1–61 in *Flora of the Venezuelan Guayana, vol. 1: introduction*, P. E. Berry, B. K. Holst, and K. Yatskievych, eds. Portland, OR: Timber Press.

Huber, O., and Berry, P. 1995. Topographical map of the Venezuelan Guayana. Packet in *Flora of the Venezuelan Guayana, vol. 1: introduction*, P. E. Berry, B. K. Holst, and K. Yatskievych, eds. Portland, OR: Timber Press.

Huxley, C. R. 1978. The ant-plants *Myrmecodia* and *Hydnophytum* (Rubiaceae), and the relationships between their morphology, ant occupants, physiology and ecology. New Phytologist 80:231–268.

Huxley, C. 1980. Symbiosis between ants and epiphytes. Biological Review 55:321–340.

Huxley, C. R. 1986. Evolution of benevolent ant-plant relationships Pp. 257–282 in *Insects and the plant surface*, B. Juniper and R. Southwood, eds. London, England: Edward Arnold.

Huxley, J. 1943. *Evolution: the modern synthesis*. New York, NY: Harper and Brothers.

Jaffe, K., Michelangeli, F., Gonzalez, J. M., Miras, B., and Ruiz, M. C. 1995. Carnivory in pitcher plants of the genus *Heliamphora* (Sarraceniaceae). New Phytologist 122:733–744.

Janzen, D. H. 1974. Tropical blackwater rivers, animals, and mast fruiting in the Dipterocarpaceae. Biotropica 6:69–105.

Juniper, B. E., Robins, R. J., and Joel, D. M. 1989. *The carnivorous plants*. New York, NY: Academic Press.

Kadereit, J. W., and Sytsma, K. J. 1992. Disassembling *Papaver*: a restriction site analysis of chloroplast DNA. Nordic Journal of Botany 12:205–217.

King, D. A. 1990. The adaptive significance of tree height. American Naturalist 135:809–828.

Knox, E. B., Downie, S. R., and Palmer, J. D. 1993. Chloroplast genome rearrangements and the evolution of giant lobelias from herbaceous ancestors. Molecular Biology and Evolution 10:414–430.

Kocher, T. D., Conroy, J. A., McKaye, K. R., and Stauffer, J. R. 1993. Similar morphologies of cichlid fish in Lakes Tanganyika and Malawi are due to convergence. Molecular Phylogenetics and Evolution 2:158–165.

Kress, W. J., Schatz, G. E., Andrianifahanana, M., and Morland, H. S. 1994. Pollination of *Ravenala madagascariensis* (Strelitziaceae) by lemurs in Madagascar: evidence for an archaic coevolutionary system? American Journal of Botany 81:542–551.

Lack, D. 1947. *Darwin's finches*. Cambridge, England: Cambridge University Press.

Liem, K. 1990. Key evolutionary innovations, differential diversity, and symecomorphosis. Pp. 147–170 in *Evolutionary innovations*, M. H. Nitecki, ed. Chicago, IL: University of Chicago Press.

Maddison, W. P., and Maddison, D. R. 1992. *MacClade: analysis of phylogeny and character evolution, vers. 3.05*. Sunderland, MA: Sinauer Associates.

Maddison, W. P., Donoghue, M. J., and Maddison, D. R. 1984. Outgroup analysis and parsimony. Systematic Zoology 33:83–103.

Maguire, B. 1955. Cerro de la Neblina, Amazonas, Venezuela. Geographic Review 45:27–51.

Maguire, B. 1970. On the flora of the Guayana Highland. Biotropica 2:85–100.

Maguire, B. 1978. The botany of the Guayana Highland 10. Memoirs of the New York Botanical Garden 29:36–62.
Martin, C. E. 1994. Physiological ecology of the Bromeliaceae. Botanical Review 60:1–81.
Mayr, E. 1963. *Animal species and evolution*. Cambridge, MA: Harvard University Press.
Mayr, E. 1970. *Populations, species, and evolution*. Cambridge, MA: Belknap Press.
Mayr, E., and Phelps, W. H., Jr. 1967. The origin of the bird fauna of the south Venezuelan highlands. Bulletin of the American Museum of Natural History 136:275–327.
McDiarmid, R. W., and Gorzula, S. 1989. Aspects of the reproductive ecology and behavior of the tepui toads, genus *Oreophynella* (Anura, Bufonidae). Copeia 2:445–451.
McWilliams, E. L. 1974. Evolutionary ecology. Pp. 40–55 in L. B. Smith and R. J. Downs, Pitcairnioideae (Bromeliaceae). Flora Neotropica 14:1–660.
Meyer, A., Kocher, T. D., Basasibwaki, P., and Wilson, A. C. 1990. Monophyletic origin of Lake Victoria cichlid fishes suggested by mitochondrial DNA sequences. Nature 347:550–553.
Meyer, A., Kocher, T. D., and Wilson, A. C. 1991. African fishes. Nature 350:467–468.
Moran, P., Kornfield, I., and Reinthal, P. 1993. Molecular systematics and radiation of the haplochromine cichlids (Teleostei: Perciformes) of Lake Malawi. Coepia 1994:274–288.
Moulton, M. P., and Pimm, S. L. 1986. The extent of competition in shaping an introduced avifauna. Pp. 80–97 in *Community ecology*, J. Diamond and T. J. Case, eds. New York, NY: Harper and Row.
Nei, M., and Li, W. 1979. Mathematical model for studying genetic variation in terms of restriction endonucleases. Proceedings of the National Academy of Sciences, USA 76:5269–5273.
Olmstead, R. G., and Palmer, J. D. 1992. A chloroplast DNA phylogeny of the Solanaceae: subfamilial relationships and character evolution. Annals of the Missouri Botanical Garden 79:346–360.
Osborn, H. F. 1902. The law of adaptive radiation. American Naturalist 36:353–363.
Owen, T. P., Benzing, D. H., and Thomson, W. W. 1988. Apoplastic and ultrastructural characterizations of the trichomes of the carnivorous bromeliad *Brocchinia reducta*. Canadian Journal of Botany 66:941–948.
Palmer, J. D. 1987. Chloroplast DNA evolution and biosystematic uses of chloroplast DNA variation. American Naturalist 130:S6–S29.
Palmer, J. D. 1990. Phylogenetic analysis of *rbc*L and other chloroplast gene sequences. American Journal of Botany 77 (suppl.):117.
Palmer, J. D., Jorgensen, R. A., and Thompson, W. F. 1985. Chloroplast DNA variation and evolution in *Pisum*: patterns of change and phylogenetic analysis. Genetics 109:195–213.
Pittendrigh, C. S. 1948. The bromeliad-*Anopheles*-malaria complex in Trinidad. I. The bromeliad flora. Evolution 2:58–89.
Pouyllau, M., and Seurin, M. 1985. Pseudo-karst dans les roches greso-quartzitiques de la Formation Roraima. Karstologia 5:45–52.
Puente, M. E., and Bashan, Y. 1994. The desert epiphyte *Tillandsia recurvata* harbors the nitrogen-fixing bacterium *Pseudomonas stutzeri*. Canadian Journal of Botany 72:406–408.
Ranker, T. A., Soltis, D. E., Soltis, P. S., and Gilmartin, A. J. 1990. Subfamilial relationships of the Bromeliaceae: evidence from chloroplast DNA restriction site variation. Systematic Botany 15:425–434.
Rickson, F. 1978. Absorption of animal tissue breakdown products into a plant system – the feeding of a plant by ants. American Journal of Botany 66:87–90.
Rieseberg, L. H., and Morefield, J. D. 1995. Character expression, phylogenetic reconstruction, and the detection of reticulate evolution. Pp. 333–353 in *Experimental and molecular approaches to plant biosystematics*, P. C. Hoch and A. G. Stephenson, eds. Monographs in Systematic Botany, Vol. 53. St. Louis, MO: Missouri Botanical Garden.
Robberecht, R., and Caldwell, M. M. 1978. Leaf epidermal transmittance of ultraviolet radiation and its implications for plant sensitivity to ultraviolet-radiation induced injury. Oecologia 32:277–287.
Robinson, H. 1969. A monograph on foliar anatomy of the genera *Connellia, Cottendorfi*a and *Navia* (Bromeliaceae). Smithsonian Contributions to Botany 2:1–41.
Rull, V. 1991. Contribucion a la paleoecologia de Pantepui y la Gran Sabana (Guayana Venezolana): clima, biogeographia, ecologia. Scientia Guaianae 2:i–xxii, 1–133.
Saghai-Maroof, M. A., Soliman, K. M., Jorgenson, R. A., and Allard, R. W. 1984. Ribosomal DNA spacer-length polymorphisms in barley: Mendelian inheritance, chromosomal location, and population dynamics. Proceedings of the National Academy of Sciences, USA 81:8014–8018.
Sanderson, M. J. 1989. Confidence limits on phylogenies: the bootstrap revisited. Cladistics 5: 113–129.
Sanderson, M. J., and Donoghue, M. J. 1994. Shifts in diversification rate with the origin of angiosperms. Science 264:1590–1593.

Schaefer, C., and Dalrymple, J. 1995. Landscape evolution in Roraima, North Amazonia – planation, paleosols and paleoclimates. Zeitschrift für Geomorphologie 39:1–28.

Schaefer, S. A., and Lauder, G. V. 1986. Historical transformation of functional design: evolutionary morphology of feeding mechanisms in loricarioid catfishes. Systematic Zoology 35:489–508.

Schluter, D. 1988a. Character displacement and the adaptive divergence of finches on islands and continents. American Naturalist 131:799–824.

Schluter, D. 1988b. The evolution of finch communities on islands and continents: Kenya vs. Galápagos. Ecological Monographs 58:229–249.

Schluter, D., and McPhail, J. D. 1992. Ecological character displacement and speciation in sticklebacks. American Naturalist 140:85–108.

Schluter, D., Price, T. D., and Grant, P. R. 1985. Ecological character displacement in Darwin's finches. Science 227:1056–1059.

Schubert, C., and Huber, O. 1989. *La Gran Sabana*. Caracas, Venezuela: Cuadernos Lagoven.

Simpson, G. G. 1944. *Tempo and mode in evolution*. New York, NY: Columbia University Press.

Simpson, G. G. 1953. *The major features of evolution*. New York, NY: Columbia University Press.

Skinner, M. W. 1988. Comparative pollination ecology and floral evolution in Pacific Coast *Lilium*. Ph.D. dissertation, Harvard University.

Smith, J. F., Sytsma, K. J., Shoemaker, J. S., and Smith, R. L. 1991. A qualitative comparison of total cellular DNA extraction protocols. Phytochemical Bulletin 23:2–9.

Smith, L. B. 1969. Botany of the Guayana Highland, Part VIII. Bromeliaceae. Memoirs of the New York Botanical Garden 18:29–32.

Smith, L. B. 1984. A new species of *Brocchinia* from Brazil. Journal of the Bromeliad Society 34:106.

Smith, L. B. 1986. Revision of the Guayana Highland Bromeliaceae. Annals of the Missouri Botanical Garden 73:689–721.

Smith, L. B., and Downs, R. J. 1974. Bromeliaceae (Pitcairnioideae). Flora Neotropica 14:1–662.

Smith, R. L., and Sytsma, K. J. 1990. Evolution of *Populus nigra* (sect. *Aigeiros*): introgressive hybridization and the chloroplast contribution of *Populus alba* (sect. *Populus*). American Journal of Botany 77:1176–1187.

Soltis, D. E., Soltis, P. S., and Doyle, J. J. (eds.). 1992. *Plant molecular systematics*. New York, NY: Chapman and Hall.

Springer, M. S., and Kirsch, J. A. W. 1991. DNA hybridization, the compression effect, and the radiation of diprotodontian marsupials. Systematic Zoology 40:131–151.

Spruce, R. 1908. Notes of a Botanist on the Amazon and Andes. Ed. A. R. Wallace.

Steyermark, J. A. 1961. *Brocchinia*. Bromeliad Society Bulletin 11:35–41.

Steyermark, J. A. 1967. Flora del Auyán-tepui. Acta Botanica Venezuelica 2(5):5–370.

Steyermark, J. A. 1984. Flora of the Venezuela Guayana, I. Annals of the Missouri Botanical Garden 71:297–340.

Steyermark, J. A. 1986. Speciation and endemism in the flora of the Venezuelan tepuis. Pp. 317–373 in *High altitude tropical biogeography*, F. Vuilleumier and M. Monasterio, eds. London, England: Oxford University Press.

Sturmbauer, C., and Meyer, A. 1992. Genetic divergence, speciation, and morphological stasis in a lineage of African cichlid fishes. Nature 358:578–581.

Swofford, D. L. 1993. *PAUP: phylogenetic analysis using parsimony, version 3.1.1*. Champaign, IL: Illinois Natural History Survey.

Sytsma, K. J. 1990. DNA and morphology: inference of plant phylogeny. Trends in Ecology and Evolution 5:104–110.

Sytsma, K. J., and Gottlieb, L. D. 1986. Chloroplast DNA evolution and phylogenetic relationships in *Clarkia* sect. *Peripetasma* (Onagraceae). Evolution 40:1248–1261.

Sytsma, K. J., and Hahn, W. J. 1994. Molecular systematics: 1991–1993. Progress in Botany 55:307–333.

Sytsma, K. J., and Hahn, W. J. 1996. Molecular systematics: 1994–1995. Progress in Botany 58:470–499.

Sytsma, K. J., and Smith, J. F. 1988. DNA and morphology: comparisons in the Onagraceae. Annals of the Missouri Botanical Garden 75:1217–1237.

Sytsma, K. J., Smith, J. F., and Berry, P. E. 1991. The use of chloroplast DNA to assess biogeography and evolution of morphology, breeding systems, and flavonoids in *Fuchsia* sect. *Skinnera* (Onagraceae). Systematic Botany 16:257–269.

Sytsma, K. J., Smith, J. F., and Gottlieb, L. D. 1990. Phylogenetics in *Clarkia* (Onagraceae): restriction site mapping of chloroplast DNA. Systematic Botany 15:280–295.

Tate, G. H. H. 1931a. Botanical results of the Tyler-Duida expedition. II. Narrative of the expedition. Bulletin of the Torrey Botanical Club 58:279–284.

Tate, G. H. H. 1931b. Botanical results of the Tyler-Duida expedition. IV. Aspects of vegetation and plant associations. Bulletin of the Torrey Botanical Club 58:287–298.

Tate, G. H. H. 1938. Auyán tepui: notes on the Phelps Venezuelan Expedition. Geographical Review 28:452–474.

Tate, G. H. H., and Hitchcock, C. B. 1930. The Cerro Duida region of Venezuela. Geographical Review 20:31–52.

Taylor, D. W. 1995. Cretaceous to Tertiary geologic and angiosperm paleobiogeographic history of the Andes. Pp. 3–9 in *Biodiversity and conservation of neotropical montane forests*, S. P. Churchhill, ed. New York, NY: New York Botanical Garden.

Thompson, J. N. 1981. Reversed animal-plant interactions: the evolution of insectivorous and ant-fed plants. Biological Journal of the Linnean Society 16:147–155.

Tilman, D. 1988. *Resource competition and community structure*. Princeton, NJ: Princeton University Press.

Tomlinson, P. B. 1969. *Anatomy of the monocotyledons: III. Commelinales - Zingiberales*. London, England: Oxford University Press.

Treseder, K. K., Davidson, D. W., and Ehleringer, J. R. 1995. Absorption of ant-provided carbon dioxide and nitrogen by a tropical epiphyte. Nature 375:137–139.

Varadarajan, G. S. 1986a. Habitats of *Brocchinia* Schultes (Pitcairnioideae, Bromeliaceae). Journal of the Bromeliad Society 36:209–216.

Varadarajan, G. S. 1986b. Taxonomy and evolution of the subfamily Pitcairnioideae (Bromeliaceae). Ph.D. dissertation, Washington State University, Pullman.

Varadarajan, G. S., and Gilmartin, A. J. 1987. Foliar scales of the subfamily Pitcairnioideae (Bromeliaceae). Systematic Botany 12:562–571.

Varadarajan, G. S., and Gilmartin, A. J. 1988. Phylogenetic relationships among groups of genera of the subfamily Pitcairnioideae (Bromeliaceae). Systematic Botany 13:283–293.

von Humboldt, A. 1818–1829. *Personal narrative of travels to the equinoctial regions of the New Continent, during the years 1799–1804*, Translated by H. M. Williams. London, England: H. G. Bohn.

Wendel, J. F., and Albert, V. A. 1992. Phylogenetics of the cotton genus (*Gossypium*): character-state weighted parsimony analysis of chloroplast-DNA restriction site data and its systematic and biogeographic implications. Systematic Botany 17:115–143.

Wilson, E. O. 1992. *The diversity of life*. Cambridge MA: Belknap Press.

9 You aren't (always) what you eat: evolution of nectar-feeding among Old World fruitbats (Megachiroptera: Pteropodidae)

John A. W. Kirsch and François-Joseph Lapointe

The claim of the Macroglossinae to stand as a distinct "subfamily" must rest chiefly, if not entirely, on these two facts, viz., that the tongue, though approached by that of certain forms of Pteropodinae, is still more highly specialized, and that all the forms included in the Macroglossinae are undoubtedly phylogenetically intimately interconnected and therefore form a perfectly natural section.

Andersen (1912: 728)

Characterem non constituere Genus, sed Genus Characterem...

Linnaeus (1751: 119)

Nectarivory is a well-defined dietary specialization in a number of avian and mammalian groups. In some cases, as in the hummingbirds, this trophic adaptation uniformly characterizes the group, making it difficult to understand the ecological and anatomical conditions that favored its development. However, among mammals and in particular bats, nectar-feeding has appeared many times and relatives with other trophic preferences are available for comparison. The Old World fruitbats, the family Pteropodidae (the sole representatives of the bat suborder Megachiroptera), provide a case in point. Within the family, Andersen (1912) recognized a distinct, albeit in some respects primitive, subfamily of nectarivores – the Macroglossinae. The subfamily consists of five genera (*Eonycteris*, *Macroglossus*, *Melonycteris*, *Notopteris*, and *Syconycteris*) found in the Indo-Australo-Pacific region and a disjunct sixth (*Megaloglossus*) limited to forested tropical Africa. Of these, *Macroglossus* is the most widespread, ranging from northern India to the Solomons. The distributions of *Syconycteris* (Australasia including the Bismarck Archipelago) and *Melonycteris* (Papua New Guinea, Bismarck Archipelago, Solomon Islands) are virtually nested within this area. *Eonycteris* ranges from India to Timor, while *Notopteris* occurs from Vanuatu to Fiji (Nowak 1991; Koopman 1994). On geographic grounds alone, one might expect *Megaloglossus* (and perhaps *Notopteris*) to be phyletically sequestered from the other macroglossine genera.

In fact, Andersen's (1912) recognition of the macroglossine clade was based on a mostly polythetic suite of putatively derived characters. These included a greatly lengthened brushy tongue, reduced dentition, narrow muzzle, and additional cranial features – characters that might be expected to converge in members of geographically disjunct groups. Nonetheless, while admitting that even the tongue was merely the hypertrophied expression of a condition found in other fruitbats, Andersen felt that the six genera were a phyletically unified group. His view has been accepted by most subsequent taxonomists (e.g., Simpson 1945; Koopman 1984, 1994; Hill in Mickleburgh et al. 1992); this despite the fact, however, that Hood (1989) demonstrated that the African *Megaloglossus* lacks several derived characters of the female reproductive system present in the other five macroglossine genera. In his analysis, these formed a monophyletic group embedded within a clade including the non-nectarivorous *Pteropus*, *Dobsonia*, and *Nyctimene*. Colgan and Flannery (1995), using RFLP analysis of nuclear ribosomal RNA cistrons, found that the same five genera were paraphyletic with respect to frugivores, and suggested that "typical" (i.e., fruit-eating) megachiropterans (Pteropodinae) might actually have evolved from a nectar-feeding ancestor; however, they did not examine *Megaloglossus* or other African fruitbats. On the other hand, Freeman (1995), on the basis of a thorough morphometric study of both micro- and megachiropteran fruit- or nectar-feeders, concluded that either trophic type in Megachiroptera could have evolved from the other. Finally, our DNA-hybridization phylogeny (Kirsch et al. 1995), which includes all six macroglossine and twelve pteropodine genera, suggests macroglossine polyphyly, grouping only two macroglossine genera together. The other four are individually associated with megachiropterans which are supposedly less specialized in their diets (Figure 9.1B). Moreover, *Megaloglossus* is part of a distinct African clade including *Lissonycteris* and *Epomophorus*; the nearest sister-groups are *Eonycteris* and *Rousettus* (some species of which are also found in Africa, while others range as far east as the Solomons).

Our phylogeny also differs from Andersen's anatomy-based tree (Figure 9.1A) in segregating the tube-nosed Nyctimeninae (*Nyctimene* and *Paranyctimene*; recognized as a subfamily by Miller [1907] and Simpson [1945], among others) from all other Megachiroptera. Some nyctimenines are at least partially – albeit probably secondarily – insectivorous (Vestjens and Hall 1977), a feature that reinforces their separation from other pteropodids. This separation is consistent with the DNA-hybridization phylogeny when rooted with several microchiropterans and non-bat mammals (Figure 6 in Kirsch et al. 1995). However, the fossil record and dietary preferences of the presumed sister-group to Megachiroptera (Microchiroptera) do not contribute much to establishing the course of trophic evolution in Old World fruitbats: frugivory and nectarivory occur among some few New World phyllostomid bats (Fleming 1993), but the oldest (Eocene or Paleocene) bats were apparently insectivorous (Beard et al. 1992; Hand 1984; Hand et al. 1994; Sigé 1991), like the majority of modern microchiropterans. And although the earliest undoubted megachiropteran, the Oligocene *Archaeopteropus transiens*, apparently had cuspidate teeth (Meschinelli 1903), this is also true of some modern non-nyctimenine fruitbats (especially *Harpyionycteris* and *Pteralopex*) and may not reflect insectivory.

Based on the reproductive and molecular data, it seems possible that nectar-feeding is not a synapomorphy of a putative Macroglossinae, but rather evolved (or was lost) several times within Pteropodidae. If so, molecular phylogenies offer the

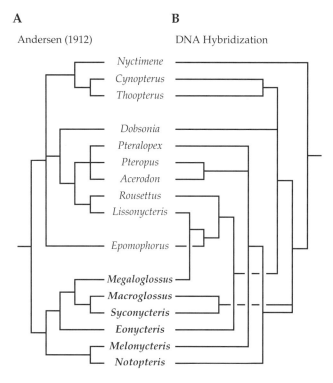

Figure 9.1. A comparison of the megachiropteran phylogeny of Andersen (1912) based on anatomy (**A**) with that of Kirsch et al. (1995) based on DNA hybridization (**B**), for sixteen genera common to both studies. Macroglossine taxa are boldfaced. Redrawn after Kirsch et al. (1995, figure 8).

opportunity to examine the conditions under which nectarivory in fruitbats is favored or abandoned, and such an exercise might prove of more general relevance to trophic evolution in other taxa – particularly those (like hummingbirds) which have no close non-nectarivorous relatives.

As a first step toward documenting the course of "evolutionary ecomorphology" in fruitbats, we undertook to determine how well the degrees of nectarivory fit Andersen's (1912) anatomical and our DNA-hybridization trees, using Andersen's phylogeny as a "control;" and also, how well anatomy, as characterized by Andersen, tracks with diet. Our expectation was that diet and the associated morphological characters would be strongly correlated on Andersen's tree because that tree was based on such features, but would not necessarily fit parsimoniously on the quite different DNA-hybridization topology. We carried out the analyses in three stages, beginning with a simple mapping of diet on the trees. During the first stage we found that diet did not track parsimoniously with **either** the anatomical or molecular phylogenies. However, because we were also interested in a possible **functional** association between diet and anatomy, we next calculated independent contrasts (Felsenstein 1985) to distinguish any such correlation from the effect of common ancestry. Filtering the effect of phylogeny as estimated by Andersen dramatically decreased the correlation among traits, but did not do so in the case of the DNA-hybridization phylogeny. Last, we generated random trees to test whether these observations differed from chance. We discovered that correlations on the DNA trees did **not** differ significantly from those on random ones, suggesting that molecular trees are not always more useful in substantiating a pattern of coevolution in characters than an unresolved or random phylogeny.

Data, methods, and results

Trees

A requirement of tree-fitting or comparative-method analyses is that the trees utilized must represent historical branching sequences. It may be questioned whether Andersen's (1912) topology, predating as it does the codification of cladistic methodology, is a true cladogram. Springer et al. (1995) carried out a cladistic-parsimony reanalysis of Andersen's data, and while some of the detailed relationships among *non*-macroglossines differ from Andersen's, Springer et al.'s trees always display a macroglossine clade – whether or not the "defining" tongue characters or Hood's (1989) reproductive features are included (Figure 9.2). We examined both Andersen's and Springer et al.'s trees for the sixteen genera common to these and the analyses based on DNA hybridization. The anatomical trees were coded with all branch lengths set equal to one, except for polytomies (i.e., they were treated as unweighted trees).

For the DNA-hybridization trees relating the same genera, we used two levels of resolution: an incompletely resolved (strict consensus) tree based on a series of analy-

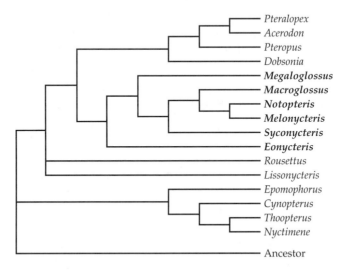

Figure 9.2. Strict consensus of two equally parsimonious trees of sixteen megachiropteran genera, based on the (informative) morphological and anatomical characters documented by Andersen (1912) and Hood (1989); CI and RI for the contributing minimum-length trees were 0.51 and 0.63, respectively. Andersen's polarities, as well as reference to *Archaeopteropus transiens*, microchiropterans, and non-bat placental mammals were utilized in constructing the likely Ancestor (which is derived in four characters). Macroglossine taxa are boldfaced. Redrawn after Springer et al. (1995, figure 2).

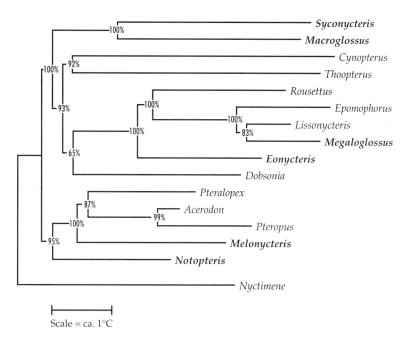

Figure 9.3. DNA-hybridization tree of sixteen megachiropterans, based on a subset of the ΔT_{mode} data presented in Kirsch et al. (1995). Numbers at nodes are bootstrap percentages (out of 100 pseudoreplicates). Macroglossine taxa are boldfaced.

ses and validations including both more and fewer than sixteen taxa, and with and without an outgroup microchiropteran included (Figure 9.1B; details are provided in Kirsch et al. 1995); and a fully resolved tree that is our best estimate of relationships for the 16 genera common to our and Andersen's study (Figure 9.3; note bootstrap percentages). We coded these two trees with internodes set to zero or one for maximum comparability with the anatomy-based topologies. However, because distance methods provide branch lengths, they are especially valuable for comparative-method analyses (Felsenstein 1985; ; Martins and Garland 1991; Garland et al. 1992); thus, we also determined character correlations on the fully resolved DNA tree with fitted distances (i.e., a weighted tree).

Dietary data

The distinction between fruit- and nectar-eating is not absolute. Many non-macroglossine fruitbats (e.g., *Pteropus, Rousettus*) eat blossoms or lap nectar and ingest pollen (Flannery 1990; Nowak 1991), but generally masticate only soft, ripe fruits. Mega- and microchiropteran nectar-feeders are similarly diverse in their diets (Wilson 1973; Fleming 1993; Freeman 1995). Only a few pteropodid species (most dramatically those of *Pteralopex* and *Harpyionycteris*) are thought to deal with tough fruits and seeds, and only these species and some *Nyctimene* show marked modifications of the generally simple, cuspless pteropodid molar structure (Andersen 1912; Hill and Smith 1984). Wilson (1973) provided semi-quantitative estimates of dietary

Table 9.1. Measurements of mandibles and molars (from Andersen [1912]), together with ratios calculated from them, and information on diet in pteropodids. Diet is percentage nectarivory (from Wilson [1973]). Abbreviations: ManL = mandible length; CorH = coronoid height (or depth); $M^1 L$ and $M^1 W$ are length and width, respectively, of the first upper molar; NI is the sum of the two ratios, independently scaled from 0 to 1.

Species	ManL	CorH	ManL / CorH	$M^1 L$	$M^1 W$	$M^1 L / M^1 W$	NI	Diet
Acerodon celebensis	50	24.5	2.04	4.85	3.7	1.31	0.38	0%
Cynopterus brachyotis	21.5	11.25	1.91	2	1.35	1.48	0.4	20%
Dobsonia inermis	34.6	18.5	1.87	4.1	2.55	1.61	0.45	20%
Eonycteris spelaea	26.5	9.45	2.8	2.3	1.2	1.92	1.15	100%
Epomophorus wahlbergi	41	17.1	2.4	3.95	2.15	1.84	0.88	50%
Lissonycteris angolensis	34	14.65	2.32	2.45	2.05	1.2	0.48	50%
Macroglossus minimus	20.35	6.75	3.01	1.25	0.65	1.92	1.28	50%
Megaloglossus woermanni	20.5	6	3.42	1.4	0.5	2.8	2	100%
Melonycteris melanops *	25	12.4	2.02	1.6	0.95	1.58	0.57	50%
Notopteris macdonaldi	25.35	8.55	2.96	1.5	0.95	1.58	1.06	50%
Nyctimene albiventer	19.2	10	1.92	1.9	1.3	1.46	0.39	0%
Pteralopex atrata	55.5	31.4	1.77	4.85	4	1.21	0.17	0%
Pteropus tonganus	51.75	24.65	2.1	5.45	3.5	1.56	0.55	0%
Rousettus amplexicaudatus	29.2	11.75	2.48	2.65	1.7	1.56	0.78	50%
Syconycteris australis	18.7	7.2	2.6	1.1	0.6	1.83	0.99	50%
Thoopterus nigrescens	27.8	14.5	1.92	2.7	2.7	1	0.14	0%

* *M. fardoulisi*, a species unknown to Andersen (1912), was used in the DNA-hybridization studies.

preferences for each chiropteran genus. According to his compilation, just two pteropodid genera – *Eonycteris* and *Megaloglossus* – are 100% nectarivorous; all other members of Macroglossinae divide their diets equally between fruit and nectar. This 50-50 division is also true for many pteropodine genera, although Wilson characterizes *Pteropus* as 0% nectarivorous, and *Cynopterus* and *Dobsonia* as 20% nectarivorous. The apparent outgroup, the nyctimenines, takes no nectar. As a means of characterizing diet, we used Wilson's (1973) compilation of percentages of nectarivory in bats (0, 20, 50, or 100%; see Table 9.1) in subsequent analyses.

Anatomical data

In order to compare the implications of Andersen's (1912) and Springer et al.'s (1995) anatomical and our DNA-based trees, all measurements or character-states were taken from Andersen's monograph, wherever possible matching the species or subspecies between his and our studies. Such matching obviates any uncertainties about generic limits as defined by Andersen, which are in any event largely not in question. Andersen does not provide tongue lengths for most species, but he does give dental dimensions for nearly all taxa included in our DNA-hybridization study (the exception being *Melonycteris fardoulisi*, a species unknown to Andersen, for which we substituted *Melonycteris melanops*). Macroglossines generally show reduction of the mandible and consequent degeneration (diminution and/or narrowing) of the teeth (Freeman 1995). To index dentition, we chose to use the length and width of the first upper molar, as it is the largest molar and therefore the one least likely to be subject to measurement error. Andersen (1912) also mentions that the coronoid process tends to be lower and more recumbent in macroglossines (Figure 9.4C), and provides mandible-length and coronoid-height measurements for pteropodids, which we also employed. In all cases, Andersen reports only minimum and maximum dimensions for a varying number of specimens; we adopted simple means of these ranges and expressed the characters as ratios (see Table 9.1). We also constructed a "nectarivory index" (NI) that combined the molar and mandibular ratios by adding them after scaling each between 0 and 1. For the mandibular ratio, the range was 1.77–3.42, while that for molars was 1.00–2.80; the composite NI covered the range 0.14–2.00. We stress that while the mandibular and molar characters may be **implicit** in Andersen's tree, and therefore their use compromises the independence of his phylogeny when those characters are correlated on it, the mandibular measurements were not included in Springer et al.'s (1995) reanalysis of Andersen's data. Those authors coded dental degeneration only in terms of the loss or relative sizes of particular teeth.

Analyses

MACCLADE – An obvious way of testing the association between phylogeny and character evolution is to trace one or more independent features onto a cladogram, using for example Maddison and Maddison's (1992) MacClade package. We did this on all four unweighted trees, coding percentage of nectarivory as both ordered and unordered states and recording the number of changes and consistency and retention indices for each character mapping.

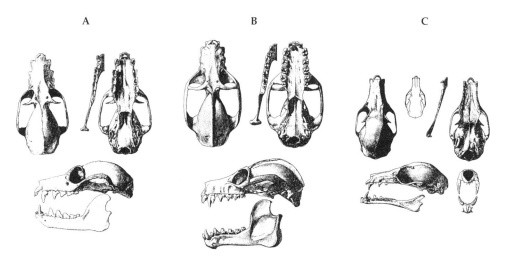

Figure 9.4. Comparisons of skulls and mandibles for three pteropodid species, showing "normal" and "extreme" states in craniodental characters. From left to right: (**A**) *Pteropus niger* (a feeder mostly on soft fruits [Andersen 1912:218]), (**B**) *Pteralopex atrata* (probably, like other members of the genus, a specialist on harder fruits and seeds [Flannery 1995]; note its multicusped canines and molariform premolars); and (**C**) *Megaloglossus woermanni* (100% nectarivorous according to Wilson [1973]). After Andersen (1912; figures 12, 21, and 68, respectively).

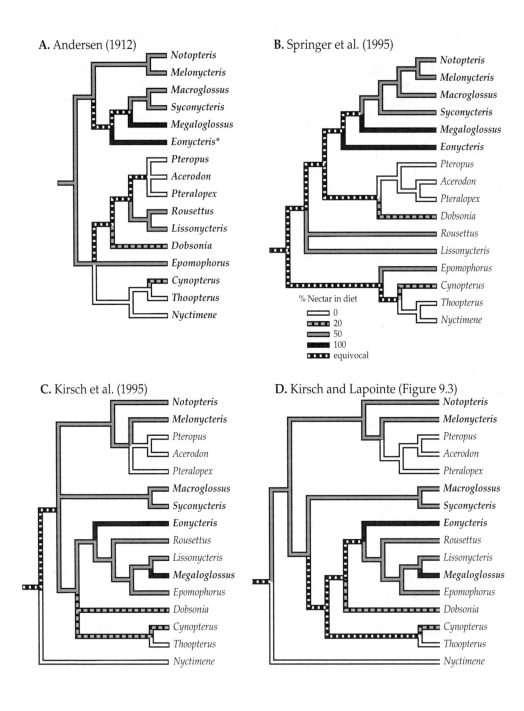

Figure 9.5. MacClade-fitting of degrees of nectarivory in sixteen megachiropteran genera (data after Wilson [1973]), coded as four unordered states (see Table 9.1), on (**A**) the tree of Andersen (1912; see Figure 9.1A), (**B**) the cladistic-parsimony reanalysis of Andersen's data by Springer et al. (1995; see Figure 9.2), (**C**) the unresolved DNA-hybridization topology of Kirsch et al. (1995; see Figure 9.1B), and (**D**) the fully resolved but unweighted DNA-hybridization tree of Figure 9.3. Macroglossine taxa are indicated in bold; tree statistics given in Table 9.2.

Table 9.2. Results of MacClade-fitting of diet on four unweighted phylogenies of sixteen megachiropteran genera, giving numbers of changes and consistency and retention indices (CI and RI, respectively) for each mapping of ordered or unordered dietary character-states.

	Andersen	Springer et al.	DNA unresolved	DNA resolved
Ordered				
No. changes	7	6	8	9
CI	0.43	0.50	0.38	0.33
RI	0.64	0.73	0.55	0.45
Unordered				
No. changes	6	6	6	7
CI	0.50	0.50	0.50	0.43
RI	0.50	0.50	0.50	0.33

The less restrictive fittings of character-state changes were those of unordered characters on less resolved trees, but even these tracings did not give particularly parsimonious results (Table 9.2). Thus, Andersen's, Springer et al.'s, and the incompletely resolved DNA-hybridization tree each required twice the minimum number of state changes (six, for a CI of 0.50 and a retention index of 0.50 for all three topologies), despite the fact that anatomical trees group the six macroglossines together (Figure 9.5). As expected, ordered states or the fully resolved DNA-hybridization tree mandate still more changes – seven and nine for ordered characters fitted to Andersen's tree and the resolved DNA phylogeny, respectively. Springer et al.'s tree with ordered characters fitted had a higher retention index than the others, presumably in part because it is one of the most resolved topologies (14 dichotomous nodes as compared with 13 for Andersen's and 12 for the unresolved DNA-hybridization tree; see Figure 9.6). It is not obvious whether any of the differences in fit are statistically significant, but certainly the results show that dietary differences are not ineluctably associated with the subfamilies defined by Andersen. In this respect, his tree does not serve very well as a "control," nor do DNA trees provide a particularly striking contrast, at least for unordered characters. Nevertheless, the specifics of the mappings are suggestive regarding the evolutionary sequences of repeated changes in dietary states, a point to which we return in the Discussion.

INDEPENDENT CONTRASTS – The MacClade tracings do not answer the question of whether there is a possibly functional – as opposed to phylogenetic – correlation between trophic and morphological characters. To do so requires a more statistical approach using quantitative measures of anatomy as well as of diet (Harvey and Pagel 1991; Garland et al. 1992). Further, one must be sure that any correlations truly represent concerted change, and are not due to relationship alone, with taxa sharing co-evolved conditions simply because they inherited them from a common ancestor (Felsenstein 1985). Comparative-method techniques allow one to make this distinction, accounting for the effect of phylogeny by providing more realistic estimates of degrees of freedom. Specifically, we employed Felsenstein's (1985) method of standardized independent contrasts based on a model of gradual evolution (FL1G), as

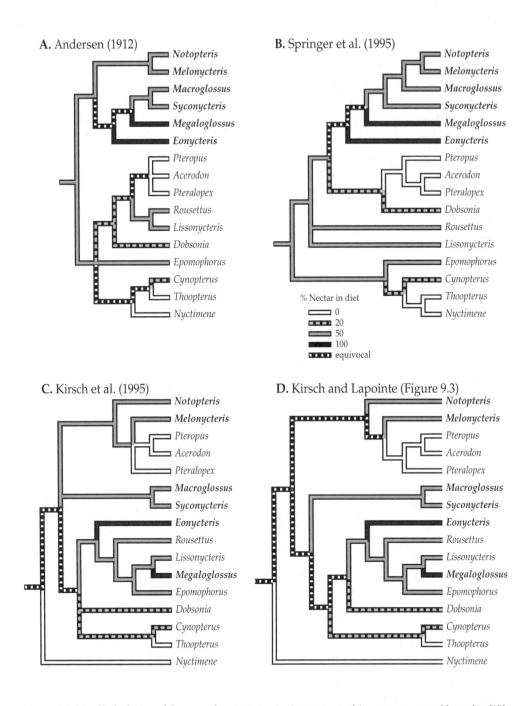

Figure 9.6. MacClade-fitting of degrees of nectarivory in sixteen megachiropteran genera (data after Wilson [1973]), coded as four ordered states (see Table 9.1), on (**A**) the tree of Andersen (1912; see Figure 9.1A), (**B**) the cladistic-parsimony reanalysis of Andersen's data by Springer et al. (1995; see Figure 9.2), (**C**) the unresolved DNA-hybridization topology of Kirsch et al. (1995; see Figure 9.1B), and (**D**) the fully resolved but unweighted DNA-hybridization tree of Figure 9.3. Macroglossine taxa are indicated in bold; tree statistics are given in Table 9.2.

implemented in Martins and Garland's (1991) CMSINGLE program. We performed correlations among all three pairwise combinations of the dietary, molar, and mandibular characters considered as continuous variables, as well as between diet and the composite NI. Again these treatments were carried out on all four unweighted trees, but also on the weighted DNA-hybridization tree. For comparison with the results of applying "standard" statistics, correlations were generated on a star phylogeny (referred to as "TIPS" in CMSINGLE), which amounts to considering phylogeny as totally unresolved. The results are listed in Table 9.3.

TIPS correlations were always significant at the 1% level (the critical value is 0.623 for $\alpha = 0.01$ with 14 degrees of freedom; Rohlf and Sokal [1981]). In contrast, none of the results of independent-contrasts analyses for correlations among diet, molar, or mandibular characters were significant at that level on either Andersen's or Springer et al.'s trees, although they were just significant for diet correlated with NI (the critical value is 0.641 in the case of independent contrasts). However, significant correlations for all comparisons were found on the three versions of the DNA tree, indicating both an association between mandibular and molar ratios and their independent as well as joint (as NI) correlation with diet.

To this point our results appear to follow expectation: molar and mandibular ratios are correlated (i.e., they should give similar results when tracking any co-evolved dietary change), and there is a significant correlation between these characters and diet when relationships are considered as unresolved; but that correlation sometimes becomes "not significant" when the effect of phylogeny is taken into account. This latter observation holds for trees constructed on the basis of *anatomy* (Andersen 1912; Springer et al. 1995), which do display one or more groups based on some extreme aspects of nectivorous specialization, notwithstanding the relatively unparsimonious results of MacClade character fitting. Correlations on the independent molecular trees, on the other hand, do not exhibit much of a decrease and remain significant when the standardized-contrasts calculations are performed. This strongly suggests that nectarivory is functionally related to anatomy and that it is not merely present in a variety of fruitbats as an accidental consequence of shared ancestry.

TREE RANDOMIZATIONS – Despite the significant correlations on the molecular trees, it occurred to us to ask a further question: what would the correlations be on a random tree? If, as seems likely from the MacClade tests, nectarivory evolved multiple times among fruitbats, perhaps **any** tree (so long as it was not generated from craniodental or lingual morphology) might give significant results. Therefore, we created a series of 99 random trees, and once more examined the three pairwise FL1G correlations among M^1, mandible, and diet, and that between NI and diet on each, comparing the resulting correlations with those initially obtained on the star and "real" unweighted trees. For separate comparisons involving the weighted DNA-hybridization tree, the actual internal and terminal branch lengths on that tree were randomly assigned to internal and terminal branches of the 99 random trees, conservatively maintaining this categorization of distances, and the FL1G tests were repeated. The null hypothesis was that correlations obtained on any of the tested phylogenies (including the star) would not be more extreme than those obtained by chance alone,

Table 9.3. Results of independent contrasts (FL1G) and randomization tests on four unweighted and one weighted phylogenies of sixteen megachiropteran genera, together with TIPS correlations for the same taxa. First line of each cell gives correlation for comparison listed at the top of column; second indicates significance (***) or non-significance (N.S.) for the parametric test (the critical values were 0.623 and 0.641 for TIPS and FL1G, respectively); third (and fourth, for TIPS) gives the probability (P) for that correlation based on the distribution of FL1G correlations among 99 random trees. Abbreviations as for Table 9.1.

	ManL/CorH: M^1L/M^1W	ManL/CorH: Diet	M^1L/M^1W: Diet	NI: Diet
TIPS	0.7785	0.8272	0.7649	0.8475
	***	***	***	***
P (w/o branch lengths)	0.48	0.61	0.50	0.61
P (w/ branch lengths)	0.55	0.58	0.58	0.64
Andersen (1912) data	0.5192	0.5684	0.5936	0.6638
	N.S.	N.S.	N.S.	***
P	1.00	1.00	0.98	1.00
Springer et al. (1995) data	0.5726	0.5979	0.6002	0.6743
	N.S.	N.S.	N.S.	***
P	1.00	1.00	0.98	1.00
DNA Hybridization (less resolved)	0.7191	0.7790	0.7252	0.8124
	***	***	***	***
P	0.85	0.84	0.75	0.78
DNA Hybridization (fully resolved)	0.7145	0.7857	0.7174	0.8135
	***	***	***	***
P	0.86	0.82	0.80	0.78
DNA Hybridization (w/branchlengths)	0.7728	0.8184	0.7324	0.8233
	***	***	***	***
P	0.58	0.66	0.79	0.79

determined from correlations on the random trees. The probability of H_0 being true is given by the ratio of the number of random-tree correlations that are larger than or equal to the value obtained from the reference phylogeny, compared to the total number of random events plus one (i.e., the correlation on the star or "real" tree in each case was included in the distribution as appropriate).

Consistent with the non-significant FL1G results on the anatomical trees, correlations on these trees were lower than virtually all those obtained for random trees (the probabilities are given in Table 9.3); Andersen's and Springer et al.'s trees are clearly outliers, and are obviously special in sequestering (some) nectar-feeders in a single clade. Moreover, despite the significant correlations for the star phylogeny, about half of the correlations on the random trees were in all cases larger than those for TIPS (showing that an unresolved phylogeny lies within – and in fact at about the midpoint of – the range of random topologies). This result suggests that the correlations are not, in fact, much inflated by failing to take phylogeny into account. Surprisingly, however, a similar finding was obtained for the DNA-hybridization phylogenies, none of which produced correlations more extreme than for random trees. The conclusion must be that anatomy and diet, while they well may be correlated with each other, are not demonstrated to be so by the molecular trees.

Discussion

If trophic specialization and anatomy were causally related in megachiropterans, one might expect a parsimonious mapping of diet on the Andersen or Springer et al. trees, as those topologies are based on characters which supposedly reflect the correlates of frugivory and nectarivory. We found instead that the MacClade results (even using unordered states) were little better for anatomy-based than for DNA-hybridization-based trees. This suggests either that morphology and diet are not coupled, or that similar diets result from combinations of characters other than (or in addition to) those perceived to be important by Andersen in arriving at his classification. We pursued the latter possibility by examining the correlations among diet and two features – molar and mandibular shapes – likely to index dietary differences, finding that there were in fact strong correlations among them on the star phylogeny. We then addressed the question of whether these correlations are independent of phylogeny by employing comparative-method analysis, specifically that of independent contrasts under the assumption of gradual evolution (FL1G): if the correlations were a consequence just of shared ancestry (even in multiple lineages), they should decrease when evolutionary relationship is accounted for. Despite the unparsimonious mappings of diet on the anatomical trees, the correlations **did** become insignificant for the Andersen and Springer et al. trees (only the composite NI showing any such association – and then a barely significant one – when phylogeny was taken into account), thus indicating a strong phyletic contribution in the context of those trees. On the other hand, correlations remained high on the DNA-based trees, arguing for a real relationship between diet and anatomy. But, given the structures of the molecular trees and the logic of the comparative method, that linkage must again be one that has appeared repetitively in the evolution of pteropodids.

The undiminished FL1G correlations on the DNA phylogenies do not, however, necessarily validate those trees as a tool for examining trophic-character correlations: we do not know that the depicted relationships are correct, only that the associations among characters are better maintained on (phylogeny-corrected) DNA-based than on anatomical trees. We therefore tested the hypothesis that the DNA trees provide better indicators of character coevolution by examining FL1G correlations on a series of random trees. If the molecular phylogeny is superior, we should observe lower correlations on the random topologies than on DNA-hybridization trees. This approach contrasts with the usual one, where **characters** are permuted or their evolution is simulated, while holding phylogeny constant (i.e., where the tree is assumed to be correct), in order to test whether the distribution of simulated or permuted characters is a result of chance or shared history. Instead, we conjectured that the character correlations were real (as they seem to be) and tested the probability that the **tree** might not be a "good" one, as reflected in the distribution of correlations over the series of random phylogenies. By assuming that the characters are "good" and phylogeny is unknown or dubious, our method is thus the inverse of maximum likelihood, which asks the question, "What is the probability of the observed data distribution, given the tree?"

In any event, the tree-randomization tests showed that **none** of the observed FL1G (or TIPS!) correlations are more extreme than those expected by chance alone, most falling toward the middle or low end of the distributions of correlations on random trees. Therefore, we conclude that molecular phylogenies may not **always** be especially useful in a comparative-method study. Random trees – or for that matter a star phylogeny – may demonstrate any correlated pattern of change in characters just as well, a possibility implicit in Martins' (1996) recent exploration of the application of comparative methods when phylogenies are unknown or poorly resolved (see also Losos [1994]).

What do these results mean for the conduct of comparative-method studies? We emphasize that we are recommending neither that DNA trees be abandoned, nor that random (or unresolved) trees be substituted in such investigations. Rather, the point is that comparative-method tests (such as independent contrasts) should be performed on **both** "real" trees and random ones, in order to avoid Type I errors (false rejection of the null hypothesis of "no correlation"). It might be objected that our findings could be anomalous, if the molecular tree used here were itself a random topology. However, the underlying data have highly significant structure (see Kirsch et al. 1995), and most comparative-method studies have not taken the additional steps we followed here. When this is done, it may well prove to be the case that the apparently significant results of many previous enquiries are also consistent with those that would be obtained using a random phylogeny.

To repeat a crucial point, our conclusion that nectarivory evolved several times among pteropodids is evident from **all** trees, even when that inference is based on simple MacClade mappings. However, the fact that convergence is documented on the Andersen and Springer et al. trees means that megachiropterans tending toward nectarivory do not always show the classic (macroglossine) suite of craniodental or lingual adaptations for such a diet; there may be other aspects of anatomy involved. Moreover, non-significant correlations obtained with global statistics may mask positive results for local portions of the trees (Bjørklund 1994). Accepting the conclusions that nectarivory evolved convergently, but that scrutiny of parts of a tree may still reveal interesting patterns of change, allows us to consider further the sequence(s) of steps by which the nectar-feeding habitus might have evolved – repeatedly – and suggests a general scenario for dietary evolution in bats. For this purpose, tracings on the cladograms using MacClade are useful.

The least-constrained mapping of the relative degree of nectarivory is that of unordered states on the unresolved DNA-hybridization tree (Figure 9.5C). It is evident from this figure that while obligate non-nectar-feeding may or may not be ancestral for all extant pteropodids, mixed feeding (50% nectarivory) would seem to have characterized the *ur*-pteropodine (i.e., the ancestor of megachiropterans other than *Nyctimene* and *Paranyctimene*); among pteropodines, both frugivory and nectarivory represent repeated dietary shifts from mixed feeding toward specialization on one or the other resource. This hypothesis is similar to Gillette's (1977) explanation of the chiropteran subordinal dichotomy (and evolution of fruit-eating or nectarivory within Microchiroptera) from multiple-resource usage, but on a more detailed level. Gillette suggested that during their adaptive radiation different lineages of bats "chose" between either of two food sources procured at the same locality – fruit and flowers or the insects living on them: the ancestral bat at first consuming both, some evolving species shifting exclusively to fruit, nectar, and pollen, others becoming obligate insectivores. Gillette's was a hypothesis later confirmed by Legendre and Lapointe (1995). We argue here that, pteropodine megachiropterans having initially specialized on the vegetation and taking both fruit and other flower products, they further diverged (repeatedly) toward ingesting *either* fruit *or* nectar and pollen (Figure 9.7).

Whether our inference of mixed feeding in the ancestral pteropodine is correct or not (the same conclusion holds for or is allowed by all mappings, and we note again Freeman's [1995] conclusion that either dietary preference could have evolved from the other), at least one emergence of 100% nectar-feeding from a more diverse trophic behavior is implied by the position of *Megaloglossus* as sister to the mixed-feeder *Lissonycteris* on the molecular trees. It is noteworthy that *Megaloglossus*, the only African macroglossine, scores highest on both molar and mandibular features, and so also has the highest NI of any megachiropteran (Table 9.1). It is interesting that the near relatives of *Megaloglossus* on the hybridization tree (*Epomophorus, Lissonycteris,* and *Rousettus*), which are exclusively or partly African, have higher-than-median mandibular – if not always molar – ratios. The only other 100% nectarivorous megachiropteran, *Eonycteris* with a mandibular ratio of 2.80, is sister to although geographically disjunct from this clade, further suggesting that mandibular characters at least are sometimes associated with the shift toward nectarivory (see Freeman [1995] for a detailed consideration of related dentary features). In this limited case, at least, history **does** matter, and apparently geography does too; but the nearest

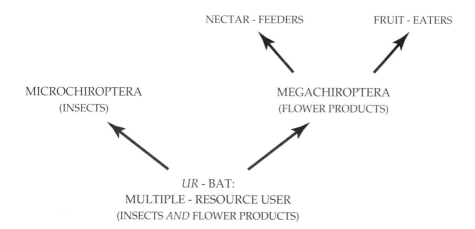

Figure 9.7. Hypothesis on the evolution of nectar-feeding in bats. After Gillette (1977, figure 2), extended to within-pteropodid phylogeny as discussed in the text.

relatives of other macroglossines such as *Melonycteris* and *Notopteris* (which are *Acerodon*, *Pteralopex*, and *Pteropus*; see Figure 9.3) do not have high mandibular (or molar) ratios. Again, nectarivory must have evolved several times among pteropodids, and not always by following the same morphological trajectory.

Conclusions

Our results show that for **these** anatomical characters and **this** DNA-hybridization phylogeny, at least, a molecular phylogeny may not be particularly useful for studying the evolution of diet in Old World fruitbats. Nevertheless, we suspect that our results may have some generality, because mammals are noted for their ability to learn and behave outside the apparent limitations of their morphology. One cannot always infer from the anatomy of any particular taxon **what** its members are able (or unable) to eat. Among marsupials, for example, the honey possum (*Tarsipes rostratus*) both is undeniably nectarivorous and shows specializations which doubtless facilitate the habit (Nowak 1991). On the other hand, of the two species of acrobatid possums, one (*Acrobates pygmaeus*) is a documented nectarivore (Turner 1984) and the other (*Distoechurus pennatus*) is known to take mostly other (animal) food (Flannery 1990); yet neither displays any obvious adaptations for nectar-feeding, and they differ anatomically mainly in that the first is volant and the other is not. A converse example is the common eutherian armadillo (*Dasypus novemcinctus*), "obviously" a myrmecophage judging from its almost featureless, peglike teeth, but in fact a notorious omnivore, at least in North America. However, a related South American species (*Euphractus sexcinctus*) has a generalist's dentition as well as habits (Smith and Redford 1990); the marsupial opossum (*Didelphis virginiana*) is a similarly generalist eater, but employs a dentition that would ordinarily be regarded as insectivorous or carnivorous (Reig et al. 1987). It is clear from our analyses that pteropodid fruitbats do not **always** require the characters noted by Andersen to facilitate the nectar-feeding (or fruit-eating) habit, but are able to consume a range of flower products using diverse craniodental apparati; yet some fruitbats clearly do have distinct food preferences. At the same time, it is not apparent which came first: dietary specialization or the anatomical characters which (sometimes) expedite it. In the larger context of evolutionary theory, the question of whether behavior drives anatomical change or if function is determined by form is at least as old as Lamarck's (1809) implicit posing of the question. It is a conundrum unlikely to be resolved by simple character mappings and correlations on phylogenies – whether anatomically or molecularly based – such as we have presented here.

Acknowledgments

Our molecular studies of fruitbats were carried out in collaboration with Drs. T. F. Flannery and M. S. Springer, and included preliminary considerations of dietary evolution; we thank our co-authors on the earlier paper for permission to paraphrase and extend those speculations here, and James Hutcheon as well as several reviewers for helpful comments on the manuscript. We also thank the National Science Foundation for repeatedly declining our grant applications, which allowed us the freedom to engage in phylogenetic terrorism without fear of reprisal through loss of financial support.

References

Andersen, K. 1912. *Catalogue of the Chiroptera in the collections of the British Museum. I. Megachiroptera.* London, England: British Museum (Natural History).
Beard, K. C., Sigé, B., and Krishtalka, L. 1992. A primitive vespertilionoid bat from the early Eocene of central Wyoming. Comptes Rendus. Académie des Sciences, Paris 314 (Série II):735–741.
Bjørklund, M. 1994. The independent contrast method in comparative biology. Cladistics 10: 425–433.
Colgan, D. J., and Flannery, T. F. 1995. A phylogeny of Indo-west Pacific Megachiroptera based on ribosomal DNA. Systematic Biology 44:209–220.
Felsenstein, J. 1985. Phylogenies and the comparative method. American Naturalist 125:1–15.
Flannery, T. F. 1990. *Mammals of New Guinea.* Carina, Queensland: Robert Brown and Associates.
Flannery, T. F. 1995. *Mammals of the South-west Pacific and Moluccan Islands.* Chatswood, New South Wales: Reed Books.
Fleming, T. H. 1993. Plant-visiting bats. American Scientist 81:460–467.
Freeman, P. W. 1995. Nectarivorous feeding mechanisms in bats. Biological Journal of the Linnean Society 56:439–463.
Garland, T., Jr., Harvey, P. H., and Ives, A. R. 1992. Procedures for the analysis of comparative data using phylogenetically independent contrasts. Systematic Biology 41:18–32.
Gillette, D. D. 1977. Evolution of feeding strategies in bats. Tebiwa 18:39–48.
Hand, S. J. 1994. Bat beginnings and biogeography: a southern perspective. Pp. 853–904 in *Vertebrate zoogeography and evolution in Australasia*, M. Archer and G. Clayton, eds. Perth, Western Australia: Hesperian Press.
Hand, S. J., Novacek, M., Godthelp, H., and Archer, M. 1984. First Eocene bat from Australia. Journal of Vertebrate Paleontology 14:375–381.
Harvey, P. H., and Pagel, M. D. 1991. *The comparative method in evolutionary biology.* Oxford, England: Oxford University Press.
Hill, J. E., and Smith, J. D. 1984. *Bats: a natural history.* Austin, TX: University of Texas Press.
Hood, C. S. 1989. Comparative morphology and evolution of the female reproductive tract in macroglossine bats (Mammalia, Chiroptera). Journal of Morphology 199:207–221.
Kirsch, J. A. W., Flannery, T. F., Springer, M. S., and Lapointe, F. -J. 1995. Phylogeny of the Pteropodidae (Mammalia: Chiroptera) based on DNA hybridisation, with evidence for bat monophyly. Australian Journal of Zoology 43:395–428.
Koopman, K. F. 1984. Bats. Pp. 145–186 in *Orders and families of recent mammals of the world*, S. Anderson and J. K. Jones, eds. New York, NY: John Wiley and Sons.
Koopman, K. F. 1994. *Chiroptera: systematics.* Berlin, Germany: Walter de Gruyter.
Lamarck, J. B. P. A. de M. de 1809. *Philosophie zoologique, ou exposition des considérations relatives a l'histoire naturelle des animaux: à la diversité de leur organization et des facultés qu'ils en obtiennent; aux causes physiques qui maintiennent en eux la vie et donnent lieu aux mouvemens qu'ils exécutent; enfin, à celles qui produisent, les unes le sentiment, et les autres l'intelligence de ceux qui en sont doués.* Paris, France: Dentu.
Legendre, P., and Lapointe, F. -J. 1995. Matching behavioral evolution to brain morphology. Brain, Behavior and Evolution 45:110–121.
Linnaeus, C. 1751. *Philosophia botanica.* Stockholm, Sweden: Godofr Kiesewetter.
Losos, J. B. 1994. An approach to the analysis of comparative data when a phylogeny is unavailable or incomplete. Systematic Biology 43:117–123.
Maddison, W. P., and Maddison, D. R. 1992. *MacClade: analysis of phylogeny and character evolution, vers. 3.1.* Sunderland, MA: Sinauer Associates.
Martins, E. P. 1996. Conducting phylogenetic comparative studies when the phylogeny is not known. Evolution 50:12–22.
Martins, E. P., and Garland, T., Jr. 1991. Phylogenetic analyses of the correlated evolution of continuous characters: a simulation study. Evolution 45:534–557.

Meschinelli, L. 1903. Un nuovo chiroptero fossile (*Archaeopteropus transiens* Meschen.) delle liquiti di Monteviale. Atti. Istituto Veneto di Scienze, Lettere, ed Arti (Venezia). Classe di Scienze Fisiche, Matematiche, e Naturali 62:1329–1344.

Mickleburgh, S. P., Hutson, A. H., and Racey, P. A. 1992. *Old World fruit bats: an action plan for their conservation.* Gland, Switzerland: International Union for the Conservation of Nature and Natural Resources.

Miller, G. S. 1907. The families and genera of bats. Bulletin of the U.S. National Museum 57:i–ixvii, 1–282.

Nowak, R. M. 1991. *Walker's mammals of the world, 5th ed., vol. I.* Baltimore, MD: The Johns Hopkins University Press.

Reig, O. A., Kirsch, J. A. W., and Marshall, L. G. 1987. Systematic relationships of the living and Neocenozoic American "opossum-like" marsupials (suborder Didelphimorphia), with comments on the classification of these and of the Cretaceous and Paleogene New World and European metatherians. Pp. 1–89 in *Possums and opossums: studies in evolution*, M. Archer, ed. Chipping Norton, New South Wales: Surrey Beatty and Sons.

Rohlf, F. J., and Sokal, R. R. 1981. *Statistical tables.* San Francisco, CA: W. H. Freeman.

Sigé, B. 1991. Rhinolophoidea et Vespertilionoidea (Chiroptera) du Chambi (Eocène inférieur de Tunisie) aspects biostratigraphique, biogéographique et paléoécologique de l'origine des chiroptères modernes. Neues Jahrbuch für Geologie und Paläontologie. Abhandlungen 182:355–376.

Simpson, G. G. 1945. The principles of classification and a classification of mammals. Bulletin of the American Museum of Natural History 85:vi–xvi, 1–350.

Smith, K. K., and Redford, K. H. 1990. The anatomy and function of the feeding apparatus in two armadillos (*Dasypoda*): anatomy is not destiny. Journal of Zoology, London 22:27–47.

Springer, M. S., Hollar, L. J., and Kirsch, J. A. W. 1995. Phylogeny, molecules versus morphology, and rates of character evolution among fruitbats (Chiroptera: Megachiroptera). Australian Journal of Zoology 43:557–582.

Turner, V. 1984. The use of Eucalyptus flowers by the Feathertail Glider, *Acrobates pygmaeus*, and its potential as a pollinator. Australian Wildlife Research 11:77–82.

Vestjens, W. J. M., and Hall, L. S. 1977. Stomach contents of forty-two species of bats from the Australian region. Australian Wildlife Research 4:25–35.

Wilson, D. E. 1973. Bat faunas: a trophic comparison. Systematic Zoology 22:14–29.

10 Leapfrog radiation in floral and vegetative traits among twig epiphytes in the orchid subtribe Oncidiinae

Mark W. Chase and Jeffrey D. Palmer

In humid cloud and rain forests around the world, a number of epiphytes grow only on the ultimate branches (2.5 cm or less in diameter) of their hosts. Some of these species, termed **obligate twig epiphytes**, are almost exclusively limited to these extreme sites (Chase 1987b). A varying degree of specialization to small branch size has been observed among obligate twig species (Catling et al. 1986); some species can occur on a wide variety of small branch sizes, whereas others are limited to only those produced in the previous one or two seasons. These sites are extreme environments – ephemeral, nutrient-poor, and often dry. They become shaded within a few years by new growth of both the host and the increasingly overlapping branches of neighboring trees, and a rapid increase in twig diameter threatens the epiphyte's attachment to its host. Among vascular plant families, twig epiphytes appear to be common in the two largest families that are composed mostly of epiphytes, cosmopolitan Orchidaceae (Chase 1987b) and neotropical Bromeliaceae (Benzing 1978, 1981). Pócs (1982) documented a series of bryophytes that also exhibits an obligate relationship to twigs.

Obligate twig epiphytism in orchids

Systematic distribution of obligate twig epiphytism

Among orchids, genera in at least three subtribes are restricted to these smallest branches (Chase 1986b, 1987b), including pantropical Angraecinae, palaeotropical Aeridinae, and neotropical Oncidiinae. Historically, orchids adapted to twigs have been viewed as closely related to genera in the same subtribe that inhabit major branches on their hosts, thus implicitly supporting the hypothesis that obligate twig epiphytism has arisen several times within groups of genera in addition to the clearly independent cases within each of these three subtribes cited above (see Dressler 1981, 1993). Both morphological and molecular data indicate that Oncidiinae, at least, are distantly related to the other two subtribes containing twig epiphytes (Dressler 1993; Chase et al. 1994). Oncidiinae thus form an ideal group in which to examine the evolutionary pathway(s) leading to this unusual life history strategy.

Obligate twig epiphytism is particularly well developed in Oncidiinae. Using data from morphology and chromosome number, Dressler and Williams (in Dressler 1981) proposed an informal set of relationships within the subtribe. Their interpretation of relationships (Dressler 1981, p. 263) placed twig epiphytes within at least five different sets of oncidioid genera, implying that twig epiphytism had evolved independently at least as many times. All previous taxonomic schemes (e.g., Pabst and Dungs

Figure 10.1. Habits of oncidioid obligate twig epiphytes (from Chase 1987b). (**A**) Three psygmoid seedlings (arrows) of *Leochilus labiatus* on *Calliandra* (Fabaceae) in Costa Rica. The seedling on the left is approximately five months old and is beginning the transformation to the adult habit (type V, Figure 10.4). Note the lack of any vegetation on the twig at the right. (**B**) One-year old psygmoid adult of *Psygmorchis pusilla* in Panama.

1977) have also split the oncidioid twig epiphytes among several groups of genera or even placed them in separate subtribes. A rigorous phylogenetic treatment of variation in floral and vegetative morphology might not support this view, but such a treatment has yet to be conducted. Previous workers have essentially concluded that oncidioid twig epiphytes represent multiple, parallel shifts to a syndrome of adaptations to life on the smallest branches of their hosts. Like a classic Vavilovian series (Vavilov 1922), independent shifts to obligate twig epiphytism within genetically closely related Oncidiinae might have produced exactly the same modifications in seed morphology, seedling habit, and velamen structure cited by Chase (1987b; Figures 10.1–10.3). To evaluate this idea, Chase and Palmer (1992) produced a molecular phylogeny for the Oncidiinae, based on parsimony analysis of chloroplast DNA (cpDNA) restriction sites. They concluded that, within the Oncidiinae, the vegetative characteristics associated with obligate twig epiphytism evolved only once, in the *Rodriguezia* clade composed entirely of obligate twig epiphytes and occupying a recently derived phylogenetic "twig" in the Oncidiinae.

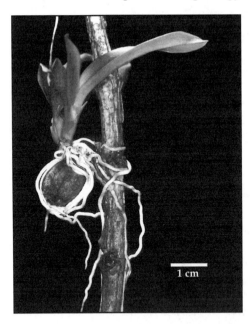

Figure 10.2. Two-year old adult of *Leochilus labiatus* growing on a branch of *Coffea* (Rubiaceae) in Costa Rica (from Chase 1987b). The seed germinated on the developing coffee fruit and managed to grow fast enough to trap the fruit by wrapping its roots around the twig. Note the withered inflorescence on the first-year growth.

Evolution of obligate twig epiphytism

Among the oncidioid twig epiphytes, several different adult habits occur (Table 10.1; Figure 10.4), but all of these share certain recurrent features. Notably, all these genera have *psygmoid* seedlings with a fan-shaped arrangement of laterally flattened, unifacial leaves (Chase 1986b). All of them also have a greatly contracted life cycle, maturing and flowering in a year from germination (except for *Psygmorchis*, which completes its development in six months). Typically, orchids require two more years to reach reproductive age. In several genera (e.g., *Macroclinium, Psygmorchis,* and *Tolumnia*), flowering plants retain their seedling habit and never develop the adult form typical of other Oncidiinae, which bear pseudobulbs (storage organs) and dorsiventrally flattened, bifacial, conduplicate leaves. Some species of *Scelochilus* begin flowering while psygmoid and later make the transition to the typical adult habit. Thus some taxa appear to be adults with juvenile habits – that is, they are *paedomorphic*. Such paedomorphosis could be achieved either by *progenesis* (early sexual development) or *neoteny* (retention of juvenile traits into adulthood; terminology from Alberch et al. 1979). In addition, some taxa (e.g., *Ionopsis satryioides*) develop terete leaves, a relatively unusual trait, overlaid on either the standard adult or juvenile habit, a further development which could be interpreted as *peramorphosis* (Table 10.1; Figure 10.4).

Such differences in habits are potentially unreliable indicators of relationships because heterochronic developmental shifts (such as neoteny or progenesis) could involve relatively minor genetic modifications, perhaps a change at a single locus (Alberch et al. 1979, among many). Closely related species often differ in habit (e.g., the two species of *Ionopsis*, one with terete leaves [Figure 10.4], the other with conduplicate leaves). Our hypothesis is that these changes in habit are correlated with adaptation to specific habitat conditions and thus do not reflect phylogenetic relationships. Regarding adaptation, T. J. Givnish (pers. comm.) has suggested that (i) steeply inclined, flattened, unifacial or terete leaves might be advantageous by reducing water loss and/or photoinhibition under brightly lit conditions and be especially adaptive for small plants with limited access to water and nutrients; (ii) the fan-shaped leaf arrangement may ensure that new leaves emerge without expanding the region of the meristem exposed to the environment; and (iii) diversion of energy to pseudobulb formation could be incompatible with rapid growth and flowering. A rapid life history is essential if reproduction is to be completed before the host's and neighboring trees' continued growth shades the twig, or the twig loosens the epiphyte's attachment through expansion and sloughing off of bark.

The adaptive value of these traits needs attention, but it is clear that none of them is uniquely associated with twig epiphytism (Chase 1987b). For example, the genus *Ornithocephalus* (Ornithocephalinae) and some species of *Maxillaria* (e.g., *M. valenzuelana*: Maxillariinae) are psygmoid but do not exhibit the seed conditions (Figure 10.3) found in obligate twig epiphytes (Chase and Pippen 1988). Furthermore, psygmoid *M. valenzuelana* grows to more than 30 cm across (enormous compared to the minute size of most twig epiphytes) and favors trunks in deep shade. Whereas individual traits associated with twig epiphytism undoubtedly have adaptive value under other

Table 10.1. Voucher data, pollinators, and habitat of the obligate twig epiphytes included in this study. Compiled from van der Pijl and Dodson (1966), Chase (1986a), Dressler (1993). See Figure 10.4 for habit illustrations.

Genus/species	Origin	Voucher	Pollinators	Habit
Capanemia superflua	Brazil	*Chase 84462* (K)	unknown	single, terete leaf, small pseudobulb
Comparettia macroplectron	Colombia	*Chase 87035* (K)	bird	standard (type V)
Goniochilus leochilinus	Costa Rica	*Chase 84191* (K)	halictid bee	standard (type V)
Ionopsis utricularioides	Costa Rica	*Chase 85039* (K)	butterfly	bifacial leaves, reduced pseudobulb
Leochilus carinatus	Mexico	*Chase 82172* (K)	halictid bee	standard (type V)
Leochilus oncidioides	Mexico	*Chase 83417* (K)	halictid bees	standard (type V)
Macradenia lutescens	Ecuador	*Chase 87176*(K)	euglossine bee	standard (type V)
Macroclinium bicolor	Mexico	*Chase 83376* (K)	euglossine bee	psygmoid, sympodial (type II)
Notylia barkeri	Mexico	*Chase 83383* (K)	euglossine bee	standard (type V)
Oncidium crispum	Brazil	*Chase 84504* (K)	no-reward flower	standard (type V)
Oncidium pubes	Brazil	*Chase 85119* (K)	oil-collecting bees	standard (type V)
Rodriguezia lanceolata	Brazil	*Chase 83002* (K)	hummingbird	standard (type V)
Psygmorchis pusilla	Ecuador	*Chase 85027* (K)	bee, deceit pollination	psygmoid, monopodial (type II)
Scelochilus ottonis	Venezuela	*Chase 85731* (K)	bee	standard (type V)
Tolumnia variegata	Puerto Rico	*Chase 83087* (K)	bee, deceit pollination	psygmoid, sympodial (type IV)
Trizeuxis falcata	Neotropical	*Chase 87277* (K)	unknown	nearly terete, psygmoid (type I)
Warmingia eugeniae	Brazil	*Chase 84460*(K)	euglossine bee	standard (type V)

circumstances, it is striking that their co-occurrence is exclusively associated with obligate twig epiphytism.

In this study, we examine patterns of floral and vegetative change among the twig epiphytes in the orchid subtribe Oncidiinae using an independent data set, cpDNA restriction sites. We wish to explore the following three hypotheses:
- The oncidioid twig epiphytes (i.e., the *Rodriguezia* clade) are a strongly supported monophyletic lineage. Chase and Palmer (1992) did not evaluate the degree of support for the *Rodriguezia* clade, and this needs to be established so that the possibility of these vegetative modifications being parallelisms can be more clearly estimated.
- Among the twig epiphytes, the differences in overall floral morphology that were responsible for previous taxonomists separating them into different groupings may be reliable indicators of relationships. Gross floral traits in the Oncidiinae are generally unreliable indicators of phylogenetic relationships (Chase 1987a; Chase and Palmer 1992), but within this group such differences may reflect a major mode of adaptive radiation within the new "adaptive zone" made accessible by the non-floral modifications associated with twig epiphytism. Different classes of pollinators are known to operate within the Oncidiinae (Table 10.1).

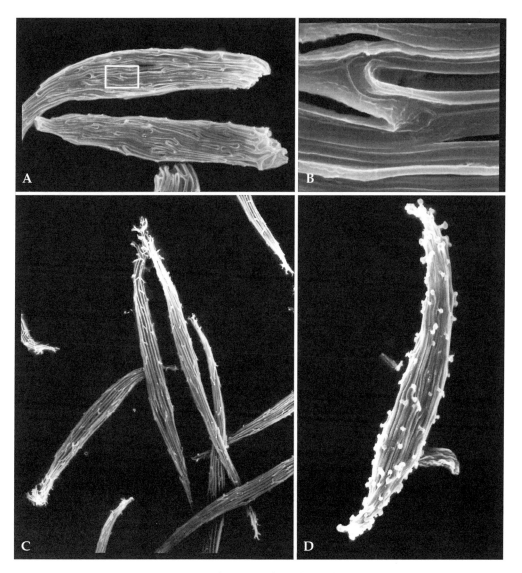

Figure 10.3. Scanning electron micrographs of the seeds of Oncidiinae (from Chase 1987b). (**A**) Seeds of *Oncidium*, typical of all genera growing on main axes of their host trees (200✕). (**B**) Magnified section of the portion of seed boxed in 3A showing the lack of raised ends of the anticlinal cell walls (2,000✕). (**C**) Seeds of the obligate twig epiphyte, *Psygmorchis pusilla*, with hooked extensions on many anticlinal walls (100✕). (**D**) Seeds of obligate twig epiphyte, *Tolumnia variegata*, with knobbed extensions to the anticlinal walls (180✕).

- The vegetative variations among the twig epiphytes are unreliable as phylogenetic indicators. Differences in growth form and leaf morphology appear to characterize major phylogenetic groups with Oncidiinae (Chase and Palmer 1992), but heterochronic shifts could take place rapidly as closely related species adapt to different habitat conditions. Floral specializations and habit differences (Table 10.1) clearly are varying independently and would provide different groupings of genera.

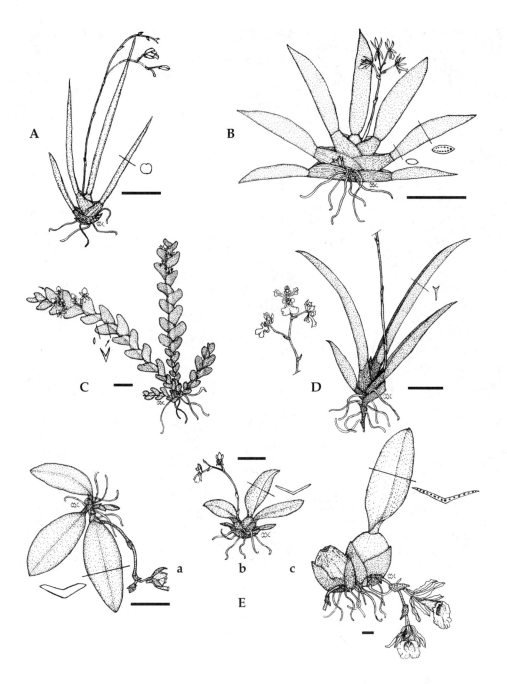

Figure 10.4. Habits and leaf cross sections of oncidioid orchids; scale bars equal 1 cm (from Chase 1986b). (**A**) Type I, *Ionopsis satyrioides*; (**B**) type II, *Macroclinium bicolor*; (**C**) Type III, *Lockhartia hercodonta*; (**D**) Type IV, *Tolumnia variegata*; (**E**) type V, *Trichocentrum panamensis* (**Ea**), *Leochilus labiatus* (**Eb**), and *Trichopilia suavis* (**Ec**). Type V is the standard type, although the pseudobulb is greatly reduced in *Trichocentrum* (all habit types for the twig epiphytes and their sister-group also indicated in Table 10.1 and Figure 10.6). The five types of habit are based on the nature of the leaves seen in cross-section, as well as the presence or absence of a pseudobulb. There is somewhat greater diversity in the Oncidiinae than is illustrated here, but the most frequent and significant variants are shown.

We will explore the idea that vegetative and floral traits are each informative at different taxonomic scales and that their relevance at these scales reflects an evolutionary pattern of "leapfrog radiation," in which the acquisition of a syndrome of vegetative traits first provided access to a new adaptive zone (in this case twigs), and then a secondary round of diversification in pollinator relationships occurred based on the offering of floral rewards (only the *Rodriguezia* clade among Oncidiinae secrete nectar). Pollinator specialization has been a hallmark of orchid evolution (e.g., van der Pijl and Dodson 1966; Dressler 1993), and specialized floral morphology has been the basis of reproductive isolation in orchids (Dodson 1962). We envision that this secondary round of adaptive radiation may have been driven by competition for pollinators and the need to stabilize novel floral structures through reproductive isolation. Through developmental heterochrony, alteration of gross vegetative format and phylogeny have become disconnected, thus making these traits unreliable estimators of relationships.

Molecular Systematics

Restriction-site mapping

Total cellular DNAs of 52 species in 34 genera were extracted according to Palmer et al. (1988) and purified on ethidium bromide-cesium chloride gradients. Generic limits are those used in the most recent family classification (Dressler 1993). More detailed information for the obligate twig epiphytes included in this study are listed in Table 10.1. Each DNA was digested with ten restriction endonucleases (six base-pair recognition sites) that cut a standard cpDNA 30–75 times: *BamH* I, *Ban* I, *Ban* II, *Bgl* II, *Cla* I, *EcoR* I, *EcoR* V, *Hind* III, *Nsi* I, and *Xba* I. Methods of Southern hybridization were those of Palmer et al. (1988). The cpDNA clone bank of *Oncidium excavatum* (Chase and Palmer 1989) was used, a species also included in this study. Hybridizations were completed for the entire chloroplast genome except for the small single copy region (*Oncidium* clones 6b, 7, and 8a), a region in which extreme numbers of insertion/deletion (indel) events were detected. Not even closely related species in the same genus shared restriction fragment patterns in the small single copy region. Hybridizations were carried out using each *Oncidium* probe individually, which made it possible to localize other length-variable regions in the large single copy region. Length variation in the cpDNA of orchids does not occur in distinct uniform segments (results not shown); with enzymes that made cuts bordering such regions, it was possible to see that each species had a unique length, thus eliminating the use of indel events as additional characters in phylogenetic analysis. We suspect that some of the changes interpreted as restriction-site changes were caused by smaller indels in other regions, but this was examined carefully by looking for consistent differences in these regions produced by several enzymes; this procedure is limited by the distance that restricted DNAs are separated on the gels prior to transfer to filters. We ran these gels only to 12.5 cm so that two such sets of gels would fit on a standard 25 cm piece of X-Ray film. It is likely that had we run gels further we could have better identified the regions in which indels were taking place and perhaps eliminated some of the unusual categories of homoplasy found in this data matrix (see below). Inability to distinguish site changes from indels, however, cannot be responsible for the unexpectedly high number of homoplasious site gains, unless independent insertions already contain restriction sites.

Mapped restriction sites rather than band absence/presence were used in analyses. *Maxillaria cucullata* was specified as the outgroup in this analysis for the same reason as in Chase and Palmer (1992); it is a member of a sister-group to Oncidiinae at some level and is definitely not a derivative of the subtribe. *Tolumnia variegata* was treated as a member of the genus *Oncidium* (as *O. variegatum*) in the earlier Chase and Palmer (1992) analysis of this same matrix, but no other taxonomic changes have been made thus far.

Phylogenetic analysis

The computer program PAUP 3.1.1 (Swofford 1993) was used to analyze the variation detected. Heuristic searches with Wagner parsimony (equal weight for restriction-site gains and losses) and successive weighting with a base weight of 1,000 applied to the re-scaled consistency index (RC; Farris 1969; Carpenter 1988) were performed. Dollo parsimony, which implements an infinitely greater weight for site gains, was not used because this extreme weighting is not biologically relevant (Albert et al. 1992). The weighting scheme of Albert et al. (1992), which applies a more moderate weight to site gains but not to their subsequent losses, was also not used because some restriction-site gains appeared unusually homoplasious (see below).

Successive weighting has not been used previously with restriction-site variation (even though the concept is older than the field of DNA systematics; Farris 1969), but its use here seems appropriate. The basic assumption of this approach relies upon a majority of characters being consistent and having little or no homoplasy in the initial Wagner analysis. Under this form of weighting, each character can be weighted individually, rather than weighting a whole class of characters consistently; thus, consistent site losses could be favored over highly homoplasious site gains. Successive weighting is a reasonable approach because it favors consistent characters regardless of their nature. Trees produced by successive weighting are never of vastly different lengths from those produced by equal weights (Wagner trees). As in this case, the trees found with successive weights are merely a subset of the Wagner trees, and they have the same Wagner length as the most parsimonious Wagner trees. In essence, successive weighting favors topologies in which homoplasy in the most consistent characters is minimized, and conversely in which additional homoplasy is permitted in characters already demonstrated to be relatively more homoplasious.

One thousand replicates of random taxon entries were made with tree bisection-reconnection (TBR) swapping and MULPARS on (holding multiple equally parsimonious trees), but limited to only ten trees per replicate. The latter strategy accelerates the rate at which random replicates accumulate due to early termination of searches that are stuck on suboptimal lengths. All trees found during these 1,000 replicates were then used as starting trees for one round of heuristic search with TBR swapping and MULPARS on but with no limits to the number of trees held in memory. All trees were swapped to completion. Successive weighting was then implemented on 10 replicates of randomly entered taxa, TBR swapping, and MULPARS, with no tree limits. Successive rounds were continued until the same weighted tree length was obtained twice, which signifies that the relative character weights have become the same in each round of re-weighting and will change no further.

Size variation in cpDNA is relatively common in orchids. In addition to our work on Oncidiinae, the only other published cpDNA restriction-site study (Yukawa et al. 1993, on Dendrobiinae) reported the same difficulties. As stated above, the small single-copy region is so variable that it is not used in this analysis, and the high level of length variation in other regions has caused us concern. We therefore tallied the types of changes each character experienced based on one of the most parsimonious Wagner trees (accelerated transformation, ACCTRAN, optimization was used throughout; Tables 10.2, 10.3). Parallel losses and gain-losses are expected to be more frequent than parallel gains and loss-gains, respectively. There is no expectation for frequencies of singular gains and losses, nor is there any expected relationship between these and other types of changes. We list in Table 10.2 the weights applied to each homoplasious character (taken from the APOLIST output of PAUP) in the last round of successive weighting (consistent characters were given a base weight of 1,000, as stated above). All unlisted characters were consistent and given a weight of 1,000.

Internal support was evaluated using the jackknife measure, based on 1,000 replicates (Farris et al. 1997). The latter program was designed for evaluating DNA nucleotide data, so we converted our (0,1) matrix into a binary nucleotide matrix (e.g., C, T). The jackknife uses unambiguous character optimization as its criterion of internal support; all supported branches must have at least one unambiguous character. The output of the jackknife program is, as with the bootstrap (Felsenstein 1985), the percentage of replicates in which a group is found, but these percentages are not equivalent to those of the bootstrap. Farris et al. (1997) considered that a jackknife value of 63% is strong support, whereas a bootstrap of 63% is only weak support. The jackknife threshold is 63%, and we indicate in Figure 10.5 only those branches for which this value has been exceeded. This method of evaluating internal support is not dependent on which character optimization is used (ACCTRAN or DELTRAN). Because use of the jackknife is relatively new, we have included bootstrap percentages (assessed using the weights applied in the final round of successive weighting) as well so that the two can be compared. The bootstrap method drops characters but replaces these with duplicates of others so that a new full-size matrix is created, whereas the jackknife drops a percentage of the characters on each replicate but does not replace them. We have indicated decay of parsimony (Bremer 1988; Mishler et al. 1991) for this same topology (Figure 10.5); this procedure discovers when each branch "decays," by determining how many extra steps beyond maximum parsimony are needed to make each branch no longer monophyletic. Unlike the bootstrap and jackknife, this method uses all the data simultaneously and relaxes parsimony to evaluate when members of a monophyletic group begin to associate with taxa from other groups.

Table 10.2. Homoplasious cpDNA restriction-site characters taken from the APOLIST output of PAUP 3.1.1. The type of homoplasy is listed for each character; the weight assigned in the last round of successive weighting is also included. All characters with a CI of 1.0 are not listed. Character numbers are the same as in the matrix included as Appendix 10.1.

Character number	Type of homoplasy	Consistency index	Weight assigned
2	parallel loss	0.50	250
4	parallel loss (or gain-loss)	0.50	250
5	parallel loss	0.50	375
6	loss-gain	0.50	455
11	loss-gain, parallel loss	0.20	100
13	parallel loss	0.33	300
14	gain-loss	0.50	444
19	parallel gain	0.33	0
21	parallel loss	0.50	333
24	parallel gain, parallel loss	0.10	10
26	parallel loss	0.50	333
28	loss-gain	0.50	250
33	parallel loss, parallel gain	0.25	175
34	parallel loss-gain	0.33	289
42	parallel loss (or loss-gain)	0.50	250
45	parallel loss	0.50	333
60	parallel gain	0.50	250
63	parallel loss	0.33	167
64	gain-loss	0.50	333
65	parallel losses	0.33	0
72	parallel loss	0.25	100
75	parallel losses, loss-gain	0.33	278
77	gain-loss	0.50	429
83	parallel loss	0.33	200
85	parallel gain	0.50	0
87	parallel gain	0.50	0
88	parallel gain, gain loss	0.33	0
92	loss-gain	0.50	438
95	parallel loss	0.50	471
96	loss-gain (or parallel gain)	0.50	375
99	gain-loss	0.50	455
106	loss-gain	0.50	458
110	parallel gains	0.33	238
111	parallel gain	0.50	438
112	parallel loss, loss-gain	0.33	111

Results

The types of molecular events occurring for each homoplasious character are presented in Table 10.2. Parallel site losses occur more often than parallel site gains, loss-gains more than gain-losses, and singular losses more than singular gains. The only one of these to deviate from expectations is the relationship of loss-gains to gain-losses, which is reversed in relative numbers. It should be noted, however, that the number of parallel gains, a molecularly unlikely event, is relatively high (Table 10.3).

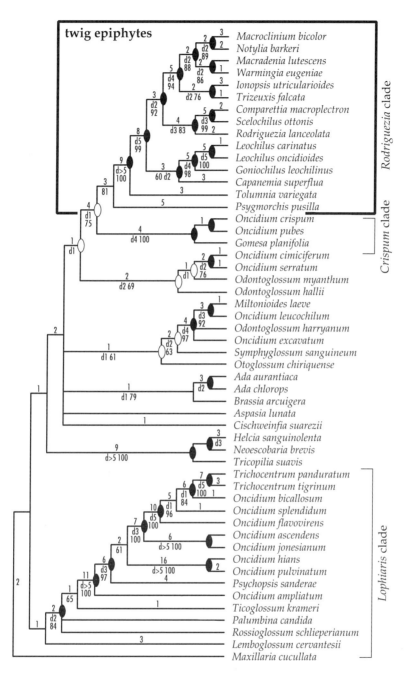

Figure 10.5. The single most parsimonious tree based on cpDNA restriction-site mapping found with successive weighting. This is also one of the 132 Wagner (equal-weighted) trees. These have 230 steps, a consistency index (CI) of 0.68, and a retention index (RI) of 0.91. Names of the clades are from Chase and Palmer (1992). Open ovals mark the branches found in all 132 Wagner trees, and solid ovals mark branches supported by the jackknife. Decay indices and bootstrap percentages are placed below the branches, but those without bootstraps indicated received percentages less than 60%; branch lengths (ACCTRAN optimization) are shown above branches. Dotted branches indicate groups not found among the 132 Wagner trees.

Table 10.3. Numbers of each category of molecular event. Numbers of loss-gains deviate from expectation by being greater than gain-losses. Both loss-gains and parallel gains should be exceedingly rare events; they are not rare in this matrix.

Type of event	Number	Weight range assigned
parallel loss	26	0–471
parallel gain	11	0–278
loss-gain	11	100–458
gain-loss	10	10–455
singular loss	47	1000
singular gains	64	1000

Under the Wagner criterion, 132 equally most parsimonious trees were found. These have 230 steps, a consistency index (CI) of 0.68, and a retention index (RI) of 0.91. After three rounds of successive weighting, a single tree with the same branch length was obtained (Figure 10.5). This successive weighted tree is among the 132 found under the Wagner criterion, and thus it has the same Wagner length, CI, and RI.

Regardless of the type of weighting used, all the oncidioid twig epiphytes, except *Psygmorchis*, form a strongly supported monophyletic group (jackknife support, 100% bootstrap, and d>5; Figure 10.5). *Psygmorchis* is unresolved in the strict consensus of the most parsimonious Wagner trees, but successive weighting places it as the sister-group of the rest of twig epiphyte clade. With the weights derived from successive weighting, *Psygmorchis* has moderate support (81%) as the sister-group of the rest of this group. Because it exhibits the morphological and life history traits typical of a twig epiphyte, it is most parsimonious to consider it a member of the *Rodriguezia* clade. The sister-group of the twig epiphytes is the Brazilian group of *Oncidium* (the *Crispum*-group, including *Gomesa*; see Chase and Palmer [1992] for a complete discussion of the other major clades identified in this analysis). It should be noted that these data do not provide strong support for the interrelationships of these major clades along the spine of the cpDNA tree; these branches are both short and relatively unreliable due to homoplasy.

The three measures of internal support, the jackknife, bootstrap, and decay, generally provide parallel estimates of internal support. The exceptions are some short branches of one to three steps; bootstrap and decay analyses sometimes identify these as more weakly supported than does jackknife analysis. For example, the branch shared by *Ionopsis* and *Trizeuxis* (Figure 10.5) is marked by only two characters, but these are free from homoplasy (CI = 1.0). The branch decays at two steps less parsimonious and has a bootstrap of only 76%, but the jackknife supports this branch. It is worth noting that on branches with five or more characters, all three methods agree. None of the branches that are only weakly to moderately supported by decay (decaying at less than three steps) and bootstrap (less than 80%) analyses, but are supported by jackknife analysis, are significant in terms of the points made in the discussion.

When life history traits, floral morphology, and habit are mapped onto the molecular phylogeny (Figures 10.5 and 10.6), it is clear that certain vegetative traits (hooked seeds, flattened seedlings, velamen modifications, rapid cycling) associated with twig epiphytism have evolved once within Oncidiinae, whereas the floral traits that suggested other relationships are unreliable. Some aspects of habitat are not good characters marking groups of genera. For example, terete leaves occur in the twig epiphytes, such as *Ionopsis*, but also in the *Lophiaris* clade in *Oncidium jonesianum* and in *Leucohyle*. Within the twig epiphytes, the situation is quite different (Figure 10.6). Here pollination syndromes and their attendant floral traits match the cpDNA tree very well with each subclade being characterized by different pollinator relationships. The first two branches within the twig clade are both deceit-pollinated taxa, which is the predominant strategy within most subtribes of Maxillarieae-Cymbidieae (*sensu* Dressler 1993). In contrast to the nearly perfect match of floral traits and cpDNA variation, the genera of the twig clade do not exhibit patterns of vegetative traits (i.e., habit) that are congruent with the cpDNA tree (Figure 10.6; Table 10.1). The overall impression is that different sets of traits are reliable at different levels in the branching pattern.

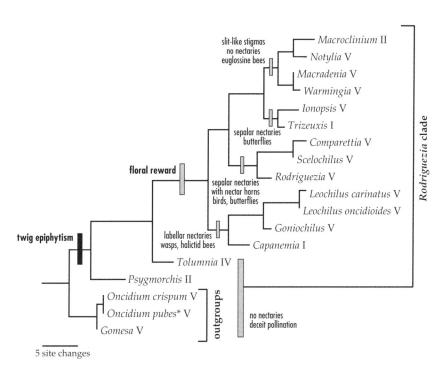

Figure 10.6. Evolution of twig epiphytism, habits, and floral traits/pollinator relationships in the oncidioid twig epiphytes. Portion of tree shown in Figure 10.5 with the twig epiphytes. Next to each clade are the general categories of pollinator relationships (van der Pijl and Dodson 1966; M. W. Chase, pers. obs.). Also indicated are the adult habit types as categorized from Table 10.1. Twig epiphytism includes evolution of rapid cycling, hooked seeds, flattened seedlings, and modifications of the velamen (Chase 1987b). *Oncidium pubes* is pollinated by oil-collecting bees and does produce on the lip callus what appears to be oil; that this is oil needs to be confirmed. This portion of the tree is strongly supported and, except for the position of *Psygmorchis*, is identical in all shortest Wagner trees.

Discussion

Homology of restriction site fragments

In these data, the observed number of parallel restriction-site gains and loss-gains is disturbingly high (Tables 10.2, 10.3), much higher than in other studies. For example, Jansen and Palmer (1988) observed 27 parallel losses, and 5 gain-losses (likely events), and only 2 parallel gains and 2 loss-gains (unlikely events) at the family level in Asteraceae. The high frequencies of these unlikely events in this oncidiod cpDNA data set may be due to our inability to distinguish between small differences in fragment length caused by slightly differently sized independent insertion/deletions (i.e., we therefore treated them as potentially homologous point substitutions). After observing the unusually high frequencies of unlikely events, we re-examined our interpretations of these on the autoradiograms, but we could not determine that our original interpretations were in error. We believe they nonetheless must have been, but this cannot be determined at the probing scale used. Yukawa et al. (1993) also reported such difficulties in their study of cpDNA restriction-site studies in Dendrobiinae (Orchidaceae). Studies of restriction-site variation are dependent on relative constancy of genome structure and size, which is typically characteristic of cpDNA in plants. Whatever the cause of the unexpectedly high number of rare events, it has serious implications for extending this technique to Orchidaceae as a whole. If the level of homoplasy uncovered in this study increases as we examine more distantly related groups of species, as would typically be expected, it will "swamp out" the signal in these data. We have in fact attempted to look at higher level relationships in orchids using restriction-site mapping and have had great difficulty in determining site homologies, even in the highly conserved inverted repeat region (M. Chase and J. Hills, unpubl. data). We believe that nucleotide sequencing offers the best alternative and are pursuing this course.

A strategy that could potentially aid in sorting out the signal in these restriction-site data is down-weighting or eliminating characters that change too frequently or that change in ways expected to be unlikely. Successive weighting appears to do this successfully here, and it should be investigated in other restriction-site studies to better evaluate its performance. Homoplasious events that should be unlikely were given the lowest weights in the last round of successive weighting (Table 10.3). Use of the relative weighting scheme of Albert et al. (1992), which weights site gains more than site losses but permits subsequent losses without penalty, produced twelve equally most parsimonious trees (Chase et al. 1992). The 12 "weighted" trees were a subset of the Wagner trees, and the successively weighted tree shown here is one of these twelve.

Evolution of floral and vegetative features

One of the reasons the study of cpDNA of Oncidiinae was initiated was to examine the evolution of whole suites of floral and vegetative features, as well as chromosome number (Chase and Palmer 1992), from the perspective of an independent data set. We ultimately hope to carry out morphological phylogenetics but felt that complex ecological and morphogenetic traits would perhaps be best viewed at present from the perspective of molecular data. Complex traits, particularly ecological adaptations, such as pollination syndromes and chromosome number, are controversially used as single characters, and this could also lead to the incorporation of correlated information if morphological traits, that are part of these syndromes, are also used. No reason existed *a priori* to believe that the modifications associated with twig epiphytism should be phylogenetically informative within Oncidiinae because (i) convergence on the syndrome is evident in several other subtribes, and (ii) the same individual traits appear to function in other ecological circumstances as well (Chase 1987b).

Gross floral traits have been the mainstay of orchid taxonomy at all levels, and the efforts of one of us (M. W. Chase) over the last few years to emphasize characters other than floral has resulted in an impasse; those who prefer to emphasize gross floral characters have continued to describe new genera that are unnatural from the perspective of floral micromorphogical details and vegetative features. The trees found here support the use of both types of characters in orchid phylogenetics. However, as happens with most characters, the appropriateness of each depends on the taxonomic level. Oncidiinae as a subtribe are well characterized by their gross floral and vegetative features (Dressler 1993), but, at the basal levels within Oncidiinae, gross floral traits appear unreliable (Chase and Palmer 1992). The floral features that Dressler and Williams (Dressler 1981) emphasized, supported relationships of subsets of the twig epiphytes to other genera with non-floral traits more typical of Oncidiinae.

The cpDNA tree for Oncidiinae suggests that the primary radiation within the subtribe was based on habitat and life-history specializations (Chase and Palmer 1992). One of these groups, the *Rodriguezia* clade (all obligate twig epiphytes), specialized on the smallest branches in the canopy and became rapid life-cyclers. Such twig epiphytes reach reproductive age in a single season, except for *Psygmorchis* which, *Arabidopsis*-like, accomplishes this in 4–6 months; the rest of Oncidiinae requires several years to reach flowering size. The features associated with this life-history innovation are probably the synapomorphies for the *Rodriguezia* clade, and these undoubtedly have adaptive value. Each of the major clades in the Oncidiinae developed non-floral specializations; those of the twig epiphytes are rapid cycling, hooked seeds, flattened seedlings, and modifications of the velamen, as discussed in Chase (1987b), and they are unique within this subtribe.

ADAPTIVE RADIATIONS IN FLORAL-POLLINATION TRAITS – Once this radiation based on life-history specializations was accomplished, adaptations to specific pollinating animals occurred. This second round of adaptive radiation leapfrogged over the life-history radiation, resulting in floral morphologies adapted for attraction and pollination by the same classes of vectors available to all plant species in neotropical habitats. Thus, genera of twig epiphytes visited by bees produced flowers that resemble those in other oncidioid clades visited by bees. Since the first workers to classify these plants had no knowledge of their life cycle, seed morphology, or seedling habit, they placed them near other genera with similar floral morphologies. An oncidioid Vavilovian series does exist, but rather than being applied to the adaptations of twig epiphytism, it is the independent shifts to the same pollinating animals within genetically closely related lineages that has produced nearly the same modifications in floral morphology.

One complex floral trait that distinguishes the twig-epiphyte clade from all others in Oncidiinae (with one exception, mentioned below) is that they offer rewards to their pollinators. When viewed from the context of the molecular phylogeny within the *Rodriguezia* clade, floral morphology/reward clearly identifies each of the subclades (Figure 10.6). The first two successive sister-groups are *Psygmorchis* and *Tolumnia*; these offer no reward and are deceit-pollinated, like nearly all other members of the subtribe (and most orchids in general [Ackerman 1986]). The most diverse groups of obligate twig epiphytes offer nectar and floral fragrances to reward their pollinators (Figure 10.6; Table 10.1). Each general class of pollinator requires specific floral traits to permit them to work the flowers, and the floral morphology of each subclade of twig epiphytes exhibits features designed to offer their reward in structures suited to their pollinators' behavior. Shallow nectaries of one group of genera are suited to halictid bees and wasps, and nectar spurs of another group are designed to permit access by only the long tongues/beaks of butterflies, some bees, and hummingbirds (Figure 10.7). The genera pollinated by fragrance-collecting male euglossine bees are a wonderful example of how a typical oncidioid floral morphology can be modified for this exclusively neotropical phenomenon. *Macroclinium* (Figure 10.4B, type II) produces small flimsy flowers that are mauled by the relatively massive euglossine male bees searching for floral fragrances to collect. A completely different morphological adaptation to this male euglossine syndrome is exhibited by the species of *Trichocentrum*, a member of the *Lophiaris*-clade (Figure 10.5E; type Va, discussed in Chase and Palmer 1992), which produces a floral spur that is devoid of nectar (M. W. Chase has observed these events in the field; they are described in van der Pijl and Dodson 1966).

ADAPTIVE RADIATIONS IN LIFE-HISTORY TRAITS – One final set of features exhibited by the oncidioid twig epiphytes is their differences in adult habits. As mentioned earlier, the distribution of these traits conflicts with the distribution of floral traits within the *Rodriguezia* clade (Table 10.1; Figure 10.6). All these taxa form psygmoid seedlings (like smaller versions of *Macroclinium*; Figure 10.4B, type II), and most of the variation in adult habit appears to involve the degree to which seedling habit is retained once reproductive age is reached. The mechanism hypothesized to be responsible for the different habits exhibited by adult individuals in the *Rodriguezia* clade, heterochronic shifts, should be phylogenetically unreliable because of the minor amounts of genetic change required to drastically alter adult habit. A hypothetical scheme for the types of heterochronic alterations involved (terminology of Alberch et al. 1979) is presented in Figure 10.8. No phylogenetic content is hypothesized for the pattern of changes proposed (i.e., the arrows do not represent cladogenesis and cannot be derived from the cpDNA trees); it is instead a model for how habits are being altered merely by shifting the timing of developmental stages.

Nearly all orchids except for the rapid cycling twig epiphytes take 2–4 years (and often more) to reach maturity, so the initial breakthrough that resulted in a new ecological niche being made available to the *Rodriguezia* clade involved a shortening of the time to sexual maturity, presumably with the retention of the standard

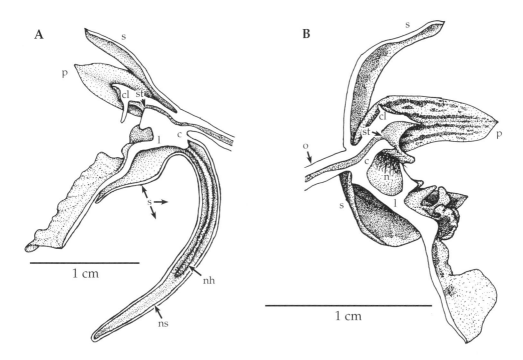

Figure 10.7. Comparison of nectary forms in two members of the *Rodriguezia*-clade (the obligate twig epiphytes; from Chase 1986a). (**A**) *Comparettia coccinea*. (**B**) *Leochilus carinatus*. *Comparettia* has a complex cavity with a nectar-secreting horn, and the species are pollinated by butterflies or hummingbirds, whereas *Leochilus* has a simple, open cavity at the base of the lip and is pollinated by generalist wasps and bees (c, column base; cl, clinandrium; l, lip base; n, nectary; nh, nectar horn; ns, nectar spur; o, ovary; p, petals; s, sepals; st, stigma).

adult habit (Figure 10.4E; type V) with pseudobulbs and conduplicate leaves. The whole maturation process, vegetative as well as sexual, simply speeded up. Genera with this pattern are represented in each pollinator category, including *Notylia*, *Rodriguezia, Comparettia, Leochilus*, and some species of *Capanemia* and *Ionopsis*. Through progenesis, *Psygmorchis* was produced; the onset of flowering was moved up to six months from germination, so that *Psygmorchis* is a sexual juvenile. Conversely, *Macroclinium* and some species of *Scelochilus* (again with different pollinator relationships) mature at the same rate as twig species with standard adult habits, but they retain varying degrees of the juvenile psygmoid habit. These species are thus neotenic rather than progenetic. Also produced are terete and triquetrous-leaved versions of both the psygmoid fan-shaped and standard type (Figure 10.5A, D, types I and IV; Figure 10.8), which could be produced by acceleration. This scenario is entirely speculative, and we offer it as an hypothesis that could be examined with current methods of molecular biology: tissue- and stage-specific mRNA activity could be used as markers to plot developmental trajectories, from which the types of heterochrony involved could be inferred.

Although habit appears reliable and gross morphology floral unreliable at the basal levels in Oncidiinae, these reverse positions within the twig-epiphyte clade. The

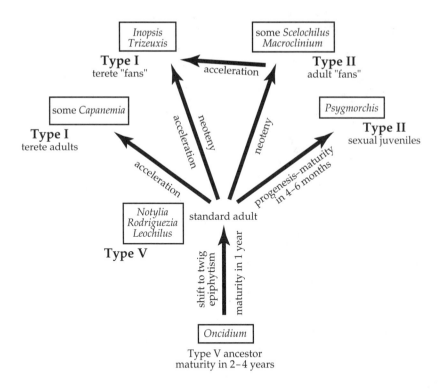

Figure 10.8. A hypothetical scheme illustrating how developmental heterochrony could be producing differences in habit that are not reflected in phylogenetic relationships. See text for discussion.

various habit types exhibited by the oncidioid twig epiphytes need to be studied carefully in the field to determine the conditions that favor one over another. The occurrence of species of *Scelochilus* that begin flowering while still psygmoid and then subsequently develop a typical habit once they have achieved a larger size, suggests that size itself may be one such factor. Being fan-shaped is adaptive at a certain small size down to seedling dimensions, but, once certain proportions are surpassed, then the standard habit appears to confer greater adaptive value.

LEAPFROGGING ADAPTIVE RADIATION – Our immediate goal in this study was to gain perspective on the relative value and appropriate taxonomic level for viewing the various characteristics exhibited by groups of genera within Oncidiinae and the twig epiphytes in particular. The general pattern suggested by this analysis is one of leapfrogging episodes of adaptive radiations, based first on vegetative features, followed by another round of radiations building upon another set of specializations in floral morphology. With time, differences in habit, which now exhibit little phylogenetic content and that vary among species in some genera or closely related genera, may be the basis for the next round of diversification. This pattern of adaptive "leapfrogging" of one suite of features after another is likely to be repeated at various hierarchical levels within most species-rich families. We hope our results will have heuristic value and help to initiate a series of evolutionary studies in Orchidaceae in general, not just

Oncidiinae. All characters appear to contribute to these radiations, and the production of molecular phylogenies is one appropriate method of formulating explicit evolutionary hypotheses that can be evaluated in studies of floral and vegetative development using both molecular and more traditional methods.

Acknowledgments

This research was supported by a National Science Foundation Post-Doctoral Fellowship in Environmental Biology to MWC. We wish also to thank T. Givnish and K. Sytsma for helpful criticisms and suggestions for improvement. Tissue samples were in some cases provided by Great Lakes Orchids (Romulus, Michigan), J & L Orchids (Easton, Connecticut), and Taylor Orchids (Monroe, Michigan); this is gratefully acknowledged.

References

Ackerman, J. D. 1986. Mechanisms and evolution of food-deceptive pollination systems in orchids. Lindleyana 1:108–113.

Alberch, P., Gould, S. J., Oster, G. F., and Wake, D. B. 1979. Size and shape in ontogeny and phylogeny. Paleobiology 5:296–317.

Albert, V. A., Mishler, B. A., and Chase, M. W. 1992. Character-state weighting for restriction site data in phylogenetic reconstruction, with an example from chloroplast DNA. Pp. 369–403 in *Molecular systematics of plants*, P. S. Soltis, D. E. Soltis, and J. J. Doyle, eds. New York, NY: Chapman and Hall.

Benzing, D. H. 1978. The life history profile of *Tillandsia cicinnata* and the rarity of extreme epiphytism among angiosperms. Selbyana 2:135–144.

Benzing, D. H. 1981 The population dynamics of *Tillandsia circinnata* (Bromeliaceae): cypress crown colonies in southern Florida. Selbyana 5:256–263.

Bremer, K. 1988. The limits of amino acid sequence data in angiosperm phylogenetic reconstruction. Evolution 42:795–803.

Carpenter, J. M. 1988. Choosing among multiple equally parsimonious cladograms. Cladistics 4:291–296.

Catling, P. M., Brownell, V. R., and Lefkovitch, L. P. 1986. Epiphytic orchids in a Belizean grapefruit orchard: distribution, colonization, and association. Lindleyana 1:194–202.

Chase, M. W. 1986a. A monograph of *Leochilus* (Orchidaceae). Systematic Botany Monographs 14:1–97.

Chase, M. W. 1986b. A reappraisal of the oncidioid orchids. Systematic Botany 11:477–491.

Chase, M. W. 1987a. Systematic implications of pollinarium morphology in *Oncidium* Sw., *Odontoglossum* Kunth, and allied genera (Orchidaceae). Lindleyana 2:8–28.

Chase, M. W. 1987b. Obligate twig epiphytes: a distinct subset of neotropical orchidaceous epiphytes. Selbyana 10:24–30.

Chase, M. W., Cameron, K. G., Hills, H. G., and Jarrell, D. J. 1994. Molecular systematics of the Orchidaceae and other lilioid monocots. Pp. 61–73 in *Proceedings of the 14th world orchid conference*, A. Pridgeon, ed. London, England: HMSO.

Chase, M. W., and Palmer, J. D. 1989. Chloroplast DNA systematics of lilioid monocots: resources, feasibility, and an example from the Orchidaceae. American Journal of Botany 76:1720–1730.

Chase, M. W., and Palmer, J. D. 1992. Floral morphology and chromosome number in subtribe Oncidiinae (Orchidaceae): evolutionary insights from a phylogenetic analysis of chloroplast DNA restriction site variation. Pp. 324–339 in *Molecular systematics of plants*, P. S. Soltis, D. E. Soltis, and J. J. Doyle, eds. New York, NY: Chapman and Hall.

Chase, M. W., and Pippen, J. S. 1988. Seed morphology in the subtribe Oncidiinae (Orchidaceae). Systematic Botany 13:313–323.

Dodson, C. H. 1962. The importance of pollination in the evolution of the orchids of tropical America. American Orchid Society Bulletin 31:525–554, 641–649, 731–735.

Dressler, R. L. 1981. *The orchids: natural history and classification.* Cambridge, MA: Harvard University Press.

Dressler, R. L. 1993. *Phylogeny and classification of the orchid family*. Cambridge, England: Cambridge University Press.
Farris, J. S. 1969. A successive approximations approach to character weighting. Systematic Zoology 18:374–385.
Farris, J. S., Albert, V. A., Källersjö, M., Lipscomb, D., and Kluge, A. G. 1997. Parsimony jackknifing outperforms neighbor-joining. Cladistics 12:99–124.
Felsenstein, J. 1985. Confidence limits on phylogenies: an approach using the bootstrap. Evolution 39:783–91.
Jansen, R. K., and Palmer, J. D. 1988. Phylogenetic implications of chloroplast DNA restriction site variation in the Mutiseae (Asteraceae). American Journal of Botany 75:751–764.
Mishler, B. D., Donoghue, M. J., and Albert, V. A. 1991. The decay index as a measure of relative robustness within a cladogram. Toronto, Canada: Hennnig X (abstract).
Pabst, G. F. J., and Dungs, F. 1977. *Orchidaceae Brasilienses*. Hildesheim, Germany: Kurt Schmerson.
Palmer, J. D., Jansen, R. K., Michaels, H. J., Chase, M. W., and Manhart, J. M. 1988. Analysis of chloroplast DNA variation. Annals of the Missouri Botany Garden 75:1180–1206.
Pijl, L. van der and Dodson, C. H. 1966. *Orchid flowers: their pollination and evolution*. Coral Gables, FL: University of Miami Press.
Pócs, T. 1982. Tropical forest bryophytes. Pp. 59–106 in *Bryophyte ecology*, A. J. E. Smith, ed. London, England: Chapman and Hall.
Swofford, D. L. 1993. *PAUP: phylogenetic analysis using parsimony, vers. 3.1.1*. Champaign, IL: Illinois Natural History Survey.
Vavilov, N. I. 1922. The law of homologous series in variation. Journal of Genetics 12:47–89.
Yukawa, T., Kurita, S., Nishida, M., and Hasebe, M. 1993. Phylogenetic implications of chloroplast DNA restriction site variation in subtribe Dendrobiinae (Orchidaceae). Lindleyana 8:211–221.

Appendix 10.1. Matrix of restriction site variation in Oncidiinae (156 characters). Zeros are used uniformly for site absence and ones for site presence.

Ada aurantiaca	0011110010110010001110001011111101011001100111000111000010000100000101100000001011000011010100100011110100010100000010101000110000001110011011010001111110
Ada chlorops	0011110010110010001110001011111101011001100111000111000010000100000101100000001011000011010100100011110100010100000010101000110000001110011011010001111110
Aspasia	0111110010110010001110001011111111011001100111000111000010000100000101101000001011000011010100100011110100000100000010101000110000001110011011010001111110
Brassia	0111110010110010001110001011111111011001100111000111000010000100000101100000001011000011010100100011110100000100000010101000110000001110011011010001111110
Capanemia	0101100101010001100111001101100111011001100111010101111000100011010000101101000001001000011010100100011111100000000001010101000110000001110011000011001110
Cischweinfia	0111110010110010001110001011111111011001100111000111000010000100000101101000001001000011101001010001111010000010000010101000110000001110011011010001111110
Comparettia	0101100101010001100111001101100111011001100110010110010001011010000110101000001011000011101001010011110000001000010101000110000001110111010000110011110
Gomesa	0111101010100010101110001011111101011001100111000111000010000110011110100000010110000110101001000110101000001000011010100011000000011100110110100011111101
Goniochilus	0101100101010001100101001001001110110011001110101011110001000101010100000110100010110100001011111000001000101001010000010100011000010100110001100010
Helcia	0111110011110010001100001111111111111001100111000111000010001011100101111001101111000001010101000111101000001000000010100011000010110001101101000011110

Ionopsis	0101100101110001100111001101100111011001100111010110000111
	0110100001001010000010110000111010010101111000001100001010
	1010001100001001111011000110000011
Lemboglossum	0111110001011011000111000101111101011001100111000111000100
	0010110001011010000010110010110101001000111010000010000010
	10100011000000011100110110100011110
Leochilus carinatus	0101100101010001100101001001100111011001100111010011100010 0
	0110100001011010000011001000011010100101011110000011001010 10
	10100011000010111011100001100111000
Leochilus oncidioides	0101100101010001100101001001100111011001100111010011100010 0
	0110100001011010000011001000011010100101011110000010001010 10
	10100011000010111011100001100111000
Macroclinium	0101100101010001100101001011001110110011001110101100000110
	1110100001001010000010010000111010010101111000001100001010
	101000110000100111001100011111100011
Macradenia	0101100101010001100111001101100111011001100111010110000110
	1110100001011010000010110000101010010101011000001100001010
	10100011000010011100110000110110 0011
Miltonioides	1111010001011000000111000101111101011001100111000111000100
	0010110001011010000010110000110100011000111101000101010000 10
	1010001100010001110011011010001111 10
Neoescobaria	0111110001110010001100001111111111110011001110001110000100
	0010110010111001101111000001010110100001111010000010000000 0
	10100011000001011000110110100011 1110
Notylia	0101110101110001100111001101100111011001100111010110000110
	1110100001011010000010110000111010010101111000001100001010
	10100011000010011100110001111110 0011
Odontoglossum hallii	0111110000110010001110001001110101011001100111000111000100
	0010110001011010000010110000110101001000111010000010000010
	1010001100000011100110110100011 1110
Odontoglossum harryanum	0111010001011000000111000101111101011001100111000111000100
	0010110001011010000010110000101100001100011110100001010000 10
	1000001100010001110011011010001111 10
Odontoglossum myanthum	0111110000110010001110001001110101011001100111000111000100
	0010110001011010000010110000110101001000111010000010000010
	10100011000000011100110110100011 1110
Oncidium ampliatum	0111110010001110001110001011111010110011001110011110000100
	0010110001011000000010110100110011100000111010000010000101 0
	1011101100000001110011011010001111 10
Oncidium ascendens	0111110110001110000111101010111100001011100111100111000100
	0010110110110001000001101001100111000000111011000100001011
	10111010100000001001001000001111 10
Oncidium bicallosum	0111110010001110011111010101010001110100111110011100001001 0
	00001101110100010000011010011001110010011100100000100001 011
	001100000010000011000001100001111 10
Oncidium cimiciferum	0111100000110010001110001011111010110011000110001110000100
	0010110001011010000010110000110101001000111010000010001001 0
	100001100000001110011011010001111 10
Oncidium crispum	0111110101010001010111000101111101011001100111000111000100
	0011100001101000001010000011010100100010101000001000011 10
	1010001100000011100110110100011 1110
Oncidium excavatum	0111010001011000000111000101111101011001100111000111000100
	0010110001011010000010110000110100011000111101000101000010
	10000011000100011100110110100011 1110
Oncidium flavovirens	0111110010001100111111010101111000110110011110011100000100
	00100101110110001000001111011100111000001110010000010000101 1
	00110000000000011000001100000111 10
Oncidium hians	0110110010001110001111101011111010000001101111001110010100
	0000110011010001000001101001101111000001111010010011000101 0
	1110110010000011001001101000111 10
Oncidium jonesianum	0111111010000011000011110101011110001011100111100111000100
	0010110110110001000001101001100111000000111011000010000101 1
	10111010100000001001001000001111 10
Oncidium leucochilum	1111010001011000000111000101111101011001100111000111000100
	0010110001011010000010110000101000011000111101000101010000 10
	1000001100010001110011011010001111 10
Oncidium pubes	0111101010100010101110001011111010110011001110011100001000
	0011100001101000001010000011010100100010101000001000001110
	1010001100000001110011011010001111 10

Oncidium pulvinatum	0010111001000111000111100101111101000000110111001110010100 0000110011010001000000110100110111100000111010010011000101 01110111001000001110010011010001111110
Oncidium serratum	0111110000110010001110010011110110110011000110001110000100 0010110001011010000010110000110101001000111101000101000100 10 1000001100000001110011011010001111110
Oncidium splendens	0111110010001110011110010101101000111010001111001110000100 0000101101101010000011101110011100100111001000001000010110 0110000000000011000011000001111110
Tolumnia variegata	0101110101010001100111011001110111101100110011101011100100 0110110001001010000010110000110101001000111100000010000101 0 101000110000000111001100101000110110
Otoglossum	0111100010110010001110001011111101100110011100011100001 00 0010110001011010000010110000110100001000111010000010000001 0 1010001100000001110011011010001111110
Palumbina	0111100011100110001110001011111010110011001110001110000100 0010110001011010000010110000110101101000111101000001000010 10 101000110000000111001101101000111110
Psychopsis	0110111001000111001111100101111110110011001100111001110000100 0010010001011000100000110101100110000011110100000100001010 1011101000000011100100110100011110
Psygmorchis	0111010101011001000111011101111101101100110001100011100000100 0011110001011010000010110000110100100010001111010000000000010 10100011000000011100110010100011110
Rodriguezia	010110010101000110011101110110011110110011001110101110000111 0110100001101010000010110000111010010101111000000100000010 10100011000000011100110000011100111100
Rossioglossum	01111100011100110001110001011111010110011001110001110000100 0010110001011010000010110000110101101000111101000001000010 10 101000110000000111001101101000111110
Scelochilus	01011001010100011001110111011001111011001100110010110010010 0110100001011010000010110000111010001010011110000000100000010 1 0100011000010011110110000110011110
Symphyglossum	011111000101100100011100010111111011001100111000111000000100 0010110001011010000010110000110100001000111100001010000010 1000001100000001110011011010001111110
Ticoglossum	0111110001110011000111000101111101011001100111000111000100 0010110001011010000010110010101011000011110100000100001010 10100011000000011100110110100011111110
Trichocentrum panduratum	0111110010000111000111100101011010001110100110110110100011100 0000110111011000100000101001100110010111100101000100001011 0011000000100000110000011000001111110
Trichocentrum tigrinum	011111100100001110001111001010110100011101101101101100100011100 0010101110110000100000101010011001011110010100010000101011 00110000001000000110000011000001111110
Trichopilia	0111100011100100011000011111111011001100111000111100000100 001011100101111000101110000010101101000111101000001000000000 10100011000001011100110110100011111110
Trizeuxis	0101100101010001100111011101100111110110011001110101100000111 0110100001011010000010110000111100100101011111000001000010100 1010001100001001110011000011000000011
Warmingia	0101100101000011001110111011001111011001100110101100001110 1110100001011010000010110000101010010010101110000011000010 1010001100001001110011000011011100011
Maxillaria	0111100011101100011100010111110111011001110011100001110000100 0010110001011010000010110000110101111000111101000001000000010 10 101000110000000111001101101000111110

11 Adaptation, Cladogenesis, and the Evolution of Habitat Association in North American Tiger Beetles: A Phylogenetic Perspective

Alfried P. Vogler and Paul Z. Goldstein

Adaptation and radiation are complex, not easily characterized phenomena. Hypotheses that simultaneously address each of these phenomena, whether formulated in general terms or with regard to a particular group of organisms, must carry specific predictions regarding the origin and frequency of adaptational events relative to speciation events if they are to be testable from within a phylogenetic framework. In this paper, we use phylogenetic information to address hypotheses that relate speciation events to adaptive shifts in the behavioral and ecological characters of North American tiger beetles (Coleoptera: Cicindelidae). Following the work of Liebherr and Hajek (1990), we will examine the taxon-cycle hypothesis of Wilson (1959, 1961) and the taxon-pulse hypothesis of Erwin (1979, 1981, 1985) for this group of organisms, and use them to illustrate both the powers and pitfalls of using phylogenies to illuminate theories of "adaptive radiation."

Views on Adaptive Radiation

Adaptation and radiation: a clarification of terms

Studies of "adaptive radiation" often suffer from confusion as to what exactly the concept embodies, and to what extent phylogenetic information can be applied in such context. At the fore are disparate interpretations of adaptation, radiation, and evolutionary "success." As Lewontin (1978) and Vrba (1983) pointed out, adaptation and radiation are quite independent phenomena that may, but need not, be coupled. One involves natural selection conferring enhanced fitness on the individual members of one or more populations, and the other a bloom of speciation events. One might well ask: Is adaptation a necessary condition of radiation, or is adaptive radiation a particular kind of radiation? Acceptance of the former would obviate the use of the word "adaptive," while the latter requires that we know something about the history of selective forces on the group under study – knowledge that, as we will discuss, cannot be directly inferred using phylogenetics (Wenzel and Carpenter 1994). Just as enhanced fitness conferred by a particular allele may ultimately precipitate fixation of that "adaptive" allele in a population (a process that need not involve speciation), some long-term macroevolutionary "trends" (*sensu* Vrba 1983) may simply be non-adaptive in the sense that newly formed species need not be more or less fit than their sister taxa. Speciation may be thought of as the accumulation of diagnostic characters which may or may not have been selected for. In short, adaptation need not be accompanied by speciation, much less radiation, or vice versa.

The phrase "adaptive radiation" is thus potentially misleading, and has been treated recently in at least two ways which differ from each other in the relative order of speciation versus adaptation events; neither is necessarily identical to Osborn's (1902) conception. In one form, perhaps most often associated with geospizine finches of the Galápagos (e.g., Lack 1947), speciation accompanies competition-mediated natural selection (finches compete for food, leading each group of finches to alter its feeding guild and subsequently or concurrently become better adapted to exploit available resources). This scenario invokes competition-based character displacement (Brown and Wilson 1956) as a mechanism for morphological diversification. The selective forces driving such adaptive (by definition) displacement are inherently microevolutionary, and thus not inferable from phylogenetic (macroevolutionary) pattern (Coddington 1988; Frumhoff and Reeve 1994; Leroi et al. 1994; Wenzel and Carpenter 1994). This kind of scenario seems to derive from Osborn's (1902) "general adaptive radiation," erected with specific reference to the parallel radiations of placental and marsupial mammals that, Osborn argued, evolved along similar paths in order to utilize analogous resources.

Several recent authors (e.g., Farrell et al. 1991; Wiegmann et al. 1993; Heard and Hauser 1995; Chapter 13) have addressed the notion of "key innovation" or "key adaptation." Clades possessing a trait hypothesized to be adaptive and derived (see Coddington 1988), so the argument goes, are expected to be more "successful" (species-rich) than their sister clades lacking the attribute, ultimately resulting in higher species numbers (Mitter et al. 1988; Farrell et al. 1991; Farrell and Mitter 1993). The equation of species diversity with "success" implies a deterministic connection between fitness and speciation. But this expectation is problematic (Cracraft 1990, p. 35; Farrell and Mitter 1993, p. 64; Heard and Hauser 1995). If a given heritable trait is adaptive, conferring enhanced fitness to the organisms that bear it, then should one necessarily expect enhanced speciation rates as well? Although workers often characterize *groups* as "successful" by virtue of their having evolved to exploit a wide array of resources, we need to caution ourselves against measuring a given clade's success by the number of species it contains. In our view, the notion of evolutionary "success" connotes different and possibly incomparable meanings when applied to single species versus higher taxa. Which is more successful: a species-rich group of organisms comprised of locally distributed habitat specialists that exhibit high extinction turnover, or a single widespread generalist species that persists over eons?

Phylogenetic approaches to the study of "adaptive radiation"

In the context of studying adaptation, phylogenetic information provides insight into the frequency, distribution, and directionality of character evolution. Thus the number of times a given attribute arose in a particular clade, as well as derived and ancestral states of that character, can be inferred from a cladogram. The unique history of selective forces that brought about modification of that character is, in our opinion, inherently unknowable, and is beyond the scope of phylogenetic testability.

However, when multiple characters are involved, phylogenetics provides the only means whereby the order of evolutionary events may be inferred (Coddington 1988; Carpenter 1989; Wenzel and Carpenter 1994).

In order to be testable in a cladistic framework, then, an hypothesis relevant to "adaptive radiation" (by any definition) must specify polarity and/or a frequency of evolutionary events. Two general hypotheses proposed to explain radiations, geographic distributions, and habitat associations in insects correspond to the two concepts of adaptive radiation outlined above: (i) the taxon-cycle hypothesis, proposed by Wilson (1959, 1961) to explain the evolutionary biogeography of Melanesian ants, may be likened operationally to the scenario in which adaptation and speciation co-occur; and (ii) the taxon-pulse hypothesis, advanced by Erwin (1979, 1981, 1985) to explain species diversification in ground beetles, corresponds to the notion of an evolutionary innovation preceding a bloom (or a "pulse") of speciation events. Drawing from the biogeographic work of Darlington (e.g., 1943, 1957; see Liebherr and Hajek 1990, p. 39), both hypotheses predict specific patterns of unidirectional niche shifts relative to speciation within a given clade of organisms. Erwin's ideas were erected with specific reference to carabid beetles, and both hypotheses have since been examined cladistically with specific regard to carabids by Liebherr and Hajek (1990).

In this paper, we use a molecular character-based hypothesis of the relationships among North American *Cicindela* to illustrate the applicability of cladistic information to studies of "adaptive radiation" with specific reference to the taxon-cycle and taxon-pulse hypotheses. Following a brief review of cicindelid biology, we use phylogenetic information to examine the evolution of habitat association as predicted by either hypothesis. Subsequently, we focus our discussion on the evolution of a single character complex in tiger beetles – phenology – as it pertains both to putative mechanisms of character displacement and to the notion of "key innovation." Finally, we examine an array of related operational issues confronting the application of phylogenetic systematics to studies of "adaptive radiation."

The taxon-cycle and taxon-pulse hypotheses

Taxon-cycle hypothesis

Wilson's (1959) taxon cycle implicated repeated, and directional, shifts in habitat association and niche (habitat) breadth in the generation of new species over the lifetime of a lineage and associated biota. The scenario is as follows: generalist species invade suboptimal or "marginal" habitats on islands, where they acquire "adaptations" to "more stable" habitats, which they subsequently invade and where they eventually become dominant, displacing the former occupants of those habitats in the process. This scenario predicts that ancestral generalist species give rise to derived species that are both distributed peripherally with respect to their ancestor (the hypothesis was proposed with specific reference to ant fauna of the Melanesian islands) and more specialized with respect to habitat association (Figure 11.1A). A "cycle" is completed when the generalist species becomes extinct due to competitive pressure, after its habitat is invaded by better specialists from other lineages or by

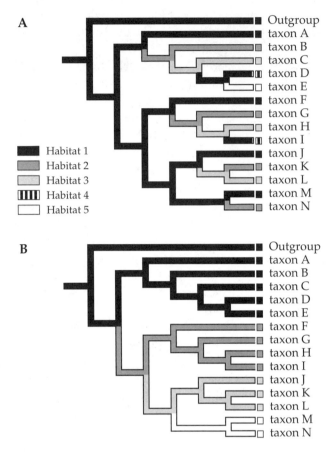

Figure 11.1. Examples of the distribution of habitat association relative to speciation under (**A**) the taxon-cycle hypothesis and (**B**) the taxon-pulse hypothesis.

new colonists. After such extinction, the processes of diversification and specialization begin anew with other groups. In each cycle, recently arriving generalist species outcompete those present in marginal areas and bring about the disappearance of the former denizens of those areas. In the jargon of Roughgarden and colleagues (Rummel and Roughgarden 1985; Roughgarden and Pacala 1989), taxon cycles in Caribbean *Anolis* lizards involve asymmetrical competition among resident and invading taxa, resulting in cyclical "coevolution-invasion turnovers." The competitive interactions of these lizards were also found to be consistent with hypotheses of character displacement (Losos 1992; see Thompson 1994, pp. 272–274 and below for a discussion of the role of character displacement in taxon cycles).

Taxon-pulse hypothesis

The taxon-pulse hypothesis of Erwin (1979, 1981, 1985) likewise specifies a relationship between speciation (cladogenesis) and the evolution of habitat association. However, the taxon-pulse hypothesis invokes ecological shifts to explain the origin of clades in which habitat association is well conserved. Under this scenario, such pulses are initiated by major ecological, geological, or paleoclimatic changes (such as the rise of a particularly effective group of predators or large-scale change in global climate [Erwin 1981, 1985]). The taxon-pulse hypothesis may be visualized as sequential speciation pulses preceded by habitat shifts, such that derived clades deviate progressively more, ecologically speaking, from their ancestors (Erwin 1985; Howden 1985; Figure 11.1B). Thus, ecological generalists would be expected to give rise to specialist and "superspecialist" species (Erwin 1985). In the specific case of lineages of ground beetles to which the taxon-pulse hypothesis was initially applied, the predicted trend in the evolution of habitat association is from continuously wet and warm habitats in the humid tropics to drier and colder climatic conditions away from the presumed ancestral habitat type (Erwin 1985).

Predictions of the two hypotheses

The taxon-cycle and taxon-pulse hypotheses result in the following predictions: (i) character-state changes in ecological parameters ("adaptive shifts" of Erwin 1985) – including changes in habitat association – are unidirectional and irreversible, with the exception of shifts that reset the taxon cycle; (ii) habitat specialists are derived with respect to habitat generalists; and (iii) derived taxa are distributed peripherally with respect to the habitats occupied by basal members of a given clade. The taxon-cycle hypothesis further predicts that inhabitants of small, isolated sink areas be derived with respect to those of larger, more "stable" source areas. The taxon-pulse hypothesis incorporates the specific prediction that tropical wetlands are the primitive condition for carabids, with "adaptive shifts" (Erwin 1985) having resulted in the subsequent colonization of drier and cooler areas. As Liebherr and Hajek (1990) pointed out, the taxon-cycle hypothesis assumes that shifts in habitat association are coincident with cladogenesis, whereas the taxon-pulse postulates that a (rare) habitat shift precedes a plume of speciation events without further habitat shifts, as shown diagramatically in Figure 11.1.

Tests of taxon-cycle and taxon-pulse hypotheses in *Cicindela*

Background: systematics and ecology of *Cicindela*

The cosmopolitan genus *Cicindela* (Coleoptera: Cicindelidae) comprises over 1,000 species. Many are brightly colored and predaceous as adults and larvae; they are commonly known as tiger beetles. Species of *Cicindela* have colonized a wide array of structurally similar habitat types and exhibit a variety of noteworthy life history traits, including pronounced habitat specialization, mate-guarding behavior, aposematism, and a variety of defensive strategies. As a result of the group's many remarkable biological features as well as its many showy species, *Cicindela* has garnered considerable attention from organismal biologists and collectors alike. The genus is thus rather well understood with respect to the alpha-taxonomy, natural history, and geographic distribution of its component species (Pearson 1988). *Cicindela* has been the focus of numerous studies of community ecology (e.g., Pearson and Mury 1979; Pearson 1986;

Schultz and Hadley 1987; Pearson and Juliano 1991; Niemela and Ranta 1993), physiological and chemical ecology (Schultz 1986; Pearson et al. 1988), character evolution (Pearson and Mury 1979; Pearson 1980; Mury-Meyer 1987), and conservation biology (Knisley and Hill 1992; Pearson and Cassola 1992; Vogler and DeSalle 1994). Because of these investigations the global distribution of cicindelid species is reasonably well known for most parts of the world (Pearson and Cassola 1992; Pearson and Juliano 1994).

Although phylogenetic (or "historical" *sensu* Pearson et al. [1988] and Pearson and Juliano [1994]) influences have been invoked to explain the distribution of certain attributes within the genus (Mooi et al. 1989; Altaba 1991), a careful phylogenetic analysis has not been carried out. Studies of these evolutionary patterns have been based on the traditional (Linnaean) classificatory scheme of Rivalier who characterized *Cicindela* from different parts of the world in a series of papers (1950–1963) (e. g., see Rivalier 1954, 1963). His groupings were based primarily on male genitalic characters. Rivalier's classification recognized over 50 subgenera and additional species groups, but no attempt was made to reconstruct hierarchical relationships among these groups.

Adult tiger beetles are typically cursorial predators of invertebrates. Larvae, in contrast, are ambush predators which position themselves at the entrance of burrows they construct in the soil or sand, from which they seize approaching prey. Tiger beetle habitats are typically early successional, sparsely vegetated, and/or disturbance-prone areas such as beaches, sand flats, dunes, water edges, tidal mud flats and open forests, savannas and grasslands (Pearson 1988). Most species are limited to a narrow range of microhabitats that appears to be restricted by various physical factors, including soil composition, moisture, and temperature (Pearson 1988). Habitat association is usually consistent throughout the geographic range of a given species (Schultz 1989), possibly because of physiological requirements of the eggs and the larvae (Willis 1967; Knisley 1984; Knisley 1987). Tiger beetle communities have been studied in some detail at several North American sites. Taxa appear to be segregated by differences in microhabitat and by differences in adult flight seasons (Pearson and Mury 1979; Knisley 1984; Mury-Meyer 1987; Schultz and Hadley 1987; Schultz 1989). Most North American species of *Cicindela* exhibit one of two phenological patterns. In species where adults are active in the spring and fall, larval development occurs over the summer, and fall-flying adults are dormant during the winter, re-emerging in the spring. Adults of such species thus show a bimodal pattern each year. In summer-active forms, adults emerge and reproduce during the middle of the summer, and early instar larvae overwinter (Pearson 1988). The total duration of the life cycle varies between one and several years.

Methodology and data base

TAXON SAMPLING – A total of 147 North American species of *Cicindela* are currently recognized (Boyd et al. 1982), grouped into 11 subgenera and some 7 named species groups (Rivalier 1954). Our sampling of taxa follows that of Pearson et al.'s (1988) ecological study addressing the distribution of traits implicated in anti-predator defense among North American cicindelids. With three exceptions (*C. horni, C. pimeriana,* and *C. ocellata* were not available), we analyzed all North American *Cicindela* of that earlier study. In addition, four species (*C. dorsalis, C. chlorocephala, C. praecisa,* and *C. unipunctata*) were included to complement groups not sampled in sufficient detail by Pearson et al. (1988). We thus studied 44 *Cicindela* species plus seven outgroups, covering the major clades of the Cicindelidae (see Vogler and Pearson 1996).

CLADOGRAM CONSTRUCTION – We used sequences from three regions of mitochondrial DNA (mtDNA) to reconstruct relationships among the 51 taxa included in our study. The data are described in more detail elsewhere (Vogler and Welsh 1997; A. P. Vogler and K. C. Kelley, submitted) but here we provide a brief summary of the data and method of analysis. The data matrix comprises a total of 1,896 positions from three regions of mtDNA. Searches for the most parsimonious trees included 10 replications of TBR branch-swapping with random addition of taxa in PAUP version 3.1.1 (Swofford 1993). Using unweighted maximum parsimony analysis, we obtained two shortest trees of 4,686 steps with CI = 0.254 (excluding uninformative characters) and RI = 0.468 (Figure 11.2). Both trees differed only in the position of the closely related *C. fulgida* and *C. splendida*. Decay indices and bootstrap values were used as a measure of support for each node. Recognizing that the work of Rivalier (1954) and Boyd et al. (1982) was explicitly non-phylogenetic, we also mapped the distribution and identity of traditional species groups onto our cladogram (Figure 11.2; see below). Several of these traditional groupings were found to be monophyletic whereas others were paraphyletic or polyphyletic in the molecular analysis, possibly as a result of different criteria for establishing groupings rather than conflicting phylogenetic signal.

CHARACTER CODING – Data on habitat associations are given by Pearson et al. (1988), who assigned a single habitat-type to each species of *Cicindela*. A total of ten such habitat-types were differentiated: sandy water edge, muddy water edge, muddy pond edge, moist forest floor, open forest floor, grassland, alpine grassland, saline flats, ocean beach, and sand dune. Habitat types were mapped on the cladograms using the "Trace character" option with both accelerated (ACCTRAN) or delayed transformation (DELTRAN) in MacClade version 3.01 (Maddison and Maddison 1992). For this analysis, "habitat" was coded as an unordered multi-state character. To further test the predictions from the taxon-cycle and taxon-pulse hypotheses, we compared these parsimonious optimizations of habitat associations as unordered character transformations with optimizations of habitat association ordered according to the predictions of each hypothesis. Following Liebherr and Hajek (1990) for this test, habitat character-states were coded as an ordered, multi-state character under Farris coding (Farris 1970) and Camin-Sokal coding (Camin and Sokal 1965), using the "ordered" and "irreversible" option, respectively, in the "Change Type" menu of MacClade (Maddison and Maddison 1992). Character-states were ordered in the sequence of their presumed level of water availability following the ranking of Altaba (1991) who characterized "sandy water edge" as the habitat type with the highest and "sand dune" with the lowest water availability in the sequence as given in Table 11.1, with the character-state "ocean beach" added by us.

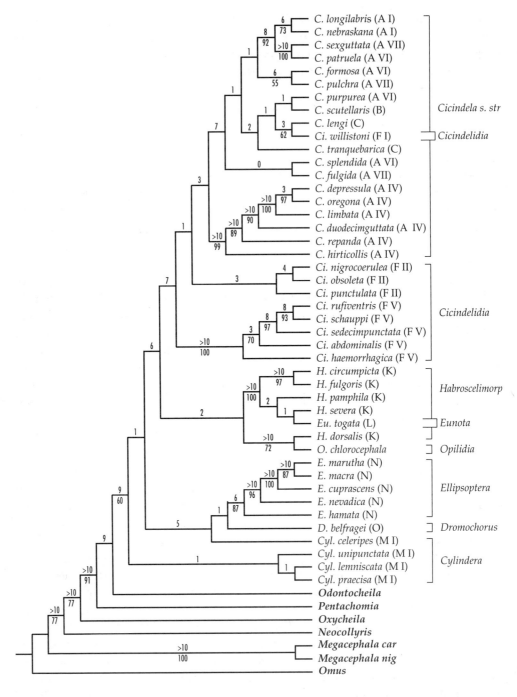

Figure 11.2. Cladogram depicting relationships of 44 North American *Cicindela* species and 7 outgroup taxa based on 1896 base pairs of mtDNA (from Vogler and Kelly 1997). The cladogram represents one of two shortest trees of 4686 steps. The numbers above branches are decay support values (Bremer 1988); below are the bootstrap support values (Felsenstein 1988) for the same node. Brackets indicate the traditionally recognized subgenera according to the classification of Rivalier (1954). The letters in parentheses after each taxon name refer to the designation of subgroupings in Pearson et al. (1988).

Table 11.1. Habitat transformation for 42 ingroup species (habitat data not available for *C. praecisa* and *C. unipunctata*). The number of taxa exhibiting a particular habitat type (Taxa), the number of nodes that are conserved (cons) for the character-state, the number of gains and losses of the character-state are given, separately under accelerated (ACCTRAN) and delayed (DELTRAN) optimization.

Habitat	Taxa	ACCTRAN			DELTRAN		
		cons	gain	loss	cons	gain	loss
sandy water edge	6	4	2	3	5	2	1
muddy water edge	1	0	1	0	0	1	0
muddy pond edge	2	0	2	0	0	2	0
moist forest floor	1	0	1	0	0	1	0
open forest floor	6	6	2	7	7	2	7
alpine grassland	1	0	1	0	0	1	0
grassland	7	4	3	1	5	2	2
saline flat	10	5	5	1	5	5	1
ocean beach	2	1	1	0	1	0	1
sand dune	4	1	3	2	0	4	0

BIOGEOGRAPHY AND PHENOLOGY – For information on geographic distribution we relied on the data included in the checklist of North American *Cicindela* (Boyd et al. 1982). This checklist provides the geographic distribution for all states in the United States and provinces in Canada and Mexico for each of 147 species of *Cicindela*. Although these data represent political rather than biogeographical entities, the listings give a rough estimate of the extent of the geographic distribution of taxa. As a preliminary estimate for which species occur at high latitudes, we list 24 in all with records from at least one province of Canada in the checklist of Boyd et al. (1982).

Phenology data for many species of *Cicindela* are available from a variety of published studies (e.g., Shelford 1908; Willis 1967; Rumpp 1977; Knisley 1979, 1984, 1987; Schultz 1989). In many cases these data are complemented by personal observations and those of colleagues (C. B. Knisley, P. Nothnagle, pers. comm.). All taxa were categorized with respect to the adult activity period as being summer or spring-fall (Pearson 1988). A third category, with adult activity periods in the spring without activity in the fall reflects minimal adult activity in *C. sexguttata* after eclosion in the fall. Adult activity periods appear to be consistent across geography and climate, although some modifications of the precise timing of key parameters – such as the onset and end of diapause, timing of larval stages – may depend on local climatic conditions. Life cycles in two species (*C. pulchra* and *C. willistoni*) are polymorphic with respect to flight season: adult activity in the desert Southwest is limited to the period of the summer monsoon, but both species are active in spring and fall in other areas. The overwintering stage in summer-active populations has not yet been investigated in either species.

Tests of predictions from taxon-cycle and taxon-pulse hypotheses

THE EVOLUTION OF HABITAT ASSOCIATIONS – Based on the phylogenetic hypothesis derived from our molecular data, we tested the predictions from the taxon-cycle and taxon-pulse hypotheses regarding the evolution of ecological characters and of biogeography. We began with predictions of directionality and frequency of changes in habitat association. We tested the postulate of the taxon-cycle/pulse hypothesis that "adaptive shifts" (Erwin 1985) such as changes in habitat association are unidirectional and irreversible but represented in different species groups. As the basis for the analysis of habitat character transformations, we mapped habitat as an unordered multi-state character on the mtDNA phylogeny (Figure 11.3).

To examine the directionality and frequency of habitat change we compiled the number of observed habitat shifts as optimized on the cladogram (Figure 11.3, Table 11.1). The distribution of the inferred character changes varied greatly for the ten recognized character states. Several habitat associations (e.g., muddy water edge, moist forest floor, alpine grassland, and ocean beach) were observed only in singleton taxa or in a single monophyletic group. Other states (e.g., saline flat, open forest floor, and sandy water edge) were found in a larger number of taxa with several character-state changes (Table 11.1). Habitat associations appear to be conserved in some subgroups of the sampled clade (Figure 11.3). Examples include the clades associated with ocean beach (*C. dorsalis* and allies), saline flats (subgenus *Habroscelimorpha* minus *C. dorsalis* clade), and grassland (*C. nigrocoerulea, C. punctulata,* and *C. obsoleta*). Other monophyletic groups include a few taxa which differ from the predominant habitat association, such as taxa occurring in sandy water edge or open forest floor (clades in the subgenus *Cicindela s. str.*). A total of 15 changes (ingroup taxa only) can be optimized unambiguously on the cladogram under the assumption of accelerated transformation (16 under delayed transformation), and only a total of seven changes were unambiguous under all resolving options. Almost all inferred changes were unique; only the change from open forest floor to muddy pond edge was observed twice (ACCTRAN only) and the change from open forest floor to sand dune observed three times (DELTRAN only). In total, the reconstruction of character-state transformations for the 44 ingroup taxa reveals 15 nodes in the cladogram for which a change of habitat association is inferred (16 nodes under DELTRAN optimization) whereas 23 cladogenetic events (27 under DELTRAN) did not coincide with an inferred change in habitat association.

We also modified the analysis of directional character change to test the proposal by Altaba (1991) who suggested a pattern of change from wet to dry habitat in North American *Cicindela*. This supposition is consistent with one of the ancillary aspects of the taxon-pulse hypothesis, which assumes an overall direction of habitat change from a wet (tropical) ancestral habitat type to more extreme drier or colder habitat types (Erwin 1985). We coded "habitat type" as an ordered character, ordering the transformation series from wet to dry habitats as proposed by Altaba (1991). The optimization of the habitat type as an ordered character (Farris 1970) when plotted on the mtDNA phylogeny for all taxa including the outgroup (Figure 11.3), requires 64 steps, substantially more than the 22 changes under unordered (Fitch) transformation. If, in accordance with the cladistic interpretation of the taxon pulse by Liebherr and Hajek (1990), the habitat transformation is optimized under Camin-Sokal coding with a fixed ancestral character state ("sandy water edge") and irreversible character transitions (Camin and Sokal 1965), a total of 109 steps are required for the optimization. The increased number of evolutionary steps required by this transformation argues

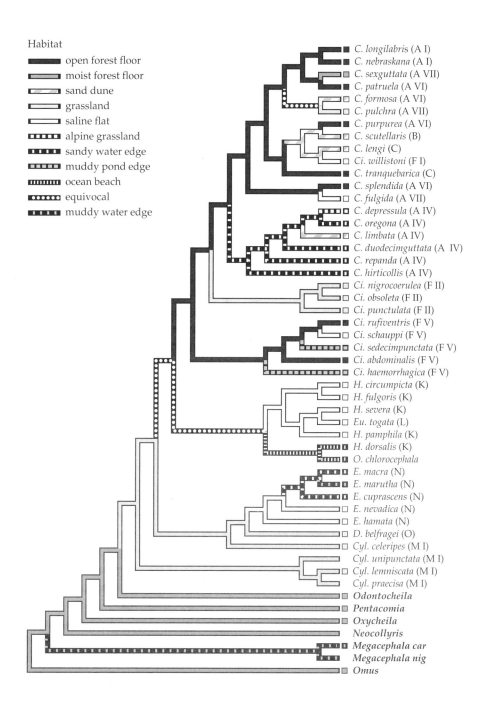

Figure 11.3. Habitat association and changes in character-states in North American *Cicindela* (ACCTRAN optimization). The habitat type is taken from Pearson et al. (1988). Habitats are ranked by water availability (highest to lowest) according to Altaba (1991), with ocean beach and alpine grassland added by us.

strongly against the proposition that habitat evolved along the transformation series proposed by Altaba (1991). We also compared the number of times in which the character transformation changed in the expected direction (from wet to dry) versus the unexpected direction with a sign test. Using only character changes in the ingroup, 8 out of the total of 15 changes under ACCTRAN (Table 11.1) were from a habitat with higher water availability to drier habitat; 11 out of 16 changes under DELTRAN were in this direction. Many of these changes were only observed under one of the two resolving options; but among the total of 7 changes which were observed in all most parsimonious optimizations, a total of 5 changes were from wet to dry habitat. This observation may indicate a predominant direction of habitat shifts from wet to drier habitat, but no stepwise character change along the proposed gradient.

In summary, our analysis of habitat change and cladogenesis suggests that numerous cladogenetic events do coincide with habitat transformation (Figure 11.1A) but in a slightly larger number of cases the habitat associations are conserved, a pattern more consistent with the taxon-pulse than the taxon-cycle hypothesis (Figure 11.1B). Thus, no clear directional and repeated pattern of habitat change was observed, contrary to predictions of either hypothesis. Similarly, the test by Liebherr and Hajek (1990) using ordered character transformations resulted in large increases of the number of mutational steps, inconsistent with the predictions of either the taxon-cycle or taxon-pulse hypothesis. However, the direction of the character change observed – from wet to dry habitat – was generally in accordance to that expected under the taxon-pulse hypothesis.

EVOLUTION OF PHENOLOGICAL SYNDROMES – Habitat associations and both latitudinal and altitudinal distributions in cicindelids and related carabids frequently have been proposed as being correlated with phenology (e.g., Thiele 1977 [pp. 248–250]; Kavanaugh 1978; Sota 1994 [and references therein]). Phenological syndromes have also been addressed in studies of resource partitioning between temporally segregated species (Pearson and Mury 1979; Pearson 1988; Pearson and Juliano 1991), invoking competition-mediated life history character displacement to explain differences in phenology (as well as feeding apparatus "adaptations" to different sized prey items; Pearson and Mury 1979; Mury-Meyer 1987). The study of phenological syndromes and inferences about their phylogenetic history can possibly provide a test to discriminate between explanations that invoke competition between populations (consistent with the taxon cycle) and unique "adaptive shifts" (consistent with the taxon pulse).

When plotted on the cladogram, those species with adult activity primarily in summer were all found to be in the basal clades of the North American *Cicindela* (Figure 11.4A). Those species with spring/fall activity and adult overwintering were exclusively found in the clade comprising the subgenera *Cicindela* (s. str.) and *C. willistoni* (classified as a member of *Cicindelidia* by Rivalier [1954]). *Cicindela sexguttata*, the species with adults active only in the spring, is situated within this clade as are two variable species (*C. pulchra* and *C. willistoni*) that exhibit summer activity in certain parts of their range and spring-fall activity in others. In summary, the overall pattern of character transformation for phenology involves a single change from summer to spring/fall adult activity, followed by modifications of the latter in certain taxa and populations (Figure 11.4A).

It is interesting that latitudinal changes are almost perfectly correlated with shifts in adult phenology (Figure 11.4B). Almost all taxa found at high latitudes show spring/fall activity. Exceptions to this pattern are the two species polymorphic for phenology (*C. pulchra* and *C. willistoni*) which are not found at higher latitudes, although they would be expected to be, based on their phylogenetic positions. Other exceptions are two species (*C. punctulata* and *C. nevadica*) with summer activity but found at higher latitude. It is noteworthy that, while one of these species is deeply nested within the summer active clade, *C. punctulata* is just basal to the spring/fall active group. This point will be addressed below.

Although there is evidence for competition between co-occurring species of tiger beetles (Pearson and Knisley 1985), no evidence for character displacement has been found in cases that have been studied in detail (Schultz and Hadley 1987). Similarly, putative displacement in mandible length was not confirmed in more recent studies (Pearson and Juliano 1991). Subsequent work on tiger beetle communities has instead emphasized "historical" (Pearson et al. 1988) factors to explain patterns of similarity and dissimilarity in tiger beetle species, recognizing that evidence for character displacement is weak (Pearson and Juliano 1994). With the availability of an explicit phylogenetic hypothesis for *Cicindela* and the reconstruction of phenology character transformation, the evidence for character displacement is even less convincing. As a conservative character, phenological shifts do not appear to be a necessary precursor or result of speciation events.

In contrast, the evolution of phenological syndromes and the apparent geographic distribution of taxa with spring/fall adult activity at higher latitudes is consistent with the notion of a taxon pulse. In its most simple formulation this pulse into novel geographic territory (area of higher latitude) could have been triggered by an adaptive shift – in this case the novel phenological syndrome that allowed persistence under more extreme climatic conditions. The evidence for a pulse of this kind in *Cicindela* is certainly better than in many other taxa for which taxon pulses have been invoked, and the presumption of such pulse is intuitively attractive as a summary description of the ecological evolution in *Cicindela*.

Caveats

We caution, however, against the whole-hearted acceptance of such a general explanation for the patterns observed in *Cicindela* and the inferred process underlying it – i.e., physiological and other adaptations facilitating the species diversification in northern latitudes. Given the several deviations from a strict correlation between seasonality and occurrence at high latitude, it is possible that this correlation is entirely coincidental. A causal correlation of both factors, however, will be more credible if similar patterns can be observed repeatedly in lineages of organisms under similar environmental conditions – i.e., if the taxon-pulse hypothesis and its presumed underlying process can be shown empirically to have generality in explaining patterns of diversity. This type of corroboration could come from convergent

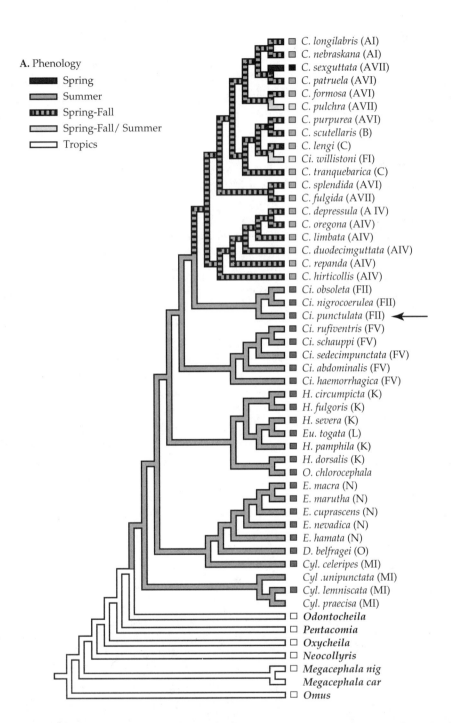

Figure 11.4. Distribution of phenological syndromes (**A**) and latitudinal distribution (**B**) on the molecular phylogeny of 47 North American *Cicindela* species. Species active in the summer generally occur at lower latitudes only; species that are active in the spring and fall have a wider distribution including areas at higher latitudes. Note that *C. punctulata* (arrow) at the base of the spring-fall active clade is a summer-active species that also occurs at high latitudes. The phenology of *C. pulchra* and *C. willistoni* varies geographically.

patterns in other subgroups of the genus *Cicindela* on other continents under similar ecological and geographic conditions, or from studies on related carabids radiating in a similar fashion. However, the evidence for a general pattern of these convergent cycles is still weak. In our estimation, the evidence for such pulses is mostly narrative, and a foundation based on a well-supported phylogenetic hypothesis is lacking in almost all instances, including the presumed pulses in carabids outlined by Erwin (1985). The unspecified physiological and life-history adaptations that have been invoked to account for many taxon pulses cannot be taken as evidence for their general existence. Each of these cycles is explained by a different event in Earth history, providing nothing more than an *ad hoc* explanation for each of these radiations.

TESTABILITY OF THE TAXON-CYCLE/TAXON-PULSE HYPOTHESES – Even if the evolution of phenological syndromes is inferred to be consistent with the expectations of a taxon pulse, it has to be clear that any explanation of historical-ecological patterns in terms of the taxon-cycle and taxon-pulse hypotheses are possibly flawed because both hypotheses make a number of hidden untestable postulates. Liebherr and Hajek (1990: p. 39) identified three of these (which they term "assumptions") required by any would-be test of either hypothesis: (i) habitat specialization is largely irreversible in a lineage, (ii) ecological specializations arise in a center of origin, and (iii) dispersal events leading to current distributions can be ascertained. Indeed there are other aspects, some of which were also identified by Liebherr and Hajek, common to both of these hypotheses that demand to be addressed by any rigorous test. Several of these assumptions center around biogeographic dynamics, specifically the inference of a center of origin, the route of dispersal away from this center, the expansion of geographic range and subsequent extinction of (unobserved) taxa, and the replacement of such unobserved taxa with invading "younger groups" (Erwin 1985: p. 440) to initiate a subsequent round of the cycle/pulse. The postulates of the taxon-cycle/ taxon-pulse hypotheses regarding the adaptive nature of the ecological changes involved are also problematic. These include (i) the assumption that "adaptations" arise (as the result of competition) in the centers of origin from where colonization of outlying regions initiates (Liebherr and Hajek 1990: p. 54), implying that the adaptations arise under conditions to which taxa have never been exposed before commencing a cycle/pulse; (ii) that the ecological conditions which were imposing such selection pressure can be inferred in retrospect; (iii) that acquired adaptations proceed in a directional fashion resulting in an increasingly narrow ecological amplitude from generalist to specialist species; and (iv) that these adaptations are the factors actually responsible for the ever increasing numbers of species in a radiation.

Of all these, only the first of Liebherr and Hajek's (1990) assumptions is legitimately addressed with phylogenetics: that is, by inspecting the most parsimonious optimization of character-states on a resolved cladogram to determine whether character-state changes in habitat are unidirectional, whether the patterns of change are repetitive (convergent), and whether the series of character changes is irreversible. The set of requirements regarding the biogeographic predictions of the taxon-cycle/pulse hypotheses is related to the inference of ancestral areas and dispersal events. The inference of ancestral areas or centers of origin is decidedly *ad hoc* (Croizat 1976), as is the invocation of extinction events of unobserved organisms. Dispersal events may be invoked to account for taxic distributions, but only when those distributions are not more easily explained by simple vicariance (*contra* Erwin 1981). Neither the taxon-cycle nor the taxon-pulse hypothesis, however, admits vicariance as an explanation of the patterns, precluding geological history as a factor to account for current distribution of taxa.

The set of assumptions of the taxon-cycle and taxon-pulse hypotheses regarding the adaptive nature of the traits implicated in the onset of cycles/pulses are untestable, because the selective forces giving rise to putatively adaptive traits cannot be identified *post facto*. Hypotheses of adaptation to particular habitats or areas lend themselves to testing when related to a specific prediction of character-state distribution. When only one character is involved, the most one can hope for is to infer the number of times each character-state evolved and in what order. Only when two characters or a character and an association are involved (ecological, biogeographic, or otherwise, as in the case of phenology and latitudinal distribution in *Cicindela*), and one hypothesized to be an adaptation to the other, can adaptation be addressed, and then only in part. If trait A is hypothesized to have evolved in response to condition X, then the inferred origin of A may not appear before that of X (Coddington 1988; Wenzel and Carpenter 1994). Of course, just because A appears after X does not in itself prove that A is an adaptation to X, merely that such a hypothesis is consistent with the observed character-state distribution.

As has also become clear from our discussion of the evolution of phenology, phylogeny provides no information on the mechanistic relationship between two characters unless their distribution across taxa differs in particular ways. It is not apparent whether the evolution of a spring/fall phenology allowed the invasion of higher latitudes, whether higher latitudes exerted selection pressure favoring spring/fall activity, or some combination of the two. Phylogenetic approaches, however, can provide a hypothesis for the order of events (see Coddington 1988). If a given trait is hypothesized to predispose its bearer to evolving a second trait, the two traits must be distributed in a particular way on the cladogram. Our phenological data constitute an example of an ambiguous order with respect to the evolution of spring/fall phenology and the invasion of high latitudes (Figure 11.4). It is worth noting, however, that despite the strong correlation of spring/fall adult activity and northern distribution, there are two cases where summer-active species persist at higher latitudes, one of which, *C. punctulata*, is at the base of the spring-fall active clade. At least under one resolving option (ACCTRAN), the character transformation to higher latitude immediately precedes the evolution of spring-fall activity cycle. This indicates, perhaps, that the spring-fall pattern is an adaptation to the occurrence at higher latitude acquired by the ancestor of the spring-fall clade after the colonization of northern habitats. (Because many of those taxa which occur at higher latitudes have in fact a very wide distribution, including large parts of the lower latitudes of the continent, it may actually have been an adaptation to the persistence under a wider range of climatic conditions.)

KEY INNOVATIONS AND THE CAUSE OF RADIATIONS – We have already examined the notion of key innovation, emphasizing that the implications of "key" for speciation rates are not necessarily well founded. Phylogenetic approaches to the investigation of key innovations, or "adaptive shifts," that initiate a taxon pulse, suffer from the same problem as the study of adaptational character changes. Namely, when a particular character-state is hypothesized to be a "key innovation" that precipitated a bloom of speciation events – by whatever mechanism – it must behave as a synapomorphy for the group in question. Unfortunately, topology is no more capable of distinguishing "key adaptation" from straightforward synapomorphy (or from "phylogenetic constraint," for that matter) than it is of partitioning adaptive from other character changes. And, as has become clear from our discussion of phenology, phylogeny provides no information on the mechanistic relationship between two characters unless their distribution across taxa differs in particular ways. In studies of taxon pulses (and adaptive radiations in general) we can identify traits when they co-occur at a given node on a cladogram. We can also identify whether a change in a particular ecological character co-occurs with a change in species richness as compared to sister taxa missing the ecological characters in question, using various null models of tree symmetry (Raup et al. 1973; Purvis 1996). But we are confronted by our inability to determine any relationship between mechanistic causation and the distribution of a character on a cladogram. The topology of the cladogram itself does not address the question of "if" and "how" an acquired feature enables a clade to undergo an increased rate of speciation (or slowed rate of extinction). This type of phylogenetic inference, however, is implicit in the presentation of the taxon-pulse hypothesis as it draws its conclusions from the co-occurrence of a particular event in Earth history and a particular adaptation – the latter inferred from a character transformation on a cladogram dated in geological time. If the presumed adaptation arose at the base of a species-rich clade it is inferred to be the causal agent of the taxon pulse.

Conclusions

Biologists approach adaptive radiation from several angles: (i) attempting to discern the selective forces and the adaptations they elicit; (ii) associating the evolution of a particular feature with species diversification (radiation); and (iii) testing the consistency of specific directional hypotheses of character evolution against phylogenetic information. For large groups, such as the *Cicindela* tiger beetles discussed in this paper, the difficulty in identifying morphological characters informative at the species level renders molecular systematic information critical to phylogeny reconstruction. While the testing of microevolutionary adaptive scenarios eludes examination by macroevolutionary pattern, species-level cladistic information at the very least provides a means of examining macroevolutionary pattern as finely as possible – i.e., in a way that suggests directions for studies of character displacement and other microevolutionary phenomena. Our data suggest that phenological "character displacement" may be less well explained by competitive and predatory interaction than by a single evolutionary event coinciding with a geographic shift toward higher latitudes. The taxon-cycle and taxon-pulse hypotheses incorporate both macro- and microevolutionary parameters, and are thus only partly addressable by cladistic pattern. However, those few aspects of these hypotheses that are supported by our data suggest that ecological and behavioral characters are rather conserved phylogenetically.

Following Cracraft (1990) and Wiegmann et al. (1993), we caution against the temptation of dubbing a conservative character (in effect, a synapomorphy) with the moniker "key innovation." Such designation is always *post hoc*, and cannot be coupled mechanistically with "radiation" *per se*. Given that most of the phylogenetic history of life on Earth comprises unique historical events that are inherently unknowable, the most we can hope for is a clear understanding of the features of adaptation and radiation that do – and do not – lend themselves to phylogenetic examination.

Acknowledgments

We wish to thank A. Brower, R. DeSalle, P. Hammond, K. Clarke-Kelley, C. B. Knisley, P. Nothnagle, D. L. Pearson, and J. N. Thompson, two anonymous reviewers and particularly K. Sytsma for helpful comments on this manuscript. We acknowledge excellent technical help by Alexandra Welsh who generated most of the DNA sequence information used for the phylogeny reconstruction. We are grateful to T. Givnish and K. Sytsma for the opportunity to contribute to this Symposium. Work on this paper was supported in part by grant GR 3 /10632 of the Natural Environment Research Council (NERC) (APV), by an NSF Graduate Research Training fellowship administered by the Dept. of Ecology and Evolutionary Biology at the University of Connecticut, by an EPA graduate fellowship and by the Department of Entomology, American Museum of Natural History (PZG).

References

Altaba, C. R. 1991. The importance of ecological and historical factors in the production of benzaldehyde in tiger beetles. Systematic Zoology 40:101–105.

Boyd, H. P. et al. 1982. *Checklist of Cicindelidae: the tiger beetles*. Marlton, NJ: Plexus Publishing, Inc.

Bremer, K. 1988. The limits of amino-acid sequence data in angiosperm phylogenetic reconstruction. Evolution 42:795–803.

Brown, W. L., and Wilson, E. O. 1956. Character displacement. Systematic Zoology 7:49–64.

Camin, J. H., and Sokal, R. R. 1965. A method for deducing branching sequences in phylogeny. Evolution 19:311–326.

Carpenter, J. M. 1989. Testing scenarios: Wasp social behavior. Cladistics 5:131–144.

Coddington, J. A. 1988. Cladistic tests of adaptational hypotheses. Cladistics 4:3–22.

Croizat, L. 1976. Biogeografia analitica y sintetica ("panbiogeografia") de las Americas. Biblioteca de la Academia de Ciencias Fisicas, Matematicas y Naturales (Caracas) 15–16:1–454, 455–890.

Erwin, T. L. 1979. Thoughts on the evolutionary history of ground beetles: hypotheses generated from comparative faunal analysis of lowland forest sites in temperate and tropical regions. Pp. 539–587 in *Carabid beetles: their evolution, natural history, and classification*, T. L. Erwin, G. E. Ball, D. R. Whitehead, and A. L. Halpern, eds. The Hague, The Netherlands: W. Junk.

Erwin, T. L. 1981. Taxon pulses, vicariance, and dispersal: an evolutionary synthesis illustrated by carabid beetles. Pp. 159–196 in *Vicariance biogeography: a critique*, G Nelson and D. E. Rosen, eds. New York, NY: Columbia University Press.

Erwin, T. L. 1985. The taxon pulse: a general pattern of lineage radiation and extinction among carabid beetles. Pp. 437–472 in *Taxonomy, phylogeny and zoogeography of beetles and ants*, G. E. Ball, eds. Dordrecht, The Netherlands: W. Junk.

Farrell, B. D., D. E. Dussard, and Mitter, C. 1991. Escalation of a plant defense: do latex and resin channels spur diversification? American Naturalist 138:881–900.

Farrell, B. D., and Mitter, C. 1993. Phylogenetic determinants of insect/plant community diversity. Pp. 253–266 in *Species diversity in ecological communities*, R. E. Ricklefs and D. Schluter, eds. Chicago, IL: University of Chicago Press.

Farris, J. S. 1970. Methods for computing Wagner trees. Systematic Zoology 19:83–92.

Frumhoff, P. C., and Reeve, H. K. 1994. Using phylogenies to test hypotheses of adaptation: a critique to some current proposals. Evolution 48:172–180.

Heard, S. B., and Hauser, D. L. 1995. Key evolutionary innovations and their ecological mechnisms. Historical Biology 10:151–173.

Howden, H. F. 1985. Expansion and contraction cycles, endemism and area: the taxon cycle brought full circle. Pp. 473–487 in *Taxonomy, phylogeny and zoogeography of beetles and ants*, G. E. Ball, eds. Dordrecht, The Netherlands: W. Junk.

Kavanaugh, D. H. 1978. Analysis of the Nearctic *Nebria* fauna. Ph. D. Dissertation. Edmonton, Alberta: University of Alberta.

Knisley, C. B. 1979. Distribution, abundance, and seasonality of tiger beetles (Cicindelidae) of the Indiana dunes region. Proceedings of the Indiana Academy of Sciences 88:125–133.

Knisley, C. B. 1984. Ecological distribution of tiger beetles (Coleoptera: Cicindelidae) in Colfax County, New Mexico. Southwest Naturalist 29:93–104.

Knisley, C. B. 1987. Habitats, food resources, and natural enemies of a community of larval *Cicindela* in southeastern Arizona. Canadian Journal of Zoology 65:1191–1200.

Knisley, C. B., and Hill, J. M. 1992. Effects of habitat change from ecological succession and human impact on tiger beetles. Virginia Journal of Science 43:133–142.

Lack, D. 1947. *Darwin's finches*. Cambridge, England: Cambridge University Press

Leroi, A. M., Rose, M. R., and Lauder, G. V. 1994. What does the comparative method reveal about adaptation? American Naturalist 143:381–402.

Liebherr, J. K., and Hajek, A. E. 1990. A cladistic test of the taxon cycle and taxon pulse hypotheses. Cladistics 6:39–59.

Losos, J. B. 1992. A critical comparison of the taxon-cycle and character-displacement models for size evolution in *Anolis* lizards in the Lesser Antilles. Copeia 23:279–288.

Maddison, W. P., and Maddison, D. R. 1992. *MacClade, ver. 3.01*. Sunderland, MA: Sinauer Associates.

Mitter, C., Farrell, B., and Wiegmann, B. 1988. The phylogenetic study of adaptive zones: has phytophagy promoted insect diversification? American Naturalist 132:107–128.

Mooi, R., Cannell, P. F., Funk, V. A., Mabee, P. M., O'Grady, R. T., and Starr, C. K. 1989. Historical perspectives, ecology and tiger beetles: an alternative discussion. Systematic Zoology 38:191–195.

Mury-Meyer, E. J. 1987. Asymmetric resource use in two syntopic species of larval tiger beetles (Cicindelidae). Oikos 50:167–175.

Niemela, J., and Ranta, E. 1993. World-wide tiger beetle mandible length ratios: was something left unmentioned? Annales Zoologici Fennici 30:85–88.

Pearson, D. L. 1980. Patterns of limiting similarity in tropical forest tiger beetles (Coleoptera: Cicindelidae). Biotropica 12:195–204.

Pearson, D. L. 1986. Community structure and species co-occurrence: a basis for developing broader generalizations. Oikos 46:419–422.

Pearson, D. L. 1988. Biology of tiger beetles. Annual Review of Entomology 33:123–147.

Pearson, D. L., Blum, M. S., Jones, T. H., Fales, H. M., Gonda, E., and White, B. R. 1988. Historical perspective and the interpretation of ecological patterns: defensive compounds of tiger beetles (Coleoptera: Cicindelidae). American Naturalist 132:404–416.

Pearson, D. L., and Cassola, F. 1992. World-wide species richness patterns of tiger beetles (Coleoptera: Cicindelidae): indicator taxon for biodiversity and conservation studies. Conservation Biology 6:376–391.

Pearson, D. L., and Juliano, S. A. 1991. Mandible length ratios as a mechanism for co-occurrence: evidence from a world-wide comparison of tiger beetle assemblages (Cicindelidae). Oikos 60:223–233.

Pearson, D. L., and Juliano, S. A. 1994. Evidence for the influence of historical processes in co-occurrence and diversity of tiger beetle species. Pp. 194–202 in *Species diversity in ecological communities*, R. E. Ricklefs and D. Schluter, eds. Chicago, IL: Chicago University Press.

Pearson, D. L., and Knisley, C. B. 1985. Evidence for food as a limiting resource in the life cycle of tiger beetles (Coleoptera: Cicindelidae). Oikos 45:161–168.

Pearson, D. L., and Mury, E. J. 1979. Character divergence and convergence among tiger beetles (Coleoptera: Cicindelidae). Ecology 60:557–566.
Purvis, A. 1996. Using interspecies phylogenies to test macroevolutionary hypotheses. Pp. 153–168 in *New uses for new phylogenies*, eds. P. H. Harvey, A Leigh Brown, J. Maynard Smith, and S. Nee, eds. New York, NY: Oxford University Press.
Raup, D. M., Gould, S. J., Schopf, T. J., and Simberloff, D. S. 1973. Stochastic models of phylogeny and the evolution of diversity. Journal of Geology 81:525–542.
Rivalier, E. 1954. Démembrement du genre *Cicindela* Linné. II. Faune americaine. Revue Française d'Entomologie 21:249–268.
Rivalier, E. 1963. Démembrement du genre *Cicindela* Linné. V. Faune australienne. Revue Française d'Entomologie 28:30–48.
Roughgarden, J., and Pacala, S. 1989. Taxon cycle among *Anolis* lizard populations: review of evidence. Pp. 403–432 in *Speciation and its consequences*, D. Otte and J. A. Endler, eds. Sunderland, MA: Sinauer Associates.
Rummel, J. D., and Roughgarden, J. 1985. A theory of faunal buildup for competition communities. Evolution 39:1009–1033.
Rumpp, N. L. 1977. Tiger beetles of the genus *Cicindela* in the Sulphur Springs Valley, Arizona, with descriptions of three new subspecies (Cicindelidae - Coleoptera). Proceedings of the California Academy of Sciences 61:169–182.
Schultz, T. D. 1986. Role of structural colors in predator avoidance by tiger beetles of the genus *Cicindela* (Coleoptera: Cicindelidae). Proceedings of the California Academy of Sciences 41:165–187.
Schultz, T. D. 1989. Habitat preference and seasonal abundances of eight sympatric species of tiger beetle, genus *Cicindela* (Coleoptera: Cicindelidae) in Bastrop State Park, Texas. Southwest Naturalist 34:468–477.
Schultz, T. D., and Hadley, N. F. 1987. Microhabitat segregation and physiological differences in co-occurring tiger beetle species, *Cicindela oregona* and *Cicidela tranquebarica*. Oecologia 73:363–370.
Shelford, V. E. 1908. Life-histories and larval habits in the tiger beetles (Cicindelidae). Zoological Journal Linnaean Society 23:157–184.
Sota, T. 1994. Variation of carabid life cycles along climatic gradients: an adaptive perspective for life-history evolution under adverse conditions. Pp. 91–112 in *Insect life-cycle polymorphism: theory, evolution, and ecological consequences for seasonality and diapause control*, H.V. Danks, ed. Dordrecht, The Netherlands: Kluwer Academic Publishers.
Swofford, D. L. 1993. *PAUP: phylogenetic analysis using parsimony, ver. 3.1*. Champaign, IL: Illinois Natural History Survey.
Thiele, H. U. 1977. *Carabid beetles in their environments*. Berlin, Germany: Springer-Verlag.
Thompson, J. N. 1994. *The coevolutionary process*. Chicago, IL: Chicago University Press.
Vogler, A. P., and DeSalle, R. 1994. Diagnosing units of conservation management. Conservation Biology 8:354–363.
Vogler, A. P., and Pearson, D. L. 1996. A molecular phylogeny of the tiger beetles (Cicindelidae): congruence of mitochondrial and nuclear rDNA data sets. Molecular Phylogenetics and Evolution 6:321–338.
Vogler, A. P., and Welsh, A. 1997. Phylogeny of North American Cicindela tiger beetles inferred from multiple mitochondrial DNA sequences. Molecular Phylogenetics and Evolution 7 (in press).
Vrba, E. S. 1983. Macroevolutionary trends: new perspectives on the roles of adaptation and incidental effect. Science 221:387–389.
Wenzel, J. W., and Carpenter, J. M. 1994. Comparing methods: adaptive traits and tests of adaptation. Pp. 79–101 in *Phylogenetics in ecology*, P. Eggleton and R. Vane-Wright, eds. London, England: Harcourt Brace.
Wiegmann, B. M., Mitter, C., and Farrell, B. 1993. Diversification of carnivorous parasitic insects: extraordinary radiation or specialized dead end? American Naturalist 142:737–754.
Willis, H. L. 1967. Bionomics and zoogeography of tiger beetles of saline habitats in the central United States (Coleoptera: Cicindelidae). University of Kansas Scientific Bulletin 47:145–313.
Wilson, E. O. 1959. Adaptive shift and dispersal in a tropical ant fauna. Evolution 13:122–144.
Wilson, E. O. 1961. The nature of the taxon cycle in the Melanesian ant fauna. American Naturalist 95:169–193.

12 MOLECULAR PHYLOGENETIC TESTS OF SPECIATION MODELS IN LAKE MALAWI CICHLID FISHES

Peter N. Reinthal and Axel Meyer

The adaptive radiations of cichlids (Teleostei: Cichlidae) in the East African Great Lakes – Victoria, Tanganyika, and Malawi (Figure 12.1) – are extremely species-rich and bear testimony to the evolutionary success of these fishes. Each lake is occupied by a species flock of at least several hundred cichlid species (Fryer and Iles 1972; Lewis et al. 1986), most of which are endemic to that lake. Each species flock is thought to be a monophyletic assemblage (Greenwood 1984) and contains a sweeping array of morphologically and behaviorally specialized fishes occupying several different ecological roles (Fryer and Iles 1972). Given the unparalleled evolutionary and ecological success of each of these adaptive radiations, their evolutionary origin and ecological maintenance have been much debated and studied (e.g. Mayr 1963, 1984; Fryer and Iles 1972; Coulter 1991; Greenwood 1991; Meyer 1993b). Yet, despite a long history of research, the precise nature of the phylogenetic relationships among endemic cichlid faunas has remained largely unresolved. Few morphological synapomorphies characterize members of each radiation; convergent and parallel evolution in form and ecology, and the masking of informative morphological characters by autapomorphies have also hindered attempts to elucidate relationships among cichlids.

Guilds of ecologically similar, specialized species – such as scale-scrapers, mollusc-crushers, algae-scrapers, or paedophages – are found in all three species flocks (e.g., Fryer and Iles 1972). Similar morphological and behavioral solutions to similar ecological problems have thus arisen in each lake (Stiassny 1981; Greenwood 1983). For example, the Victoria endemic *Macropleurodus bicolor* and the Malawi endemic *Chilotilapia rhoadesii* both exhibit similar, highly derived dentitions, jaw structures, and feeding behaviors whereby they prey on gastropods, crushing their shells with their oral jaws (Greenwood 1983).

The similarity in morphological and behavioral specializations in separate basins gives rise to the much debated question of their origin through common ancestry or convergence/parallelism. If each kind of specialization arose only once in East African cichlids, it would indicate the polyphyly of the species flocks, with each of several different lineages having a geographic distribution that extends beyond the boundaries of a single lake (Stiassny 1981; Greenwood 1983). The remarkable morphological and behavioral similarities among guild members from different lake basins have led authors to suggest that, if such species were found in the same basin, they would be placed in the same genus (Fryer and Iles 1972; Greenwood 1983). Such interpretations, if correct, would mean that each species flock is polyphyletic.

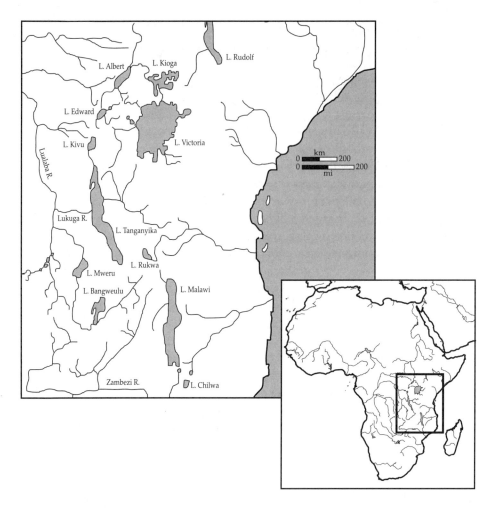

Figure 12.1. Geography of East Africa, including major river systems and the Great Lakes of the rift zone (redrawn after Fryer and Iles 1992).

The lack of morphological synapomorphies that characterize members of each species flock has called into question the utility of even the most careful of phylogenetic reconstructions based on morphology (see Stiassny 1991 for discussion), and highlights the need for other kinds of characters with which to reconstruct the evolutionary history of the East African cichlids. Variation in DNA sequences has proven to be a useful source of phylogenetic information, independent of homoplasy-ridden morphology (Meyer et al. 1990). Based on such data, it now appears well established that cichlid species flocks from Lake Victoria and Lake Malawi are each monophyletic and arose quite recently within the confines of their lakes in a true intralacustrine manner (Meyer et al. 1990, 1991; Sturmbauer and Meyer 1992; Meyer 1993b; Meyer et al. 1994). DNA studies on the question of the between-flock relationships (Meyer et al. 1990; Kocher et al. 1993) demonstrated that the striking morphological

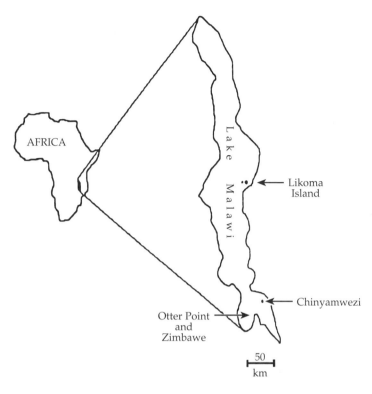

Figure 12.2. Lake Malawi, showing locations where cichlid specimens were collected. See Table 12.1 for a list of the ecologically equivalent pairs of species at Likoma Island and Otter Point.

similarities between certain members of flocks in different lakes (e.g., *Macropleurodus/Chilotilapia*) did, in fact, evolve convergently.

Lake Tanganyika contains the oldest and genetically most diverse cichlid species flock (Nishida 1991). This flock is composed of several morphologically and genetically distinct lineages (Nishida 1991; Sturmbauer and Meyer 1992, 1993; Sturmbauer et al. 1994). The monophyly of the Lake Tanganyika flock has not been clearly established (Nishida 1991; Sturmbauer and Meyer 1992; Kocher et al. 1993; Meyer 1993b; Moran et al. 1994; Sturmbauer et al. 1994; Sültmann et al. 1995; Meyer et al. 1996). Several ancestral lineages are likely to have colonized the Tanganyika basin and relationships among its lineages are still unknown (Sturmbauer and Meyer 1992, 1993; Kocher et al. 1993; Meyer 1993; Sturmbauer et al. 1994; Meyer et al. 1994; Sültmann et al. 1995). It is known, however, that several lineages of Tanganyika cichlids have been able to leave the confines of that lake's basin. For example, a small number of lamprologine cichlids – a lineage otherwise found only in Lake Tanganyika – live in the Zaire river. These lamprologines had previously been assumed to be representatives of basal lineages that, early in the evolution of the Tanganyika flock, colonized that lake and gave rise to about 30% of all cichlid species now found in it. However, mitochondrial DNA data (Sturmbauer et al. 1994) suggest that riverine lamprologine species are recently derived, and are likely to have left the lake when the Lukaga River began to flow from Lake Tanganyika into the Zaire River.

It is apparent from the phylogeny of the three species flocks (reviewed by Meyer 1993b and Meyer et al. 1994) that the endemic haplochromine cichlids of Lake Tanganyika are closely related to the lineages of East African riverine haplochromines that first colonized Lake Malawi and more recently Lake Victoria (Nishida 1991; Sturmbauer and Meyer 1992). The haplochromine species flocks endemic to Lakes Malawi and Victoria are much younger than the Lake Tanganyika species flock, and therefore genetically more similar to each other than either is to the Tanganyika flock. There are several East African haplochromine lineages that are not endemic to lakes (e.g., *Astatoreochromis, Astatotilapia, Serranochromis*) which are more closely related to the

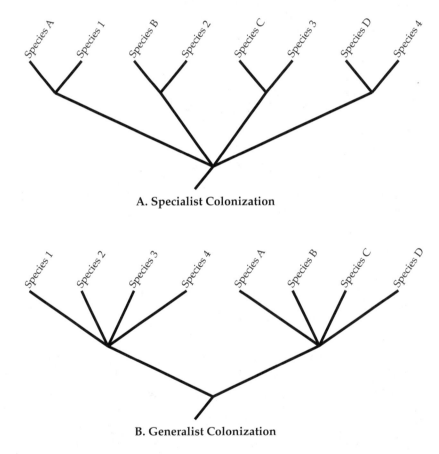

Figure 12.3. Alternative phylogenies expected under two alternative speciation models. Species A – D represent those from Otter Point, while species 1 – 4 represent those from Likoma Island from the middle of Lake Malawi. Species 1 and A, 2 and B, 3 and C, and 4 and D are the ecologically equivalent pairs of species. (**A**) Under the specialist-colonization hypothesis, ecological and morphological specialization precedes colonization and geographically distinct specialists would be more closely related to each other than they are to other members found at the same locality. (**B**) Under the generalist colonization hypothesis, all species found at a locality would be descended from the same generalist ancestor, and hence be more closely related to each other than to corresponding specialists from other localities.

haplochromines endemic to Lakes Malawi and Victoria than to those endemic to Lake Tanganyika (Meyer et al. 1990, 1991; Meyer et al. 1994; Sültmann et al. 1995).

Possibility of sympatric speciation and radiation

Molecular systematics can not only provide insights into relationships among broad groups of organisms, but can also permit tests of alternative mechanisms of speciation that may have generated closely related taxa within individual flocks. Speciation is a central issue in evolutionary biology (Slatkin 1987; Lynch 1989; Otte and Endler 1989; Coyne 1992) but only few phylogenetic tests of models of speciation have been conducted (e.g., Otte and Endler 1989; Harrison 1991; Coyne 1992; Mayden 1992). The intralacustrine origins of the haplochromine species flocks does not necessarily imply that they each arose through sympatric speciation. All three large East African lakes underwent repeated lake-level fluctuations that could have fragmented single basins into several separate or satellite lakes that persisted for thousands of years. These major lake-level changes shaped the current distribution of genetic variation and are likely to have been a major influence on speciation in these fishes by restricting gene flow to individual basins (Sturmbauer et al. 1994; Johnson et al. 1996; Meyer et al. 1996; Verheyen et al. 1996). In this paper, we present a case study involving an endemic group of Lake Malawi cichlids to demonstrate how phylogenetic information might be used to test conflicting theories for speciation and the origin of convergence.

Lake Malawi contains an estimated 500 to 1,000 endemic cichlid species, and hosts more species of fish than any other lake in the world (Fryer and Iles 1972; Ribbink et al. 1983; Lewis et al. 1986). One monophyletic lineage of rock-dwelling cichlids known as "mbuna" is a diverse assemblage of more than 200 species; they are restricted to rocky habitats, have localized geographic distributions and very limited dispersal (Hert 1992), and are believed to have evolved extremely rapidly (Fryer 1959, 1977; Kornfield 1978; Ribbink et al. 1983; Lewis et al. 1986; Owen et al. 1990). The eight mbuna species examined in this paper (Table 12.1) are species belonging to the *Pseudotropheus tropheops* species complex. This complex consists of at least 38 known species and is expected to be assigned to a new genus (D. S. Lewis and P. N. Reinthal, unpubl. data). The informal species names used here are those of Ribbink et al. (1983). The status of these taxa as distinct species is supported by genetic, morphological, behavioral, colorational, and

Table 12.1. Species pairs from the *Pseudotropheus tropheops* species complex from Otter Point and Likoma Island, with a summary of the ecological strategies that differentiate sympatric species and unite pairs of allopatric species.

Species Pairs		Ecological Strategy		
Otter Point	Likoma Island	Depth	Sediment	Habitat
P. t. "orange chest"	P. t. "red cheek"	Shallow	Free	Rock surfaces
P. t. "broad mouth"	P. t. "yellow chin"	Shallow	Rich	Rock/Sand
P. cf. *gracilior*	P. t. "dark"	Deep	Medium	Rock surfaces
P. cf. *microstoma*	P. t. "membe"	Deep	Rich	Rock/Sand

ecological differences (Ribbink et al. 1983; P. N. Reinthal, unpubl. data). Although presumptive data have been presented for hybridization between separate mbuna genera elsewhere (Stauffer et al. 1996), hybridization has never been observed in the field and species translocated within Lake Malawi remain distinct and do not interbreed (Ribbink et al. 1983; P. N. Reinthal, unpubl. data).

Each of the eight species examined has a very restricted geographic distribution, limited to a small part of the shore of Lake Malawi. This kind of distribution is typical of the vast majority of endemic cichlids in all three lakes, especially among rock-dwelling taxa (Reinthal 1993). Only open-water species tend to occur throughout a whole lake basin (e.g., Fryer and Iles 1972; Coulter 1991). At each of two localities from Lake Malawi – Likoma Island in the north, and Otter Point about 200 km to the south (Figure 12.2) – four species were found to be the major representatives of the *Pseudotropheus tropheops* species complex. The four species at a given locality show some ecological differentiation. More important, however, each species corresponds to an ecologically equivalent species at the other locality. Both members of each pair have extremely similar ecological habits (Table 12.1) and morphological adaptations (P. N. Reinthal, unpubl. data). At least two evolutionary models, involving initial colonization by a generalist or specialist, could account for the presence of species with equivalent ecological strategies at separate locations (Figure 12.3).

Allopatric differentiation

Like other East African lakes, Lake Malawi experienced major and minor lake-level fluctuations during its geological history (Sturmbauer and Meyer 1992; Johnson et al. 1996; Sturmbauer et al., submitted). Most recently between 1 and 25 thousand years ago, a large drop in lake level severely disrupted mbuna habitat availability and use (Fryer 1977; Scholz and Rosendahl 1988; Owen et al. 1990). The specialist colonization model (Figure 12.3A) assumes that speciation and ecological specialization occurred in an ancestral lake community during periods of low water levels, when species were concentrated in smaller areas and competition was intensified (McKaye and Marsh 1983). During these high competition situations, habitat preferences and morphological specializations might have been established (Fryer 1959). Ecotypic and morphologically specialized species would colonize newly available appropriate habitat in different localities as lake level rose again. Parallel specialists (sister species under this model) at disjunct localities might later differentiate through breeding preferences as a result of limited dispersal capabilities (imposed by stenotopy, philopatry, and mouth brooding) and sexual selection or genetic drift in populations isolated by non-rocky habitat (Kosswig 1947; Lande 1981; Dominey 1984; McKaye et al. 1990; Ribbink 1991). Male coloration appears to be an important factor in both male-male competition for territories and female mate choice. Both of these factors would be consistent with rapid evolution under models of sexual selection. If sexual selection were to result in different color patterns in allopatric populations, ecological equivalents would be classified as distinct taxa. The same scenario would be true if coloration were to change rapidly as a result of founder events. Both sexual selection and drift are consistent with micro-allopatric models of speciation, but require that ecological specializations evolve prior to colonization events. In either case, the testable prediction of the specialist colonization model is that allopatric pairs of species with similar ecological and morphological specializations would be more closely related to each other than to other members of their respective local communities (Figure 12.3A).

Sympatric differentiation

Alternatively, the generalist colonization model (Figure 12.3B) assumes that habitat and open niches became available as the lake level rose, and that ancestral colonists of different new communities were not specialists but generalists. We define a generalist colonist as a species that utilized a wider range of habitats than is now observed for individual taxa. Ecological specialization, morphological differentiation, and speciation would then have occurred subsequent to colonization via sympatric specialization to different microhabitats. Different communities would evolve in parallel in separate localities. This model would result in the formation of similar communities of sympatric but ecologically distinct species. Under this "generalist colonization model," a species' closest relatives would be found at the same locality irrespective of their particular ecological strategy or morphological adaptations (Figure 12.3B). The ecologically equivalent species at separate geographic localities within the lake would be relatively distantly related to each other, having evolved parallel adaptations. The generalist colonization model predicts that genetic variation correlates with locality rather than morphological or ecological similarity between species and would suggest repeated convergent evolution within a single lake basin.

Striking examples of convergence between endemic species of cichlid flocks from different East African lakes have been demonstrated before based on mitochondrial DNA-based phylogenetic analyses at the between-lake level, but not before at the within-lake level (see Meyer et al. 1990; Kocher et al. 1993; Meyer 1993b). In this paper we examine the possibility that ecological and morphological specializations in mbunas have evolved sympatrically at very local scales.

Materials and methods

Populations sampled, ecological and morphological data

Eight species of the *Pseudotropheus tropheops* species group were studied at Likoma Island and Otter Point in Lake Malawi (Figure 12.2). Underwater observations and transect studies revealed that each species has distinct and restricted patterns of habitat use and resource utilization; each species had an apparent equivalent at the other locality (Table 12.1). Ecological preferences characterized also included depth, habitat type, and sediment restrictions. Stomach contents analyses demonstrated that different species at each locality eat algal species in different proportions, and thus are ecologically separated along this axis as well (P. N. Reinthal, in prep.).

Amplification and sequencing of mtDNA

Nucleotide sequences of the mitochondrial genome (mtDNA) were obtained for 29 individuals of nine species of the *P. tropheops* group. We analyzed four species from

Otter Point, four species with equivalent ecological strategies from Likoma Island, and the only *P. tropheops* species found at Chinyamwezi Rock, a location in the southeast arm of Lake Malawi (Figure 12.2). In addition, *P. tropheops* "orange chest" is also found at Zimbawe Rock, an island about 7 km from Otter Point; individuals from both sites were examined to determine geographic variation within a species. Representatives of two other mbuna genera (*Labeotropheus fuelleborni, Labidochromis* sp.) and another Malawi haplochromine (*Chilotilapia rhoadesi*) were sequenced as outgroup taxa. A 427 base-pair segment coding for part of the threonine transfer RNA gene, all of the proline tRNA gene, and the most variable part of the control region were sequenced after amplification, using the primer sequences and protocol of Kocher et al. (1989) with minor modifications (Meyer et al. 1990; Sturmbauer and Meyer 1992). Total DNAs were extracted from frozen or ethanol (75%) preserved muscle tissues following standard procedures (Kocher et al. 1989; Meyer et al. 1990).

Table 12.2. Ten variable mtDNA sites, showing genetic variability for the species examined from the *Pseudotropheus tropheops* species complex from Otter Point, Zimbawe, and Chinyamwezi in southern Lake Malawi and Likoma Island in the north. The informal species names are those of Ribbink et al. (1981); all individuals (number shown) of a species from a given site were genetically identical.

Locality	Species	No.	Site 64 / 1	73 / 2	105 / 3	185 / 4	253 / 5	268 / 6	275 / 7	276 / 8	348 / 9	356 / 10
Otter Point	*P. t.* "broad mouth"	2	T	A	C	C	C	T	C	T	T	C
	P. cf. *gracilior*	2	T	A	C	C	C	T	C	T	T	C
	P. cf. *microstoma*	2	T	G	C	C	C	T	T	T	T	T
	P. t. "orange chest"	4	T	A	C	C	C	T	C	T	T	C
Zimbawe	*P. t.* "orange chest"	3	T	A	C	C	C	T	C	T	T	T
Chinyamwezi	*P. t.* "chinyamwezi"	3	T	A	C	T	T	T	C	C	C	T
Likoma Island	*P. t.* "red cheek"	3	T	A	C	C	C	T	C	T	T	C
	P. t. "yellow chin"	3	T	A	T	T	C	T	C	C	C	T
	P. t. "dark"	3	C	A	C	T	T	C	C	C	C	T
	P. t. "membe"	4	T	A	C	T	T	T	C	C	C	T

Phylogenetic analysis

The genealogical relationships among the mtDNA sequences were estimated under maximum parsimony using PAUP 3.0s (Swofford 1991). Given the high degree of similarity among taxa, there were no ambiguities in the alignment. Exhaustive searches were conducted. Geographic and ecological characters were superimposed on the resulting tree using MacClade version 3.05 (Maddison and Maddison, 1992). The robustness of the phylogenetic estimates was tested by bootstrap analysis (Felsenstein 1985) using PAUP with 100 bootstrap replications and the support decay index (Bremer 1988).

Results

Ecology

Observations on the ecology and morphology of the *Pseudotropheus tropheops* species complex revealed differentiation among sympatric members of a community and similarities between communities at different localities (P. N. Reinthal, in prep.). From the surface to the depth at which rocky bottoms are replaced by sand (the entire depth range for the species in question), two of the four species at each locality were restricted to shallow water, and two were restricted to deeper water. At Otter Point, *P. t.* "broad mouth" and *P. t.* "orange chest" were shallow-water species, with individuals defending territories at mean depths of 2.0 ± 0.6 m (s.d.) and 2.9 ± 1.3 m, respectively. The deep-water species, *P. gracilior* and *P. microstoma*, showed mean territory depths of 5.2 ± 1.2 m and 4.5 ± 1.1 m. The maximum depth of utilizable habitat at this site is 7.6 m; no overlap in depth was observed between shallow- and deep-water species.

At Likoma Island, we also found two shallow-water and two deep-water species. The mean depth of territories for *P. t.* "yellow" and *P. t.* "red cheek" were 3.0 ± 2.3 m and 3.9 ± 2.7 m; *P. t.* "dark" and *P. t.* "membe" showed mean depths of 9.7 ± 2.3 m and 12.1 ± 0.9 m, respectively. Utilizable rocky habitat at Likoma extended from the surface to a depth of more than 30 m.

Sympatric species found at roughly the same depth differed by microhabitat utilization. In shallow water at Otter Point, *P. t.* "orange chest" defended territories centered over large boulders and rock surfaces. *P. t.* "broad mouth" was found over sand patches among rocks. In deeper water, *P. gracilior* defended large, sediment-covered rocks; *P. microstoma* occupied the rock/sand interface. Similar ecological differentiation by microhabitat occurred at Likoma Island. Shallow *P. t.* "red cheek" and deep *P. t.* "dark" were found over rock surfaces; shallow *P. t.* "yellow chin" and deep *P. t.* "membe" at the rock/sand interface.

The diets of each species reflect this habitat partitioning. The diets of species found over rocks in shallow water were dominated by attached algae (*Cladophora*, *Calothrix*), while the diets of species found over rocks in deep water were dominated by diatoms and other pelagic algae that settle on rock surfaces. Species found over rock/sand interfaces in shallow water were omnivores; similar species in deep water had the greatest dietary proportion of invertebrates.

Trophic morphological traits also showed differentiation among sympatric species. Characters such as neurocrania shape, mouth shape, tooth shape, and intestinal morphology can all be used as predictors of ecological strategy. There are many apparent convergences in morphology between ecologically similar species at the two different sites, but these are beyond the scope of the current paper (P. N. Reinthal, in prep.).

Molecular systematics

We found a remarkably high level of genetic similarity within the *Pseudotropheus tropheops* complex in Lake Malawi. In nine *P. tropheops* species examined, only 10 of 427 bases exhibited base substitutions. All of the observed substitutions are transitions

(Table 12.2), further supporting the notion that these species are very closely related (Brown et al. 1982). Only one transition involves a purine; pyrimindine transitions tend to be more frequent than purine transitions in fishes (Meyer 1993a).

Five substitutions were phylogenetically informative, and five were autapomorphic (sites 1, 2, 3, 6, and 7) (Table 12.2). Three of the five informative substitutions (sites 4, 8, and 9) supported the hypothesis that all species from a locality are more closely related to each other. A mutation at site 5 united three species from Likoma Island; another mutation at site 10 was shared by three species from Otter Point. No intraspecific variation was found in any species. It is interesting that retention of ancestral polymorphisms of mitochondrial DNA has been proposed for Malawi cichlids of the *Pseudotropheus zebra* species group (Moran and Kornfield 1993, 1995). However, if the results presented here were due to ancestral polymorphism, the polymorphism would be expected to be observed in at least one of the two localities. Given the number of individuals examined here (29), the probability of the data showing the biogeographic pattern observed here would be exceedingly low. This is not to say that the polymorphism did not become alternatively fixed at the two distinct localities or there was a possibilty of mtDNA capture. Until the presence of a polymorphism is detected in these taxa or variation in nuclear DNA is examined, the phylogenetic signal of the data presented here should not be ignored.

Conspecifics of *P. tropheops* "orange chest" from Zimbawe Island and Otter Point (Figure 12.2) differed only at site 10. The exceedingly small amount of genetic variation within and between communities supports the idea of an exceptionally recent origin of the Lake Malawi species flock, and agrees with other estimates of extremely limited genetic variation found in species of the Malawi and Victoria species flocks and in some Neotropical cichlids (Kornfield 1978; Owen et al. 1990; Meyer et al. 1990; Sturmbauer and Meyer 1992; Moran and Kornfield 1993, 1995; Sültmann et al. 1995; but see Klein et al. 1993 and Ono et al. 1993 for MHC variation in Lake Malawi cichlids). Because mtDNA evolves 5 to 10 times faster than a typical single-copy nuclear-protein-coding gene and the portion sequenced is the fastest evolving region of the entire mitochondrial genome (Meyer 1993a) the amount of DNA variation detected in this study would be unlikely to be increased by further sequencing of other portions of mtDNA.

Even though Otter Point and Likoma Island species differed from each other by only about 0.7% sequence divergence, we are able to use mtDNA sequences to assess the evolutionary relationships of these taxa. A parsimony analysis without outgroups results in a single most parsimonious tree (CI = 1.0, tree length = 10 steps) and reveals that all Otter Point species are genetically more closely related to each other than they are to any Likoma Island species and vice versa. This result supports the generalist-colonization model (Figure 12.3B) and documents the parallel evolution of extensive ecological and morphological similarity within a single species flock. A bootstrap analysis with no outgroup supports the Likoma-Otter Point distinction at levels over 95%. The central branch of an unrooted tree that differentiates the species from the two localities has a decay index of three. If the tree is constrained to support the specialist-colonization model, our data require a tree 23 steps in length. This would represent an additional 13 steps from the most parsimonious tree.

Genetic separation of Otter Point and Likoma species is supported by parsimony using representatives of two other mbuna genera (*Labeotropheus fuelleborni* or *Labidochromis* sp.) or another Malawi haplochromine cichlid species (*Chilotilapia rhoadesii*) as outgroup taxa. For example, with *Chilotilapia rhoadesii* as outgroup, we find three equally parsimonious trees (length = 13 steps; CI = 0.85; RI = 0.87). The strict consensus tree separates the Likoma species (with *Pseudotropheus tropheops* "red cheek," *P. t.* "dark" and *P. t.* "membe" as an unresolved trichotomy) from the Otter Point species (with *P. t.* "orange chest," *P. t.* "broad mouth," and *P. gracilior* as an unresolved trichotomy); it never supports the specialist-colonization model by placing ecologically equivalent species as sister taxa.

Including the Zimbawe *Pseudotropheops tropheops* "orange chest" population and *P. t. íchinyamwezíí* in a parsimony analysis, using PAUP with *Chilotilapia rhoadesii* as an outgroup, results in nine most parsimonious trees (length = 14 steps; CI = 0.79; RI = 0.84). The strict consensus tree unites all four species from Likoma Island as sister taxa, but leaves those from Otter Point, Zimbawe, and Chinyamwezi as unresolved basal taxa. Three of the species – *P. t.* "red cheek," *P. t.* "dark," and *P. t.* "membe" – form an unresolved trichotomy sister to *P. t.* "yellow chin." These results also support the generalist-colonization model.

Discussion

Generalist-colonization model supported

The genetic data clearly support the model of *in situ* parallel evolution at separate geographic localities (Figure 12.3B) because we found that sympatric species are more closely related to each other than they are to their allopatric ecologically equivalent species. The most parsimonious explanation of these results is that colonization occurred, possibly by a generalist, at different parts of the lake and preceded ecological specialization and speciation events. These results make it unlikely that ecological specialization and speciation predated colonization, and instead support the hypothesis that speciation and morphological diversification occurred sympatrically on repeated, independent occasions at geographically separated sites in Lake Malawi. Interestingly, at localities where only a single *P. tropheops* species is found (e.g., Chinyamwezi, Figure 12.2), it expands its ecological range to encompass habitat used by all four species at Likoma Island or Otter Point (P. N. Reinthal, unpubl. data).

This finding tends to implicate ecological specialization as a driving force behind speciation and morphological differentiation at any one locality. However, divergence of trophic morphology need not necessarily precede or initiate speciation. If, for example, differences in coloration arose through sexual selection it could reproductively isolate populations in different habitats (Lande 1981; Dominey 1984; McKaye, Louda, and Stauffer 1990; McKaye 1991; Ribbink 1991; McElroy and Kornfield 1991), so that morphological and ecological divergence could then evolve secondarily (Reinthal 1990).

Parallel evolution of morphological traits within a single lake basin

Ecologically equivalent species from Likoma Island and Otter Point are very similar morphologically; there appears to be as much morphological variation within a species of *Pseudotropheops tropheops* at a given site as there is between it and the corresponding species from the other site. Morphometric analyses of neurocrania, body shapes, mouth shapes, and intestines revealed features that are more similar between ecologically equivalent species from different localities than between sympatrically occurring species (P. N. Reinthal, unpubl. data). Because ecologically equivalent species are not closely related genetically, it appears that morphological parallelism and convergence are important in cichlid evolution. Such homoplasy has previously been found to obstruct phylogenetic inference based on morphology at broader taxonomic scales (Meyer et al. 1990; Kocher et al. 1993). Convergence involving cichlids from different lakes has been identified using molecular techniques (Meyer et al. 1990; Meyer et al. 1991; Kocher et al. 1993; Meyer 1993b), including instances where it had been incorrectly interpreted as an indication of common recent ancestry (Greenwood 1983). This study presents the first clear example of within-lake parallel evolution in cichlid morphology and ecology, and suggests that speciation and adaptive radiation may have occurred sympatrically, at least in some instances.

Trophic morphology and phylogeny

Much of the morphological variation found in the family Cichlidae is related to trophic structures (particularly the oral jaw) which appear subject to strong selection pressures and convergence, so care must be taken in using trophic morphology as a basis for reconstructing cichlid phylogeny. There is a tendency for autapomorphies in trophic structure to mask synapomorphic variation in such traits as well. Previous work on phenotypic plasticity (e.g., Meyer 1987; Wimberger 1991) and trophic polymorphisms (Meyer 1990) demonstrated that trophic structures are highly labile and often easily influenced by environmental factors. Molecular phylogenetic analyses have also revealed that some cichlid species of Lake Tanganyika show large, discontinuous divergence in trophic structures even if they are closely related to each other (Sturmbauer and Meyer 1992, 1993; Verheyen et al. 1996).

Models of speciation

Our findings suggest that a significant amount of local differentiation takes place *in situ*. New species could potentially arise via a sympatric model in which differential habitat utilization and small genetic differences would promote reproductive isolation (Kondrashov and Mina 1986; Maynard Smith 1986). Sympatric speciation has been suggested as being responsible for the co-occurrence of a small number of cichlid species in two crater lakes in Cameroon (Schliewen et al. 1994). Alternatively, our results are also consistent with a micro-allopatric speciation model, in which a relatively small area of incompatible habitat (e.g., open water, or sand between rocks) forms an effective barrier to dispersal (Fryer 1977; Sturmbauer and Meyer 1992; Moran and Kornfield 1995). Species isolated over short distances could differentiate rapidly. Short-term fluctuations in lake level would re-unite differentiated taxa (i.e., sister species) and competition for shared, preferred resources would result in divergent selection for use of different microhabitats. Different patterns of microhabitat (e.g., depth) and resource utilization might lead to morphological differentiation between reproductively isolated taxa (as a result of differing selection pressures, phenotypic plasticity, or a combination of both) and reinforce reproductive isolation. The resulting distribution of genetic variation would have all members of a particular location more closely related to each other than to ecological equivalents from other more distant locations.

Several different speciation models have been suggested to account for the evolutionary origin of the adaptive radiation of cichlid fishes (Kosswig 1947; Fryer 1959; Fryer and Iles 1972; Fryer 1977; Kornfield 1978; Ribbink et al. 1983; Lewis et al. 1986; Meyer et al. 1990; Owen et al. 1990; Dominey 1991; Greenwood 1991; Sturmbauer and Meyer 1992; Meyer 1993b). Our data support the notion of true intralacustrine speciation. Speciation occurred (and is occurring) at many sites within Lake Malawi simultaneously and in parallel, probably through the mechanism of sympatric or micro-allopatric speciation. The availability of new habitat or open niches for these fishes is largely influenced by physical factors such as lake-level fluctuation (Ribbink et al. 1983; Scholz and Rosendahl 1988; Owen et al. 1990; Sturmbauer and Meyer 1992; Meyer et al. 1996; Verheyen et al. submitted; Sturmbauer et al. submitted). These abiotic changes contribute to intralacustrine speciation since gene flow seems to be extremely restricted and only newly available habitats allow for the expansion of species (Sturmbauer and Meyer 1992; Bowers et al. 1994; Moran and Kornfield 1995). We believe that these factors, in combination with tight ecological specialization and the combined synergistic effects of mating systems and sexual selection, are largely responsible for the evolutionary origin of the extraordinary species assemblages of cichlid fishes in Lake Malawi.

Conclusions

The remarkable adaptive radiations of cichlid fishes of the African Great Lakes are well known for their "explosive" rates of speciation and extensive ecological and morphological differentiation. The debate about which mechanisms of speciation underlie the formation of these adaptive radiations has not, however, been settled. The species flocks of cichlids in each lake provide model systems for testing different speciation scenarios for the origin of these adaptive radiations. Four pairs of cichlid species of the *Pseudotropheus tropheops* complex at two localities (separated by roughly 200 km) in Lake Malawi were found to have equivalent ecological strategies and morphological adaptations. To distinguish between two alternative models to account for this pattern, a 427-bp stretch of the fastest evolving portion of the mitochondrial genome was sequenced from 31 individuals of 12 species. Hypotheses based on an assumption of an initial colonization of each site by (i) several lineages of specialists or (ii) a generalist that then diversified *in situ* were tested by combining quantitative data on ecology and morphology with a molecular phylogeny based on DNA sequences. Cladistic analyses of the sequence data indicated that all species from the same locality – irrespective of morphological or ecological specialization – share an immediate common ancestor, and thus evolved in a true intralacustrine manner. These results suggest that species evolved *in situ*, via sympatric or microallopatric speciation, and that convergent morphological diversification and ecological specialization evolved repeatedly and in parallel following colonizations of geographically separate areas within the basin of Lake Malawi.

Acknowledgments

We thank J. Coyne, J. Endler, D. Futuyma, E. Mayr, L. McDade, G. Orti, M. Stiassny, C. Sturmbauer, P. Wimberger, and G. Kling for insightful comments on earlier drafts and B. Curtsinger, J. Whitman, and R. Jennings for help during fieldwork in Malawi. We thank the Government of Malawi for permission to conduct the fieldwork and I. Kornfield and P. Moran for some purified mitochondrial DNAs. Part of this work was done while PNR was a Kalbfleisch Research Fellow at the American Museum of Natural History and AM an Alfred P. Sloan Postdoctoral Fellow in the laboratory of the late Allan C. Wilson at the University of California, Berkeley. Support came from the American Museum of Natural History and the National Geographic Society to PNR. Support came from the National Science Foundation (BSR-9107838) and the Alfred P. Sloan Foundation to AM and the National Science Foundation and the National Institutes of Health to the late Allan C. Wilson.

References

Bowers, N., Stauffer, J. R., and Kocher, T. D. 1994. Intra- and interspecific mitochondrial DNA sequence variation within two species of rock-dwelling cichlids (Teleostei: Cichlidae) from Lake Malawi, Africa. Molecular Phylogenetics and Evolution 3:75–82.

Bremer, K. 1988. The limits of amino acid sequence data in angiosperm phylogenetic reconstruction. Evolution 42:795–803.

Brown, W. M, Prager E. M., and Wang, A. 1982. Mitochondrial DNA sequences of primates: tempo and mode of evolution. Journal of Molecular Evolution 18:225–239.

Coulter, G. W. 1991. Zoogeography, affinities and evolution with special regard to the fishes. Pp. 275–305 in *Lake Tanganyika and its life*, G.W. Coulter, ed. London, England: Oxford University Press.

Coyne, J. A. 1992. Genetics and speciation. Nature 355:511–515.

Dominey, W. J. 1984. Effects of sexual selection and life history on speciation: species flocks in African cichlids and Hawaiin *Drosophila*. Pp. 231–254 in *Evolution of fish species flocks*, A. A. Echelle and I. Kornfield, eds. Orono, ME: University of Maine Press.

Felsenstein, J. 1985. Confidence limits on phylogenies. Evolution 39:783–791.

Fryer, G. 1959. The trophic interrelationships and ecology of some littoral communities of Lake Nyasa with especial reference to the fishes, and a discussion of the evolution of a group of rock-frequenting Cichlidae. Proceedings of the Zoological Society of London 132:153–281.

Fryer, G. 1977. Evolution of species flocks of cichlid fishes in African lakes. Zeitschrift für Zoologische Systematik und Evolution-forschungen 15:141–165.

Fryer. G., and Iles, T. D. 1972. The cichlid fishes of the great lakes of Africa. Edinburgh, Scotland: Oliver and Boyd.

Greenwood, P. H. 1983. On *Macropleurodus, Chilotilapia* (Teleostei, Cichlidae) and the interrelationships of African cichlid species flocks. Bulletin of the British Museum of Natural History (Zoology) 45:209–231.

Greenwood, P. H. 1984. African cichlids and evolutionary theories. Pp. 141–154 in *Evolution of fish species flocks*, A. A. Echelle and I. Kornfield, eds. Orono, ME: University of Maine Press.

Greenwood, P. H. 1991. Speciation. Pp. 86–102 in *Cichlid fishes: behaviour, ecology and evolution*, M. H. Keenleyside, ed. London, England: Chapman and Hall.

Harrison, R. G. 1991. Molecular changes at speciation. Annual Review of Ecology and Systematics 22:281–308.

Hert, E. 1992. Homing and home-site fidelity in rock-dwelling cichlids (Pisces: Teleostei) of Lake Malawi, Africa. Environmental Biology of Fishes 33:229–237.

Klein, D., Ono, H., O'Huigin, C., Vincek, V., Goldschmidt, T., and Klein, J. 1993. Extensive MHC variability in cichlid fishes of Lake Malawi. Nature 364:330–334.

Kocher, T. D., Conroy, J. A., McKaye, K. R., and Stauffer, J. R. 1993. Similar morphologies of cichlid fish in Lakes Tanganyika and Malawi are due to convergence. Molecular Phylogenetics and Evolution 2:158–165.

Kocher, T. D., Thomas, W. K., Meyer, A., Edwards, S.V., Pääbo, S., Villablanca, F. X., and Wilson, A. C. 1989. Dynamics of mitochondrial DNA evolution: amplification and sequencing with conserved primers. Proceedings of the National Academy of Sciences, USA 86:6196–6200.

Kondrashov A. S., and Mina, M. V. 1986. Sympatric speciation: when is it possible? Biological Journal of the Linnean Society 27:201–223.

Kornfield, I. 1978. Evidence for rapid speciation in African cichlid fishes. Experientia 34:335–336.

Kosswig, C. 1947. Selective mating as a factor for speciation in cichlid fish of East African Lakes. Nature 159:604–605.

Lande, R. 1981. The minimum number of genes contributing to quantitative variation between and within populations. Proceedings of the National Academy of Sciences, USA 78:3721–3725.

Lewis, D. S. C., Reinthal, P. N., Trendall, J. 1986. *A guide to the fishes of Lake Malawi national park*. Gland, Switzerland: World Wildlife Fund.

Lynch, J.D. 1989. The gauge of speciation: on the frequencies of modes of speciation. Pp. 527–553 in *Speciation and its consequences*, D. Otte and J. A. Endler, eds. Sunderland, MA: Sinauer Associates.

Maddison, W. P,., and Maddison, D. R. 1992. *MacClade: Analysis of phylogeny and character evolution, version 3.05.* Sunderland, MA: Sinauer Associates.

Mayden, R. L. 1992. *Systematics, historical ecology and North American freshwater fishes*. Palo Alto, CA: Stanford University Press.

Maynard Smith, J. 1966. *Sympatric speciation*. American Naturalist 100:637–650.

Mayr, E. 1963. *Animal species and evolution*. Cambridge, MA: Harvard University Press

Mayr, E. 1984. Evolution of fish species flocks: a commentary. Pp. 3–11 in *Evolution of fish species flocks*, A. A. Echelle and I. Kornfield, eds. Orono, ME: University of Maine Press.

McElroy, D. M., and Kornfield, I. 1991. Coloration in African cichlids: diversity and constraints in Lake Malawi endemics. Netherlands Journal of Zoology 41:250–268.

McKaye, K.R. 1991. Sexual selection and the evolution of the cichlid fishes of Lake Malawi. Pp. 241–257 in *Cichlid fishes: behaviour, ecology and evolution*, M. H. Keenleyside, ed. London, England: Chapman and Hall.

McKaye, K. R., Louda, S., and Stauffer, J. 1990. Bower size and male reproductive success in a cichlid fish lek. American Naturalist 135:597–613.

McKaye, K. R., and Marsh, A. C. 1983. Food switching by two specialized algae-scraping cichlid fishes in Lake Malawi, Africa. Oecologia 56:245–248.

Meyer, A. 1987. Phenotypic plasticity and heterochrony in *Cichlasoma managuense* (Pisces, Cichlidae) and their implications for speciation in cichlid fishes. Evolution 41:1357–1369.

Meyer, A. 1990. Ecological and evolutionary aspects of the trophic polymorphism in *Cichlasoma citrinellum* (Pisces: Cichlidae). Biological Journal of the Linnean Society 39:279–299.

Meyer, A. 1993a. Evolution of mitochondrial DNA in fishes. Pp. 1–38 in *Biochemistry and molecular biology of fishes, vol. 2*, P.W. Hochachka and T. P. Mommsen, eds. Amsterdam, The Netherlands: Elsevier.

Meyer, A. 1993b. Phylogenetic relationships and evolutionary processes in East African cichlid fishes. Trends in Ecology and Evolution 8:279–284.

Meyer, A., Knowles, L., and Verheyen, E. 1997. Widespread geographic distribution of mitochondrial haplotypes in Lake Tanganyika rock-dwelling cichlid fishes. Molecular Ecology (in press).

Meyer, A., Kocher, T. D., Basasibwaki, P., and Wilson, A. C. 1990. Monophyletic origin of Lake Victoria cichlid fishes suggested by mitochondrial DNA sequences. Nature 347:550–553.

Meyer, A., Kocher, T. D., and Wilson, A. C. 1991. African fishes. Nature 351:467–468.

Meyer, A., Montero, C., and Spreinat, A. 1994. Evolutionary history of the cichlid fish species flocks of the East African great lakes inferred from molecular phylogenetic data. Advances in Limnology 44:409–425.

Moran, P., and Kornfield, I. 1993. Retention of an ancestral polymorphism in the mbuna species flock (Pisces: Cichlidae) of Lake Malawi. Molecular Biology and Evolution 10:1015–1029.

Moran, P., and Kornfield, I. 1995. Were population bottlenecks associated with the radiation of the mbuna species flock (Teleostei: Cichlidae) of Lake Malawi? Molecular Biology and Evolution 12:1085–1093.

Moran, P., Kornfield, I., and Reinthal, P. 1993. Molecular systematics and radiation of the haplochromine cichlids (Teleostei: Perciformes) of Lake Malawi. Copeia 1994:274–288.

Nishida, M. 1991. Lake Tanganyika as an evolutionary reservoir of old lineages of East African cichlid fishes: inferences from allozyme data. Experientia 47:974–979.

Ono, H., O'Huigin, C., Tichy, H., and Klein, J. 1993. Major-histocompatibility-complex variation in two species of cichlid fishes from Lake Malawi. Molecular Biology and Evolution 10:1060–1072.

Otte, D., and Endler, J. A. (eds.). 1989. *Speciation and its consequences*. Sunderland, MA: Sinauer Associates.

Owen, R. B., Crossley, R., Johnson, T. C., Tweddle, D., Kornfield, I., Davidson, S., Eccles, D. H., and Engstrom, D. E. 1990. Major low levels of Lake Malawi and their implications for speciation rates in cichlid fishes. Proceedings of the Royal Society of London, Series B 240:519–553.

Reinthal, P. N. 1990. Morphological analysis of the neurocranium of a group of rock-dwelling cichlid fishes (Cichlidae: Perciformes) from Lake Malawi, Africa. Zoological Journal of the Linnean Society 98:123–139.

Reinthal, P. N. 1993. Evaluating biodiversity and conserving Lake Malawi's fish fauna. Conservation Biology 7:211–219.

Ribbink, A. J. 1991. Distribution and ecology of the cichlids of the African Great Lakes. Pp. 36–59 in *Cichlid fishes: behaviour, ecology and evolution*, M. H. Keenleyside, ed. London, England: Chapman and Hall.

Ribbink, A. J., Marsh, B. A., Marsh, A. C., Ribbink, A. C., and Sharp, B. J. 1983. A preliminary survey of the cichlid fishes of rocky habitats in Lake Malawi. South African Journal of Zoology 18:149–310.

Schliewen, U. K., Tautz, D., and Pääbo, S. 1994. Sympatric speciation suggested by monophyly of crater lake cichlids. Nature 368:629–632.

Scholz, C. A., and Rosendahl, B. R. 1988. Low lake stands in Lake Malawi and Tanganyika, East Africa, delineated with multifold seismic data. Science 240:1645–1648.

Slatkin, M. 1987. Gene flow and the geographic structure of natural populations. Science 236:787–792.

Stauffer, J. R., Jr., Bowers, N. J., Kocher, T. D., and McKaye, K. R. 1996. Evidence of hybridization between *Cynotilapia afra* and *Pseudotropheus zebra* (Teleostei: Cichlidae) following an intralacustrine translocation in Lake Malawi. Copeia 1996:203–207.

Stiassny, M. L. J. 1981. Phylogenetic versus convergent relationships between piscivorous cichlid fishes from Lake Malawi and Tanganyika. Bulletin of the British Museum of Natural History (Zoology) 40:67–101.

Stiassny, M. L. J. 1991. Phylogenetic intrarelationships of the family Cichlidae: an overview. Pp. 1–35 in *Cichlid fishes: behaviour, ecology and evolution*, M. H. Keenleyside, ed. London, England: Chapman and Hall.

Sturmbauer, C., and Meyer, A. 1992. Genetic divergence, speciation and morphological stasis in a lineage of African cichlid fishes. Nature 358:578–581.

Sturmbauer, C., and Meyer, A. 1993. Mitochondrial phylogeny of the endemic mouthbrooding lineages of cichlid fishes from Lake Tanganyika in Eastern Africa. Molecular Biology and Evolution 10:751–768.

Sturmbauer, C., Verheyen, E., and Meyer, A. 1994. Mitochondrial phylogeny of the Lamprologini, the major substrate spawning lineage of cichlid fishes from Lake Tanganyika in Eastern Africa. Molecular Biology and Evolution 11:691–703.

Sültmann, H., Mayer, W. E., Figueroa, F., Tichy, H., and Klein, J. 1995. Phylogenetic analysis of cichlid fishes using nuclear DNA markers. Molecular Biology and Evolution 12:1033–1047.

Swofford, D. 1991. *PAUP: phylogenetic analyses using parsimony, vers. 3.0s*, Champaign, IL: Illinois Natural History Survey.

Verheyen, E., Rüber, L., Snoeks, J., and Meyer, A. 1996. Mitochondrial phylogeography of rock-dwelling cichlid fishes reveals evolutionary influence of historical lake level fluctuations of Lake Tanganyika, Africa. Philosophical Transactions of the Royal Society of London, Series B 351:797–805.

Wimberger, P. 1991. Plasticity of jaw and skull morphology in the Neotropical cichlid *Geophagus brasiliensis* and *G. steindachneri*. Evolution 45:1545–1564.

13 Rapid Radiation Due to a Key Innovation in Columbines (Ranunculaceae: *Aquilegia*)

Scott A. Hodges

Adaptive radiations can occur through two primary mechanisms. First, they may be triggered through extrinsic causes due to new environmental circumstances. The chance dispersal of taxa to newly formed islands or lake systems has resulted in many groups that have speciated to fill a wide diversity of ecological roles (e.g., Darwin's Finches [Grant 1986], the Hawaiian silversword alliance [see Chapter 3; Baldwin et al. 1991], and the African cichlid fishes [see Chapter 12; Meyer et al. 1990]). Second, radiations may occur due to intrinsic characters of organisms; the evolution of a key innovation can allow a taxon to utilize existing niche space in a novel manner. For example, the evolution of flight in birds has been proposed as a key innovation that provided the means to occupy many niches previously unavailable to the direct ancestors of the birds.

Unfortunately, it has been difficult in the past to directly test these hypotheses on the causes of adaptive radiations. For instance, on island chains there are some lineages that have produced species with a startling array of ecological associations, but others do not show such diversity. The simple explanation to these divergent patterns is that those taxa that first arrived on the islands were able to speciate and utilize the open niches while those taxa that arrived subsequent to the occupation of the niches could not radiate. However, we often lack information on the timing of island colonization. Thus explicit tests of extrinsic causes of species diversification can be difficult.

Similarly, there have been strong arguments against hypotheses of key innovations as causes of adaptive radiations. In particular, hypotheses of key innovations that result in increased rates of species diversification have been challenged (Cracraft 1990). It is often difficult to understand how a proposed key innovation actually affects diversification rates. Increased diversification results from either an increase in speciation rates and/or a decrease in extinction rates. Thus it is imperative to understand how a proposed key innovation can affect these processes. Ideally, there will be experimental studies to test how a proposed key innovation can directly affect processes thought to be important in speciation and extinction. In addition, statistical tests of changes in diversification rates are needed. In the early 1970's Raup et al. (1973), using computer simulations, showed that patterns of cladogenesis that would traditionally have been attributed to key innovations match patterns that can be generated through random processes. Thus it is necessary to test patterns of diversity against random models to determine if diversification rates have actually changed.

Phylogenetic analyses are essential for testing hypotheses of adaptive radiations. Such analyses can test for monophyly – a basic assumption of adaptive radiation. For example, the spider genus *Tetragnatha* in Hawaii was thought to be an example of a single adaptive radiation. However, Gillespie et al. (1994), using both morphological and mitochondrial sequence data, found this group to be polyphyletic in origin. In contrast, monophyly was established for the Hawaiian silversword alliance (Baldwin et al. 1991) and the African cichlid fishes (Meyer et al. 1990). In addition, phylogenetic analyses can aid in the identification of sister taxa. Because sister taxa form a monophyletic group, they are by definition of equal age. Therefore, disparities in the species diversity between sister taxa can indicate differences in diversification rates.

Here I show how molecular phylogenetic analyses can provide data to identify a radiation, to delineate the timing of the evolution of proposed key innovations, and to identify the sister taxa that are imperative in analyses of diversification. Specifically, I show that the evolution of the columbines (Ranunculaceae: *Aquilegia*) probably involved a rapid radiation due to the key innovation of floral nectar spurs. I further show how variation in nectar spurs can adapt species to pollination by different pollinators and, therefore, how floral spurs may affect processes important in species diversification. Finally, using phylogenetic analyses, I show that the evolution of floral spurs is temporally linked to the radiation of *Aquilegia*, and that multiple, independent origins of floral spurs throughout the angiosperms are correlated with increased rates of species diversification.

Rapid radiation in *Aquilegia*

Aquilegia has a high diversity of floral morphologies and color that correspond to different pollination syndromes. In addition, columbines can be found in diverse habitats from the high alpine to desert springs. Despite the high morphological and ecological diversity in the genus, the species are largely interfertile suggesting that they may be of recent origin (Clausen et al. 1945). However, other authors (Stebbins 1950; Prazmo 1965; Grant 1994a) have suggested that the genus is of at least mid-Tertiary age, based on the widespread distribution of the columbines in temperate regions of the Northern Hemisphere.

While rapid adaptive radiations are generally associated with small geographic regions, phylogenetic analyses can establish that widespread taxa have also radiated rapidly (Hodges and Arnold 1994a). A recent divergence among taxa will result in lower levels of sequence divergence than more ancient radiations. Among species of *Aquilegia*, very low levels of sequence divergence were found in comparison with the levels of sequence divergence in the closely related genera, *Isopyrum* and *Thalictrum* (Hodges and Arnold 1994a). Furthermore, this pattern of comparatively low sequence variation in *Aquilegia* was found in both a nuclear DNA region (the ITS region of the rDNA) and a chloroplast DNA region (the spacer between the *atp*B and *rbc*L genes) (Figure 13.1). The finding of similar patterns in both a nuclear and chloroplast DNA region suggests that this pattern is not simply an anomaly of a particular gene region but is indicative of a genome-wide phenomenon (Avise 1994).

This pattern of low nucleotide variation among columbines could be explained by either a rapid radiation or a drastic reduction in the rate of nucleotide substitutions in the *Aquilegia* clade. However, it seems unlikely that this pattern is due to a reduction in nucleotide substitutions because the overall branch lengths for the columbine clade are not shorter than the branch lengths for their close relatives (Figure 13.1; Hodges and Arnold 1994a). Thus, it is likely that the columbines have experienced a rapid and recent radiation.

The radiation of the columbines is likely to have occurred via a key innovation rather than via invasion of a newly formed habitat with few competing species. The columbines are distributed throughout the mountainous regions of the Northern Hemisphere as are its close relatives *Isopyrum,* and *Thalictrum*. As such, the columbines do not occupy a geographic range that is substantially different from their close

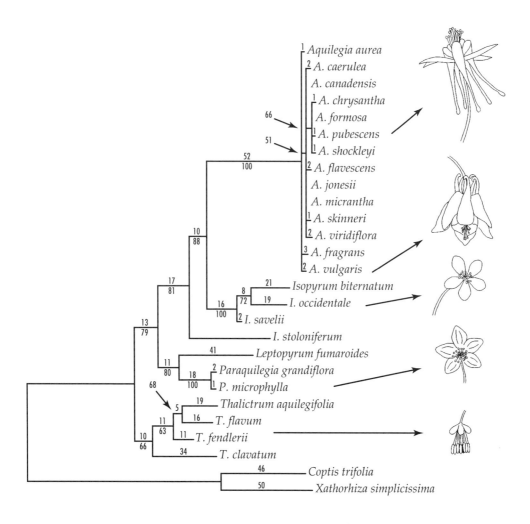

Figure 13.1. Phylogram from maximum parsimony analysis (Swofford 1993) of ITS and cpDNA sequence data for *Aquilegia* and its close relatives (truncated to show only branches with bootstrap values over 50%). A single most parsimonious tree resulted from the analysis; percent bootstrap values are indicated below branches (or by arrows) and inferred branch lengths are given above branches. Line drawings (not to scale) depict the floral form of some species. Note the specialized nectar spurs of *Aquilegia* and the open radiate forms of its relatives. Redrawn from Hodges and Arnold (1994a).

relatives that show no signs of having experienced a rapid radiation (Figure 13.1; Hodges and Arnold 1994a). Therefore, it does not appear that *Aquilegia* has dispersed into a new habitat that its close relatives were unable to invade.

The columbines are clearly different from their close relatives in their modes of pollination. Species of *Aquilegia* differ strikingly in their floral morphologies and associated pollinators while their close relatives have simple radiate flowers that do not restrict the types of floral visitors (or are even wind pollinated as in many *Thalictrum*) (Figures 13.1, 13.2). Thus Hodges and Arnold (1994a) proposed that the evolution of nectar spurs was a key innovation for this group that allowed specialization to different pollinators and promoted diversification.

Adaptive significance of nectar spurs

While the pattern of species diversification as revealed through the molecular phylogenetic analysis outline above suggests that the columbines experienced a rapid radiation due to a key innovation, identifying and testing a specific key innovation hypothesis is much more difficult. Imperative to testing a key innovation hypothesis is determining how the key innovation may affect processes important in species diversification (Cracraft 1990; Skelton 1993; Heard and Hauser 1995; Hodges and Arnold 1995). Increased species diversification can result from increased speciation and/or decreased extinction. By explicitly stating how a proposed key innovation may affect processes of speciation or extinction, we can test these hypotheses directly.

Hodges and Arnold (1994a) suggested that the evolution of nectar spurs could have been the key innovation that spurred the columbine radiation. An underlying assumption of most species concepts is the necessity for reproductive isolation (Dobzhansky 1937; Mayr 1942; Grant 1963). Therefore, characters that can promote reproductive isolation may increase speciation rates and thus species diversification. Nectar spurs are likely to facilitate reproductive isolation between species because variation in nectar spur length, shape, color and orientation can facilitate the visitation of some types of pollinators while restricting the visitation of others (Grant 1952). As a result, if populations of a species with nectar spurs become specialized for different pollinators while in allopatry, then even if these populations come into secondary contact the floral morphology differences would provide a prezygotic reproductive isolating mechanism that would prevent the taxa from merging (Grant 1952; Hodges and Arnold 1994a, 1995). Thus, in contrast to non-spurred relatives, taxa with spurs can become specialized on different pollinator types which increases reproductive isolation and possibly speciation.

There are several studies that support the hypothesis that nectar spur morphology influences reproductive success and isolation among species. For instance, in the orchid genus *Plantanthera*, Nilsson (1988) showed that experimentally reducing the length of nectar spurs had a pronounced effect on both the insertion and removal of pollinia by pollinators. Nilsson (1988) further showed that individual plants with short spurs had significantly lower fruit set than longer spurred individuals. Thus, spur length is intimately tied to reproductive success among individuals of *Plantanthera*.

The length of nectar spurs among populations of a species and among species of a single genus are correlated with pollinator morphology. In *Aconitum columbianum* (Ranunculaceae), populations vary nearly three-fold in the length of the nectar spurs and this variation is significantly correlated with the tongue lengths of the bee species that pollinate plants in these populations (Brink 1980). Spur morphology has also been shown to be correlated with pollinator morphology both among populations within a species and among species of *Diascia* (Scrophulariaceae) (Steiner and Whitehead 1990, 1991). In *Diascia*, there are two spurs per flower and these spurs contain oils that members of the wasp genus *Rediviva* collect by inserting their forelegs into the spurs (Steiner and Whitehead 1990, 1991). The species of *Diascia* vary in spur length and when Steiner and Whitehead collected *Rediviva* on these different species, they found that there was a very strong correlation between spur length and the foreleg length of *Rediviva*. Again, these results suggest that spur morphology is intimately associated with pollinator type and therefore with reproductive isolation.

There also is evidence that spur morphology and color are important in reproductive isolation in *Aquilegia*. Miller (1981) established that variation in both length and color of nectar spur among populations of *Aquilegia caerulea* was correlated with the types of pollinators (bees versus hawkmoths) present. In mixed populations that had plants that produced either blue or white flowers, Miller (1981) showed that in years of high bumblebee abundance, blue-flowered individuals had higher seed set than white-flowered individuals and the reverse was true in years of high hawkmoth abundance. In addition, Miller suggested that the unusually long spurs of *A. caerulea* var. *pinetorum* are a result of selection due to hawkmoths with longer tongues rather than to the hawkmoths that occur in the rest of the range of *A. caerulea*. These patterns suggest that pollinator type can strongly influence spur morphology and color even within a single species of *Aquilegia*.

The features that affect reproductive isolation between the two columbine species, *Aquilegia formosa* and *A. pubescens*, have been used as a paradigm for the ability of floral characters to confer reproductive isolation (Grant 1952). In a classic study, Grant (1952) suggested that differences in floral morphology and pollinators were a major factor in keeping *A. formosa* and *A. pubescens* reproductively isolated. *A. formosa* has short, pendent, red-spurred flowers predominately visited by hummingbirds, while *A. pubescens* has long, upright, white- or yellow-spurred flowers predominately visited by hawkmoths (Grant 1952; Grant 1993; Grant 1994b). Grant suggested that these differences in floral morphology and pollinator preference were important in maintaining these columbine species as different taxa. However, Chase and Raven (1975) challenged this conclusion and suggested that ecological factors, particularly soil type, were the major factors maintaining the species and that the differences in floral morphology had little, if anything, to do with reproductive isolation.

As a test of whether floral morphology – in particular nectar spur morphology – affects reproductive isolation between *A. formosa* and *A. pubescens*, Hodges and Arnold (1994b) measured the transition of species-specific characters across hybrid zones between these two species. Alternative characters that are selected for on opposite sides of a hybrid zone were implicated in causing reproductive isolation.

Selection of alternative characters on opposite sides of a hybrid zone should result in sharp transitions while traits that are neutral would be expected to show a decay in clinal variation over time (Endler 1986; Barton and Hewitt 1989). Length and orientation of nectar spurs have very sharp transitions across both an elevational and a habitat transect between *Aquilegia formosa* and *A. pubescens* (Figure 13.2). The sharp transition of spur characters is in contrast with the transition of random regions of DNA (RAPDs, randomly amplified polymorphic DNAs) that are presumably neutral with respect to plant fitness (Figure 13.2; Hodges and Arnold 1994b). Thus these data suggest that introgression of alternative nectar spur characters is selected against in these species. This selection presumably occurs due to the foraging behavior of hummingbirds on *A. formosa*-like individuals and by hawkmoths on *A. pubescens*-like individuals. An alternative explanation of this pattern is that all genes that affect nectar spur morphology are closely linked with other traits, for example physiological traits, that are the actual targets of selection (Chase and Raven 1975).

The debate on the ability of differences in nectar spur morphology in providing a means of reproductive isolation in *Aquilegia* is illustrative of the need of researchers to explicitly state how a proposed key innovation may affect processes of diversification. As such it is important to determine how spurs influence reproductive isolation. While it seems unlikely that all floral characters in *Aquilegia formosa* and *A. pubescens* are tightly linked with ecological characters important in differentiating these species, this hypothesis can be tested. For instance, direct tests of the influence of spur morphology on reproduction could be made by either experimentally altering spur morphology or by transplanting species and their hybrids to increase the range of variation found in nature. In addition, in the future, it may be possible to locate both floral morphology traits and physiological traits on linkage maps of the columbine genome. This would provide a test of the linkage hypothesis of Chase and Raven (1975).

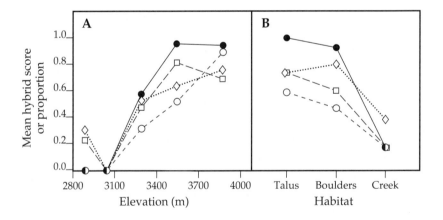

Figure 13.2. Clinal variation in both nectar spur length (●) and RAPD markers (open symbols) for (**A**) an elevational transect from a low elevation *A. formosa* population to a high elevation *A. pubescens* population and (**B**) a habitat transect along one side of a lake. Symbols indicate the mean hybrid score for nectar spur length or the proportion of individuals with a particular RAPD marker in each population. Redrawn from Hodges and Arnold (1994b).

Together, the data on floral spur morphology across multiple species suggest that relatively small changes in spur morphology can influence pollinator behavior and therefore act as a premating isolation barrier between species. Thus, if populations of species become isolated and specialized to different pollinators, then even if these species come into secondary contact they will maintain their reproductive isolation and increase the rate of species diversification.

Tests of differential diversification

Providing a plausible explanation for how a proposed key innovation may affect species diversification is not, by itself, sufficient to test a key-innovation hypothesis (Cracraft 1990; Skelton 1993; Heard and Hauser 1995; Hodges and Arnold 1995). It is important to establish that changes in species diversification rates are correlated with the evolution of proposed key innovations. Because diversification rate is defined as the net increase in taxa per unit time, it is necessary to compare the number of taxa in a group with a proposed key innovation with the number of taxa in a similar group of equivalent age. Because sister taxa are by definition of equal age, comparisons of species numbers between them are an obvious test for changes in diversification rates. Therefore it is essential to have an unbiased estimate of which taxa are sister-groups and thus cladistic analyses that include all possible sister taxa are ideal. Furthermore, nearly all sister-clade comparisons will have unequal numbers of taxa and therefore it is important to determine if there has actually been a change in diversification rate between sister taxa. Recently, several null models have been proposed as tests for changes in diversification rate and for specific key innovation hypotheses (Slowinski and Guyer 1989, 1993; Sanderson and Donoghue 1994; see Sanderson and Donoghue 1995 for recent review).

In the cladistic analysis for *Aquilegia* and relatives shown in Figure 13.1, there is a long branch leading to the rapid radiation of the columbines. This long branch implies that the ancestor to the columbines did not diversify for a relatively long period of time, and that then an event happened that caused a rapid radiation. To determine if this event may have been the evolution of nectar spurs and to identify the sister-group of the columbines, Hodges and Arnold (1995) included two additional species to their molecular phylogenetic analysis, *Aquilegia ecalcarata* and *Semiaquilegia adoxoides*. These two species lack nectar spurs and had been postulated as the ancestors to the spurred columbines (Munz 1946). As such, these taxa could help define both the timing of the evolution of nectar spurs and the appropriate sister-group to utilize in measures of diversification.

Inclusion of the two spurless species *A. ecalcarata* and *S. adoxoides* in the molecular phylogenetic analysis produced a single most parsimonious tree that placed *A. ecalcarata* among most of the other European and Asian columbines and *S. adoxoides* as sister to the columbines (Figure 13.3; Hodges and Arnold 1995). The relationship of *A. ecalcarata* to the rest of the columbines is ambiguous due to the lack of informative characters; it may itself be basal to the rest of the columbines (and thus their sister-group) or it may be derived within the columbines (and thus have secondarily lost nectar spurs). In fact, the most parsimonious tree resulted in the placement of *A. ecalcarata* within the columbines. In contrast, the placement of *S. adoxoides* is clearly basal to the columbines and therefore this species is the most appropriate taxa to use for sister-group comparisons with *Aquilegia*.

The placement of *Semiaquilegia adoxoides* close to and basal to the columbines also suggests that the evolution of spurs occurred near the time of the columbine radiation. By identifying *Semiaquilegia* as closely related and basal to the columbines, the most parsimonious placement of the origin of nectar spurs is placed on a short branch very near the columbine radiation (Figure 13.3). Therefore, the timing of the evolution of nectar spurs and the radiation are apparently closely linked and supports a key innovation hypothesis (Hodges and Arnold 1995).

Once sister taxa are identified, comparisons with null models of diversification can be used to test whether changes in diversification are likely to have occurred in taxa with a proposed key innovation. Sanderson and Donoghue (1994), using a model of random speciation, developed a maximum likelihood approach to test for a correlation between increased diversification rate and a proposed key innovation. The test compares the diversification rate of the sister-group lacking the key innovation and the diversification rate for the two basal lineages that possess the proposed key innovation. Sanderson and Donoghue (1994) point out that a change in diversification should be associated with the branch where the key innovation evolved and, therefore, that both of the clades that possess the key innovation should exhibit increased diversification. Thus, if neither of the clades with the key innovation or only one of them shows increased diversification, then this would falsify a simple key innovation hypothesis. The maximum likelihood model therefore tests whether particular patterns of diversification match the observed diversities of the taxa being tested (Sanderson and Donoghue 1994).

The columbines have a basal split that largely separates the Old World from the New World species (Figure 13.3) and therefore these groups were compared with *Semiaquilegia* using the Sanderson and Donoghue model (Figure 13.4; Hodges and Arnold 1995). All models that are inconsistent with the key innovation hypothesis were rejected and the two simplest models consistent with a key innovation hypothesis were not (Figure 13.4). Thus, under this model of diversification, the key innovation hypothesis for *Aquilegia* is supported.

Despite the correlation of the evolution of nectar spurs and increased diversification in *Aquilegia*, for every clade there will be many synapomorphies and therefore many possible characters that could be a key innovation. To further test a key innovation hypothesis it is essential to utilize comparative tests to determine if increased diversification is generally associated with the evolution of the proposed key innovation (Skelton 1993; Slowinski and Guyer 1993). For instance, Mitter et al. (1988) determined that 12 out of 16 groups of plants that had evolved some sort of resin canals showed greater numbers of species than their suspected sister-groups. Thus, using a simple sign test, Mitter et al. (1988) showed that the evolution of resin canals was indeed associated with greater species diversity. Linking the evolution of a particular trait with increased diversity across many groups provides strong support for a key innovation hypothesis.

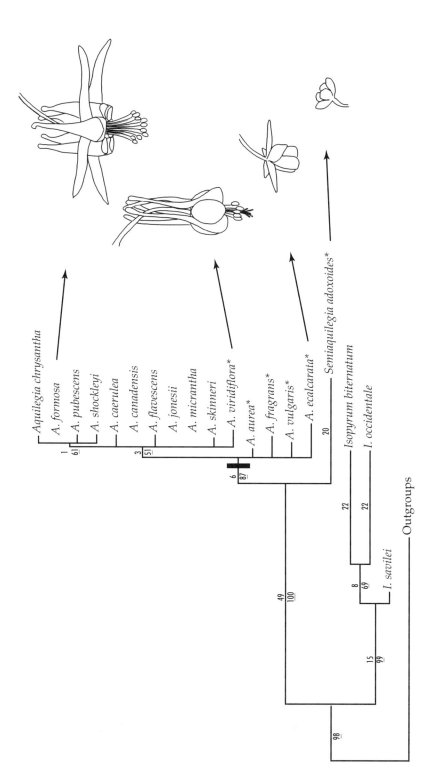

Figure 13.3. Phylogram (truncated to show only branches with bootstrap values over 50%) from a maximum parsimony analysis of the species included in Figure 13.1 as well as *Aquilegia ecalcarata* and *Semiaquilegia adoxoides*. A single most parsimonious tree resulted from the analysis. Percent bootstrap values are underlined and below or to the left of branches; inferred branch lengths are indicated above branches. The black bar indicates the most parsimonious position for the evolution of nectar spurs (all *Aquilegia* species except *A. ecalcarata* have nectar spurs). Species from Europe or Asia are indicated by an asterisk. Line drawings (not to scale) represent the floral forms for some of the species. Redrawn from Hodges and Arnold (1995).

Many groups of plants have independently evolved floral spurs and thus these groups provide the means to test if spurs are generally associated with increases in diversification rate. There are at least 15 separate instances where floral spurs have evolved (Table 13.1). To identify the sister taxa of these groups, the literature was searched for large cladistic analyses that included all of the likely sister taxa for each of the groups in Table 13.1. Once sister-groups are identified, then a sign test of the species diversity in the key innovation group in comparison with their sister-groups can be made (e.g., Mitter et al. 1988). Of the eight groups where sister-groups can be identified (Table 13.2), seven have more species in the clade with floral spurs (sign test, $P < 0.05$).

In addition to the sign test, Slowinski and Guyer (1993) proposed a test for increased diversification between sister taxa. Using a model of random speciation and extinction, they showed that, in a single clade of n species which splits into two subclades, all combinations of subclade sizes (r and n-r species) are equally likely. Slowinski and Guyer (1993) then showed that the probability of observing a given species diversity in a clade with a proposed key innovation or an even greater difference can be calculated as (n-r)/(n-1) where r is the number of species in the clade with the proposed key innovation. Of the seven spurred taxa where the diversity of their sister-group has been identified (Table 13.2), six have significant or nearly significant deviations from expected. A combined test that regards each sister-group relationship as a separate test of the same phenomenon (Slowinski and Guyer 1993)

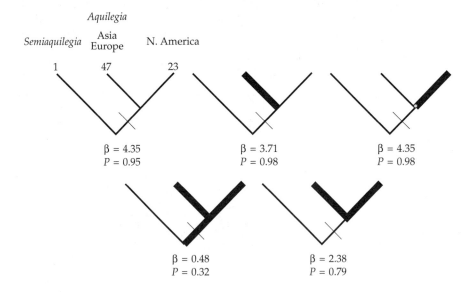

Figure 13.4. Five models of change in diversification rate (indicated by different line thickness). These models were used to test if the evolution of nectar spurs in *Aquilegia* (slash line) is correlated with a change in the rate of diversification (see Sanderson and Donoghue 1994). The log-likelihood ratios (β) and corresponding probability values (P) are given for each model (P values ≥ 0.95 reject a given model). The top three models are inconsistent with a key innovation hypothesis, and are rejected; the bottom two models are consistent with such an hypothesis, and are not rejected. Redrawn from Hodges and Arnold (1995).

Table 13.1. Taxa that have independently evolved floral spurs.

Balsaminaceae	Scrophulariaceae
(except *Hydrocera*)	*Chaenorhinum*
Fumariaceae	*Cymbalaria*
Gentianaceae	*Diascia*
Halenia	*Kickxia*
Geraniaceae	*Linaria*
Pelargonium	*Nuttallanthus*
Lentibulariaceae	Tropaeolaceae
Leguminosae	Violaceae
Amherstia	*Anchietea*
Bauhinia	*Corynostylis*
Orchidaceae	*Noisettia*
Angraecum	*Viola*
Habenaria	Vochysiaceae
many other genera	(except *Amphilochia*
Ranunculaceae	and *Euphronia*)
Aconitum	
Aquilegia	
Delphinium	

results in a highly significant association between the evolution of nectar spurs and increased species diversity ($\chi^2 = 45.24$ or 43.46, $P < 0.001$ for 14 df).

One case serves to make the point of the influence of using a particular analysis used to identify sister taxa. Three analyses of the Ranunculaceae (morphological data, cpDNA sequence data, and cpDNA restriction-site data) support two different sister taxa for *Aconitum* and *Delphinium* (Table 13.2). If *Nigella* alone is the sister-group then a significant result is found ($P = 0.039$); however, if *Nigella*, *Actaea*, and *Cimicifuga* are the sister-group then this comparison is non-significant ($P = 0.095$). Additionally, in a preliminary analysis of *rbc*L sequence data for the Violaceae, *Noisettia* and *Viola* form a clade of floral spur-bearing taxa and their sister-group is a subset of the genus *Hybanthus* (S. A. Hodges et al., unpubl. data). Because the relatively large genus of *Hybanthus* (150 spp.) was found to be polyphyletic in this analysis, the actual number of species in the sister-group for *Noisettia* and *Viola* must be less than 150 species – but presently it is impossible to determine how much less. However, it is possible to state that there are more species in the spur-bearing clade than its sister clade. Thus, the analyses that determine sister-group comparisons are critical for tests of key innovations.

One group, *Pelargonium* (Geraniaceae), does not fit the overall pattern in Table 13.2. There are several reasons to expect that not all groups that evolve a particular key innovation will show increased species diversity as compared to their sister-group. First, the sister-group may have evolved its own key innovation allowing it also to diversify. Second, key innovations are likely to be context-specific – an increase in diversification will occur only under particular ecological conditions. For instance, Liem (1973) suggests that the evolution of the cichlid pharyngeal jaw allowed rapid radiation upon invasion of the impoverished African rift lakes. Third, there may be genetic constraints to the modification of the key innovation. For instance in *Pelargonium*, the spur is fused

Table 13.2. Species diversity of taxa that have independantly evolved floral nectar spurs and their sister-groups.

Taxa with nectar spurs	Species	Sister taxa	Species	P	References
Aquilegia	70	*Semiaquilegia*	1	0.014	Hodges and Arnold 1995
Delphinium, Aconitum	350	*Nigella*	14	0.039	Hoot 1991, 1995
Delphinium, Aconitum	350	*Nigella, Actaea, Cimicifuga*	37	0.095	Johansson and Jansen 1993
Fumariaceae	450	*Hypercoum*[†]	15	0.032	Hoot and Crane 1995
Tropaeolaceae	88	Akaniaceae and Bretschneideraceae	2	0.022	Chase et al. 1993
Anchiectia, Corynostylis	12	*Agatea*	1	0.083	Hodges et al. unpubl. data
Noisettia, Viola	401	subset of *Hybanthus*	<150	?	Hodges et al. unpubl. data
Lentibulariaceae	245	Byblidaceae	2	0.008	Olmstead et al. 1993
Pelargonium	280	*Geranium, Erodium, Nonsonia* and *Sarcocaulon*	399	0.588	Price and Palmer 1993

[†] S. Hoot, personal comm. In an analysis of *rbc*L sequences from members of the Papaverales, *Hypercoum* was found to be the sister-group of Fumariaceae.

to the pedicel of the flower, rendering it "hidden" (Endress 1994). Thus, any modification of spur length also requires modification of the length of the pedicel. This fusion of floral parts thus could constrain developmentally the ease of diversification within this group. It is interesting that several other groups of flowering plants also have hidden spurs (e.g. *Bauhinia, Epidendrum,* and *Dactyladenia*) and thus the generality of this hypothesis could be tested with sister-group analyses.

A second possible exception to the overall pattern found in Table 13.2 is the family Vochysiaceae. In a recent analysis of the Myrtales (Conti et al. 1996), the Vochysiaceae (210 spp.) were found to be the sister-group of the Myrtaceae (3,850 spp.) which lack spurs. As such, this relationship would argue against the hypothesis that the evolution of nectar spurs can increase species diversification. However, while the consensus tree of the 100 shortest trees obtained from the analysis exhibited this sister-group arrangement, there was less than 50% bootstrap support for this relationship (Conti et al. 1996). If only branches with ≥ 50% bootstrap support are considered, then the Myrtaceae and Vochysiaceae form a polytomy with the families Heteropyxidaceae and the Psiloxylaceae (4 species total) (Conti et al. 1996). Therefore, it is possible that these two small families are the actual sister-group to the Vochysiaceae. Furthermore, there are members of the Vochysiaceae that lack nectar spurs (*Euphronia* and *Amphilochia,* Table 13.1). If these taxa are basal to the family then they would be the actual sister-group to the clade that has evolved nectar spurs. Once again, this possible ambiguity in sister-group relationships illustrates the sensitivity of the analysis to the identification of sister taxa and thus the importance of utilizing studies that include all possible sister taxa.

Finally, it is possible that a character correlated with nectar spurs is the actual causal agent for the pattern of increased diversification found in Table 13.2. While this is possible, at present I know of no other character that is correlated with nectar spurs across all of the groups analyzed thus far. Clearly there are additional groups that have independently evolved nectar spurs that are not included in the present analysis (Table 13.1). As sister-groups for these taxa are identified, additional tests of the role of the evolution of floral spurs on species diversification rates will be possible.

Conclusions

Understanding why some groups of organisms are particularly diverse while others are not has long spurred the curiosity of biologists. With the ability to rapidly obtain data on genetic differences among taxa we have seen an explosion of studies that provide reconstructions of the phylogeny of organisms. Clearly molecular phylogenetic analyses have provided much of the recent data for reconstructing phylogenies of organisms. As shown here, these data are imperative to our understanding of the processes that control species diversity. Both the identification of a recent radiation in *Aquilegia* and much of the tests of the causal agent of this diversification were dependent on molecular analyses. However, in addition to phylogenetic analyses, population level studies are needed to understand how traits influence the processes of speciation and extinction and ultimately clade diversification. By combining phylogenetic, population, and comparative studies a clearer understanding of the biology of diversity will be possible.

References

Avise, J. C. 1994. *Molecular markers, natural history and evolution*. New York, NY: Chapman and Hall.
Baldwin, B. G., Kyhos, D. W., Dvorak, J., and Carr, G. D. 1991. Chloroplast DNA evidence for a North American origin of the Hawaiian silversword alliance (Asteraceae). Proceedings National Academy of Sciences, USA 88:1840–1843.
Barton, N. H., and Hewitt, G. M. 1989. Adaptation, speciation and hybrid zones. Nature 341:497–503.
Bremer, B., Olmstead, R. G., Struwe, L., and Sweere, J. A. 1994. *rbc*L sequences support exclusion of *Retzia, Desfontainia,* and *Nicodemia* from the Gentianales. Plant Systematics and Evolution 190:213–230.
Brink, D. E. 1980. Reproduction and variation in *Aconitum columbianum* (Ranunculaceae), with emphasis on California populations. American Journal of Botany 67:263–273.
Chase, M. W., Soltis, D. E., Olmstead, R. G., Morgan, D., Les, D. H., Mishler, B. D., Duvall, M. R., Price, R. A., Hells, H. G., Qiu, Y.-L., Kron, K. A., Rettig, J. H., Conti, E., Palmer, J. D., Manhart, J. R., Sytsma, K. J., Michaels, H. J., Kress, W. J., Karol, K. G., Clark, W. D., Hedren, M., Gaut, B. S., Jansen, R. K., Kim, K.-J., Wimpee, C. F., Smith, J. F., Furnier, G., Strauss, S. H., Xiang, Q.-Y., Plunkett, G. M., Soltis, P. S., Swensen, S. M., Williams, S. E., Gadek, P. A., Quinn, C. J., Eguiarte, L. E., Golenberg, E., Geralk H. Learn, J., Graham, S. W., Barrett, S. C. H., Dayanandan, S., and Albert, V. A. 1993. Phylogenetics of seed plants: an analysis of nucleotide sequences from the plastid gene *rbc*L. Annals of the Missouri Botanical Garden 80:528–580.
Chase, V. C., and Raven, P. H. 1975. Evolutionary and ecological relationships between *Aquilegia formosa* and *A. pubescens* (Ranunculaceae), two perennial plants. Evolution 29:474–486.
Clausen, J., Keck, D. D., and Hiesey, W. M. 1945. *Experimental studies on the nature of species. II.* Washington, DC: Carnegie Institute Washington Publ.
Conti, E., Litt, A., and Sytsma, K. J. 1996. Circumscription of Myrtales and their relationships to other rosids: evidence from *rbc*L sequence data. American Journal of Botany 83:221–233.
Cracraft, J. 1990. The origin of evolutionary novelties: pattern and process at different hierarchical levels. Pp. 21–46 in *Evolutionary innovations*, M. H. Nitecki, ed. Chicago, IL: University of Chicago Press.
Dobzhansky, T. 1937. *Genetics and the origin of species*. New York, NY: Columbia University Press.
Endler, J. A. 1986. *Natural selection in the wild*. Monographs in Population Biology, R. M. May, ed. Princeton, NJ: Princeton University Press.

Endress, P. K. 1994. *Diversity and evolutionary biology of tropical flowers*. Cambridge, England: Cambridge University Press.

Gillespie, R. G., Croom, H. B., and Palumbi, S. R. 1994. Multiple origins of a spider radiation in Hawaii. Proceedings National Academy of Sciences, USA 91:2290–2294.

Grant, P. R. 1986. *Ecology and evolution of Darwin's finches*. Princeton, NJ: Princeton University Press.

Grant, V. 1952. Isolation and hybridization between *Aquilegia formosa* and *A. pubescens*. Aliso 2:341–360.

Grant, V. 1963. *The origin of adaptations*. New York, NY: Columbia University Press.

Grant, V. 1993. Origin of floral isolation between ornithophilous and sphingophilous plant species. Proceedings National Academy of Sciences, USA 90:7729–7733.

Grant, V. 1994a. Historical development of ornithophily in the western North American flora. Proceedings National Academy of Sciences, USA 91:10407–10411.

Grant, V. 1994b. Modes and origins of mechanical and ethological isolation in angiosperms. Proceedings National Academy of Sciences, USA 91:3–10.

Heard, S. B., and Hauser, D. L. 1995. Key evolutionary innovations and their ecological mechanisms. Historical Biology 10:151–173.

Hodges, S. A., and Arnold, M. L. 1994a. Columbines: a geographically wide-spread species flock. Proceedings National Academy of Sciences, USA 91:5129–5132.

Hodges, S. A., and Arnold, M. L. 1994b. Floral and ecological isolation between *Aquilegia formosa* and *Aquilegia pubescens*. Proceedings National Academy of Sciences, USA 91:2493–2496.

Hodges, S. A., and Arnold, M. L. 1995. Spurring plant diversification: are floral nectar spurs a key innovation? Proceedings Royal Society of London. Series B 262:343–348.

Hoot, S. B. 1991. Phylogeny of the Ranunculaceae based on epidermal microcharacters and macromorphology. Systematic Botany 16:741–755.

Hoot, S. B. 1995. Phylogeny of the Ranunculaceae based on preliminary *atp*B, *rbc*L and 18S nuclear ribosomal DNA sequence data. Plant Systematics and Evolution [Suppl.] 9:241–251.

Johansson, J. T., and Jansen, R. K. 1993. Chloroplast DNA variation and phylogeny of the Ranunculaceae. Plant Systematics and Evolution 187:29–49.

Liem, K. F. 1973. Evolutionary strategies and morphological innovations: cichlid pharyngeal jaws. Systematic Zoology 22:425–441.

Mayr, E. 1942. *Systematics and the origin of species*. New York, NY: Columbia University Press.

Meyer, A., Kocher, T. D., Basasibwaki, P., and Wilson, A. C. 1990. Monophyletic origin of Lake Victoria cichlid fishes suggested by mitochondrial DNA sequences. Nature 347:550–553.

Miller, R. B. 1981. Hawkmoths and the geographic patterns of floral variation in *Aquilegia caerulea*. Evolution 35:763–774.

Mitter, C., Farrell, B., and Wiegmann, B. 1988. The phylogenetic study of adaptive zones: has phytophagy promoted insect diversification? American Naturalist 132:107–128.

Munz, P. A. 1946. *Aquilegia*: The cultivated and wild columbines. Gentes Herbarum 7:1–150.

Nilsson, L. A. 1988. The evolution of flowers with deep corolla tubes. Nature 334:147–149.

Olmstead, R. G., Michaels, H. J., Scott, K. M., and Palmer, J. D. 1993. Monophyly of the Asteridae and identification of their major lineages inferred from DNA sequences of *rbc*L. Annals of the Missouri Botanical Garden 80:700–722.

Prazmo, W. 1965. Cytogenetic studies on the genus *Aquilegia* IV. Fertility relationships among the *Aquilegia* species. Acta Societatis Botanicorum Poloniae 34:667–685.

Price, R. A., and Palmer, J. D. 1993. Phylogenetic relationships of the Geraniaceae and Geraniales from *rbc*L sequence comparisons. Annals of the Missouri Botanical Garden 80:661–671.

Raup, D. M., Gould, S. J., Schopf, T. J. M., and Simberloff, D. S. 1973. Stochastic models of phylogeny and the evolution of diversity. Journal of Geology 81:525–542.

Rodman, J., Price, R. A., Karol, K., Conti, E., Sytsma, K. J., and Palmer, J. D. 1993. Nucleotide sequences of the *rbc*L gene indicate monophyly of mustard oil plants. Annals of the Missouri Botanical Garden 80:686–699.

Sanderson, M. J., and Donoghue, M. J. 1994. Shifts in diversification rate with the origin of angiosperms. Science 264:1590–1593.

Skelton, P. W. 1993. Adaptive radiation: definition and diagnostic tests. Pp. 45–58 in *Evolutionary patterns and processes*, D. R. Lees and D. Edwards, eds. London, England: Academic Press.

Slowinski, J. B., and Guyer, C. 1989. Testing the stochasticity of patterns of organismal diversity: an improved null model. American Naturalist 134:907–921.

Slowinski, J. B., and Guyer, C. 1993. Testing whether certain traits have caused amplified diversification: an improved method based on a model of random speciation and extinction. American Naturalist 142:1019–1024.

Stebbins, G. L. 1950. *Variation and evolution in plants*. New York, NY: Columbia University Press.

Steiner, K. E., and Whitehead, V. B. 1990. Pollinator adaptation to oil-secreting flowers – *Rediviva* and *Diascia*. Evolution 44:1701–1707.

Steiner, K. E., and Whitehead, V. B. 1991. Oil flowers and oil bees: further evidence for pollinator adaptation. Evolution 45:1493–1501.

Swofford, D. L. 1993. *PAUP: phylogenetic analysis using parsimony, version 3.1.1*. Champaign, IL: Illinois Natural History Survey.

14 Origin and Evolution of *Argyranthemum* (Asteraceae: Anthemideae) in Macaronesia

Javier Francisco-Ortega, Daniel J. Crawford, Arnoldo Santos-Guerra, and Robert K. Jansen

Macaronesia comprises five Atlantic archipelagos (Azores, Desertas-Madeira, Selvagens, Canaries, and Cape Verde) off the western coasts of Europe and Africa, situated between latitudes 15° and 40° N. The region has 24 major islands that exhibit a broad range of variation both in their ecology and geology. Geological ages vary between 21 million years (My) for Fuerteventura to 0.8 My for El Hierro (Rothe 1982; Mitchell-Thomé 1985; Galopim de Carvalho and Brandão 1991; Boekschoten and Manuputty 1993; Carracedo 1994). The combination of latitudinal gradients and northeastern trade winds has produced a number of distinct ecological zones (Bramwell 1972; Figure 14.1). Three of these zones (humid lowland scrub, laurel forest, and heath belt) are under the direct influence of the northeastern trade winds and are mainly situated on northern slopes of the islands between 400 and 1,200 m. The other four ecological zones (coastal desert, arid lowland scrub, pine forest, and high altitude desert) are much more arid because they do not receive the direct influence of these moisture-laden winds. The great habitat diversity and insular isolation are the main factors responsible for the rich flora of Macaronesia; at least 831 species and 40 genera are endemic to the region (Humphries 1979; Hansen and Sunding 1993; La Roche and Rodríguez-Piñero 1994).

Argyranthemum (Asteraceae: Anthemideae) is the largest endemic plant genus found on volcanic islands in the Atlantic Ocean (Howard 1973; Hansen and Sunding 1993). The genus includes 24 species and 15 subspecies that are endemic to the archipelagos of Madeira, Selvagens, and Canaries (Humphries 1976a,b; Borgen 1980; Rustan 1981; Figure 14.2). Each taxon is restricted to a specific ecological zone (Table 14.1, Figure 14.2). *Argyranthemum* has radiated into all major habitats of these islands, with at least one taxon occurring in each of the ecological zones of Macaronesia. The two ecological zones with the highest number of taxa are coastal desert (14 taxa) and humid lowland scrub (12 taxa).

The extent of morphological variation among taxa is so great that recent ecogeographical surveys suggest that, in addition to the 24 described species, new distinct morphological forms are found in the pine forests of southern Tenerife and La Palma, the lowland arid scrub of southern Tenerife, the high altitude desert of La Palma, and the coastal deserts of northern and eastern La Palma and southeastern El Hierro (Santos-Guerra 1983; Santos-Guerra et al. 1993; Francisco-Ortega et al. 1996a; Table 14.1; Figure 14.3).

The ability of *Argyranthemum* to exploit the many ecological zones in Macaronesia and to colonize multiple islands is unique among genera endemic to Atlantic archipelagos. This radiation is comparable only to that seen in other non-endemic genera

Figure 14.1. Ecological zones in Macaronesia. (**A**) Pine forest in the Caldera de Taburiente National Park, La Palma; (**B**) Cloud belt in northern Tenerife; this zone receives the direct influence of the northeastern trade winds; (**C**) Coastal zone in Desertas Island; the island of Bugio can be seen in the background; (**D**) High altitude desert in the island of Tenerife, plants of *Echium wildpretii* shown in the foreground; (**E**) Laurel forest in Cubo de La Galga in northern La Palma.

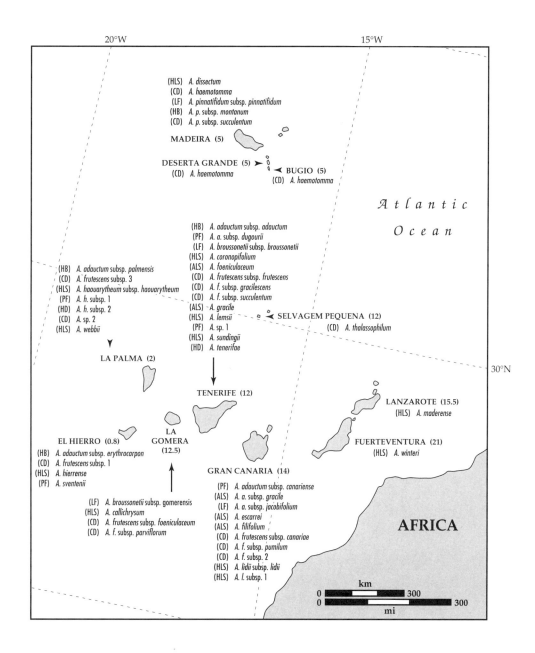

Figure 14.2. Distribution of *Argyranthemum* in the Canaries, Selvagens, and Madeira archipelagos. Ecological zones are coded in Table 14.1. Geological age (My) of each island is also indicated (based on Galopim de Carvalho and Brandão 1991 and Carracedo 1994).

Table 14.1. Distribution of *Argyranthemum* in Macaronesia (after Francisco-Ortega et al. 1996a,b). Ecological zones are as follows: CD, coastal desert; ALS, arid lowland scrub; HLS, humid lowland scrub; LF, laurel forest; HB, heath belt; PF, pine forest; HD, high altitude desert. Dash indicates that a particular ecological zone is absent. Geographical data compiled from Carlquist (1974), Humphries (1979), Galopim de Carvalho and Brandão (1991), and Carracedo (1994). Each taxon is endemic to a particular zone and a particular island, the exception being *A. haemotomma*, which is found in Madeira, Deserta Grande, and Bugio. Taxa found in each island are given in Figure 14.1. The table also includes unpublished taxa from the coastal desert of El Hierro (1), coastal desert of La Palma (2), coastal desert of Gran Canaria (1), humid lowland scrub of Gran Canaria (1), pine forest of Tenerife (1), pine forest of La Palma (1), and high altitude desert of La Palma (1).

Island	Altitude (m)	Area (km)	Number of taxa per ecological zone							
			CD	ALS	HLS	LF	HB	PF	HD	Total
Canary Islands										
El Hierro	1,520	307	1	0	1	0	1	1	–	4
La Palma	2,423	789	2	0	2	0	1	1	1	6
La Gomera	1,484	425	2	0	1	1	0	–	–	4
Tenerife	3,714	2,355	3	2	3	1	1	2	1	13
Gran Canaria	1,950	1,625	3	3	2	1	0	1	–	10
Fuerteventura	807	1,717	0	0	1	–	–	–	–	1
Lanzarote	670	717	0	0	1	–	–	–	–	1
Madeira Islands										
Madeira	1,861	730	2	0	1	1	1	–	–	5
Deserta Grande	479	20	1	0	0	–	–	–	–	1
Bugio	384	7	1	0	0	–	–	–	–	1
Selvagens Islands										
Selvagem Pequena	49	0.5	1	–	–	–	–	–	–	1
Total			14*	5	12	4	4	5	2	47

*Indicates that *Argyranthemum haemotomma* has been added only once.

such as *Aeonium* (Crassulaceae), *Echium* (Boraginaceae), *Euphorbia* (Euphorbiaceae), *Sideritis* (Lamiaceae), and *Lotus* (Fabaceae). Thus, *Argyranthemum* is an ideal system for understanding the origin, radiation, and patterns of evolution of the flora of Macaronesia. The goal of this paper is to summarize the evolutionary implications for *Argyranthemum* of recent macromolecular studies, including comparisons of isozymes, chloroplast DNA (cpDNA) restriction sites and DNA sequences of the internal transcribed spacer (ITS) regions of the nuclear ribosomal repeat. These data will be discussed within the framework of the results of prior studies on the genus.

Natural history of *Argyranthemum*

The Macaronesian species of *Argyranthemum* exhibit a broad range of diversity in habit. The high altitude endemic *A. tenerifae* has a dwarf, dome-shaped growth form. All subspecies of *A. adauctum* in the laurel forest and heath belt zones are prostrate. Most species in the lowland scrub zone are decumbent, but some (e.g., *A. foeniculaceum*) occasionally develop candelabra-like branching (Humphries 1976).

Considerable variation in leaf and cypsela morphology exists. Leaves of species from arid areas show a reduction in lobe size and number (Figure 14.3A; Humphries 1979). The fruits (cypselas) are extremely variable in shape, size, color, pappus and wing characteristics (Figure 14.3B). In addition, at least 10 taxa (*A. adauctum, A. broussonetii* subsp. *gomerensis, A. callichrysum, A. escarrei, A. filifolium, A. haemotoma, A. hierrense, A, sventenii, A. thalassophilum, A. webbii*) have cypselas fused into groups, a unique feature in the Chrysantheminae. Most species of *Argyranthemum* have white ray florets; however, *A. maderense* and some populations of *A. callichrysum* have yellowish ray florets (Figure 14.3C). The flowers of *A. haemotoma* have a distinct color pattern with reddish ray florets and disk florets that expand deeply red but turn yellow as the stamens are exserted (Figure 14.3e).

Most taxa of *Argyranthemum* are only locally abundant and appear to be adapted to very specific ecological conditions. There are usually considerable numbers of populations and individuals within the distribution of each taxon. However, eleven taxa (*A. adauctum* subspp. *erythrocarpon, jacobifolium,* and *palmensis, A. broussonetii* subsp. *gomerensis, A. coronopifolium, A. haemotoma, A. lidii, A. sundingii, A. thalassophilum, A. webbii,* and *A. winteri*) are on the verge of extinction with only one population (*A. thalassophilum*) or very few populations (the other ten taxa) of several individuals remaining. In contrast, ten taxa are widespread (*A. adauctum* subspp. *canariense, dugourii,* and *gracile,* and seven subspecies of *A. frutescens*). *Argyranthemum frutescens* from Tenerife is considered a weed because it aggressively invades the range of other taxa (Humphries 1976a,b; Brochmann 1984, 1987).

Argyranthemum is diploid ($2n = 18$; Humphries 1975). The genus has been regarded as self-compatible, although results concerning percentage of autogamy have not been reported (e.g., Humphries 1975; Borgen 1976). However, our preliminary greenhouse studies (J. Francisco-Ortega, unpubl. data) and plant breeding studies conducted in Australia (Cunneen 1995) indicate that self-compatibility is very limited. Isozyme data from populations of five taxa also suggest that *Argyranthemum* is

Figure 14.3. Morphological variation in *Argyranthemum*. (**A**) Leaf shape variation in (from top to bottom and from left to right) *A. lidii*, *A. sventenii*, *A. adauctum* subsp. *dugourii*, *A. foeniculaceum*, *A. frutescens* subsp. *succulentum*, *A. coronopifolium*, *A. gracile*, *A. dissectum*, and *A.* sp. nov. from coastal zones of northern La Palma. (**B**) Cypsela variation, upper row: species with fused cypselas (from left to right: *A. broussonetii* subsp. *gomerensis*, *A. webbii*, *A. adauctum* subsp. *adauctum*, *A. thalassophilum*, *A. adauctum* subsp. *dugourii*, and *A. callichrysum*), lower row: species with unfused cypselas (from left to right: *A. broussonetii* subsp. *broussonetii*, *A. coronopifolium* (ray cypsela), *A. coronopifolium* (disk cypsela), *A. maderense*, *A. lemsii*, *A. frutescens* subsp. *frutescens*). (**C**) Inflorescences with yellow ligules of *A. maderense*. (**D**) *Argyranthemum sundingii*, a species endemic in Tenerife, which has been suggested to be of hybrid origin between *A. broussonetii* subsp. *broussonetii* and *A. frutescens* subsp. *frutescens* (Brochmann, 1987). (**E**) Ray florets of *A. haemotomma* are reddish in color, the disk florets change from red to yellow as the florets open and the stamens exert.

allogamous because the total genetic diversity among populations (G_{ST}) ranges between 0.115 and 0.385 (J. Francisco-Ortega, unpubl. data). Therefore, we believe that *Argyranthemum* should be considered an outbreeder. Breeding systems for its continental relatives have been reported for *Ismelia* (Jain and Gupta 1960; Chaudhuri et al. 1976), *Chrysanthemum coronarium* (Chaudhuri et al. 1976), and *C. segetum* (Howarth and Williams 1972). These studies also indicated strong self-incompatibility mechanisms. Thus, *Argyranthemum* follows the pattern of the Hawaiian silversword alliance (Carr et al. 1986; Baldwin and Robichaux 1995) and could be regarded as another exception to Baker's rule: self-compatibility is the most advantageous breeding system for a colonizer to establish and evolve after long-distance dispersal (Baker 1955, 1967).

Post-zygotic barriers between *Argyranthemum* species are weak and reproductive isolation is mainly ecological or geographical. Artificial interspecific hybrids are easy to produce and hybrid swarms have been reported for *A. adauctum-A. filifolium* in Gran Canaria (Borgen 1976), *A. coronopifolium-A. frutescens* (Humphries 1976a,b; Brochmann 1984) in Tenerife, and *A. broussonetii-A. frutescens* (Figure 14.3d; Brochmann 1987) in Tenerife. The breakdown of ecological barriers due to human disturbance has enhanced the ability of *Argyranthemum* to produce hybrid swarms. If this process continues, it is possible that rare species may be assimilated by the more aggressive species (Levin et al. 1996). In some areas of the Teno peninsula in Tenerife, the number of pure populations of the rare *A. coronopifolium* have diminished because of extensive hybridization with the widespread *A. frutescens* (Humphries 1976a,b; Brochmann 1984; Bramwell 1990).

Origin of *Argyranthemum*

Non-molecular systematic evidence

Previous morphological and phytochemical data supported a close relationship between *Argyranthemum* and three other continental genera: *Chrysanthemum*, *Ismelia* and *Heteranthemis* (Greger 1977; Christensen 1992; Bremer and Humphries 1993). *Ismelia* is a monotypic genus endemic to Morocco. *Heteranthemis* is also monotypic with a disjunct distribution in southern Iberia and northern Morocco. The two species of *Chrysanthemum* have a worldwide distribution, but they are considered to have a Mediterranean origin (Heywood 1976; Heywood and Humphries 1977).

Morphological cladistic studies (Bremer and Humphries 1993) suggested that *Argyranthemum-Ismelia-Heteranthemis-Chrysanthemum* form a monophyletic group, which was recognized as the subtribe Chrysantheminae. The possibility of close affinities between *Argyranthemum* and the South African genus *Cymbopappus* (Hutchinson 1917; Bramwell 1976; Takhtajan 1986) or the Japanese genus *Nipponanthemum* (Kitamura 1978) has also been proposed.

Two alternative hypotheses have been suggested for the origin of *Argyranthemum*. According to the first, the genus is a relict of a flora that existed in the Mediterranean basin before the Pleistocene (Bramwell 1972, 1976; Humphries 1976a,b, 1979; Aldridge 1979). Two lines of evidence support this hypothesis: (i) the woody perennial habit of *Argyranthemum*, which is unique in the Chrysantheminae, has been considered the ancestral state; and (ii) the presence of European Tertiary fossils from at least six genera that do not currently occur in Europe, even though all have extant representatives in Macaronesia (reviewed by Sunding 1979). Thus, in this model, species of *Argyranthemum* have been derived from a relictual continental ancestor that became extinct in the mainland following major climatological changes in the late Tertiary (Humphries 1976a). These endemics – referred to as active epibiotics – are suggested to have gone through a secondary evolutionary episode from a relict group (Bramwell 1972, 1976).

The second hypothesis suggests that *Argyranthemum* originated very recently and that its woody habit represents a common insular adaptation seen in lineages endemic to Pacific and Atlantic volcanic islands. Woodiness has been considered to be a common adaptation to the uniformity of the climate of insular environments (Carlquist 1965, 1974). Thus, in this model, woodiness in *Argyranthemum* would not represent an ancestral character-state within the Chrysantheminae.

Molecular systematic evidence

Our macromolecular studies of *Argyranthemum* have focused on four questions regarding the origin of the genus: (i) What is the closest continental relative? (ii) Is *Argyranthemum* a relict of an older Mediterranean flora? (iii) When did the genus originate? and (iv) Is *Argyranthemum* monophyletic, thus suggesting a single colonization of Macaronesia?

IDENTIFICATION OF MAINLAND RELATIVES – Cladistic analyses of cpDNA restriction-site data (Francisco-Ortega et al. 1995a) from a sample of seven genera of the subtribes Chrysantheminae, Leucantheminae, Anthemidinae, and Achilleinae strongly support a close relationship of *Argyranthemum* with the three other genera of the Chrysantheminae (Figure 14.4). The monophyly of these four genera is supported by a bootstrap value of 100%. An expanded cpDNA restriction-site phylogeny including all the taxa of *Argyranthemum* (Francisco-Ortega et al. 1996b) provides strong support (100% bootstrap value) for the monophyly of the genus (Figure 14.5).

Phylogenetic analyses of ITS sequence data for 32 genera from 8 subtribes of the Anthemideae have been completed (Francisco-Ortega et al. 1997; Figure 14.6). The ITS tree is congruent with the cpDNA tree in supporting a close relationship of *Argyranthemum* with *Chrysanthemum* and *Ismelia*. The two trees differ in that *Heteranthemis* groups with five other Mediterranean genera (*Hymenostemma*, *Lepidophorum*, *Leucanthemopsis*, *Lonas*, and *Prolongoa*) in the ITS tree, rather than with the other genera of the Chrysantheminae. The position of *Heteranthemis* is problematic in the ITS tree because of low bootstrap support and because considerable rate heterogeneity in ITS sequences is seen in all pairwise comparisons involving *Heteranthemis* and the other genera of the Chrysantheminae. The ITS tree suggests that the continental *Ismelia carinata* and *Chrysanthemum coronarium* are the sister-group of *Argyranthemum*. However, support for this relationship is also very low.

Previous hypotheses of a close relationship of *Argyranthemum* with the South African flora through *Cymbopappus* or with the Far Eastern flora through the Japanase

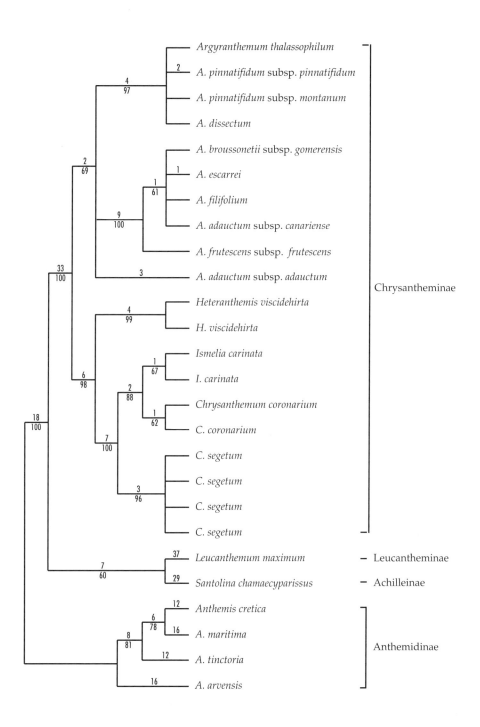

Figure 14.4. Single most parsimonious tree of selected genera of the Anthemideae based on cpDNA restriction site data (modified from Francisco-Ortega et al. 1995b). The tree has a length of 240 steps and CI' = 0.72 excluding autapomorphies. Bootstrap values (100 replicates) are shown below each branch; the number of steps are shown above each branch. The four species of *Anthemis* were designated as outgroups.

genus *Niponanthemum* is not supported by the ITS data. All the Mediterranean genera (except *Phalacrocarpum* and *Anacyclus*) form a weakly supported clade (33% bootstrap value) that includes *Argyranthemum*. Furthermore, a 16 bp deletion in the ITS2 region of *Argyranthemum* and all other Mediterranean genera is not present in taxa from the Far East or South Africa. Thus, a Mediterranean origin for *Argyranthemum* is also supported by the ITS data.

POSITION OF *ARGYRANTHEMUM* IN THE CHRYSANTHEMINAE – Phylogenetic analysis of ITS data shows that *Argyranthemum* is in a relatively derived position among the Mediterranean genera of the Anthemideae (Figure 14.6). However, weak support for most basal nodes of this lineage indicates uncertainty about its position within the Mediterranean clade. In contrast, cpDNA restriction-site data strongly support *Argyranthemum* as the sister-group to the three continental genera of the Chrysantheminae (Figure 14.4). A combined analysis of the ITS and cpDNA restriction-site data also supports the monophyly of the Chrysantheminae (Figure 14.7) and is concordant with the cpDNA restriction-site data alone in showing *Argyranthemum* as sister to the monophyletic group *Heteranthemis-Ismelia-Chrysanthemum*. Thus, *Argyranthemum* is as old as the other genera of the Chrysantheminae.

Isozyme data also indicated that *Argyranthemum* may be an old member of the Chrysantheminae. Genetic diversity values and the number of unique alleles were higher in *Argyranthemum* than in the other three genera (Francisco-Ortega et al. 1995b). This is concordant with the cpDNA restriction-site data. Intergeneric comparisons within the subtribe show that cpDNA restriction-site divergence is as high within *Argyranthemum* as between the other three genera (Francisco-Ortega et al. 1995a).

Figure 14.5. (Opposite) (**A**) One of the two equally parsimonious trees from weighted parsimony analysis of cpDNA restriction-site variation in *Argyranthemum* (weight of 1.01 in favor of site gains; modified from Francisco-Ortega et al. 1996b). Arrow indicates the branch which collapsed in the second tree. Dashed lines indicate branches that collapse in the strict consensus tree of 54 equally parsimonious trees. Branch lengths are shown above each branch. Bootstrap values (100 replicates) greater than 50% are shown below branches. The distribution of the species among the five recognized sections (Humphries 1976b) of *Argyranthemum* (Arg), *Monoptera* (Mon), *Preauxia* (Pre), *Sphenismelia* (Sph), *Stigmatotheca* (Sti) is given. (**B**) Geographic distribution among islands and ecological habitats for each species are overlain on the cpDNA restriction site tree. Ecological habitats are coded as in Table 14.1. Solid circles indicate ecological habitats not influenced by the trade winds. Taxa of *Argyranthemum* are coded as follows: ada (*A. adauctum* subsp. *adauctum*), ada can (*A. adauctum* subsp. *canariense*), ada dug (*A. adauctum* subsp. *dugourii*), ada ery (*A. adauctum* subsp. *erythrocarpon*), ada gra (*A. adauctum* subsp. *gracile*), ada jac (*A. adauctum* subsp. *jacobifolium*), ada pal (*A. adauctum* subsp. *palmensis*), bro (*A. broussonetii* subsp. *broussonetii*), bro gom (*A. broussonetii* subsp. *gomerensis*), cal (*A. callichrysum*), cor (*A. coronopifolium*), dis (*A. dissectum*), esc (*A. escarrei*), fil (*A. filifolium*), foe (*A. foeniculaceum*), fru (*A. frutescens* subsp. *frutescens*), fru can (*A. frutescens* subsp. *canariae*), fru foe (*A. frutescens* subsp. *foeniculaceum*), fru gra (*A. frutescens* subsp. *gracilescens*), fru par (*A. frutescens* subsp. *parviflorum*), fru pum (*A. frutescens* subsp. *pumilum*), fru suc (*A. frutescens* subsp. *succulentum*), fru 1 (*A. frutescens* subsp. nov. 1), fru 2 (*A. frutescens* subsp. nov. 2), gra (*A. gracile*), hae (*A. haemotomma*), hao (*A. haouarytheum* subsp. *haouarytheum*), hao 1 (*A. haouarytheum* subsp. nov. 1), hao 2 (*A. haouarytheum* subsp. nov. 2), hie (*A. hierrense*), lem (*A. lemsii*), lid (*A. lidii* subsp. *lidii*), lid 1 (*A. lidii* subsp. nov. 1), mad (*A. maderense*), pin (*A. pinnatifidum* subsp. *pinnatifidum*), pin mon (*A. pinnatifidum* subsp. *montanum*), pin suc (*A. pinnatifidum* subsp. *succulentum*), sun (*A. sundingii*), sve (*A. sventenii*), ten (*A. tenerifae*), tha (*A. thalassophilum*), web (*A.webbii*), win (*A. winteri*), A 1 (*A.* sp. nov. 1), and A 2 (*A.* sp. nov. 2). Taxa of the outgroup are coded as follows: Chr cor (*Chrysanthemum coronarium*), Chr seg (*C. segetum*), Het vis (*Heteranthemis viscidehirta*), Ism car (*Ismelia carinata*).

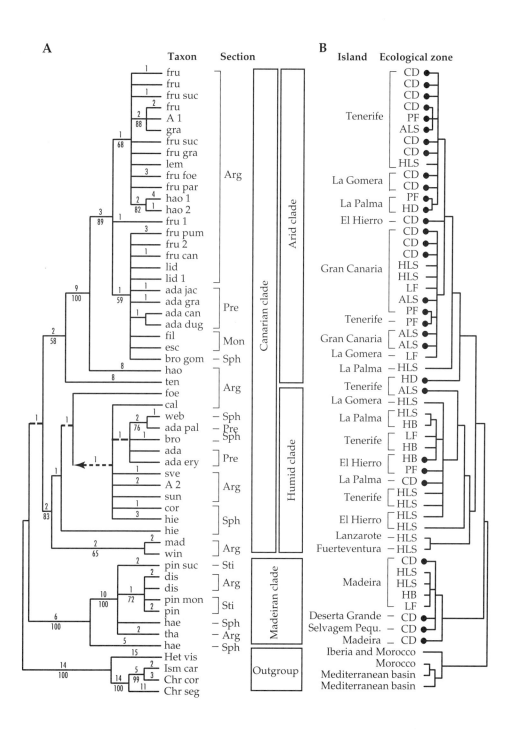

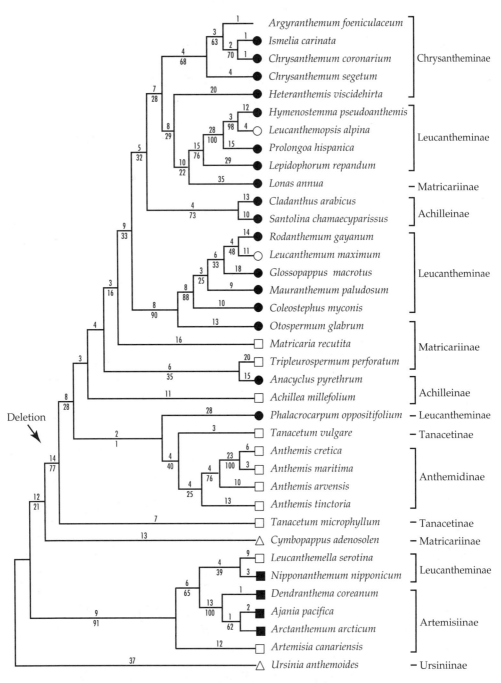

Figure 14.6. Single most parsimonious tree from the ITS nucleotide sequence data of 37 taxa of 32 genera of the Anthemideae (adapted from Francisco-Ortega et al. 1997). Tree length is 744 steps; CI' = 0.466 (excluding autapomorphies); RI = 0.632. The number of nucleotide changes is indicated along each branch. Bootstrap values (100 replicates) are shown below each branch. Distribution: Mediterranean (●); predominantly Mediterranean (○); predominantly in Europe and East Asia (□); predominantly in Europe and Eastern Asia (■); South Africa (Δ). Arrow indicates clade which has a ca. 17 bp deletion in ITS2.

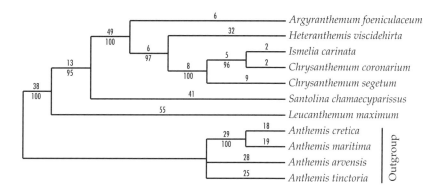

Figure 14.7. Strict consensus tree of the two most parsimonious trees obtained from combined analysis of ITS nucleotide sequences and cpDNA restriction sites (adapted from Francisco-Ortega et al. 1997). The tree has a length of 223 steps; CI' = 0.695 (excluding autapomorphies); RI = 0.848. Bootstrap values > 50% (100 replicates) are shown below each branch. Number of changes are given above each branch.

Other factors could perhaps account for the high genetic diversity within *Argyranthemum*. The large population size of many of the taxa (Humphries 1976a, 1976b; Francisco-Ortega et al. 1996a), the existence of natural hybrid swarms (Borgen 1976), and a breeding system which appears to favor outcrossing might have contributed to the high allozyme diversity of *Argyranthemum*, as suggested for *Schiedea* (Caryophyllaceae) and *Alsinidendron* in the Hawaiian Islands (Weller et al. 1996). However, the continental relatives are also highly allogamous (Jain and Gupta 1960; Howarth and Williams 1972; Chaudhuri et al. 1976), and tend to form large populations, at least for *Chrysanthemum segetum* and *C. coronarium* (Howarth and Williams 1972; Heywood 1976).

The position of *Argyranthemum* in the Chrystantheminae can be explained by three possible evolutionary scenarios for the subtribe: (i) *Argyranthemum* originated in the mainland, but colonized Macaronesia before it became extinct in the continent; (ii) the Chrystantheminae originated in the Macaronesian Islands followed by a subsequent colonization from one of the islands onto the mainland and the formation of three distinct genera; and (iii) two simultaneous and independent radiations from a common extinct ancestor occurred in both the Mediterranean basin and Macaronesia – the first producing the three continental genera of the Chrystantheminae and the second involving radiation of *Argyranthemum* within the Macaronesian archipelagos.

Argyranthemum does not occur on the mainland; however, there are two predominantly Macaronesian groups of Asteraceae (*Sonchus* subg. *Dendrosonchus* and *Cheirolophus*) that also have representatives in the western Mediterranean (Peltier 1973; Boulos 1974; Dostal 1976; Nyffeler 1992). These distributions suggest that *Argyranthemum* may have occurred in both Macaronesia and the mainland, and that the genus became restricted to Madeira and the Canaries following extinction on the continent. Recent molecular phylogenetic analysis of *Aeonium* (Crassulaceae) indicates that the Macaronesian species are basal to the African taxa, even though the genus has a derived position within the subfamily Sempervivoideae (Mes 1995). This relationship strongly suggests that the mainland taxa of *Aeonium* were derived from island species. Thus, we cannot exclude the possibility of dispersal of *Argyranthemum* from Macaronesia to the Mediterranean basin during the evolutionary history of the Chrysantheminae. The low levels of isozyme variation of the continental genera compared with those of *Argyranthemum* argue in favor of this hypothesis. Following such a colonization of the continent, there may have been a severe genetic bottleneck that led to the paucity of levels of genetic variation of the mainland genera of the Chrysantheminae.

Limited population sampling in isozyme studies of the continental genera (Francisco-Ortega et al. 1995b), however, may account for the low genetic variation and few unique alleles in *Chrysanthemum, Ismelia,* and *Heteranthemis*. This also could explain the unexpected high G_{ST} value reported for *C. coronarium*, a species that has a strong self-incompatibility system (Chaudhuri et al. 1976). However, the fact that phylogenies based on cpDNA restriction-site data (Figures 14.2, 14.5) and *ndh*F sequences (L. Watson, unpubl. data) strongly support the position of *Argyranthemum* as the sister-group to the other three genera of the Chrysantheminae suggests that the high isozyme diversity may be due to the fact that the genus is old relative to the continental genera in the subtribe.

TIME OF ORIGIN – Estimates of divergence times from isozyme data suggest that *Argyranthemum* diverged from the rest of the Chrysantheminae between 2.5 and 3.0 million years ago (Mya) (Francisco-Ortega et al. 1995b). Similarly, estimates from cpDNA restriction-site data indicate that divergence took place between 1.5 and 3.0 Mya (Francisco-Ortega et al. 1997). These values place the origin of *Argyranthemum* in the late Tertiary, when all of the Canary Islands were already formed except El Hierro (Carracedo 1994; Figure 14.2). The ITS data were also used to estimate divergence times; however, relative rate tests showed heterogeneity of ITS sequences in the Anthemideae (Francisco-Ortega et al. 1997). Statistically significant differences in rates within the Chrysantheminae were detected in comparisons between *Heteranthemis* and the three other genera. Therefore, estimates of divergence times between *Argyranthemum* and its continental relatives did not include *Heteranthemis*. The values obtained (0.26–3.23 Mya) were generally lower than those obtained from isozymes and cpDNA restriction-site data, although there was considerable overlap. The molecular clock was rejected in almost half of the pairwise comparisons in the Anthemideae, which may explain why times of divergence from the ITS data are not in complete agreement with those from the two other molecular markers.

Evolution within *Argyranthemum*

Three factors have been implicated in the evolution of *Argyranthemum* in Macaronesia: adaptive radiation, ecological vicariance (dispersal to similar ecological zones followed by divergence), and hybridization (Bramwell 1972; Borgen 1976, 1984; Humphries 1976a, 1979; Brochmann 1984, 1987; Marrero-Rodríguez 1992). Two species (*A. sundingii* and *A. escarrei*) have been postulated to originate through hybridization (Borgen 1976, 1980; Brochmann 1987; Figure 14.3d). Borgen (1976, 1984) argued that *Argyranthemum* could provide the best example of the importance of hybridization in the evolution of the Macaronesian flora.

We have addressed four major questions regarding the evolution of *Argyranthemum*: (i) Has radiation of *Argyranthemum* been recent? (ii) What has been the relative role of ecological vicariance and insular radiation in species diversification? (iii) Are there any correlations between the ecology and molecular phylogenies? and (iv) Has hybridization been important in the evolution of the genus?

RECENT RADIATION – Divergence estimates from both isozyme and cpDNA restriction-site data support a recent radiation of *Argyranthemum*. Studies of genetic variation using 17 isozyme loci are in agreement with the pattern reported for other oceanic island groups. The mean genetic identity between species of *Argyranthemum* is 0.895, and pairwise comparisons range between 0.685 and 1.000 (Francisco-Ortega et al. 1996c). The high genetic identity values are concordant with those reported in other oceanic island endemics from the Pacific Ocean (reviewed by Crawford 1990) and Macaronesia (e.g., *Chamaecytisus proliferus* [Fabaceae], Francisco-Ortega et al. 1992). Isozyme data indicate recent speciation in *Argyranthemum*, a result also obtained for island endemics from Pacific Ocean archipelagos (Crawford 1990).

The average levels of genetic diversity in *Argyranthemum* species (H_T = 0.230) are much higher than the mean value (H_T = 0.064) reported for island endemics from the Pacific region (DeJoode and Wendel 1992). However, these values are consistent with the high levels of genetic diversity reported for other Macaronesian endemics such as *Avena canariensis* (Poaceae) (Morikawa and Leggett 1990), *Chamaecytisus proliferus* (Francisco-Ortega et al. 1992), *Dactylis smithii* (Poaceae) (Sahuquillo and Lumaret 1995), and *Lolium canariense* (Poaceae) (Oliveira et al. 1995). These data suggest a recent radiation of the genus into a number of morphological forms in Macaronesia and that, despite the high allozyme diversity and the strong insular and ecological isolation of the species, interspecific genetic identities still remain very high in most pairwise comparisons.

Nucleotide sequence divergence values based on cpDNA restriction-site data are also very low for most of the comparisons (0.122% ± 0.034) and support the hypothesis of recent speciation for *Argyranthemum* in Macaronesia. Estimates of time of divergence using cpDNA restriction-site data indicate that radiation in *Argyranthemum* occurred ca. 1.2 Mya. A recent radiation ca. 0.7 Mya for *Argyranthemum* is also supported by isozyme data (Francisco-Ortega et al. 1995b). Both markers indicate that speciation in *Argyranthemum* began in the middle Quaternary.

ADAPTIVE RADIATION AND INTER-ISLAND COLONIZATION – Sequencing of the ITS region for 16 taxa of *Argyranthemum* identify only eight variable positions, three of which are phylogenetically informative (Francisco-Ortega et al. 1997). At this level the ITS sequences are not sufficient for inferring phylogenetic relationships – only two resolved nodes with low bootstrap support are evident in the ITS trees.

Phylogenetic analyses of cpDNA restriction-site data identify two major clades in *Argyranthemum* (Figure 14.5). The first comprises all the taxa endemic to the northern archipelagos of Madeira, Desertas, and Selvagens. The second lineage includes only the Canary Island taxa. The existence of two major lineages in *Argyranthemum*

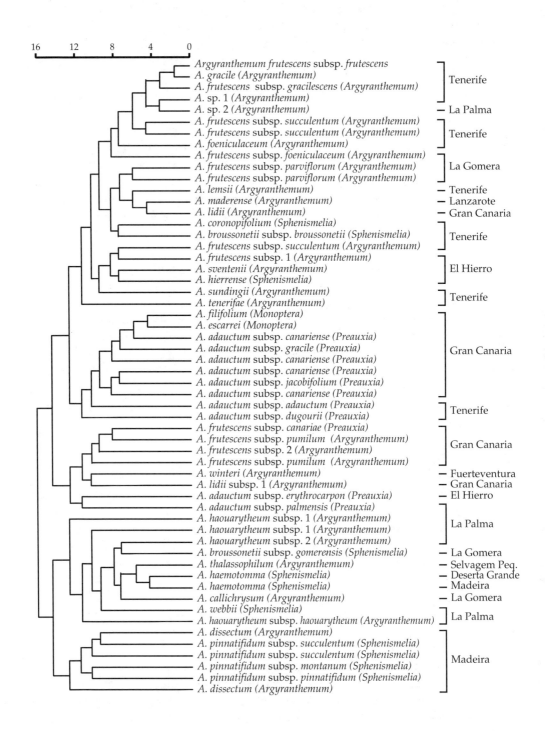

Figure 14.8. Neighbor-joining tree based on Roger's genetic distance obtained from allozyme frequency data from 56 populations of *Argyranthemum* (modified from Francisco-Ortega et al. 1996c).

is also supported by isozyme data. Thirty of the 67 allozymes are unique to the Canaries, whereas the northern archipelagos (Madeira, Desertas, and Selvagens) have only eight unique alleles. The four archipelagos involved in the study share only 29 allozymes. The fact that 57% of the alleles are not shared by the northern archipelagos and the Canary Islands supports the existence of two major geographical lineages for *Argyranthemum* in Macaronesia as seen with the cpDNA analysis.

A phenogram based on isozyme data does not identify two major groups for Macaronesia (Figure 14.8). However, all taxa from Madeira, except *A. haemotomma*, group together. The two other species from the northern archipelagos (*A. haemotomma* and *A. thalassophilum*) are part of a group which comprises species from La Gomera and La Palma.

None of the taxa on single islands form monophyletic groups in the cpDNA phylogeny (Figure 14.5) suggesting that insular radiation has not been of great importance in the evolutionary history of *Argyranthemum*. However, the phylogeny of the taxa from the northern archipelagos (Madeira, Desertas, and Selvagens) suggests that ecological shifts within a single island have occurred at least once in *Argyranthemum*. The best example of insular radiation is on the island of Madeira. All four major ecological zones of this island have been colonized by five different taxa of *Argyranthemum*, and three of them (*A. dissectum, A. pinnatifidum* subsp. *pinnatifidum*, and *A. pinnatifidum* subsp. *montanum*) form a monophyletic group. The basal position of the coastal desert taxa suggests that radiation on this island initiated in this ecological zone. Likewise, insular radiation may have been important in Gran Canaria. All taxa from this island are members of an unresolved (and weakly supported) clade with three taxa from La Gomera and Tenerife (Figure 14.5).

The taxa endemic to the remaining islands (El Hierro, La Palma, La Gomera, and Tenerife) do not form monophyletic groups in the cpDNA tree. Therefore, inter-island colonization rather than insular radiation probably has been the primary manner by which *Argyranthemum* has exploited the rich ecological diversity in Macaronesia. Ecological shifts between the two major climate zones of the Canaries have not occurred very often. Horizontal colonization between habitats appears to be the primary route for establishing new populations in the various ecological zones of the islands. Replication of these zones on different islands provided unique opportunities for the establishment of taxa that were already pre-adapted to these zones.

Molecular phylogenies of other Macaronesian plant groups including the *Sonchus* alliance (Kim et al. 1996), *Crambe* sect. *Dendrocrambe* (Brassicaceae) (Francisco-Ortega et al. 1996d), and *Aeonium* (Mes 1995) also indicate that divergence following inter-island colonization has been much more important than divergence associated with the occupation of new ecological zones in an island. The only example in which molecular data suggests an important role for insular radiation in morphological differentiation of a Macaronesian plant group is the *Chamaecytisus proliferus* complex in the Canary Islands. Both isozyme (Francisco-Ortega et al. 1992) and seed protein data (Sánchez-Yelamo et al. 1995) suggest a dispersal route from the eastern to the western islands with insular radiation into two morphological forms in Gran Canaria, Tenerife, and La Palma. It is noteworthy, however, that taxa from the pine forest of

Gran Canaria and Tenerife are convergent in floral morphology (namely the reflexed standard petal), that may reflect optimization of pollination mechanisms in this particular ecological zone (Francisco-Ortega et al. 1993).

The polyphyly of *Argyranthemum haouarytheum* in the cpDNA phylogeny (Figure 14.5) provides the strongest support for our hypothesis that there has been recent and extensive inter-island colonization following adaptive radiation within a single island. The three populations of *A. haouarytheum* from La Palma are part of the Arid clade from the Canary Islands. Ecogeographical studies (Francisco-Ortega et al. 1996a) suggest that there are two distinct groups in this island. The first comprises two morphological forms from the arid zones of the pine forest and high altitude desert. The second group includes only the morphological form from the humid lowland scrub. The cpDNA phylogeny places these two groups in distinct lineages. The first lineage of *A. haouarytheum* from the humid lowland scrub is basal to a strongly supported clade that comprises more than half of taxa endemic to the Canary Islands. The island of La Palma (formed 1.5 Mya) is the second youngest in the Canaries (Figure 14.2). The basal position of this endemic from La Palma supports the hypothesis that approximately half of the speciation events in *Argyranthemum* in the Canary Islands occurred after most of the islands were already formed. We assume that *A. haouarytheum* originated in La Palma, and that it did not exist in an older island on which it would now be extinct. This assumption seems reasonable because none of the extant Canarian taxa of *Argyranthemum* occur on more than one island. Two possible explanations could account for the polyphyly of *A. haouarytheum*. The first involves hybridization between *A. haouarytheum* and a species adapted to arid zones dispersed from Tenerife or La Gomera (Francisco-Ortega et al. 1996b). This scenario could explain the presence of two morphological forms of *A. haouarytheum* from La Palma in the clade that mainly comprises *A. frutescens* from Tenerife and La Gomera. However, low bootstrap values for this clade (68%) and the clade including its sister-group (59%) suggest that a second hypothesis is possible. This scenario involves an initial radiation of *A. haouarytheum* from the humid lowland scrub into the arid zones of the pine forest and the high altitude desert of La Palma. These forms later colonized other islands with similar environments, providing an example of horizontal colonization between similar ecological zones.

Molecular phylogenies for Macaronesian endemic lizards (Lacertidae: *Gallotia galloti*, Thorpe et al. 1994; Thorpe 1996) and beetles (Tenebrionidae: *Pimelia*, Juan et al. 1995; *Hegeter*, Juan et al. 1996) are not concordant with these evolutionary patterns in *Argyranthemum*. In these animal groups, it appears that insular radiation accounts for the morphological differentiation found within a particular island. A similar pattern – where insular radiation has been the prevalent evolutionary mode – has been reported in molecular phylogenetic studies of Hawaiian plant groups: the silversword alliance (Chapter 3; Baldwin et al. 1990, Baldwin and Robichaux 1995), and Hawaiian lobeliads (Givnish et al. 1995).

Loss of dispersal mechanisms has been suggested as one of the most important consequences of evolution in oceanic islands (Carlquist 1996a, 1996b; Cody and Overton 1996). Reduction in dispersability would be a major obstacle for inter-island colonization and would enhance speciation events on single islands. However, the prevalence of inter-island colonization, suggested by the cpDNA phylogeny for *Argyranthemum*, indicates that dispersal between islands has occurred frequently in the evolution of this genus.

A comparison of continental and insular genera of the Chrysantheminae can provide interesting insights into dispersability in this group. Continental species of *Chrysanthemum* and *Ismelia* have small, usually winged seeds which make them suitable for wind dispersal. Seeds of *Heteranthemis* are three times larger than those of the other continental genera, but they have a pappus with three prominent awns. In contrast, the Macaronesian taxa exhibit a broad range of variation in cypsela morphology (Figure 14.3b). All 21 taxa of sects. *Argyranthemum* and *Stigmatotheca*, with the exception of *A. callichrysum*, *A. sventenii*, and *A. thalassophilum*, have small cypselas with prominent wings and coriaceous pappus. These sections include approximately half of the taxa of *Argyranthemum*. Thus, it appears that the genus has not experienced a significant reduction of dispersability compared with the continental genera. However, taxa from other sections have large cypselas (sect. *Sphenismelia*) or cypselas that are fused into groups (sects. *Monoptera* and *Preauxia*) (Figure 14.3b). These taxa would appear to have a more limited dispersal ability. *Argyranthemum adauctum* fits in this latter group; however, this species has been very succcessful in colonizing different islands as six subspecies occur in four of the Canary Islands. A similar situation has been reported for *Bidens* (Asteraceae) in the Hawaiian islands where *B. menziesii*, a species with reduced awns, is widespread on three islands (Hawaii, Maui, and Molokai). It is noteworthy that six of the seven taxa with large cypselas (*A. broussonetii* subsp. *gomerensis*, *A. coronopifolium*, *A. haemotomma*, *A. thalassophilum*, and *A. webbii*) have a reduced number of populations and also a reduced number of individuals per population. A similar situation is found in two species of sect. *Argyranthemum* (*A. sundingii* and *A. winteri*).

CORRELATIONS BETWEEN ECOLOGY AND PHYLOGENY – The two major clades of *Argyranthemum* from the Canaries appear to have an ecological interpretation. In ecological zones not directly influenced by the northeastern trade winds (high altitude desert, pine forest, arid lowland scrub, and coastal desert), 88% of the taxa are in the Arid clade (Figure 14.5). In contrast, most of the taxa restricted to areas that are influenced by these winds (heath belt, laurel forest, and humid lowland scrub) are in the Humid clade (Clade B in Figure 14.5). This correlation between ecology and phylogeny supports our hypothesis concerning the importance of horizontal colonization between similar ecological zones as a major factor in the evolution of *Argyranthemum*.

Distinct isozyme alleles are not found for either of the ecological clades from the Canaries. The only islands with unique alleles are Lanzarote (1), Tenerife (2), and Gran Canaria (8). It appears that during the evolution of *Argyranthemum* in the Canaries there was virtually no fixation of unique allozymes. Most changes for allozymes involved population frequencies. Ecological isolation had a more pronounced effect, however, on the chloroplast genome in terms of the accumulation of unique changes.

The importance of colonization between similar ecological zones in the evolution of the Macaronesian flora is exhibited in other groups including *Crambe*, *Taeckholmia*

(Asteraceae), *Limonium* (Plumbaginaceae), and *Cheirolophus* (Asteraceae). These genera have most of their taxa occurring in the same habitat but on different islands (Bramwell 1972, 1976; Marrero-Rodríguez 1992).

A correlation between molecular phylogeny and ecology in oceanic islands has also been suggested for the endemic Hawaiian genus *Cyanea* (Givnish et al. 1995). Two major clades (characterized by orange vs. purple fruits) can be recognized in *Cyanea*. Most species in sunnier microsites are restricted to the purple-fruited clade. Orange-fruited taxa occupy shaded microsites, have longer leaves at a given height, produce longer flowers, and frequently bear thornlike prickles and divided leaves in the juvenile state. In *Argyranthemum*, however, there was no correlation between the cpDNA phylogeny and morphology.

HYBRIDIZATION – Hybridization barriers between *Argyranthemum* species are weak and mainly the result of ecological or geographical isolation. Several studies have revealed extensive natural hybridization in areas of sympatry (Borgen 1976; Brochmann 1984, 1987). Introgression of neutral isozyme alleles and lineage sorting could account for the mosaic-like patterns of variation depicted in the phenogram of genetic distances (Francisco-Ortega et al. 1996c; Figure 14.8). The role of hybridization in the evolution of *Argyranthemum* is also evident in the cpDNA restriction-site phylogeny (Francisco-Ortega et al. 1996b; Figure 14.5).

The best example of reticulate evolution was identified in the *Argyranthemum adauctum* species complex. The seven subspecies form a coherent morphological group that is recognized as the distinct section *Preauxia*. Subspecies from the pine forest were part of the Arid clade in the cpDNA tree, whereas taxa from the laurel forest and heath belt (with the exception of *A. adauctum* subsp. *jacobifolium* from Gran Canaria) were part of the Humid clade (Figure 14.5). The three subspecies of *A. adauctum* from Gran Canaria may represent a case of radiation because no other *Argyranthemum* species on this island is restricted to the laurel forest. Unless it is assumed that there has been extensive parallel evolution of morphology in this group, the most likely explanation for the polyphyly of the *A. adauctum* in the cpDNA phylogeny would be hybridization. There are at least 21 restriction-site changes differentiating the two cpDNA groups of the *A. adauctum* complex. Therefore, it is unlikely that convergent restriction-site changes or lineage sorting of ancestral polymorphisms could account for the separation of the subspecies of *A. adauctum* into these two clades. We believe that early colonizers from this complex could have become sympatric with the endemics *A. webbii* in La Palma, *A. broussonetii* in Tenerife, or *A. hierrense* in El Hierro, all of these species thriving in areas influenced by the northeastern trade winds. Hybridization would have given new opportunities for *A. adauctum* to adapt to the ecology of the heath belt of these islands. Isozyme data largely agree with the hybridization hypothesis because all taxa of this complex (except *A. adauctum* subsp. *palmensis* and *A. adauctum* subsp. *erythrocarpon*) form a single cluster with *A. filifolium* and *A. escarrei* in the isozyme tree. Morphological studies also suggest a close relationship between these two species and the *A. adauctum* complex (Humphries 1976b).

There are three other instances (*A. broussonetii*, *A. frutescens*, and *A. hierrense*) in which replicates of the same species group in distinct cpDNA clades. The two subspecies of *A. broussonetii* from the laurel forest of Tenerife and La Gomera do not form a monophyletic group. Two morphological characters are shared by these two subspecies: large, usually red-brown cypselas and broad lobed leaves. However, plants from La Gomera have wingless cypselas that are always fused in groups of 2 to 5 (vs. unfused cypselas with prominent wings), their pappus is absent or reduced to a coriaceous rim (vs. prominent pappus), and their leaves are petiolate (vs. leaf blade decurrent on the stems). Therefore, the polyphyly of *A. broussonetii* is apparently not the result of hybridization or convergent evolution. Instead, we believe that these two taxa are not conspecific. The polyphyly of the two other taxa (*A. hierrense* and *A. frutescens*) is supported weakly. However, replicates of these two taxa are part of two distinct clades that include taxa from several islands. This supports our hypothesis that recent horizontal colonizations between similar ecological zones rather than hybridization are likely to account for the polyphyly of these species.

Our studies indicate that hybridization has been an important factor in the evolution of some Macaronesian endemics, as suggested for the Hawaiian archipelago by Gillett (1972) in *Bidens* (Asteraceae), *Gouldia* (Rubiaceae), *Pipturus* (Urticaceae), and *Scaevola* (Goodeniaceae) and Smith et al. (1996) in *Cyrtandra* (Gesneriaceae). This view of the importance of hybridization in insular habitats is also supported by studies of the Hawaiian silversword alliance (Chapter 3; Baldwin et al. 1990; Carr 1995).

Hybridization in island groups such as *Argyranthemum* or the silverswords could be important in two ways. First, introgression might generate genetic diversity in otherwise uniform island plants. Second, hybridization could possibly lead to new species by stabilization of hybrid recombinants. The importance of these two processes has been demonstrated by Carr (1995) in his elegant study of intergeneric hybridization in the silversword alliance. After just two generations of backcrossing between *Argyroxiphium sandwicense* and *Dubautia menziesii*, there was little segregation for the morphological characters that distinguish these two taxa. In *Argyranthemum*, our data do not resolve whether interspecific hybridization has been involved in the origin of *A. sundingii* (between *A. frutescens* subsp. *frutescens* and *A. broussonetii* [Borgen 1980; Brochmann 1987]) or of *A. escarrei* (between *A. adauctum* subsp. *canariense* and *A. filifolium* [Borgen 1976]). However, it seems likely that hybrid speciation was involved in the origin of some taxa.

Conclusions

We have used the endemic genus *Argyranthemum* as a model system to examine the origin and patterns of evolution of the flora of the Macaronesian Islands. Our studies have included three independent sets of macromolecular data, isozymes, cpDNA restriction sites, and ITS sequences. The results are consistent with a previous hypothesis that some members of the endemic flora of Macaronesia may be ancient in relation to their continental relatives. Furthermore, the primary avenue of taxonomic diversification is extensive inter-island colonization between similar ecological zones. A correlation between the cpDNA phylogeny of *Argyranthemum* and the two climatic regions caused by the trade winds provides strong evidence for the importance of this phenomenon. Incongruence between morphology, isozyme, and cpDNA data supports previous hypotheses (e.g., Borgen 1976; Brochmann 1984, 1987, Marrero-Rodríguez 1992) that hybridization has also played an important role in the evolution of *Argyranthemum*.

Future work on Macaronesian plants is heading in two directions. First, we are expanding our molecular comparisons to the other genera of the Asteraceae (the *Sonchus* alliance, Kim et al. 1996a, 1996b; *Pericallis*, *Cheirolophus*, endemic Inuleae, and the Gonosperminae, Francisco-Ortega et al., unpubl. data), and to genera in other families (Brassicaceae: *Crambe*, Francisco-Ortega et al. 1996d; Malvaceae: *Lavatera*, Fuertes-Aguilar et al. 1996) to determine if the processes detected in *Argyranthemum* are operating for the Macaronesian flora as a whole. Early results from these studies suggest that some elements of the flora may be relictual and that inter-island colonization between similar ecological zones is very important in the Macaronesian Islands. The second direction, in collaboration with D. Levin, is focusing on the role of hybridization on species extinction (Levin et al. 1996) using both cpDNA and RAPD markers. We believe that these ongoing studies will provide a much clearer understanding of the evolution of the rich endemic flora of Macaronesia.

Acknowledgments

This research was supported by a personal grant (Ministerio de Educación y Ciencia, Spain; PF92 42044506) and a specific contract from Instituto Canario de Investigaciones Agrarias (through the advice of M. Fernández-Galván) to JFO and from the National Science Foundation (DEB-9318279) to RKJ. We thank D. Levin for critically reading this paper. The senior author is deeply grateful to T. Delevoryas, D. Levin, B. Simpson, and B. Turner for their friendship, enthusiasm, and encouragement during a postdoctoral leave at the University of Texas.

References

Aldridge, A. E. 1979. Evolution within a single genus: *Sonchus* in Macaronesia. Pp. 279–291 in *Plants and islands*, D. Bramwell, ed. London, England: Academic Press.

Baker, H. G. 1955. Self-compatibility and establishment after "long distance" dispersal. Evolution 9:347–349.

Baker, H. G. 1967. Support for Baker's law – as a rule. Evolution 12:853–856.

Baldwin, B. G., Kyhos, D. W., and J. Dvorák, J. 1990. Chloroplast DNA evolution and adaptive radiation in the Hawaiian silversword alliance (Asteraceae - Madiinae). Annals of the Missouri Botanical Garden 77:96–109.

Baldwin, B. G., and Robichaux, R. H. 1995. Historical biogeography and ecology of the Hawaiian silversword alliance (Asteraceae): new molecular phylogenetic perspectives. Pp. 259–287 in *Hawaiian biogeography: evolution on a hot spot archipelago*, W. L. Wagner and V. A. Funk, eds. Washington, DC: Smithsonian Institution Press.

Boekschoten, G. J., and Manuputty, J. A. 1993. The age of the Cape Verde Islands. Courier des Forschungsinstitutes Senckenberg 159:3–5.

Borgen, L. 1976. Analysis of a hybrid swarm between *Argyranthemum adauctum* and *A. filifolium* in the Canary Islands. Norwegian Journal of Botany 23:121–137.

Borgen, L. 1980. A new species of *Argyranthemum* (Compositae) from the Canary Islands. Norwegian Journal of Botany 27:163–165.

Borgen, L. 1984. Biosystematics of Macaronesian flowering plants. Pp. 477–496 in *Plant biosystematics*, V. F. Grant, ed. Toronto, Canada: Academic Press.
Boulos, L. 1974. Révision systématique du genre *Sonchus* s. l. V. Sous-genre 2. *Dendrosonchus*. Botaniska Notiser 127:7–37.
Bramwell, D. 1972. Endemism in the flora of the Canary Islands. Pp. 141–159 in *Taxonomy, phytogeography and evolution*, D. H. Valentine, ed. London, England: Academic Press.
Bramwell, D. 1976. The endemic flora of the Canary Islands; distribution, relationships and phytogeography. Pp. 207–240 in *Biogeography and ecology in the Canary Islands*, G. Kunkel, ed. The Hague, The Netherlands: Dr W. Junk.
Bramwell, D. 1990. Conserving biodiversity in the Canary Islands. Annals of the Missouri Botanical Garden 77:28–37.
Bremer, K., and Humphries, C. J. 1993. Generic monograph of the Asteraceae-Anthemideae. Bulletin of the Natural History Museum of London, Botany 23:71–177.
Brochmann, C. 1984. Hybridization and distribution of *Argyranthemum coronopifolium* (Asteraceae - Anthemideae) in the Canary Islands. Nordic Journal of Botany 4:729–736.
Brochmann, C. 1987. Evaluation of some methods for hybrid analysis, exemplified by hybridization in *Argyranthemum* (Asteraceae). Nordic Journal of Botany 7:609–630.
Carlquist, S. 1965. *Island life: a natural history of the islands of the world*. New York, NY: Natural History Press.
Carlquist, S. 1966a. The biota of long-distance dispersal. I. Principles of dispersal and evolution. Quarterly Reviews in Biology 41:247–270.
Carlquist, S. 1966b. The biota of long-distance dispersal. II. Loss of dispersability in the Pacific Compositae. Evolution 20:30–48.
Carlquist, S. 1974. *Island biology*. New York, NY: Columbia University Press.
Carr, G. D. 1995. A fully fertile intergeneric hybrid derivative from *Argyroxiphium sandwicence* ssp. *macrocephalum* x *Dubautia menziesii* (Asteraceae) and its relevance to plant evolution in the Hawaiian Islands. American Journal of Botany 82:1574–1581.
Carr, G. D., Powell, E. A., and Kyhos, D. W. 1986. Self-incompatibility in the Hawaiian Madiinae (Compositae): an exception to Baker's rule. Evolution 40:430–434.
Carracedo, J. C. 1994. The Canary Islands: an example of structural control on the growth of large oceanic-island volcanoes. Journal of Volcanology and Geothermal Research 60:225–241.
Chaudhuri, B. K., Chaudhuri, S. K., Basak, S. L., and Dana, S. 1976. Cytogenetics of a cross between two species of annual *Chrysanthemum*. Cytologia 41:111–121.
Christensen, L. P. 1992. Acetylenes and related compounds in Anthemideae. Phytochemistry 31:7–49.
Cody, M. L., and Overton, J. M. 1996. Short-term evolution of reduced dispersal in island plant populations. Journal of Ecology 84:53–61.
Crawford, D. J. 1990. *Plant molecular systematics*. New York, NY: John Wiley and Sons.
Cunneen, T. M. 1995. Breeding for improvement of the Marguerite Daisy (*Argyranthemum* spp.). Acta Horticulturae 420:101–103.
DeJoode, D. R., and Wendel, J. F. 1992. Genetic diversity and origin of the Hawaiian Islands cotton, *Gosyppium tomentosum*. American Journal of Botany 79:1311–1319.
Dostal, J. 1976. *Cheirolophus* Cass. Pp. 249–250 in *Flora Europaea, vol. 4*, T. G. Tutin, V. H. Heywood, N. A. Burges, D. M. Moore, D. H. Valentine, S. M. Walters, and D. A. Webb, eds. Cambridge, England: Cambridge University Press.
Francisco-Ortega, J., Jackson, M. T., Catty, J. P., and Ford-Lloyd, B. V. 1992. Genetic diversity in the *Chamaecytisus proliferus* complex (Fabaceae: Genisteae) in the Canary Islands in relation to in situ conservation. Genetic Resources and Crop Evolution 39:149–158.
Francisco-Ortega, J., Jackson, M. T., Santos-Guerra, A., and Ford-Lloyd, B. V. 1993. Morphological variation in the *Chamaecytisus proliferus* (L.fil.) complex (Fabaceae: Genisteae) in the Canary Islands. Botanical Journal of the Linnean Society 112:187–202.
Francisco-Ortega, J., Jansen, R. K., Crawford, D. J., and Santos-Guerra, A. 1995a. Chloroplast DNA evidence for intergeneric relationships of the Macaronesian endemic genus *Argyranthemum* (Asteraceae). Systematic Botany 20:413–422.
Francisco-Ortega, J., Crawford, D. J., Santos-Guerra, A., and Sa-Fontinha, S. 1995b. Genetic divergence among Mediterranean and Macaronesian genera of the subtribe Chrystheminae (Asteraceae). American Journal of Botany 82:1321–1328.

Francisco-Ortega, J., Santos-Guerra, A., Mesa-Coello, R., González-Feria, E., and Crawford, D. J. 1996a. Genetic resource conservation of the endemic genus *Argyranthemum* Sch. Bip. (Asteraceae: Anthemideae) in the Macaronesian Islands. Genetic Resources and Crop Evolution 43:33–39.

Francisco-Ortega, J., Jansen, R. K., and Santos-Guerra, A. 1996b. Chloroplast DNA evidence of colonization, adaptive radiation, and hybridization in the evolution of the Macaronesian flora. Proceedings of the National Academy of Sciences, USA 93:4085–4090.

Francisco-Ortega, J., Crawford, D. J., Santos-Guerra, A., and Carvalho, J. A. 1996c. Isozyme differentiation in the endemic genus *Argyranthemum* (Asteraceae: Anthemideae) in the Macaronesian Islands. Plant Systematics and Evolution 202:137–152.

Francisco-Ortega, J., Fuertes-Aguilar, J., Kim, S. C., Crawford, D. J., Santos-Guerra, A., and Jansen, R. K. 1996d. Molecular evidence for the origin, evolution, and dispersal of *Crambe* (Brassicaceae) in the Macaronesian Islands. P. 41 in *Abstracts 2nd symposium fauna and flora of the Atlantic islands*. Las Palmas de Gran Canaria: Universidad de Las Palmas de Gran Canaria.

Francisco-Ortega, J., Santos-Guerra, A., Hines, A., and Jansen, R. K. 1997. Molecular evidence for a Mediterranean origin of the Macaronesian endemic genus *Argyranthemum* (Asteraceae). American Journal of Botany (in press).

Fuertes-Aguilar, J., Ray, M. F., Francisco-Ortega, J., and Jansen, R. K. 1996. Systematics and evolution of the Macaronesian endemic Malvaceae based on morphological and molecular evidence. P. 51 in *Abstracts 2nd symposium fauna and flora of the Atlantic islands*. Las Palmas de Gran Canaria: Universidad de Las Palmas de Gran Canaria.

Galopim de Carvalho, A. M., and Brandão, J. E. 1991. *Geologia do Archipélago da Madeira*. Lisbon, Portugal: Museu Nacional de História Natural (Mineralogia e Geologia) da Universidade de Lisboa.

Gillett, G. W. 1972. The role of hybridization in the evolution of the Hawaiian flora. Pp. 205–219 in *Taxonomy, phytogeography, and evolution*, D. H. Valentine, ed. London, England: Academic Press.

Givnish, T. J., Sytsma, K. J., Smith, J. F., and Hahn, W. J. 1995. Molecular evolution, adaptive radiation, and geographic speciation in *Cyanea* (Campanulaceae), the largest plant genus endemic to Hawaii. Pp. 288–337 in *Hawaiian biogeography: evolution on a hot spot archipelago*, W. L. Wagner and V. A. Funk, eds. Washington, DC: Smithsonian Institution Press.

Greger, H. 1977. Anthemideae - chemical review. Pp. 899–941 in *The biology and chemistry of the Compositae, vol. 2*, V. H. Heywood, J. B. Harborne, and B. L. Turner, eds. London, England: Academic Press.

Hansen, A., and Sunding, P. 1993. Flora of Macaronesia. Checklist of vascular plants, 4th revised ed. Sommerfeltia 17:1–296.

Heywood, V. H. 1976. *Chrysanthemum* L. Pp. 168–169 in *Flora Europaea, vol. 4*, T. G. Tutin, V. H. Heywood, N. A. Burges, D. M. Moore, D. H. Valentine, S. M. Walters, and D. A. Webb, eds. Cambridge, England: Cambridge University Press.

Heywood, V. H., and C. J. Humphries. 1977. Anthemideae - systematic review. Pp. 851–898 in *The biology and chemistry of the Compositae, vol. 2*, V. H. Heywood, J. B. Harborne, and B. L. Turner, eds. London, England: Academic Press.

Howard, R. A. 1973. The vegetation of the Antilles. Pp. 1–38 in *Vegetation and vegetational history of northern Latin America*, A. Graham, ed. Amsterdam, The Netherlands: Elsevier.

Howarth, S. E., and Williams, J. T. 1972. *Chrysanthemum segetum* L. Journal of Ecology 60:573–584.

Humphries, C, J. 1975. Cytological studies in the Macaronesian genus *Argyranthemum* (Compositae: Anthemideae). Botaniska Notiser 128:239–255.

Humphries, C. J. 1976a. Evolution and endemism in *Argyranthemum* Webb ex Schultz Bip. (Compositae: Anthemideae). Botánica Macaronésica 1:25–50.

Humphries, C. J. 1976b. A revision of the Macaronesian genus *Argyranthemum* Webb ex Schultz Bip. (Compositae - Anthemideae). Bulletin of the Natural History Museum of London, Botany 5:145–240.

Humphries, C. J. 1979. Endemism and evolution in Macaronesia. Pp. 171–199 in *Plants and islands*, D. Bramwell, ed. London, England: Academic Press.

Hutchinson, J. 1917. Notes on African Compositae: IV. *Matricaria* Linn. and *Chrysanthemum*, DC. Kew Bulletin 1917:111–118.

Jain, H. K., and Gupta, S. B. 1960. Genetic nature of self-incompatibility in annual *Chrysanthemum*. Experientia 16:364–365.

Juan, C., Oromi, P., and Hewitt, G. M. 1995. Mitochondrial DNA phylogeny and sequencial colonization of Canary Islands by darkling beetles of the genus *Pimelia* (Tenebrionidae). Proceedings of the Royal Society of London, Series B 261:173–180.

Juan, C., Oromi, P., and Hewitt, G. M. 1996. Phylogeny of the genus *Hegeter* (Tenebrionidae, Coleoptera) and its colonization of the Canary Islands deduced from Cytochrome Oxidase I mitchondrial DNA sequences. Heredity 76:392–403.

Kim, S. C., Crawford, D. J., Francisco-Ortega, J., and Santos-Guerra, A. 1996. A common origin for woody *Sonchus* and five related genera in the Macaronesian Islands: molecular evidence for extensive radiation. Proceedings of the National Academy of Sciences, USA 93:7743–7748.

Kitamura, S. 1978. *Dendranthema* and *Nipponanthemum*. Acta Phytotaxonomica et Geobotanica, Kyoto 29:165–170.

La Roche, F., and Rodríguez-Piñero, J. C. 1994. Aproximación al número de táxones de la flora vascular silvestre de los archipiélagos macaronésicos. Revista de la Academia Canaria de Ciencias 6:77–98.

Levin, D. A., Francisco-Ortega, J., and Jansen, R. K. 1996. Hybridization and the extinction of rare plant species. Conservation Biology 10:10–16.

Marrero-Rodríguez, A. 1992. Evolución de la flora canaria. Pp. 55–92 in *Flora y vegetación del Archipiélago Canario. Tratado florístico, 1a. parte*, G. Kunkel, ed. Las Palmas de Gran Canaria: Edirca.

Mes, T. H. M. 1995. Origin and evolution of the Macaronesian Sempervivoideae (Crassulaceae). Ph.D. dissertation. Utrecht, Netherlands: University of Utrecht.

Mitchell-Thomé, R. C. 1985. Radiometric studies in Macaronesia. Boletim do Museu Municipal do Funchal 37:52–85.

Morikawa, T., and Leggett, J. M. 1990. Isozyme polymorphism in natural populations of *Avena canariensis* from the Canary Islands. Heredity 64:403–411.

Nyffeler, R. 1992. A taxonomic revision of the genus *Monanthes* Haworth (Crassulaceae). Bradleya 10:49–82.

Oliveira, K. A., Arbones, E., and Bregu, R. 1995. Diversidad genética en poblaciones naturales de *Lolium canariense*. Pp. 21–24 in *Actas 35th reunión científica de la Sociedad Española para el Estudio de Pastos*. La Laguna, Tenerife: Centro de Investigación y Tecnología Agrarias and Universidad de La Laguna.

Peltier, J. R. 1973. Endémiques macaronésiènnes au Maroc. Inventaire bibliographique et problémes taxonomiques. Monographiae Biologicae Canariensis 4:134–142.

Rothe, P. 1982. Zur geologie der Kapverdischen Inseln. Courier des Forschungsinstitutes Senckenberg 52:1–9.

Rustan, Ø. H. 1981. Infraspecific variation in *Argyranthemum pinnatifidum* (Lowe) Lowe. Bocagiana 55:2–18.

Sahuquillo, E., and Lumaret, R. 1995. Variation in the subtropical group of *Dactylis glomerata* L. 1. Evidence from allozyme polymorphism. Biochemical Systematics and Ecology 23:407–418.

Sánchez-Yelamo, M. D., Espejo-Ibáñez, M. C., Francisco-Ortega, J., and Santos-Guerra, A. 1995. Electrophoretical evidence of variation in populations of the fodder legume *Chamaecytisus proliferus* from the Canary Islands. Biochemical Systematics and Ecology 23:53–63.

Santos-Guerra, A. 1983. *Vegetación y flora de La Palma*. Santa Cruz de Tenerife, Canary Islands: Interinsular Canaria.

Santos-Guerra, A., Francisco-Ortega, J., and González-Feria, E. 1993. Contribution to the knowledge of genus *Argyranthemum* Webb ex Schultz Bip. (Compositae) in the Canary Islands. P. 27 in *Abstracts 1st symposium fauna and flora of Atlantic islands*. Funchal, Madeira: Museu Municipal do Funchal.

Smith, J. F., Burke, C. C., and Wagner, W. L. 1996. Interspecific hybridization in natural populations of *Cyrtandra* (Gesneriaceae) on the Hawaiian Islands – evidence from RAPD markers. Plant Systematics and Evolution 200:61–77.

Sunding, P. 1979. Origins of the Macaronesian flora. Pp. 13–40 in *Plants and islands*, D. Bramwell, ed. London, England: Academic Press.

Takhtajan, A. 1986. *Floristic regions of the world* (translation by C. Jeffrey). Edinburgh, Scotland: Oliver and Boyd.

Thorpe. R. S. 1996. The use of DNA divergence to help determine the correlates of evolution of morphological characters. Evolution 50:524–531.

Thorpe, R. S., McGregor, D. P., Alastair, M. C., and Jordan, W. C. 1994. DNA evolution and colonization sequence of island lizards in relation to geological history: mtDNA RFLP, Cytochrome b, Cytochrome Oxidase, 12S rRNA sequence, and nuclear RAPD analysis. Evolution 48:230–240.

Weller, S. G., Sakai, A. K., and Straub C. 1996. Allozyme diversity and genetic identity in *Schiedea* and *Alsinidendron* (Caryophyllaceae: Alsinoideae) in the Hawaiian Islands. Evolution 50:23–34.

15 Plant-Pollinator Interactions and Floral Radiation in *Platanthera* (Orchidaceae)

Jeffrey R. Hapeman and Ken Inoue

The Orchidaceae, with ca. 20,000 species (Dressler 1993), is perhaps the largest family of flowering plants. Given the spectacular floral diversity of the orchids, plant-pollinator interactions are thought to be one of the primary selective forces driving the remarkable diversification of this family (Benzing 1987; Dressler 1993). The significance of floral specialization in the radiation of the orchids was first recognized by Charles Darwin (1877) in his book *The Various Contrivances by which Orchids are Fertilized by Insects*. Darwin described the importance of orchid floral structure in attracting pollinators and the relation of floral structure to pollinator morphology, as well as the important role of the pollinators in orchid reproduction. Written at a time when many in the scientific community felt that floral structure had little or no relation to pollination, his book convincingly demonstrated the adaptive nature of orchid floral structure. In particular, Darwin perceptively described how floral morphology and pollinator morphology could interact, leading to the possibility of what would now be termed plant-pollinator coevolution.

The research of Darwin and those who followed him (including Dodson 1962; van der Pijl and Dodson 1966; Dressler 1981, 1993; Williams 1982; Nilsson 1992) has continued to demonstrate the importance of floral differentiation in orchid diversification. The spectacular variation in floral form and color, the production of ethereal scents to lure oil-collecting bees, and the tremendous variation in the placement of pollinia on pollinators have been studied and attempts have been made to relate them to orchid phylogeny. Yet much of this research has been plagued by one nagging problem: orchid phylogeny is based on a systematic scheme derived primarily from floral morphology, the same characters whose adaptive significance is typically under study (Chase and Palmer 1992). In order to avoid bias or circularity (Maddison and Maddison 1992), an independent source of characters is needed when studying orchid floral evolution (contra Brooks and McLennan 1994). Historically, this has been difficult to achieve using morphology, given that vegetative characters are fairly conserved, particularly at the generic level (Benzing 1987; Dressler 1993). The advent of molecular techniques of phylogenetic inference, with the possibility of a wealth of variable characters (Hillis 1987; Clegg 1993), has provided an excellent solution to the problem of character bias (Sytsma 1990; Givnish et al. 1995) and has opened up the possibility of investigating floral evolution in the orchids in an unbiased fashion (Chase and Palmer 1992).

We have begun a molecular systematic study of the genus *Platanthera* (Orchidaceae) and related orchidoid genera to investigate the role of floral divergence and pollinator diversification (and the interaction between the two) in orchid diversification. *Platanthera*, with ca. 85 species, is the largest genus of north temperate terrestrial orchids

Table 15.1. Distribution, floral syndrome, and pollination ecology of *Platanthera* species under study

Section/Group Taxon	Distribution	Color	Pollination time	Pollinia placement	Pollinator	References
Blephariglottis*						
blephariglottis	E U.S.	white	both	eye	moths, butterflies	Smith & Snow 1976
ciliaris	E U.S.	orange	diurnal	eye	*Papilio* spp.	Robertson & Wyatt 1990
cristata	SE U.S.	orange	diurnal	eye?	butterflies, bees?	Folsom 1979, Hapeman unpubl.
integrilabia	SE U.S.	white	both	eye	hawkmoths, butterflies	Zettler 1996
grandiflora	E U.S.	purple	diurnal	eye	*Hemaris* spp., butterflies	Stoutamire 1974, Moldenke 1949
lacera	E U.S.	green/white	both	proboscis	noctuid moths, *Hemaris* spp.	Duckett 1983, Stoutamire 1974
leucophaea	E U.S.	white	nocturnal	proboscis	hawkmoths	Sheviak & Bowles 1986
peramoena	E U.S.	purple	diurnal	eye	*Hemaris thysbe*	Hapeman 1997
praeclara	Central U.S.	white	nocturnal	eye	hawkmoths	Cuthrell 1994
psycodes	E U.S.	purple	both	proboscis	*Hemaris* spp., butterflies, moths	Stoutamire 1974
Gymnadeniopsis†						
clavellata	E U.S.	green	selfing	proboscis	selfing	Catling 1991
integra	SE U.S.	yellow	diurnal	proboscis	bees, butterflies, selfing	Morong 1893, Luer 1975
nivea	SE U.S.	white	?	proboscis	butterflies?	
Platanthera?						
bifolia	palearctic	green/white	nocturnal	proboscis	hawkmoth	Nilsson 1983
metabifolia	E palearctic	white	nocturnal	proboscis	hawkmoth	Inoue 1983
orbiculata	nearctic	green/white	nocturnal	eye	noctuid moths	Stoutamire unpubl.
okuboi	Japan	green/white	nocturnal	proboscis	hawkmoths	Inoue 1983
obtusata	nearctic, Sweden	green/white	nocturnal	eye	mosquitoes, pyralid moths	Thien and Utech 1972, Voss and Riefner 1983
hookeri	E nearctic	green/yellow	nocturnal	eye	noctuid moths?	
macrophylla	E nearctic	green/white	nocturnal	eye	noctuid moths or hawkmoths?	
Boninensis						
boninensis	Bonin Isl.	cream	nocturnal	proboscis	noctuid moths	Inoue 1983

Taxon	Location	Color	Activity	Attachment	Pollinator	Reference
Tipuloides*						
tipuloides ssp. *nipponica*	Japan	green	nocturnal	proboscis	noctuid moths	Inoue 1983
tipuloides ssp. *sororia*	Japan	green	nocturnal	proboscis	hawkmoths, noctuid moths	Inoue 1983
Sachalinensis						
sachalinensis	E Asia	green	nocturnal	proboscis	noctuid moths	Inoue 1983
Mandarinorum*						
mandarinorum ssp. *ophrydioides*	Japan	green	nocturnal	eye	noctuid & geometrid moths	Inoue 1983
mandarinorum ssp. *hachijoensis*	Japan	green	nocturnal	eye	hawkmoths	Inoue 1983
amabilis	Japan	green	nocturnal	proboscis	noctuid moths?	Inoue 1983
minor	E Asia	green	xnocturnal	eye		Inoue 1983
Limnorchis						
dilatata var. *leucostachys*	NW U.S.	white	nocturnal	proboscis	noctuid moths	Kipping 1971
hyperborea var. *hyperborea*	nearctic	green	both	proboscis	bees, butterflies, noctuid moths	Catling & Catling 1989
hyperborea var. *viridiflora*	Japan, Aleutia	green	?	proboscis	noctuid and/or pyralid moths?	
stricta	NW. U.S.	green	both	proboscis	geometrid moths, bees, flies	Patt et al. 1989
Japonica						
japonica	E Asia	white	nocturnal	proboscis	hawkmoths?	
Tulotis						
fuscescens	E Asia, Aleutia	green	nocturnal	proboscis	noctuid moths?	
sonoharae	Ryukyu Isl.	green	nocturnal	proboscis	noctuid moths?	
ussuriensis	E Asia	green	nocturnal	proboscis	pyralid moths	Inoue 1983

* polyphyletic
† paraphyletic
? not enough resolution to determine

(Ohwi 1965; Inoue 1983; Mabberley 1987; Wood et al. 1993). Species of *Platanthera* occur in North America, Asia, Europe, North Africa, as well as Borneo and Sarawak. Major centers of diversity are found in North America and East Asia. Compared to other genera in the tribe Orchideae, *Platanthera* has apparently undergone a tremendous radiation in floral form and pollination syndrome (Plate 3). Flowers range from green and white to bright purple or orange, and may have elaborate dissections of the labellum and petals, lending the flowers a "fringed" appearance (Plate 3). All species have a nectar spur on the lip, whose length may exceed 4 cm in some taxa. The majority of species are pollinated by noctuid and pyralid moths, but there are species pollinated by beetles, butterflies, hawkmoths, bumblebees, flies and even mosquitoes (see Table 15.1). The diversity of pollination syndromes found in *Platanthera* thus represents nearly all of the non-deceptive pollination syndromes found in the Orchidaceae (van der Pijl and Dodson 1966). As a consequence, *Platanthera* serves as a small-scale model for understanding the role of pollinators and floral specialization in the adaptive radiation of the Orchidaceae as a whole.

This paper will examine preliminary results from our research on floral evolution in *Platanthera*. Questions addressed include:
- What have been the pathways to different floral syndromes?
- How have important traits such as pollinia placement and flower color evolved? Is there evidence of repeated independent evolution (i.e., convergence) of similar floral morphology as a result of sharing similar pollinators?
- How useful is floral morphology for inferring phylogeny?

To address these questions, we overlay floral morphological traits and pollination data on the molecular phylogeny to determine the most parsimonious states for the origin and evolution of these traits (Kocher et al. 1993; Crisp 1994; Givnish et al. 1995).

Systematic Considerations

Platanthera is a member of the tribe Orchideae, which contains approximately 1,300 species of terrestrial orchids in 57 genera. Relationships between the genera are poorly understood, and indeed, even the generic boundaries are unclear (Dressler 1993). Traditionally, American botanists have taken a broad taxonomic view by including all North American species of *Platanthera* in the genus *Habenaria* (Ames 1910; Correll 1950). More recently, however, specialists on the Orchidaceae have recognized *Platanthera* as distinct (Luer 1975; Case 1987; Smith 1993; Homoya 1994); Dressler (1993) goes so far as to place *Platanthera* and *Habenaria* in different subtribes of the Orchideae. As part of our search for appropriate outgroup taxa, we are investigating intergeneric relationships in the Orchideae (in collaboration with colleagues at the Royal Botanic Gardens, Kew). As we will report elsewhere, we have found that *Platanthera* is clearly monophyletic and rather distantly removed from *Habenaria* s.s. Thus, the current trend to recognize *Platanthera* as distinct appears to be well justified; however, a precise definition of synapomorphic morphological characters that unite the various species of *Platanthera* is still lacking. Currently, *Pla-*

tanthera is defined by (i) the relatively broad anther; (ii) the presence of fusiform or elongate root-tuberoids (or tuberoids lacking), as opposed to spheroid or globose tuberoids; and (iii) a stigma that lacks processes and is united into one large receptive surface. The inadequacy of these characters in defining *Platanthera* is demonstrated by the fact that several species (*P. nivea*, *P. clavellata*, and *P. integra*) have stigmatic processes, although they are smaller than those typically found in species of *Habenaria* s.s.

Ecology and Morphology of *Platanthera*

Platanthera grow in a variety of habitats ranging from acid, peaty savannas and bogs to circumneutral floodplains and prairies, and can also be found in both temperate and boreal woodlands. A few species grow in the tropical montane rain forests of Borneo (Luer 1975; Case 1987; Smith 1993; Wood et al. 1993; Homoya 1994).

Vegetative characters shared by *Platanthera* and other taxa in the tribe Orchideae include soft, entire, and sometimes fleshy leaves. The leaves are cauline in most species, although some are characterized by basal leaves.

Flowers of *Platanthera* are typical of those of other orchidoid genera. The single fertile anther (a synapomorphy uniting all orchid taxa above the primitive apostasioids and cypripedioids) has been secondarily divided into two structures, variously called pollinia, pollinaria, or hemipollinaria (hereafter referred to as pollinia) (Dressler 1993). The pollinia are composed of pollen masses held together by elastic threads, and are drawn out at one end into a caudicle. The caudicle is capped at its apex by a sticky disk of tissue, the viscidium, which serves to attach the pollinium to the pollinator (Figure 15.1). Since most of the pollinators of *Platanthera* are lepidopterans and thus are hairy/scaly over most of their body, the only smooth surfaces where the sticky viscidia can effectively attach to the pollinators are the eyes and proboscis. Indeed, the pollinia attach to one of these two locations in the pollinators of all species of *Platanthera*. The labellum is well developed, resupinate, and bears a nectar spur which varies in length from about 2 mm in *P. stricta* to more than 40 mm in *P. praeclara* (Patt et al. 1989; Cuthrell 1994). The nectar spur varies in shape from short and saccate (*P. stricta*), to elongate (*P. praeclara*), to somewhat S-shaped

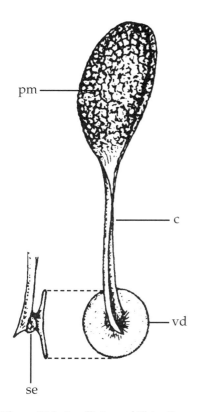

Figure 15.1. A pollinium of *Platanthera chlorantha*. **pm**=pollen mass, **c**=caudicle, **vd**=viscidium, **se**=stipe (from Nilsson 1978, used with permission).

(*P. peramoena*). Inoue (1983) has suggested that spur curvature may be functionally related to pollinator class. The petals and labellum are typically entire but are occasionally dentate or deeply divided, the latter giving the flowers a "fringed" appearance. The labellum may bear fleshy tubercles or projections. Flowers may be yellow, orange, purple, white or green, although most are green or whitish-green (see Plate 3). Some flowers (such as those of *P. leucophaea* and *P. chlorantha*) are strongly scented, while others (such as *P. peramoena*) are at most weakly scented. Scent production is often strongly periodic. In many species the scent is only released in the early hours of the evening when pollinators are most active, while at other times the flowers are virtually scentless (Nilsson 1978; Sheviak and Bowles 1986).

Apart from these general characteristics, many species express a variety of additional specializations. In a number of species, such as *Platanthera dilatata* and *P. hyperborea*, the labellum is "trapped" in an upright position for the first few days of anthesis, blocking access to the stigma and thus making the flowers functionally protandrous (Plate 3 J, M). A tubercle on the labellum in section *Tulotis* serves to deflect the proboscis of the pollinator, allowing removal of only one pollinium at a time (Plate 3 G) (Stoutamire 1971; Inoue 1983). The flowers of *P. hookeri* are strongly "hook-shaped" in profile which may force pollinators to approach the flower from one side at a time, serving a similar function to the tubercle in *Tulotis* (Catling and Catling 1991).

There is a tremendous diversity of pollination syndromes within *Platanthera*, involving apparent adaptations to attract and ensure fertilization mediated by specific groups of insects (Table 15.1). The majority of taxa are pollinated by noctuid and pyralid moths, and hawkmoths are also common pollinators. One or two taxa are pollinated by bumblebees, several by butterflies, and one by beetles (*P. chorisiana*, not included in this analysis). Several taxa are pollinated by both moths and mosquitoes (Thien and Utech 1970; Voss and Riefner 1983), and at least one species is pollinated by flies, bees, and moths (Patt et al. 1989). Some species, such as *P. hyperborea*, appear to be pollinated by several different types of pollinators. Detailed studies, however, usually reveal that one class of pollinator or one class of proboscis length predominates, although a number of insects may act as pollinators of a given species (Inoue 1983; Nilsson 1978, 1983; Patt et al. 1989). Catling and Catling (1991) have suggested that the number of pollinator species varies inversely with spur length in *Platanthera*.

Pollination data were primarily collected from the literature. The pollination biology of *Platanthera* has been well studied, primarily through the efforts of Warren Stoutamire in the United States, Ken Inoue in Japan and eastern Asia, and L. Anders Nilsson in Europe. The senior author has supplemented this with field observations of several critical taxa during the past two years.

Molecular Systematics

Fresh or silica gel-dried plant material was collected for 19 taxa during numerous trips throughout the eastern United States by the senior author. Ken Inoue collected leaf material for all of the East Asian taxa included in the analysis, and Mark

Chase kindly provided DNA for additional U.S. taxa. Leaf material for a number of other taxa was obtained by the senior author through correspondence with various colleagues. Leaf material was obtained from all currently recognized sections of the genus, usually with more than one taxon representing each section. Voucher specimens for DNA samples are deposited at the University of Wisconsin Herbarium (WIS) and the Shinshu University Herbarium, Japan (SHIN).

DNA extraction

Total DNAs were extracted from fresh, silica gel-dried or −80° C frozen leaf tissue collected from individual plants using either a 2% CTAB (Doyle and Doyle 1987) or a modified 6% CTAB procedure (Smith et al. 1991). Leaf tissue was precooled in liquid nitrogen to maximize DNA yield. Some DNA samples were further purified on CsCl gradients.

Amplification and sequencing

The Internal Transcribed Spacer (ITS) region of nuclear ribosomal DNA (nrDNA) was amplified from total genomic DNA via the polymerase chain reaction (PCR) (Oste 1988; Ehrlich 1989) using the primers "ITS4" and "ITS5" (White et al. 1990) and the protocol detailed in Baum et al. (1994). The ITS region has been shown to evolve rapidly, and is useful for resolving phylogenetic relationships at fairly low taxonomic levels (Baldwin 1992; Baum et al. 1994; Baldwin et al. 1995; Yuan and Küpfer 1995).

PCR-amplified DNA was cleaned with the GeneClean II™ kit (Bio 101) and then directly sequenced via the dideoxy method (Sanger et al. 1977), using the Sequenase™ 2.0 kit (US Biochemicals). Primers for sequencing included the two primers used for amplification, as well as "ITS2" and "ITS3B" (White et al. 1990). A modified double-stranded sequencing protocol described in Baum et al. (1994) was followed. Some taxa were sequenced with the AmpliTaq FS dye-terminator cycle sequencing kit (Perkin-Elmer) and analyzed on an ABI Model 377 automated sequencer (Applied Biosystems Inc.). PCR-amplified DNA for these taxa was cleaned using Qiaquick spin columns (Qiagen Inc.).

Sequence alignment and phylogenetic analysis

Sequences of ITS1 and ITS2 were manually aligned using SeqApp 1.9 (Gilbert 1993) and analyzed using Fitch parsimony in PAUP 3.1.1 (Swofford 1993). Alignment required the insertion of a number of gaps; however, most alignments were fairly unambiguous. When several alignment alternatives were found, the alignment that yielded the fewest informative characters was chosen. This is the most conservative approach and minimizes the *a priori* creation of informative characters (Baum et al. 1994). Due to the large size of the data set, the heuristic search strategy of PAUP was implemented. The random addition sequence and steepest descent options were utilized to increase the probability of finding all possible "islands" of trees (Maddison 1991). In addition to unweighted parsimony, character-state step-matrix weighting of transitions-transversions (Albert and Mishler 1992) and coding of informative indels (i.e., those shared by two or more taxa) were explored (for a discussion of gap coding see Baldwin 1993; Baum et al. 1994). The phylogenetic "signal" of the sequences was assessed by the CI (Kluge and Farris 1969) and the g_1 statistic (Hillis and Huelsenbeck 1992) (the latter by generating 10,000 random trees using the Random Trees option in PAUP).

Due to concerns about the monophyly of *Platanthera*, and in order to select the proper outgroups, 16 taxa representing 13 genera from the tribe Orchideae were included in the data set which was then analyzed using global parsimony (Maddison et al. 1984). To assess the monophyly of *Platanthera*, the taxa of *Platanthera* were not constrained to the ingroup. Based on a larger *rbc*L study of the Orchidaceae, two species of *Orchis* and *Ophrys* were chosen as super-outgroups (K. Cameron et al. unpubl. data). In an effort to gauge the strength of each clade in the strict consensus tree, both bootstrap (Felsenstein 1985) and decay analyses (Bremer 1988) were conducted. While bootstrapping and decay analyses may have their limitations (see Sanderson 1989, 1995), we feel they give at least some relative measure of internal support for the resulting clades.

Evolutionary analyses

Patterns of floral morphological evolution and the evolution of pollination syndromes were investigated by overlaying the appropriate morphological and ecological characters onto the molecular phylogeny with MacClade 3.05 (Maddison and Maddison 1992). Analyses in MacClade were conducted using the ACCTRAN option, a more conservative approach which minimizes convergence and parallelisms in character-states (Maddison and Maddison 1992). This procedure required that the *Platanthera ciliaris-blephariglottis-cristata* polytomy be resolved. All three possible branching patterns were investigated. The order of branching had no effect on character evolution, with the exception of floral color, where two arrangements agreed with each other but differed from the third. Consequently, the basal floral color for that group was coded as equivocal. The trees were redrawn with the polytomy present. Morphological and ecological characters were obtained both through the study of specimens (both live and herbarium) and through a detailed review of the literature on *Platanthera* and its relatives. The results of individual character analyses were compared to investigate the relations between phylogeny, morphology, and ecology.

Results and Discussion

Sequence alignment and analysis

The lengths of both ITS regions were within the ranges reported for other angiosperms and were similar for all the species investigated; differences in length were primarily the result of small indels. Alignment required the insertion of 39 gaps, 18 in ITS1 and 21 in ITS2; the gaps ranged from one to 14 base pairs in length. Seven of the gaps were potentially informative. Using the method of Baum et al. (1993), these gaps were coded and added to the data set. There was a total of 447 sites, of which 254 were potentially informative.

Sequence divergence (treating gaps as missing) was fairly high, ranging from 0.3% to 38.0% for ITS1 and 0.0% to 37.7% for ITS2. While these high values may imply high levels of "noise," comparison of the g_1 statistic from our data with the published data of Hillis and Huelsenbeck (1992) indicated a highly significant (p<0.01) phylogenetic signal (i.e., a significant amount of non-random structure in the data set). In addition, the CI value of the resulting trees indicated a relatively low level of homoplasy given the number of taxa and characters under consideration (see Chapter 2).

Phylogenetic analysis

Analysis of the data yielded two most parsimonious trees of 1123 steps (excluding uninformative characters, CI = 0.540); addition of the indels yielded the same two trees, and simply increased the length of several branches (1178 steps, CI = 0.546). Weighting the transversions over the transitions 2:1 yielded one shortest tree, which was identical to one of the two trees found in the unweighted analysis. Bootstrap and decay analyses were conducted on this tree, which was chosen as a working model for floral character evolution in *Platanthera* (Figure 15.2).

All analyses supported a monophyletic *Platanthera*, with *Galearis* as the sister-group. Contrary to the beliefs of many systematists, *Platanthera* was rather far removed from *Habenaria s.s.* In all analyses, five major clades could be recognized within *Platanthera*: Blephariglottis, Lacera, Limnorchis, Platanthera, and Tulotis (Figure 15.2). These informal clade names are derived from traditional sectional names where the clades overlap significantly with the sections. The Lacera clade represents a group of species not previously recognized at the sectional level, and is named for one species of the clade, *P. lacera*. While the clades overlap traditional sections to some degree, the monophyly of few sections of *Platanthera* was well supported (Table 15.1). It should be noted, however, that while relationships within the clades are rather well supported, the relationships between the clades are only weakly supported (at least by such measures as bootstrap support).

Floral evolution in *Platanthera*

Any discussion of large-scale patterns of floral evolution in *Platanthera* must be accompanied by a cautionary note: Relationships between clades of *Platanthera* are only weakly supported by our ITS data, even though relationships within clades are generally well supported. However, given the lack of other suitable phylogenies with which to test our hypotheses, the models we present are currently the best estimates of floral evolution in the genus. While it may be possible to strengthen the relationships between the clades by adding more characters to the data set, it is likely that the major clades of *Platanthera* rapidly diversified; if this is the case, it is possible that no amount of sequencing will allow us to confidently resolve intersectional relationships in this genus (Sytsma 1990).

POLLINATION SYNDROME – The basal condition in *Platanthera* appears to involve generalized settling moth pollination (i.e., noctuids and pyralids) (Figures 15.3, 15.4). This corresponds with the assessment of previous workers (van der Pijl and Dodson 1966; Nilsson 1983; Dressler 1993), but is inconsistent with Inoue (1983) who sug-

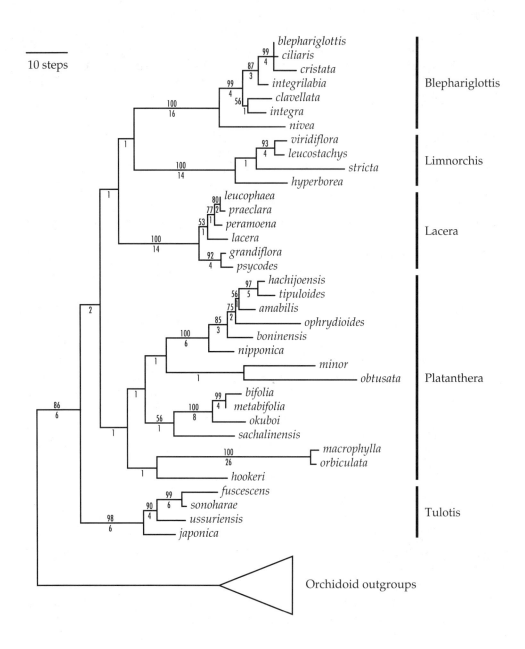

Figure 15.2. Phylogram of *Platanthera* based on a weighted parsimony analysis of nuclear ribosomal ITS sequence data. This is the only tree produced by the weighted analysis, and is identical to one of the two trees from the unweighted analysis. Bootstrap values greater than 50% are listed above the branches; decay values are below the branches. The outgroups are represented by a wedge; relationships among the outgroups will be discussed elsewhere.

Floral Diversification in Platanthera

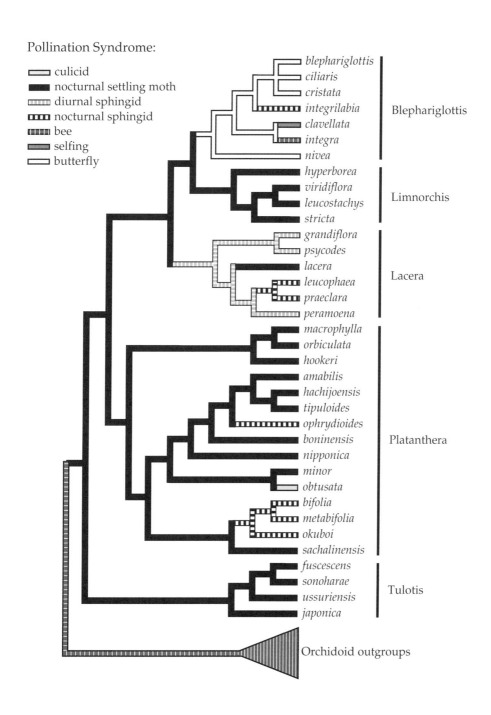

Figure 15.3. Cladogram illustrating pollination syndrome evolution in *Platanthera*, drawn from an ACC-TRAN-optimized analysis of the pollination data on the ITS-derived phylogeny using MacClade 3.05 (Maddison and Maddison 1992).

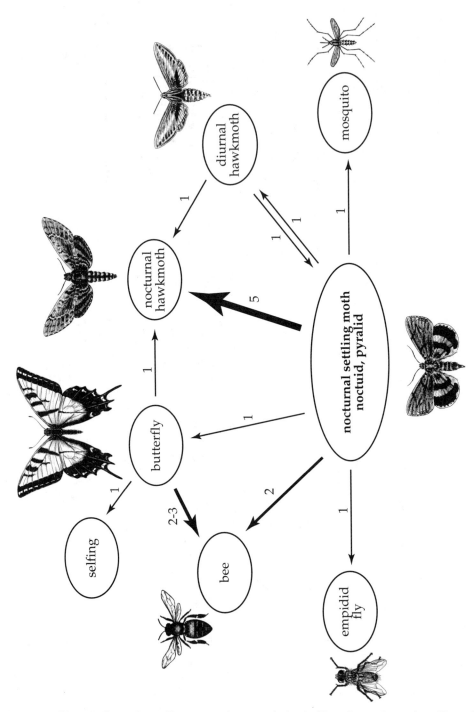

Figure 15.4. Diagram illustrating pollination syndrome evolution in *Platanthera*, redrawn from Figure 3 to demonstrate more clearly the evolutionary pathways to individual syndromes. The arrows indicate the direction of pollination syndrome evolution. The width of the arrows (and the numbers next to them) indicate the number of independent gains (i.e., >1 = convergence) of the given syndrome. (Figures of insects modified from Swann and Papp 1972, used with permission.)

gested that either bee or beetle pollination was the basal condition. Diurnal hawkmoth pollination has arisen only once, in the Lacera clade, and diurnal butterfly pollination has also arisen only once or at most twice. These syndromes apparently require a change in color (to purple and orange or yellow, respectively), in addition to numerous other morphological modifications to insure proper placement of the pollinia and the presence of the appropriate attractive stimuli. In contrast, hawkmoth pollination has arisen convergently five times, and requires only slight modifications from the generalized green to greenish-white floral syndrome of most species of *Platanthera*, such as increased scent production, a shift to whiter flowers and perhaps an increase in spur length. Thus, most species of *Platanthera* may be somewhat preadapted for hawkmoth pollination, facilitating the frequent shifts to this syndrome.

Dipteran (mosquito and/or fly) pollination has arisen two times, each time in a different clade of *Platanthera* (Figures 15.3, 15.4). Species pollinated by flies and mosquitoes are very similar morphologically to those pollinated by small moths, and apparently all species that are pollinated by mosquitoes and flies are pollinated by moths as well (Thien and Utech 1970; Voss and Riefner 1983; , Patt et al. 1989; Catling and Catling 1991). A switch from moths to dipterans may require only slight changes, perhaps in spur length and odor. As with hawkmoth pollination, the convergent shifts to dipteran pollination may also reflect some degree of preadaptation to this syndrome.

Finally, bumblebee pollination has arisen either two or three times, in two different clades (Table 15.1). The three species for which there are sufficient pollination records by bumblebees are apparently also pollinated by moths and/or butterflies. Therefore, the presence of a "bee pollination" syndrome in *Platanthera* is questionable at this point.

FLORAL COLOR – The primitive floral color is green (Figure 15.5). A number of species have shifted to greenish-white coloration, while pure white flowers have convergently arisen five times involving four sections. It should be noted that all hawkmoth-pollinated species have white flowers, but not all white-flowered species are hawkmoth-pollinated. Thus, while hawkmoths might exert some selective pressure for the development of white flowers, other pollinators apparently do so as well (Table 15.1). Brightly colored flowers have arisen in two clades: yellow and orange in the Blephariglottis clade and purple in the Lacera clade (Figure 15.5).

The adaptive significance of floral color has been investigated by many pollination ecologists (e.g., Grant and Grant 1965; Faegri and van der Pijl 1966; Kevan 1983; Scogin 1983). The colors of flowers of *Platanthera* fit the classical syndromes: white flowers are typically pollinated by hawkmoths or noctuids; green and greenish-white flowers are pollinated by moths (noctuids and pyralids); yellow and orange species are primarily pollinated by butterflies (Table 15.1). The purple-flowered species are all pollinated, at least in part, by day-flying hawkmoths in the genus *Hemaris*. Observations on adult nectar hosts of *Hemaris thysbe* (Fleming 1970) indicate that this species has a distinct preference for purple to purplish-pink flowers. Thus, selection by these diurnal hawkmoths may have led to the evolution of purple flowers in *Platanthera*.

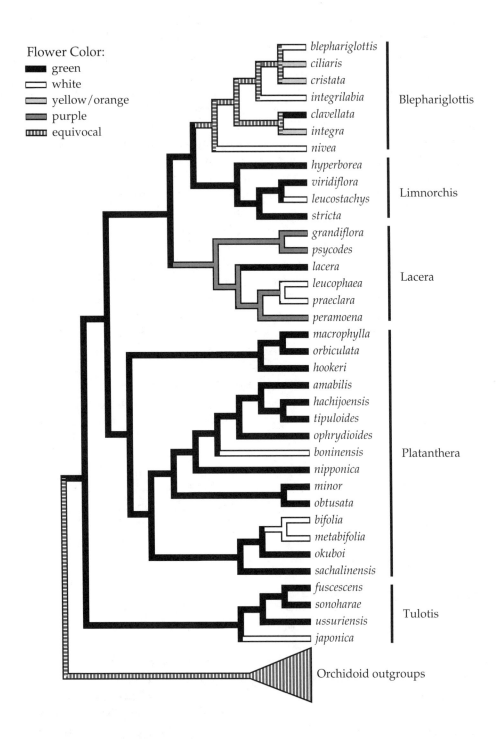

Figure 15.5. Cladogram illustrating floral color evolution in *Platanthera*, drawn from an ACCTRAN-optimized analysis of the pollination data on the ITS-derived phylogeny using MacClade 3.05 (Maddison and Maddison 1992).

Figure 15.6. Cladogram illustrating the evolution of pollinia placement (and hence column morphology) in *Platanthera*, drawn from an ACCTRAN-optimized analysis of the pollination data on the ITS-derived phylogeny using MacClade 3.05 (Maddison and Maddison 1992).

POLLINATION TIME – While only a few species are diurnally pollinated, diurnal pollination has arisen twice: once in the yellow, orange or white species of the Blephariglottis clade and once in the purple-flowered species of the Lacera clade. Comparisons of pollination time and color show that only the diurnally pollinated species are brightly colored; floral color and pollination time are clearly closely linked.

POLLINIA PLACEMENT – While the primitive condition is proboscis placement of the pollinia, eye placement has arisen independently at least seven times, involving all of the clades except Tulotis (Figure 15.6). Pollinia placement is a particularly homoplasious character; in the Platanthera clade alone, eye placement has convergently arisen at least three times. Placement of the pollinia on the eyes enforces a strong floral specificity, as both the distance between the viscidia and the length of the spur must match, respectively, the eye spacing and proboscis length of the pollinator. Figures 15.7 and 15.8 illustrate the functional constraints imposed by eye placement of the pollinia. With proboscis placement, eye spacing is irrelevant and the proboscis need only be long enough to reach the nectar (Nilsson 1978, 1983, 1988). Given this, it would seem that placement of pollinia on the eyes might result in fewer successful pollinations than placement on the proboscis. However, the frequent convergent gains of eye placement argue for some selective advantage to this trait. Further research on the pollination biology of *Platanthera* will be necessary to determine what may be the selective advantage(s) of this trait.

The switch from tongue to eye placement (or perhaps vice versa) of the pollinia has apparently been important in speciation in *Platanthera*. Three species-pairs (*P. psycodes-P. grandiflora, P. leucophaea-P. praeclara*, and *P. bifolia-P. chlorantha*) are differentiated primarily by placement of the pollinia. Two of these pairs are found in the Lacera clade. In each of these pairs, the orchids share at least one pollinator species and are sympatric in some areas of their respective ranges. The shift in column morphology allows the plants to effectively "partition" the pollinators. These pairs may have speciated sympatrically through bi-directional selection on column morphology from an ancestrally intermediate or polymorphic column type. Studies of the developmental morphology of the column may provide insight as to how this process might have occurred.

Pollinator partitioning via pollinia placement has been the subject of a great deal of research in the neotropical Catasetinae and Gongorae (Dressler 1968; Williams 1982). Apparently this study is the first to investigate pollinator partitioning via pollinia placement in an unbiased phylogenetic context, in part due to the previous difficulties in clearly determining the phylogenetic relationships of the species in question. In many cases, the differences in column structure, and thus pollinia placement, are used in separating species and in assessing their relationships. The homoplasious nature of column structure in *Platanthera* demonstrates that caution in character selection is necessary when studying floral evolution in groups, such as the orchids, where runaway sexual selection is expected.

FRINGING – Fringing of the labellum and petals has arisen only twice, in the Blephariglottis and Lacera clades. The convergent nature of this character is of particular interest, as fringing has previously been considered a synapomorphy uniting the species of section *Blephariglottis*. The degree of floral dissection has been shown to be

Floral Diversification in Platanthera 449

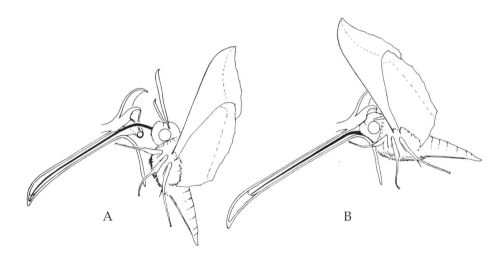

Figure 15.7. Constraints placed on spur length by eye placement of pollinia in *Platanthera*. In **A**, the spur is shorter than the distance from the tip of the moth's tongue to the base of its eyes; consequently, the eyes do not contact the viscidia, the pollinia are not removed, and the moth is effectively a nectar thief. In **B**, the spur is longer than the distance from the tip of the moth's tongue to the base of its eyes, and pollination is effected. (Taken from Nilsson 1978, used with permission.)

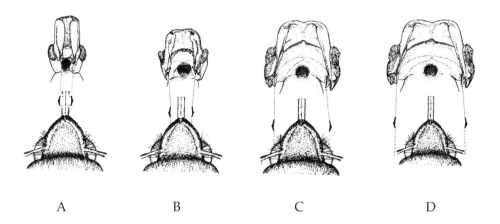

Figure 15.8. Column morphology and pollinia placement in *Platanthera*. **A** and **C** demonstrate proboscis and eye placement, respectively. In **B**, the moth's head is wider than the distance between the viscidia, which are also too widely placed to contact the moth's proboscis. Consequently, the viscidia will contact the scaly face of the moth, where they are unlikely to stick. **D** illustrates a moth that is too small to effect efficient pollination; the moth's head fits between the viscidia without contacting them. While this moth may occasionally serve as a pollinator by removing one pollinium at a time, it would not be as effective as the moth in **C**. **p**=pollinium, **v**=viscidium. (Taken and adapted from Nilsson 1983, used with permission.)

related to the frequency of pollination in the hawkmoth-pollinated *Viola cazorlensis* Gand. (Violaceae) (Herrera 1993). In fact, seven of the nine fringed species of *Platanthera* are pollinated at least in part by hawkmoths. If fringed flowers confer an advantage in pollination frequency, the selective advantage of fringing for *Platanthera* would be great, as several species of *Platanthera* are apparently pollinator-limited (Inoue 1983; Cuthrell 1994).

NECTAR SPURS – The general trend has been from shorter to longer nectar spurs. There have been several reversals to shorter spurs, however, in species pollinated by flies, mosquitoes and bees. All of these insects have shorter mouthparts than most lepidoptera. Very long spurs (>2.5 cm) appear to have evolved independently several times in response to hawkmoth and papilionid butterfly pollination.

OTHER TRAITS – A thickened callus of tissue, or tubercle, is found on the labella of all species of the Tulotis clade (with the exception of *P. japonica*) (Plate 3 G) and on the labella of *P. limosa* and *P. hologlottis* (not included in this study). The tubercle appears to function as a barrier preventing a forward approach to the flowers, forcing the pollinator to insert its proboscis from one side of the flower at a time. As a result, the pollinator can remove only one pollinium at a time. This may allow the plant to increase the odds of outcrossing, particularly if the pollinia are removed by different pollinators, as would be expected (Stoutamire 1971; Catling and Catling 1991). Because *P. limosa* and *P. hologlottis* are more closely related to the Limnorchis clade, the tubercle they share with the Tulotis clade is convergent (J. Hapeman, unpubl. data).

Other species also exhibit specializations that ensure that pollinia are removed individually. Several other species of the Limnorchis clade have the labellum "trapped" in an upright position initially, which allows removal of only one pollinium at a time (perhaps increasing outcrossing) and also apparently makes the flowers functionally protandrous by blocking access to the stigma (Plate 3 I, J, M). The ice tong-shaped flowers of *P. hookeri* also have been suggested to serve in manipulating pollinators into removing only one pollinium (Catling and Catling 1991).

Conclusions

Several broad conclusions can be drawn from our results. First, molecular data provide an adequate and non-biased foundation for studying floral evolution. Sequence data supply a large number of homologous, variable characters which are independent of the floral characters. Without this independent data set, it would have been difficult (if not impossible) to accurately study floral evolution in this fascinating and diverse genus. In *Platanthera*, most of the taxonomy is based on floral morphology, as the plants are vegetatively fairly uniform. The poly- or paraphyletic nature of many sections of *Platanthera* is an indication of how convergent floral morphology has historically led to misinterpretation of phylogenetic relationships in this genus.

Second, given the frequency of convergence and reversals in floral morphology, it is clear that floral characters are unsuitable for inferring phylogeny in *Platanthera*. Much of floral morphology appears to be rather homoplasious; that is, it is related more to ecology than to phylogeny. Shifts in column structure (= pollinia placement) as a result of pollinator partitioning is a particularly clear example of this. A second example would be the convergent evolution of "fringed" flowers in hawkmoth-pollinated lineages. While some features of floral morphology may be useful for inferring phylogeny, it is difficult or impossible to determine *a priori* which characters are suitable and which are homoplastic. This is similar to the findings of Chase and Palmer (1992) for the Oncidiinae (Orchidaceae). These two results taken together hold potentially large repercussions for much of orchid systematics, which historically has been based heavily on floral morphology.

Finally, pollinators have been important in shaping the evolution of *Platanthera*, affecting a diversity of changes in floral color and morphology. Hawkmoths have been involved in the evolution of long nectar spurs. Diurnal hawkmoths and butterflies have driven the evolution of brightly colored flowers. Partitioning of pollinators via differential pollinia placement has led to speciation through shifts in column morphology. Thus, if *Platanthera* is an adequate model for the Orchidaceae, then floral diversification as a result of pollinator-mediated selection has been an important driving force in the radiation of this fascinating and diverse family.

Acknowledgments

This research was made possible by a grant from the American Orchid Society Research and Education Committee to JRH and Thomas J. Givnish, a grant from the University of Wisconsin Davis Fund and a National Science Foundation Graduate Research Fellowship to JRH and a Grant-in-Aid from the Ministry of Education, Science and Culture of Japan (No. 06640897) to KI. Special thanks go to Mark Chase for providing critical DNAs and to Michael E. Abrams, Paul Kores and Mia Molvray, John Pleasants, Milo Pyne, and Jon Titus for providing leaf material.

Warren Stoutamire provided useful discussions on pollination and evolution in *Platanthera*, and was kind enough to provide unpublished personal observations on pollination. Tom Givnish and Ken Sytsma provided valuable discussion and guidance while Molly Nepokroeff and Tim Evans provided training and assistance in the lab. Warren Stoutamire, L. Anders Nilsson, Tom Givnish, Ken Sytsma, and Michelle Zjhra provided constructive comments on earlier drafts of this manuscript. Special thanks go to Carolyn Hapeman for invaluable field assistance and comments on the earliest drafts of this manuscript. I would also like to acknowledge Carlyle Luer who, with his spectacular books on North American terrestrial orchids, first inspired my love of orchids as a whole, and *Platanthera* in particular.

References

Albert, V. A., and Mishler, B. D. 1992. On the rationale and utility of weighting nucleotide sequence data. Cladistics 8:73–83.

Ames, O. 1910. *Illustrations and studies of the family Orchidaceae Volume IV: The genus Habenaria in North America*. Stanfordville, NY: Earl L. Coleman.

Baldwin, B. G. 1992. Phylogenetic utility of the internal transcribed spacers of nuclear ribosomal DNA in plants: an example from the Compositae. Molecular Phylogenetics and Evolution 1:3–16.

Baldwin, B. G. 1993. Molecular phylogenetics of *Calycadenia* (Compositae) based on ITS sequences of nuclear ribosomal DNA in plants: chromosomal and morphological evolution reexamined. American Journal of Botany 80:222–238.

Baldwin, B. G., Sanderson, M., Porter, J. M., Wojciechowski, M. F., Campbell, C. S., and Donoghue, M. J. 1995. The ITS region of nuclear ribosomal DNA: a valuable source of evidence on angiosperm phylogeny. Annals of the Missouri Botanical Garden 82:247–277.

Baum, D. A., Sytsma, K. J., and Hoch, P. C. 1994. The phylogeny of *Epilobium* L. (Onagraceae) based on nuclear ribosomal ITS sequences. Systematic Botany 19:363–388.

Benzing, D. A. 1987. Major patterns and processes in orchid evolution: a critical synthesis. Pp. 33–78 in *Orchid biology: reviews and perspectives, IV,* J. Arditti, ed. Ithaca, NY: Cornell University Press.

Bremer, K. 1988. The limits of amino acid sequence data in angiosperm phylogenetic reconstruction. Evolution 42:795–803.

Brooks, D. R., and McLennan, D. A. 1994. Historical ecology as a research programme: scope, limitations and the future. Pp. 1–27 in *Phylogenetics and ecology,* P. Eggleton and R. I. Vane-Wright, eds. London, England: Academic Press.

Case, F. W. 1987. *Orchids of the western Great Lakes region.* Bloomfield, MI: Cranbook Institute of Science.

Catling, P. M., and Catling, V. R. 1989. Observations of the pollination of *Platanthera huronensis* in southwestern Colorado. Lindleyana 4:78–84.

Catling, P. M., and Catling, V. R. 1991. A synopsis of breeding systems and pollination in North American orchids. Lindleyana 6:187–210.

Chase, M. W., and Palmer, J. D. 1992. Floral morphology and chromosome number in subtribe Oncidiinae (Orchidaceae): evolutionary insights from a phylogenetic analysis of chloroplast DNA restriction site variation. Pp. 324–339 in *Molecular systematics of plants,* P. S. Soltis, D. E. Soltis, and J. J. Doyle, eds. New York, NY: Chapman and Hall.

Chase, M. W., Soltis, D. E., Olmstead, R. G., Morgan, D., Les, D. H., Mishler, B. D., Duvall, M. R., Price, R. A., Hills, H. G., Qiu, Y., Kron, K. A., Rettig, J. H., Conti, E., Palmer, J. D., Manhart, J. R., Sytsma, K. J., Michaels, H. J., Kress, W. J., Karol, K. G., Clark, W. D., Hedrén, M., Gaut, B. S., Jansen, R. K., Kim, K.- J., Wimpee, C. F., Smith, J. F., Furnier, G. R., Strauss, S. H., Xiang, Q.- Y., Plunkett, G. M., Soltis, P. S., Swensen, S. M., Williams, S. E., Gadek, P. A., Quinn, C. J., Eguiarte, L. E., Golenberg, E., Learn, G. H., Graham, S. W., Barrett, S. C. H., Dayanandan, S., and Albert, V. A. 1993. Phylogenetics of seed plants: an analysis of nucleotide sequences from the plastid gene *rbc*L. Annals of the Missouri Botanical Garden 80:528–580.

Clegg, M. T. 1993. Chloroplast gene sequences and the study of plant evolution. Proceedings of the National Academy of Science, USA 90:363–367.

Cole, F. R., and Firmage, D. H. 1984. The floral ecology of *Platanthera blephariglottis.* American Journal of Botany 71:700–710.

Correll, D. S. 1950. *Native orchids of North America north of Mexico.* Stanford, CA: Stanford University Press.

Crisp, M. D. 1994. Evolution of bird-pollination in some Australian legumes (Fabaceae). Pp. 1–27 in *Phylogenetics and Ecology,* P. Eggleton and R. I. Vane-Wright, eds. London, England: Academic Press.

Cuthrell, D. L. 1994. Insects associated with the prairie fringed orchids, *Platanthera praeclara* Sheviak and Bowles and *P. leucophaea* (Nuttall) Lindley. M. S. thesis, North Dakota State University.

Darwin, C. 1877. *The various contrivances by which orchids are fertilized by insects.* 2nd ed., revised. Chicago, IL: University of Chicago Press.

Doyle, J. J., and Doyle, J. L. 1987. A rapid DNA isolation procedure for small quantities of fresh leaf tissue. Phytochemistry Bulletin 19:11–15.

Dressler, R. L. 1993. *Phylogeny and classification of the orchid family.* Portland, OR: Dioscorides Press.

Duckett, C. 1983. Pollination and seed production of the ragged fringed orchis, *Platanthera lacera* (Orchidaceae). Honor's thesis, Brown University, Providence, RI.

Ehrlich, H. A. 1989. *PCR technology: principles and applications for DNA amplification.* New York, NY: Stockton Press.

Faegri, K., and Pijl, van der L. 1966. *The principles of pollination ecology.* New York, NY: Pergamon Press.

Felsenstein, J. 1985. Confidence limits on phylogenies: an approach using the bootstrap. Evolution 39:783–791.

Fernald, M. L. 1950. *Gray's manual of botany.* 8th ed. New York, NY: D. Van Nostrand.

Fleming, R. C. 1970. Food plants of some adult sphinx moths (Lepidoptera: Sphingidae). The Michigan Entomologist 3:17–23.

Folsom, J. P. 1984. A reinterpretation of the status of relationships of taxa of the yellow-fringed orchid complex. Orquidea 9:320–345.
Gilbert, D. 1993. SeqApp version 1.9, a multiple sequence editor for Macintosh computers. Published electronically on the Internet, available via anonymous ftp from ftp.bio.indiana.edu.
Givnish, T. J., Sytsma, K. J., Smith, J. F., and Hahn, W. J. 1995. Molecular evolution, adaptive radiation, and geographic speciation in *Cyanea* (Campanulaceae). Pp. 288–337 in *Hawaiian biogeography: evolution on a hot spot archipelago*, W. L. Wagner and V. A. Funk, eds. Washington, DC: Smithsonian Institution Press.
Grant, V., and Grant, K. A. 1965. *Flower pollination in the phlox family*. New York, NY: Columbia University Press.
Hapeman, J. R. 1997. Pollination and floral biology of *Platanthera peramoena* (A. Gray) A. Gray (Orchidaceae) Lindleyana (in press).
Herrera, C. M. 1993. Selection on complexity of corolla outline in a hawkmoth-pollinated violet. Evolutionary Trends in Plants 7:9–13.
Hillis, D. M. 1987. Molecular versus morphological approaches to systematics. Annual Review of Ecology and Systematics 18:23–42.
Hillis, D. M., and Huelsenbeck, J. P. 1992. Signal, noise, and reliability in molecular phylogenetic analyses. Journal of Heredity 83:189–195.
Homoya, M. A. 1993. *Orchids of Indiana*. Bloomington, IN: Indiana University Press.
Inoue, K. I. 1983. Systematics of the genus *Platanthera* (Orchidaceae) in Japan and adjacent regions with special reference to pollination. Journal of the Faculty of Science, the University of Tokyo Section III, 13: 285–374.
Kevan, P. G. 1983. Floral colors through the insect eye: What they are and what they mean. Pp. 3–30 in *Handbook of experimental pollination biology*, C. E. Jones and R. J. Little, eds. New York, NY: Van Nostrand Reinhold.
Kipping, J. L. 1971. *Pollination studies of native orchids*. M.S. Thesis, San Francisco State College.
Kluge, A. G., and Farris, J. S. 1969. Quantitative phyletics and the evolution of anurans. Systematic Zoology 18:1–32.
Kocher, T. D., Conroy, J. A., McKaye, K. R., and Stauffer, J. R. 1993. Similar morphologies of cichlid fishes in Lakes Tanganyika and Malawi are due to convergence. Molecular Phylogenetics and Evolution 2:158–165.
Luer, C. A. 1975. *The native orchids of the United States and Canada*. New York, NY: New York Botanical Garden.
Mabberley, D. J. 1987. *The plant book: a portable dictionary of the higher plants*. New York, NY: Cambridge University Press.
Maddison, W. P., Donoghue, M. J., and Maddison, D. R. 1984. Outgroup analysis and parsimony. Systematic Zoology 33:83–103.
Maddison, W. P., and Maddison, D. R. 1992. *MacClade: analysis of phylogeny and character evolution, version 3.05*. Sunderland, MA: Sinauer Associates.
Moldenke, H. N. 1949. *American wildflowers*. New York, NY: Van Nostrand.
Morong, T. 1893. A new species of *Listera*, with notes on other orchids. Bulletin of the Torrey Botanical Club 20:31–39.
Nilsson, L. A. 1978. Pollination ecology and adaptation in *Platanthera chlorantha* (Orchidaceae). Botaniska Notiser 131:35–51.
Nilsson, L. A. 1983. Processes of isolation and introgressive interplay between *Platanthera bifolia* (L.) Rich. and *P. chlorantha* (Custer) Reichb. (Orchidaceae). Botanical Journal of the Linnaean Society 87:325–350.
Nilsson, L. A. 1992. Orchid pollination biology. Trends in Ecology and Evolution 7:255–259.
Ohwi, J. 1965. *Flora of Japan*. Washington, DC: Smithsonian Institution Press.
Oste, C. 1988. Polymerase chain reaction. Biotechniques 6:162–167.
Patt, J. M., Merchant, M. W., Williams, D. R. E., and Meeuse, B. J. D. 1989. Pollination biology of *Platanthera stricta* (Orchidaceae) in Olympic National Park, Washington. American Journal of Botany 76:1097–1106.
Pijl, L. van der, and Dodson, C. H. 1966. *Orchid flowers: their pollination and evolution*. Miami, FL: Fairchild Tropical Garden and the University of Miami Press.
Robertson, J. L., and Wyatt, R. 1990. Evidence for pollination ecotypes in the yellow-fringed orchid, *Platanthera ciliaris*. Evolution 44:121–133.
Sanderson, M. J. 1989. Confidence limits on phylogenies: the bootstrap revisited. Cladistics 5:113–129.
Sanderson, M. J. 1995. Objections to bootstrapping phylogenies: a critique. Systematic Biology 44:299–320.

Sanger, F., Nicklen, S., and Coulson, A. R. 1977. DNA sequencing with chain terminating inhibitors. Proceedings of the National Academy of Science, USA 74:5463.

Scogin, R. 1983. Visible floral pigments and pollinators. Pp. 160–172 in *Handbook of experimental pollination biology*, C. E. Jones and R. J. Little, eds. New York, NY: Van Nostrand Reinhold.

Sheviak, C. J., and Bowles, M. L. 1986. The prairie fringed orchids: a pollinator-isolated pair. Rhodora 88:267–290.

Smith, G. W., and Snow, G. E. 1976. Pollination ecology of *Platanthera ciliaris* and *P. blephariglottis* (Orchidaceae). Botanical Gazette (Crawfordsville) 137:133–140.

Smith, J. F., Sytsma, K. J., Shoemaker, J. S., and Smith, R. L. 1991. A qualitative comparison of total cellular DNA extraction protocols. Phytochemical Bulletin 23:2–9.

Smith, W. R. 1993. *Orchids of Minnesota*. Minneapolis, MN: University of Minnesota Press.

Stoutamire, W. P. 1971. Pollination in temperate American orchids. Pp. 233–243 in *Proc. 6th world orchid conference*, M. J. G. Corrigan, ed. Sydney, Australia: Halstead Press.

Stoutamire, W. P. 1974. Relationships of the purple-fringed orchids *Platanthera psycodes* and *P. grandiflora*. Brittonia 26:42–58.

Swofford, D. L. 1993. *PAUP: phylogenetic analysis using parsimony, version 3.1.1*. Champaign, IL: Illinois Natural History Survey.

Sytsma, K. J. 1990. DNA and morphology: inference of plant phylogeny. Trends in Ecology and Evolution 5:104–110.

Thien, L. B., and Utech, F. 1970. The mode of pollination in *Habenaria obtusata* (Orchidaceae). American Journal of Botany 57:1031–1035.

Voss, E. G., and Riefner, R. E. 1983. A pyralid moth (Lepidoptera) as pollinator of blunt-leaf orchid. Great Lakes Entomologist 16:57–60.

White, T. J., Birns, T., Lee, S., and Taylor, J. 1990. Amplification and direct sequencing of fungal ribosomal RNA genes for phylogenetics. Pp. 315–322 in *PCR protocols: a guide to methods and applications*, M. Innis, D. Gelfand, J. Sninsky, and T. White, eds. San Diego, CA: Academic Press.

Wood, J. J., Beaman, R. S., and Beaman, J. H. 1993. *The plants of Mount Kinabalu: 2. orchids*. London, England: Royal Botanic Gardens, Kew.

Yuan, Y., and Küpfer, P. 1995. Molecular phylogenetics of the subtribe Gentianinae (Gentianaceae) inferred from the sequences of internal transcribed spacers (ITS) of nuclear ribosomal DNA. Plant Systematics and Evolution 196:207–226.

Zettler, L. W., Ahuja, N. S., and McInnis, T. M. 1996. Insect pollination of the endangered monkey-face orchid *(Platanthera integrilabia)* in McMinn County, Tennessee – one last glimpse of a once common spectacle. Castanea 61:14–24.

16 Phylogenetic Perspectives on the Evolution of Dioecy: Adaptive Radiation in the Endemic Hawaiian Genera *Schiedea* and *Alsinidendron* (Caryophyllaceae: Alsinoideae)

Ann K. Sakai, Stephen G. Weller, Warren L. Wagner, Pamela S. Soltis, and Douglas E. Soltis

The biota of the Hawaiian Islands has long been noted for many spectacular examples of adaptive radiation (Carlquist 1974; Wagner et al. 1990). As a result of moist northeasterly tradewinds and mountains over 3,000 m above sea level, these islands harbor a diversity of habitats that include dry deserts in rain-shadows with less than 1,250 mm precipitation per year, wet rain forests with more than 7,000 mm precipitation per year, and seasonally cold alpine habitats (Wagner et al. 1990). The landscape is even more diverse because of the effects of island age, with the erosion, subsidence, and eventual disappearance of islands formed for the past 70 million years by the Hawaiian volcanic "hot spot" (Wagner et al. 1990). Islands have been available for colonization far longer than the age of the current major Hawaiian Islands, which range in age from Kaua`i, formed about 5.7 million years ago (Mya) to the island of Hawai`i, formed less than 0.5 Mya (Macdonald et al. 1983). The great isolation of the archipelago (ca. 4,000 km from the nearest large land mass, North America) has sharply limited immigration, and the native flora of the Hawaiian Islands is the result of only 291 presumed colonists (Sakai et al. 1995a). Although two-thirds of the colonists are represented by only a single species (Sakai et al. 1995a), other colonizations led to large species-rich lineages, and the eight largest lineages (2.7% of colonists) account for 31.4% of the 971 current species. Because speciation far exceeded rates of immigration and emigration, 89% of native flowering plant species are found only in the Hawaiian Islands. In many cases, these lineages are associated with speciation into different habitats and to different islands in the archipelago, with changes in vegetative habit, floral morphology, pollinator vector, and breeding system (Sakai et al. 1995b).

One of the most striking features of the native Hawaiian flora is the high incidence of dioecy: 14.7% of the flora is dioecious (with separate male [staminate] and female [pistillate] plants in populations), the highest of any known flora worldwide. One-fifth (20.7%) of the native Hawaiian flora is dimorphic, including species that are dioecious, gynodioecious (with females and hermaphrodites), subdioecious (with males, females, and rare hermaphrodites), and polygamodioecious (with plants with male and hermaphroditic flowers or with female and hermaphroditic flowers) (Sakai et al. 1995a). Hypotheses on selective forces promoting the evolution of dioecy include the classic argument that dioecy evolved as a mechanism to avoid inbreeding depression. Because of the separation of sexes on different plants, dioecy is a particularly effective mechanism to prevent selfing and thus expression of inbreeding depression in selfed progeny. Other hypotheses suggest that resource allocation, sexual selection, and ecological factors may be of primary importance in the evolution of dioecy (e.g., Givnish 1980; Bawa 1980; Thomson and Brunet 1990).

Surveys of floras have shown associations between several ecological factors and dioecy, but these analyses have been hard to interpret because it is difficult to infer whether selection has acted directly on these traits or instead on correlated traits. A very serious problem in many of these analyses is that species with common ancestors may not represent independent events in the evolution of dioecy (Felsenstein 1985; Donoghue 1989; Thomson and Brunet 1990; Sakai et al. 1995b). A large number of dioecious species within the same lineage may represent multiple independent origins of dioecy or a single change to dioecy in a common ancestor. Phylogenetic considerations are particularly important in the Hawaiian Islands where the flora is the result of relatively few lineages of varying sizes (Sakai et al. 1995a).

Although over half of current dimorphic species in the Hawaiian Islands are in lineages arising from dimorphic colonists, one third of current dimorphic species are in lineages arising from monomorphic (primarily hermaphroditic or monoecious) colonists (Sakai et al. 1995a). This autochthonous evolution of dimorphism occurred in at least 12 lineages, and it is within these lineages that phylogenetic analysis and comparison of conditions associated with dioecious and hermaphroditic species may be especially relevant to determine causal factors in the evolution of dioecy. Mapping of ecological conditions and breeding system traits in a phylogenetic context can be used to distinguish conditions that may have been associated with the origin of a trait (e.g., dimorphism), in contrast to those conditions that may have appeared after the trait was well established. In addition, because of the recent origin and speciation of many lineages in the Hawaiian Islands, extant species with intermediate states (e.g., gynodioecy) may well represent unstable transitory conditions and provide further evidence for factors promoting the evolution of dioecy. The recent origin of many Hawaiian lineages also has implications for the types of data most useful in phylogenetic reconstruction. More ancient lineages may show significant differences among taxa in both morphological and molecular characters. In more recent lineages, rapid speciation into diverse habitats may have occurred with relatively few molecular changes.

The 25 species of *Schiedea* and four species of *Alsinidendron* (Caryophyllaceae: Alsinoideae) form a monophyletic group as evidenced by synapomorphic shifts to apetaly and in nectary morphology (Weller et al. 1990; Harris and Wagner 1993; Wagner et al. 1995). The lineage is endemic to the Hawaiian Islands, found on all the major islands, and has radiated into a number of diverse habitats. Species include rain forest vines, woody shrubs in dry and mesic habitats, and perennial herbs on windswept sea cliffs (Table 16.1, Figure 16.1). Dioecy has arisen autochthonously within the lineage from a presumed hermaphroditic ancestor (Wagner et al. 1995; Weller et al. 1995), and ten species of *Schiedea* have dioecious, subdioecious, or gynodioecious breeding systems. Associations of habitat, pollination vector, floral morphology, and breeding system are evident. Hermaphroditic species are found primarily in mesic or wet forest and are apparently insect-pollinated or autogamous (Weller and Sakai 1990). Dimorphic species occur in dry habitats, and all are

Table 16.1. Breeding system, growth form, habitat, and distribution for species of *Alsinidendron* and *Schiedea*, based on herbarium, greenhouse, or field observations. After Weller et al. (1995), with modification.

Species	Breeding system	Habit	Habitat	Distribution
Alsinidendron H. Mann				
lychnoides (Hillebr.) Sherff	hermaphroditism, facultative autogamy	vine	wet forest	Kaua`i
obovatum Sherff	hermaphroditism, facultative autogamy	subshrub	diverse mesic forest	O`ahu
trinerve H. Mann	hermaphroditism, cleistogamy	subshrub	wet forest, diverse mesic forest	O`ahu
viscosum (H. Mann) Sherff	hermaphroditism, facultative autogamy	vine	wet forest, diverse mesic forest	Kaua`i
Schiedea Cham. & Schlechtend.				
adamantis St. John	gynodioecy	shrub	dry shrubland	O`ahu
amplexicaulis H. Mann	hermaphroditism	?	?	Kaua`i (extinct)
apokremnos St. John	gynodioecy	shrub	dry cliffs	Kaua`i
attenuata W. L. Wagner, Weller & Sakai	hermaphroditism	shrub	diverse mesic forest pockets on cliffs	Kaua`i
diffusa A. Gray	hermaphroditism, facultative autogamy	vine	wet forest	East Maui, Moloka`i, Hawai`i
globosa H. Mann	subdioecy	suffruticose herb	dry coastal cliffs	O`ahu, Maui, Moloka`i
haleakalensis Degener & Sherff	dioecy	shrub	dry subalpine cliffs	East Maui
helleri Sherff	hermaphroditism	vine	wet forest cliffs	Kaua`i
hookeri A. Gray	hermaphroditism	subshrub	diverse mesic forest	O`ahu
implexa (Hillebr.) Sherff	hermaphroditism	subshrub	mesic forest?	Maui (extinct)
kaalae Wawra	hermaphroditism	perennial herb	diverse mesic forest, wet forest	O`ahu
kealiae Caum & Hosaka	subdioecy	subshrub	dry forest	O`ahu
ligustrina Cham. & Schlechtend.	dioecy	shrub	dry shrubland, often cliffs	O`ahu
lydgatei Hillebr.	hermaphroditism	shrub	dry shrubland	Moloka`i
mannii St. John	subdioecy	shrub	dry ridges in diverse mesic forest	O`ahu
membranacea St. John	hermaphroditism	perennial herb	diverse mesic forest	Kaua`i
menziesii Hook.	hermaphroditism	shrub	dry shrubland	Lana`i and West Maui
nuttallii Hook. var. *nuttallii*	hermaphroditism	subshrub	diverse mesic forest	O`ahu
nuttallii Hook. var. *pauciflora* Degener & Sherff	hermaphroditism	subshrub	diverse mesic forest	Kaua`i
pubescens Hillebr.	hermaphroditism	vine	diverse mesic forest	O`ahu, Moloka`i, Lana`i and Maui
salicaria Hillebr.	gynodioecy	shrub	dry shrubland	West Maui
sarmentosa Degener & Sherff	gynodioecy	shrub	dry forest and shrubland	Moloka`i
spergulina A. Gray	dioecy	shrub	cliffs in dry shrubland	Kaua`i
stellarioides H. Mann	hermaphroditism	subshrub	diverse mesic forest	Kaua`i
verticillata F. Brown	hermaphroditism	perennial herb	soil pockets and cracks on dry coastal cliffs	Nihoa
sp. nov. (Perlman 13448)	hermaphroditism	subshrub	mesic forest	Kaua`i

apparently wind-pollinated. Weller and Sakai (1990) hypothesized that the shift to the dry, windy cliffs and ridges resulted in a loss of insect pollinators, and thus led to increased selfing and expression of inbreeding depression. These latter conditions provided a favorable environment for the spread of male-steriles (females) into these populations. Interpreting these associations in a phylogenetic context may help to reveal the importance of these ecological shifts as causal factors in the evolution of dioecy. If dimorphism evolved many times independently and always in association with these ecological shifts, then these ecological shifts in habitat and pollination vector may be important as causal factors in the evolution of dimorphism. This is particularly true if shifts in habitat preceded shifts in breeding system (Donoghue 1989). Alternatively, if dimorphism evolved only once early within the lineage but led to several dimorphic species, then the association of habitat and breeding system may be coincidental and may not reflect a causal relationship.

In this chapter, we present an overview and synthesis of the recent literature that places the evolution of breeding systems of the Hawaiian Alsinoideae in a phylogenetic context (Wagner et al. 1995; Weller et al. 1995; Soltis et al. 1996). Using a morphological data set with all taxa, we show that significant differences in interpretation of the evolution of dioecy can arise depending upon the procedures used in character mapping, particularly how a trait such as breeding system is coded, and how equivocal placement of mapped characters is resolved (Weller et al. 1995). Results from biogeographic analyses (Wagner et al. 1995; Soltis et al. 1996) are briefly summarized. We also discuss the results and relative importance of morphological data, molecular data, and a combined data set in phylogenetic reconstruction of this lineage.

Materials and methods

Outgroup selection

Floral nectaries are a key distinctive feature delineating the endemic Hawaiian Alsinoideae as a monophyletic group. Although the tips of the nectaries in *Schiedea* are tubular and those of *Alsinidendron* are flaplike, developmental studies clearly indicate that these structures are homologous in the two taxa (Harris and Wagner 1993; Wagner et al. 1995). The subtropical Hawaiian Alsinoideae are sufficiently divergent from mainland species that it has been difficult to identify a mainland sister-group. *Schiedea* and *Alsinidendron* are clearly aligned with the exstipulate Alsinoideae because of the presence of exstipulate leaves, capsules splitting into as many valves as styles, distinct sepals, and development of homologous nectaries. In analyses of morphological data, a hypothetical subfamily Alsinoideae ancestor was used as the outgroup. Two species of *Minuartia* L. (Alsinoideae) were used as additional outgroups because of similarities in nectary structure and type of capsule dehiscence (Weller et al. 1995). Because material of *Minuartia* was not available for molecular work, *Moehringia lateriflora,* also in subfamily Alsinoideae, and the endemic Hawaiian *Silene struthioloides* (Caryophyllaceae: Silenoideae) were used as outgroups for molecular analyses.

Character selection and coding

MORPHOLOGICAL DATA – Characters were scored using herbarium specimens of all taxa as well as greenhouse-grown flowering material of all but four species (Wagner et al. 1995; Weller et al. 1995). Two species (*S. amplexicaulis*, known from only two herbarium collections and with much missing data; and *S. implexa*) are apparently extinct. Forty-three floral and vegetative characters were coded (see Weller et al. 1995 appendix for data matrix), where (0) represents the plesiomorphic (primitive) state, and (1) or higher represents apomorphic (derived) states. Autapomorphic characters were not included in the matrix, and multi-state characters were unordered because it was difficult to determine the most likely transition series. Characters included traits based on roots, stems, habit, several leaf characters (shape, size, texture, venation, pubescence), inflorescence orientation and pubescence, sepals (width, orientation, texture, shape, pubescence), several floral characteristics (nectary shape, pollen color, etc.), and seed traits (shapes of cells on seed surfaces). Because breeding system was a major trait of interest in character mapping, breeding system and characters that appeared to be correlated with breeding system evolution (inflorescence condensation, characters 19, 20; staminal filament/sepal length ratio, character 35; breeding system, character 43) were excluded from analyses. Previous results (Weller et al. 1995) showed that inclusion of these characters had little effect on the topology of the trees.

MOLECULAR DATA – Methods for cpDNA and rDNA restriction-site analysis described in Soltis et al. (1996) are summarized here. Most material was obtained from greenhouse plants (University of California, Irvine) raised from seeds collected in the field. For a few taxa, field-collected leaves were mailed directly to Washington State University for molecular analysis (vouchers of specimens at US, collection numbers reported in Soltis et al. 1996).

High-molecular-weight total DNAs were isolated using either a mini-prep protocol of Doyle and Doyle (1987) or a large-scale DNA isolation procedure (Rieseberg et al. 1988; Soltis et al. 1991). DNA was extracted successfully from 20 of the 21 taxa with available material. DNAs were digested with the following 28 endonucleases: *Acc*I, *Apa*I, *Ava*I, *Ava*II, *Ban*I, *Ban*II, *Bgl*I, *Bgl*II, *Bst*EII, *Bst*NI, *Bst*XI, *Cla*I, *Cfo*I, *Eco*RI, *Eco*RV, *Hae*II, *Hind*III, *Hpa*II, *Nci*I, *Pvu*II, *Pst*I, *Smn*I, *Ssp*I, *Sac*I, *Sac*II, *Sal*I, *Xba*I, and *Xho*I. DNA fragments were separated in 1.0% agarose gels, denatured, and transferred to nylon membranes following the general methods of Palmer (1986). Heterologous cpDNA probes from lettuce (Jansen and Palmer 1987) and petunia were labeled using the Random Primed DNA Labeling Kit (U. S. Biochemical Corporation, Cleveland, Ohio) and hybridized to the membrane-bound DNA fragments. To analyze rDNA variation, filters were probed with pGMr-1, a plasmid containing a single 18S-26S rDNA repeat from *Glycine max*. Restriction sites were scored as present (1) or absent (0). Missing data were scored as question marks; 2.4% of the data matrix cells were scored as missing (see Soltis et al. 1996 for full data matrix).

Phylogenetic analysis

Morphological data, molecular data, and the combined data were analyzed using PAUP 3.1.1 (Swofford 1993) with MULPARS on, TBR branch-swapping, and "unweighted" character-state changes in which gains and losses are weighted equally. Breeding system characters were not used in the phylogenetic reconstruction, and character-states were unordered. In the combined data set, only taxa for which there were both molecular and morphological data (20 of the 30 taxa) were used. The random addition sequence option was used to increase the likelihood of obtaining all equally most parsimonious trees (Maddison 1991), with 10 replicate tree searches using the morphological data and 100 replicate tree searches for the molecular data and combined data. Bootstrap analysis (Felsenstein 1985) was conducted to estimate reliability for monophyletic groups, with 1,000 replicates for morphological and combined data and 100 replicates for the molecular data. To evaluate the nonrandom structure of the combined cpDNA/rDNA data set, the skewness test (Hillis 1991; Huelsenbeck 1991; Hillis and Huelsenbeck 1992) was performed. For the skewness analyses, the RANDOM TREES feature of PAUP was used to generate 10,000 random trees and to calculate the g_1 statistic based on the distribution of the lengths of these trees. Robustness of major clades also was assessed using decay indices (Bremer 1988; Donoghue et al. 1992).

Because the position in a tree where a change in character-state (e. g., from hermaphroditism to dimorphism) might have occurred can be ambiguous, breeding system was mapped in two different ways on the tree based on morphological data. With the ACCTRAN (accelerated transformation) option of PAUP, the earliest possible occurrence of the transition is used. With this option, transitions to dimorphism occur at more interior nodes (earlier in the evolutionary lineage) and some instances of hermaphroditism are more likely to appear as reversals. With the DELTRAN (delayed transformation) option, these transitions occur later in the tree. In this case, transitions to dimorphism may appear as cases of parallel evolution, and reversals to hermaphroditism will be hypothesized less frequently.

We also used the tree based on morphological data to map the breeding system character coded in two different ways, varying the number of character-states. Breeding system was either coded with dioecy, subdioecy, gynodioecy, and hermaphroditism each as a separate state, or with binary coding, lumping all dimorphic breeding systems together (0 = hermaphroditism, 1 = dimorphism). This latter coding system assumes that the critical step in the evolution of dioecy may be the initial appearance of females in the population.

Results and discussion

Phylogenetic reconstruction

MORPHOLOGICAL DATA – Analysis of the morphological data set resulted in six equally most parsimonious trees with 132 steps, showing four major clades within the Hawaiian Alsinoideae (Weller et al. 1995). The six trees differed in the placement of the extinct *Schiedea amplexicaulis*, for which many characters were missing, and the relationship of four species (*S. hookeri*, *S. kealiae*, *S. menziesii*, and *S. sarmentosa*) within the *S. globosa* clade.

The four clades are defined primarily by leaf characters (Figure 16.1). The *S. membranacea* clade is characterized in general by the occurrence of broad, multi-nerved

Adaptive Radiation in Hawaiian Alsinoideae 461

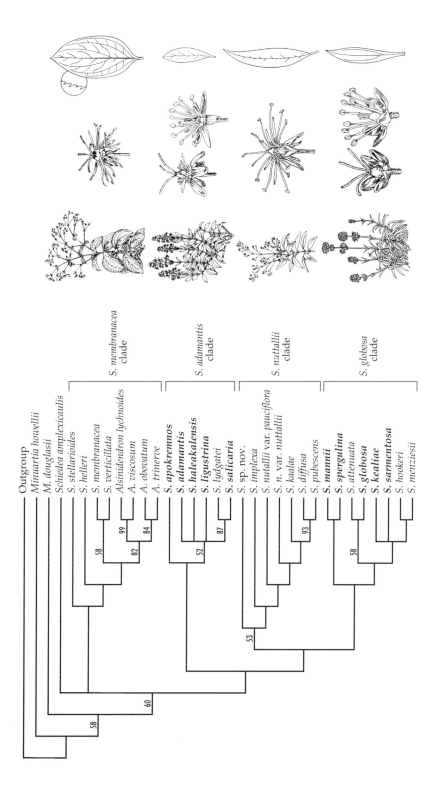

Figure 16.1. Strict consensus of six equally most parsimonious trees resulting from analysis of morphological data showing the presence of four major clades. Also shown are results of bootstrap analysis of morphological data, based on 1000 replicates (only clades supported >50% are shown). One thousand trees were saved for each replication. Boldface indicates species with dimorphic breeding systems. Pictured (top to bottom) to illustrate the four clades are the inflorescences, flowers, and leaves of *S. membranacea*, *S. adamantis*, *S. nuttallii*, and *S. globosa*. After Weller et al. (1995), with modification.

leaves and leaf margins that are ciliate or toothed. The four species of *Alsinidendron* found within this clade are well defined as a group by several synapomorphies (campanulate calyx, petaloid sepal texture, flaplike nectary extensions, black nectar, gray pollen). The Kaua`i species of *Alsinidendron* (*A. lychnoides* and *A. viscosum*) are vines with papery sepals in fruit. The O`ahu species (*A. obovatum* and *A. trinerve*) have fleshy, dark purple sepals in fruit. The species pairs within *Alsinidendron*, as well as the clade of four *Alsinidendron* species, are well supported by bootstrap analysis and decay analysis (Figure 16.1; Weller et al. 1995). The *S. membranacea* clade as a whole is not well supported in bootstrap analysis and decays in three steps when relaxing parsimony. All of the species in this clade are hermaphroditic, and all but *S. verticillata* occur in mesic or wet habitats. Most species in this clade occur on Kaua`i; two *Alsinidendron* species occur on O`ahu, and *S. verticillata* occurs on Nihoa, the southernmost of the northwestern Hawaiian Islands, 200 km northwest of Kaua`i (Figure 16.2).

The *S. adamantis* clade is characterized by leaves that are broadest above the midpoint and (in general) by papillate seeds and acute margins on the epidermal cells of the seeds (Figure 16.1). The clade (excluding *S. apokremnos*) is supported at the 52% level by bootstrap analysis but decays if parsimony is relaxed by two steps. The clade is not well resolved, with the exception of the *S. lydgatei-S. salicaria* sister-group (supported by a bootstrap value of 87% and decay index of 2; Weller et al. 1995). All species in this clade occur in dry habitats, and all but *S. lydgatei* are dimorphic. Five of the six species occur on O`ahu and the Maui Nui complex (Maui, Moloka`i, and Lana`i, treated together because of their recent geological connections, Figure 16.2).

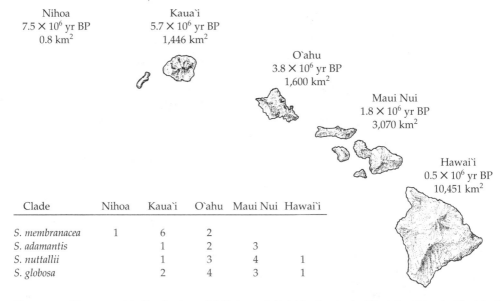

Clade	Nihoa	Kaua`i	O`ahu	Maui Nui	Hawai`i
S. membranacea	1	6	2		
S. adamantis			1	2	3
S. nuttallii		1	3	4	1
S. globosa		2	4	3	1

Figure 16.2. Biogeographic distribution of *Schiedea* and *Alsinidendron* in the Hawaiian Islands. The Maui Nui complex includes the islands of Maui, Moloka`i, Kaho`olawe, and Lana`i as one area because the four islands were connected during the Pleistocene. The numbers of species on each island from each clade (see Figure 16.1) are indicated. Note that some species occur on more than one island (see Table 16.1). yr BP = years before present.

The *S. nuttallii* clade (Figure 16.1) is characterized in general by the presence of fleshy stems, large leaves with a single vein, and attenuate to caudate, strongly reflexed sepals. It is supported at the 53% level by bootstrap analysis and decays by relaxation of parsimony in two steps. Within the clade, *S. diffusa* and *S. pubescens* form a strongly supported sister-group. Both species are vines with pendent flowers and pendent, non-glandular inflorescences with purple pubescence. All species within this clade are hermaphroditic and grow in wet or mesic habitats. Species are widely distributed on six of the major islands (Figure 16.2).

The *S. globosa* clade is the most weakly supported of the four major clades in the morphological data set. It is delimited only by the presence of long, attenuate leaf tips (Figure 16.1), and has a bootstrap value < 50% and a decay index of 1. This clade contains both hermaphroditic species in mesic habitats as well as dimorphic species in dry habitats, primarily on O`ahu, Maui, and Moloka`i (Figure 16.2). The *S. globosa* and *S. nuttallii* clades are linked in all trees but only by the presence of slightly asymmetric leaves. These two clades are linked to the *S. adamantis* clade by the presence of woody tissue.

MOLECULAR DATA – Although 46 cpDNA restriction-site mutations were detected, only 18 of these were shared by two or more species and thus phylogenetically informative within *Schiedea* and *Alsinidendron* (Soltis et al. 1996). In addition, 13 restriction site mutations differentiated the outgroup species from the *Schiedea-Alsinidendron* complex. The remaining restriction-site mutations were found only in one taxon (autapomorphies) and thus were not useful in phylogenetic reconstruction. All three restriction-site changes in rDNA that were detected were phylogenetically informative within the *Schiedea-Alsinidendron* complex (see Soltis et al. 1996 for data matrix).

The combined cpDNA/rDNA data set showed considerable non-random structure in the data (skewness test on combined cpDNA and rDNA data set: $g_1 = -1.690$, $P < 0.01$, Hillis and Huelsenbeck 1992). The rDNA mutations in general complemented relationships suggested by the cpDNA analysis. One rDNA mutation lent further support to the monophyly of *Alsinidendron*. Another gave further support to the strong relationship between the two O`ahu species of *Alsinidendron*. The third rDNA mutation suggested a close relationship among *S. pubescens, S. nuttallii, S. diffusa,* and *S. kaalae,* species whose relationships were unresolved using cpDNA data. Because of the low level of cpDNA and rDNA restriction-site divergence found in *Schiedea* and *Alsinidendron,* parsimony analysis resulted in a large number (870) of most parsimonious trees, each of 53 steps (CI = 0.895 excluding autapomorphies, RI = 0.947). Many of the taxa had very similar restriction-site profiles, and many of the detected clades were supported by only one restriction-site mutation.

Using the 50% majority-rule tree for the combined cpDNA/rDNA data set (Figure 16.3A), the monophyly of the *Schiedea-Alsinidendron* lineage relative to the outgroups is strongly supported (100% bootstrap value). Several clades are defined within these two genera. The three *Alsinidendron* species formed a monophyletic group (97% bootstrap value) and the sister-group status (96% bootstrap value) of the two O`ahu species, *A. obovatum* and *A. trinerve,* was consistent with the morphological results. Based on molecular data, the placement of *S. membranacea* is problematic.

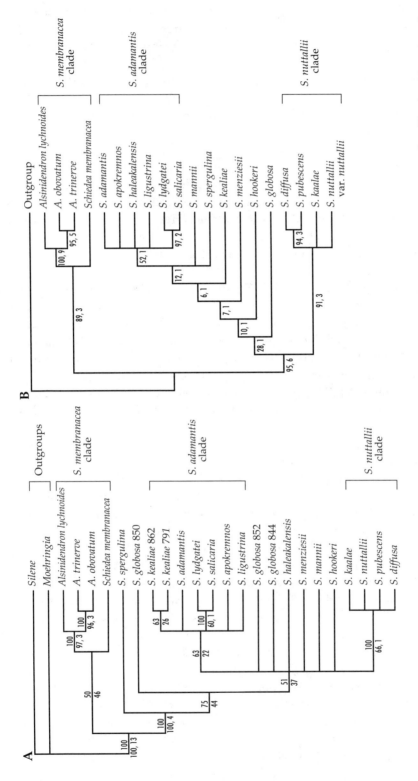

Figure 16.3. (A) Majority-rule consensus of 870 most parsimonious trees resulting from analysis of cpDNA/rDNA restriction-site data. Numbers above branches are the percentage of these 870 trees that support the branches; numbers below branches are bootstrap percentages, followed by decay indices. The species forming the three clades denoted in this figure are similar to those in Figure 16.1 (morphologically based phylogeny) with the following exceptions: in the morphologically based phylogeny, *S. kealiae* is placed in the *S. globosa* clade and *S. haleakalensis* is placed in the *S. adamantis* clade. Molecular material was not available for *A. viscosum, S. stellarioides, S. helleri, S. verticillata, S. sp. nov.*, and *S. implexa*. (B) Strict consensus of 20 most parsimonious trees resulting from analysis of the combined molecular and morphological data set. Numbers below branches are bootstrap percentages followed by decay indices. After Soltis et al. (1996), with modification.

One cpDNA mutation links this species with *Alsinidendron*, similar to the topology suggested by morphological data. A different cpDNA mutation is shared by *S. membranacea* and all other *Schiedea* species. Thus, half of the trees show *S. membranacea* as sister to all other species of *Schiedea*, making *Schiedea* a monophyletic group rather than a paraphyletic group that includes *Alsinidendron*, as suggested by the morphological data and the other cpDNA mutation.

The other species of *Schiedea* form a well supported monophyletic group (100% bootstrap value), but relationships within this clade are poorly resolved. The *S. adamantis* and *S. nuttallii* clades defined by morphological data are weakly supported by molecular data. One clade in the molecular data set (22% bootstrap value) is similar to the *S. adamantis* clade, but includes *S. kealiae* and lacks *S. haleakalensis*. The close relationship of *S. salicaria* and *S. lydgatei* suggested in the morphological data is also supported by molecular data (60% bootstrap value). A second clade defined by one rDNA restriction-site mutation and supported by a bootstrap value of 66%, includes all four species in the molecular analysis that are in the *S. nuttallii* clade defined by morphological data.

COMBINED MOLECULAR AND MORPHOLOGICAL DATA – The strict consensus tree derived from the combined molecular and morphological data (including only those species scored for both molecular and morphological data) is very similar to those obtained in separate analyses of morphological and molecular data, but is more fully resolved (Figure 16.3B). Phylogenetic analysis resulted in 20 most parsimonious trees, each 166 steps long (CI = 0.610 excluding autapomorphies, RI = 0.725; skewness test with significant non-random structure: g_1 = -1.245, $P < 0.01$, Hillis and Huelsenbeck 1992). As expected from the separate molecular and morphological analyses, the four species analyzed from the *S. membranacea* clade continue to form a clade (including *S. membranacea*), and the four species in the *S. nuttallii* clade remain as a clade. The *S. adamantis* clade is consistent with the morphological analysis, including *S. haleakalensis* and excluding *S. kealiae*. Unlike the separate analyses, species in the *S. globosa* clade defined by morphological data are part of a larger clade that includes the *S. adamantis* clade nested within it.

Character-state evolution

Here we consider the consensus tree based on morphological data first because it includes all taxa, and then turn to the consensus tree derived from the combined data set, given that it is both largely congruent with and better resolved and supported than the tree based on DNA data alone.

BREEDING SYSTEMS – Using the tree based on morphological data, the inferred number of independent origins of dioecy depends upon whether breeding system was coded with binary states (hermaphroditism, dimorphism) or with four different states (hermaphroditism, gynodioecy, subdioecy, dioecy). Separate origins for dimorphism occur in two different areas of the tree. If dimorphism is coded as four states, one unequivocal origin occurs at the base of the *S. adamantis* clade (Figure 16.4). One to five origins of dimorphism occur in the *S. globosa* clade because the character shift is equivocal (it can be mapped on trees at different points without altering the total

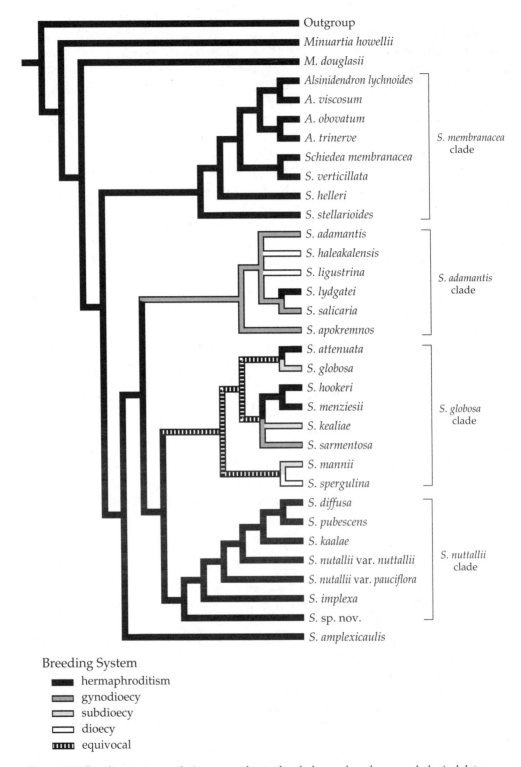

Figure 16.4. Breeding system evolution mapped onto the phylogeny based on morphological data.

number of changes). With equivocal character shifts, accelerated transformations – occurring in the most basal positions possible on the tree – may lead to very different views of character-state evolution than delayed transformations, which occur as close to the tips of the tree as possible. If transitions are accelerated, one origin of dimorphism occurs in the *S. globosa* clade; if transitions are delayed, five origins result in this clade. The equivocal results for the *S. globosa* clade result from the topology of the clade and the particular relationships of hermaphroditic and dimorphic species. The clade is weakly supported and an entirely different picture of the evolution of dimorphism in this clade might emerge if the taxa were realigned.

If breeding system is considered as a two-state character (hermaphroditic or dimorphic), dimorphism evolves only once if placement is at the earliest possible equivocal position or sometimes twice if changes to dimorphism are delayed (Weller et al. 1995). Even more possibilities in the number of origins of dioecy occur if breeding system characters are included in construction of the phylogeny and characters are coded as ordered (Weller et al. 1995).

Using the combined tree, two origins of dimorphism are implied – one in *S. globosa*, and one in the ancestor of the large clade with the remaining eight dimorphic species. With this interpretation, only a single reversal to hermaphroditism occurs (*S. lydgatei*). Only one origin of dimorphism occurs with an interpretation that is one step longer, involving the origin of dimorphism in the ancestor of the sister-group of the *S. nuttallii* clade. This interpretation requires three reversals to hermaphroditism (*S. hookeri* and *S. menziesii* in addition to *S. lydgatei*).

HABITAT SHIFTS – All ten dimorphic species occur in dry habitats, and 16 of the 19 hermaphroditic species occur in mesic or wet habitats, suggesting the importance of dry habitats for the evolution of dimorphism in this group. Using the phylogeny based on morphological data, within the *S. adamantis* clade five species are dimorphic but a reversal to hermaphroditism has occurred in *S. lydgatei* without a reversal to a more mesic habitat (Figure 16.5). Within the hermaphroditic *S. membranacea* clade of eight species, *S. verticillata* has maintained a hermaphroditic breeding system but apparently has shifted to a very dry coastal habitat. This latter shift suggests that colonization of dry habitats may occur without changes in the reproductive system. In the *S. globosa* clade of three hermaphroditic species and five dimorphic species, the relationship of habitat shifts and breeding system shifts is difficult to discern, given that both habitat and breeding system transitions are equivocal within the clade. Hermaphroditism in the sister species *S. menziesii* (in dry habitat) and in *S. hookeri* (in more mesic habitat) may represent a reversal from dimorphism.

Because of the absence from the combined analysis of several key species (*S. verticillata* and *S. attenuata*) important for interpreting shifts in habitat and breeding system, conclusions based on the combined data set must be interpreted with caution. With this data set, a single shift to dry habitats is suggested in the ancestor of the large clade comprising species of the *S. adamantis* and *S. globosa* clades defined using morphological data. A shift to more mesic habitat then occurs with a shift to hermaphroditism in *S. hookeri*. Both the *S. nuttallii* clade and the *S. membranacea* clade occur in mesic or wet environments. Unfortunately, no molecular

Figure 16.5. Habitat mapped onto the morphological phylogeny.

data were available for *S. verticillata*, a hermaphroditic member of the *S. membranacea* clade of particular interest because it occurs in dry coastal habitats. The habitat of the ancestral members of the complex is uncertain. Although the more basal *S. membranacea* and *S. nuttallii* clades occur in mesic and wet habitats, the closest relatives of *Schiedea* and *Alsinidendron* may be continental dry-site species (Wagner et al. 1995; Weller et al. 1995).

BIOGEOGRAPHY – The complex biogeography of *Schiedea* and *Alsinidendron*, including the initial point of colonization of the Hawaiian Islands, numbers and age-related patterns of inter- and intra-island colonizations and speciation events, the association of habitat shifts with inter- and intra-island colonizations, and the relative specialization of colonizing species, is discussed in detail in Wagner et al. (1995). Based on phylogenetic analyses of morphological data as well as the combined data sets, the endemic Hawaiian Alsinoideae originated from colonists on older islands, with radiations from older to younger islands (Figure 16.2). Phylogenetic analysis of morphological data suggests a Kaua`i origin for the *S. membranacea, S. adamantis,* and *S. nuttallii* clades, with an O`ahu origin for the *S. globosa* clade (Wagner et al. 1995). Analysis of a more limited number of species based on combined data supports this hypothesis but points to a possible origin on O`ahu as well (Soltis et al. 1996). Wagner et al. (1995) suggest that the center of distribution of the more basal *S. membranacea* clade on Kaua`i may result from either colonization directly to Kaua`i or to islands that are now vanished beneath the ocean through erosion or subsidence, with the Kaua`i and Nihoa species the sole survivors of these various earlier clades.

This interpretation of an origin on older islands is supported not only by the biogeographical distribution of the species but also by the more specialized morphological features in the more basal *S. membranacea* clade (particularly *Alsinidendron*) and the greater number of molecular mutations that differentiate this clade from others (Soltis et al. 1996). Based on both the morphological and combined data, colonization of new islands appears to be by more recent, less specialized species. Most cpDNA and rDNA restriction-site mutations differentiate *Schiedea* from *Alsinidendron*. The lack of many restriction-site mutations in *Schiedea* suggests either that most species of *Schiedea*, unlike *Alsinidendron*, are the result of recent rapid speciation, or alternatively, that there has been a slowdown in the rate of cpDNA evolution (Soltis et al. 1996). Unfortunately, the only *Schiedea* species from the older islands of Kaua`i or Nihoa with material for the molecular analysis was *S. membranacea*; further studies of other *Schiedea* species from these older islands may result in a revised interpretation.

Conclusions

The endemic Hawaiian Alsinoideae, with 26 species of *Schiedea* and four species of *Alsinidendron*, is one of the six largest monophyletic lineages of native Hawaiian flowering plants (Wagner et al. 1990). Species occur on all the major Hawaiian Islands and have radiated into a number of different habitats with significant changes in vegetative and floral morphology. This lineage is one of the most diverse in breeding system expression, with hermaphroditic, gynodioecious, subdioecious, and dioecious species.

Based on both morphological and molecular data, the clade including *Alsinidendron* (*S. membrancea* clade) is the sister-group to the remaining species of *Schiedea*. An apparently ancient split separates the *S. membranacea* clade from the clade comprising the majority of *Schiedea* species; the latter clade has apparently radiated relatively recently and rapidly. Because of the presumed relatively recent origin of much of the lineage, relatively few restriction-site mutations in either cpDNA or rDNA were phylogenetically informative within *Schiedea*, resulting in a weakly resolved molecular phylogeny. Similarly, the phylogeny of *Schiedea* species based on morphological characters was not strongly supported. Although weak, the resulting trees based on molecular data and on morphological data were largely consistent, and a combined analysis of the morphological and molecular data sets provided greater resolution and stronger support for many parts of the tree than either analysis alone. Some results of the combined data set should still be interpreted with caution because of missing taxa. Although two-thirds of known species were included in both the molecular and combined data sets, phylogenetic analysis of the more complete morphological data set suggests that interpretations might change if species with missing molecular data were to be included. Combined molecular and morphological data may be particularly important in reconstruction in lineages where recent adaptive radiation among some species has resulted in morphological diversity but little molecular diversification. Results for *Schiedea* and *Alsinidendron* show that the morphological and molecular data sets, though providing relatively weakly supported or unresolved estimates of phylogeny when used independently, are largely congruent and provide more robust estimates when combined.

Phylogenetic analysis suggests that colonization has been from older to younger islands, and speciation has resulted from both inter-island colonizations as well as intra-island colonizations often associated with habitat shifts. Examination of breeding system evolution in a phylogenetic context suggests that dimorphism (gynodioecy, subdioecy, dioecy) has arisen from 1 to 6 times, depending on how breeding system is coded and how equivocal placements of changes are resolved. Dimorphism most likely arose either once (at the base of a large clade including all the dimorphic species) or twice (at the base of each of the two clades including dimorphic species). Reversals to hermaphroditism in some species are indicated in all phylogenetic analyses. Unfortunately, the least resolved clade in both the morphological and molecular analyses is one with both dimorphic and hermaphroditic species. Shifts to dimorphism are coincident with shifts to dry habitats or occur in clades where the origin of both features is equivocal, and as a consequence, it is difficult to determine the causal relationship between these factors using phylogenetic approaches. Three hermaphroditic species in different clades (*S. lydgatei* with a reversal to hermaphroditism; *S. menziesii* with a potential reversal to hermaphroditism; and *S. verticillata*) do occur in dry habitats, indicating that shifts to dry habitat do not require a shift to dimorphism. The occurrence of all dimorphic species in dry environments, however, indicates that this habitat shift strongly favors the evolution of dimorphism. Weller and Sakai (1990) have suggested that shifts in habitat to dry windy environments may often be associated with a loss of insect pollinators and shift to wind pollination. All dimorphic species are members of the two clades (*S. adamantis* and *S. globosa* clades) with condensed or globose inflorescences, morphology more consistent with wind pollination. In addition, dimorphic species generally occur in larger populations than hermaphroditic species (Weller et al. 1995), resulting in higher conspecific densities, another factor that may be important in the evolution of wind pollination. Initial loss of insect pollinators may lead to increased selfing rates and increased expression of inbreeding depression. Inbreeding depression studies of several species of *Schiedea* have all shown high levels of inbreeding depression (*S. salicaria* and *S. globosa* [Sakai et al. 1989], *S. lydgatei* [Norman et al. 1995], *S. adamantis* [Sakai et al. 1997]), supporting this hypothesis. This combination of high levels of inbreeding depression and high selfing rates may favor the spread of females in these populations (Charlesworth and Charlesworth 1978). Coincident shifts to wind pollination may occur, resulting in the observed pattern of dimorphic, wind-pollinated species in dry environments and hermaphroditic, insect-pollinated species in mesic and wet environments. Additionally, changes in resource allocation may occur (Sakai et al. 1997), although preliminary data suggest this occurs after the introduction of females into the population.

Despite the occurrence of convergent evolution associated with the evolution of dimorphic breeding systems, two distinct clades with dimorphic species were recognized in the phylogeny based on morphological data. Similar changes have occurred in several dimorphic species in dry environments, which presumably have favored wind pollination, more condensed inflorescences, and smaller leaves. The presence of these two clades suggests that morphological traits can be phylogenetically informative in the face of similar selective pressures. The phylogeny based on molecular data was largely congruent with the phylogeny based on morphology, but dimorphic species occur in only one clade, although the molecular data were limited by the paucity of informative molecular characters, and the lack of molecular data from several species.

Further resolution of phylogenetic relationships of species as well as characterization of habitat differences and studies of selfing rates and levels of inbreeding depression in all clades of this lineage may clarify patterns. In addition, experimental approaches to breeding system evolution as well as studies of the capacity of these populations to respond to selection on breeding systems may elucidate the mechanisms resulting in the remarkable breeding system diversity within this group.

References

Bawa, K. S. 1980. Evolution of dioecy in flowering plants. Annual Review of Ecology and Systematics 11:15–40.

Bremer, K. 1988. The limits of amino acid sequence data in angiosperm phylogenetic reconstruction. Evolution 42:795–803.

Carlquist, S. 1974. *Island biology*. New York, NY: Columbia University Press.

Charlesworth, B., and Charlesworth, D. 1978. A model for the evolution of dioecy and gynodioecy. American Naturalist 112:975–997.

Donoghue, M. J. 1989. Phylogenies and the analysis of evolutionary sequences, with examples from seed plants. Evolution 43:1137–1156.

Donoghue, M. J., Olmstead, R. G., Smith, J. F., and Palmer, J. D. 1992. Phylogenetic relationships of Dipsacales based on *rbc*L sequences. Annals of the Missouri Botanical Garden 79:333–345.

Doyle, J. J., and Doyle, J. L. 1987. A rapid DNA isolation procedure for small amounts of fresh leaf tissue. Phytochemical Bulletin 19:11–15.

Felsenstein, J. 1985. Phylogenies and the comparative method. American Naturalist 125:1–15.

Givnish, T. J. 1980. Ecological constraints in the evolution of breeding systems in seed plants: dioecy and dispersal in gymnosperms. Evolution 34:959–972.

Harris, E. M., and Wagner, W. L. 1993. Using floral ontogeny to track evolution in *Schiedea* and *Alsinidendron* (Caryophyllaceae). American Journal of Botany (Supplement) 80:25.

Hillis, D. M. 1991. Discriminating between phylogenetic signal and random noise in DNA sequences. Pp. 278–294 in *Phylogenetic analysis of DNA sequences*, M. M. Miyamoto and J. Cracraft, eds. Oxford, England: Oxford University Press.

Hillis, D. M., and Huelsenbeck, J. P. 1992. Signal, noise, and reliability in molecular phylogenetic analyses. Journal of Heredity 83:189–195.

Huelsenbeck, J. P. 1991. Tree-length distribution skewness: an indicator of phylogenetic information. Systematic Zoology 40:257–270.

Jansen, R. K., and Palmer, J. D. 1987. Chloroplast DNA from lettuce and *Barnadesia* (Asteraceae): structure, gene localization, and characterization of a large inversion. Current Genetics 11:553–564.

Macdonald, G. A., Abbott, A. T., and Peterson, F. L. 1983. *Volcanoes in the sea: the geology of Hawaii, 2nd ed.* Honolulu, HI: University of Hawaii Press.

Maddison, D. R. 1991. The discovery and importance of multiple islands of most parsimonious trees. Systematic Zoology 40:315–328.

Norman, J. K., Sakai, A. K., Weller, S. G., and Dawson, T. E. 1995. Inbreeding depression in morphological and physiological traits of *Schiedea lydgatei* (Caryophyllaceae) in two environments. Evolution 49:297–306.

Palmer, J. D. 1986. Isolation and structural analysis of chloroplast DNA. Methods in Enzymology 118:167–186.

Rieseberg, L. H., Soltis, D. E., and Palmer, J. D. 1988. A molecular reexamination of introgression between *Helianthus annuus* and *H. bolanderi* (Compositae). Evolution 42:227–238.

Sakai, A. K., Karoly, K., and Weller, S. G. 1989. Inbreeding depression in *Schiedea globosa* and *S. salicaria* (Caryophyllaceae), subdioecious and gynodioecious Hawaiian species. American Journal of Botany 76:437–444.

Sakai, A. K., Wagner, W. L., Ferguson, D. M., and Herbst, D. R. 1995a. Origins of dioecy in the Hawaiian flora. Ecology 76:2517–2529.

Sakai, A. K., Wagner, W. L., Ferguson, D. M., and Herbst, D. R. 1995b. Biogeographical and ecological correlates of dioecy in the Hawaiian flora. Ecology 76:2530–2543.

Sakai, A. K., Weller, S. G., Chen, M.-L., Chou, S.-Y., and Tasanont, C. 1997. Evolution of gynodioecy and maintenance of females: the role of inbreeding depression, outcrossing rates, and resource allocation in *Schiedea adamantis* (Caryophyllaceae). Evolution 51 (in press).

Soltis, D. E., Soltis, P. S., Collier, T. G., and Edgerton, M. L. 1991. Chloroplast DNA variation within and among genera of the *Heuchera* group (Saxifragaceae): evidence for chloroplast transfer and paraphyly. American Journal of Botany 78:1091–1112.

Soltis, P. S., Soltis, D. E., Weller, S. G., Sakai, A. K., and Wagner, W. L. 1996. Molecular phylogenetic analysis of the Hawaiian endemics *Schiedea* and *Alsinidendron* (Caryophyllaceae). Systematic Botany 21:365–379.

Swofford, D. L. 1993. *PAUP: phylogenetic analysis using parsimony, vers. 3.1.1.* Champaign, IL: Illinois Natural History Survey.

Thomson, J. D., and Brunet, J. 1990. Hypotheses for the evolution of dioecy in seed plants. Trends in Ecology and Evolution 5:11–16.

Wagner, W. L., Herbst, D. R., and Sohmer, S. H. 1990. *Manual of the flowering plants of Hawai`i.* Honolulu, HI: University of Hawaii Press and Bishop Museum Press.

Wagner, W. L., Weller, S. G., and Sakai, A. K. 1995. Phylogeny and biogeography in *Schiedea* and *Alsinidendron* (Caryophyllaceae). Pp. 221–258 in *Hawaiian biogeography: evolution on a hot spot archipelago*, W. L. Wagner and V. A. Funk, eds. Washington, DC: Smithsonian Institution Press.

Weller, S. G., and Sakai, A. K. 1990. The evolution of dicliny in *Schiedea* (Caryophyllaceae), an endemic Hawaiian genus. Plant Species Biology 5:83–95.

Weller, S. G., Sakai, A. K., and Straub, C. 1996. Allozyme diversity and genetic identity in *Schiedea* and *Alsinidendron* (Caryophyllaceae: Alsinoideae) in Hawai`i. Evolution 50:23–34.

Weller, S. G., Sakai, A. K., Wagner, W. L., and Herbst, D. R. 1990. Evolution of dioecy in *Schiedea* (Caryophyllaceae: Alsinoideae) in the Hawaiian Islands: biogeographical and ecological factors. Systematic Botany 15:266–276.

Weller, S. G., Wagner, W. L., and Sakai, A. K. 1995. A phylogenetic analysis of *Schiedea* and *Alsinidendron* (Caryophyllaceae: Alsinoideae): Implications for the evolution of breeding systems. Systematic Botany 20:315–337.

17 Ecological and Reproductive Shifts in the Diversification of the Endemic Hawaiian *Drosophila*

Michael P. Kambysellis and Elysse M. Craddock

The Drosophilidae endemic to the isolated Hawaiian archipelago provide one of the most spectacular examples of insular diversification known to evolutionary biologists (Zimmerman 1958; Carson et al. 1970; Williamson 1981). From one or at most two original founders from a distant continental source (Throckmorton 1966; Thomas and Hunt 1991; DeSalle 1995; Kaneshiro et al. 1995), more than 800 contemporary species have evolved after an initial colonization more than 30 Mya (Beverley and Wilson 1985; DeSalle 1992). This explosive speciation has resulted in forms that vary widely in size and morphology, with different species displaying bizarre modifications of the head, forelegs, wings, or mouthparts, especially in males (Hardy 1965; Hardy and Kaneshiro 1981). Such morphological divergence largely reflects the development of extraordinary secondary sexual characteristics correlated with male mating behavior (Spieth 1966, 1974, 1982). Significantly, the morphological and behavioral diversity among Hawaiian taxa far exceeds that of drosophilids anywhere else in the world. Given the small area, remoteness, and comparative geological youth of the current Hawaiian Islands (< 5.3 My), it is astounding that ca. 25% of the world's species of *Drosophila* are found on the six major islands of this archipelago. Most species are single-island endemics, and are often restricted to a single volcano, or even a single "kipuka" (a patch of forest isolated by lava flows). The Hawaiian drosophilids form a spectacular sexual radiation, and have been a frequent target for studies of geographic speciation and the evolution of mating barriers (Carson 1983, 1986; Kaneshiro 1983, 1989; Kaneshiro and Boake 1987; Carson et al. 1990; DeSalle 1995; Kaneshiro et al. 1995).

Although most often noted for their morphological and behavioral diversity, the Hawaiian drosophilids are also ecologically quite diverse. They occupy a wide range of altitudes (50 to 3,000 m) and moisture regimes that range from dry open forests to rain forests receiving more 8,000 mm of annual precipitation. Within this broad spectrum of habitats, different drosophilid species exploit a diverse array of ecological substrates for breeding, including sap exudates (fluxes) and decaying parts of several different plant species (Heed 1968, 1971; Montgomery 1975). Previous treatments have placed little emphasis on the role of ecological characteristics in the evolution of the Hawaiian drosophilids, with much greater attention being given to the role of sexual selection (Carson 1978, 1986; Kaneshiro 1983, 1989) and founder events (Carson 1971; Carson and Templeton 1984) in driving divergence and species proliferation. This chapter instead focuses on ecological differentiation in the Hawaiian drosophilids and associated adaptive differentiation, particularly with respect to the female reproductive system. Using a phylogenetic approach, we reconstruct the sequence of ecological shifts in the evolution of the group and show how a phylogenetic analysis of ecological, morphological, developmental, physiological, and behavioral aspects of female reproduction in the endemic Hawaiian *Drosophila* can illuminate our understanding of the adaptive radiation of this remarkable group of organisms.

Today, roughly 30 species groups and 55 subgroups of Hawaiian drosophilids are recognized, based largely on morphology (K. Kaneshiro, pers. comm.). Given that systematic studies are still incomplete, the number of these groups may well increase. Clearly, the Hawaiian drosophilid radiation has been complex, likely resulting from the operation of multiple forces rather than just a single one. What was the adaptive basis of divergence in this group? These flies have evolved adaptations to the intraspecific sexual milieu (Carson 1978) as well as to the external environment. In the first arena, the driving force is sexual selection; in the second, environmental adaptation is driven by natural selection.

The exaggerated development of a variety of male secondary sexual characteristics, many of which provide key characters for identifying species groups, is clear evidence for the significance of sexual selection in the evolution of the endemic Hawaiian drosophiloids. These traits are generally directly involved in the elaborate courtship and mating behaviors of these flies (Spieth 1966, 1974, 1982). For example, during the "head-under-wing" stage of courtship the enlarged and erect antennae of males of the *antopocerus* species group articulate with the female wing vanes and protect the male's head during his vigorous lunging movements (Spieth 1968a). In the modified-mouthparts group, the male labellum is variously modified in shape and hairiness (Hardy and Kaneshiro 1975); males of this group use this structure to grasp the female's genital area during courtship (Spieth 1966). Although members of individual species groups share particular morphological and behavioral modifications indicative of their phylogenetic affinity, nonetheless each species displays a unique courtship pattern and unique morphology. Intraspecific coadaptation of male and female behaviors constitutes a specific mate recognition system distinct from that of related species.

Classical models of sexual selection accent the intermale competition for mates and the role of epigamic selection by females in a runaway process that generates extreme forms of particular male characters (Fisher 1930; Lande 1981). Given that the sexual system is dynamic and highly labile with a variety of mating types segregating among both males and females, any destabilization, as during a founder event, may be followed by a shift of balance and a readjustment of the intersexual interactions, leading to a novel mating system (Carson 1986; Kaneshiro 1989). Thus sexual selection may be pivotal to the speciation process and the radiation of a group of organisms such as the Hawaiian drosophilids. Although most models of sexual selection assume that female preferences for certain male secondary sexual characters are counterbalanced by natural selection constraining the runaway process, Kaneshiro (1987) suggests an alternative mechanism whereby secondary sexual characters are not necessarily subject to direct selection, but rather originate and are maintained pleiotropically or via genetic linkage to some other feature of the mate recognition system.

Whereas the divergent morphologies and mating behaviors of Hawaiian drosophilids are readily explained as a consequence of sexual selection, this force cannot account for the other prominent feature of the group, namely their ecological divergence. Drosophilids are saprophages and the Hawaiian species have radiated from the original founder to breed in a diverse array of substrates that includes fungi, rotting flowers, leaves, roots, stems, bark, and tree fluxes from some 40 families of endemic Hawaiian plants (Heed 1968, 1971; Montgomery 1975). A few Hawaiian drosophilids are even parasitic on spiders' eggs (Hardy 1965; Heed 1968). Exploitation of such diverse breeding substrates has succeeded because natural selection has molded various aspects of the female reproductive system to adapt each species to its particular breeding niche and its local environment. Herein lies the real basis of the adaptive radiation of the endemic Hawaiian drosophilids.

It can be envisaged that saturation of the initial breeding substrates and the accompanying competition, coupled with the availability of open niches as new islands and habitats were colonized, led rather quickly to an ecological radiation, with natural selection driving the adaptation of the flies to novel breeding niches. Breeding substrates of the Hawaiian flies vary widely in predictability and nutrient reserves for larval growth. At the same time, patterns of egg production and distribution can vary dramatically: some Hawaiian species produce just one egg at a time, whereas others mature and oviposit hundreds of eggs at a time (Kambysellis and Heed 1971). This wide range of female reproductive strategies and the concomitant ecological divergence raise the following questions which provide the focus of this chapter.

- What is the relationship of female reproductive traits to the nature of the larval substrate?
- How are these traits (e.g., ovariole number per ovary, number of mature eggs per ovariole, length of egg respiratory filaments) adapted to features of the breeding substrate?
- What was the ancestral larval substrate for the Hawaiian drosophilids, and what has been the subsequent pattern of shifts in substrate?
- Do species utilizing a particular kind of substrate (e.g., plant stems) form a monophyletic group? If not, how many times has each substrate been invaded?
- Where there have been multiple invasions of the same substrate, have the female reproductive traits shown correlated shifts? And
- What is the relationship, if any, between variation in ecology and female reproductive traits and variation in male mating behavior and sexual characteristics, and what are the implications of such a relationship for evolution and speciation in the Hawaiian *Drosophila*?

Before addressing these questions and analyzing the evolution of female reproductive biology in these flies in relation to their ecological diversification in breeding substrate and host plant use, we present some further background on the biology and taxonomy of the Hawaiian *Drosophila*. This information provides the necessary context for understanding the adaptive radiation of the group.

Natural History of the Hawaiian Drosophilidae

The endemic Hawaiian drosophilids comprise an extremely diverse and species-rich assemblage, quite distinct from continental forms and, as a paradigm of the process of adaptive radiation on oceanic islands, a group worthy of intensive research. Initial multi-disciplinary studies conducted under the auspices of the Hawaiian *Drosophila* Project (Spieth 1981) accumulated a wealth of information on the morphology and systematics of the group, their polytene chromosomal inversion differences, karyotypes, ecology, behavior, and allozymic differentiation (see reviews by Carson et al. 1970; Carson and Kaneshiro 1976). These early studies suggested that, despite their extreme morphological divergence, the Hawaiian flies are genetically quite closely related and form a cohesive group. More recent molecular analyses have, in fact, confirmed that the Hawaiian drosophilids are monophyletic (Thomas and Hunt 1991; DeSalle 1992, 1995; Kambysellis et al. 1995).

Systematics

Although the initial taxonomic treatment distinguished nine genera of Hawaiian drosophilids (Hardy 1965), it is now clear that there are only two evolutionary lineages, the drosophiloids and the scaptomyzoids, with the majority of species belonging to either *Drosophila* or *Scaptomyza* (Kaneshiro 1976). The existence of forms that are in some respects intermediate between drosophiloids and scaptomyzoids suggests that both lineages may have derived from one original founder, with the genus *Scaptomyza* evolving in Hawai`i, and spreading from there to the rest of the world (Throckmorton 1966; DeSalle 1992; Kambysellis et al. 1995).

Systematically, the Hawaiian *Drosophila* are more closely allied to subgenus *Drosophila* (Thomas and Hunt 1991; DeSalle 1992, 1995; Kambysellis et al. 1995) than subgenus *Sophophora*, to which the well-studied species *D. melanogaster* belongs. However, there is no single continental form that can be designated as the sister to the Hawaiian lineages (DeSalle 1995). Whereas the majority of Hawaiian *Drosophila* are placed in subgenus *Drosophila*, a group of six or more species are instead placed in subgenus *Engiscaptomyza* (Kaneshiro 1969; Hardy and Kaneshiro 1981). Members of this group (represented in this study by *D. crassifemur*) are somewhat intermediate between drosophiloids and scaptomyzoids; they were retained in the genus *Drosophila* on the basis of external morphology, but display *Scaptomyza*-like internal anatomy and mating behavior.

The Hawaiian scaptomyzoids alone include more than 200 species, more than in the rest of the world combined. Most belong to the genus *Scaptomyza*, but there are also three other scaptomyzoid genera with from one to eleven species (Hardy and Kaneshiro 1981). Overall, the scaptomyzoids are less diverse morphologically than the more numerous drosophiloids, although their male genitalia are well differentiated (Takada 1966). Remarkably, they appear to be ecologically more diverse than the drosophiloids, utilizing several novel breeding substrates in addition to those utilized by the drosophiloids (K. Kaneshiro, pers. comm.). Analysis of the evolution of ecological shifts in the adaptive radiation of the scaptomyzoids must, however, await systematic and phylogenetic analysis of this enigmatic group.

Male sexual behavior

Whereas the Hawaiian drosophiloids are characterized by extremely elaborate species-specific courtship patterns and marked morphological differentiation of the males, Hawaiian *Scaptomyza* have a more uniform morphology and simplified courtship behavior, with basically an "assault-type" mating (Spieth 1966). These traits correlate with a lack of lek behavior and suggest that sexual selection may be relatively unimportant in the scaptomyzoids. This contrasts with the morphological and behavioral divergence of the drosophiloids, and the prominent role of sexual selection in the majority of the members of this lineage. Intermale competition is especially marked in members of the picture-winged group where males display a high level of agonistic behavior and often engage in ritualized fighting (Spieth 1982). Furthermore, body size of these species is generally very large, and mature males advertise their sexual readiness by characteristic displays on their leks.

Significantly, courtship and mating in Hawaiian drosophiloids take place only on the leks, not on the feeding and oviposition sites (Spieth 1966, 1968b). This spatial separation between mating and oviposition sites contrasts with the usual drosophilid pattern and suggests that the processes of sexual selection among males and natural selection on female traits exerted by the breeding substrate may be independent.

Reproductive isolation and speciation

Sympatric species are behaviorally isolated by their unique courtship repertoires. Strong premating barriers essentially prevent interspecific hybridization in the field, although in a few exceptional instances limited hybridization between closely related species has been recorded (Kaneshiro and Val 1977; Carson et al. 1989). Allopatric species typically show lower levels of premating isolation; laboratory hybridizations can therefore be used to assess the extent of postmating barriers (Yang and Wheeler 1969; Craddock 1974a,b). These range from limited F_1 hybrid male sterility to complete hybrid inviability, depending on the genetic distance between species.

Clearly, much of the great proliferation of Hawaiian drosophilid species has been triggered by founder events (Carson 1971; Carson and Templeton 1984). Chance colonization of a new island or volcano by a single fertilized female has frequently resulted in genetic divergence from the ancestral population and subsequent speciation. Genetic drift in the initial founder population must be central to founder-effect speciation, but it should be recognized that the new population will usually find itself in a novel environment and selective regime, and will likely experience a shift in sexual selection as well (Kaneshiro 1989). Thus drift, natural selection, and sexual selection may jointly operate to lead to speciation following founder events.

Some speciation events in the radiation of the Hawaiian *Drosophila* have apparently occurred independently of inter- or intra-island dispersal events. In some instances, shifts in sexual selection may have been responsible (Carson 1986; Kaneshiro 1987). Such speciation might be considered non-adaptive with respect to the external environment, but adaptive with respect to the intraspecific sexual environment. In other cases, however, it is germane to ask what role the environment and natural selection may have played in speciation and subsequent adaptation.

Ecological diversification of the Hawaiian *Drosophila*

Colonization of the Hawaiian Islands by drosophilids has been accompanied by radiation into a wide variety of altitudes and ecological niches in both wet and dry montane forests that vary markedly in ecological complexity. Moreover, the habitats these flies occupy are floristically diverse, providing opportunities for the evolution of a range of fly-host plant interactions. The adult flies are quite mobile and feed at a variety of sites throughout the forest; however, the larvae of a particular species are extremely localized in their distribution, typically being restricted to one or a few specific substrates, chiefly in decaying plant material. This is due to the specificity of female behavior in selecting sites for oviposition. Presumably, their behavior is guided by olfactory and other chemosensory cues, as well as tactile cues perceived by sensory hairs on the ovipositor, which they use to probe prospective substrates prior to actual oviposition. Female behavior is thus a primary factor in the specificity of host plant use for breeding, and behavioral divergence and evolution of particular female preferences must be considered as one of the components in the ecological radiation of these flies.

Knowledge of the breeding sites of Hawaiian drosophilids began with the records of Perkins (1913) in the *Fauna Hawaiiensis*, but is mainly due to the careful field work of Heed (1968, 1971) and Montgomery (1975). Although host plants are only known for about 20% of the endemic drosophilids, some 40 of the 114 families of vascular plants represented in Hawai`i (Wagner 1991) have been recorded as substrates, including a number of native ferns. In addition, species in the white-tip scutellum group use fungi. Although relatively little is known of the chemistry of endemic Hawaiian plants (Kircher and Heed 1970), it can be surmised that the plant families utilized provide a chemical environment tolerated by the developing larvae.

Ecological divergence in the breeding niche is not restricted to the type of endemic Hawaiian plant used, but also involves the particular part of the plant selected for oviposition. Virtually all parts of a decaying plant may be used. Individual species are remarkably specific in their choice of substrate and will breed in decaying leaves, flowers, fruits, stems, bark, or roots, but usually not in several or all of these. Other Hawaiian drosophilids use fluxes or sap exudates of particular trees as their breeding substrate, while yet others use fungi. Notwithstanding the specificity of the interaction between a fly species and its particular host plant and the part of the plant utilized, it must be pointed out that the plant is not the primary nutritive source for the larvae or the adults. Drosophilids are saprophagous, and as the larvae mine the substrate, they feed on the microbial fauna associated with the decaying plant material. Although most continental *Drosophila* feed on yeasts (see Begon 1982), both larvae and adults of the Hawaiian species feed primarily on bacteria (Robertson et al. 1968; M. Kambysellis, unpubl. data). Thus the ecological interaction is quite complex with a three-way interplay between bacteria, plants, and flies. We have initiated a project on identification and analysis of the bacteria found in the gut and fecal material of *Drosophila* species from different substrates and different habitats, and the bacteria associated with various host plants, but the data are complex and will not be discussed further here. Suffice it to say that the microflora is an important component of Hawaiian habitats and may be a critical factor in the ecological interactions between *Drosophila* and their decaying breeding substrates. Selection of a particular plant for oviposition may depend as much on the associated microflora as on the chemistry of the plant. It may well be the combined odors released by the plant and associated bacteria that make a particular substrate attractive to a searching female and trigger her oviposition behavior.

Methods

Molecular systematics

The 39 Hawaiian taxa examined in this study are included in Table 17.1. Species groups and subgroups are largely based on morphological traits, following the arrangement of Hardy and Kaneshiro (1981) and Kaneshiro et al. (1995). Five continental drosophilids were used as outgroups, including one scaptomyzoid (*Scaptomyza adusta*) and four drosophiloids, representing the subgenera *Drosophila* (*D. virilis* and *D. buzzatii*) and *Sophophora* (*D. birchii* and *D. melanogaster*).

DNA was extracted from adult flies of each species, and a 1 kb segment of the *Yp1* yolk protein gene was PCR-amplified and sequenced (Kambysellis et al. 1995). The sequenced region encompassed the two introns and 64% of the coding sequences of the *Yp1* gene. The sequence for *D. melanogaster* was provided by Hung and Wensink (1981).

Global parsimony was used to construct a phylogeny based on the *Yp1* nucleotide sequence data, using an unweighted analysis implemented in PAUP version 3.1.1 (Swofford 1993). Given the large number of taxa involved, we conducted heuristic searches, using the random stepwise addition option for generating a different tree to begin each of ten replicate searches. The resulting topology of the ingroup taxa was independent of whether the outgroup chosen was *D. melanogaster*, both Sophophorans, or all four continental *Drosophila*. The level of support for each branch of the single most parsimonious tree was assessed via bootstrap analysis (Felsenstein 1985, 1988) and decay analysis (Bremer 1988; Donoghue et al. 1992). Bootstrap values from 1,000 replications were calculated using the random-input file option of MEGA (Kumar et al. 1993), and decay indices obtained by recursively saving trees one step longer than the previous trees and calculating the strict consensus until all the branches collapsed.

Since our *Yp1* phylogeny focuses on the picture-winged species and includes few early-divergent Hawaiian drosophilids (only 5 species from two lineages), we made use of another molecular phylogeny of 15 Hawaiian drosophilids, kindly supplied by Rob DeSalle, that includes 7 additional species. This maximum parsimony tree includes five groups of non-picture-wings (represented by two species each) and five picture-winged species, and is based on nucleotide sequences of fragments of four mitochondrial and four nuclear genes, totalling ~2.5 kb in all (DeSalle 1992; Baker and DeSalle 1997).

Analysis of character evolution

The evolution of ecological and female reproductive characters pertinent to the adaptive radiation of the Hawaiian Drosophilidae was analyzed using MacClade 3.04

Table 17.1. Genera and species groups of Hawaiian Drosophilidae analyzed in this study and their distinguishing features.

Group	Number of species	Distinguishing features	Taxa analyzed
Scaptomyza	200+	assault-type mating	*S. albovittata*
Drosophila			
Subg. *Engiscaptomyza*	6	intermediate morphology	*D. crassifemur*
Subg. *Drosophila*			
White-tip scutellum	~80	slender iridescent body	*D. longipedis* *D. iki*
Modified tarsi	~80	modifications of first and/or second tarsomere of male foreleg	
Bristle tarsi		clump of heavy bristles on front basitarsus	*D. petalopeza*
Spoon tarsi		second tarsomere flattened and concave	*D. waddingtoni*
antopocerus	~17	large porrect antennae in males	*D. adunca* *D. tanythrix* *D. yooni*
Modified mouthparts	~110	dense hairs, bristles or spines on apical lobes of male labella	*D. mimica* *D. soonae* *D. infuscata*
Picture wings	110	maculations on wings	
primaeva	2		*D. primaeva*
adiastola	16		*D. ornata* *D. truncipenna* *D. spectabilis* *D. setosimentum* *D. adiastola*
planitibia	17		*D. picticornis* *D. setosifrons* *D. substenoptera* *D. nigribasis* *D. oahuensis* *D. obscuripes* *D. melanocephala* *D. cyrtoloma* *D. hemipeza* *D. planitibia* *D. differens* *D. silvestris* *D. heteroneura*
glabriapex	34		*D. pilimana* *D. fasciculisetae* *D. lineosetae* *D. macrothrix* *D. punalua*
grimshawi	39		*D. mulli* *D. sproati* *D. silvarentis* *D. heedi* *D. hawaiiensis* *D. disjuncta* *D. bostrycha* *D. grimshawi*[1] *D. pullipes*

[1]The molecular phylogeny includes both the Kaua`i and Maui populations of *D. grimshawi*. These may be incipient species.

(Maddison and Maddison 1992) to overlay character-states on the molecular phylogeny and determine the most parsimonious scenario for the origin of those states. We applied both accelerated and delayed transformation optimization options; the character reconstructions presented, however, use ACCTRAN, which minimizes parallel evolution. Records of host plant use (Heed 1968, 1971; Montgomery 1975) formed the basis for mapping the ecological shifts from one plant family to another, from monophagy to oligophagy and polyphagy, and from one breeding substrate to another. The evolution of female reproductive types was analyzed using available data on the egg/ovarian types of Hawaiian drosophilids (Kambysellis and Heed 1971; M. Kambysellis and E. Craddock, unpubl. data). The island distributions of the Hawaiian species were used in a similar manner to display geographic patterns of speciation in the group.

Results and Discussion

Molecular phylogeny of the Hawaiian Drosophilidae

Figure 17.1A presents the single most parsimonious tree of 39 Hawaiian *Drosophila* species based on a cladistic analysis of *Yp1* DNA sequences. This phylogeny (length L = 1,550; CI = 0.605; RI = 0.706) includes 34 picture-winged species, representing all five recognized species groups and 12 of the 14 recognized species subgroups (Kaneshiro et al. 1995), as well as two members of the modified-mouthparts group, and three members of the *antopocerus* group. Most of the previously recognized species groups and subgroups are resolved as monophyletic clades with strong statistical support (Figure 17.1A). The placements of two species (*D. mulli* and *D. punalua*) in our molecular phylogeny differ from their previously recognized morphological affinities. *Drosophila mulli* had been classified with *D. sproati* in the *grimshawi* group (Kaneshiro et al. 1995), but aligns with the *glabriapex* group in the *Yp1* phylogeny. *Drosophila punalua* had been placed in the *glabriapex* group, but our molecular data place it in the *grimshawi* group. A more significant discrepancy is that the *picticornis* subgroup does not form a monophyletic clade with the two other recognized subgroups (*cyrtoloma* and *planitibia*) of the *planitibia* species group. Although resolution of the basal branches is weak, other data also suggest that the taxonomic affinity of the *picticornis* subgroup needs to be reevaluated (Russo et al. 1995).

Figure 17.1B shows another molecular phylogeny of the Hawaiian Drosophilidae, provided by Baker and DeSalle (1997). This phylogeny is the single most parsimonious tree based on nuclear and mitochondrial DNA sequences (L = 1,946; CI = 0.50; RI = 0.49) and is quite robust, as indicated by the bootstrap values and decay indices. It includes representatives of the more primitive Hawaiian groups (*antopocerus*, spoon tarsi, bristle tarsi, white-tip scutellum, modified-mouthparts), the more advanced picture-winged group (5 spp.), and the more divergent scaptomyzoids (represented by *D. crassifemur* [subgenus *Engiscaptomyza*] and *Scaptomyza albovittata*). The two outgroups used are the continental species *D. melanogaster* and *D. mulleri*. This tree of 15 species provides a broad overview of the relationships among some of the major species groups of Hawaiian flies.

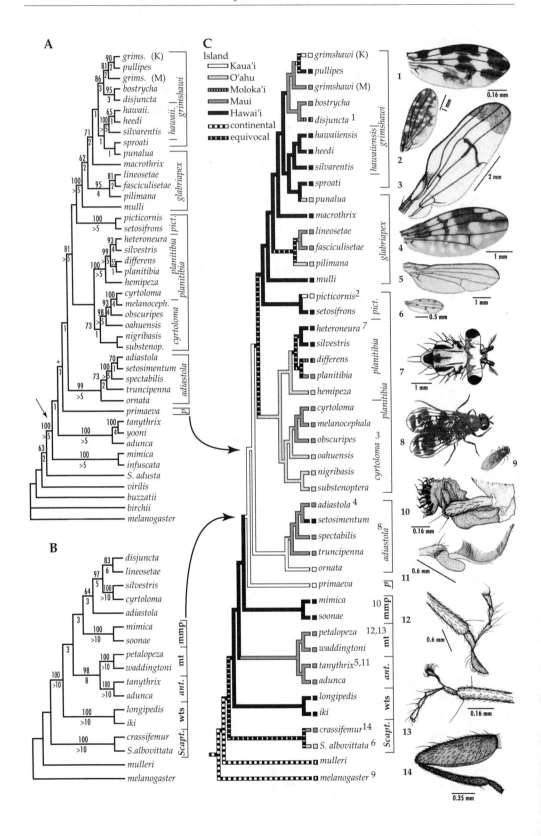

The two independent molecular phylogenies (Figure 17.1A,B) are congruent in indicating that the Hawaiian drosophilids are monophyletic, that the genus *Scaptomyza* diverged between the continental and Hawaiian *Drosophila*, and that the picture-winged flies comprise the most recently derived of the Hawaiian species-groups. The trees differ, however, in their placement of the *antopocerus* and modified-mouthparts groups. Because the topology of these two groups is weak in the *Yp1* tree (Figure 17.1A; Kambysellis et al. 1995), we favor the systematic arrangement arising from DeSalle's data (Figure 17.1B), which places the *antopocerus* group as sister to the modified-mouthparts group, with the latter being sister to the picture-winged group. Morphological similarities in the mouthparts of modified-mouthparts species and some of the *adiastola* species group – one of the more primitive picture-winged groups (Hardy 1965) – provide further support for this topology.

There is considerable overlap in the taxa included in the two analyses: the five picture-winged species, the two *antopocerus* species, and one of the two modified-mouthpart species in the Baker-DeSalle tree are included in our *Yp1* tree. However, there is no overlap in the sequence data sets used to derive the two phylogenies, which precludes combining the sequences in a single analysis. So, for illustrative purposes only, we have simply inserted the phylogeny of the 10 primitive, non-picture-winged Hawaiian drosophilids resolved in the Baker-DeSalle analysis into our *Yp1* phylogeny of the 34 species of the more derived picture-winged group to form a composite phylogeny (Figure 17.1C). This sampling of 44 Hawaiian species represents the broad diversity of Hawaiian drosophilids, based on the best available molecular data. For efficiency, we use this composite phylogeny as the basis for subsequent analyses of the evolution of ecological and reproductive characters.

Some of the features which distinguish species groups, as well as the distributions of the Hawaiian species, are also shown in Figure 17.1C. For the better sampled groups of picture-wings, the composite phylogeny indicates that dispersal has generally

Figure 17.1. (Opposite) Molecular phylogenies of the endemic Hawaiian Drosophilidae. (**A**) The single most parsimonious tree of 39 Hawaiian *Drosophila* species and 5 continental drosophilids based on nucleotide sequences of the *Yp1* gene. Bootstrap values from 1,000 replications are shown above the branches; decay indices are shown below. The arrow indicates the base of the Hawaiian *Drosophila*; the asterisk indicates the base of the picture-winged group. (**B**) The single most parsimonious tree of 15 Hawaiian drosophilids with two continental outgroups based on nucleotide sequences from 4 mtDNA and 4 nDNA gene regions (Baker and DeSalle 1997). (**C**) Composite tree, showing the phylogenetic relationships of 10 non-picture-winged Hawaiian species from (**B**) and 34 picture-winged species from (**A**) (as arrowed). Abbreviations for the morphologically defined species groups (in bold) are as follows: **Scapt** – scaptomyzoids; **wts** – white-tip scutellum; **ant** – *antopocerus*; **mt** – modified tarsi; **mmp** – modified mouthparts; ***p*** – *primaeva*. Picture-winged species subgroups (italics) are as shown; *pict* represents the *picticornis* subgroup. Island distributions of the 44 species are overlaid on the tree assuming accelerated transformation. To the right of the tree are shown distinctive morphological features of representative species and species groups (drawings taken from Hardy 1965). These include (1–6) the wings of six species approximately to scale; (7) the head of a male *D. heteroneura*; (8) a male *D. hamifera*, with (9) *D. melanogaster* for scale; and representative species with modified (10) mouthparts, (11) antennae, (12, 13) foretarsi, and (14) femurs. The numbers also identify the species whose traits are illustrated in the molecular phylogeny. Species illustrated but not in the phylogeny are as follows: (3) *D. neoperkinsi* of the *cyrtoloma* subgroup; (8) *D. hamifera* of the *adiastola* group; (10) *D. scolostoma* of the modified-mouthparts group; (12) *D. clavata* and (13) *D. attenuata*, both of the modified-tarsi group.

proceeded from older to younger islands in the Hawaiian chain. For example, in the *adiastola* group the Kaua`i species *D. ornata* is basal, with subsequent evolution of the Maui species, and finally the species on the youngest island of Hawai`i. Carson (1970, 1981), using chromosomal banding patterns, was the first to trace inter-island migrations in *Drosophila* down the Hawaiian chain. Among the 106 picture-winged species analyzed, he inferred a minimum of 45 inter-island colonization events, with eight founders from the currently oldest high island (Kaua`i) serving as progenitors of the 94 species found on the younger, more southeasterly islands in the chain (Carson 1992). Similar patterns have been inferred more recently, based on both chromosomal banding patterns (Kaneshiro et al. 1995) and DNA sequences (DeSalle 1995). Our molecular phylogeny (Figure 17.1A), although incomplete, confirms the general pattern of island hopping down the chain from older to younger islands. However, our molecular data imply fewer instances of back-migrations than are suggested by the chromosomal data.

Ecological character analysis

Table 17.2 summarizes data on oviposition sites known for the endemic Hawaiian *Drosophila*; Figures 17.2 and 17.3 display the inferred evolutionary patterns of ecological shifts in these character-states.

HOST PLANT ASSOCIATION – With respect to the taxonomic breadth of plants used as breeding substrates, the majority of Hawaiian drosophilids (81%) are monophagous; each monophagous species uses plants belonging to a single plant family. Oligophagy (the use of two to four plant families) is restricted to some of the more derived species. Only two of the 44 species analyzed show polyphagy (use of ≥ 5 plant families); in both cases, polyphagy is recently and independently derived (Figure 17.2A). Our unambiguous finding that specialization on specific plant hosts is the ancestral condition in Hawaiian *Drosophila* (a conclusion consistent under both ACCTRAN and DELTRAN) is contrary to the longstanding idea that specialization is a derived condition (Futuyma and Moreno 1988). Even where there have been shifts from one plant family to another (Figure 17.2B), most such changes have been from specialist to specialist; in only a few cases have they involved increased generalization.

It is interesting that allopatric island populations of *D. grimshawi* include both specialists and generalists (Montgomery 1975). This species is exceptional, in that it is distributed on all of the high islands except Hawai`i, which is inhabited by a very closely related species (*D. pullipes*) that differs only in the color of the legs and pleurae (Hardy and Kaneshiro 1972). Whereas *D. pullipes* and its sister taxa, the Kaua`i (Figure 17.2A) and O`ahu populations of *D. grimshawi*, are restricted to breeding on *Wikstroemia* of the plant family Thymelaeaceae (Figure 17.2B), the separate clade of *D. grimshawi* populations from Maui Nui (Maui, Moloka`i and Lana`i, interconnected during the Pleistocence due to lower sea levels) are polyphagous, using Liliaceae, Urticaceae, and eight other endemic families (but not Thymelaeaceae), as well as two introduced plants. By analyzing segregation in crosses between specialist and generalist forms and selection for opposite ovipositional behaviors, Ohta (1989) showed that only a few genes may regulate this behavior, the genetic variance being highly

Table 17.2. Summary of recorded oviposition preferences of endemic Hawaiian Drosophila with respect to utilization of plant families and specific breeding substrates. Except for the first column, all values shown are percentages.

Species group or subgroup	No. of species	% Monophagy	Use of major plant families/genera[1]				Breeding Substrates[2]							
			Araliaceae	Cheirodendron	Campanulaceae	Clermontia	leaves	stems	bark	flux	fruit	flowers	fungi	ferns
Non-picture-wings[3]														
antopocerus	11	91	91	91	0	0	100	0	0	0	0	0	0	0
Modified tarsi														
Fork tarsi	11	91	82	82	0	0	100	0	0	0	0	0	0	0
Bristle tarsi	16	94	63	56	6	6	100	0	0	0	0	0	0	0
Spoon tarsi	14	71	79	64	14	14	100	0	0	0	0	0	0	0
Ciliated tarsi	17	71	53	53	18	12	59	18	0	0	24	0	6	12
Modified-mouthparts	21	86	0	0	48	48	43	43	0	0	38	10	24	0
Picture-wings[3,4]														
primaeva group	1	100	100	100	0	0	0	0	100	0	0	0	0	0
adiastola group	11	82	0	0	73	55	18	73	55	0	18	27	0	18
planitibia group														
picticornis subgroup	2	50	50	50	0	0	0	0	50	50	0	0	0	0
cyrtoloma subgroup	6	100	100	83	0	0	0	17	100	0	0	0	0	0
planitibia subgroup	5	40	40	40	100	100	0	100	60	0	0	0	0	0
glabriapex group	22	82	5	5	0	0	5	77	14	5	0	0	0	0
grimshawi group[5]	26	85	54	8	19	19	15	46	46	19	19	4	8	0
hawaiiensis	9	67	11	0	0	0	0	22	11	100	0	0	0	0
Total picture-wings	82	79	32	15	22	20	8	52	38	19	8	5	2	2
Total, others	90	83	54	51	18	17	79	13	0	0	13	2	7	2
Grand total	172	81	44	34	20	18	45	32	17	9	11	3	5	2

[1]Cheirodendron is a genus in the plant family Araliaceae; Clermontia belongs to the family Campanulaceae.
[2]Some groups sum to more than 100% because certain species use more than one substrate.
[3]Ecological data from Heed (1968).
[4]Ecological data from Montgomery (1975).
[5]The values shown do not include the hawaiiensis subgroup which is listed separately.

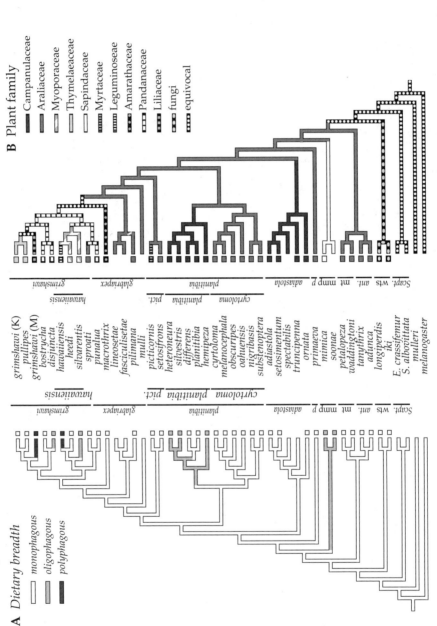

Figure 17.2. Evolution of host plant use by Hawaiian *Drosophila*, assuming ACCTRAN. Ecological data are from Heed (1968) and Montgomery (1975); no ecological data are available for the species that lack a coding box. (**A**) Degree of specialization in larval host plant. Most species are monophagous with respect to choice of the plant family used by the larvae; oligophagy and polyphagy are derived conditions. (**B**) Plant family used by larvae. Abbreviations for the drosophilid species groups and subgroups are those given in Figure 17.1.

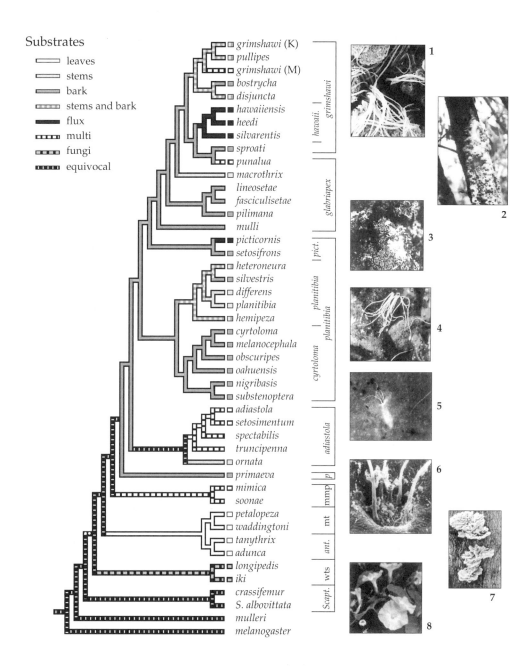

Figure 17.3. Evolution of larval substrate in Hawaiian *Drosophila*. Species are grouped into groups and subgroups as in Figure 17.1. To the right are shown representative oviposition substrates: (1) SEM of eggs of *D. silvarentis* oviposited in bark near a tree flux; (2) *Myoporum* tree flux; (3) *D. heedi* eggs oviposited in soil flux; (4) eggs oviposited in decaying *Sapindus* bark; (5) portion of decaying *Cheirodendron* leaf with a *D. adunca* egg; (6) SEM of *D. adunca* egg oviposited in *Cheirodendron* leaf; (7) fungi; (8) flowers of *Ipomoea*, a scaptomyzoid substrate.

additive. It appears that the dominant mutations favoring generalist ovipositional behavior arose in populations on Maui Nui, although the taxonomic relationships of these *grimshawi* taxa are ambiguous, and ecologically differentiated populations may well represent incipient species (Piano et al. 1997).

The family of host plants shown for oligo- and polyphagous species in Figure 17.2B is the one most frequently recorded for each drosophilid. Excluding the clade of fungus-breeding taxa, two plant families – Araliaceae and Campanulaceae – are used for breeding sites by a large majority of Hawaiian *Drosophila*. It is interesting that these two families are rather closely related (Chase et al. 1993; K. Sytsma, pers. comm.), which suggests that their members might bear some physical and chemical similarity. Two endemic genera in these families – *Cheirodendron* (Araliaceae) and *Clermontia* (Campanulaceae) – stand out as the predominant host plants, supporting the reproduction of 52% of the 172 *Drosophila* species reported (see Table 17.2). Although *Cheirodendron* is widespread in mesic to wet forests, there are only five species native to Hawai`i (Wagner et al. 1990). *Clermontia* is also widespread in moist lowland and montane forests, but has 22 endemic species and 9 heteronymic subspecies (Lammers 1995), making it the sixth largest of the 216 flowering plant genera endemic to Hawai`i (Wagner et al. 1990). The fact that larvae of so many Hawaiian drosophilids develop in the rotting tissues of *Cheirodendron* and *Clermontia* suggests that these genera may contain chemical attractants or that drosophilid larvae readily tolerate their secondary plant compounds. Kircher and Heed (1970) suggested that the physical characteristics of *Cheirodendron* leaves – which rot slowly under moist conditions – might make them a good breeding substrate.

Excluding the primitive white-tip scutellum group whose larvae develop in fungi, the primitive drosophiloids breed on Araliaceae (Figure 17.2B), primarily *Cheirodendron* (Table 17.2). The Araliaceae was apparently the first plant family invaded by Hawaiian drosophilids, as inferred using either ACCTRAN or DELTRAN reconstructions; this host is retained by many of the picture-winged species (Figure 17.2B). Based on our analysis, two independent shifts to the Campanulaceae appear to have occurred, both within the picture-winged group. The first involves the *adiastola* group; the second, the *planitibia* subgroup of the *planitibia* species group. In both instances, *Clermontia* is the principal genus utilized, but *Cyanea* – an even larger lobelioid genus of wet forest interiors and the largest plant genus endemic to Hawai`i (Givnish et al. 1995) – is also a frequent host. Shifts to the Campanulaceae have also occurred in some species of the modified-tarsi, modified-mouthparts, and *grimshawi* groups (Table 17.2), but the individual fly species involved were not included in our phylogenetic analysis.

The greatest host plant diversity is shown by the terminal *grimshawi* species group, which has undergone numerous host plant shifts into a taxonomically diverse group of plant families. Because of the frequency of oligophagy and polyphagy in this species group, the extreme diversity of host plants, and the fact that our molecular phylogeny includes only a partial sampling of *grimshawi* group species, we remain cautious as to the exact order and pattern of host plant shifts in this group. Members of this group evolved late and were among the last drosophilid lineages to colonize Hawaiian forests; thus, it is tempting to speculate that interspecific competition and saturation of favored host plants may have favored the adoption of several new hosts by the *grimshawi* group. Alternatively, Carson and Ohta (1981) suggested that generalism in this group evolved in response to the presence of a series of unfilled niches.

BREEDING SUBSTRATE – An even more complex evolutionary pattern emerges when the actual substrates used for breeding are considered (Figure 17.3; Table 17.2), particularly those used by more recently derived groups. The scaptomyzoids, although poorly studied, are ecologically diverse (K. Kaneshiro, pers. comm.). Substrates for the two scaptomyzoids included in the molecular phylogeny are unknown, but similar species breed in flowers (Heed 1968), an ecological resource that is also utilized by a few of the modified-mouthpart species (Table 17.2). It is likely that the scaptomyzoids have undergone an ecological radiation in breeding substrate that is at least as broad as that of the drosophiloids. However, the poor representation of scaptomyzoids in our molecular phylogeny, as well as limited ecological data, precludes any meaningful analysis of the scaptomyzoid radiation at this time.

Confining our attention to the Hawaiian drosophiloids, the molecular phylogeny suggests that their ancestral breeding substrate was fungi, the substrate of the most primitive white-tip scutellum group – the so-called fungus feeders (Figure 17.3). A shift to breeding in decaying leaves then occurred, involving primarily the leaves of *Cheirodendron* (Heed 1968), which are widely used by the *antopocerus* and modified-tarsi groups (Table 17.2). Evolution of the modified-mouthparts group initiated diversification into a greater array of ecological substrates, including decaying stems and fruits (Table 17.2). This large group of over 100 species is poorly represented in the molecular phylogeny, so detailed analysis must await more data.

Evolution of the highly derived picture-winged species (represented by the branch to the ancestral Kaua`i species *D. primaeva*) was associated with a shift into another novel substrate, decaying bark. The relatively late invasion of this breeding site indicated by the phylogenetic mapping refutes Spieth's (1982) assertion that the original ancestor of the Hawaiian fauna was a bark-breeder. The shift to bark-breeding in the picture-winged species is by no means complete; many taxa are known to use decaying stems. Indeed, bark and stems are the predominant substrates used by 90% of the ecologically well-sampled picture-wings (Montgomery 1975). The distinction between breeding on bark vs. stems may be somewhat arbitrary; the term stem is applied to branches ca. 2 cm or less in diameter (Montgomery 1975). Nevertheless, there appears to be some evolutionary basis for this distinction, in that substrate choice is generally consistent within a species group. For example, although two species of the *adiastola* group are shown as using multiple substrates (Figure 17.3), the group as a whole is predominantly stem-breeding (Table 17.2), whereas the *cyrtoloma* subgroup is predominantly bark-breeding. Because some species are reported to use both substrates, we coded substrate use as bark exclusively, stems exclusively, or stems as well as bark.

Another substrate used by picture-winged species is tree flux. Shifts to flux-breeding occurred twice in this group, in *D. picticornis* and the *hawaiiensis* subgroup of the *grimshawi* group. In each case, the shift was from breeding in decaying bark to breeding in the moistened bark of living trees. In fact, some of these species can also use decaying bark; nonetheless, the use of fluxes is reported as a distinct ecological specialization. Adaptive aspects of this specialization and variation in this ecological way of life are discussed later.

Although use of a single substrate is the rule, one modified-mouthparts species and four of the 34 species of picture-wings included in our phylogeny are multi-substrate users; two of the latter are also oligophagous and one (*D. grimshawi* of Maui) is polyphagous. This broader array of breeding niches and generalist tendencies suggests that these species are opportunistic, or that competition from sympatric drosophilids has forced them to oviposit in substrates to which they are maladapted. Rearing only one or a few adults from some of the rarer substrates – as compared with these species' potential fecundity and the much larger numbers of adults reared from more typical substrates (Montgomery 1975) – suggests that oviposition is sometimes incidental in these polyphagous species. The assumption underlying the observations and interpretations of most evolutionary biologists is that endemic organisms are well (if not perfectly) adapted to their particular habitats. This assumption may need to be questioned in some instances. A possible case in point involves drosophilids found to breed in plants exotic to Hawai`i (e.g., the picture-winged *D. grimshawi* and *D. crucigera* [Montgomery 1975], and the *Exalloscaptomyza* that breed in flowers of the morning glory *Ipomoea*). On the other hand, adaptation in insular organisms can occur in a matter of a few generations (e.g., Grant and Grant 1993), so host shifts from endemic to introduced plants may have been rapidly followed by adaptation. The ecological niche of drosophilid larvae and adults includes the array of bacteria and yeasts associated with plant decomposition. It is premature to make any judgment about the complex interactions involved, but our current and future research may reveal the significance of the microflora in the ecological specificity of the endemic Hawaiian flies.

Female reproductive diversification in the Hawaiian *Drosophila*

Patterns of ovarian development in the endemic Hawaiian Drosophilidae reflect a diverse array of reproductive strategies (Kambysellis and Heed 1971). Potential fecundity varies widely as a result of differences in the structure and function of the ovaries. In different species, these may produce just one egg or hundreds of eggs at a time. This variation is due to genetic differences specifying the anatomical structure of the ovaries, especially the number of ovarioles per ovary, which vary from one to more than 50 among the Hawaiian species. Control of the process of egg maturation within an ovariole is also variable, affecting the maximum number of mature eggs per ovariole, which varies from one to three or more. Species also differ in whether their ovaries show synchronous development (with all ovarioles maturing one or more eggs simultaneously) or asynchronous development. The number of functional ovarioles may vary from one to all of those present.

Egg morphology also differs among species at the ultrastructural level (Kambysellis 1993) as well as the gross level (Throckmorton 1966). The size and shape of the egg, the absolute and relative length of the respiratory filaments, and their number vary remarkably among the Hawaiian species. Whereas most continental *Drosophila* have short respiratory filaments, many of the Hawaiian species, particularly the bark breeders, have extraordinarily long filaments, up to three or four times the length of the egg. For example, in *D. sejuncta*, the egg is 0.97 ± 0.02 mm long, while the two posterior filaments are 3.80 ± 0.11 mm long (Kambysellis and Heed 1971). Yet other Hawaiian drosophilids possess only rudimentary filaments or lack them entirely.

Eggs of the Hawaiian species also vary in the structure and the thickness of the eggshell, or chorion (Kambysellis 1993). Whereas the outer endochorion is very thin in all the continental drosophilids studied thus far, in Hawaiian species the thickness varies from very thin in the scaptomyzoids to more than tenfold thicker in the picture-wings. It is interesting that there is some evidence that greater eggshell thickness is due to an increased level of chorion gene amplification during egg development (J. C. Martínez-Cruzado, pers. comm.). Structurally, there are two distinctive but variable chorion features in Hawaiian drosophilid eggs. The dorsal ridge – a structure absent from continental species, as well as the primitive scaptomyzoids and white-tip scutellum flies – is present in rudimentary form in the *antopocerus* and modified-tarsi groups, but is well developed in eggs of modified-mouthparts and picture-winged species. The collar, a structure at the anterior end of the egg near the micropyle, is absent from the scaptomyzoids and the white-tip scutellum groups but well formed in all the more derived Hawaiian lineages (Kambysellis 1993).

Another extraordinarily variable female character is the ovipositor (Throckmorton 1966), the structure at the posterior end of the abdomen used in egg laying. This structure varies enormously among the Hawaiian species in shape and in length, the longest ovipositors being found in the bark-breeding species (Franchi et al. 1997). It appears that the length of the ovipositor correlates with the depth to which the egg is inserted in the substrate during oviposition.

EGG/OVARIAN TYPES – Kambysellis and Heed (1971) recognized four discrete reproductive types among Hawaiian drosophilids, based on suites of ovarian and egg characters. Here we expand the number of these types, based on additional data; the characteristics of these types are summarized in Table 17.3. The number of ovarioles and number of mature eggs per fly provide a reliable indicator of potential lifetime fecundity (David 1970), because egg production is continuous throughout adult life once females become reproductively mature. Potential fecundity varies over an extraordinarily wide range in the Hawaiian drosophilids, from a very low value in species with Type Ia ovaries that mature only one egg at a time, to an extremely high value in species with Type IIIb ovaries, which are characterized by very high ovariole numbers and several mature eggs per ovariole.

Type Ia eggs have a smooth chorion but completely lack a dorsal ridge and respiratory filaments; Type Ib eggs have a chorion with a pattern of follicle imprints, a modified dorsal ridge in some species, and may possess rudimentary respiratory filaments (Figure 17.4). Type IIIb eggs differ from those of Type IIIa in having significantly longer respiratory filaments (Figure 17.4), a greater number of mature eggs per ovariole, and a clustered vs. solitary deposition of eggs in the substrate (Table 17.3).

Table 17.3. Characteristics of female reproductive types in endemic Hawaiian drosophilids.

Type	Ovarian traits				Egg traits			Chorion traits			Oviposition
	Ovarioles per fly	Functional ovarioles	Maximum # mature eggs per ovariole	Ovarian development	Number of respiratory filaments	Filament[1] length	Length/ width	Endochorion thickness	Dorsal ridge	Collar	Egg positioning
Ia	2–7	1–2	1	alternate ovarioles	0	—	1.4–2.9	very thin	absent	absent	singly on surface
Ib[2]	11	3–6	1	alternate ovarioles	0	—	2.5	thin	modified	absent	singly on surface
II	8–20	all	1	alternate ovarioles	4	short (0.7–1.1)	3.2–3.8	thin	rudimentary	present	inserted singly
IIIa	24–101	all	2	asynchronous	4	short (0.7–1.2)	3.2–4.0	thick, solid	well-formed	present	inserted singly
IIIb	28–87	all	4	asynchronous	4	long (1.5–3.9)	3.4–4.6	thick, solid	well-formed	present	clusters into substrate
IV[3]	38–45	all	1 rarely 2	synchronous	2	very short (0.3)	5.3	?	rudimentary	present	singly (lab observation)
V[4]	37–42	all	5	synchronous	4	very short (0.2)	2.8	open, meshlike	absent	absent	singly into substrate

[1]The values in parentheses represent the range of ratios of the length of the posterior pair of filaments to the length of the egg.
[2]Description of reproductive traits based on the laboratory stock of *Scaptomyza albovittata*.
[3]Based on three individuals of the sole representative of this class, the endangered species *D. mulli*.
[4]Based on one species, *D. nigra*.

Female reproductive types IV and V are identified here for the first time. Type IV is distinguished by synchronous ovarian development and unusually elongate eggs with two very short filaments, the anterior pair being missing (Figure 17.4). Exceptionally short filaments also characterize Type V eggs; these have a more typical shape and the usual four filaments, but possess a quite differently structured chorion with a unique, open meshlike construction distinct from the solid chorion of all the other Hawaiian species (Kambysellis 1993).

EVOLUTION OF FEMALE REPRODUCTIVE TYPES – Overlaying the seven egg/ovarian types on our molecular tree (Figure 17.4) demonstrates that such types are generally conserved within a lineage, but that there has also been a series of reproductive shifts in the evolution of the Hawaiian drosophilids. Lineages of the more primitive, non-picture-winged species display greater diversity in female reproductive strategies (five egg/ovarian types represented) than the picture-winged species with three egg/ovarian types. The poorly sampled scaptomyzoids exhibit the lowest fecundity (Types Ia and Ib); further study, however, may reveal greater ovarian diversity. Among the non-picture-winged drosophiloids, female reproduction shifts from Type V in the most primitive lineage, the white-tip scutellum flies, to Type II in the *antopocerus* and modified-tarsi groups, and then to Type IIIa in the modified-mouthparts group; the last type has a much higher reproductive potential than the others because it entails a greater number of ovarioles, all of which are functional.

The most primitive picture-winged species retain the Type IIIa pattern, but more derived species evolved types IIIb and IV. Significantly, species possessing Type IIIa ovaries and those bearing Type IIIb ovaries are each found in two disjunct portions of the tree (Figure 17.4). This implies that there must have been convergent evolution and/or evolutionary reversals in reproductive type. However, the exact pattern of female reproductive shifts and the inferred ancestral states in this portion of the phylogeny are sensitive to the assumptions made. Under ACCTRAN and disregarding the evolution of Type IV, there appear to have been four changes of reproductive type (Figure 17.4): two independent gains of Type IIIa (in the modified-mouthparts and *glabriapex* groups), and two "losses" of Type IIIa, with convergent shifts to Type IIIb occurring in the *planitibia* and the *grimshawi* groups. Under DELTRAN, Type IIIa arose only once (in the modified-mouthparts group) but there were three independent shifts to Type IIIb, on the branches leading to the *cyrtoloma* and *planitibia* subgroups, to the *picticornis* subgroup, and to the *grimshawi* species group, respectively. Although *D. punalua* may be misplaced in the molecular phylogeny, like other members of the *glabriapex* group it exhibits the Type IIIa pattern.

One picture-winged species (*D. mulli*) displays a unique egg/ovarian pattern, Type IV. Morphologically, *D. mulli* is somewhat anomalous in that it lacks the wing pigmentation characteristic of all the picture-winged species, yet external male genitalia and other traits clearly place it in the *grimshawi* species group (Kaneshiro et al. 1995). This phylogenetic affinity is confirmed by the molecular phylogeny (Figure 17.1A) although the subgroup designation does not agree with its morphological placement. Independent of this inconsistency, it is clear that there have been several discrete reproductive shifts in the evolution of the endemic Hawaiian drosophilids.

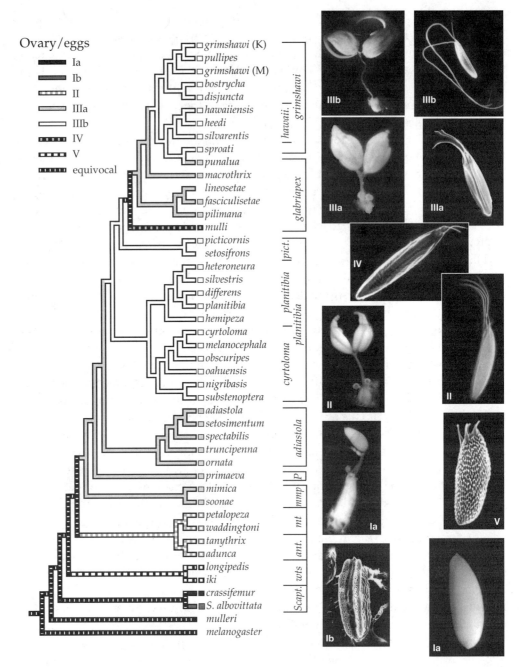

Figure 17.4. Evolution of ovarian types in Hawaiian drosophilids, assuming ACCTRAN. Specific egg/ovarian types generally characterize entire clades, indicating that reproductive shifts occurred early in their evolution. Ovaries of four types are shown on the near right, together with an egg of *S. albovittata* at the bottom. The ovaries are from *D. sejuncta* (Type IIIb), *D. punalua* (Type IIIa), *D. waddingtoni* (Type II), and *D. crassifemur* (subgenus *Engiscaptomyza*) (Type Ia). Note the larva in the vagina of *D. crassifemur*. To the far right are representative eggs of the various egg/ovarian types: *D. claytonae* (Type IIIb); *D. truncipenna* (Type IIIa); *D. mulli* (Type IV); *D. waddingtoni* (Type II); *D. longipedis* (Type V); and *D. crassifemur* (Type Ia).

The question is, what is the adaptive significance of the female reproductive differentiation observed among lineages of this group of flies?

RELATIONSHIP OF FEMALE REPRODUCTIVE STRATEGIES TO BREEDING SUBSTRATE – Although the patterns of evolutionary divergence in host plant and larval substrate (Figure 17.3) and in female reproductive strategy (Figure 17.4) are complex, comparison of these two patterns suggests that female reproductive strategies broadly correlate with host plants and larval substrates. Among the non-picture-winged drosophiloids, the early radiation into three successive breeding niches – fungi, leaves, and then a more diverse range of substrates including stems and fruits – was accompanied by evolution of three distinct reproductive patterns, involving ovarian types V, II, and IIIa, respectively. Larval substrates have not been recorded for the two scaptomyzoids included in the molecular phylogeny, but other scaptomyzoids of subgenus *Exalloscaptomyza* with the Type Ia pattern of low fecundity use decaying flowers (Kambysellis and Heed 1971).

The ecology of the picture-wings is considerably better analyzed (Montgomery 1975), but the breeding substrates of some of the sequenced species in the *adiastola* and *glabriapex* groups, and of *Drosophila mulli* remain unknown (Figure 17.3). The unique Type IV reproductive pattern of the latter species suggests that field studies may reveal a novel breeding substrate. Reproductive patterns of the remaining picture-winged species fall into either Types IIIa or IIIb. Type IIIa (moderate to high ovariole numbers, resulting in higher fecundity) appears to have first evolved in the modified-mouthparts lineage, and then was retained by the more primitive *primaeva* and *adiastola* groups of the picture-winged species; it also characterizes the more derived *glabriapex* group (Figure 17.4).

Unfortunately, the ecology of species with this reproductive type is rather poorly known. *Drosophila ornata* of the *adiastola* group and *D. macrothrix* of the *glabriapex* group are clearly stem breeders. Although they use other substrates as well, *D. punalua*, *D. adiastola*, and *D. setosimentum* oviposit predominantly in stems. Based on preliminary data, ovarian Type IIIa thus appears to be associated with the stem-breeding habit. One of the two apparent exceptions to this relationship, the record of a single individual of *D. pilimana* reared from bark, may be misleading. The other exception, the bark-breeding habit of *D. primaeva*, is inconsistent with the general association between reproductive type and breeding substrate.

Egg/ovarian Type IIIb (with the highest fecundity) appears to have evolved several times in the picture-winged group with at least two independent reproductive shifts from Type IIIa to Type IIIb in the *planitibia* group and the *grimshawi* group (Figure 17.4). Type IIIb is well correlated with the bark-breeding habit (e.g., the *cyrtoloma* subgroup), but is also associated with the use of decaying stems. Multiple records of stem and bark-breeding in Type IIIb species from the *planitibia* subgroup and the *grimshawi* group raise the question of whether there is a significant distinction between decaying stems and bark for ovipositing females.

A few picture-winged species use tree fluxes as a breeding substrate; all have females with Type IIIb ovaries. In both evolutionary shifts to flux-breeding, the ancestor was a bark-breeder (Figure 17.3) with Type IIIb ovaries (Figure 17.4), suggesting that this ovarian type was pre-adapted to breeding in fluxes. Although no major reproductive shifts took place in the adaptation to this novel niche, some changes in the female reproductive system are associated with different types of fluxes, as we discuss next.

ECOLOGICAL AND REPRODUCTIVE DIVERGENCE IN A PAIR OF SYMPATRIC SPECIES, *D. silvarentis* AND *D. heedi* – The preceding discussion has detailed the broad phylogenetic patterns of ecological and reproductive shifts in the evolution of the endemic Hawaiian *Drosophila*, but another perspective on evolution and adaptation can be gained by a close analysis of related but sympatric species, and ultimately, analysis of entire Hawaiian communities of endemic drosophilids, plants, and bacteria. Detailed studies of a pair of sympatric species endemic to the island of Hawai`i illustrate how a more careful analysis can lead to a fuller understanding of the ecological relationships among species and the adaptive aspects of their resource use. The two species, *D. silvarentis* and *D. heedi*, are closely related members of the *hawaiiensis* subgroup; both are flux breeders (Figure 17.3) and have Type IIIb ovaries (Figure 17.4). Nevertheless, they are ecologically and reproductively differentiated, demonstrating that the broad classifications of substrate and ovarian types may obscure some of the biologically significant aspects of the adaptation of these flies.

This pair of recently evolved sympatric species has been hailed as a striking example of precise niche partitioning (Kaneshiro et al. 1973). According to the chromosomal phylogeny of Carson (1981), they are sister species derived from a common founder from Maui. They are also the only picture-winged species that inhabit the arid, poorly vegetated area of the high-altitude saddle between Mauna Kea and Mauna Loa. They both depend on the flux drippings of one of the two trees in the area, *Myoporum sandwicense* (Myoporaceae). Based on cytological identification of larvae from natural substrates, Kaneshiro et al. (1973) determined that *D. silvarentis* oviposits only on tree fluxes, whereas *D. heedi* breeds exclusively on ground fluxes – that is, the soil moistened by the flux dripping from overhanging branches. Kaneshiro et al. (1973) argued that the few *D. silvarentis* larvae found in ground fluxes had dropped to the ground for pupation after having completed development in the branches overhead. Thus, the two species appeared to be significantly separated by the females' choice of oviposition site, the novel shift to breeding in soil fluxes presumably being favored because it reduced competition with the congener *D. silvarentis* (Kaneshiro et al. 1973).

Morphological differences in the eggs of these two species (Figure 17.5) allowed us to test more carefully the hypothesis that ecological divergence between these closely related species is actually due to a shift in the oviposition site. Eggs of *D. silvarentis* have the very long respiratory filaments (Figure 17.5B) typical of flies with Type IIIb ovaries. Although ovaries of *D. heedi* are also of the IIIb type, their eggs have shorter respiratory filaments and are much smaller than eggs of *D. silvarentis* (Figure 17.5E). Our field data (Table 17.4) confirm that *D. heedi* oviposits exclusively in soil fluxes, but show that *D. silvarentis* oviposits in both soil and tree fluxes, although preferentially in the tree fluxes. The oviposition sites of the two species are thus not completely separate, the overlap being due to the fact that *D. silvarentis* females will also oviposit in soil fluxes, particularly when they are fresh. This may have been expected given that the stimulus for oviposition is probably the odor arising from the flux drippings and/or from the

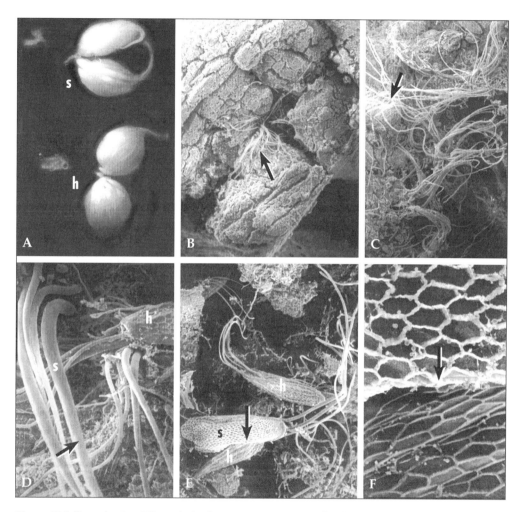

Figure 17.5. Reproductive differentiation between two sympatric flux-breeding species from Hawai`i. (**A**) Mature ovaries of *D. silvarentis* (s) and *D. heedi* (h) showing differences in ovariole numbers and filament length at the anterior end of each ovary (to the right). (**B-F**) SEMs of oviposited eggs in bark (**B**), and in soil fluxes (**C-F**). Arrows indicate respiratory filaments; eggs of *D. silvarentis* (s) and *D. heedi* (h) are identified. The arrow in (**E**) highlights a pair of overlapping eggs, which are shown at higher magnification in (**F**), emphasizing the distinctive chorion ultrastructures of the two species.

bacteria associated with the flux. The older, more established ground fluxes probably facilitate propagation of elements of the soil microflora, which then make the ground flux attractive to *D. heedi* females but less attractive to *D. silvarentis* females (Table 17.4). The "tree/soil fluxes" provide an intermediate niche suitable to both species, these substrates at the base of a vertical tree trunk being distinct from the soil fluxes which are moistened by the drippings from a horizontal branch overhead.

Although there is ovipositional overlap in the ground fluxes, the two species are not equally successful in ground fluxes. Table 17.5 shows that the relative success of *D. heedi* increases with time since oviposition; *D. heedi* clearly is better adapted to

Table 17.4. Reproductive divergence between two sympatric and ecologically similar species of Drosophila on the island of Hawai`i. Means and standard deviations of thorax length and ovariole number based on 32 field-collected females for *D. silvarentis*, and on 52 field-collected females for *D. heedi*.

Species	Thorax length (mm)	Ovarian development	Number of ovarioles per fly	Number of oviposited eggs in fluxes			
				tree	new soil	old soil	tree/soil
D. silvarentis	3.1 ± 0.03	asynchronous	39 ± 0.6	2,300	92	988	65
D. heedi	2.5 ± 0.04	synchronous	60 ± 0.7	0	2	3,136	48

conditions in the ground flux and appears to outcompete *D. silvarentis*, which shows high mortality. The reduced survival of *D. silvarentis* in ground fluxes may be due to the nutrient poverty of this substrate; the larvae of this species are much larger than those of *D. heedi* (see Table 17.4). Few *D. silvarentis* larvae apparently reach the critical size to enter pupation in ground fluxes. Larvae of the two species in the same ground flux would compete directly for nutrients, and it appears that selection against *D. silvarentis* is stronger, selecting against soil oviposition by this species.

From our molecular phylogeny (Figure 17.1A), it appears that *D. heedi* diverged from its sister-group more recently than *D. silvarentis*. Furthermore, *D. heedi* is the only member of the *hawaiiensis* subgroup to use soil fluxes – all others use tree fluxes. Regardless of the precise ancestor of *D. heedi*, it appears likely that this species diverged from breeding in tree fluxes to avoid competition, which can be quite severe on this substrate. Tree fluxes are quite rare, and eggs laid directly in the flux source usually die before completing embryogenesis, the only ones surviving being those laid in a restricted area surrounding the source (M. Kambysellis and E. Craddock, unpubl. data). Although soil fluxes are even rarer than tree fluxes, *D. heedi* has adapted to using this breeding resource effectively by making smaller eggs, by reducing the time required for larval development, and by reducing the length of the respiratory filaments; smaller eggs and smaller larvae at pupation result in smaller adults (Table 17.4). To optimize use of this unpredictable resource, *D. heedi* has evolved ovaries with more ovarioles and synchronous egg development (Table 17.4). When, after a prolonged search, a female does find one of the very rare soil fluxes in appropriate condition, she

Table 17.5. Proportions of *D. heedi* and *D. silvarentis* at different life stages in samples from soil fluxes. The fraction of *D. heedi* increases significantly (P < 0.001) between oviposition and the adult stage.

Life history stage	Sample size	Ratio of *heedi/silvarentis*	Reference
Oviposited eggs	4,124	3.2 : 1	Kambysellis and Craddock (unpubl.)
Third instar larvae	15	4.0 : 1	Kaneshiro et al. (1973)
Adults reared from flux	225	8.8 : 1	Kaneshiro et al. (1973)
Adults reared from flux	279	8.6 : 1	Kambysellis and Craddock (unpubl.)

can lay a large number of eggs on that flux simultaneously. The ecological and reproductive data presented here demonstrate that several reproductive traits have evolved in an adaptive direction in a relatively short time period (< 0.4 My), providing a clear example of natural selection resulting in character displacement. As demonstrated experimentally by Schluter (1994; see Chapter 18), interspecific competition can drive adaptive divergence. As the above example makes clear, resource competition is one of the forces that fostered the adaptive radiation of the Hawaiian *Drosophila* and led to the exploitation of novel niches for oviposition and breeding.

Adaptive significance of female reproductive strategies in Hawaiian *Drosophila*

Female reproductive potential in Hawaiian flies varies from the production and oviposition of one egg at a time to clusters of more than a hundred eggs. Coincident with this variation in fecundity is the exploitation of a variety of larval substrates. We believe that the differences in several features of female reproductive strategy – ovariole number, ovariole function, length of the ovipositor, length of the respiratory filaments, chorionic ultrastructure, and oviposition behavior – are adaptive and have arisen in response to strong selection exerted by the larval substrate. Most of these traits are largely genetically determined. Ovariole number is also affected by environmental factors, particularly larval nutrition (Kambysellis and Heed 1971). However, the genetic component of control is also strong, as evidenced by the retention of ovarian structure and function when larval imaginal discs are transplanted into species with very different ovarian types (Kambysellis 1970; Kambysellis and Heed 1971).

In species in which eggs are oviposited deeply within the breeding substrate – as in bark-breeders – ovipositors are very long; by contrast, scaptomyzoids that simply drop their eggs on the surface have extremely short ovipositors (Throckmorton 1966; Franchi et al. 1997). The length of the respiratory filaments of the egg also appears to be adaptive. The eggs of bark-breeding species (Type IIIb) have extraordinarily long respiratory filaments (1.5–3.9 mm) that are 1.5 to 4 times the length of the egg itself (Kambysellis and Heed 1971; Table 17.3; Figure 17.4). These long filaments act as "snorkels" facilitating respiratory exchange for the deeply buried developing embryo (Margaritis 1983) in the potentially anaerobic environment of the rotting substrate. The upper layers of bark substrates are generally dry and hard, and thus unsuitable for embryonic development, so the eggs must be inserted deeply. This has selected for long, heavily sclerotized ovipositors and eggs with long respiratory filaments (Type IIIb) in the invasion of the bark niche by picture-winged drosophilids of the *planitibia* and *grimshawi* groups (see Kambysellis and Heed 1971; Franchi et al. 1997).

Scaptomyzoid eggs (Types Ia and Ib), which are simply deposited on the surface of the breeding substrate, are devoid of respiratory filaments; the eggs of leaf-breeders (Type II) and stem-breeders (Type IIIa) are inserted just below the substrate surface (Figure 17.3), and have short filaments that are less than or equal to the length of the egg (see Figure 17.4). Among flux-breeding species (Type IIIb), the respiratory filaments are very long for those using tree fluxes, but considerably shorter for those using soil fluxes (Figure 17.5). Thus, species differences in the lengths of ovipositors and egg respiratory filaments appear adapted to enhance embryo survival under different substrate conditions.

The thickness of the egg chorion also appears to show adaptive variation. Continental drosophilids – and the more primitive Hawaiian drosophilids that oviposit in decaying flowers and leaves – have a very thin chorion. The eggs of stem- and bark-breeders, however, have a thick solid chorion, probably to help withstand the mechanical pressure of the substrate (Kambysellis 1993). In fungus-breeders, the chorion has a unique, scaffold-like structure with several open layers of interconnecting pillars, permitting respiration and extensive exchange of the ammoniacal gases that abound in this hostile breeding substrate.

Following embryogenesis and hatching, drosophilid larvae face a new set of selective pressures. They must obtain adequate nutrition to grow and develop, but must also withstand the chemical environment in which they find themselves. This may be quite noxious in fungal substrates (Jaenike et al. 1983) or even in rotting plant tissues. To survive, the larvae must be able to detoxify many of the chemicals they encounter. Shifts to breeding in new genera and families of plants that are chemically different probably required enzymatic adaptation to additional compounds.

The longevity of the substrate and its rate of deterioration also impose selective constraints on larval development – and hence, on adult size. For example, the ephemeral nature of decaying flowers puts a premium on brief development, selecting for rapid embryogenic and larval development; in the extreme case of *Exalloscaptomyza*, the females often larviposit (Figure 17.4) a first instar larva on morning-glory flowers (Kambysellis and Heed 1971)! Rapid larval development of flower-breeding species results in very small flies. Slowly decaying bark, on the other hand, permits extended larval development, resulting in large larvae and adults; in fact, the Hawaiian bark-breeding picture-wings are the giants of the *Drosophila* world (Figure 17.1). The extreme range of variation in body size among the Hawaiian drosophilids is partly a consequence of their varied developmental histories which are related, in turn, to variation in oviposition site.

Perhaps the most remarkable outcome of varied selection regimes in the larval niche is the interspecific variation in fecundity, which ranges from one egg per day to several hundred. The phylogenetic trend in the Hawaiian drosophilids has been toward greater fecundity (Table 17.3; Figure 17.4), and appears to be driven by the exploitation of substrates that provide greater food resources. The increase in the physical volume of the substrate (from flowers, to leaves, to stems, branches, and tree trunks) should result in an increase in its larval carrying capacity. Females that responded to this opportunity by maturing and laying more eggs would have been selectively favored, leading to an increase in the frequency of alleles that underlie ovarian traits (e.g., ovariole number, number of mature eggs per ovariole) that contribute to higher fecundity.

Implications for Hawaiian *Drosophila* evolution

The preceding phylogenetic analysis of the reproductive ecology of the endemic Hawaiian Drosophilidae, taken with previously published accounts of the behavior and biology of the group (Carson 1978, 1986; Carson et al. 1970; Speith 1982; Kaneshiro and Boake 1987), provides a new perspective on the evolution and adaptive radiation of this extraordinary group of organisms. The ecological differentiation of the group into lineages using different larval substrates (fungus-breeders, leaf-breeders, bark-breeders, flux-breeders) is a major component of the phylogenetic diversification of Hawaiian drosophilids (Figure 17.3). Given the long-standing emphasis on the role of sexual selection in the speciation of Hawaiian *Drosophila*, it is important to ask how natural selection (on female reproductive strategies) and sexual selection (on male secondary sexual characteristics, female choice) might interact (T. J. Givnish, Chapter 1). We are interested in the evolutionary diversification of the whole mating system, and therefore need to consider the interactions between males and females, as well as between both sexes and their environment.

Mating systems in *Drosophila* are highly variable (Markow 1996), but nowhere more than among the endemic Hawaiian species, with their varied male morphologies, male courtship repertoires, and patterns of female reproduction. What is the relationship between the reproductive ecology of various species groups and the mating systems they display? The main parameters to evaluate (Markow 1996) are age at reproductive maturity, remating frequency, number of sperm transferred and stored per mating, and ovariole number. Among Hawaiian species, there is great variation in all of these parameters, as well as in the incidence of lek behavior by males – and hence, in the intensity of sexual selection. Although male behavioral data are incomplete for the more primitive groups, the following extreme patterns are evident.

Among the majority of the picture-winged group, female reproductive maturation and the onset of receptivity to insemination are significantly delayed following adult eclosion, often for several weeks (Kambysellis and Craddock 1991), which implies that females will generally have dispersed far from their larval substrate by the time they mate. Males generally mature much more rapidly (Boake and Adkins 1994); the differential in maturation rates between sexes would promote outcrossing. Males of these species invest a great deal of time and effort in defending space on their leks and in courtship displays, but only rarely is a male successful in mating with a sexually receptive female who has been attracted to a lek. Sexual selection among males of these species is intense (Spieth 1966, 1982). Remating is extremely rare in the picture-wings, although a few multiple inseminations have been detected (Craddock and Johnson 1978). Sperm remain viable in the female's sperm storage organs for months, and in a single copulation a male can transfer thousands of sperm (Kambysellis and Craddock 1991), enough to fertilize a female's eggs for almost a year (Carson et al. 1970) – that is, for her entire reproductive life. In such lek species, body sizes are large and ovariole numbers high (Types IIIa and IIIb), leading to females of high fecundity. Breeding substrates of these species – decaying stems, bark, and fluxes – are unpredictable, yet nutritionally rich and relatively long-lasting. Once located, these substrates provide abundant oviposition opportunities and nutritional support for large numbers of larvae.

Thus, in this reproductive strategy where the lekking males have limited opportunities to mate, there is a premium on male mating success, with strong selection for increased size, and for unique combinations of behavioral and morphological traits, and in the females, selection for high fecundity, for a capacity to store abundant sperm, and to accurately select a productive substrate for oviposition, as well as selection on ovipositor and egg traits that will ensure embryonic survival and thus reproductive success. Thus sexual selection on male traits and natural selection on female traits operate concurrently to engender a successful reproductive strategy that may have first evolved in the modified-mouthparts group and then been retained (with modifications) by the more derived picture-wings.

At the other end of the spectrum, there is an alternative reproductive strategy characterized by a more rapid rate of reproductive maturation, lower fecundity, few sperm transferred per mating, and a high incidence of remating, with many more opportunities for males to mate and much less intense sexual selection. Such a strategy seems to apply to the scaptomyzoids and the fungus-breeders, where first insemination takes place early in the female's reproductive maturation, and only a few sperm (from 10 to about 200) are transferred per mating (Kambysellis and Craddock 1991). In most *Drosophila*, sperm are mainly stored in the paired spermathecae and only temporarily in the seminal receptacle, but among 170 field-collected females from seven species of fungus-breeders, sperm were found only in the seminal receptacle (Kambysellis and Craddock 1991), the spermathecae apparently being nonfunctional (Throckmorton 1966). It appears that each copulation provides only enough sperm to inseminate one synchronously matured batch of eggs (Table 17.3), and that following oviposition, the females must remate to fertilize the following clutch of eggs. Lek behavior has not been observed in the white-tip scutellum group, and is definitely absent from the scaptomyzoids which have an assault-type mating. It is unknown whether courtship takes place on the breeding substrate, but in any event, these groups display a very different mating system from the lekking picture-wings which utilize quite different breeding substrates. Further observations on all these reproductive and ecological aspects of the biology of the more primitive non-picture-winged lineages are required in order to correlate the environmental, behavioral, and physiological features of these flies with the evolutionary forces that have led to their diversification.

Although sexual selection may be less important in some drosophilid lineages than in the picture-wings, nonetheless the biology of all groups and all species has been molded by natural selection. Indeed, ecological constraints and competition for resources seem to have been primary in Hawaiian drosophilid evolution. The short branches at the base of the molecular phylogeny suggest an early radiation into separate evolutionary lineages that are ecologically distinct (Kambysellis et al. 1995). This initial ecological differentiation was rapidly followed by adaptation of the female reproductive system to each particular substrate, accomplishing the reproductive shifts and the correlations between reproductive type and breeding substrate detailed above. In certain lineages, sexual selection operating alone or in conjunction with founder events (Kaneshiro 1989) was instrumental in the species proliferation of each lineage that followed. Natural selection continued to hone the adaptation of each particular species to its individual niche, and in some instances directional selection brought about subsequent ecological shifts, as have occurred in the two shifts to flux-breeding within the picture-winged group, in the shift of *D. heedi* to utilizing soil fluxes, and the shift of *D. mulli* to some other as yet unknown substrate. Thus natural selection has clearly been significant in the evolution and adaptive radiation of the Hawaiian drosophilid fauna, and in molding the rich array of ecological and female reproductive variations described in this chapter.

Conclusions

The adaptive radiation of the endemic Hawaiian *Drosophila* has been accompanied by divergence in multiple female reproductive traits (including ovipositor length, number of ovarioles per ovary, number of mature eggs per ovary, egg chorion thickness, and length of the egg respiratory filaments) that collectively enhance adaptation to the particular breeding substrate that characterizes each of the several evolutionary lineages. Tracing the evolution of several ecological and reproductive characters on an independently derived molecular phylogeny (based on mtDNA and nDNA sequences) demonstrates that female reproductive strategies broadly correlate with ecological divergence in host plant and larval substrate. Female reproductive Type I (low fecundity) is associated with flower-breeders; Type II, with leaf-breeders; Type IIIa, with stem-breeders; Type IIIb (highest fecundity), with bark- and flux-breeders; and Type V, with fungus-breeders. The breeding substrate of Type IV is unknown.

Our phylogenetic analysis has confirmed that the Hawaiian drosophilids are monophyletic, supporting their origin from a single ancestor, with subsequent divergence into two lineages of Hawaiian drosophilids, the scaptomyzoids and the drosophiloids, as previously indicated by the morphological analysis. Mapping substrate use on the molecular phylogeny indicates that the most primitive Hawaiian drosophiloids were fungus-breeders. Ecological shifts then occurred from fungi to decaying leaves, stems and fruit, bark, and finally to tree fluxes, with the latter two substrates being invaded by the most derived picture-winged group. There have been two independent shifts to flux-breeding, and apparently two or three shifts from stem-breeding to bark-breeding, although the distinction between these two substrates may depend mainly on the depth to which eggs are inserted during oviposition. Female reproductive traits have shown correlated shifts in the parallel shifts to new breeding substrates, validating the adaptive nature of the female reproductive variation.

Following the use of fungi by primitive Hawaiian drosophiloids, the first plant family invaded was Araliaceae. There have been several independent shifts to Campanulaceae, as well as shifts to numerous other endemic Hawaiian plant groups. Most Hawaiian *Drosophila* are monophagous at the family level; polyphagy is uncommon and seems to be a derived state. Natural selection exerted by the breeding substrate has been a major factor in Hawaiian *Drosophila* evolution, which, together with sexual selection on male behavior and morphology, has contributed to the evolution of a diversity of mating systems. In the early phyletic diversification of this group, adaptive shifts to new breeding substrates appear to have been most important, but such shifts are also a feature of more recent speciation events; this is best exemplified by the adaptation of the species *D. heedi* to soil fluxes, in response to a relatively recent shift from the tree flux niche of its immediate ancestor.

Although our analyses of the adaptive radiation of the Hawaiian *Drosophila* should be considered preliminary because of the limited ecological and molecular data available, nonetheless, we believe that our findings reflect the major trends in this group, because the molecular phylogeny includes representatives of most morphologically defined groups, and because ecological and reproductive patterns are generally consistent within groups. It will, however, be important to investigate further the ecologically diverse but currently poorly analyzed scaptomyzoids and the modified-mouthparts and white-tip scutellum groups if we are to better understand the adaptive radiation of the remarkable Hawaiian Drosophilidae.

Acknowledgments

We thank Rob DeSalle for sharing his molecular phylogeny of the Hawaiian drosophilids (reproduced here as Figure 17.1B) prior to publication. Jacob Cohen was particularly helpful to us in the preparation of the figures; we also acknowledge personnel of the NYU Computer Center for their assistance. Thanks also to the editors and reviewers for their constructive comments which helped improve the manuscript. The work of the authors was supported by NSF Grant BSR 89-18650.

References

Baker, R., and DeSalle, R. 1997. Multiple sources of character information and the phylogeny of Hawaiian drosophilids. Systematic Biology (in press).
Begon, M. 1982. Yeasts and *Drosophila*. Pp. 345–384 in *The genetics and biology of Drosophila, vol. 3b*, M. Ashburner, H. L. Carson, and J. N. Thompson, Jr., eds. London, England: Academic Press.
Beverley, S. M., and Wislon, A. C. 1985. Ancient origin for Hawaiian Drosophilinae inferred from protein comparisons. Proceedings of the National Academy of Sciences, USA 82:4753–4757.
Boake, C. R. B., and Adkins, E. 1994. Timing of male physiological and behavioral maturation in *Drosophila silvestris* (Diptera: Drosophilidae). Journal of Insect Behavior 7:577–583.
Bremer, K. 1988. The limits of amino-acid sequence data in angiosperm phylogenetic reconstruction. Evolution 42:795–803.
Carson, H. L. 1970. Chromosome tracers of the origin of species. Science 168:1414–1418.
Carson, H. L. 1971. Speciation and the founder principle. Stadler Genetics Symposium 3:51–70.
Carson, H. L. 1978. Speciation and sexual selection in Hawaiian *Drosophila*. Pp. 93–107 in *Ecological genetics: the interface*, P. F. Brussard, ed. New York, NY: Springer-Verlag.
Carson, H. L. 1981. Chromosomal tracing of evolution in a phylad of species related to *Drosophila hawaiiensis*. Pp. 286–297 in *Evolution and speciation: essays in honor of M. J. D.White*, W. R. Atchley and D. S. Woodruff, eds. Cambridge, England: Cambridge University Press.
Carson, H. L. 1986. Sexual selection and speciation. Pp. 391–409 in *Evolutionary processes and theory*, S. Karlin and E. Nevo, eds. London, England: Academic Press.
Carson, H. L. 1992. Inversions in Hawaiian *Drosophila*. Pp. 407–439 in Drosophila *inversion polymorphism*, C. B. Krimbas and J. R. Powell, eds. Boca Raton, FL: CRC Press.
Carson, H. L., Hardy, D. E., Spieth, H. T., and Stone, W. S. 1970. The evolutionary biology of the Hawaiian Drosophilidae. Pp. 437–543 in *Essays in evolution and genetics in honor of Theodosius Dobzhansky*, M. K. Hecht and W. C. Steere, eds. New York, NY: Appleton-Century-Crofts.
Carson, H. L., and Kaneshiro, K. Y. 1976. *Drosophila* of Hawaii: systematics and ecological genetics. Annual Review of Ecology and Systematics 7:311–345.

Carson, H. L., Kaneshiro, K. Y., and Val, F. C. 1989. Natural hybridization between the sympatric Hawaiian species *Drosophila silvestris* and *Drosophila heteroneura*. Evolution 43:190–203.
Carson, H. L., and Ohta, A. T. 1981. Origin of the genetic basis of colonizing ability. Pp. 365–370 in *Evolution today*, G. G. E. Scudder and J. L. Reveal, eds. Pittsburgh, PA: Carnegie-Mellon University Press.
Carson, H. L., and Templeton, A. R. 1984. Genetic revolutions in relation to speciation phenomena: the founding of new populations. Annual Review of Ecology and Systematics 15:97–131.
Chase, M. W., Soltis, D. E., Morgan, D., Les, D. H., Duvall, M. R., Price, R., Hills, H. G., Qiu, Y., Kron, K. A., Rettig, J. H., Conti, E., Palmer, J. D., Clegg, M. T., Manhart, J. R., Sytsma, K. J., Michaels, H. J., Kress, W. J., Donoghue, M. J., Clark, W. D., Hedren, M., Gaut, B. S., Jensen, R. K., Kim, K., Wimpee, C. F., Smith, J. F., Furnier, G.R., Strauss, S. H., Xiang, Q., Plunkett, G. M., Soltis, P. M., Eguiarte, L. E., Learn, G. H., Graham, S., and Albert, V. A. 1993. Phylogenetics of seed plants: an analysis of nucelotide sequences from the plastid gene *rbc*L. Annals of the Missouri Botanical Garden 80:528–580.
Craddock, E. M. 1974a. Degrees of reproductive isolation between closely related species of Hawaiian *Drosophila*. Pp. 111–139 in *Genetic mechanisms of speciation in insects*, M. J. D. White, ed. Sydney, Australia: Australia and New Zealand Book Co.
Craddock, E. M. 1974b. Reproductive relationships between homosequential species of Hawaiian *Drosophila*. Evolution 28:593–606.
Craddock, E. M., and Johnson, W. E. 1978. Multiple insemination in natural populations of *Drosophila silvestris*. Drosophila Information Service 53:138–139.
David, J. R. 1970. Le nombre d'ovarioles chez la *Drosophila*: Relation avec la fecondité et valeur adaptative. Archives de Zoologie Experimentale et Generale 111:357–370.
DeSalle, R. 1992. The origin and possible time of divergence of the Hawaiian Drosophilidae: evidence from DNA sequences. Molecular Biology and Evolution 9:905–916.
DeSalle, R. 1995. Molecular approaches to biogeographic analysis of Hawaiian Drosophilidae. Pp. 72–89 in *Hawaiian biogeography: evolution on a hot spot archipelago*, W. L. Wagner and V. A. Funk, eds. Washington, DC: Smithsonian Institution Press.
Donoghue, M. J., Olmstead, R. G., Smith, J. F., and Palmer, J. D. 1992. Phylogenetic relationships of Dipsacales based on *rbc*L sequences. Annals of the Missouri Botanical Garden 79:333–345.
Felsenstein, J. 1985. Confidence limits on phylogenies: an approach using the bootstrap. Evolution 39:783–791.
Felsenstein, J. 1988. Phylogenies from molecular sequences: inference and reliability. Annual Review of Genetics 22:521–565.
Fisher, R. A. 1930. *The genetical theory of natural selection*. Oxford, England: Clarendon Press.
Franchi, L., Francisco, P., Kambysellis, M. P., and Craddock, E. M. 1997. Morphological variation and adaptive evolution of the ovipositor in the endemic Hawaiian *Drosophila*. Journal of Morphology (in press).
Futuyma, D. J., and Moreno, G. 1988. The evolution of ecological specialization. Annual Review of Ecology and Systematics 19:207–233.
Givnish, T. J., Sytsma, K. J., Smith, J. F., and Hahn, W. J. 1995. Molecular evolution, adaptive radiation, and geographic speciation in *Cyanea* (Campanulaceae, Lobelioideae). Pp. 288–337 in *Hawaiian biogeography: evolution on a hot spot archipelago*, W. L. Wagner and V. A. Funk, eds. Washington, DC: Smithsonian Institution Press.
Grant, B. R., and Grant, P. R. 1993. Evolution of Darwin's finches caused by a rare climatic event. Proceedings of the Royal Society of London, Series B 251:111–117.
Hardy, D. E. 1965. *Insects of Hawaii, vol. 12*. Honolulu, HI: University of Hawaii Press.
Hardy, D. E., and Kaneshiro, K. Y. 1972. New picture-winged *Drosophila* from Hawaii. Part III Drosophilidae, Diptera. University of Texas Publications 7213:155–162.
Hardy, D. E., and Kaneshiro, K. Y. 1975. Studies in Hawaiian *Drosophila*, modified mouthparts species No. 1: *mitchelli* subgroup. Proceedings of the Hawaii Entomological Society 22:57–64.
Hardy, D. E., and Kaneshiro, K. Y. 1981. Drosophilidae of Pacific Oceania. Pp. 309–347 in *The genetics and biology of Drosophila, vol. 3a*, M. Ashburner, H. L. Carson, and J. N. Thompson, Jr., eds. London, England: Academic Press.
Heed, W. B. 1968. Ecology of the Hawaiian Drosophilidae. University of Texas Publications 6818:387–419.
Heed, W. B. 1971. Host plant specificity and speciation in Hawaiian *Drosophila*. Taxon 20:115–121.
Hung, M.-C., and Wensink, P. C. 1981 The sequence of the *Drosophila melanogaster* gene for yolk protein, I. Nucleic Acids Research 9:6407–6419.
Jaenike, J., Grimaldi, D. A., Sluder, A. E., and Greenleaf, A. L. 1983. α-amanitin tolerance in mycophagous *Drosophila*. Science 221:165–167.

Kambysellis, M. P. 1970. Compatibility in insect tissue transplantations. I. Ovarian transplantations and hybrid formation between *Drosophila* species endemic to Hawaii. Journal of Experimental Zoology 175:169–180.

Kambysellis, M. P. 1993. Ultrastructural diversity in the egg chorion of Hawaiian *Drosophila* and *Scaptomyza*: ecological and phylogenetic considerations. International Journal of Insect Morphology and Embryology 22:417–446.

Kambysellis, M. P., and Craddock, E. M. 1991. Insemination patterns in Hawaiian *Drosophila* species (Diptera: Drosophilidae) correlate with ovarian development. Journal of Insect Behavior 4:83–100.

Kambysellis, M. P., and Heed, W. B. 1971. Studies of oogenesis in natural populations of Drosophilidae. I. Relation of ovarian development and ecological habitats of the Hawaiian species. American Naturalist 105:31–49.

Kambysellis, M. P., Ho, K.-F., Craddock, E. M., Piano, F., Parisi, M., and Cohen, J. 1995. Pattern of ecological shifts in the diversification of Hawaiian *Drosophila* inferred from a molecular phylogeny. Current Biology 5:1129–1139.

Kaneshiro, K. Y. 1969. The *Drosophila crassifemur* group of species in a new subgenus. University of Texas Publications 6918:79–83.

Kaneshiro, K. Y. 1976. A revision of generic concepts in the biosystematics of Hawaiian Drosophilidae. Proceedings of the Hawaii Entomological Society 22:255–278.

Kaneshiro, K. Y. 1983. Sexual selection and direction of evolution in the biosystematics of Hawaiian Drosophilidae. Annual Review of Entomology 28:161–178.

Kaneshiro, K. Y. 1987. The dynamics of sexual selection and its pleiotropic effects. Behavior Genetics 17:559–569.

Kaneshiro, K. Y. 1989. The dynamics of sexual selection and founder effects in species formation. Pp. 279–296 in *Genetics, Speciation and the Founder Principle*, L. V. Giddings, K. Y. Kaneshiro, and W. W. Anderson, eds. New York, NY: Oxford University Press.

Kaneshiro, K. Y., and Boake, C. R. B. 1987. Sexual selection and speciation: issues raised by Hawaiian *Drosophila*. Trends in Ecology and Evolution 2:207–212.

Kaneshiro, K. Y., Carson, H. L., Clayton, F. E., and Heed, W. B. 1973. Niche separation in a pair of homosequential *Drosophila* species from the island of Hawaii. American Naturalist 107:766–774.

Kaneshiro, K. Y., Gillespie, R. G., and Carson, H. L. 1995. Chromosomes and male genitalia of Hawaiian *Drosophila*: tools for interpreting phylogeny and geography. Pp. 57–71 in *Hawaiian biogeography: evolution on a hot spot archipelago*, W. L. Wagner and V. A. Funk, eds. Washington, DC: Smithsonian Institution Press.

Kaneshiro, K. Y., and Val, F. C. 1977. Natural hybridization between a sympatric pair of Hawaiian *Drosophila*. American Naturalist 111:897–902.

Kircher, H. W., and Heed, W. B. 1970. Phytochemistry and host plant specificity in *Drosophila*. Pp. 192–209 in *Recent advances in phytochemisty, vol III*, C. Steelink and K. C. Runeckles, eds. New York, NY: Appleton-Century-Crofts.

Kumar, S., Tamura, K., and Nei, M. 1993. *MEGA: molecular evolutionary genetics analysis, vers. 1.01*. University Park, PA: Pennsylvania State University.

Lammers, T. G. 1995. Patterns of speciation and biogeography in *Clermontia* (Campanulaceae, Lobelioideae). Pp. 338–362 in *Hawaiian biogeography: evolution on a hot spot archipelago*, W. L. Wagner and V. A. Funk, eds. Washington, DC: Smithsonian Institution Press.

Lande, R. 1981. Models of speciation by sexual selection on polygenic traits. Proceedings of the National Academy of Sciences, USA 78:3721–3725.

Maddison, W. P., and Maddison, D. R. 1992. *MacClade: analysis of phylogeny and character evolution, vers. 3*. Sunderland, MA: Sinauer Associates, Inc.

Margaritis, L. H. 1983. Structure and physiology of the egg-shell. Pp. 153–230 in *Comprehensive insect biochemistry, physiology and pharmacology, vol. 1*, G.A. Kerkut and L.I. Gilbert, eds. Oxford, England: Pergamon Press.

Markow, T. A. 1996. Evolution of *Drosophila* mating systems. Pp. 73–106 in *Evolutionary biology, vol. 29*, M. K. Hecht et al., eds.. New York, NY: Plenum Press.

Montgomery, S. L. 1975. Comparative breeding site ecology and the adaptive radiation of picture-winged *Drosophila* (Diptera: Drosophilidae) in Hawaii. Proceedings of the Hawaii Entomological Society 22:65–103.

Ohta, A. T. 1989. Coadaptive changes in speciation via the founder principle in the *grimshawi* species complex of Hawaiian *Drosophila*. Pp. 315–328 in *Genetics, speciation and the founder principle*, L. V. Giddings, K. Y. Kaneshiro, and W. W. Anderson, eds. New York, NY: Oxford University Press.

Perkins, R. C. L. 1913. Introduction. Pp. xv–ccxxvii in *Fauna Hawaiiensis, vol. 16,* D. Sharp, ed. Cambridge, England: Cambridge University Press.

Piano, F., Craddock, E. M., and Kambysellis, M. P. 1997. Phylogeny and taxonomic status of the island populations of the Hawaiian *Drosophila grimshawi* complex: evidence from combined data. Molecular Phylogenetics and Evolution (in press).

Robertson, F. W., Shook, M., Takei, G., and Gaines, H. 1968. Observations on the biology and nutrition of *Drosophila disticha* Hardy, an indigenous Hawaiian species. University of Texas Publications 6818:279–299.

Russo, C. A. M., Takezaki, N., and Nei, M. 1995. Molecular phylogeny and divergence times of drosophilid species. Molecular Biology and Evolution 12:391–404.

Schluter, D. 1994. Experimental evidence that competition promotes divergence in adaptive radiation. Science 266:798–801.

Spieth, H. T. 1966. Courtship behavior of endemic Hawaiian *Drosophila.* University of Texas Publications 6615:245–314.

Spieth, H. T. 1968a. Evolutionary implications of the mating behavior of the species of *Antopocerus* Drosophilidae in Hawai`i. University of Texas Publications 6818:319–333.

Spieth, H. T. 1968b. Evolutionary implications of sexual behavior in *Drosophila.* Pp. 157–193 in *Evolutionary biology, vol. 2* , M. K. Hecht and W. C. Steere, eds. New York, NY: Plenum Press.

Spieth, H. T. 1974. Mating behavior and evolution of the Hawaiian *Drosophila.* Pp. 94–101 in *Genetic mechanisms of speciation in insects,* M. J. D. White, ed. Sydney, Australia: Australia and New Zealand Book Co.

Spieth, H. T. 1981. History of the Hawaiian *Drosophila* project. Drosophila Information Service 56:6–14.

Spieth, H. T. 1982. Behavioral biology and evolution of the Hawaiian picture-winged species group of *Drosophila.* Pp. 351–437 in *Evolutionary biology, vol. 14,* M.K. Hecht, B. Wallace, and G.T. Prance, eds. New York, NY: Plenum Press.

Swofford, D. L. 1993. *PAUP: phylogenetic analysis using parsimony, vers. 3.1.1.* Champaign, IL: Illinois Natural History Survey.

Takada, H. 1966. Male genitalia of some Hawaiian Drosophilidae. University of Texas Publications 6615:335–396.

Thomas, R. H., and Hunt, J. A. 1991. The molecular evolution of the alcohol dehydrogenase locus and the phylogeny of Hawaiian *Drosophila.* Molecular Biology and Evolution 8:687–702.

Throckmorton, L. H. 1966. The relationships of the endemic Hawaiian Drosophilidae. University of Texas Publications 6615:335–396.

Wagner, W. L. 1991. Evolution of waif floras: A comparison of the Hawaiian and Marquesan archipelagoes. Pp. 267–284 in *The unity of evolutionary biology,* E. C. Dudley, ed. Portland, OR: Dioscorides Press.

Wagner, W. L., Herbst, D. R., and Sohmer, S. H. 1990. *Manual of the flowering plants of Hawai`i.* Honolulu, HI: University of Hawai`i Press and Bishop Museum Press.

Williamson, M. H. 1981. *Island populations.* Oxford, England: Oxford University Press.

Yang, H. Y., and Wheeler, M. 1969. Studies on interspecific hybridization within the picture-winged group of endemic Hawaiian *Drosophila.* University of Texas Publications 6918:133–170.

Zimmerman, E. C. 1958. Three hundred species of *Drosophila* in Hawaii? A challenge to geneticists and evolutionists. Evolution 12:557–558.

18 History of ecological selection in sticklebacks: uniting experimental and phylogenetic approaches

Eric B. Taylor, John Donald McPhail, and Dolph Schluter

Adaptive radiation is the diversification of a clade into a number of species exploiting different resource types and differing in the morphological and physiological traits used to exploit those resources (Huxley 1942; Futuyma 1986; Schluter 1996a; see Chapter 1). Its study has at least two fundamental objectives. The first is to gain knowledge of the histories of taxa that have undergone adaptive radiation, which includes the elucidation of evolutionary relationships among descendant species, their geographical patterns of origination (e.g., whether in sympatry or allopatry), and the sequence of character changes at different stages of differentiation. Comparisons of these attributes between taxa will also determine whether patterns of diversification are idiosyncratic and clade-specific, or whether there are broad consistencies among clades, such as in showing increased resource specialization through time, or repeatable sequences of resource partitioning.

The second goal is to understand the forces that drive adaptive radiation – the processes that cause new species to form and become differentiated in morphology, physiology, behavior, and ecology. The major ecological forces are thought to be divergent natural selection caused by differences between populations in the resource environments they experience, and by resource competition between sympatric populations (Fisher 1930; Dobzhansky 1937; Huxley 1942; Mayr 1942; Simpson 1944, 1953; Lack 1947; Grant 1986; Schluter 1996a). Non-ecological forces affecting adaptive radiation include genetic drift, founder events, and fixation of alternative advantageous genes in populations facing similar selection pressures (Wright 1940; Muller 1940; Mayr 1954). Sexual selection may play a part in speciation by ecological or non-ecological forces depending on the mechanisms underlying divergence in mate preferences (i.e., drift or divergent natural selection).

The chief domains of the first of these objectives are phylogenetics, biogeography, and paleontology, whereas experimental and observational tests on natural populations are needed to accomplish the second. Partly because different methodologies are needed to address them, the two goals are rarely realized simultaneously for any single lineage. Consequently, most adaptive radiations remain poorly understood.

In this report we present preliminary results of a joint molecular and experimental inquiry into an adaptive radiation apparently in its very early stages: ecological and morphological divergence in the threespine sticklebacks (*Gasterosteus* spp.) in small postglacial lakes, streams, and nearby coastal waters of southwestern British Columbia (Schluter and McPhail 1992; McPhail 1994). Our ultimate goal is to piece together the history of environmental selection pressures to which populations and species have been subjected throughout diversification, and their roles in speciation and divergence. Here we focus on the sympatric species pairs that occur in small lakes of the Strait of Georgia region (Schluter and McPhail 1992; McPhail 1994) (Figure 18.1). We attempt to answer three related questions:

- With what combination of sympatry and allopatry have species originated?
- Have ecologically and morphologically similar pairs of species pairs evolved independently in different lakes? And
- Did competitive interactions contribute to the evolution of differences between species after sympatry was achieved?

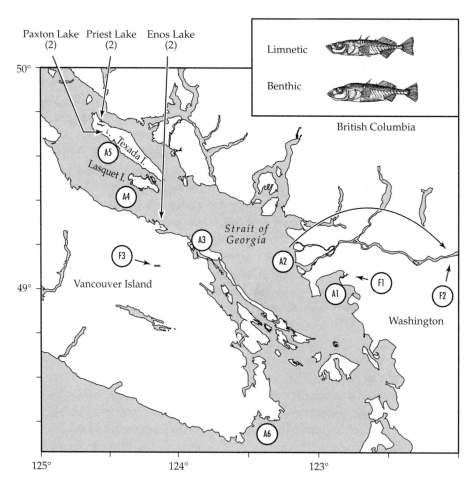

Figure 18.1. Map of the study area showing populations sampled and locations of Paxton, Priest, and Enos lakes. Each lake contains two species, a Limnetic and a Benthic (see inset). Sample size for each species in each of these three lakes was N = 14 to 16. Other permanent freshwater populations sampled are Little Campbell River (F1), Salmon River (F2), and Nanaimo Lake (F3). Anadromous populations sampled are Little Campbell River (A1), Salmon River (A2), Nanaimo River (A3), French Creek (Vancouver Island; A4), Cranby Creek (Texada Island; A5), and Witty's Lagoon (Vancouver Island; A5). Sample sizes for the latter populations were, respectively, N = 14, 27, 14, 15, 29, 15, 13, 17, 20. Illustrations are of the Benthic and Limnetic species from Paxton Lake.

We begin with a summary of our study system and then outline alternative scenarios for the origin and differentiation of species under a hypothesis of ecological character displacement. We then describe results of an initial phylogenetic analysis of species pairs based on mitochondrial DNA restriction sites. Finally, we summarize an ecological experiment designed to test the character displacement hypothesis, by measuring whether natural selection favors divergence between sympatric, morphologically similar species. We conclude with notes on the potential advantages of joint molecular and experimental studies.

Biology of sympatric sticklebacks

The threespine stickleback (*Gasterosteus aculeatus*) occurs in coastal waters throughout much of the northern hemisphere, where it is an open water planktivore. Most marine populations are anadromous and migrate to the lower sections of rivers and streams to spawn before returning to the sea (Wooton 1976; Bell and Foster 1994). The species, however, has independently given rise to permanent freshwater populations many times (perhaps hundreds) throughout its current range (Bell and Foster 1994; Ortí et al. 1994). Several of these freshwater populations are reproductively isolated from the marine (anadromous) form (e.g., Hay and McPhail 1975; Borland 1986; Zuiganov et al. 1987).

In the Strait of Georgia region of southwestern British Columbia, freshwater forms include sympatric pairs of species that occur in small lakes at low elevation (McPhail 1984, 1992, 1994; Schluter and McPhail 1992). These pairs occur in four unconnected drainages on three adjacent islands: Texada (Paxton and Priest/Emily lakes), Vancouver (Enos Lake), and Lasqueti (Hadley Lake, where both species are now extinct). The species are not yet formally described and we refer to them as Benthics or Limnetics, according to the habitats within the lakes that they mainly use. Similar, nearby lakes contain only a single species, which we hereafter refer to as Solitary species. The lakes of this region are only about 12,000 years old, and were formed when submerged coastal areas were uplifted following deglaciation at the end of the Pleistocene (McPhail 1994). This puts an upper limit on the duration of sympatry of species pairs. The lakes are presently inaccessible from the sea.

Each species pair consists of a Benthic and a Limnetic. The Limnetic is smaller, has a more slender body, a narrower gape, and longer, more numerous gill rakers than the Benthic species. Gill rakers are protuberances along the gill arches that assist in prey retention, and act either as a physical sieve of ingested prey or, when very densely packed, to direct water currents and particle movement within the buccal cavity (Sanderson et al. 1991). The morphological differences between sympatric species are associated with differences in habitat use and diet. Limnetics are mainly planktivorous and forage offshore in open water, especially in the non-breeding season. Benthics on the other hand are confined to the littoral zone and deeper areas of open sediment where they consume invertebrates inhabiting the sediment or attached to vegetation (Bentzen and McPhail 1984; Schluter and McPhail 1992; Schluter 1993). Both species breed in the littoral zone where males build and defend nests on the sediments.

Several pieces of evidence indicate that sympatric sticklebacks are good biological species rather than morphs of a single species. Morphological differences between them persist for at least two generations in the laboratory (McPhail 1984, 1992; Hatfield 1995), indicating a genetic basis. F_1 and F_2 hybrids are morphologically intermediate (McPhail 1984, 1992; Hatfield 1995). Means and variances of quantitative traits in hybrids and backcrosses indicate that species differences in at least two traits, gill raker number and length, are polygenic (Hatfield 1996). Some morphological traits (e.g., gill raker length) are developmentally plastic, such that differences between the species are reduced when their natural diets are reversed (Day et al. 1994). This effect, however, is minor or absent in most traits (e.g., gill raker number), and in no trait is the difference between species eliminated by diet. Limnetics and Benthics mate assortatively in the laboratory (Ridgway and McPhail 1984; Nagel 1994; Hatfield 1995; Hatfield and Schluter 1996), and females of the two species show a reduced level of preference for F_1 hybrid males (Hatfield and Schluter 1996; Vamosi 1996). Hybridization is also rare in the wild, as indicated by a low frequency of F_1 hybrids there (identified by morphology; McPhail 1984, 1992) and by significant differences between the species in allozyme frequencies (McPhail 1984, 1992). Nevertheless, the species are young: sympatric forms differ in allele frequencies at only 2 or 3 of 25 allozyme loci screened (Nei's [1978] D = 0.02; McPhail 1984, 1992).

Morphological differences between the sympatric species are probably adaptive, as suggested by the following evidence. First, species differences in size and shape are correlated with differences in habitat use and diet; morphologically intermediate species are also ecologically intermediate (Schluter and McPhail 1992). Second, the species differ greatly in their efficiencies of prey exploitation in the two main habitats. In open water (i.e., large aquaria with natural prey densities) Limnetics capture plankton three times faster than Benthics, whereas Benthics are more successful than Limnetics at obtaining food in the littoral zone, mainly because they can ingest larger prey (Schluter 1993). Hybrids tend to be intermediate in both habitats. Third, growth rates measured in a transplant experiment in the wild showed the same pattern of differences as foraging efficiency: Limnetics grow at about twice the rate of Benthics in open water, whereas the Benthics hold an equivalent advantage in the littoral zone (Schluter 1995a). Hybrids tend to grow at intermediate rates in both habitats. Such steep trade-offs indicate that morphological attributes that improve foraging success in one habitat reduce feeding efficiency in the second habitat. The suite of traits that distinguish Limnetics and Benthics are heritable within populations, and are genetically intercorrelated (Baumgartner 1986; Lavin and McPhail 1987; Schluter 1996b).

Comparative tests suggested that morphological differences between species in sympatry are also the result of ecological character displacement (defined here as morphological evolution resulting from resource competition between species). Stickleback species occurring alone in small lakes (Solitary) tend to be morphologically intermediate between Limnetics and Benthics (Figure 18.2) and exploit both habitats (Schluter and McPhail 1992). In the one Solitary population studied in detail, in Cranby Lake on Texada Island (next to Paxton Lake but in a separate drainage),

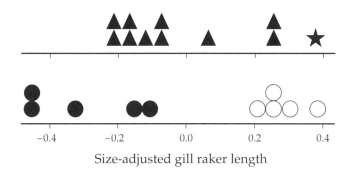

Figure 18.2. Morphological pattern suggesting ecological character displacement in sticklebacks. Circles indicate mean size-corrected gill raker lengths of males in sympatric (lower) and allopatric (upper) freshwater populations. Means of Solitary populations (▲) are intermediate between Benthics (●) and Limnetics (○). Star (★) indicates mean gill raker length of the marine form, the ultimate ancestor of all freshwater populations. Similar patterns are seen in other morphological traits (Schluter and McPhail 1992). The five species-pairs included in this comparison are from Paxton, Enos, Hadley, Priest, and Emily lakes. Details on measurements and size correction are given by Schluter and McPhail (1992). After Schluter (1996a) and Schluter and McPhail (1993).

different individuals were found to exploit either mainly plankton or mainly littoral invertebrates according to slight differences in their morphology (Schluter and McPhail 1992). This reinforces the conclusion that different combinations of traits are favored for efficient foraging in the two habitats. It also confirms that both habitats are present in a lake containing only a single species.

Hypotheses of species origination and differentiation

Consideration of the above data led to the development of two working hypotheses for the origin and differentiation of sympatric stickleback species (Schluter and McPhail 1992; McPhail 1994). In both, freshwater species are ultimately the result of colonization by the marine (anadromous) form (Figure 18.3). This assumption is justified by the obvious recency of all freshwater populations in the region and the great age of sticklebacks within the Pacific basin (e.g., Ortí et al. 1994). We also assume that the present-day marine species is morphologically similar to the original colonists. This is supported by the strong resemblance in morphology of old lineages within the Pacific to one another (Ortí et al. 1994) and to Miocene fossils (Bell 1994).

The first hypothesis posits a single invasion to each lake followed by sympatric speciation (Figure 18.3). The mechanism that would most likely favor this process is frequency- and density-dependent disruptive selection for resource specialization ("competitive speciation" – Rosenzweig 1978; Pimm 1979; Wilson and Turelli 1986). Under this scenario, premating isolation could evolve as a by-product of morphological divergence (cf. Rice and Hostert 1993) and/or reinforcement (e.g., Liou and Price 1994; Noor 1995), and might involve sexual selection (Liou and Price 1994; Hatfield and Schluter 1996; Vamosi 1996). A potential weakness of the sympatric speciation hypothesis is that the lakes containing species pairs are confined to a small

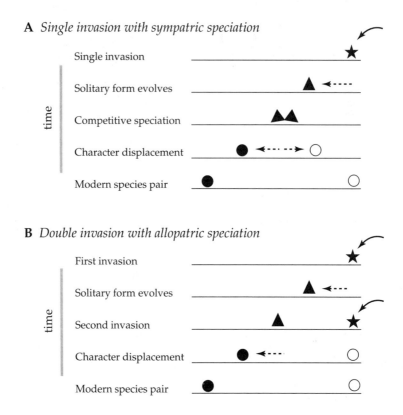

Figure 18.3. Alternative scenarios for the origin and morphological differentiation of sympatric species that compete for resources. Symbols represent population means in a morphological variable (e.g., gill raker length; for explanation of symbols, see Figure 18.2). Under the single-invasion hypothesis (**A**) the marine ancestor (a planktivore) evolved to an intermediate phenotype following colonization, exploiting both open-water plankton and littoral invertebrates. Sympatric speciation occurred subsequently, probably because of frequency- and density-dependent selection for resource specialization ("competitive speciation"). Ecological and morphological differences between the new species were maintained or enhanced by ecological character displacement. In the double-invasion hypothesis (**B**) the Solitary form resulting from the first invasion was displaced toward a Benthic phenotype in response to competition for plankton by the second invader, which ultimately became the Limnetic species.

geographic region, whereas single-species lakes occur throughout the northern hemisphere (Schluter and McPhail 1992).

The second model is double invasion, in which lakes were colonized twice by the same marine ancestor at different dates (Figure 18.3). This hypothesis was partly inspired by geological evidence that sea levels within this area of the Georgia Strait rose a second time (i.e., following the first postglacial rebound of coastal land), enabling access once again to these lakes from the sea (Schluter and McPhail 1992, McPhail 1994). Sea-level fluctuations were apparently confined to this area, providing a possible explanation for the limited geographical distribution of Benthic-Limnetic pairs. Under this scenario we expect that some morphological evolution resulting in enhanced exploitation of the littoral zone (e.g., toward the intermediate body form characteristic of lakes that today have only a single species) occurred in the first invader by the time of the second invasion, and was subsequently enhanced by divergent character displacement (Figure 18.3).

This scenario supposes that the present-day Benthic species is descended from the first invader and the Limnetic from the second (see below). A possible variant of the double invasion hypothesis is that each lake was colonized simultaneously by two stickleback lineages as it formed: the marine species plus a freshwater species from elsewhere that was itself the result of an earlier colonization by the marine form (Kassen et al. 1995).

Thus far, evidence in favor of the double invasion hypothesis includes the (slightly) greater allozyme resemblance of marine sticklebacks to Limnetics than to Benthics (McPhail 1984, 1992), and the lower hatching success of Benthic eggs in saltwater relative to marine and Limnetic eggs (Kassen et al. 1995). Both lines of evidence also indicate that if the double invasion scenario is correct, then the Benthic species is probably descended from the first invader and the Limnetic from the second. However, these data are not conclusive evidence for double invasion because allozyme differences and salinity tolerances might be by-products of adaptive ecological differentiation. The question also arises as to whether separate species pairs are independently derived. Did present-day ecological differences between Limnetics and Benthics arise only once in one of the lakes, after which the pair of species invaded other lakes together?

Alternative scenarios are possible that do not incorporate competition-induced character displacement. For example, greater differences in sympatry than allopatry may be the result of differential extinction of species pairs or of resource differences among lakes. Our scenarios also do not yet incorporate other ecological interactions such as predation, whose potential contribution to divergence remains obscure. Details of other mechanisms also remain to be worked out, such as the roles of sexual selection (Hatfield and Schluter 1996; Vamosi 1996) and phenotypic plasticity (Day et al. 1994).

As described next, we inferred the phylogeny of stickleback populations in different lakes using mtDNA RFLP's to test between the single- and double-invasion hypotheses, and to determine if the Benthics and Limnetics in different lakes reflect a single divergence event followed by dispersal, or whether multiple, independent divergences have occurred in different lakes. We also conducted a field experiment to test whether competitive interactions for food between closely related, morphologically similar species favor divergence between them.

mtDNA phylogeny of populations

Methods

Sticklebacks were collected from several lake and river systems within the Strait of Georgia region (Figure 18.1). Samples included three Benthic-Limnetic species pairs from Enos, Paxton, and Priest lakes; one Solitary species from Nanaimo Lake; two parapatric anadromous-stream resident populations from Salmon and Little

Campbell rivers; and four allopatric anadromous populations from Cranby Creek, French Creek, Nanaimo River, and Witty's Lagoon. Sample sizes were 14 to 16 for sympatric species pairs and between 14 to 30 for other populations (Figure 18.1). Fish were collected with minnow traps or pole seines. Liver, spleen, heart, and muscle tissues were extracted and placed in 95% ethanol and stored at $-4°C$.

Genomic DNA was obtained using Pronase digestion and phenol/chloroform extraction (Taylor et al. 1996). Typically, 2 to 5 µg of genomic DNA was digested overnight with 12 endonucleases: three multi-pentameric (*BstN* I, *Nci* I, *Sty* I), five multi-hexameric (*Ava* I, *Ava* II, *Ban* I, *Hae* II, *Hinc* II), and four hexameric (*Dra* I, *Hind* III, *Pst* I, *Pvu* II) enzymes. Digestion conditions were as recommended by the manufacturer (New England Biolabs). Restricted DNAs were electrophoresed on 0.8–1.2% agarose gels, and Southern blotted under vacuum as detailed by Taylor et al. (1996).

Mitochondrial DNA restriction fragment variation was assayed by hybridization of membrane-bound genomic DNA with digoxigenin-labeled threespine stickleback mtDNA cloned into puc19. Hybridization conditions (58° C, final wash stringency, 2x SSC/0.1% SDS at 58° C) and detection of probe-stickleback DNA hybrids by chemiluminesence are given by Taylor et al. (1996).

All restriction fragment length polymorphisms (RFLPs) in *Gasterosteus* could be accounted for by single or double restriction-site changes. Consequently, a presence/absence restriction-site matrix was constructed for each RFLP observed for each enzyme and was given a single capital letter code (e.g. *Ava* I A, B, C, etc.). Each fish was then characterized by a 12-letter composite haplotype code; each letter represented the restriction-site code for each of the 12 enzymes (e.g. AAAAAAAAAAAA, ABAAAAAAAAAA, etc). A site matrix for each of the composite haplotypes resolved was constructed using the program REAP (McElroy et al. 1992). The composite haplotype restriction-site matrix so generated formed the basis for subsequent analyses of haplotype and nucleotide diversity within populations, and of divergence among populations, using REAP. The restriction-site matrix was used to generate a phylogenetic tree using Wagner parsimony. Support for each branch in the haplotype tree was assessed by bootstrap resampling of the restriction-site matrix with 100 replications, using the routines SEQBOOT, MIX, and CONSENSE in PHYLIP 3.5 (Felsenstein 1993).

Evolutionary divergence among haplotypes was estimated as the number of nucleotide substitutions per site (d of Nei and Miller 1990), or percentage divergence ($d \times 100$). Haplotype frequency distributions within each population and the associated d values among haplotypes were used to estimate nucleotide diversity within populations (π of Nei and Tajima 1981). Nucleotide divergences between all pairs of populations were then estimated by calculating total nucleotide diversity between each pair and extracting that portion not explained by diversity within populations, using the routine DA of REAP. Estimates of nucleotide divergence (Nei and Miller 1990) among populations were used to estimate population relationships using the neighbor-joining algorithm (Saitou and Nei 1987) implemented in the routine NEIGHBOR of PHYLIP 3.5. Because populations were highly polymorphic, bootstrap confidence in the topology of the population tree could not be assessed using available computer packages.

Molecular systematic results

A total of 113 restriction sites representing some 600 base pairs was assayed in our survey of *Gasterosteus* mtDNA. This sampling of the genome resolved 59 composite haplotypes distinguished by one to six restriction-site changes. Overall, nucleotide divergence among haplotypes was fairly low, with a range of 0.17 to 0.83%. An unrooted neighbor-joining analysis based on nucleotide divergence distinguished two main clusters of haplotypes (hereafter designated A and B) arranged in a dumbbell pattern; relationships within each cluster approximated a star (Figure 18.4). The foci of the A and B groups (haplotypes 28 and 22) differ by about 0.3% (two restriction-site differences). Average difference in sequence between haplotypes from the two groups is 0.48%. Wagner parsimony analysis of the haplotypes also organized haplotypes into two groups almost identical to A and B.

The existence of the A and B groups is tentative, as nucleotide divergence between them is small. The groups were not clearly resolved in a bootstrap analysis (haplotype 60 was deleted for this analysis, as it falls between groups; see Figure 18.4). Indeed, the topology of the entire tree is uncertain, because only six

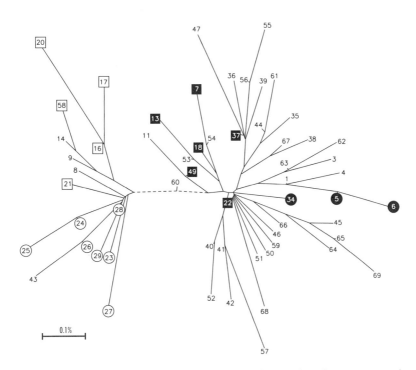

Figure 18.4. Unrooted neighbor-joining tree of 59 mtDNA haplotypes, based on estimates of nucleotide divergence. Dashed line separates the A lineage on the right from the B lineage on the left. Symbols indicate haplotypes found in one or both sympatric species in the three lakes, Paxton (●), Enos (○), and Priest (■ = haplotypes in the A lineage; □ = B haplotypes). Haplotypes 7, 37, and 49 (Priest Lake), and haplotype 28 (Enos Lake) were also found in one or more of the anadromous populations; haplotype 22 (in one individual from Priest Lake) was found in all anadromous and both stream resident populations (Salmon and Little Campbell rivers).

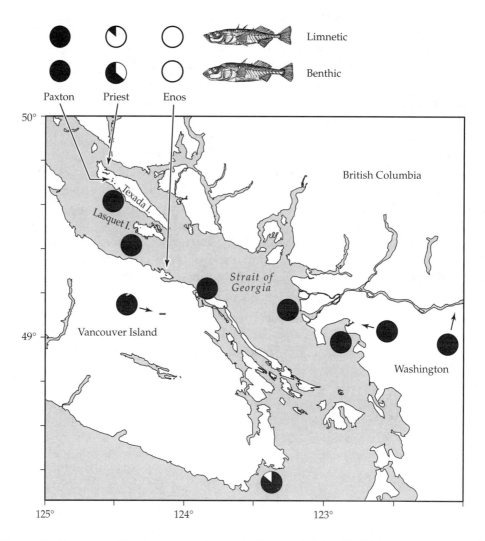

Figure 18.5. Frequency of the two putative haplotype lineages A (●) and B (○) in sample populations. Freshwater populations are indicated by arrows; remaining populations are anadromous. Sample populations are identified in Figure 18.1.

branches were supported at the 50% level or better in 100 bootstrap replicates. All six branches subtend pairs of haplotypes, of which five grouped haplotypes present in the same body of water (e.g., haplotypes 5 and 6; Figure 18.4). Topology of the Wagner parsimony tree was equally uncertain in a bootstrap analysis – other than a few terminal branches, no branches were supported at greater than the 50% level.

Most of the stickleback populations examined are dominated by group A mtDNA haplotypes (Figure 18.5). This was true of both species in Paxton Lake, which are exclusively from the A group. In contrast, Limnetics and Benthics from Enos Lake are composed entirely of haplotypes from the B group. Haplotypes A and B were present both in Benthics and Limnetics from Priest Lake and, to a lesser extent, Witty's Lagoon. Such differences between separate species pairs in the major classes of haplotype represented are the first indication that the morphologically similar populations in different lakes do not each have a common origin. A second indication is that the haplotypes within each lake tend to fall into distinct clusters in which each haplotype is connected to the others by a small number of steps (Figure 18.4).

Under the double-invasion hypothesis outlined earlier (Figure 18.3), Benthics should be more different genetically from the contemporary anadromous species than Limnetics. In contrast, under the single invasion hypothesis, both species should be roughly equidistant from the marine species. Estimates of nucleotide divergence among populations averaged 0.15% (range 0.00002 to 0.56%). Limnetic sticklebacks from Paxton Lake were roughly equally divergent from the various anadromous populations than were Benthic sticklebacks from the same lake ($d \times 100 = 0.18 \pm 0.004$ [SE] and 0.21 ± 0.005, respectively). Limnetics and Benthics in Enos Lake were also roughly equidistant from the anadromous form (0.25 ± 0.009 and 0.23 ± 0.008, respectively). In contrast, Priest Lake Limnetics were more distant from anadromous populations than were Benthics (0.21 ± 0.008 and 0.07 ± 0.005). These data do not support the double-invasion hypothesis.

Nucleotide divergence was lower between Benthic and Limnetic species within each lake than between either population and its morphological counterpart in other lakes. Consequently, the neighbor-joining population phylogeny based on these distances showed the members of a given species pair to be closely allied (Figure 18.6). The high degree of resemblance between species from the same lake is the result of their sharing haplotypes not found in other lakes or in the marine form (Figure 18.4). Moreover, the unique haplotypes found in the two species within each lake tend to be closely related to one another (Figure 18.4), except in Priest Lake where the unique haplotypes fell into two groups. Finally, sympatric pairs of lake species (in Enos, Paxton, and Priest lakes) tended to be more divergent from one another and from neighboring anadromous and stream populations than the latter populations were from one another (Figure 18.6).

These results appear to rule out the double-invasion hypothesis and are much more consistent with a single invasion into each lake followed by sympatric speciation. In the Discussion (see below), we consider a further possibility – a modified double-invasion hypothesis – in which persistent gene flow occurred following secondary contact, obscuring phylogenetic origins. In the interim, note that morphological and ecological divergence between species following secondary contact is more difficult to achieve in the presence of gene flow than in its absence, and may require forces similar to those needed for full sympatric speciation (i.e., frequency- and density-dependent natural selection). This is not only because gene flow counteracts selection but because it also causes existing levels of premating isolation to decay – particularly in the present case where hybrid viability and fertility is high (McPhail 1984, 1992; Hatfield 1995). Our phylogenetic results, therefore, suggest that the role of ecological forces is even greater than first envisioned under strict double invasion. In the following section we summarize results of an experiment to measure the strength of one of these forces: competition-induced selection for divergence in morphology and ecology.

Experimental studies on character displacement

Resource competition was a major component of adaptive radiation theory that developed in the first half of this century (Huxley 1942; Simpson 1944, 1953; Lack 1947; Amadon 1950). According to this theory divergent natural selection is the ultimate cause of adaptive radiation, and resource competition is one of the most important agents of such selection. Two influences on patterns of adaptive radiation in nature were envisioned. On the one hand, preemption of resources by species in other taxa was seen as suppressing adaptive radiation in a given lineage: freedom from such competition is the presumed reason for enhanced scope and speed of diversification in novel environments such as remote archipelagoes and newly formed lakes. On the other hand, competition between sympatric species within a radiating lineage was thought to drive their divergence onto novel resource types where species would then experience novel selection pressures.

Most evidence for the importance of competition in adaptive radiation is indirect. This evidence consists mainly of comparative tests of ecological character displacement (Brown and Wilson 1956), whereby ecological and morphological differences between closely related species in sympatry are compared with differences in allopatry (e.g., Figure 18.2) (reviewed in Arthur 1987; Taper and Case 1992) or with differences generated from a null model (reviewed in Gotelli and Graves 1996). The number of convincing cases remains few, although a large number of suggestive cases have not been studied in detail (e.g., Schluter and McPhail 1993; Robinson and Wilson 1994). A single experimental simulation over nine generations, involving two competing strains of bean beetles on novel resources, failed to detect any changes in resource preferences (Taper and Case 1992). This scarcity of evidence makes it difficult to argue that competition's role in adaptive radiation has been verified beyond doubt.

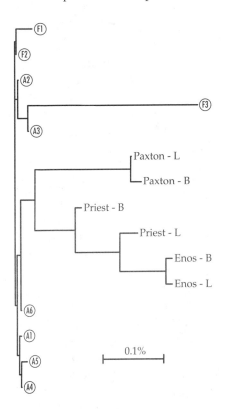

Figure 18.6. Unrooted neighbor-joining tree of populations, based on their estimated mtDNA nucleotide divergences. A1–A6 are anadromous populations, F1 and F2 are populations resident in freshwater streams, and F3 is a Solitary lake population (locations given in Figure 18.1). -B and -L indicate the Benthic and Limnetic species, respectively, in Paxton, Priest, and Enos lakes.

Here we summarize an experimental test of competition-induced divergence between sticklebacks (Schluter 1994). The goal of the experiment was to test whether natural selection favored divergence between two sympatric species (semi-species if reproductive isolation was weaker in the past than presently). Morphological differences between the experimental sympatric forms used were about half that seen between present-day Limnetics and Benthics, representing a relatively early stage of sympatry under both the single- and double-invasion hypotheses (Figure 18.3).

Methods

The experiment was carried out in two experimental ponds at the University of British Columbia. Each pond is 23 m × 23 m and slopes gradually to a depth of 3 m. The ponds were constructed in 1991 and seeded with plants and invertebrates from Paxton Lake. Ponds are sand-lined and edged with limestone (also from near Paxton Lake). All invertebrates found in the diets of experimental fish were characteristic of the species in the wild. Fish predators of sticklebacks were absent but insect and avian predators were common. Densities of fish were roughly those which, after an initial period in which half to two thirds inevitably die, produce growth rates similar to that seen in the wild. We feel, therefore, that pond conditions were realistic. The experiment was carried out in summer of 1993.

The design incorporated two treatments: a control in which individuals from a Solitary species (Cranby Lake) were placed alone, and an experimental in which the Solitary species was placed with a planktivore (the Limnetic from Paxton Lake) (top two panels in Figure 18.7). Each pond was divided in two with an impermeable plastic sheet and one of the treatments was assigned to each side. The prediction was that individuals in the Cranby population most similar morphologically and ecologically to the Limnetic competitor would suffer disproportionately in its presence, generating natural selection toward a more benthic lifestyle. The experiment was carried out within a single generation: fish were introduced as fry in spring and retrieved in autumn as subadults. Growth rate was the component of fitness compared among phenotypes across this time interval (see Schluter 1994 also for data on survival).

A potential problem with two-treatment designs of this form is that one cannot simultaneously control for differences between treatments in the total density of fish and in the relative frequencies of the two types (Schluter 1995b). Given this constraint, the preferred design held the density of the target (Cranby) species constant between pond sides, such that addition of Limnetics to the treatment side also increased the total density of fish there (Figure 18.7). One possible alternative design would keep density constant on the treatment side by reducing the number of Cranby individuals to compensate for the addition of Limnetics (Bernardo et al. 1995). This alternative design is the poorer choice because any responses by the Cranby form to treatment would not necessarily be the result of sympatry with Limnetics (i.e., a reduced density of Cranby fish might be the sole cause). The strength of the preferred design is that interspecific competition alone is varied between treatments (see the classic discussion of this issue in Maiorana 1977 and Werner and Hall 1977). Moreover, the hypothesis of character displacement makes an explicit prediction about which phenotypes

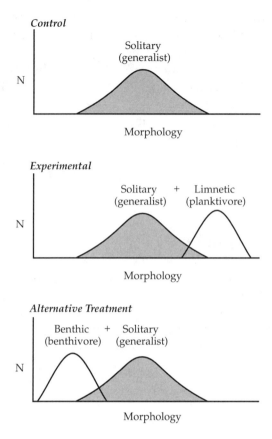

Figure 18.7. (**Top two panels**) Experimental design to test the hypothesis of character displacement. The goal was to measure whether natural selection on a Solitary species favors divergence from a closely related, morphologically similar Limnetic). Curves indicate frequency distributions of a morphological trait (e.g., gill raker number); individuals on the left are more Benthic-like, whereas individuals on the right are more Limnetic-like. The variance of the Solitary species was inflated by hybridization to allow a more sensitive measurement of changing selection pressures. In this design both treatments have the same number of individuals of the target (Solitary) species, so that interspecific competition is fully controlled, necessitating an increase in total fish density in the experimental treatment. An additional treatment (**third panel**) was not included, but could be used to test how selection on the target changes when a competitor for the alternative niche is added instead.

in the Cranby population should be most affected when Limnetics are added: growth depression should be greatest in phenotypes closest to the Limnetic and become attenuated in phenotypes progressively more different from the competitor (Schluter 1994). The experiment is therefore a valid test of this hypothesis, which is supported if the prediction is upheld. A third treatment, in which the Benthic is added instead of the Limnetic (third panel of Figure 18.7), would have been ideal to additionally test the effect of competitor species of different morphology and niche use. This was not possible because of constraints on the number of fish that could be raised.

The experiment involved one other manipulation: the target (Cranby) population was hybridized to two other species (the Limnetic and Benthic from Paxton Lake) prior to introduction, to increase levels of phenotypic variation and thereby achieve a more sensitive measurement of natural selection (Schluter 1994). The Cranby target population on both treatment and control sides was a mixture of equal numbers of offspring from three cross types: Cranby × Benthic, Cranby × Cranby, and Cranby × Limnetic. This manipulation is similar in principal to perturbations of clutch size or tail length in experimental studies of selection in birds, except that hybridization varies a suite of traits simultaneously. It is also like the whole-body manipulations of hatchling size in lizards (Sinervo et al. 1992) except that the present manipulation is genetic rather than phenotypic. Limnetics and Benthics hybridize naturally and no intrinsic reductions in F_1 hybrids have yet been detected for growth rate, viability, and fertility (McPhail 1984, 1992; Hatfield 1995). Any effects of hybridization independent of treatment are controlled for by design. Finally, if hybrids are particularly sensitive to competition, for unknown reasons, irrespective of their morphology or that of the competitor then both hybrid types (i.e., Cranby × Benthic and Cranby × Limnetic) should be affected equally when Limnetics are added, and produce a pattern of stabilizing selection on morphology rather than directional selection.

Hybridization also allowed us to overcome a potentially serious problem in selection studies where individuals cannot be measured prior to release: shifts in phenotype distribution between treatments may be caused by developmental plasticity rather than by selection. Growth and development of important trophic traits such as gill raker length and gape width were altered by diet (Day et al. 1994) but gill raker number and body armor were not. Since the Cranby species and Paxton Limnetics and Benthics differ in gill raker number and body armor, hybridization allowed us to employ a composite index of these traits that was correlated with underlying variation among individuals in the developmentally more plastic trophic traits, and could therefore be substituted for them when testing for selection (Schluter 1994).

Results

The prediction of the character displacement hypothesis was upheld (Figure 18.8). In both experimental units (ponds) growth was significantly depressed in those Cranby phenotypes closest to the Limnetic competitor in morphology and diet (Schluter 1994). In contrast, no relationship between growth and morphology was detected in the controls. This effect did not appear to be confined to individuals on the far right of the phenotype distribution (chiefly Cranby × Limnetics), but rather it diminished gradually with increasing morphological distance from the Limnetic (Schluter 1994). No significant non-linearities were detected in the form of natural selection. Moreover, the growth differential was steepest in the pond where final limnetic density was highest (due to differences in mortality rates between ponds), matching the trend between sides of each separately (Schluter 1994). The experiment thus strongly supported the hypothesis of divergent character displacement.

Controlling for interspecific competition meant that total fish densities were necessarily different between treatments (Figure 18.7). It would not have been possible to predict the results *a priori* from variation in total density *per se*, but now it is worthwhile to ask whether density differences might nevertheless be their cause. If only

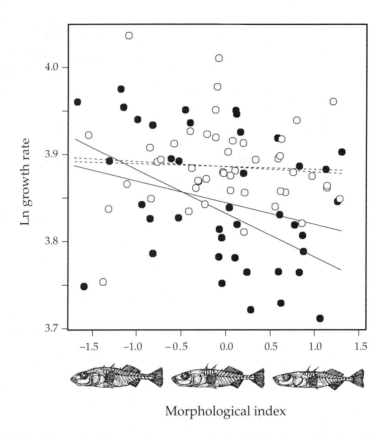

Figure 18.8. Pooled results of the pond experiment. Symbols indicate growth rates of Cranby phenotypes in the presence (filled) and absence (open) of the Limnetic species. Growth rate is measured as mm/90 days. Growth differentials (lines) are regressions of ln[growth rate] on morphology in each half-pond in the presence (solid) and absence (dotted) of the Limnetics. Growth differentials have zero slope in both controls, and slopes of −0.25 and −0.50 in the treatments. The morphological variable is a composite index representing the suite of trophic traits that distinguish sympatric Limnetic and Benthic species; more Benthic-like individuals (including Cranby × Benthic hybrids) are on the left and more Limnetic-like individuals (including Cranby × Limnetic hybrids) are on the right. For clarity, each symbol is an average of three adjacent points. From Schluter (1996a) after Schluter (1994).

the added density of a competitor mattered (and not its phenotype) then we would expect to see the same result regardless of the phenotype of the added competitor – for example, a competitor that exploits more benthos and less plankton than the Limnetic (e.g., the Cranby form, or even the Benthic [Figure 18.7]). While we cannot exclude this possibility with present data, we are unable to suggest a realistic mechanism and consider it unlikely. Growth suppression of Cranby phenotypes was directly proportional to their similarity in diet to the added Limnetic (Schluter 1994), suggesting that competition with a planktivore was the cause. Also, juvenile Limnetics are not aggressive – in the wild they forage in schools of many thousands of individuals – suggesting that food depletion rather than interference is the likely mechanism of competition.

Discussion

Molecular phylogeny of sticklebacks

The threespine sticklebacks of southwestern British Columbia bear all the classic signatures of an adaptive radiation in its very early stages. Speciation has been rapid, morphological differences are large and correlated with ecological changes, and divergence is accentuated in sympatry. Species diversity in each lake is admittedly low compared to, say, the cichlids of Lake Victoria, but this simplicity is a great advantage for ecological study. Any general theory to explain how and why adaptive radiation occurs should also apply to the sticklebacks; hence, they serve as a tractable model for testing mechanistic theories of adaptive radiation. Here we have attempted to estimate the historical (phylogenetic) relationships among populations and thereby test the single- and double-invasion models of divergence. Secondarily, we used ecological experiments to determine whether resource competition was one of the dominant ecological factors promoting divergence in sympatry.

Our mtDNA phylogeny showed that the Benthic-Limnetic species pairs of Enos, Paxton, and Priest lakes each comprises a distinct assemblage of haplotypes. Moreover, sympatric stickleback species are more similar genetically to each other than each species is to its morphological counterpart in other lakes (Figure 18.6). These results argue that divergence into Benthic and Limnetic morphologies and life-styles has occurred independently many times, and that the several sympatric species pairs have diverse origins. This conclusion is bolstered by the observation that levels of mtDNA divergence between Benthics and Limnetics vary between lakes (cf. Taylor and Bentzen 1993). The close affinity between Limnetics and Benthics from the same lake is underscored by the presence of lake-specific haplotypes possessed by both, and suggests that the two species within a lake share a more recent common ancestor than do Limnetics or Benthics from different lakes.

These results are incompatible with the double-invasion hypothesis and favor the single-invasion alternative (Figure 18.3). The lack of a clearly greater resemblance between the Limnetic and the marine (anadromous) sticklebacks than between the Benthic and the marine also weighs against the double-invasion scenario. Thus, the species pairs of *Gasterosteus* join a growing list of organisms, particularly fish, in which phylogenetic (mtDNA) evidence favors sympatric rather than allopatric speciation (Meyer et al. 1990; Taylor and Bentzen 1993; Schliewen et al. 1994; Taylor et al. 1996; Bernatchez et al. 1996; see Chapter 12).

An alternative explanation for these results, however, is that speciation in *Gasterosteus* reached an advanced stage in allopatry and that gene flow occurred subsequently following secondary contact within each lake. Under this modified double-invasion scenario, mtDNA transfer between species within each lake plus random haplotype extinction (lineage sorting) has swamped evidence of two invasions, and yielded a pattern of monophyly within lakes that is more apparent than real.

Hybridization, although uncommon today, occurs between sympatric stickleback species (McPhail 1984, 1992). Mitochondrial DNA introgression following secondary contact is also known in other postglacial fishes (Bernatchez et al. 1995) and in other animals (references in Knight et al. 1996). An instructive example is the case of limnetic-benthic species pairs of lake whitefish (*Coregonis clupeaformis* complex) occurring in three Yukon lakes (Bernatchez et al. 1996). The region is an area of overlap between two major mtDNA haplotype lineages, one endemic to Beringia (II) and the other with a Eurasian origin (III) (Bernatchez et al. 1996). In one lake the benthic species is entirely group III whereas the planktivore is predominantly group II, suggesting allopatric speciation with a trickle of gene flow from benthic to limnetic. In a nearby lake in the same tributary (Squanga Creek) of the Yukon River, however, both species are dominated by group III haplotypes. This difference between the lakes matches differences in rates of hybridization inferred from morphology (low in the first lake and moderate in the second). The third lake is in the Alsek River basin which was connected to the Yukon during the Pleistocene. All fish of both limnetic and benthic species are of group III (however, hybridization rates inferred from morphology are low in this lake).

The whitefish pattern is reminiscent of our A and B haplotype lineages (Figure 18.5). Substantial difference between species in the frequencies of these two groups was found only in Priest Lake (the Limnetic species is primarily B whereas the Benthic is mainly A). If the patterns are indeed the result of secondary contact between an A and a B lineage, one would need to imagine that the primary direction of gene flow differed among lakes, such that A haplotypes went extinct in Enos Lake and B haplotypes went extinct in Paxton.

We are unsure whether gene flow following secondary contact is a likely explanation for the mtDNA relationships. Distinguishing the gene flow hypothesis from sympatric speciation may also be difficult. Possibly, interspecific transfer of neutral genes is slowed in the nucleus if nuclear genes are more strongly linked to those coding for niche-specific adaptations (e.g., this could explain the conflict between the mtDNA results and the earlier allozyme results). In this case double invasion, if correct, might be expected to have left traces in the nucleus. We are currently exploring this possibility using microsatellites.

Competition and adaptive radiation

Our molecular results suggest that the morphological differences between Limnetics and Benthics evolved largely in sympatry and that this divergence occurred despite gene flow. Furthermore, results of the pond experiment (Figure 18.8) support earlier conclusions made from comparative data (Figure 18.2) that competition for resources is largely responsible for ecological and morphological differentiation in the species pairs. These data from sticklebacks are among the strongest evidence to date in support of the venerable hypothesis that character displacement is essential to adaptive radiation (Simpson 1944, 1953; Lack 1947; Grant 1986; Schluter 1988). The work also emphasizes the feasibility of experimental studies of natural selection and the promise they hold for testing the ecological causes of adaptive radiation in general.

With experimental measurement of growth differentials, it is also possible to approximate expected rates of evolution (Lande 1979) from estimates of additive genetic variance in the suite of morphological traits (see data in Schluter 1996b).

Under the double-invasion scenario morphological differences between the sympatric species were initially half that between modern Limnetics and Benthics (Figure 18.3). If fitness is proportional to growth rate, then even the lesser of the two slopes (Figure 18.8) is sufficient to double this initial difference and generate the present-day forms in about 250 years (Schluter 1994). This calculation is rough because it assumes that additive genetic variance in morphology remains constant, that selection intensity does not diminish with time, and that gene flow is absent. The calculation is nevertheless helpful in suggesting that the duration of sympatry (12,000 yrs) was sufficiently long for differences between Limnetics and Benthics to have evolved by character displacement. It is not yet possible to calculate whether 12,000 yrs is sufficiently long to produce Limnetics and Benthics under a single-invasion scenario. Such a calculation would require greater knowledge of the consequences of gene flow between mating types and the genetics of assortative mating.

The phylogenetic results also raise an interesting question about mechanisms of speciation, for which the experimental findings suggest a possible solution. What selective forces maintained species integrity in the face of gene flow (modified double-invasion scenario) or favored the evolution of premating reproductive isolation between Benthic-like and Limnetic-like morphological extremes within a Solitary ancestral population (single-invasion scenario)? Our results hint that the answer may lie in the mechanisms of "competitive speciation": strong trade-offs in the feeding efficiencies of different genotypes between resource environments plus resource depletion (Rosenzweig 1978; Pimm 1979; Wilson and Turelli 1986). Together these processes allow the stable coexistence of genotypes specialized for alternative resources despite hybrid competitive inferiority. Strong interspecific trade-offs are known in the sticklebacks (Schluter 1993, 1995a) and intraspecific trade-offs are known from other fish species exploiting similar benthic and limnetic environments (Robinson et al. 1996). Character displacement implies both resource depletion and competitive inferiority of intermediate phenotypes (Figure 18.8). Hybrids between Limnetics and Benthics are phenotypically intermediate (McPhail 1984, 1992; Hatfield 1995) and suffer a consumption disadvantage (Schluter 1993; Hatfield 1995).

This scenario still leaves open the question of how these ecological forces favored the origin and maintenance of premating isolation, particularly in the face of gene flow. Reasonable mechanisms include divergence of mate preference as a by-product of divergent selection (reviewed in Rice and Hostert 1993), divergent selection directly on the sensory apparatus (reviewed in Endler 1989), or reinforcement of premating isolation – particularly if disruptive sexual selection on mate preferences is involved (Liou and Price 1994). None of these seems implausible in the sticklebacks. Assortative mating is based in part on the adaptive morphological traits that distinguish the species, especially body size but probably also shape (Schluter and Nagel 1995; Nagel 1995). Mate preferences are divergent, such that F_1 hybrid males are unattractive to females of the two parental species, with the intensity of sexual selection against them depending critically on divergent mating habitat preferences (Hatfield and Schluter 1996; Vamosi 1996). The precise contributions of these various forces remains to be worked out.

Molecular and experimental studies of adaptive radiation

Our study of the stickleback adaptive radiation is far from complete, but the results thus far provide some insights into the ecological forces at work, the speed of diversification, and the likely pattern of speciation. Though preliminary, we also hope that they illustrate some of the benefits to be gained from a joint molecular and field-experimental approach. The many uses of phylogenetics in correlative (comparative) studies and in ancestor reconstruction are now well known (e.g., Harvey and Pagel 1991). Comparative study begins with assumptions about evolutionary relationships, and a molecular phylogeny can exclude many seemingly sensible scenarios for diversification. The importance of phylogenies in framing experimental studies are less well celebrated. Yet understanding the forces that cause adaptive radiation requires that stages earlier in the history of species be recreated so that forces thought to have been present in the past can be measured and tested. By narrowing the range of feasible histories of a lineage, phylogenetic methods help in the design of realistic and relevant experiments.

The reverse is also true: information on the strength of selective forces, measured experimentally, can help in the reconstruction of character-state changes through time. The task of generating a phylogeny from a set of molecular data is unlikely to be informed by field experimental results, although estimated trees are often sufficiently uncertain as to be consistent with several alternative historical trajectories, in which case understanding of mechanisms might help identify the more likely scenarios. Clearly, our phylogeny (Figure 18.6) which suggests sympatric speciation or, at the very least, secondary contact with gene flow, makes little sense in the absence of the ecological context that would favor or allow it. In this way historical hypotheses can benefit from an understanding of the ecological forces at work in nature.

Conclusions

We present preliminary results of a joint molecular and experimental study of an adaptive radiation in its early stages: the threespine sticklebacks of southwestern British Columbia. The goal of the study is to piece together the history of changes leading up to the diversity of modern forms, and to understand the ecological forces that drove them. We focus on the multiple sympatric pairs of species in small lakes. In each case, one species (the Benthic), consumes mainly prey from the littoral zone whereas the other (the Limnetic) is planktivorous.

Our phylogeny showed that each Benthic-Limnetic species pair comprises a distinct set of mtDNA haplotypes. Sympatric sticklebacks are more similar genetically to each other than each species is to its ecological and morphological counterparts in other lakes. These results suggest multiple independent origins of Benthics and Limnetics. They are also more consistent with sympatric than allopatric speciation, although an alternative hypothesis that gene flow followed secondary contact might also account for the data. A full analysis of morphological changes through time awaits a more complete phylogeny that includes more Solitary populations. Nevertheless, the findings favor the surprising conclusion that ecological and morphological divergence between sympatric species took place largely in sympatry and is maintained despite gene flow, suggesting a history of strong divergent selection.

Previous study has shown that these sympatric species are more divergent than stickleback species occurring alone in lakes (Solitary), suggesting that ecological character displacement has occurred. Our pond experiment tested directly whether competition for food favored divergence between sympatric species at a relatively early stage of diversification. The target of the experiment was a Solitary stickleback intermediate between Limnetic and Benthic in morphology and microhabitat. Addition of a planktivorous Limnetic differentially depressed growth of phenotypes in the target species most similar to it in diet and morphology, generating natural selection in favor of a more benthic life-style. This is the first experimental support for the hypothesis that competition is critical to adaptive radiation. Competition's role in morphological and ecological divergence now seems clear, and there are hints that it played a role also in speciation.

We argue that an integration of molecular-phylogenetic and field-experimental approaches is crucial to understanding adaptive radiation. Phylogenetic and historical evidence identifies the histories of lineages and helps frame realistic experiments to measure ecological forces at different stages of diversification. Experimental studies are needed to test alternative possible causes of change, and to identify the feasibility of mechanisms suggested by alternative historical scenarios.

Acknowledgments

We are grateful to P. O'Reilly for providing cloned stickleback mtDNA. N. Grabovac, T. Hatfield, L. Hummelbrunner, C. Kam, R. Kassen, J. McLean, J. Pritchard, and P. Troffe assisted in the lab and field. N. Scott and H. Diggon graciously provided access to lakes. We thank B. W. Robinson and T. J. Givnish for their comments on the manuscript. This work was funded by grants from NSERC (Canada) to all three authors.

References

Amadon, D. 1950. The Hawaiian honeycreepers (Aves, Drepaniidae). Bulletin of the American Museum of Natural History 95:157–268.

Arthur, W. 1987. *The niche in competition and evolution.* Chichester, England: Wiley.

Baumgartner, J. V. 1986. Phenotypic and genetic aspects of morphological differentiation in the threespine stickleback, *Gasterosteus aculeatus.* Ph.D. Dissertation, State University of New York, Stony Brook.

Bell, M. A. 1977. A late Miocene marine threespine stickleback, *Gasterosteus aculeatus aculeatus,* and its zoogeographic and evolutionary significance. Copeia 1977:277–282.

Bell, M. A. 1994. Paleobiology and evolution of threespine stickleback. Pp. 438–471 in *Evolutionary biology of the threespine stickleback,* M. A. Bell and S. A. Foster, eds. Oxford, England: Oxford University Press.

Bell, M. A., and Foster, S. A. 1994. Introduction to the evolutionary biology of the threespine stickleback. Pp. 1–27 in *Evolutionary biology of the threespine stickleback,* M. A. Bell and S. A. Foster, eds. Oxford, England: Oxford University Press.

Bentzen, P., and McPhail, J. D. 1984. Ecology and evolution of sympatric sticklebacks *Gasterosteus*: specialization for alternative trophic niches in the Enos Lake species pair. Canadian Journal of Zoology 62:2280–2286.

Bernardo, J., Resetarits, W. J. Jr., and Dunham, A. E. 1995. Criteria for testing character displacement. Science 268:1065–1066.

Bernatchez, L., Glémet, H., and Danzmann, R. G. 1995. Introgression and fixation of Arctic charr (*Salvelinus alpinus*) mitochondrial genome in an allopatric population of brook trout (*Salvelinus fontinalis*). Canadian Journal of Fisheries and Aquatic Sciences 52:179–185.

Bernatchez, L., Vuorinen, J. A., Bodaly, R. A., and Dodson, J. J. 1996. Genetic evidence for reproductive isolation and multiple origins of sympatric trophic ecotypes of whitefish (*Coregonus*). Evolution 50:624–635.

Borland, M. 1986. Size–assortative mating in threespine sticklebacks from two sites on the Salmon River, British Columbia. M.Sc. Dissertation, University of British Columbia, Vancouver.

Brown, W. L., Jr., and Wilson, E. O. 1956. Character displacement. Systematic Zoology 5:49–64.

Day, T., Pritchard, J., and Schluter, D. 1994. Ecology and genetics of phenotypic plasticity: a comparison of two sticklebacks. Evolution 48:1723–1734.

Dobzhansky, T. 1937. *Genetics and the origin of species*. New York, NY: Columbia University Press.

Endler, J. A. 1989. Conceptual and other problems in speciation. Pp 625–648 in *Speciation and its consequences*, D. Otte and J. A. Endler, eds. Sunderland, MA: Sinauer Associates Inc.

Felsenstein, J. 1993. *PHYLIP 3.5*. Distributed by the author.

Fisher, R. A. 1930. *The genetical theory of natural selection*. Oxford, England: Oxford University Press.

Futuyma, D. J. 1986. *Evolutionary biology, 2nd ed.* Sunderland, MA: SinauerAssociates Inc.

Gotelli, N. J., and Graves, G. R. 1996. *Null models in ecology*. Washington, DC: Smithsonian Instution Press.

Grant, P. R. 1986. *Ecology and evolution of Darwin's finches*. Princeton, NJ: Princeton University Press.

Harvey, P. H., and Pagel, M. D. 1991. *The comparative method in evolutionary biology*. Oxford, England: Oxford University Press.

Hatfield, T. 1995. Speciation in sympatric sticklebacks: hybridization, reproductive isolation and the maintenance of diversity. Ph.D. Dissertation, University of British Columbia, Vancouver.

Hatfield, T. 1996. Genetic divergence in adaptive characters between sympatric species of stickleback. American Naturalist: submitted.

Hatfield, T., and Schluter, D. 1996. A test for sexual selection on hybrids of two sympatric sticklebacks. Evolution 50:2429–2434.

Hay, D. E., and McPhail, J. D. 1975. Mate selection in the threespine sticklebacks *Gasterosteus*. Canadian Journal of Zoology 53:441–450.

Huxley, J. 1942. *Evolution, the modern synthesis*. London, England: Allen and Unwin.

Kassen, R., Schluter, D., and McPhail, J. D. 1995. Evolutionary history of threespine sticklebacks (*Gasterosteus* spp.) in British Columbia: insights from a physiological clock. Canadian Journal of Zoology 73:2154–2158.

Knight, A., Batzer, M. A., Stoneking, M., Tiwari, H. K., Scheer, W. D., Herrera, R. J., and Deininger, P. L. 1996. DNA sequences of *Alu* elements indicate a recent replacement of the human autosomal genetic complement. Proceedings of the National Academy of Sciences, USA 93:4360–4364.

Lack, D. 1947. *Darwin's finches*. Cambridge, England: Cambridge University Press.

Lande, R. 1979. Quantitative genetic analysis of multivariate evolution, applied to brain:body size allometry. Evolution 33:402–416.

Lavin, P. A., and McPhail, J. D. 1987. Morphological divergence and the organization of trophic characters among lacustrine populations of the threespine stickleback (*Gasterosteus aculeatus*). Canadian Journal of Fisheries and Aquatic Sciences 44:1820–1829.

Liou, L. W., and Price, T. D. 1994. Speciation by reinforcement of premating isolation. Evolution 48:1451–1459.

Maiorana, V. C. 1977. Density and competition among sunfish: some alternatives. Science 195:94.

Mayr, E. 1942. *Systematics and the origin of species*. New York, NY: Columbia University Press.

Mayr, E. 1954. Change of genetic environment and evolution. Pp. 157–180 in *Evolution as a process*, J. Huxley, A. C. Hardy and E. B. Ford, eds. London, England: Allen and Unwin.

McElroy, D., Moran, P., Bermingham, E., and Kornfield, I. 1992. REAP: An integrated environment for the manipulation and phylogenetic analysis of restriction data. Journal of Heredity 83:157–158.

McPhail, J. D. 1984. Ecology and evolution of sympatric sticklebacks (*Gasterosteus*): morphological and genetic evidence for a species pair in Enos Lake, British Columbia. Canadian Journal of Zoology 62:1402–1408.

McPhail, J. D. 1992. Ecology and evolution of sympatric sticklebacks (*Gasterosteus*): evidence for a species pair in Paxton Lake, Texada Island, British Columbia. Canadian Journal of Zoology 70:361–369.

McPhail, J. D. 1994. Speciation and the evolution of reproductive isolation in the sticklebacks (*Gasterosteus*) of southwestern British Columbia. Pp. 399–437 in *Evolutionary biology of the threespine stickleback*, M. A. Bell and S. A. Foster, eds. Oxford, England: Oxford University Press.

Meyer, A., Kocher, T. D., Basasibwaki, P., and Wilson, A. C. 1990. Monophyletic origin of Lake Victoria cichlid fishes suggested by mitochondrial DNA sequences. Nature 347:550–553.

Muller, H. J. 1940. Bearings of the *Drosophila* work on systematics. Pp. 185–268 in *The new systematics*, J. S. Huxley, ed. Oxford, England: Clarendon Press.

Nagel, L. M. 1994. The parallel evolution of reproductive isolation in threespine sticklebacks. M.Sc. Dissertation, University of British Columbia, Vancouver.

Nei, M. 1978. Estimation of average heterozygosity and genetic distance from a small number of individuals. Genetics 89:583–90.

Nei, M., and Miller, J. C. 1990. A simple method for estimating average number of nucleotide substitutions within and between populations from restriction data. Genetics 125:873–879.

Nei, M., and Tajima, F. 1981. DNA polymorphism detectable by restriction endonucleases. Genetics 97:145–163.

Noor, M. A. 1995. Speciation driven by natural selection in *Drosophila*. Nature 375: 674–675.

Ortí, G., Bell, M. A., Reimchen, T. E., and Meyer, A. 1994. Global survey of mitochondrial DNA sequences in the threespine stickleback: evidence for recent migrations. Evolution 48:608–622.

Pimm, S. 1979. Sympatric speciation: a simulation model. Biological Journal of the Linnean Society 11:131–139.

Rice, W. R., and Hostert, E. E. 1993. Laboratory experiments on speciation: what have we learned in 40 years? Evolution 47:1637–1653.

Ridgway, M. S., and McPhail, J. D. 1984. Ecology and evolution of sympatric sticklebacks (*Gasterosteus*): mate choice and reproductive isolation in the Enos Lake species pair. Canadian Journal of Zoology 62:1813–1818.

Robinson, B. W., and Wilson, D. S. 1994. Character release and displacement in fishes: a neglected literature. American Naturalist 144:596–627.

Robinson, B. W., Wilson, D. S., and Shea, G. O. 1996. Trade-offs of ecological specialization: an intraspecific comparison of pumpkinseed sunfish phenotypes. Ecology 77:170–178.

Rosenzweig, M. L. 1978. Competitive speciation. Biological Journal of the Linnean Society 10:274–289.

Saitou, N., and Nei, M. 1987. The neighbor-joining method: a new method for reconstructing phylogenetic trees. Molecular Biology and Evolution 4:406–425.

Sanderson, S. L., Cech, J. J., and Patterson, M. R. 1991. Fluid dynamics in suspension-feeding blackfish. Science 251:1346–1348.

Schliewen, U. K., Tautz, D., and Pääbo, S. 1994. Sympatric speciation suggested by monophyly of crater lake cichlids. Nature 368:629–632.

Schluter, D. 1988. Character displacement and the adaptive divergence of finches on islands and continents. American Naturalist 131:799–824.

Schluter, D. 1993. Adaptive radiation in sticklebacks: size, shape and habitat use efficiency. Ecology 74:699–709.

Schluter, D. 1994. Experimental evidence that competition promotes divergence in adaptive radiation. Science 266: 798–801.

Schluter, D. 1995a. Adaptive radiation in sticklebacks: trade-offs in feeding performance and growth. Ecology 76:82–90.

Schluter, D. 1995b. Criteria for testing character displacement. Science 268:1066–1067.

Schluter, D. 1996a. Ecological causes of adaptive radiation. American Naturalist 148(Suppl.):40–64.

Schluter, D. 1996b. Adaptive radiation along genetic lines of least resistance. Evolution 50:1766–1774.

Schluter, D., and McPhail, J. D. 1992. Ecological character displacement and speciation in sticklebacks. American Naturalist 140:85–108.

Schluter, D., and McPhail, J. D. 1993. Character displacement and replicate adaptive radiation. Trends in Ecology and Evolution 8:197–200.

Schluter, D., and Nagel, L. M. 1995. Parallel speciation by natural selection. American Naturalist 146:292–301.

Simpson, G. G. 1944. *Tempo and mode in evolution*. New York, NY: Columbia University Press.

Simpson, G. G. 1953. *The major features of evolution*. New York, NY: Columbia University Press.

Sinervo, B., Doughty, P., Huey, R. B., and Zamudio, K. 1992. Allometric engineering: a causal analysis of natural selection on offspring size. Science 258:1927–1930.

Taper, M. L., and Case, T. J. 1992. Coevolution among competitors. Evolutionary Biology 8:63–109.

Taylor, E. B., and Bentzen, P. 1993. Evidence for multiple origins and sympatric divergence of trophic ecotypes of smelt (*Osmerus*) in northeastern North America. Evolution 47:813–832.

Taylor, E. B., Foote, C. J., and Wood, C. C. 1996. Molecular genetic evidence for parallel life-history evolution in a Pacific salmon (sockeye salmon and kokanee, *Oncorhynchus nerka*). Evolution 50:401–416.

Vamosi, S. 1996. Postmating isolation mechanisms between sympatric populations of three-spined sticklebacks. M.Sc. Dissertation, University of British Columbia, Vancouver.

Werner, E. E., and Hall, D. J. 1977. Density and competition among sunfish: some alternatives. Science 195:94–95.

Wilson, D. S., and Turelli, M. 1986. Stable underdominance and the evolutionary invasion of empty niches. American Naturalist 127:835–850.

Wooton, R. J. 1976. *The biology of the sticklebacks*. New York, NY: Academic Press.

Wright, S. 1940. The statistical consequences of Mendelian heredity in relation to speciation. Pp. 161–163 in *The new systematics*, J. S. Huxley, ed. Oxford, England: Clarendon Press.

Zuiganov, V. V., Golovatjuk, G. J., Savvaitova, K. A., and Bugaev, V. F. 1987. Genetically isolated sympatric forms of threespine stickleback, *Gasterosteus aculeatus*, in Lake Azabachije (Kamchatka peninsula). Environmental Biology of Fishes 18:241–247.

19 PHYLOGENETIC STUDIES OF CONVERGENT ADAPTIVE RADIATIONS IN CARIBBEAN *ANOLIS* LIZARDS

Todd Jackman, Jonathan B. Losos, Allan Larson, and Kevin de Queiroz

A clade may be called an adaptive radiation if its constituent lineages have diversified to utilize a number of different aspects of the environment (Simpson 1953). Contrary to some recent discussions of adaptive radiation, the relevant criterion is not the number of species, but the adaptive disparity among these species; even clades with relatively few species can constitute an adaptive radiation if the species demonstrate considerable ecological and morphological disparity (see Chapters 1 and 8). Of course, many of the classic cases of adaptive radiation (e.g., Australian marsupials [Chapter 4], African rift lake cichlids [Chapter 12]) are clades that exhibit extensive amounts of both adaptive diversification and speciation.

Caribbean anoline lizards present a particularly interesting example of adaptive radiation. They entail not one, but four separate instances of radiation, having diversified largely independently on each island of the Greater Antilles. Multiple independent radiations within a clade associated with different geographic regions are not uncommon (e.g., cichlids in each of the African Rift Lakes, or marsupials in Australia and South America); what is striking about the Caribbean anoles is that these independent evolutionary theaters have produced extraordinarily similar radiations, both in terms of the current diversity of adaptive forms and the apparent evolutionary pathway to those forms.

Molecular systematics has contributed greatly to our understanding of the anole radiations. Subsequent to Etheridge's (1960) osteological study (the first modern treatment of anoles), almost every newly developed technique has been applied to this group, including methods based on karyotypes (Gorman and Atkins 1969; Gorman and Stamm 1975; Williams 1989), allozymes (Yang et al. 1974; Gorman and Kim 1976; Gorman et al. 1980a, 1983; Burnell and Hedges 1990), immunological distance (Gorman et al. 1980b; Wyles and Gorman 1980a,b; Shochat and Dessauer 1981; Gorman et al. 1984; Hass et al. 1993), and DNA sequences (Hass et al. 1993). In spite of these studies, a robust phylogenetic hypothesis for the entire group has proven elusive. Indeed, higher-level anoline relationships were left largely unresolved by osteological, karyotypic, and immunological work conducted through 1989 (Figure 19.1; see Cannatella and de Queiroz 1989 and Williams 1989). However, new phylogenetic methods and new data – especially DNA sequences – have recently begun to clarify some of the problematical aspects of anoline phylogeny.

This chapter will attempt to illustrate the utility of molecular systematics as part of an integrative research program aimed at unraveling the patterns and processes

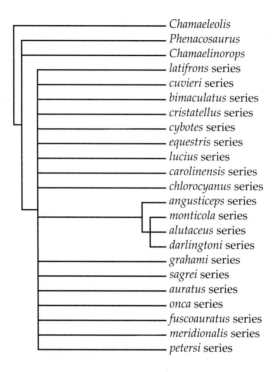

Figure 19.1. Consensus tree based on taxonomic congruence of studies of anoline relationships based on osteological, immunological, and karyotypic data (Cannatella and de Queiroz 1989). Non-capitalized names represent *Anolis* species series. Assignment of West Indian *Anolis* to species groups is not controversial; the relationships among these groups are disputed.

operating in the anoline radiation. In particular, we hope to demonstrate how molecular techniques can provide the historical phylogenetic framework upon which all studies of adaptation must be based (see Lauder 1981; Greene 1986; Coddington 1988; Baum and Larson 1991; Arnold 1994; Larson and Losos 1996). We first discuss recent inferences regarding anole phylogeny based on new DNA data. Then we address three issues in anoline evolution into which molecular systematics provides powerful insights: patterns of adaptive evolution, the significance of key innovations, and patterns of community evolution.

Natural history of Caribbean anoles

In terms of species diversity and sheer biomass, *Anolis* is one of the two dominant vertebrate groups in the Caribbean (the other is the frog genus *Eleutherodactylus*). Currently, 139 species of Caribbean anoles are recognized (Powell et al. 1996) and new species are being discovered every year (e.g., Estrada and Hedges 1995; Diaz et al. 1996). The vast majority (111) of these species occur on the islands of the Greater Antilles (Cuba, Hispaniola, Jamaica, and Puerto Rico; Figure 19.2). On each of these islands, lineages have adaptively diversified, resulting in species occupy-

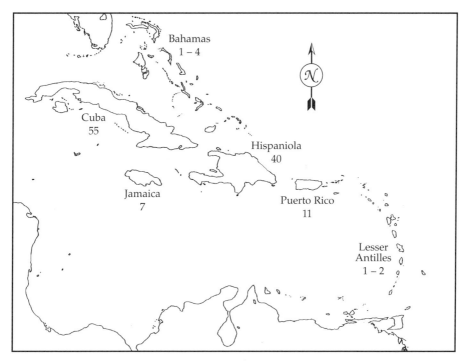

Figure 19.2. Number of anole species (including *Chamaeleolis* and *Chamaelinorops*) per island in the Caribbean.

ing a wide variety of ecological niches and as many as twelve species occurring sympatrically.

Studies of anoline assemblages have played an important role in the development of theories of community ecology (e.g., Schoener 1968, 1974; Roughgarden 1974). A wealth of data suggests that interspecific competition among anoles plays an important role in structuring anoline assemblages (see review by Losos 1994). In particular, co-occurring species almost invariably differ in either *body size* (correlated with prey size [Roughgarden 1974; Schoener 1968; Schoener and Gorman 1968]), *structural habitat* (e.g. perch height and diameter), or *microclimate* (shade vs. open sun) (Rand 1964, 1967; Schoener and Schoener 1971a,b; Moermond 1979a,b). For example, eight species of anoles occur at the Luquillo Experimental Forest in Puerto Rico. Three species occur high in the trees, of which one is small-bodied, one intermediate, and one large. Two other species perch on tree trunks and forage on the ground; one occurs in open, sunlit habitats, the other in closed forest. Of the two species that use grassy habitats and shrubs, one is found usually in the open and the other in shaded microhabitats. The last species is very small and is found primarily on twigs, probably at all heights.

Both experimental and unintentional manipulations, as well as observations of natural systems, indicate that species alter their behavior and habitat use depending on whether another species is present. In Jamaica, for example, *Anolis opalinus* perches higher when *A. lineatopus* is present than when it is absent (Jenssen 1973). Similarly,

Table 19.1. Morphological characteristics of anole ecomorph classes. Data from Losos (1990a, 1992, unpubl.); all lengths are in mm. Morphological differences are discussed in Glossip and Losos (1997) and Beuttell and Losos (unpubl.). Lamellae tabulated are those underlying the third and fourth phalanges of pedal digit IV.

Ecomorph	Snout-vent length	Shape	Hindlimb length	Forelimb/hindlimb length ratio	Tail length	Color	Number of lamellae
Crown-giant	>120	intermed.	intermed.	intermed.	intermed.	green or brown	28-42
Grass-bush	35-50	slender	long	low	very long	brown with yellow stripe	13-21
Trunk	40-60	intermed.	mod. long	mod. high	short	gray	17-18
Trunk-crown	45-80	intermed.	intermed.	intermed.	intermed.	green	19-30
Trunk-ground	50-75	stocky	long	low	long	brown or gray	15-21
Twig	35-85	very thin	very short	high	short	white	15-28

Schoener (1975) showed that the four most geographically widespread anoles in the Caribbean – *A. carolinensis, A. distichus, A. grahami,* and *A. sagrei* – all adjust their perch height depending on the other species with which they co-occur at a given locality.

The evolutionary effects of such habitat shifts become apparent when populations differing in habitat use are compared. Habitat differences among populations of *A. carolinensis* and *A. sagrei* are correlated with differences in the number of subdigital lamellae on the toepad (Lister 1976) and hindlimb length (Losos et al. 1994). Larger differences of the same kind are associated with habitat differences among sympatric species in the Greater Antilles (see below).

When comparing the anoles found on the Greater Antilles, the most striking observation is that the same set of ecological types – termed "ecomorphs" (Rand and Williams 1969; Williams 1972) – occur on each of these islands. Six ecomorph classes have been identified (Table 19.1). Each is composed of a set of morphologically and behaviorally similar species that utilize a similar micro-environment. The ecomorphs are named for the part of the arboreal or terrestrial environment they most frequently use: trunk-ground, trunk, trunk-crown, crown-giant, twig, and grass-bush (Plate 4). Morphometric analyses indicate that members of the same ecomorph class cluster in morphological space (Figure 19.3; Losos 1992; Losos and de Queiroz 1997). Several studies show that members of a given ecomorph class are similar also in ecology and behavior (Losos 1990 a,b; Irschick and Losos 1996).

Williams (1972) proposed that the ecomorphs evolved as a result of competition-driven resource partitioning. This study, two decades ahead of its time, used a phylogeny for Puerto Rican anoles based on osteological and karyotypic data to infer patterns of morphological and ecological evolution. Williams suggested that the original colonizing species on Puerto Rico was morphologically unspecialized. As new species were added to the island by speciation, competition led to resource partitioning, which in turn led to the evolution of the different ecomorphs. A re-analysis of this hypothesis conducted 20 years later (Losos 1992) and discussed below, confirmed most of Williams' phylogenetic interpretations.

Three other important aspects of the Caribbean anoline radiation must be mentioned. First, not all ecomorphs are represented on each island. Cuba and Hispaniola have all six ecomorph classes, Puerto Rico has all but the trunk anoles, and Jamaica has all but the grass-bush and trunk anoles. Thus, sets of ecomorphs on the different islands are nested, with the smaller islands having subsets of the ecomorphs found on the larger ones. In addition, not all Greater Antillean species belong to one of the six ecomorph classes. However, ecomorph species are numerically dominant in

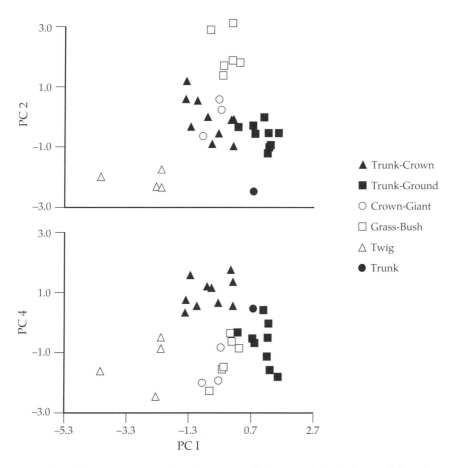

Figure 19.3. The position of the Greater Antillean ecomorphs in a morphological space defined by a principal components analysis (PCA). Variables in the analysis were body mass, hind- and forelimb length, tail length, and number of lamellae underlying the third and fourth phalanges of pedal digit IV. The effect of size was removed from each of these variables by using residuals from regressions against snout-vent length (SVL); these regressions used species means. In addition, SVL was also included in the PCA. PC 3 loads only for SVL and is not shown; only crown-giants are clearly distinguished from other ecomorphs on the basis of this axis. Figure from Losos and de Queiroz (1997).

almost all sites within the Greater Antilles; in fact, at many localities, particularly in the lowlands (and in all of Puerto Rico), only species categorized as ecomorphs can be found. Relatively few Greater Antillean species (21 of 111, including species in Cuban *Chamaeleolis* and Hispaniolan *Chamaelinorops*) cannot be placed into one of the standard ecomorph classes. These non-ecomorph species occupy distinctive ecological niches (e.g., streams) and are morphologically distinct, and do not have parallels on other islands (Williams 1983).

Second, the anoline adaptive radiation entails diversification not only in structural habitat, but in thermal microclimate as well. Some species have adapted to cool montane climates, whereas others occur in hot and arid habitats (e.g., Ruibal 1961; Huey and Webster 1976; Hertz 1980, 1981). Divergence in microclimate has occurred independently of adaptation to structural habitats, but is more common in species of certain ecomorph classes (e.g., trunk-ground anoles, grass-bush anoles) than others (Losos 1994 and references therein). Adaptation to different microclimates apparently permits members of the same ecomorph class to occur sympatrically through a partitioning of different thermal conditions (Schoener and Schoener 1971a,b; Huey and Webster 1976); indeed, this thermal partitioning is so well defined that individuals from two species may occupy the same tree at different times of the day, depending upon whether the tree is in the shade or in the sun (Schoener 1970; Huey and Webster 1976).

Third, in addition to *Anolis*, three other genera of anoles are commonly recognized, though only two of them occur in the West Indies. *Phenacosaurus* contains a small number of South American species that are similar to Greater Antillean twig anoles (Miyata 1983). Members of the Cuban genus *Chamaeleolis* (4 recognized spp.) are also superficially similar to twig anoles, but much larger (Hass et al. 1993). By contrast, *Chamaelinorops* (1 sp.) from Hispaniola is distinct from all other anoles in morphology, ecology, and behavior (Forsgaard 1983; Jenssen and Feely 1991; Flores et al. 1994). Etheridge (1960) concluded that all three of these genera are part of *Anolis*, but recent workers have placed some or all of them outside of *Anolis* (e.g., Guyer and Savage 1986, 1992; Williams 1989). Recent molecular studies have cast doubt on the latter hypothesis, suggesting that *Chamaeleolis* and *Chamaelinorops* fall within *Anolis* (Wyles and Gorman 1980a; Burnell and Hedges 1990; Hass et al. 1993; but see Case and Williams 1987).

Analysis of the anole radiation

Molecular systematics

We investigated anoline phylogeny by examining 1,455 base pairs of DNA encoding the mitochondrial *ND2* gene and five mitochondrial transfer RNAs (Figure 19.4; Jackman et al. [unpubl.] provide details of the sequencing protocol). We included 53 anoles as well as 2 outgroups, including all recognized anoline genera; although we do not follow Guyer and Savage (1986, 1992) in splitting *Anolis* into four genera, we have included representatives of all of their taxa. Although we included only 31 of 90 ecomorph species in this analysis, the phylogenetic affinities of the remaining ecomorph species not included in this study are clear. For example, the *A. alutaceus* series

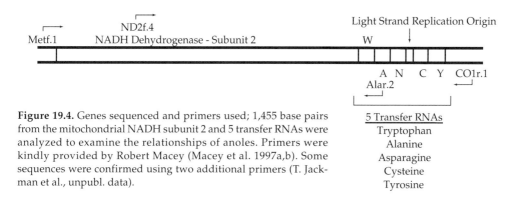

Figure 19.4. Genes sequenced and primers used; 1,455 base pairs from the mitochondrial NADH subunit 2 and 5 transfer RNAs were analyzed to examine the relationships of anoles. Primers were kindly provided by Robert Macey (Macey et al. 1997a,b). Some sequences were confirmed using two additional primers (T. Jackman et al., unpubl. data).

5 Transfer RNAs
Tryptophan
Alanine
Asparagine
Cysteine
Tyrosine

of Cuba contains 15 very similar species, all of which are grass anoles; two members of this series, *A. alutaceus* and *A. vanidicus*, are included in our study. Similarly, the *A. cybotes* series of Hispaniola contains eight trunk-ground anoles that are morphologically very similar; two of these (*A. marcanoi*, *A. strahmi*) are included in our analysis. Although these groups were initially defined on the basis of morphological evidence, immunological and electrophoretic studies generally support the close relationships of member species (Wyles and Gorman 1980a,b; Hass et al. 1993). Only one ecomorph series (the Hispaniolan *A. sheplani* lineage of twig anoles) is not included in our sample; all other ecomorph species are represented by a closely related species belonging to the same ecomorph class. Hence, our analysis of patterns of ecomorph evolution is unlikely to be compromised by partial sampling of species.

To evaluate specific phylogenetic hypotheses in a statistical framework, we used the Wilcoxon signed-ranks test (Templeton 1983). This test asks whether an alternative phylogenetic hypothesis is significantly less parsimonious than the most parsimonious topologies. The trees are compared character-by-character to investigate whether differences in the number of steps required for each character differs between the alternative trees. The differences in the numbers of steps between alternative phylogenetic hypotheses are then ranked and these rankings are used as the basis for the test (see Larson 1994 for a detailed description). For example, to test the hypothesis that *Anolis* is monophyletic, we compared the most parsimonious tree recovered from our data to the shortest tree produced from our data in which *Anolis* was monophyletic.

Figure 19.5 presents the most parsimonious tree derived from our DNA sequence data, using an unweighted analysis. Both *a priori* and *a posteriori* tests for saturation of substitutions (multiple substitutions occurring at the same site between lineages being compared) were employed (T. Jackman et al., unpubl. data). *A priori* tests consisted of comparisons of transition-to-transversion ratios at different site categories (Holmquist 1983), as well as examination of maximum likelihood distances plotted against the number of differences (Moritz et al. 1992). *A priori* tests suggested that only silent transitional changes may be saturated with mutations. An *a posteriori* test for differences in homoplasy (de Queiroz 1989; Larson 1994) showed a significant

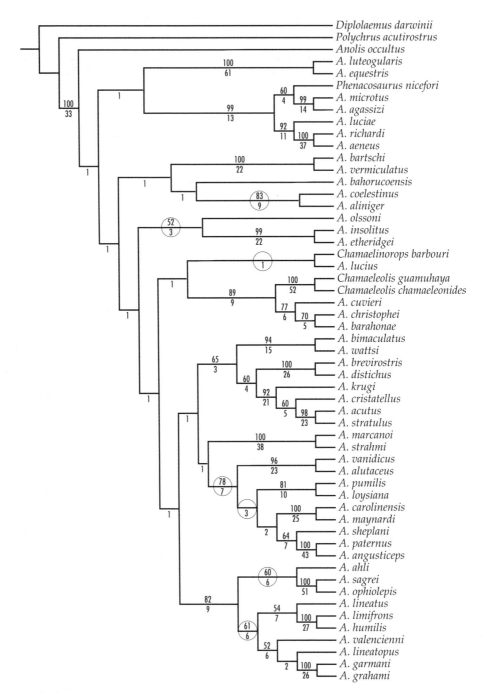

Figure 19.5. The most parsimonious tree based on an unweighted maximum parsimony analysis of DNA sequences (T. Jackman et al., unpubl. data). Bootstrap values (Felsenstein 1985a) greater than 50% are above and to the left of the nodes; decay index values (Bremer 1988) are below and to the left of the nodes. The length of the tree is 8,889 steps, based on 866 informative characters. The consistency index is 0.38. Circles indicate nodes for which support increases (≥ 5% for bootstrap analyses, ≥ 2 steps for decay analysis) when silent transitions are omitted.

increase in the number of inferred homoplasies for silent transitions. The removal of silent transitions significantly affected the tree only by improving support for the nodes circled in Figure 19.5. Consequently, we concluded that the signal-to-noise ratio in this data set is fairly high, and that the high levels of noise in silent transitions do not have a major effect on tree topology.

Several significant conclusions follow from the most parsimonious tree (Figure 19.5). First, *Anolis* is paraphyletic with respect to the other three anole genera, and we are able to reject the hypothesis that *Anolis* is monophyletic (Table 19.2). We were also able to reject the hypotheses that *Anolis* is monophyletic with respect to the other two Caribbean genera alone (Table 19.2).

Second, considerable inter-island movement has occurred during the evolution of *Anolis*. If we use parsimony to reconstruct the distribution of ancestral forms using MacClade (Maddison and Maddison 1992), Hispaniola appears as the ancestral location for many of the deepest nodes in the phylogeny (Figure 19.6). Thirteen lineages have engaged in inter-island dispersal: six to Cuba, three to Puerto Rico, one to Jamaica (not including a recent colonization of Jamaica by the Cuban *A. sagrei* species group [Williams 1969]), two to Hispaniola (one ancient and one relatively recent), and one to Central America. Contrary to previous discussions (e.g., Williams 1969; Guyer and Savage 1986), our analysis indicates that Central America was reinvaded by a lineage of Caribbean anoles.

Table 19.2. Tests of phylogenetic hypotheses using the Wilcoxon signed-rank test.

Hypothesis	T_s^1	n^2	z^3	Probability
Anolis is monophyletic[4]	10,545	234	3.78	< 0.001
Anolis is monophyletic relative to *Chamaelinorops* and *Chamaeleolis* – alpha and beta subsections monophyletic[5]	7,398	207	2.04	< 0.05
Phenacosaurus is sister to the other anoline genera	10,210	210	1.53	NS
All ecomorph classes are monophyletic	10,184	375	11.93	< 0.001
Species on each island in the Greater Antilles form a monophyletic radiation[6]	9,501	233	4.01	< 0.001

[1]Wilcoxon signed-ranks statistic.
[2]Number of characters that differed in numbers of changes on the two trees.
[3]Normal approximation when n > 100 (Zar 1984).
[4]The scenario of Guyer and Savage (1986).
[5]The scenario of Williams (1969); all *Anolis* are traditionally placed in the alpha or beta subsections.
[6]This hypothesis can also be rejected for each island individually except Jamaica.

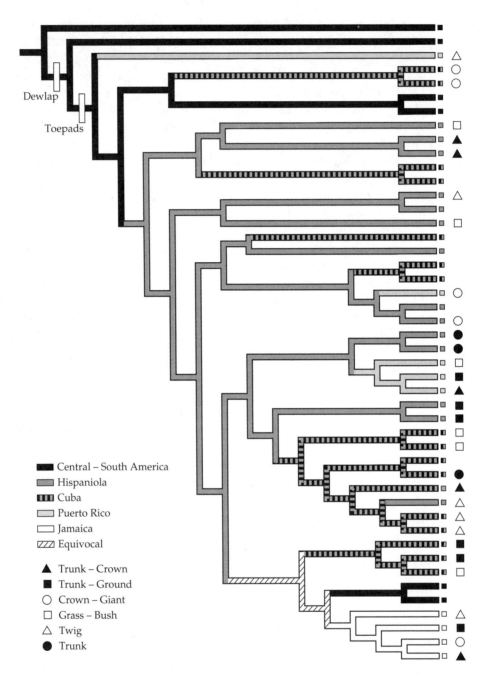

Figure 19.6. Geographic relationships and ecomorph evolution in Caribbean anoles. Phylogeny is the same as in Figure 19.5, but taxa from small islands (e.g., northern and southern Lesser Antilles, St. Croix, Malpelo) are omitted. Species names are replaced by ecomorph symbols. Ancestral states for geographic location were reconstructed using parsimony; character-states (= geographic locations) were considered unordered. Omission of some taxa does not alter the reconstruction because all small islands represent autapomorphic character-states. *Anolis carolinensis* was treated as if it were from Cuba (it actually occurs in Florida, the Bahamas, and elsewhere) because it is closely related to and surely derived from the Cuban species *A. porcatus* (Buth et al. 1980). Evolution of the anole dewlap and expanded subdigital toepads, as inferred by parsimony, is indicated by open bars. Both features have been secondarily lost in several taxa (not shown).

Given the apparent evolutionary lability of geographic distribution, perhaps one should not give too much credence to the particular reconstruction based on our present molecular tree. Alternative geographic scenarios account for current species distributions nearly as well. For example, a reconstruction in which Cuba is the ancestral location for all Caribbean anoles and has never been re-invaded requires only three additional steps. Nevertheless, our molecular phylogeny can rule out one extreme hypothesis, that each island represents a monophyletic radiation (Table 19.2). Furthermore, although the geologic history of the Caribbean is still controversial (Hedges 1996; Crother and Guyer 1996), it is difficult to see how a simple vicariance scenario could explain the distribution of lineages among islands. Explaining patterns such as this will require complex scenarios that may incorporate elements of dispersal, extinction, and restricted ancestral geographic ranges. The extent to which these scenarios will also include a vicariance component remains to be determined (see Hedges 1996; Crother and Guyer 1996).

Third, the tree reveals a number of poorly supported nodes at the deepest levels of the phylogeny. Lack of resolution could result from substitutional saturation that obscures phylogenetic signal, or from a rapid successive branching of lineages. Many evolutionary biologists have suggested that species colonizing depauperate regions may diversify rapidly (e.g., Simpson 1953; Rensch 1959); in this situation, branching of numerous lineages would occur nearly simultaneously on an evolutionary time scale.

For several reasons, we believe the anoline situation most likely reflects rapid, successive branching of lineages (T. Jackman et al., unpubl. data). Within the anoles, 866 informative characters exist (772 with silent transitions removed). As noted above, saturation of substitutions can be shown only for silent transitions. Moreover, short reconstructed branches characterize the deep nodes in the tree regardless of the methods used to reconstruct the branches or whether silent sites are included or excluded. Furthermore, branch lengths from trees constructed using only amino acid and tRNA transversion characters (the most conservative sets of characters) were similarly short. Tests involving subsampling of taxa also demonstrate that a rapid radiation occurred between the major anoline lineages.

Other data sets that have been used to examine anoline relationships suffer from a similar lack of resolution deep in the tree (Cannatella and de Queiroz 1989; Burnell and Hedges 1990; Hass et al. 1993). A combined analysis of species in common between our data set, 16S DNA data (Hass et al. 1993), allozyme data (Burnell and Hedges 1990), and morphological data produced a phylogenetic hypothesis almost entirely congruent with the trees presented here (T. Jackman et al., unpubl.). An immunological data set (Hass et al. 1993) could not be included in the combined analysis because the data are in the form of distances rather than discrete characters. Nonetheless, the phylogeny produced in the analysis by Hass et al. (1993), involving eight taxa, agreed with our DNA study in the close relationships of several taxa (e.g., *Chamaeleolis* and *A. cuvieri*, and members of the *A. cristatellus* series), as well as in indicating many short internal branches. The immunological analysis also suggested some discrepant relationships among the shorter branches of the tree, but these relationships were not robust when the immunological data were analyzed using different methods (Hass et al. 1993).

Adaptation

Integral to the study of adaptive radiation is the study of adaptation itself; a clade constitutes an adaptive radiation only if lineages within the clade have adapted to different ecological roles. Here we follow Gould and Vrba (1982, p. 5) in defining an adaptation as a feature that was "built by natural selection for the function it now performs."

Investigating the adaptive status of a feature thus requires information on its evolutionary history, which necessitates a phylogenetic perspective. The hypothesis of adaptation implies that (1) a trait currently provides an advantage at some task (such as locomotion or feeding) that itself promotes survival or fecundity; and that (2) this trait and its associated advantage in performance arose in a lineage experiencing a selective regime that would also have favored evolution of the trait. If such a trait instead arose in a lineage whose descendants only later entered a selective regime favoring that trait, then the trait would constitute an "exaptation" (Greene 1986; Coddington 1988; Baum and Larson 1991; Arnold 1994; Larson and Losos 1996).

Studies on anoline adaptation have focused on two traits: subdigital toepads (Figure 19.7) and the length of the hindlimbs. Because toepads are widely distributed in anoles and are absent in *Polychrus* and all more distant outgroups, we may conclude that toepads evolved in the ancestral anole (Figure 19.6), although they have been secondarily reduced or lost in some taxa (Peterson and Williams 1981). Because *Polychrus*, like *Anolis*, is arboreal, we may conclude also that toepads initially evolved in an arboreal species. Laboratory studies indicate that toepads provide clinging ability on smooth surfaces, similar to those of leaves and the bark of some trees, and that species with larger pads have greater clinging ability (Irschick et al. 1996). Hence, the conclusion that toepads evolved as an adaptation to enhance arboreal capabilities seems reasonable. This proposition is further bolstered by the observation that similar pads have evolved convergently in two other clades of arboreal lizards, geckos, and praesinohaemid skinks (discussed in Larson and Losos 1996).

Toepad characteristics vary among anoles, especially in the size of the pad and the number of transversely expanded scales (termed lamellae) that form the pad. Based on the functional studies cited above (Irschick et al. 1996), larger pads probably confer greater clinging ability. It seems likely that species with more lamellae may have greater ability to mold the toepad to narrow or irregular surfaces, but this hypothesis requires additional functional data.

Comparisons among ecomorph classes indicate that they differ systematically in pad size and number of lamellae, even after the effect of body size is removed (Table 19.3). Because members of the same ecomorph class are not necessarily closely related (see below), these results indicate that similar morphological features have evolved independently in species occupying similar habitats. Such parallel evolution has been considered strong evidence of adaptation; this conclusion is bolstered when the functional consequences of the trait are understood as well.

Variation among species in hindlimb length relative to body size also appears adaptive. Species in different ecomorphs differ systematically in relative hindlimb length (Table 19.3). The functional consequences of differences in limb length are well understood: species with longer limbs have greater running capabilities on broad surfaces (Losos 1990b). On narrow surfaces, long-legged species run no faster than short-legged species but stumble more frequently (Losos and Sinervo 1989; Losos and Irschick 1996).

Hence, one might hypothesize that long legs are an adaptation for moving rapidly on broad surfaces, whereas short legs are adaptive for moving without difficulty on narrow surfaces. In support of this hypothesis, behavioral studies indicate that longer-legged species run more frequently (Table 19.4) and run faster in nature than shorter-legged species (D. Irschick and J. Losos, unpubl. data). Comparisons among species indicate a relationship between relative hindlimb length and both mean perch diameter and mean perch height (Table 19.4); the relationship is negative in the height-hindlimb regression. In a multiple regression involving both perch diameter and height, more of the variance in relative hindlimb length is explained and both variables are individually more significant (Table 19.4).

We interpret this comparative finding as demonstrating that species that use broader surfaces tend to have longer legs, as predicted from the functional studies. Terrestrial species tend to have longer legs than more arboreal species, presumably because they frequently use the broadest surface of all, the ground. This difference in relative

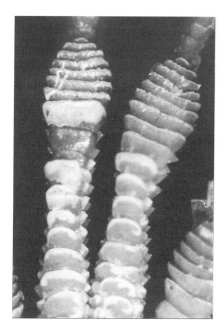

Figure 19.7. Toepads of *A. sagrei* from Cuba. Each toepad consists of a number of transverse scales, termed lamellae. Each lamella is covered with millions of microscopic hairs, termed setae. Adhesion is thought to occur as a result of the intermolecular forces generated between electrons on the setae and on the surface (Cartmill 1985; Irschick et al. 1996 and references therein).

Figure 19.8. Male of *A. lineatopus* extending dewlap.

hindlimb length is perhaps most vivid between twig anoles and grass-bush anoles (see Table 19.1), which utilize similar mean perch diameters (d_{tw} = 2.0 cm vs. d_{gb} = 2.4 cm) but differ strikingly in their use of broad surfaces. Twig anoles usually perch at 2 m or more in height and rarely come to the ground; grass-bush anoles do much of their foraging and interacting on the ground and run considerably more frequently (31% of movements vs. 9% in twig anoles; see Losos 1990a, 1992, unpubl. data). Hence, we conclude that limb length has evolved adaptively with respect to habitat use in anoles. The probable functional explanation underlying this adaptive evolution relates to the differing demands for movement on surfaces of different breadths.

Key innovations

A key innovation can be viewed as a character that permits members of a clade to interact with the environment in a manner unavailable to members of its sister taxon that lacks that key character (Larson et al. 1981). If the key innovation provides evolutionary access to many adaptive zones, then it may lead to an adaptive radiation, but the two are not necessarily linked: adaptive radiation can occur in the absence of the evolution of a key innovation, as when a species colonizes an environment with many unutilized resources; conversely, a key innovation can occur

Table 19.3. Phylogenetically structured analyses of variance for toepad area, number of lamellae, and hindlimb length across ecomorphs.[1] The effect of body size has been removed by analyzing the residuals after regressing each variable against snout-vent length, using mean values for each species. Data are drawn from Losos (1990a, 1992, unpubl. data), Glossip and Losos (in press), and K. Beuttell and J. Losos (unpubl. data). Degrees of freedom (d.f.) differ because data for toepad area are available for fewer species.

Variable	F	d.f.	Non-phylogenetic P[2]	No. of simulations producing F values greater than observed	
				Speciational model	Gradual model
Pad area	38.79	5,16	< 0.001	0/1000	0/1000
# lamellae	3.07	5,23	0.029	53/1000	65/1000
Hindlimb length	12.65	5,23	< 0.001	0/1000	0/1000

[1]Because closely related species may be phenotypically similar, statistical analyses such as these should be conducted within a phylogenetic context. Garland et al. (1993) provided a means of conducting phylogenetically structured analyses of variance by simulating character evolution on the phylogeny to establish a null distribution of F-statistics, with which to compare observed ANOVA values. Using this method, we find that an F-statistic as great as that seen in the real data is rarely achieved in the simulations for any of the variables; we thus conclude that statistical differences exist among the ecomorphs for these variables. We assumed two modes of character evolution: (1) the **speciational model** (*sensu* Garland et al. 1993) assumes that the amount of character change along each branch is drawn from the same distribution; (2) the **gradual model** assumes that the expected amount of change along on a branch is proportional to the length of that branch in time. To calculate branch lengths, we fitted the DNA data to a molecular clock model (T. Jackman et al., unpubl. data).

[2]Probabilities from ordinary, phylogenetically unstructured ANOVA.

Table 19.4. Analyses of covariation in run frequency and hindlimb length with perch height and diameter, using the method of independent contrasts (Felsenstein 1985b; Garland et al. 1992). Phylogenies for each analysis were generated from Figure 19.5 by deleting taxa for which data were unavailable. Each analysis was performed twice, using speciational and gradual models of evolutionary change (see Table 19.3). Contrast standardization was investigated using the method of Garland et al. (1993); all contrasts were found to be appropriately standardized. All regressions were conducted through the origin (Garland et al. 1992).

Dependent variable	Independent variable(s)	r^2	F	d.f.[1]	P
Gradual model:					
Run frequency	Hindlimb length	0.48	12.80	1,14	0.003
Hindlimb length	Perch diameter	0.12	4.16	1,30	0.050
Hindlimb length	Perch height	0.13	4.41	1,30	0.044
Hindlimb length[2]	Perch diameter, height	0.27	6.03	2,29	0.006
Speciational model:					
Run frequency	Hindlimb length	0.52	15.13	1,14	0.002
Hindlimb length	Perch diameter	0.14	4.81	1,30	0.036
Hindlimb length	Perch height	0.19	7.09	1,30	0.012
Hindlimb length[2]	Perch diameter, height	0.34	8.20	2,29	0.002

[1] Degrees of freedom differ because ecological data are available for more species than are locomotor behavior data. In the ecological analyses, data for *A. porcatus* were substituted for those from *A. carolinensis* and data for *A. guazuma* were substituted for *A. paternus*. In the locomotor behavior analyses, substitutions were *A. cybotes* for *A. marcanoi* and *A. chlorocyanus* for *A. coelestinus*. Because these species pairs are closely related and belong to the same ecomorph class, substitution of these species in the phylogenies used in the statistical analyses is appropriate. Data from Losos (1990a, 1992, unpubl. data) and Irschick and Losos (1996).
[2] In the multiple regressions, the significance values for the independent variables in the gradual and speciational analyses, respectively, are: 0.013 and 0.003 for perch height, and 0.014 and 0.009 for perch diameter.

without leading to adaptive radiation, as when members of a lineage interact with the environment in a new manner, but the lineage does not diversify. In the case of anoles, toepads may constitute a key innovation linked to adaptive radiation. The toepads of anoles allow them to use a variety of arboreal niches in a manner not possible in ancestral arboreal lizards lacking pads.

A concept often confounded with key innovation is the idea that some characters may promote speciation or retard extinction and thus be responsible for the species richness (or poverty) of a clade (see discussion in Heard and Hauser [1994] and references therein). With more than 300 described species, the anoline clade is large relative to its sister taxon *Polychrus* and to most other recognized lizard clades. Exceptions include a number of clades (e.g., Gekkonidae, Scincidae) that appear to be considerably older; this latter statement must be evaluated cautiously at present, however, because we lack phylogenetic and temporal information on many of the most relevant clades, particularly within the Iguania. One might ask whether the evolution of any particular character is responsible for the striking species richness of anoles.

The dewlap, an extensible fold of skin located on the throat, also characterizes the clade containing anoles plus *Polychrus* (see Plate 4, Figure 19.8). Male anoles (and females of some species) display their dewlaps in many contexts, including aggression, courtship, and predator deterrence (Leal and Rodríguez-Robles 1997). Species differ in the size, color, and patterning of their dewlaps (see Williams and Rand 1977). Observations (Rand and Williams 1970; Echelle et al. 1971; Williams and Rand 1977) and experiments (Losos 1985) suggest that attributes of the dewlap are important in species recognition. Consequently, factors that promote evolutionary change in dewlap appearance could have the incidental effect of causing populations to speciate (cf. Endler 1992; Marchetti 1993). One such possibility is that, when anoline populations occupy new habitats, they may experience different light environments (Fleishman 1992). Because some colors are seen more effectively than others in particular situations (e.g., open sun, deep shade [Endler 1993]), selection might favor evolutionary change in dewlap color in the new habitat; this change, in turn, might lead to reproductive isolation from the parental species. In this way, the evolution of the dewlap may have led to increased rates of speciation and thus be responsible, at least in part, for the great species richness of this clade.

We conclude that ecological differentiation and accelerated rates of speciation in anoles are likely to have their causal bases in different characters. Expanded toepads – while permitting evolution of arboreality – seem unlikely to accelerate rates of speciation, unless partitioning of arboreal habitats per se explains the high rates of anoline speciation (Rice and Hostert [1993] summarize experimental evidence that habitat differentiation can lead to higher rates of speciation). Likewise, dewlaps appear causally associated with speciation but are not required for the invasion of arboreal habitats – unless they provide a system of communication essential in arboreal environments. Further investigations may reveal that evolutionary interactions between toepads and dewlap underlie both the ecological disparity and taxonomic diversity of anoles.

Evolution of community structure

The presence of four of the six ecomorph classes on all four of the Greater Antilles could be explained in many ways: (1) they might each have arisen once and then dispersed to the other islands; (2) they might each have arisen independently on each island; or (3) they might have had a complex history, involving different numbers of independent origins and subsequent dispersal events in each case.

In agreement with previous studies, our DNA data clearly contradict the first hypothesis. Rather, each ecomorph class has evolved independently on each island (Figure 19.7). Indeed, with the exception of the crown-giants of Hispaniola (*A. barahonae*) and Puerto Rico (*A. cuvieri*), and the twig anoles of Cuba and Hispaniola, no members of the same ecomorph on different islands are even closely related. The Templeton test indicates that the most parsimonious tree (Figure 19.5) is significantly shorter than the shortest tree in which each of the ecomorph classes is monophyletic (Table 19.2). Although it is possible that one ecomorph may be the ancestral type from which the others evolved multiple times, anoline radiations clearly occurred independently on each island of the Greater Antilles.

Given that the ecomorphs have evolved mostly independently on each island, one might then ask whether the sequence of ecomorph evolution has been the same on all four islands. Previous phylogenies for the anoles of Jamaica and Puerto Rico have been used to reconstruct the evolution of ecomorphs based on parsimony reconstructions of morphological evolution (Losos 1992). Use of parsimony assumes that rates of evolution have not been so high that parallelism or reversal for ecomorphological characteristics would occur extensively within individual islands (Larson and Losos 1996).

The historical trajectory of ecomorphological evolution appears to have been remarkably similar on Jamaica and Puerto Rico (Figure 19.9). On both islands, the two-ecomorph stage contained a twig anole and a generalist (a morphological intermediate not corresponding to any of the extant ecomorphological classes). At the three-ecomorph stage, both islands had a crown-giant and a twig anole; in addition, Puerto Rico had a trunk-crown anole, whereas Jamaica had a species living in the crown, but not clearly identifiable as either a crown-giant or a trunk-crown anole. At the four-ecomorph stage, the islands were again identical. Finally, the fifth ecomorph to evolve on Puerto Rico was the grass-bush anole, the one that is absent from Jamaica.

Three shortcomings with this earlier study must be noted. First, the Jamaican phylogeny (Hedges and Burnell 1990) was constructed using methods whose assumptions may be violated by the data (Irschick et al., unpubl. data). Second, the Puerto Rican phylogeny was an amalgamation of several studies that used osteological, karyotypic, allozymic, and immunological data (Losos 1990a). Because different taxa were included and data were of different types (discrete characters, distances), the various data could not be combined explicitly to produce the phylogeny. Although several of the relationships in Figure 19.5 do not agree with the phylogenetic hypotheses previously used to reconstruct ecomorph evolution, these relationships should be considered tentative until more species are added.

Third, the anoles on Puerto Rico, rather than representing a monophyletic radiation like those on Jamaica (except the recent colonist *A. sagrei*), consist of three clades: *A. cuvieri*, *A. occultus*, and the members of the *A. cristatellus* species group. Due to lack of information on phylogeny of the entire genus, the Puerto Rican anoles were treated as monophyletic, with the proviso that future information might alter our understanding; an alternative possibility is that one of these clades had evolved its particular ecomorph prior to arriving in Puerto Rico. Examination of the DNA phylogeny (Figure 19.5) fails to resolve this question. None of the twig anoles is closely related to *A. occultus*; hence, the *A. occultus* lineage most likely evolved into a twig anole independently. The hypothesis that the twig ecomorph represents the ancestral condition among anoles is non-parsimonious. By contrast, the *A. cuvieri* clade is closely related (but not the sister taxon) to the Hispaniolan crown-giants; consequently, the ancestor of *A. cuvieri*, which may not have occurred on Puerto Rico, may have already become a crown-giant by the time it existed independently on Puerto Rico. Hence, although the two- and three-species stages in the Jamaican and Puerto Rican radiations may have been very similar in ecomorph composition (assuming that the phylogenies used in these analyses withstand further examination), we cannot eliminate the possibility that some of the ecomorphs evolved prior to occurrence on Puerto Rico.

Interpretation of the sequence of ecomorph evolution on Cuba and Hispaniola is hindered because it is difficult to determine which lineages specialized *in situ* and which were already specialized by the time they first occupied their present islands. Nonetheless, inspection of the DNA phylogeny (Figure 19.6) suggests that Cuba and Hispaniola may not have followed the same sequence of ecomorph evolution as that suggested for the other two islands. In particular, the oldest clades (as determined both by topological position of the clade in the phylogenetic tree and by estimated time of divergence based on number of DNA changes on each branch [T. Jackman et al., unpubl.

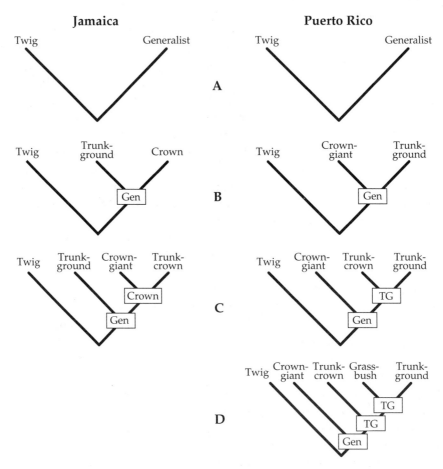

Figure 19.9. Reconstruction of ecomorph evolution on Jamaica and Puerto Rico at the (a) two-, (b) three-, (c) four-, and (d) five-ecomorph stages. Ancestral states were estimated by reconstructing ancestral values for scores on principal components axes (see Figure 19.4). Once the value of PCA axes 1-4 had been estimated, the positions of ancestors in PCA space were plotted. If an ancestor fell within the boundaries defined by extant members of an ecomorph class, then the species was considered to be a member of that ecomorph class. If the ancestor did not fall within the boundaries of any of the ecomorphs and fell in an intermediate position between the ecomorph spaces, then it was considered a generalist. Randomization analyses indicate that parsimony does not necessarily reconstruct generalists at deep nodes within the tree and that the congruence between the two reconstructions is greater than expected by chance (Losos, 1992). Morphology of basal ancestors was not reconstructed because of lack of firm knowledge of the sister taxa to these radiations. (Figure from Losos [1992] with permission.)

data]) do not include twig anoles, which were the first ecomorphs inferred to have evolved on Puerto Rico and Jamaica. However, our analysis omitted one clade of twig anoles from Hispaniola. Hence, this conclusion will need to be re-evaluated when this lineage is included. In addition, grass-bush anoles are relatively early diverging lineages and crown-giants may be relatively late diverging lineages in Hispaniola, contrary to the pattern on Jamaica and Puerto Rico. Consequently, we conclude that available evidence rejects the hypothesis that the sequence of ecomorph evolution has been identical on all four islands, even though the evolutionary outcome has been very similar.

Conclusions

Despite more than 30 years of systematic studies of the anoline radiation, much still remains to be resolved. In addition to clarifying phylogenetic relationships within and between the taxa on each of the Greater Antilles, molecular studies will be useful for studying diversification in the little studied continental anolines. Within the Caribbean, more research (both ecological and phylogenetic) needs to be done on the larger islands of Cuba and Hispaniola. Although the relatively modest radiations of Puerto Rico and Jamaica have been well studied, the large number of anoline species and ecological roles on Cuba and Hispaniola require further investigation.

Nonetheless, the combination of systematic, field, and laboratory research makes clear that anoles are a paradigmatic example of adaptive radiation – four times over! Independently on each of the islands of the Greater Antilles, anoles have adaptively diversified to occupy a similar spectrum of disparate ecological roles. Given historical contingencies and ecological differences among islands, one might have expected radiations on the islands to have resulted in striking differences in ecology and morphology of the component species, but this prediction is not realized; rather, the same roles have been filled repeatedly. Further ecological research should investigate why the outcome of anoline radiation is so predictable, in contrast to most other radiations examined. Gould (1989, p. 51) has argued that historical contingency plays such an important role in evolutionary diversification that a "replay of the tape [of evolutionary diversification] would lead evolution down a pathway radically different from the road actually taken." We suggest that, at least for anoles of the Greater Antilles, the tape has been replayed, four times, and the outcome has been substantially the same each time.

Acknowledgments

We thank T. Givnish for useful comments and exhortations and B. Crother and M. Leal for providing access to unpublished manuscripts. This work was supported by the National Science Foundation (DEB-9318642) and the David and Lucille Packard Foundation.

References

Arnold, E. N. 1994. Investigating the origins of performance advantage: adaptation, exaptation and lineage effects. Pp. 123-168 in *Phylogenetics and Ecology,* P. Eggleton and R. Vane-Wright, eds. London, England: Academic Press.

Baum, D. A., and Larson, A. 1991. Adaptation reviewed: a phylogenetic methodology for studying character macroevolution. Systematic Zoology 40:1-18.

Bremer, K. 1988. The limits of amino-acid sequence data in angiosperm phylogenetic reconstruction. Evolution 42:795-803.

Burnell, K. L., and Hedges, S. B. 1990. Relationships of West Indian *Anolis* (Sauria: Iguanidae): an approach using slow-evolving loci. Caribbean Journal of Science 26:7-30.

Buth, D. G., Gorman, G. C., and Lieb. C. S. 1980. Genetic divergence between *Anolis carolinensis* and its Cuban progenitor, *Anolis porcatus.* Journal of Herpetology 14:279–284.

Cannatella, D. C., and de Queiroz, K. 1989. Phylogenetic systematics of the anoles: is a new taxonomy warranted? Systematic Zoology 38:57–68.

Cartmill, M. 1985. Climbing. Pp. 73–88 in *Functional Vertebrate Morphology,* M. Hildebrand, D. M. Bramble, K. F. Liem, and D. B. Wake, eds. Cambridge, MA: Belknap Press.

Case, S. M., and Williams, E. E. 1987. The cybotoid anoles and *Chamaelinorops* lizards (Reptilia: Iguanidae): evidence of mosaic evolution. Zoological Journal of the Linnean Society 91:325–341.

Coddington, J. A. 1988. Cladistic tests of adaptational hypotheses. Cladistics 4:3–22.

Crother, B. I., and Guyer, C. 1996. Caribbean historical biogeography: Was the dispersal-vicariance debate eliminated by an extraterrestrial bolide? Herpetologica 52:440–465.

de Queiroz, K. 1989. *Morphological and biochemical evolution in the sand lizards.* Ph.D. Dissertation, University of California, Berkeley.

Diaz, L. M., Estrada, A. R., and Moreno, L. V. 1996. A new species of *Anolis* (Sauria: Iguanidae) from the Sierra de Trinidad, Sancti Spiritus, Cuba. Caribbean Journal of Science 32:54–58.

Echelle, A. A., Echelle, A. F., and Fitch, H. S. 1971. A comparative analysis of aggressive display in nine species of Costa Rican *Anolis.* Herpetologica 27:271–288.

Endler, J. A. 1992. Signals, signal conditions, and the direction of evolution. American Naturalist 139:S125–153.

Endler, J. A. 1993. The color of light in forests and its implications. Ecological Monographs 63:1–27.

Estrada, A. R., and Hedges, S. B. 1995. A new species of *Anolis* (Sauria: Iguanidae) from eastern Cuba. Caribbean Journal of Science 31:65–72.

Etheridge, R. 1960. *The relationships of the anoles (Reptilia: Sauria: Iguanidae): an interpretation based on skeletal morphology.* Ph.D. Dissertation, University of Michigan, Ann Arbor.

Felsenstein, J. 1985a. Confidence limits on phylogenies: an approach using the bootstrap. Evolution 39:783–791.

Felsenstein, J. 1985b. Phylogenies and the comparative method. American Naturalist 125:1–15.

Fleishman, L. J. 1992. The influence of sensory system and the environment on motion patterns in the visual displays of anoline lizards and other vertebrates. American Naturalist 139: S36–61.

Flores, G., Lenzycki, J. H., and Palumbo, J., Jr. 1994. An ecological study of the endemic Hispaniolan anoline, *Chamaelinorops barbouri* (Lacertilia: Iguanidae). Breviora 499:1–23.

Forsgaard, K. 1983. The axial skeleton of *Chamaelinorops.* Pp. 284–295 in *Advances in herpetology and evolutionary biology: essays in honor of Ernest E. Williams,* A. G. J. Rhodin and K. Miyata, eds. Cambridge, MA: Museum of Comparative Zoology, Harvard University.

Garland, T., Jr., Dickerman, A. W., Janis, C. M., and Jones, J. A. 1993. Phylogenetic analysis of covariance by computer simulation. Systematic Biology 42:265–292.

Garland, T., Jr., Harvey, P. H., and Ives, A. R. 1992. Procedures for the analysis of comparative data using phylogenetically independent contrasts. Systematic Biology 41:18–32.

Glossip, D., and Losos, J. B. 1997. Ecological correlates of number of subdigital lamellae in anoles. Herpetologica (in press).

Gorman, G. C., and Atkins, L. 1969. New karyotypic data for 16 species of *Anolis* (Sauria: Iguanidae) from Cuba, Jamaica, and the Cayman Islands. Herpetologica 24:13–21.

Gorman, G. C., Buth, D. G., Soulé, M., and Yang, S. Y. 1980a. The relationships of the *Anolis cristatellus* species group: electrophoretic analysis. Journal of Herpetology 14:269–278.

Gorman, G. C., Buth, D. G., Soulé, M., and Yang, S. Y. 1983. The relationships of the Puerto Rican *Anolis*: electrophoretic and karyotypic studies. Pp. 626–642 in *Advances in herpetology and evolutionary biology: essays in honor of Ernest E. Williams,* A. G. J. Rhodin and K. Miyata, eds. Cambridge, MA: Museum of Comparative Zoology, Harvard University.

Gorman, G. C., Buth, D. G., and Wyles, J. S. 1980b. *Anolis* lizards of the eastern Caribbean: a case study in evolution. III. A cladistic analysis of albumin immunological data, and the definition of species groups. Systematic Zoology 29:143–158.

Gorman, G. C., and Kim, Y. S. 1976. *Anolis* lizards of the eastern Caribbean: a case study in evolution II. Genetic relationships and genetic variation of the *bimaculatus* group. Systematic Zoology 20:167–185.

Gorman, G. C., Lieb, C. S., and Harwood, R. H. 1984. The relationships of *Anolis gadovi:* albumin immunological evidence. Caribbean Journal of Science 20:145–152.

Gorman, G. C., and Stamm, B. 1975. The *Anolis* lizards of Mona, Redonda, and La Blanquilla: Chromosomes, relationships, and natural history notes. Journal of Herpetology 9:197–205.

Gould, S. J. 1989. *Wonderful life: The Burgess Shale and the nature of history.* New York, NY: W. W. Norton.

Gould, S. J., and Vrba, E. S. 1982. Exaptation – a missing term in the science of form. Paleobiology 8:4–15.

Greene, H. W. 1986. Diet and arboreality in the emerald monitor, *Varanus prasinus,* with comments on the study of adaptation. Fieldiana Zool. New Series 31:1–12.

Guyer, C., and Savage, J. M. 1986. Cladistic relationships among anoles. Systematic Zoology 35:509–531.

Guyer, C., and Savage, J. M. 1992. Anole systematics revisited. Systematic Biology 41:89–110.

Hass, C. A., Hedges S. B., and Maxson L. R. 1993. Molecular insights into the relationships and biogeography of West Indian anoline lizards. Biochemical Systematics and Ecology 21:97–114.

Heard, S. B., and D. L. Hauser. 1995. Key evolutionary innovations and their ecological mechanisms. History of Biology 10:151–173.

Hedges, S. B. 1996. The origin of West Indian amphibians and reptiles. Pp. 95–128 in *Contributions to West Indian herpetology: a tribute to Albert Schwartz,* R. Powell and R. W. Henderson, eds. Ithaca, NY: Society for the Study of Amphibians and Reptiles.

Hedges, S. B., and Burnell, K. L. 1990. The Jamaican radiation of *Anolis* (Sauria: Iguanidae): An analysis of relationships and biogeography using sequential electrophoresis. Caribbean Journal of Science 26:31–44.

Hertz, P. E. 1980. Responses to dehydration in *Anolis* lizards sampled along altitudinal transects. Copeia 1980:440–446.

Hertz, P. E. 1981. Adaptation to altitude in two West Indian anoles (Reptilia: Iguanidae): field thermal biology and physiological ecology. Journal of Zoology 195:25–37.

Holmquist, R. 1983. Transitions and transversions in evolutionary descent: an approach to understanding. Journal of Molecular Evolution 19:134–144.

Huey, R. B., and Webster, T. P. 1976. Thermal biology of *Anolis* lizards in a complex fauna: the cristatellus group on Puerto Rico. Ecology 57:985–994.

Irschick, D. J., Austin, C. C., Petren, K., Fisher, R. N., Losos, J. B., and Ellers, O. 1996. A comparative analysis of clinging ability among pad-bearing lizards. Biological Journal of the Linnean Society 59:21–35.

Irschick, D. J., and Losos, J. B. 1996. Morphology, ecology, and behavior of the twig anole, *Anolis angusticeps.* Pp 291–301 in *Contributions to West Indian herpetology: a tribute to Albert Schwartz,* R. Powell and R. W. Henderson, eds. Ithaca, NY: Society for the Study of Amphibians and Reptiles.

Jenssen, T. A. 1973. Shift in the structural habitat of *Anolis opalinus* due to congeneric competition. Ecology 54:863–869.

Jenssen, T. A., and Feely, P. C. 1991. Social behavior of the male anoline lizard *Chamaelinorops barbouri,* with a comparison to *Anolis.* Journal of Herpetology 25:454–461.

Larson, A. 1994. The comparison of morphological and molecular data in phylogenetic systematics. Pp. 371–390 in *Molecular ecology and evolution: approaches and applications,* B. Schierwater, B. Streit, G. P. Wagner, and R. DeSalle, eds. Basel, Switzerland: Birkhäuser Verlag.

Larson, A., and Losos, J. B. 1996. Phylogenetic systematics of adaptation. Pp. 187–220 in *Adaptation,* M. R. Rose and G. V. Lauder, eds. New York NY: Academic Press.

Larson, A., Wake, D. B., Maxson, L. R., and Highton, R. 1981. A molecular phylogenetic perspective on the origin of morphological novelties in the salamanders of the tribe Plethodontini (Amphibia, Plethodontidae). Evolution 35:405–422.

Lauder, G. V. 1981. Form and function: structural analysis in evolutionary morphology. Paleobiology 7:430–442.

Leal, M., and Rodríguez-Robles, J. A. 1997. Signalling displays during predator-prey interactions in a Puerto Rican anole, *Anolis cristatellus.* Animal Behavior (in press).

Lister, B. C. 1976. The nature of niche expansion in West Indian *Anolis* lizards II. Evolutionary components. Evolution 30:677–692.

Losos, J. B. 1985. An experimental demonstration of the species recognition role of *Anolis* dewlap color. Copeia 1985:905–910.

Losos, J. B. 1990a. Ecomorphology, performance capability, and scaling of West Indian *Anolis* lizards: an evolutionary analysis. Ecological Monographs 60:369–388.

Losos, J. B. 1990b. The evolution of form and function: morphology and locomotor performance in West Indian *Anolis* lizards. Evolution 44:558–569.

Losos, J. B. 1992. The evolution of convergent structure in Caribbean *Anolis* communities. Systematic Biology 41:403–420.

Losos, J. B. 1994. Integrative approaches to evolutionary ecology: *Anolis* lizards as model systems. Annual Review of Ecology and Sytematics 25:467–493.

Losos, J. B., and de Queiroz, K. 1997. Ecological release and the evolution of specialization in Caribbean *Anolis* lizards. Biological Journal of the Linnean Society (submitted).

Losos, J. B., and Irschick, D. J. 1996. The effect of perch diameter on escape behaviour of *Anolis* lizards: laboratory predictions and field tests. Animal Behavior 51:593–602.

Losos, J. B., Irschick, D. J., and Schoener, T. W. 1994. Adaptation and constraint in the evolution of specialization of Bahamian *Anolis* lizards. Evolution 48:1786–1798.

Losos, J. B., and Sinervo, B. 1989. The effect of morphology and perch diameter on sprint performance of *Anolis* lizards. Journal of Experimental Biology 145:23–30.

Macey, J. R., Larson, A. Ananjeva, N. B., Fang, Z., and Papenfuss, T. J. 1997a. Two novel gene orders and the role of light-strand replication in rearrangement of the vertebrate mitochondrial genome. Molecular Biology and Evolution 14:91–104.

Macey, J. R., Larson, A., Ananjeva, N. B., and Papenfuss, T. J. 1997b. Evolutionary shifts in three major structural features of the mitochondrial genome among iguanian lizards. Journal of Molecular Evolution (in press).

Maddison, W., and Maddison, D. 1992. *MacClade 3: analysis of phylogenetic and character evolution.* Sunderland, MA: Sinauer Associates, Inc..

Marchetti, K. 1993. Dark habitats and bright birds illustrate the role of the environment in species divergence. Nature 362:149–152.

Miyata, K. 1983. Notes on *Phenacosaurus heterodermus* in the Sabana de Bogotá, Colombia. Journal of Herpetology 17:102–105.

Moermond, T. C. 1979a. The influence of habitat structure on *Anolis* foraging behavior. Behaviour 70:147–167.

Moermond, T. C. 1979b. Habitat constraints on the behavior, morphology, and community structure of *Anolis* lizards. Ecology 60:152–164.

Moritz, C., Schneider, C. J., and Wake, D. B. 1992. Evolutionary relationships within the *Ensatina eschscholtzii* complex confirm the ring species interpretation. Systematic Biology 41: 273–291.

Peterson, J. A., and Williams, E. E. 1981. A case history in retrograde evolution: the *onca* lineage in anoline lizards. II. Subdigital fine structure. Bulletin of the Museum of Comparative Zoology 149:215–268.

Powell, R., Henderson, R. W. , Adler, K., and Dundee, H. A. 1996. An annotated checklist of West Indian amphibians and reptiles. Pp. 51–93 in *Contributions to West Indian herpetology: a tribute to Albert Schwartz*, R. Powell and R. W. Henderson, eds. Ithaca, NY: Society for the Study of Amphibians and Reptiles.

Rand, A. S. 1964. Ecological distribution in anoline lizards of Puerto Rico. Ecology 45:745–752.

Rand, A. S. 1967. The ecological distribution of anoline lizards around Kingston, Jamaica. Breviora 272:1–18.

Rand, A. S., and Williams, E. E. 1969. The anoles of La Palma: aspects of their ecological relationships. Breviora 327:1–19.

Rand, A. S., and Williams, E. E. 1970. An estimation of redundancy and information content of anole dewlaps. American Naturalist 104:99–103.

Rensch, B. 1959. *Evolution above the species level.* New York, NY: Columbia University Press.

Rice, W. R., and Hostert, E. E. 1993. Laboratory experiments on speciation: what have we learned in forty years? Evolution 47:1637–1653.

Roughgarden, J. 1974. Niche width: biogeographic patterns among *Anolis* lizard populations. American Naturalist 108:429–442.

Ruibal, R. 1961. Thermal relations of five species of tropical lizards. Evolution 15:98–111.

Schoener, T. W. 1968. The *Anolis* lizards of Bimini: resource partitioning in a complex fauna. Ecology 49:704–726.

Schoener, T. W. 1970. Nonsynchronous spatial overlap of lizards in patchy habitats. Ecology 51:408–418.

Schoener, T. W. 1974. Resource partitioning in ecological communities. Science 185:27–39.

Schoener, T. W. 1975. Presence and absence of habitat shift in some widespread lizard species. Ecological Monographs 45:233–258.

Schoener, T. W., and Gorman, G. C. 1968. Some niche differences in three Lesser-Antillean lizards of the genus *Anolis*. Ecology 49:819–830.

Schoener, T. W., and Schoener, A. 1971a. Structural habitats of West Indian *Anolis* lizards. I. Jamaican lowlands. Breviora 368:1–53.

Schoener, T. W., and Schoener, A. 1971b. Structural habitats of West Indian *Anolis* lizards. II. Puerto Rican uplands. Breviora 375:1–39.

Shochat, D., and Dessauer, H. C. 1981. Comparative immunological study of the albumins of *Anolis* lizards of the Caribbean Islands. Comparative Biochemistry and Physiology 68A:67–73.

Simpson, G. G. 1953. *The major features of evolution.* New York, NY: Columbia University Press.

Templeton, A. 1983. Phylogenetic inference from restriction endonuclease cleavage site maps with particular reference to the evolution of humans and the apes. Evolution 37:221–244.

Williams, E. E. 1969. The ecology of colonization as seen in the zoogeography of anoline lizards on small islands. Quarterly Review of Biology 44:345–389.

Williams, E. E. 1972. The origin of faunas – evolution of lizard congeners in a complex island fauna: a trial analysis. Evolutionary Biology 6:47–89.

Williams, E. E. 1983. Ecomorphs, faunas, island size, and diverse end points in island radiations of *Anolis*. Pp. 326–370 in *Lizard ecology: studies of a model organism,* R. B. Huey, E. R. Pianka, and T. W. Schoener, eds. Cambridge, MA: Harvard University Press.

Williams, E. E. 1989. A critique of Guyer and Savage 1986: cladistic relationships among anoles (Sauria: Iguanidae): are the data available to reclassify the anoles? Pp. 433–477 in *Biogeography of the West Indies: past, present, and future,* C. A. Woods, ed. Gainesville, FL: Sandhill Crane.

Williams, E. E., and Rand, A. S. 1977. Species recognition, dewlap function, and faunal size. American Zoologist 17:261–270.

Wyles, J. S., and Gorman, G. C. 1980a. The albumin immunological and Nei electrophoretic distance correlation: a calibration for the saurian genus *Anolis* (Iguanidae). Copeia 1980:66–71.

Wyles, J. S., and Gorman, G. C. 1980b. The classification of *Anolis:* conflict between genetic and osteological interpretation as exemplified by *Anolis cybotes*. Journal of Herpetology 14:149–153.

Yang, S. Y., Soulé, M., and Gorman, G. C. 1974. *Anolis* lizards of the eastern Caribbean: a case study in evolution. I. Genetic relationships, phylogeny, and colonization sequence of the roquet group. Systematic Zoology 23:387–399.

Zar, J. H. 1984. *Biostatistical analysis, 2nd ed.* Englewood Cliffs, NJ: Prentice-Hall.

20 Molecular and morphological evolution during the post-Palaeozoic diversification of echinoids

Andrew B. Smith and D. T. J. Littlewood

Echinoids – sea urchins, sand dollars, and their allies – are an ancient and diverse group of benthic marine invertebrates that can be found in a wide range of habitats, from intertidal sediments and rock platforms to the abyssal plains. They fall broadly into two major groups: the regular echinoids, which are epifaunal and feed using their dental apparatus for grazing and rasping sessile organisms, and the irregular echinoids, which are deposit feeders and primarily infaunal. Echinoids have an extensive, though patchy, fossil record. They first appeared in the late Ordovician, approximately 450 million years ago (Mya), and diversified during the Palaeozoic. Like many other groups, however, echinoids were significantly affected by the mass extinction at the end of the Permian; only two closely related lineages survived into the Triassic (Smith and Hollingworth 1990). Consequently the modern fauna of some 900 species is derived almost entirely from a second major bout of diversification that started in the Triassic and early Jurassic, approximately 200 Mya.

The classification of echinoids is based largely on the morphology of their complex endoskeleton. This dependence on morphology not only provides a wealth of phylogenetically informative characters, but also ensures full integration of extant and extinct taxa, both being classified on the same suite of characters. Furthermore, the reasonably good fossil record of echinoids means that their morphological diversification can be reconstructed against an absolute time scale. Their fossil record demonstrates that evolutionary change has proceeded erratically, with some clades showing periods of relatively rapid morphological innovation over short time intervals while others have remained morphologically little changed over long time intervals (e.g., Kier 1982; Smith and Wright 1990).

The close correspondence between structure and function in the echinoid endoskeleton makes it reasonably easy to deduce the palaeoecology of extinct forms (Smith 1984). Based on this approach, late Palaeozoic echinoids are all inferred to have had a very similar life-style, distinct from that pursued by the great majority of echinoids alive today. All were undoubtedly vagile and epifaunal and, although they had by that time evolved a fully functional dental apparatus, it was much less robust than that seen in today's shallow-water species that graze on hard substrates. Indeed, the earliest grazing traces produced by echinoids feeding on hard substrates do not appear in the fossil record until some 50 My after the end of the Permian, supporting the view that Palaeozoic echinoids lacked the ability to rasp with their lantern and were presumably generalist scavengers. The earliest infaunal deposit feeders appear even later, after a further 20 My (Smith 1984). Late Palaeozoic echinoids thus occupied only a very limited ecological spectrum in comparison to today's fauna.

Other aspects of echinoid evolution are less well documented, but also imply a major post-Palaeozoic diversification in form. For example, all taxa have a complex life cycle. Adults are benthic; to reproduce, most shed their gametes into the water where fertilization takes place. The fertilized eggs typically develop into pelagic larvae (plutei) which live as plankton for several weeks or months. A dramatic metamorphosis ensues, during which almost every larval organ is resorbed and lost, including the gut and nervous system. A rudiment forms laterally from which the adult develops, and the remaining larval organs are resorbed as it settles to the bottom. There is considerable morphological diversity among these planktonic stages, suggesting a complex evolutionary history. Furthermore, whereas most echinoids pass through a pelagic larval stage, some groups undergo a more direct development that omits the pluteus stage – a process known as lecithotrophic development (Emlet et al. 1987) . But with virtually no fossil record of echinoid larvae, the evolution of larval life history can only be pieced together indirectly by placing the known larvae of extant forms in a phylogenetic context (e.g., Wray 1992; Smith et al. 1995b).

The evolution of echinoids has traditionally been reconstructed through comparisons of the adult morphology and anatomy of present-day forms, combined with data from the fossil record. Recently, however, ribosomal genes have been sequenced for several echinoids and Smith et al. (1992) and Littlewood and Smith (1995) have constructed morphological, molecular, and total evidence phylogenies for the major clades of echinoids. Furthermore, morphological data have been compiled for larvae of echinoids and placed in a phylogenetic context (Wray 1992, 1994, 1996; Smith et al. 1995a). Investigations of other echinoderm groups have begun based on a similar approach, but taxonomic coverage is, as yet, less extensive. Lafay et al. (1995) present morphological and molecular phylogenies of asteroids; Smith et al. (1995b) have done the same for ophiuroids.

These combined morphological and molecular studies are now providing a more secure phylogenetic basis from which to examine broad evolutionary questions. For the symposium we were asked to address a number of themes (see Chapter 1). Here we investigate the following questions using echinoids:

- Do phylogenies based on molecular data differ substantially from those based on morphological data and materially affect our interpretation of relationships?
- Do major ecological innovations within a clade appear to have evolved only once or do they show multiple origins?
- What is the relationship between genetic and morphological divergence?

A working phylogenetic hypothesis for echinoids

If we are to make any headway in understanding evolutionary diversification, we must first construct a reliable phylogenetic hypothesis for the taxa in question. Use of a single data base of any kind – morphological or molecular – has the potential to generate incorrect relationships. Bootstrapping single data sets is not sufficient to tell us whether our results are reliable, because bootstrapping only indicates whether there is sufficient data to justify phylogenetic statements (sister-group relationships) as more than just chance pairings, and does not tell us whether the data themselves are reliable (e.g., see Nei 1991). This has been highlighted by Lecointre et al. (1993) and Philippe and Douzery (1994), who have derived conflicting topologies for the same three taxa, sometimes each with more than 95% bootstrap support, simply by changing the representative species. Therefore, confidence in the correctness of a phylogenetic hypothesis must come primarily from verification based on independent data sets.

Identifying the same set of relationships using multiple independent sources of data (both morphological and molecular) undoubtedly provides the strongest evidence that we are approaching the correct phylogeny (Hillis 1987, 1995; Miyamoto and Fitch 1995). In addition, topological congruence of the inferred tree with the stratigraphic record of first occurrences in the fossil record can lend further support (Smith and Littlewood 1994). Sadly, congruence among data sets seems to be the exception rather than the rule (Patterson et al. 1994). How much incongruence exists is difficult to quantify, because few comparative studies have included exactly the same taxa, making computation of character incongruence indices difficult (Omland 1994). Problems of homoplasy and systematic bias in characters means that different data sets generally identify topologies that differ to a greater or lesser degree. Each phylogenetic analysis should therefore be thought of as providing an independent estimate of the true phylogeny but with some unknown amount of error included (Larson 1994).

There are currently four independent data sets from which to construct a phylogeny of echinoids, based respectively on adult morphology, larval morphology, complete small subunit ribosomal RNA (SSU rRNA) gene sequence, and partial large subunit ribosomal RNA (LSU rRNA) gene sequence (see Table 20.1 for sources). Our analysis here deals with 29 extant species which includes representatives from most of the major lineages of extant echinoid (Table 20.2). Although the data sets are not

Table 20.1. Summary of morphological and molecular data for echinoids used in this analysis.

Adult morphology: 163 characters (34 autapomorphies). Data for all 28 taxa are listed in Appendix 1 of Littlewood and Smith (1995).

Larval morphology: 48 characters (8 autapomorphies). Data for 26 taxa are listed in Appendix 1 of Smith et al. (1995a).

SSU rRNA gene sequences: 270 characters (186 autapomorphies). Data for 22 taxa are given in Appendix 2 of Littlewood and Smith (1995). All 1,800 bases of the gene were sequenced; aligned sequences have been deposited with EMBL under accession number DS19161 and are available via ftp.

LSU rRNA gene sequences: 92 characters (56 autapomorphies). Data for 13 taxa are presented in Appendix 3 of Littlewood and Smith (1995). The first 400 bases from the 5' end of the gene were sequenced; aligned sequences have been deposited with EMBL under accession number DS19161 and are available via ftp.

Table 20.2. Higher classification and relevant ecology of echinoid taxa used in this study. Ecological habit: e = epifaunal; i = infaunal. Feeding strategy: o = lantern feeder, omnivore; g = lantern feeder, specialist grazer; d = deposit feeder (db = bulk sediment swallower; de epipsammic grazer; df = sediment siever with food-groove system; dp = selective deposit feeder using phyllopodes). Larval development: p = planktotrophic; l = lecithotrophic; fp = facultative planktotrophic; (p/l) = both planktotrophic and lecithotrophic development found within this order.

	Ecological habit	Feeding strategy	Larval development
"Regular" echinoids			
Subclass Cidaroidea			
Order Cidaroida			(p/l)
Eucidaris	e	o	p
Cidaris	e	o	p
Subclass Euechinoida			
Order Echinothurioida			(l)
Phormosoma	e	o	l
Asthenosoma	e	o	l
Order Diadematoida			(p)
Diadema	e	o	p
Centrostephanus	e	o	p
Order Arbacioida			(p)
Arbacia	e	g	p
Order Phymosomatoida			(p)
Stomopneustes	e	g	p
Glyptocidaris	e	g	p
Order Temnopleuroida			(p/l)
Temnopleurus	e	g	p
Mespilia	e	g	p
Salmacis	e	g	p
Order Echinoida			(p/l)
Echinus	e	g	p
Paracentrotus	e	g	p
Psammechinus	e	g	p
Strongylocentrotus	e	g	p
Colobocentrotus	e	g	p
Heliocidaris	e	g	p/l
Sphaerechinus	e	g	p
Tripneustes	e	g	p
Lytechinus	e	g	p
"Irregular" echinoids			
Order Cassiduloida			(p/l)
Cassidulus	i	db	l
Echinolampas	i	db	p
Order Clypeasteroida			(p/fp)
Fellaster/Clypeaster	i	df	p/fp
Echinocyamus	i	de	p
Echinodiscus	i	df	p
Encope	i	df	p
Order Spatangoida			(p/l)
Spatangus	i	dp	p
Echinocardium	i	dp	p
Meoma	i	dp	p
Brissopsis	i	dp	p

fully complementary, many of the same taxa appear in all four analyses. Each has been analyzed cladistically to generate four independent estimates of the true phylogeny. Three of the four topologies show considerable overall similarity, but differ to a greater or lesser extent in their detailed structure; the topology derived from larval morphology is, at first sight, very different (Figure 20.1). A further complexity is that, depending upon whether fossil taxa are included or excluded, two slightly differing topological arrangements emerge based on adult morphology (Littlewood and Smith 1995).

Faced with alternative cladograms, our problem is to decide which represents the best estimate of true echinoid phylogeny. Where real conflict exists between morphological and molecular data sets, it is generally impossible to decide whether one or the other is correct, or whether both in fact are misleading. This also strongly influences how we interpret the results. A molecular biologist might argue that such conflict demonstrates that phylogenies based on molecular data do materially affect our perception of the evolutionary tree, whereas a morphologist might consider that the result demonstrates systematic bias in the molecular data set. This, for example, has been the underlying gist of the arguments concerning mammal-bird relationships (e.g., Hedges et al. 1990; Marshall 1992).

However, an alternative explanation for a mismatch might be that the conflict is more apparent than real, and that the observed differences could simply have arisen by chance. Two data sets could be producing estimates of the same underlying phylogenetic signal, but with imprecision due to sampling errors associated with small data matrices. If this were the case, then the best option would be to combine all the data and re-analyze it. This would find the best supported topology on the basis of overall congruence among the set of all characters (Kluge 1989; Eernisse and Kluge 1993; Kluge and Wolf 1993).

So we have two possibilities: constructing a consensus tree of the rival cladograms to highlight those areas of agreement, or combining data and re-analyzing. If independent lines of evidence produce non-congruent topologies in which alternate groupings each show high levels of support, then the conflict is deep-seated and one or other data set is positively misleading. In such cases, consensus methods are more appropriate (de Queiroz 1993; Miyamoto and Fitch 1994; Larson 1994). However, a consensus approach in this case is very uninformative (Figure 20.2, left), because of the variable position of *Arbacia* in the LSU-derived topology and the strongly divergent structure of the tree derived from larval morphology. Even a consensus tree for adult morphology, SSU rRNA sequences, and LSU rRNA sequences with *Arbacia* removed, though showing somewhat more structure, still includes many polytomies (Figure 20.2, right).

On the other hand, if the independent lines of evidence are providing estimates of the same underlying phylogeny within acceptable limits of random error, then combining data is the best option. A simple test of whether topologies can be considered suboptimal estimates of one another or represent significantly different topologies is provided by Templeton's Test (Templeton 1983; Larson 1994). When Templeton's Test is applied to the echinoid data, the topologies derived from all but

Adult morphology: semi-strict consensus of 13 trees

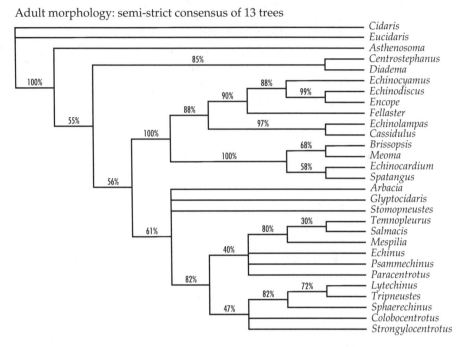

SSU rRNA data: semi-strict consensus of 2 trees

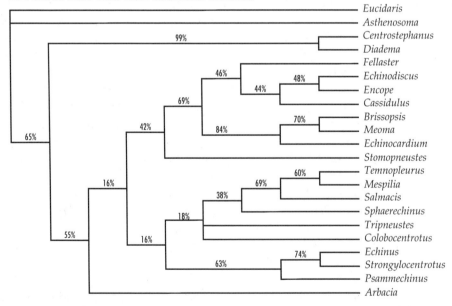

Figure 20.1 (above and facing page). Cladograms derived from the four individual data sets using parsimony in PAUP 3.1.1 (Swofford 1993). In each case, all characters were equally weighted; ten replicate heuristic searches were conducted using randomly assembled starting trees. **Adult morphology:** 163 characters, tree length = 260 steps, CI (consistency index excluding invariant positions) = 0.70, RI (retention index excluding invariant positions) = 0.89, and RC (rescaled consistency index) = 0.65. The tree was rooted using the late Palaeozoic fossil echinoid *Archaeocidaris* (not shown) as an outgroup. **Larval morphology:**

Larval morphology: semi-strict consensus of 13 trees

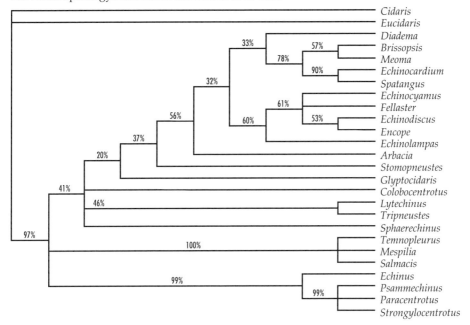

LSU rRNA data: semi-strict consensus of 8 trees

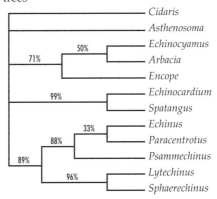

48 characters, tree length = 91 steps, CI = 0.55, RI = 0.82, and RC = 0.48. The two cidaroids (*Cidaris* and *Eucidaris*) were used to root the tree, in accord with the morphological and LSU rRNA analyses. **SSU rRNA sequences:** 278 characters, tree length = 440, CI = 0.50, RI = 0.59, and RC = 0.44. The cidaroid *Eucidaris* was used to root the tree, in accord with the morphological and LSU rRNA analyses. **LSU rRNA sequences:** 91 characters, tree length = 132, CI = 0.69, RI = 0.77, and RC = 0.64. Rooting was by reference to invariant positions in the sequences of other echinoderm classes (see Smith et al. 1991 for details). Bootstrap values are based on 1,000 replicates.

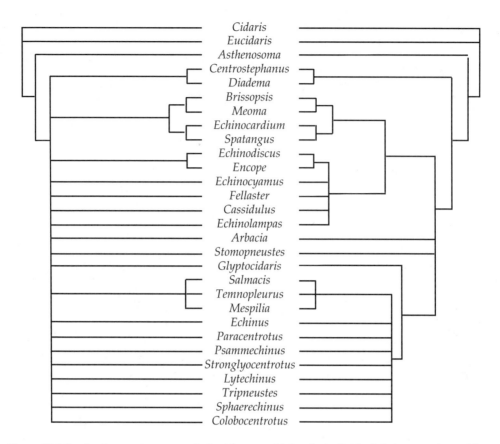

Figure 20.2. Semi-strict consensus trees derived from combining the individual cladograms derived from adult morphology, larval morphology, SSU rRNA and LSU rRNA gene sequences. The right-hand tree excludes the larval data set and the LSU rRNA gene sequence data for *Arbacia*.

the LSU rRNA data set differ significantly (Table 20.3). Topologies derived from adult morphology and SSU data sets show much overall similarity but differ in a number of places where one or both data sets have no strong signal in support (Figure 20.3, top). The larval data set produces a topology very much at variance with those produced by adult morphology and SSU rRNA sequences, but the bootstrap values again show that few of the conflicting branches are well supported (Figure 20.3, bottom). Only the topology derived from LSU rRNA sequence data shows non–significant differences in structure. For LSU rRNA, problems arise only from the placement of a single taxon (*Arbacia*), whose position is very different from that observed in trees derived from morphology or SSU rRNA sequences. Other aspects of the LSU rRNA cladogram are either congruent or unresolved.

A phylogenetic analysis of the four data sets combined identifies two equally parsimonious trees which differ in the placement of three closely related taxa (*Echinus, Psammechinus, Paracentrotus*). Identical results are obtained irrespective of whether *Arbacia* is included or excluded from the LSU rRNA data set. A semi-strict consensus of these two trees (Figure 20.4) differs from all four trees based on the separate data sets. Templeton's Test, however, shows that adult, SSU, and LSU topologies can all be considered as suboptimal estimates of this combined data tree (Table 20.3). Only larval data continue to show significant conflict in topological structure.

So there is sufficient congruence among three of our data sets for them to be considered as suboptimal estimates of the total evidence topology. The larval data set is the most incongruent, but also contains the fewest phylogenetically informative characters. Analysis with or without larval data has an effect on relationships within one clade (Camarodonta) but does not affect inferred relationships elsewhere in the tree. We therefore take the strict consensus tree of the combined data as our best working hypothesis of echinoid phylogeny, although we recognize that the relationships among camarodonts may be incorrect. Indeed, the inclusion of a wider range of taxa, including fossils (Littlewood and Smith 1995), results in a slightly different arrangement of Camarodonta. Bootstrap support is high for almost all branches of the combined data tree, indicating that the topology is stable under subsampling of the characters involved (Figure 20.4).

Do molecular data affect our interpretation of echinoderm relationships?

As demonstrated above, molecular data can generate phylogenies different from those based on traditional morphological evidence, but how this affects our interpretation of echinoderm relationships depends on how much reliance is placed upon

Table 20.3. Comparison of tree topologies using Templeton's Test. Significant levels of P indicate that the two topologies being compared cannot be considered suboptimal estimates of each other under Templeton's Test. In all cases data sets on the left are optimized onto topologies derived from data sets in columns. n.s. = no significant difference ($P > 0.10$).

	Adult morphology	Larval morphology	SSU rRNA	LSU rRNA (including *Arbacia*)	Combined data
Adult morphology		$P < 0.01$	$P < 0.01$	$P < 0.01$	n.s.
Larval morphology	$P < 0.01$		n.s.	n.s.	$P < 0.05$
SSU rRNA	$P < 0.01$	$P < 0.01$		$P < 0.01$	n.s.
LSU rRNA (including *Arbacia*)	n.s.	n.s.	n.s.		n.s.

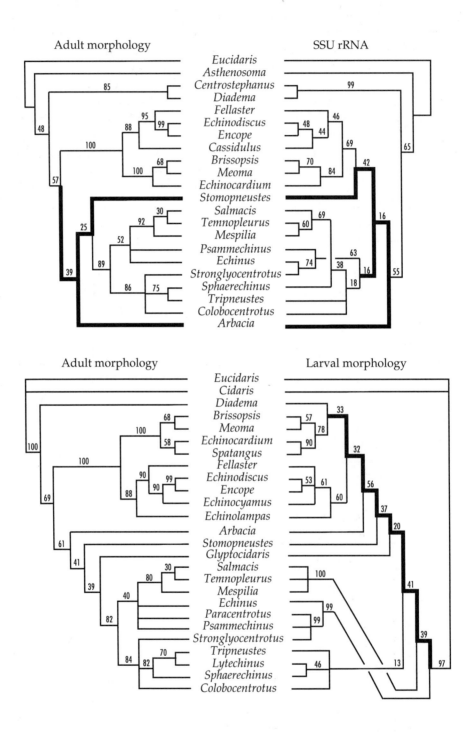

Figure 20.3. Comparison of topologies derived from adult morphology with those based on SSU rRNA sequences and larval morphology. Bootstrap support values are based on 1,000 replicates. Poorly supported branches in conflict are highlighted in bold.

individual data sets. In the case of echinoids, if we simply took our LSU rRNA phylogeny as "the truth" then we would be forced to reinterpret our ideas drastically about the evolutionary diversification of echinoids. Irregular echinoids would be diphyletic, not monophyletic, and the regular stirodont *Arbacia* would fall within the morphologically very divergent order Clypeasteroida (sand dollars and sea biscuits). Similarly, if we took the larval data alone as providing the "true phylogeny" then there would be even more severe repercussions for the adult morphology-based phy-

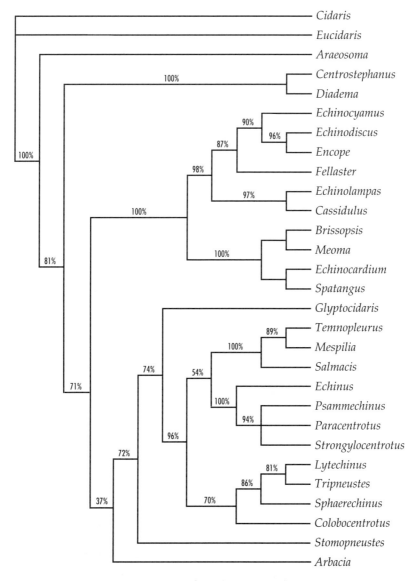

Figure 20.4. A semi-strict consensus constructed from the two trees found under unweighted parsimony analysis of the combined data. Both trees had a length of 952 steps, CI = 0.58, RI = 0.78, and RC = 0.55. Bootstrap values are based on 1,000 replicates.

logeny (see also Wray 1996). But such radical reinterpretations are unjustified in the light of all available evidence.

The important question then is how well single data sets capture what we believe to be our best estimate of the correct phylogeny, and whether combining data sets materially *improves* our interpretation of phylogenetic relationships. To answer this question, we must assess the effect of including/excluding each data set on the total evidence tree, our best and most reliable phylogeny. To quantify the effect of individual data sets we use two simple metrics, involving (i) topological compatibility and (ii) the number of additional steps required in the combined tree.

Index of topological compatibility

The compatibility between the topology found using single data sets and the combined data consensus – considered the best estimate of the "correct" phylogeny – gives an indication of how close each data set comes to estimating the correct topology. The proportion of clades common to two topologies is easily calculated and ranges from 0 (no compatibility) to 1.0 (complete compatibility) (Bledsoe and Raikow 1990; Swofford 1991). Table 20.4 lists the indices for each pairwise comparison of data sets. From such comparisons, it is clear that adult morphology estimates tree topology most accurately relative to the total evidence tree. SSU sequence data and larval morphology generate topologies that are somewhat less accurate than the morphological data, while the LSU sequence data give a rather poor estimate. The estimate provided by LSU data once *Arbacia* is pruned is much more accurate, but includes only half as many clades.

Comparing each topology with the total evidence topology (Figure 20.5) shows that, in most cases of conflict, the total evidence tree is resolved in favor of the

Table 20.4. Indices of topographical congruence for the cladograms constructed from adult morphology (adult), larval morphology (larva), SSU rRNA and LSU rRNA gene sequences, and all four data sets combined (total). T_s and T_i are measures determined from the proportion of clades in common between the two topologies (see Bledsoe and Raikow 1990). N_1 = number of clades in common between the two trees; N_2 = number of clades present in the first tree but absent from the second; N_3 = number of clades that appear in the first tree but are unresolved in the second, and thus are potentially congruent. T_s = (number of clades in common)/(total number of clades); T_i = (number of clades in common, including possible compatible groupings where one topology leaves relationships unresolved)/(total number of clades).

	N_1	N_2	N_3	T_s	T_i
Adult vs. larva	9	8	5	0.53	0.64
Adult vs. SSU	10	8	0	0.56	0.56
Adult vs. LSU	3	2	3	0.60	0.75
Larva vs. SSU	6	9	2	0.40	0.47
Larva vs. LSU	3	3	1	0.50	0.57
SSU vs. LSU	3	2	3	0.60	0.75
Total vs. adult	18	3	3	0.86	0.88
Total vs. larva	11	6	4	0.65	0.71
Total vs. SSU	12	6	0	0.67	0.67
Total vs. LSU	4	3	1	0.57	0.62

topology supported by morphological data. It is interesting that combining the SSU rRNA and adult morphology data sets generated two trees in which the relationships among camarodonts were much closer to those found under a full morphological analysis with both fossil and recent taxa included (Littlewood and Smith 1995, Figure 1a). However, there is enough signal from larval morphology and SSU rRNA sequences to overturn relationships implied from adult morphology concerning the position of *Strongylocentrotus*.

Number of extra steps

When characters from one subset of the data are optimized onto the "correct" topology based on the combined data, the number of implied character-state changes increases in comparison to the most parsimonious topology based on that subset of data alone. Each increase in tree length implies that there is hidden homoplasy in one of the characters, relative to the other characters included in the analysis. The greater the increase in tree length for a given data set, the less reliable it is for estimating the "correct" topology (see also Chapter 2). As a proportion of the total number of character-state changes, morphology and SSU sequences perform equally well, each requiring an additional 2.3% homoplasy when optimized onto the combined-data tree (Table 20.5). LSU sequence data, even excluding *Arbacia*, shows slightly greater levels of homoplasy (3.8% increase in tree length) than either the adult morphology

Table 20.5. Comparison of tree statistics based on a single data set (a) vs. optimization of that data set onto the combined-data topology (b) or the morphology-based tree (c). Difference in tree lengths (D) is a measure of the inferred amount of homoplasy in a data set.

	Tree length	CI	RI	RC
Adult morphology (all 29 taxa):				
a	260 steps	0.70	0.89	0.65
b	266 steps	0.69	0.88	0.63
D_{ab} = 6 steps, 2.3% increase in inferred homoplasy				
Larval morphology (26 taxa):				
a	91 steps	0.55	0.82	0.48
b	99 steps	0.50	0.77	0.42
D_{ab} = 8 steps, 8.8% increase in inferred homoplasy				
SSU rRNA sequence data (22 taxa):				
a	440 steps	0.50	0.59	0.44
b	450 steps	0.48	0.55	0.40
D_{ab} = 10 steps, 2.3% increase in inferred homoplasy				
c	460 steps	0.46	0.51	0.37
D_{ac} = 20 steps, 4.5% increase in inferred homoplasy				
LSU rRNA sequence data (13 taxa):				
a	132 steps	0.70	0.77	0.64
b	137 steps	0.65	0.71	0.57
D_{ab} = 5 steps, 3.8% increase in inferred homoplasy				

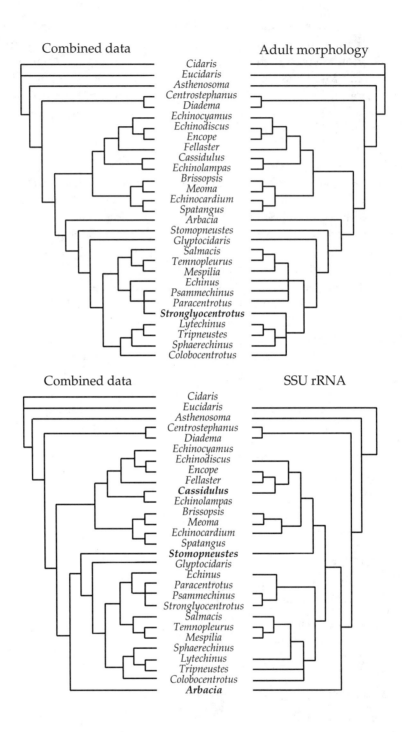

Figure 20.5 (above and facing page). Comparison of the topologies of trees derived from single data sets with that derived from the analysis of the combined data. Taxa whose position is significantly altered have their names in bold.

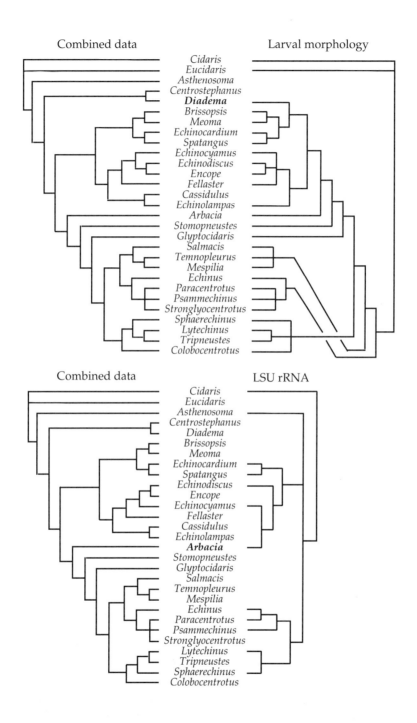

or SSU sequence data, whereas the larval morphology data perform significantly worse, requiring an additional 8.8% increase in tree length when optimized onto the "correct" topology.

In the case of echinoids, our best estimate of the "correct" topology is most strongly influenced by the adult morphological data. Nevertheless, relationships within parts of the tree based on adult morphology are discordant with the combined-data tree. Our best estimate of echinoid phylogeny is thus materially improved by the addition of molecular data, though in a small way. Comparing the topologies derived from adult morphology and SSU rRNA data shows that, where the two do not match, the arrangement supported by morphological data usually has much higher bootstrap support values than the one supported by sequence data (Figure 20.4). SSU rRNA sequences are therefore basically supportive of relationships based on morphology or, where in conflict, lack sufficient signal to alter significantly the relationships inferred from adult morphology. Adult morphology does provide a considerably greater number (136) of phylogenetically informative characters than either set of molecular data (84 for SSU rRNA sequences; 36 for LSU rRNA sequences). But contrary to the assertion by Hedges and Sibley (1994), the number of characters per se is not important, it is whether they provide a phylogenetic signal that is internally congruent. A large data set of nearly random data may have many potentially informative characters, but most of these will conflict, resulting in virtually no signal, whereas a small but consistent data set can generate a strong phylogenetic signal (see also Chapter 2). The greater phylogenetic signal in adult morphology is indicated by consistency and retention indices that are considerably higher than those associated with data matrices of comparable size for larval morphology, LSU rRNA sequences, and SSU rRNA sequences (see Table 20.5).

For the LSU rRNA data, the presence of one rogue sequence (*Arbacia*) brings the LSU rRNA-derived tree into conflict with both the morphological and SSU rRNA-derived trees. However, the strength of signal derived from the SSU rRNA and morphological data sets is sufficient to swamp the conflicting (and presumably erroneous) signal from the LSU rRNA data concerning the placement of *Arbacia*.

Areas of phylogenetic uncertainty based on morphology are also difficult to resolve using molecular data. Such areas often involve the order of branch divergence that occurred over a relatively short period in the distant past, resulting in short internodes and long terminal branches. This kind of topology is known to generate problems for all methods of phylogenetic reconstruction (Felsenstein 1978; Penny and Hendy 1986; Lanyon 1988). Whereas morphological data can often be sought from early fossil members of each terminal branch to reduce the problems of convergence (Smith 1994), there is no such simple solution for molecular data.

The levels of bootstrap support for individual branches of the trees based on single data bases (Figure 20.3) provide an estimate of how robustly each data set supports each clade. The signal from adult morphology is weakest for the relationship of *Arbacia* and *Stomopneustes* to the irregular echinoids and camarodont echinoids (Figure 20.3), with these branches appearing in only 25 and 39% of replicates, respectively. The SSU rRNA data are no improvement – the nodes associated with these four taxa show bootstrap support values of only 16 and 42%. Thus, the most weakly supported nodes in the morphological analysis are also weakly supported in the SSU rRNA sequence analysis.

Similarly, the precise relationship among camarodont echinoids receives only weak bootstrap support from adult morphology, particularly with respect to the placement of Echinidae and Temnopleuridae (Figure 20.3). The key branches are supported by only 47 and 52% bootstrap replicates. The SSU rRNA data lead to a different topology, but the key branches are again supported by only 16, 18, and 38% of bootstrap replicates. Clearly, both morphology and SSU rRNA sequences contain no strong phylogenetic signal with which to resolve these particular relationships.

Conversely, the two best supported branches in the SSU rRNA topology are those uniting the three spatangoids (*Brissopsis, Meoma, Echinocardium:* 84% bootstrap support) and the two diadematoids (*Centrostephanus, Diadema*: 99% support). These branches are also very strongly supported by adult morphology (100% and 85% bootstrap support, respectively). In general, branches well supported by adult morphology are also well supported by SSU sequences.

Ecological innovations

Having constructed a phylogeny for echinoids that is robust and (we hope) reliable, it is possible to ask how frequently ecological innovations have occurred in the history of a clade. This is done by selecting particular ecological traits of interest and optimizing them onto the total evidence cladogram. Given the coarse resolution of our analysis, we are rather limited in the number of meaningful ecological traits that we can compare. However, our analysis serves as a pilot study and we hope that with denser sampling – particularly among the irregular echinoids – it will be possible to tackle a broader range of ecological attributes in the near future.

Within echinoids, some ecological traits appear to have evolved only once while others appear to have evolved independently many times. None of these ecological attributes was used explicitly as a phylogenetic character; however, given the strong correlation between structure and function, virtually all morphological characters are correlated to a greater or lesser extent with an echinoid's biology. Although this might introduce an element of circularity, the fact that a phylogeny based on SSU rRNA sequences has almost the same branching pattern as the morphological tree suggests that such circularity is not a significant problem. At present, there seems to be no way in which generalizations can be drawn, and each ecological trait needs to be treated independently. Two contrasting examples are given below.

Infaunal vs. epifaunal habit

One of the most obvious mechanisms of resource partitioning by echinoids is by infaunal vs. epifaunal habit. Functional analyses leave little doubt that Palaeozoic echinoids were epifaunal and that the infaunal habit has been secondarily acquired (Smith 1984; McNamara 1990). The switch to living and feeding on unconsolidated sediment appears to have occurred in the Jurassic and triggered (or was associated with) a wide range of functional and structural changes, such as the development of

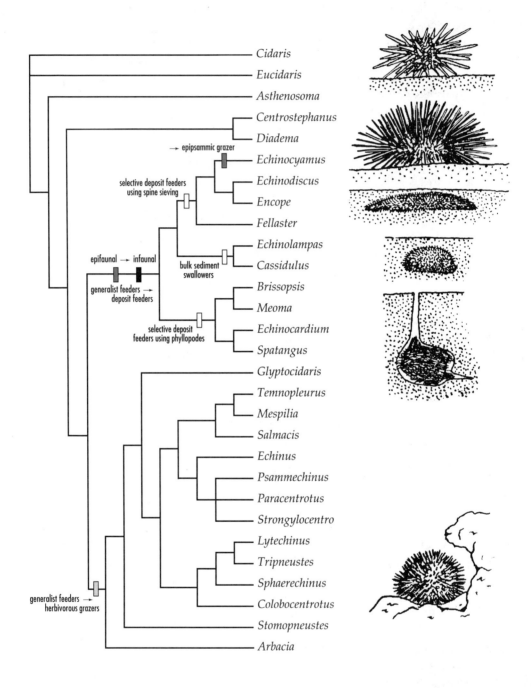

Figure 20.6. The best working phylogenetic hypothesis for echinoids based on the analysis of the combined data, with key ecological innovations related to feeding and mode of life optimized onto the topology.

a dense, uniform aboral spine canopy to create and maintain a sediment-free envelope surrounding the test, and the adoption of unidirectional locomotion allowing functional differentiation of spines for burrowing. The monophyletic origin of an infaunal mode of life (Figure 20.6) is supported by the combined data as well as by adult morphology and SSU rRNA data separately.

The shift to unconsolidated sediments was also accompanied by a change in feeding strategy. All regular echinoids possess a dental apparatus known as Aristotle's lantern, which was present in late Palaeozoic echinoids prior to the diversification of the modern fauna. The great majority of these lantern-bearing echinoids are grazers, feeding on plants, algae, and sessile invertebrates. They can also feed by using their lantern to scoop up bottom sediment, but this mode of feeding is not common except in mud swallowing deep-sea echinothuriods.

Irregular echinoids, by contrast, are all specialist deposit feeders using adoral tubefeet to manipulate particles into the peristome. The majority have completely lost their lantern apparatus. Cassiduloids are primitively bulk-sediment swallowers, passing large quantities of sediment through their guts. Sediment is collected by oral tubefeet and passed to the mouth. Because they rely on suckered tubefeet for harvesting food particles, they appear to be restricted to sand- or gravel-size sediments. Spatangoids, on the other hand, feed using enlarged penicillate tubefeet (phyllopodes) which are brushlike and use sticky mucus for capturing particles. Probably because of this morphological innovation, spatangoids have been very successful in exploiting fine-grained sediments that have proved inaccessible to cassiduloids (Smith 1984). At the ordinal level of our analysis (Figure 20.6), feeding strategy of echinoids thus shows remarkable consistency. Even taking into account taxa not covered in our analysis, there is little evidence of convergent evolution in broad feeding strategies among echinoids.

Lecithotrophic vs. planktotrophic development

Most echinoids show planktotrophic development, in which they pass through a free-swimming larval stage that must feed and grow in the plankton before metamorphosizing into adults. However, some taxa show lecithotrophic development, in which the larvae do not feed but develop directly from a relatively large, yolk-rich egg. In many cases, lecithotrophy is associated with egg brooding. There are a few taxa with facultatively planktotrophic larvae, in which the larvae live for a short time in the plankton but do not have to feed in order to undergo metamorphosis to the adult stage.

It is clear from the distribution of larval types (Figure 20.7) that evolution has consistently been from planktotrophic to lecithotrophic development (Strathmann 1985; Emlet, 1990). This transition involves an increase in larval size and a reduction/loss of organs associated with flotation and particle capture. The transformation from planktotrophic to non-planktotrophic larva thus involves a significant simplification of the larval body plan, achieved largely through the retention of character-states present in the early developmental stages of planktotrophic larvae (i.e., arms rudimentary or absent; skeletal basket of rods absent). Such gross simplification through retarded development appears at least five times independently in the echinoid

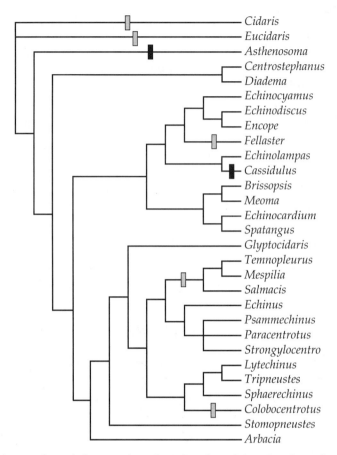

Figure 20.7. The best working phylogenetic hypothesis for echinoids based on the analysis of all data combined, with larval life-history strategies optimized onto the topology. Black bars = lecithotrophic development; gray bars = lecithotrophic development known in other members of the same family.

phylogeny (for detailed discussion see Wray 1996), representing a major change in development and larval ecology that has evolved independently many times.

Genetic vs. morphological divergence

Our combined-evidence phylogeny can be used to compare rates of morphological and molecular evolution over geological time. Optimizing character transformations onto the combined-evidence tree provides a working estimate of the numbers of morphological and molecular changes that have occurred along each branch (see Figure 20.8). For this analysis, we focused on the ten taxa common to all four data sets so that comparisons would not be biased by differences in sampling. Optimization was carried out manually for each character and characters were partitioned equally among branches where any doubt existed as to their correct position (i.e., where using the DELTRAN or ACCTRAN option in PAUP suggested alternative character optimizations). Multiple consecutive deletions or insertions were treated as single events (apomorphies). Then, using the fossil record of first occurrences, the cladogram was calibrated to derive an evolutionary tree based on geological time. For each branch, we calculated estimates of the numbers of transformations in adult and larval morphological character-states, the numbers of fixed point mutations in SSU and LSU rRNA gene sequences, and the lengths of branch durations. Direct comparisons of the rates of morphological and molecular evolution over geological time are then possible (Smith et al. 1995a,b).

This approach involves two major assumptions. First, divergence in the molecular sequences chosen for study must not be approaching saturation. As saturation levels increase, multiple hits become sufficiently common to generate substantial underestimates of the number of mutations along all but the most recent branches. To minimize this problem, we removed all regions of the molecule where there is any ambi-

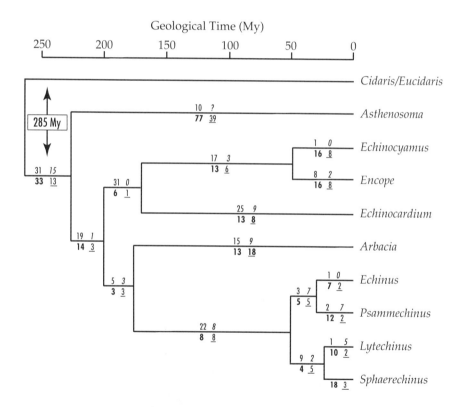

Figure 20.8. Evolutionary tree constructed for the ten taxa common to all four data sets. Branch lengths are calibrated against geological time based on the first known appearance of one or the daughter taxa in the fossil record. Estimated durations of branches can be read directly off the time scale at the top, except for the basal dichotomy where both sides are combined. Above each branch are the numbers of morphological character-state changes assigned to that branch: the first figure (roman) is based on adult morphology, the second (*italic*) on larval morphology. Below each branch are the numbers of molecular character-state changes assigned to that branch: the first figure (**bold**) is based on SSU rRNA sequence data, the second (underlined) on LSU rRNA sequence data.

guity concerning alignment. Second, we must assume that our set of morphological characters are effectively a random sample of all possible characters that could be defined for echinoids. There is no objectivity in defining morphological characters, and characters range from trivial to fundamental in nature, so we must simply assume that the characters used are not biased in any way. As in previous studies, we employ the nonparametric Spearman rank-correlation test to determine how strongly our variables are correlated (Smith et al. 1992, 1995; Omland 1994).

Spearman rank-correlation coefficients for all paired variables are plotted in Figure 20.9. From these we draw the following conclusions:

1. All four kinds of morphological and molecular characters have undergone divergence with comparable levels of correlation with estimated branch duration. Because evolutionary changes accrue over time to both perceived morphology (through character-state transformations) and gene sequences (through fixed point mutations), on average there should be a positive correlation between time and the number of observed evolutionary changes. As a result, the amount of morphological and molecular divergence should also be positively correlated, as has indeed been found previously for echinoids (Smith et al. 1992) and ducks (Omland 1994). The accrued changes in molecular data sets are no more strongly correlated with elapsed time than are those in morphological data sets.

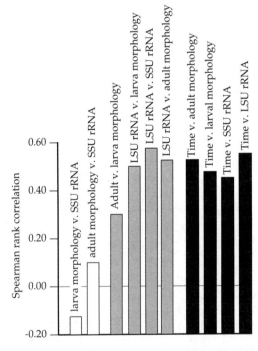

Figure 20.9. Spearman rank-correlation coefficients derived from pairwise comparisons of numbers of morphological and molecular changes assigned to branches on the combined-data topology. All comparisons involving temporal duration are based on the data presented in Figure 20.8. Other comparisons are based on the full set of taxa available for comparison.

2. The estimated numbers of character-state transformations in morphological data show little or no overall correspondence with those for molecular data. Although morphological and molecular changes are both positively correlated with time, the stochastic nature of evolutionary change has meant that morphological and molecular evolution have proceeded more or less independently. There is no correspondence whatsoever between rates of SSU rRNA sequence evolution and morphological evolution. Branches where morphological changes are concentrated are not usually branches along which gene sequence mutations are concentrated, or vice versa. For LSU rRNA sequence data, there is evidence of a positive correlation between rates of morphological and molecular evolution, but no greater than might be expected from the fact that both are very roughly correlated with elapsed time.

3. The correlation between morphological and molecular change is considerably weaker for more recent branches (< 65 Mya) – associated with the diversification of spatangoids, clypeasteroids, and camarodonts – than it is for older branches. This is almost entirely due to the fact that a few terminal branches of relatively recent origin show a substantial number of point mutations (*Temnopleurus* among the camarodonts, and *Encope* and *Echinodiscus* among the irregular echinoids). The reason for this is unknown, but may have to do with saturation effects, or with the operation of the central limit theorem on morphological and molecular changes over long periods (see Lewontin 1974).

4. Changes in larval morphology have occurred independently of those affecting adult morphology. There is no concerted evolution affecting both adult and larval stages (see Smith et al. 1995a for a more detailed account). This is hardly surprising given that larva and adult lead such totally different modes of life (planktotrophic suspension-feeders vs. vagile, benthic deposit-feeders or grazers).

5. There is a positive correlation between rates of LSU and SSU sequence evolution, but it is only a little greater than that shown between molecular data and estimated branch duration. SSU and LSU rRNA genes exist on eukaryotic chromosomes in multiple copies, separated by 5.8S genes and transcribed spacers. These DNA arrays evolve in a concerted fashion (Hillis and Dixon 1991) such that we might expect rates of SSU and LSU evolution to be related. Why we have no evidence for stronger correlation is a puzzle. Perhaps our LSU data set, which is derived from less than a tenth of the full gene sequence, is not representative of the entire gene. Alternatively, the stochastic nature of evolutionary change may combine with relatively small sequence samples to produce a simple explanation for the low correlation between rates of LSU and SSU rRNA sequence evolution.

Our results show that not only has the morphological evolution of echinoids proceeded erratically over geological time, but so has their molecular evolution. Furthermore, larval life history appears to have evolved more or less independently of adult morphology in the echinoids. Finally, and perhaps most surprisingly, we found little evidence of correspondence between the rates of sequence evolution in two ribosomal genes, or between the amounts of molecular and morphological evolution, at least over the past 250 million years.

References

Bledsoe A. H., and Raikow, R. J. 1990. A quantitative assessment of congruence between molecular and nonmolecular estimates of phylogeny. Journal of Molecular Evolution 30:247–259.

de Queiroz, A. 1993. For consensus (sometimes). Systematic Biology 42:368–372.

Eernisse, D. J., and Kluge, A. G. 1993. Taxonomic congruence versus total evidence, and the phylogeny of amniotes inferred from fossils, molecules and morphology. Molecular Biology and Evolution 10:1170–1195.

Emlet, R. B. 1990. World patterns of developmental mode in echinoid echinoderms. Pp. 329–335 in *Advances in invertebrate reproduction*, M. Hoshi and O. Yamishita, eds. Amsterdam, The Netherlands: Elsevier.

Emlet, R. B., McEdward, L. R., and Strathmann, R. R. 1987. Echinoderm larval ecology viewed from the egg. Pp. 55–136 in *Echinoderm studies 2*, M. Jangoux and J. M. Lawrence, eds. Amsterdam, The Netherlands: A. A. Balkema.

Felsenstein, J. 1978. Cases in which parsimony or compatibility methods will be positively misleading. Systematic Zoology 27:401–410.

Felsenstein, J. 1988. Phylogenies from molecular sequences: inference and reliability. Annual Review of Genetics 22:521–565.

Hedges, S. B., Moberg, K. D., and Maxson, L. R. 1990. Tetrapod phylogeny inferred from 18S and 28S ribosomal RNA sequences and a review of the evidence for amniote relationships. Molecular Biology and Evolution 7:607–633.

Hedges, S. B., and Sibley, C. G. 1994. Molecules vs. morphology in avian evolution: the case of the "pelecaniform" birds. Proceedings of the National Academy of Sciences, USA 91:9861–9865.

Hillis, D.M. 1987. Molecular versus morphological approaches to systematics. Annual Review of Ecology and Systematics 18:23–42.

Hillis, D. M. 1995. Approaches for assessing phylogenetic accuracy. Systematic Biology 44:3–16.

Hillis, D. M., and Dixon, M.T. 1991. Ribosomal DNA: molecular evolution and phylogenetic inference. Quarterly Review of Biology 66:411–453.

Kier, P. M. 1982. Rapid evolution in echinoids. Palaeontology 25:1–9.

Kluge, A. G. 1989. A concern for evidence and a phylogenetic hypothesis of relationships among *Epicrates* (Boidae, Serpentes). Systematic Zoology 38:7–25.

Kluge, A. G., and Wolf, A. J. 1993. Cladistics: what's in a word? Cladistics 9:183–199.

Lafay, B., Smith, A. B., and Christen, R. 1995. A combined morphological and molecular approach to the phylogeny of asteroids (Asteroidea: Echinodermata). Systematic Biology 44:190–208.

Lanyon, S. M. 1988. The stochastic model of molecular evolution: what consequences for systematic investigations? Auk 105:565–573.

Larson, A. 1994. The comparison of morphological and molecular data in phylogenetic systematics. Pp. 372–390 in *Molecular ecology and evolution: approaches and applications*, B. Schierwater, B. Streit, G. P. Wagner, and R. De Salle, eds. Basel, Switzerland: Birkhäuser Verlag.

Lecointre, G., Philippe, H., Lê, H. V. L., and Le Guyader, H. 1993. Species sampling has a major impact on phylogenetic inference. Molecular Phylogeny and Evolution 2:205–224.

Lewontin, R. C. 1974. *The genetic basis for evolutionary change*. New York, NY: Columbia University Press.

Littlewood, D. T. J., and Smith, A. B. 1995. A combined morphological and molecular phylogeny for sea urchins (Echinoidea: Echinodermata). Philosophical Transactions of the Royal Society of London, Series B 347:213–234.

Marshall, C. R. 1992. Substitution bias, weighted parsimony and amniote phylogeny as inferred from 18S rRNA sequences. Molecular Biology and Evolution 9:370–373.

McNamara, K. J. 1990. Echinoids. Pp. 205–231 in *Evolutionary trends*, K. J. McNamara, ed. London, England: Belhaven Press.

Miyamoto, M. M., and Fitch, W. M. 1995. Testing species phylogenies and phylogenetic methods with congruence. Systematic Biology 44:64–76.

Nei, M. 1991. Relative efficiencies of different treemaking methods for molecular data, Pp. 90–128 in *Phylogenetic analysis of DNA sequences*, M. M. Miyamoto and J. Cracraft, eds. Oxford, England: Oxford University Press.

Omland, K.E. 1994. Character congruence between a molecular and a morphological phylogeny for dabbling ducks (*Anas*). Systematic Biology 43:369–386.

Patterson, C., Williams, D. M., and Humphries, C. J. 1994. Congruence between molecular and morphological phylogenies. Annual Review of Ecology and Systematics 24:153–188.

Penny, D., and Hendy, M. D. 1986. Estimating the reliability of evolutionary trees. Molecular Biology and Evolution 3:403–417.

Philippe, H., and Douzery, E. 1994. The pitfalls of molecular phylogeny based on four species, as illustrated by the Cetacea/Artiodactyla relationships. Journal of Mammalian Evolution 2:133–152.

Smith, A. B. 1984. *Echinoid palaeobiology.* London, England: Allen and Unwin.

Smith, A. B. 1994. *Systematics and the fossil record: discovering evolutionary patterns.* Oxford, England: Blackwells Scientific Publications.

Smith, A. B., and Hollingworth, N. T. J. 1990. Tooth structure and phylogeny of the Upper Permian echinoid *Miocidaris keyserlingi.* Proceedings of the Yorkshire Geological Society 48:47–60.

Smith, A. B., Lafay, B., and Christen, R. 1992. Comparative variation of morphological and molecular evolution through time: 28S ribosomal RNA versus morphology in echinoids. Philosophical Transactions of the Royal Society of London, Series B 338:365–382.

Smith, A. B., and Littlewood, D. T. J. 1994. Paleontological data and molecular phylogenetic analysis. Paleobiology B20:259–273.

Smith, A. B., Littlewood, D. T. J., and Wray, G. A. 1995a. Comparing patterns of evolution: larval and adult life history stages and ribosomal RNA of postPalaeozoic echinoids. Philosophical Transactions of the Royal Society of London, Series B 349:11–18.

Smith, A. B., Paterson, G. L. J., and Lafay, B. 1995b. Ophiuroid phylogeny and higher taxonomy: morphological, molecular and palaeontological perspectives. Zoological Journal of the Linnean Society 114:213–243.

Smith, A. B., and Wright, C. W. 1990. British Cretaceous echinoids. Part 2, Echinothurioida, Diadematoida and Stirodonta (1, Calycina). Palaeontographical Society Monographs, pp. 101–198, pls 33–72.

Strathmann, R. R. 1985. Feeding and nonfeeding larval development and life history evolution in marine invertebrates. Annual Review of Ecology and Systematics 16:339–361.

Swofford, D. L. 1991. When are phylogeny estimates from molecular and morphological data incongruent? Pp. 295–333 in *Phylogenetic analysis of DNA sequences,* M. M. Miyamoto and J. Cracraft, eds. Oxford, England: Oxford University Press.

Swofford, D. L. 1993. *PAUP: Phylogenetic analysis using parsimony, vers. 3.1.1.* Champaign, IL: Illinois Natural History Survey.

Templeton, A. 1983. Phylogenetic inference from restriction endonuclease cleavage site maps with particular reference to the evolution of humans and the apes. Evolution 37:221–244.

Wray, G. A. 1992. The evolution of larval morphology during the postPaleozoic radiation of echinoids. Paleobiology 18:258–287.

Wray, G. A. 1994. Larval morphology and echinoid phylogeny. P. 921 in *Echinoderms through time,* B. David, A. Guille., J. P. Feral, and M. Roux, eds. Rotterdam, The Netherlands: A. A. Balkema.

Wray, G. A. 1996. Parallel evolution of nonfeeding larvae in echinoids. Systematic Biology 45:308–322.

21 HOW FAST IS SPECIATION? MOLECULAR, GEOLOGICAL, AND PHYLOGENETIC EVIDENCE FROM ADAPTIVE RADIATIONS OF FISHES

Amy R. McCune

Species flocks of fishes in freshwater lakes – including the cichlids of African rift lakes (Fryer and Iles 1972; Greenwood 1981; Meyer et al. 1990) and the cyprinodontids of Lake Titicaca (Parenti 1984; Parker and Kornfield 1995) – involve some of the most dramatic instances of adaptive radiation and explosive speciation on earth (Brooks 1950; Echelle and Kornfield 1984; Martens et al. 1994). Like oceanic islands, newly formed lakes are often relatively isolated from sources of potential colonists. Newly formed lakes and islands are biotically impoverished and provide rich ecological opportunities to early colonists (e.g., see Carlquist, 1974; McCune 1990; Givnish et al. 1995). Thus, evolutionary biologists have long been interested in both lakes and islands as hotbeds of ecological diversification and rapid speciation.

Despite our presumption that adaptive radiations of fishes in lakes are rapid, or even "explosive," data on how long speciation takes in these lacustrine radiations of fishes have not been compared explicitly, and assessments of what constitutes rapid speciation are quite varied throughout the evolutionary literature. For example, Futuyma (1986) gives several examples of rapid speciation which include the origins of (1) five species of cichlids in the 4,000-year lifetime of Lake Nabugabo in East Africa; (2) several species of Death Valley pupfish in 20,000 to 30,000 years; and (3) two genera of mammals – *Thalarctos* (polar bear) and *Microtus* (vole) – during the Pleistocene, over a period of ca. 1.64 million years (My). Examples of rapid speciation in the paleontological literature involve time spans ranging from 0.01 to 0.04 My for snails in Lake Turkana in Kenya (Williamson 1981), 0.2 My for trilobites (Cisne et al. 1980), and 0.5 to 1.4 My for horses (Hulbert 1993).

Comparative data on speciation rate (SR, species My^{-1}) can yield a relative assessment of how long speciation takes, but the inverse of SR approximates a more direct measure of the time required for speciation (TFS, My species^{-1}). In this paper, I describe several measures of SR and TFS and discuss the role of phylogenetic topology in the estimation of these measures. Because estimates of both SR and TFS depend critically on estimating the time at which members of a clade began to diversify, I begin by discussing the advantages and limitations of different types of geological or genetic dating. I then summarize data on average SR and TFS for adaptive radiations of fishes within lakes, reviewing the pertinent literature and explaining the source of geological or genetic estimates for the origin of each adaptive radiation. To provide a broader context, I compare estimates of SR and TFS for fish radiations in lakes to several widely known radiations of animals on oceanic islands. I conclude by discussing how mode of speciation might account, in part, for the apparently more rapid speciation of endemic fishes in lakes than for animals on islands.

Methods

Literature survey and taxa studied

I have tried to include all examples of endemic adaptive radiations of fishes in lakes, but I do not claim that my review is exhaustive. Sources included the literature on species flocks of fishes, recent symposia (e.g., Echelle and Kornfield 1984; Martens et al. 1994) on this topic, and computer keyword searches on speciation, fishes, and speciation rates. In general, criteria for inclusion of a study were existence of reasonable estimates for (1) the number of endemic taxa involved in a radiation, and (2) the age of such endemics or the radiation as a whole, based either on geological dating or genetic divergence. For genetic divergence, I generally based my calculations on DNA sequence data, but in one case where such data were unavailable I used protein electrophoretic data for comparison with geologically estimated rates. Though not terribly restrictive, these criteria forced the omission of some notable cases, such as the cottoid fishes in Lake Baikal, where sequence data are only available for a few species (e.g., Grachev et al. 1992) and there is no consensus about even the geological age of the basin (Mazepova 1994), let alone the age of the lake. By focusing on adaptive radiations, I also excluded informative studies on pairs of species (e.g. Kirkpatrick and Selander 1979; Schluter and McPhail 1992).

Accurate dates of clade origins are critical to computation of SR and TFS, so I have verified dates from original sources and explained the empirical basis for the customarily cited geological dates for each lake. Computation of speciation rates also depends on accurate estimates of species number. For taxonomically more diverse or less intensively studied groups (e.g., cichlids, Hawaiian drosophilids, Titicaca cyprinodontids, Newark semionotids), species number is an estimate which may include known but undescribed species, and excludes extinct or undiscovered taxa. I have used data on species numbers uncritically, because it was impractical – and, in most cases, beyond my expertise – to critique species-level taxonomy of various groups. Fortunately, the effects of changes in species number on estimates of speciation rates are likely to be small relative to the effects of variation in geological or genetic estimates of age.

To provide a context for the fish data, I selected additional data for several well-known examples of animal speciation on islands, but made no attempt to be comprehensive. For the purpose of calculating speciation rates, island and lake faunas share three important characteristics. First, the formation of oceanic islands via volcanism and the formation of lakes via rifting, volcanism, or glaciation are relatively discrete events that can often be dated geologically with precision. Second, many oceanic islands and isolated lakes have produced one or more endemic radiations in some of the groups that reached them. Finally, the time of origin of such radiations can be associated with that of the islands/lakes themselves. In particular, the maximum possible age of an endemic radiation can reasonably be estimated as the time of origin of an island or lake; this maximum age may be close to the actual age of a radiation, to the extent that radiation is most likely in groups originating from one or a few colonists that dispersed to an island/lake soon after its origin (e.g., see Givnish et al. 1995). For island radiations, the time of clade origin was simply assumed to be the time of island origin.

Sources of dating, accuracy, and assumptions

All methods of dating species divergence require certain assumptions. In this section I review the assumptions inherent in the geological dating of lakes, as well as the assumptions necessary to estimate time of divergence based on genetic divergence data.

GEOLOGICAL DATING OF LAKES – At best, any geologically derived estimate for the origin of an adaptive radiation by dating the origin of its lake can only produce a maximum age for the radiation, assuming the radiation occurred *in situ*. If an adaptive radiation did indeed begin early in the history of a lake, then this geological estimate is useful. However, initial colonization of a newly formed lake may be long delayed by the absence of watershed connections or deficiencies in environmental features, such as water chemistry or structural habitat. Use of geological dates for the origin of lakes also entails the assumption that the taxa in question did not evolve in previously existing lakes. Determining whether there were previously existing lakes depends on stratigraphic study of (often inaccessible) sediments underlying an existing lake, or an interpretation of a possible relationship of exposed sediments to inaccessible deposits. Lake Victoria, for example, was long considered to be several million years older than the current published estimate of 0.25 to 0.75 My because nearby older lacustrine sediments – now known to be unrelated to Lake Victoria – were once considered to be a previous incarnation of the lake (see Temple 1969a).

When dating the origin of a lake, it is important to recognize the difference between a date of origin for the lake basin (i.e., the topographic basin that can hold water) and the time this basin filled with water. Traditional geologic dating generally dates the origin of the basin, based on biostratigraphic methods and/or radiometric dates for volcanic rocks associated with basin formation. But to determine the date of origin of an endemic radiation, the most useful date is the beginning of the most recent episode in which the basin held water. The distinction between the origin of the topographic basin and the origin of a lake filling that basin is extremely important. For example, while the basin occupied by Lago Chichancanab in Mexico has been isolated since the early Pleistocene (ca. 1.64 million years ago [Mya] according to Humphries 1984, source of geological data not cited), the earliest known lacustrine sediments from cores are ^{14}C-dated at 28,000 years (Covich and Stuiver 1974; Hodell et al. 1995). Furthermore, the age of earliest lake sediments is not always the most relevant date for origin of an endemic radiation. While a lake basin may persist for millions of years, it may be filled with a habitable lake for only part of its history. In Lago Chichancanab, although the oldest lacustrine sediments are 28,000 years old, sediment cores show a layer of terrestrial sediments, containing terrestrial fossils, was deposited across the entire basin only 8,000 years ago (Covich and Stuiver 1974; Hodell et al. 1995), suggesting that the most appropriate geologically derived date for the origin of the endemic fauna is 8,000 years.

The difference between the times of origin of lake basins and habitable lakes is most clearly illustrated when the entire basin history is preserved, as in the late Triassic lacustrine deposits of the Newark Basin in eastern North America. The Newark Basin formed as a rift valley about 225 Mya (Olsen 1986). For the following 35 My, Newark Basin sediments show repeating sequences of lake formation and evaporation: deep-lake sediments are followed by sediments deposited in progressively shallowing water; followed by terrestrial deposits containing fossil soils, dinosaur footprints, and roots preserved *in situ*; followed by sediments deposited in progressively deepening water; followed by deep-lake sediments. The fish faunas in successive units of deep-lake units (separated stratigraphically by terrestrial deposits) can be quite distinct, giving further evidence that successive sedimentary cycles were deposited in distinct lakes that existed at different times (Olsen 1980, 1988). As remarkable as it may seem, these cycles of lake formation and evaporation occurred in deep rift valley lakes that were probably as large as the east African rift lakes (Olsen 1984).

GEOLOGICAL DATING OF ISLANDS – As with lakes, the geological dating of islands can provide an estimate for the age of an endemic adaptive radiation, but similar assumptions must be made. The age of an island's origin is the maximum age for an endemic radiation, but provides an overestimate if colonization occurred well after the island was formed. Perhaps less obvious is that islands may be younger than the initial origin of their endemic biota if older, now submerged, islands were the site of initial colonization and diversification. For example, recent studies of the Galapagós Islands suggest that submerged seamounts nearby are former islands and are 4 My older than the oldest exposed island (Christie et al. 1992). Similarly, while the oldest of the emergent Hawaiian Islands was formed 5.6 Mya, an extensive submerged chain of seamounts to their northwest indicates that this archipelago formed 65 Mya (Carson and Clagne 1995).

ACCURACY OF GEOLOGICAL DATES – The reliability and precision of geological dating varies widely with the types of data on which a date is based, and also depends on when the date was estimated. Dates for basins are generally based on stratigraphic data and/or radiometric dates for volcanic rocks associated with the formation of a topographic basin (e.g., a lava flow damming a valley). While the accuracy and range of radiometric dates have improved continuously since they were first produced in the early 1900s (Press and Siever 1986; Harland et al. 1990), basin formation may substantially precede initial lake formation and subsequent periods of lake drying. Sediment cores provide detailed information about the history of lakes, but in the largest lakes (e.g., Tanganyika and Malawi), extraordinary lake depth and sediment thickness have prevented paleolimnologists from sampling much of the sedimentary record. As a result, although numerous cores have been taken in the shallow portions of the African great lakes, it has not been feasible to core the oldest sediments in the deepest parts of the basins. Recent study of the history of Lake Tanganyika, for example, has relied on seismic profiles to determine the depth of the sediment basement, and then the age of the basement has been estimated by extrapolating sedimentation rate from ^{14}C-dated cores of the youngest sediments (Tiercelin and Mondeuer 1991; Cohen et al. 1993), which assumes a constant average rate of sedimentation.

DATING LINEAGES BY PROTEIN ELECTROPHORESIS – Vast amounts of electrophoretic data have been amassed for various groups of organisms over the past 30 years. The severe limitations of electrophoretic data for estimating divergence times have been reviewed previously (Avise and Aquadro 1982) and will not be repeated here.

DATING LINEAGES BY DNA SEQUENCE DIVERGENCE – Percent divergence of DNA sequence data has been used, where available, to estimate time of divergence. Absolute calibration of these data is problematic, given that the assumption of constant average rates is controversial and that – even if DNA sequence evolution is reasonably clocklike – different genes in the same taxa and the same genes in different taxa evolve at different rates (see Li and Grauer 1991; Avise 1994). Furthermore, all such calibrations ultimately depend on geologically derived dates. For example, calibration of mitochondrial DNA (mtDNA) data for Lake Victoria cichlids was based on rates of divergence in cytochrome b (cyt b) used for mammals (Meyer et al. 1990), but the actual calibration was derived from the fossil record of ungulates (Irwin et al. 1991). Using this calibration for cichlids assumed that (i) the rate in ungulates was generalizable to other vertebrates, and that (ii) the cyt b rate was generalizable to other mtDNA genes, given that the variation Meyer et al. (1990) studied occurred only in the control region. Recent work suggests that body size, generation time, and metabolic rate may influence rates of divergence in mtDNA (Martin and Palumbi 1993). Martin and Palumbi (1993) argue that the rates of mtDNA divergence in mammals may be as much as 2.5 to 5.0 times greater than those in bony fishes, based on RFLP data for salmon. Thus, where calibration of fish DNA sequence data was based on mammalian rates, I also provide an alternative calibration using the slower rate determined for salmon (Smith 1992). However, calibrating mtDNA sequences in other bony fishes using the salmonid data – which involve RFLPs (restriction fragment length polymorphism) for the entire mtDNA molecule – may not be an improvement over using a calibration based on mammalian cyt b, because there is accumulating evidence that RFLP data for the entire mtDNA molecule is not a reliable estimator of the rate of base substitution for cyt b or certain other regions of the molecule (see McVeigh et al. 1991; Bernatchez and Danzmann 1993).

INTERPRETING VARIATION IN DATING – As discussed above, each method of dating involves certain assumptions and/or limitations. Given the variation in the dates of origin produced, which kind of dating is likely to provide the most accurate estimate of the actual date of origin? Traditional geological dates given for the formation of a basin are the least useful, simply because they can be a very poor estimate of *lake* origin, which has central biological relevance for the radiation of fish. Sedimentological data, when available back to the time of lake origin or the most recent episode of complete drying, are thus preferred over dates of basin origin for the purpose of estimating time of origin on an endemic lineage. However, I know of no reason to prefer paleolimnological (sediment cores or seismic) data over DNA sequence data or vice versa. In several cases, reviewed below, agreement between them is considerable.

Calculation of speciation rate and time for speciation

Speciation rates (SR) were calculated using both linear and logarithmic models. Both estimate a net average rate of speciation, equal to the difference between the gross rates of speciation and extinction. Hence, SR underestimates the gross rates of speciation.

The linear model for estimating SR assumes a comb-shaped tree (Figure 21.1A), in which bifurcations occur asymmetrically in only one lineage. The logarithmic model assumes a more symmetrical or balanced tree (Figure 21.1B), in which bifurcations occur with the same frequency in all lineages.

Let n = the number of known species in a monophyletic clade, t = the age of that clade, and d = the number of episodes of speciation, corrected for those that occur simultaneously. Under the linear model, the net speciation rate would be:

$$SR_{lin} = \frac{n}{t} \quad (21.1)$$

The time required for speciation would thus be:

$$TFS_{lin} = \frac{t}{d} = \frac{t}{n-1} \quad (21.2)$$

Under the logarithmic model, the net speciation rate would be:

$$SR_{ln} = \frac{\ln n}{t} \quad (21.3)$$

Implicit in this model is sequential splitting of 1 species into 2, 4, 8, ..., 2^d species after d episodes of speciation at equal time intervals, each equal to TFS. Given that $n = 2^d$, we obtain

$$d = \frac{\ln n}{\ln 2} \quad (21.4)$$

so that

$$TFS_{ln} = \frac{t}{d} = \frac{t \ln 2}{\ln n} \quad (21.5)$$

These calculations provide only approximations of the average net rate of speciation and time required for speciation. Actual rates will vary through time and depend on the phylogeny of the groups in question (see Discussion).

Results

Data for the numbers of endemic species in different clades of lacustrine fishes and the estimated dates of origin of those clades are listed in Table 21.1. The basis for each estimated age is given and explained in more detail below.

General patterns

Independent estimates of the age of a clade based on different types of data can vary by as much as two orders of magnitude. For example, the origin of the Lago

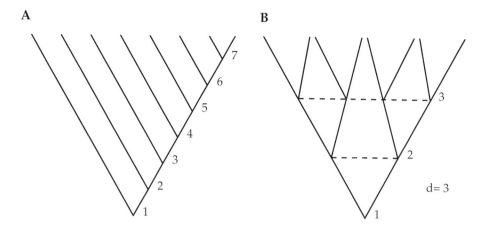

Figure 21.1. Two phylogenetic topologies, corresponding to linear vs. logarithmic models used to calculate speciation rate (SR) or time required for speciation (TFS). (**A**) Comb-shaped tree, assumed in linear estimates of SR and TFS; (**B**) Balanced tree, assumed in logarithmic estimates. Most actual phylogenies probably lie somewhere between these extremes. Sets of simultaneous divergences are numbered on both trees. In (**A**) speciation events occur sequentially, so there are seven episodes of divergence (d=7; see text); in (**B**) there are three sets of simultaneous divergences, involving 1, 2, and 4 cladogenetic events (d=3).

Chichancanab basin is 1.64 My, yet the last episode of complete drying in the basin is thought to be only 8,000 years ago (Table 21.1). In this case, the geological date for basin origin is clearly inappropriate and can be dismissed in preference for the one based on sedimentological data. In other cases, such as the Tanganyikan cichlids, it is more difficult to disregard any of the estimates, which vary by 10-fold (1.2–12 My for Tropheini). For most radiations, however, the variation in age estimates is far less, and in the notable cases of Lake Chichancanab *Cyprinodon* and Barombi Mbo cichlids, there is close agreement between age estimates based on sedimentology and DNA sequence divergence.

A plot of the logarithm of the number of species within a clade vs. age shows the expected tendency for there to be more species in older clades (Figure 21.2). Even this pattern is not without exception, however, given that the two very species-rich clades of Lake Malawi cichlids (◇) and Lake Victoria cichlids (○) are also very young. Despite the considerable variation in geologically and genetically estimated dates of origin for many clades of fishes (Table 21.1), estimates for the origins of Lake Tanganyikan cichlids stand out as being so variable (those for Ectodini range from 2.4–12 My) as to be uninformative. Perhaps the most striking pattern apparent in Figure 21.2 is that lacustrine fish clades are generally younger (≤ 3 My) than those tabulated of island birds and arthropods (≥ 5 My).

Calculated values of SR and TFS, based on both linear and logarithmic models, are summarized in Table 21.2. This table includes data on fishes and some of the best studied groups of terrestrial animals that have speciated on island archipelagos. For the fishes surveyed, the more conservative measure of how long speciation takes (TFS_{ln}) generally falls between about 1,500 to 300,000 years, with the exception that

Table 21.1. Ages of clades of endemic fishes in lakes, estimated from geological and genetic data. Reference for species number is given when it differs from source of age data. All endemic complexes thought to be monophyletic, except where noted. For greater detail about dating and evaluating the reliability of age estimates, see text.

Lake	Taxon	Number of endemic species	Age (My)	Source of age estimate
Lake Victoria, East Africa	cichlids	300 (Meyer 1993)	.25–.75	geological date based on traditional stratigraphic, paleontological, and archeological evidence (Bishop 1969; Temple 1969a)
			.2	803 bp of mtDNA sequence data for 14 species (Meyer et al. 1990), including 363 bp cyt b calibrated by rate of divergence in mammals (2.5% My^{-1})
			.5–.56	mtDNA sequence data from Meyer et al (1990), using slower calibration of 0.9–1.0% My^{-1} (Martin & Palumbi 1993); calibration based on RFLP data for mtDNA divergence in salmon (from Smith 1992, used by Martin and Palumbi 1993)
			.012	data from 1994 field season includes seismic profiles and sediment cores from the deepest part of the basin showing a terrestrial deposit across the entire basin (Johnson et al. 1996)
Lake Nabugabo, East Africa	cichlids	5 (Fryer and Iles 1972; monophyly not demonstrated)	.004	geological date; ^{14}C date for a water-rolled piece of charcoal from a strand line of L. Victoria; at the level of this strand line, lakes Victoria and Nabugabo would have been connected (Bishop 1969; Temple 1969b)
Lake Malawi, East Africa	cichlids	400 (Meyer 1993)	1–2	geological date cited by Meyer (1993) from Fryer & Iles (1972) who cite Fryer (1959), who based date on biostratigraphic studies of Dixey (1941) and Hopwood (pers. comm.)
			0.7	803 bp of mtDNA sequence data from Meyer et al. (1990) for 24 species, using calibration of 2.5% My^{-1} for cyt b divergence in mammals (Meyer 1993)
			1.75–2.1	mtDNA sequence data as above, recalibrated by 0.9–1.0 % My^{-1} derived from RFLP data for mtDNA in salmon (Smith 1992 in Martin and Palumbi 1993)

Location	Fish group	# species	Age (My)	Notes
Lake Tanganyika, East Africa	cichlids	171 (Sturmbauer and Meyer 1993); not monophyletic	9–12	sedimentological date for earliest lake sediments deposited in basin, based on depth of lake's sedimentary basement from reflection seismic data; sediment thickness calibrated by sedimentation rates determined by radiocarbon dating of recent sediment cores (Cohen et al. 1993; Tiercelin and Mondeguer 1991)
	cichlids – Tropheini only	6	1.2–2.4	842 bp of mtDNA sequence data for all six species; 2.5% My^{-1} calibration implicit (Sturmbauer et al. 1994)
			3.0–6.0	recalibrated using 1% My^{-1} (Smith 1992 in Martin and Palumbi 1993)
	cichlids – Ectodini only	30	2.4–3.5	852 bp mtDNA sequence data for 12 species of the 12 Ectodini genera; authors report that Ectodini are twice as old as the Malawi flock and six times older than the Victoria flock; 2.5% My^{-1} calibration implicit (Sturmbauer et al. 1994)
			6.0–8.75	recalibrated using 1% My^{-1} (Smith 1992 in Martin and Palumbi 1993)
	cichlids – Lamprologini only	85	2.4–3.5	402 bp of cyt b from 16 species and 452 bp control region and tRNA genes from 25 species; 2.5% My^{-1} calibration implicit (Sturmbauer et al. 1994)
			6.0–8.75	recalibrated using 1% My^{-1} (Smith 1992 in Martin and Palumbi 1993)
Barombi Mbo, Cameroon	cichlids	11 (Schliewen et al. 1994)	2.8–3.1	2.8% divergence (uncorrected) from 340 bp of cyt b sequence (Schliewen et al. 1994); calibrated using 0.9–1.0% My^{-1} (RFLP mtDNA data for salmon from Smith 1992 used by Martin and Palumbi 1993)
			1.1	2.8% divergence, as above, calibrated using 2.5% My^{-1}
			1	K/Ar date for basaltic lava flows on lake shore date origin of crater (Maley et al. 1990)

Location	Taxon	# species	Age (My)	Notes
Lake Lanao, Philippines	cyprinids	18 (Kornfield and Carpenter 1984)	3.6–5.5	K/Ar date for lava flow damming the lake; cited as a pers. comm. from Frey in Lewis (1978)
			.091–2.06	electrophoretic data: 20 loci for the 3 extant species (others became extinct ca. 1960–1980) from Kornfield and Carpenter (1984); calibration of Nei's D using range of rates (c = 0.8–18.0 My) given in Avise and Aquadro (1982)
Lake Titicaca, Peru and Bolivia	cyprinodontids *Orestias*	~22 (Parenti 1984; A. Parker, pers. comm.)	2.8	geological: there have been lakes in the Andean Altiplano lakes for 2.8 My (Dejoux 1994)
			.02–.8	geological: trough of L.Titicaca itself formed ca. 0.8 Mya and significant low lake levels occurred ca. 0.02–0.024 Mya (Dejoux 1994)
			.15	~400 bp control region sequence data (A. Parker, in prep.) being analyzed currently; rough calculation of divergence used calibration of 0.9% My^{-1}, adjusted for control region being non-coding
			.06	~400 bp control region sequence data (A. Parker, in prep.); rough conversion to 2.5% My^{-1} calibration
Lago Chichancanab, Mexico	*Cyprinodon*	5	1.64	geological: date given for age of basin as Early Pleistocene (Humphries 1984); source of geological data unknown
			.028	paleolimnological: ^{14}C dating of oldest lacustrine sediments in basin (Covich and Stuiver 1974; Hodell et al. 1995)
			.008	paleolimnological: ^{14}C dating of the most recent layer of terrestrial sediments containing terrestrial fossils which is continuous throughout the entire basin as shown from multiple sediment cores (Covich and Stuiver 1974)
			.008	mtDNA sequence, calibrated using average genetic distance for Tanganyikan cichlids from Sturmbauer and Meyer (1993); 2.5% My^{-1} implied (Strecker 1996)
Lake P4, Newark Basin New Jersey	*Semionotus*	6 monophyly not demonstrated	.005–.008	sediment thickness, calibrated by sedimentary cycle duration determined by K/Ar data, number of cycles, average sedimentation rates and varve counts (Olsen 1986; McCune 1996)

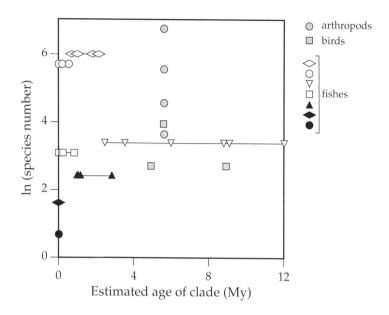

Figure 21.2. Logarithm of species number as a function of the estimated age of a clade. Lines connecting like symbols show the range of age estimates for a given clade. Most clades of fishes (except Tanganyikan cichlids) are less than 3 My old. Estimates for the three Tanganyikan clades are similar and highly variable (see Tables 21.1 and 21.2), and thus uninformative, so only the range (2.4–12 My) for the representative Ectodini is shown here. Island radiations of birds and arthropods are all estimated to be ≥ 5 My old.

all three Tanganyikan cichlid lineages fall well outside that range (0.47 to 4.60 My). The estimates for the ages of Tanganyikan cichlids, however, are too variable to be useful. Estimates of TFS_{ln} for island arthropods and birds are generally higher than those for fishes, falling between 0.6 and 1.3 My (Table 21.2; Figure 21.3).

Review of age estimates for individual case studies

LAKE VICTORIA CICHLIDS – The geological date generally cited for the origin of Lake Victoria is 250,000 to 750,000 years. This is based largely on the stratigraphic and paleontological research of W. W. Bishop in the 1950s and 1960s, as interpreted by Temple (1969a) for a biological audience. More recently, analysis of cores from Lake Victoria has shown that, about 14,000 years ago, the maximum depth of Lake Victoria was at least 75 m below its current depth, leaving a lake much reduced in area or even dry (Kendall 1969; Livingstone 1980). Discovery of such a dramatically reduced lake level led to speculation about possible implications for the origin and evolution of the endemic cichlid fauna (McCune et al. 1984; but see Stager et al. 1986), but without cores from the deepest areas of the lake, it has not been possible to determine the degree of reduction until very recently. During their 1994 field season, Johnson et al. (1996) obtained high resolution seismic profiles and 7 piston cores sampling even the deepest parts of Lake Victoria. Both cores and seismic data show a terrestrial layer, at about 12,000 years ago, which appears to be continuous across the entire basin. These data suggest that Lake Victoria was completely dry then, and

Table 21.2. Speciation rate (SR) and time required for speciation (TFS) for animals on islands and endemic fishes within lakes. Sources of data for island arthropods and birds are cited in the table. For fishes, estimates are based on maximum and minimum ages from Table 21.1. Electrophoretic data are excluded because they are too variable for absolute dating (Avise and Aquadro 1982). Where sedimentological data are available for dating lake origins, stratigraphic dates for basin origins are excluded because the age of the lake, not the basin, is the relevant age, as described in the text. Where sedimentological and DNA sequence data agreed, these dates are given preference over other genetically or sedimentologically derived dates. Where the number of endemics is marked by an asterisk, monophyly has not been demonstrated, so these complexes are treated as individual lineages (i.e., 5 species originating in 4,000 years is treated as each of 5 species having originated in 4,000 years).

Lake or Island	Taxon	Number of endemic species	Age (My)	SR_{lin} (sp/My)	SR_{ln} (My)	TFS_{lin} (My/sp)	TFS_{ln} (My/sp)
Hawaiian Islands	crickets	250 Shaw 1995	5.6	45	.99	.022	.703
Hawaiian Islands	drosophilids	860 Otte 1994	5.6	153	1.21	.006	.574
Hawaiian Islands	honeycreepers	47 Tarr and Fleischer 1995	5.6	8.4	.69	.119	1.008
Hawaiian Islands	spiders (*Tetragnatha*)	100? Gillespie and Croom 1995	5.6	15	.82	.056	.843
Hawaiian Islands	plant bugs (*Sarona*)	40 Asquith	5.6	7	.66	.140	1.052
Galapagos Islands	finches	14 Grant 1986	5–9 Grant 1986; Christie et al. 1992	2–3	.29–.52	.357–.642	1.3–2.4
Lake Victoria	cichlids	300	.012–.2	1500–25,000	28–475	.000044–.001666	.002–.024
Lake Nabugabo	cichlids	5*	.004	250	–	.004	.004
Lake Malawi	cichlids	400	.7–2.0	190–571	3–9	.002–.005	.081–.231
Lake Tanganyika	cichlids–Tropheini	6	1.2–12.0	2.5–5.0	.75–1.49	.200–2.0	.464–4.642
Lake Tanganyika	cichlids–Ectodini	30	2.4–12.0	2.5–12.5	.28–1.42	.3–2.0	.489–2.446
Lake Tanganyika	cichlids–Lamprologini	85	2.4–12.0	7–35.4	.37–1.85	.103–.705	.374–1.87
Barombi Mbo	cichlids	11	1.0–1.1	11	2.2–2.4	.091–.100	.289–.318
Lake Titicaca	cyprinodontids	~22	.020–.15	28–293	3.8–155	.001–.036	.004–.179
Lago Chichancanab	cyprinodontids	5	.008	625	201	.002	.0034
Lake P4, Newark Basin	semionotids	6*	.005–.008	125–200	–	.005–.008	.005–.008

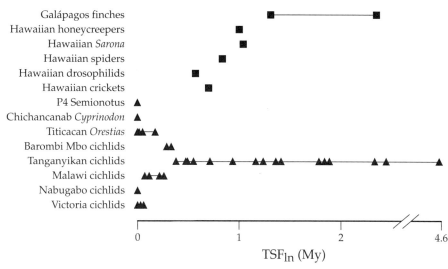

Figure 21.3. Estimates of TFS_{ln} based on a full range of geological and genetic estimates for clade age. For geological estimates, sedimentological dates (where available) for lake origin were used in preference over stratigraphic and radiometric dates for basin formation, just as DNA sequence data are used over electrophoretic data. Thus, dates of basin formation listed in Table 21.1 were excluded for lakes Chichancanab, Titicaca, and Victoria, but retained for Barombi Mbo and Malawi, for which sedimentological dates are not available. Horizontal lines connect varying estimates of TFS_{ln} for the same clade, derived from different estimates of clade age. Data for fishes are represented as triangles while data for island arthropods and birds are represented by squares. Estimates of TFS_{ln} for the three monophyletic lineages of cichlids from Lake Tanganyika are lumped into one line because they are nearly identical and highly variable because of the vast range of age estimates for these lineages. Note that, with the exception of Lake Tanganyikan cichlids, TFS_{ln} is lower for fishes in lakes than for animals on islands. In general, TFS_{ln} for fishes is < 0.3 My, while for island arthropods and birds, TFS_{ln} ranges from about 0.6 to 1.3 My.

imply that the endemic cichlids of Lake Victoria originated as recently as 12,000 years ago.

Meyer et al. (1990) presented evidence for monophyly of the Victoria cichlid flock and, based on mtDNA sequence calibrated by mammalian rates of cyt *b* divergence (2.5% My^{-1}), dated the origin of the Victoria flock at 200,000 years. Martin and Palumbi (1993) have suggested that rates of mtDNA sequence divergence may be slower in fishes than mammals, and that a more appropriate calibration based on salmon might be 0.9–1.0% My^{-1} (data from Smith 1992 in Martin and Palumbi 1993). Using that calibration, the age of the Victoria flock would be estimated as 500,000 years. With either calibration of the genetic data, agreement with the traditional geological date is reasonable. However, neither date based on sequence divergence is consistent with the newest paleolimnological data.

LAKE NABUGABO CICHLIDS – Cichlids from Lake Nabugabo are often cited as an example of rapid speciation in animals (Stanley 1979; Futuyma 1986). Lake Nabugabo is a satellite lake of Lake Victoria, thought to have been separated by declining lake level about 4,000 years ago. Five species of cichlids, whose closest relatives are taxa from Lake Victoria, inhabit Nabugabo, suggesting that all five evolved in no more than 4,000 years. The date for separation of the two lakes is based on a ^{14}C date (3,720

± 120 years; see Bishop 1969 and Temple 1969b) for a piece of water-rolled charcoal found in a cave at a former strand line of Lake Victoria. Lake Nabugabo is at a higher elevation than Lake Victoria, and at the elevation of this strand line, the two lakes would have been connected. Because the closest relatives of each of the Nabugabo cichlids appear to be Victoria endemics, the Nabugabo cichlids themselves are not thought to be monophyletic (Greenwood 1965). Thus, calculations of SR and TFS given in the tables are based on the conservative assumption that each speciation event took a full 4,000 years.

LAKE MALAWI CICHLIDS – The geological age of origin widely reported for Lake Malawi is 1–2 My (Fryer and Iles 1972). This age appears to be based on fossils dated as early to mid-Pleistocene in the underlying Chiwondo beds (Hopwood, pers. comm. to Dixey 1941 in Fryer 1959), but it is not certain that the Chiwondo beds should be considered part of the Malawi lacustrine sequence (A. Cohen, pers. comm.). Based on seismic data, there is evidence of continuous lacustrine sedimentation for at least the last 1 My and there could be Malawi lake sediments as old as 5–6 My (C. Scholz, pers. comm.).

Meyer et al. (1990) sequenced mtDNA for 24 species of Malawi cichlids and they estimated, using the mammalian cyt b calibration of 2.5% My^{-1}, an age of 700,000 years for the Malawi flock. Using a 1% My^{-1} calibration (Martin and Palumbi 1993), the estimated age of Malawi cichlids increases to 1.75–2.1 My.

Estimates of the age of origin of Malawi cichlids based on geological and sequence data (calibrated by the salmonid rate) roughly agree, but paleolimnological studies also show that Lake Malawi has experienced substantial fluctuations in lake level over its history. Using seismic reflection profiles, Scholz and Finney (1994) detected a "major and prolonged lowstand before approximately 78,000 years bp" in which Lake Malawi dropped at least 350 meters. Another major reduction in lake level about 25,000 years ago reduced the surface area of the lake by about 50% (Scholz and Rosendahl 1988). Although neither reduction would have been sufficient to have separated the two deep basins within Lake Malawi, both would have had a major impact on the fishes and may have ramifications for estimating rates of speciation in Malawi cichlids. Owen et al. (1990) argued that Lake Malawi was 120–150 m lower than today between 1390 and 1860 AD, based on sediment coring, radiometric dating, and mapping the bottom topography of the southern arms of the present-day lake. Owen et al. (1990) claimed that a drop in lake level of this magnitude would leave rocky shores around islands dry. Many of these islands are apparently now inhabited by "mbuna" species endemic to only one or two islands (see Chapter 12), and Owen et al. (1990) suggested that these 80+ endemics (of unknown relationships) have evolved in the less than 200 years since the end of the last lake recession.

LAKE TANGANYIKA CICHLIDS – Lake Tanganyika has been much studied by geologists and limnologists in recent years. Using the reflection seismic-radiocarbon method (RSRM), the lake's sedimentary basement has been identified and dated by extrapolation of sedimentation rates derived from radiometrically dated cores of more recent sediments (Cohen et al. 1993). The extrapolation of sedimentation rates involves a similar assumption to that underlying a molecular clock: it is widely recognized that rates of sedimentation vary substantially over short time intervals, and although there are likely to be episodes of erosion and/or non-deposition, average sedimentation rates are reasonably "constant" if estimated over a large enough interval. The RSRM method dates the earliest lacustrine sediments in Lake Tanganyika at 9–12 My (Cohen et al. 1993). However, recent studies of the northern basin of Lake Tanganyika show a complicated history of changing basin morphology and substantial lake level fluctuations (Cohen et al., in prep; Lezzar et al. 1996). These authors document 5 major lake level lows (ranging from 160–700 m below present level) during the last million years. The largest of these (at about 1 Mya) was sufficient to separate the northern and southern basins of the lake (Lezzar et al., in press), but it is clear that there has been at least one major body of water continuously in the Tanganyikan basin for at least the last million years (A. Cohen, pers. comm.). It is certainly also biologically significant that the morphology of Lake Tanganyika has changed dramatically through its history. The flat, pre-rift surface of the region from 3.6–7.4 Mya did not attain the current rift basin morphology until about 0.36–0.39 Mya (Lezzar et al. 1996). It remains unclear how these vast lake-level fluctuations and the changing basin morphology affected the origin and evolution of the Tanganyikan flock, and which dates are relevant for estimating the origin of Tanganyikan endemics.

There is no genetically derived date for the origin of Tanganyikan cichlids because the endemic cichlids are apparently polyphyletic. However, Sturmbauer et al. (1994) sequenced approximately 850 bp of mtDNA for a sample of each of three monophyletic lineages of Tanganyikan endemics, including all 6 species of Tropheini, 12 out of the 30 species in the endemic tribe Ectodini, and 16 of the 85 endemic Lamprologini. They reported that the Tropheini is two times older than the whole Lake Malawi flock and six times older than the Victoria flock, or about 1.2–1.4 My old, and that the Lamprologini and Ectodini are of similar age, about 2.4–3.5 My. Implicit in these age estimates is a 2.5% My^{-1} mammalian calibration; if recalibrated using the 1% My^{-1} suggested by Martin and Palumbi (1993), the age estimates would increase to maxima of 6 My and 8.75 My, respectively (Table 21.1).

BAROMBI MBO CICHLIDS – Using sequence data from 790 bp of mtDNA (cyt b and control region), Schliewen et al. (1994) found the 11 species of cichlids inhabiting Barombi Mbo in Camaroon are monophyletic and have diverged little from each other genetically. The maximum divergence in cyt b (uncorrected, calculated from their Table 1) for any pair of species is 2.8%, suggesting an origin 1.1 Mya, in good agreement with the postulated origin of the Barombi Mbo crater 1 Mya (Maley et al. 1990) if the divergence data are calibrated at 2.5% My^{-1}.

LAKE LANAO CYPRINIDS – Neither the geological nor genetic divergence dates are satisfactory for the Lanao cyprinids. The K/Ar date for the lava flow damming Lake Lanao is based on verbal communication (Frey in Lewis 1978), and using a date for basin formation to estimate age of a clade within a lake in that basin entails several assumptions, previously enumerated. There are electrophoretic data for only 3 of the 18 endemic species, because the other 15 species have recently become extinct, presumably due to human activities (Kornfield and Carpenter 1984). The range of possible rates of divergence for such electrophoretic data varies by two orders of magnitude (Avise and Aquadro 1982), making any conclusion difficult. Unless preserved specimens can be used, sequence data are not likely to be forthcoming. Given the lack of substantiated geological information and sequence data, this case was excluded from Table 21.2 and Figures 21.2 and 21.3.

LAKE TITICACA CYPRINODONTIDS – At present, there are thought to be about 22 species of the cyprinodontid *Orestias* endemic to Lake Titicaca (Parenti 1984; A. Parker, pers. comm.). The age of the oldest lake on the Andean Altiplano has been cited as Pliocene (Parenti 1984; Dejoux 1994). However, beginning about 3 Mya there were at least five distinct lacustrine deposits on the Altiplano, with the maximum extent of each lake occurring at the end of a glaciation (Dejoux 1994). Apparently, the tectonic formation of the present-day Titicaca trough occurred about 0.8 Mya (Lavenu 1992). Analysis of pollen and ostracod distributions in sediment cores suggests a very low lake level about 20,000 to 24,000 years ago (Wirrmann et al. 1992; Ybert 1992). Lack of water in a lake basin is clearly devastating for fish, but substantially lower levels may also devastate lake biotas due to changes in water chemistry or elimination of structural habitats. While the data apparently do not exist to determine the age of Lake Titicaca that would be relevant to the origin of *Orestias*, it seems likely that it would have been after the tectonic origin of the trough at 0.8 Mya and before the identified low at 0.02 Mya.

About 400 bp of mtDNA control region sequence data has been obtained from many of the endemic *Orestias* and is currently being analyzed (A. Parker, University of Maine, pers. comm.). Although the analysis is very preliminary, a rough calculation of maximum divergence (using Martin and Palumbi's [1993] salmon data for calibration) suggests an origin for the clade between 150,000 and 300,000 years ago (A. Parker, pers. comm.), or as little as 60,000 years ago using the 2.5% My^{-1} calibration, both of which are well within the range of geological dates discussed above.

LAGO CHICHANCANAB *CYPRINODON* – As discussed earlier, Lago Chichancanab clearly illustrates the difference between geological age of a basin, the age of the lake within that basin, and the age of a habitable lake. The biostratigraphic age of the basin is early Pleistocene (Humphries 1984), about 1.64 Mya. The earliest lacustrine sediments in Lago Chichancanab are only 28,000 years old (Covich and Stuiver 1974; Hodell et al. 1995), and cores show a hiatus in lacustrine sedimentation about 8,000 years ago. At this time, the sediments and preserved fossils are unquestionably terrestrial (Covich and Stuiver 1974), suggesting that the origin of the five endemic species of *Cyprinodon* (the status of some of these taxa is questionable [U. Strecker, pers. comm.]) occurred within the last 8,000 years. Sequence divergence data are remarkably consistent with this paleolimnological age (Strecker et al., in press). Even electrophoretic data are as consistent with the more recent age for Lago Chichancanab as those data could be. In a survey of 13 loci for 13 populations representing 5 species, the maximum Nei's D for any pair of species is 0.01. Judged against the full range of values for the calibration constant for fishes (c = 0.8–18.0 My [Avise and Aquadro 1982]), the electrophoretic data imply a maximum divergence time between any pair of species of 8,000 to 180,000 years.

NEWARK SEMIONOTIDS – The fishes from a single early Jurassic lake deposit (Lake P4) in the Newark Basin of New Jersey are the subject of ongoing studies (McCune et al. 1984; McCune 1987a,b, 1990, 1996). Based on a broad geographic and stratigraphic survey of species distributions, at least 6 of the 21 semionotids found in the P4 lake deposit are endemic species (McCune 1996). Dating the duration of this former lake is less difficult than dating many modern lakes because the complete sedimentary sequence is exposed, instead of being buried at the bottom of a lake. From its initial formation through maximum lake stand and subsequent evaporation, the duration of Lake P4 is estimated to be about 21,000–24,000 years (Olsen 1980; McCune 1996); all six endemics appear during the first 5,000–8,000 years of lake history. Because these endemic species have not been shown to be monophyletic, calculations of SR and TFS conservatively assume the full 5,000 to 8,000 years for the formation of each species.

Estimation of the complete sequence of lake formation and evaporation is based on studies of hundreds of similar lake formation-evaporation cycles in the Lockatong and Passaic formations in the Newark Basin, derived from (i) extrapolations from counts of varves (annual couplets of sediment layers); (ii) K/Ar dates bracketing total sediment thickness; and (iii) a Fourier analysis of the distribution of sediment types ranked to reflect deposition in varying depths of water (Olsen 1980; 1986). While the P4 lake deposit of the Towaco Formation was not subject to the same intensity of study as the Lockatong and Passaic lake cycles, there are compelling reasons why these data are generalizable to Towaco lake cycles (Olsen 1986; see also McCune 1996).

Discussion

Estimating speciation rate and time for speciation

Both SR_{lin} and SR_{ln} are estimates of net average speciation rate, that is the difference between gross speciation rate and gross extinction rate. Similarly, TFS_{lin} and TFS_{ln} are calculated without taking into account extinction rates. If extinction rates are similar across groups, comparisons of SR and TFS can be made, but if some groups have actually experienced greater proportional extinction, values of SR will be somewhat lower and values of TFS will be somewhat higher. It is difficult to see how a correction for extinction rate could be introduced for the cases described in this paper, though I have included known extinct species (as in the case of Hawaiian honeycreepers) in the tally of species number. Given that SR and TFS are based on net speciation, one must be cautious about interpreting differences in SR and TFS. A lower SR may be a result of either lower speciation or higher extinction.

As noted earlier, SR_{lin} and TFS_{lin} assume a comblike phylogeny (Fig. 21.1A); SR_{ln} and TFS_{ln} assume a fully balanced dichotomous phylogeny (Fig. 21.1B). Most phylogenies lie somewhere between these two extremes, with published phylogenies tending to be more imbalanced than expected under an equal-rate, random-speciation model (Guyer and Slowinkski 1991; Heard 1992, submitted). Which measure of rate is more appropriate will depend on the topology of speciation in a particular clade. To the extent that speciation rates within a clade vary (with the resulting tree being relatively asymmetrical or imbalanced [Heard, submitted]) or that some lineages are better "speciators" than others (e.g., Vrba 1980), a linear measure may better reflect the actual speciation rate. To the extent that all members of a clade were equally likely to speciate (with the resulting tree being symmetrical or balanced), a logarithmic measure would be more appropriate.

The most accurate estimate of TFS must incorporate information about the actual phylogenetic topology of a group. Consider a hypothetical case of 8 endemic species having arisen in 1 million years. TFS_{actual} can vary from 0.14 My to 1.00 My for some simple phylogenetic topologies (Figure 21.4). For example, if the phylogeny is a true polytomy or "star radiation" (Figure 21.4A) – involving simultaneous speciation in 8 allopatric populations (perhaps inhabiting 8 islands or lakes) – TFS_{actual} would achieve its maximum value of 1 My. Of the fully resolved phylogenies illustrated, TFS_{actual} would be greatest (0.33 My) for the balanced tree assumed by TFS_{ln} (Figure 21.4B) and shortest (0.14 My) for the comblike tree assumed by TFS_{lin} (Figure 21.4F). Intermediate topologies (Figure 21.4C,D,E) give intermediate values of TFS_{lin}, ranging from 0.17 to 0.25 My.

Speciation in fish appears to be exceptionally fast

The time required for speciation (TFS) appears to be much less in fishes (excluding Tanganyikan cichlids, for which age estimates are so variable as to be uninformative) than in the groups of island birds and arthropods selected for comparison (Table 21.2; Figure 21.3). Using the more conservative estimates from the logarithmic model, TFS_{ln} in fishes ranges from about 0.0015 to 0.3 My per species, whereas TFS_{ln} in birds and arthropods ranges from about 0.6 to 1.0 My (Table 21.2; Figure 21.3). What could explain such a difference in speciation rate? Below, I discuss two possible artifactual explanations relating to the nature of available data, as well as the possibility that the higher rate of speciation in fishes might reflect different modes of speciation.

SPECIATION APPEARS FASTER WHEN THE INTERVAL OVER WHICH IT IS MEASURED IS SHORTER – While fishes display the lowest values for TFS_{ln}, their values are also estimated over the shortest intervals (Table 21.2; Figure 21.3). Excluding the Tanganyikan cichlids, the fish clades are thought to have arisen within the last 2 My, while the bird and arthropod clades are estimated to be 5 to 6 My old. Rate comparisons should be made over the same time interval (Gingerich 1983; Gould 1984); TFS will increase with age t, other things being equal (see eqs. 21.1–21.5), because the longer the time interval over which speciation is measured, the greater the probability of including episodes of *non-speciation* in the average, thereby reducing estimated SR and increasing estimated TFS. Averaging periods of speciation and non-speciation affects the measured speciation rate in any taxon, living or fossil, but the effect is particularly graphic in the following example from the fossil record.

In Newark semionotids, the measured TFS_{ln} for fishes in the P4 lake cycle is quite low – 5,000 to 8,000 years per species (Tables 21.1 and 21.2). However, there are only about 40 species of semionotids known from Newark Supergroup lacustrine sediments overall, and only one species is known from the first 25 My; all others first appear in the subsequent 2 My. Clearly, the high SR_{ln} was not sustained throughout the approximately 35 million years of Newark sedimentary record or there would

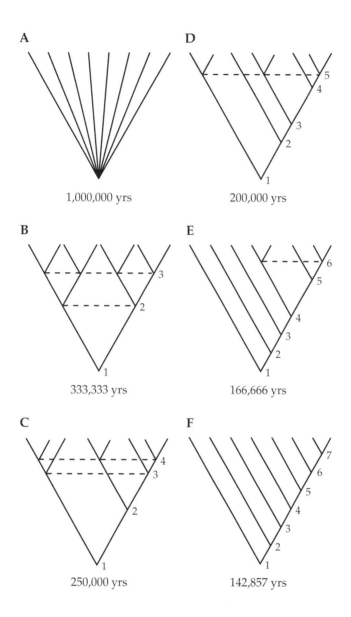

Figure 21.4. Variation in TFS$_{ln}$ with phylogenetic topology. For a hypothetical clade of eight species differentiating in 1 My, TFS$_{ln}$ varies from 0.14 to 1.0 My, depending on phylogenetic topology. Simultaneous episodes of divergence are indicated by a dotted line and numbered, and the estimated TFS$_{ln}$ is given below each cladogram. (**A**) In the case of an unresolved polytomy, speciation may have occurred simultaneously in eight lineages and thus, TFS$_{ln}$ is less than or equal to the maximum value of 1 My. (**B**) For a fully symmetric tree, there are three episodes of divergence, yielding an estimated TFS$_{ln}$ of 0.33 My per speciation event. (**C**) Four episodes of divergence yield a TFS$_{ln}$ of 0.25 My. (**D**) Five episodes of divergence yield a TFS$_{ln}$ of 0.20 My. (**E**) Six episodes of divergence yield a TFS$_{ln}$ of 0.17 My. (**F**) The comblike phylogeny gives the shortest estimated TFS$_{ln}$ of approximately 0.14 My.

have been 4,375 species produced during Newark history. Why was semionotid speciation not sustained throughout Newark history? First, only a fraction of the Newark Basin fossil record reflects lacustrine habitat. About half to two-thirds of each cycle of lake formation and evaporation comprises marshy to terrestrial sediments; while these sediments were being deposited, intralacustrine speciation was not possible. Second, even the lacustrine portions of many Newark sedimentary cycles have not yielded any fishes, perhaps because the lakes in which these sediments were deposited were never colonized by fishes. Thus, in the absence of lacustrine conditions and/or fishes, semionotid speciation would have been impossible throughout much of Newark history. For these fossil fishes – or any taxon, living or fossil – the highest estimates of speciation rate will often be measured over the shortest intervals of time, because they are most likely to isolate periods of speciation from non-speciation. Given that the sedimentary record of lakes often will provide a more detailed chronology than geological studies of oceanic islands, there is a possibility that speciation simply appears to be faster in lacustrine fishes than island arthropods and birds simply because the historical record of lakes is better resolved than that of islands.

TIME SINCE DIVERGENCE DOES NOT NECESSARILY ESTIMATE TIME FOR SPECIATION – Another reason that estimates of speciation rate will be faster over shorter intervals is that any estimate of divergence time, whether molecular or paleontological, will confound divergence during speciation with divergence after speciation. Hence, estimates of speciation rate made over shorter intervals may better isolate divergence during speciation from that of subsequent genetic divergence. Both paleontological and molecular data can help distinguish these two possibilities. Sturmbauer and Meyer (1993), for example, found that although divergence in mtDNA sequences indicated that the Tanganyikan cichlid tribe Ectodini is about five times older than the entire cichlid flock from Lake Malawi, seven of the nine Ectodini lineages were equally divergent. They interpreted their finding as evidence that these seven lineages arose approximately simultaneously, early in the history of Lake Tanganyika. In my own studies of semionotids from the Newark Basin, I found that novel variation – later incorporated into morphologies of new species – appeared first during the earliest quarter (ca. 5,500 years) of lake history (McCune 1990), and that all six endemic semionotid species to appear in Lake P4 initially appeared during the first 5,000 to 8,000 years of the sedimentary record (McCune 1996). Thus, both inference based on mtDNA divergence of Tanganyikan cichlids and direct study of variation and species appearances of fossil semionotids suggest that those radiations did not occur at a constant pace. Instead, speciation appears to have been concentrated in the early phases of each lake's history.

HAVE I INADVERTENTLY COMPARED RAPID SYMPATRIC SPECIATION IN FISHES TO SLOWER ALLOPATRIC SPECIATION IN BIRDS AND ARTHROPODS? – An intriguing possibility is that the more rapid speciation in fishes than in island birds and arthropods may be due to differences in their respective modes of speciation. For sexually reproducing animals, theory suggests that when sympatric speciation occurs, it should proceed more rapidly than allopatric speciation (Kondrashov and Mina 1986). Sympatric speciation must involve both disruptive selection on a polymorphic trait (e.g., preferred habitat or dietary preference) and non-random mating, and thus is at least partly a result of selection acting directly on mating systems (Kondrashov and Mina 1986). In contrast, there is no special mechanism leading to allopatric speciation. Given a geographic barrier, drift and the process of adaptation to local environments will eventually and indirectly lead to reproductive isolation as a by-product of differences accumulated through selection and drift. Thus, because sympatric speciation involves selection acting directly on mating systems, and allopatric speciation is an indirect result of local population processes, we would expect sympatric speciation, when it occurs, to happen more rapidly than allopatric speciation.

For the Hawaiian and Galapagós Islands, the likelihood of allopatric speciation giving rise to sister taxa on separate islands seems high. But given the restriction of entire monophyletic groups of fishes to individual lakes, the role of allopatric speciation is less obvious. For example, while the diversification of cichlids has clearly occurred within lakes such as Barombi Mbo, Malawi, or Victoria (Meyer et al. 1990; Schliewen et al. 1994), it has long been debated whether such "intralacustrine" speciation is sympatric or allopatric (Brooks 1950; Fryer and Iles 1972; Echelle and Kornfield 1984; Martens et al. 1994 and references therein). For the Malawi cichlids, proponents of allopatric speciation argue that cichlids have strong habitat preferences and populations must have become isolated by alternating patches of rocky shore and sandy substrate, leading to "micro-allopatric" speciation (Fryer and Iles 1972). However, the habitat specificity and/or trophic specializations of cichlids, together with a propensity for assortative mating (Dominey 1984), satisfy important prerequisites for sympatric speciation as well (Kondrashov and Mina 1986).

Could the apparently very fast process of speciation in fishes reported here simply be a result of studying groups that have undergone rapid sympatric speciation? This hypothesis is being tested by comparing divergence of fishes that probably have speciated allopatrically to divergence of endemic lacustrine fishes that *may* have speciated sympatrically (McCune and Lovejoy 1997). Percent sequence divergence in cytochrome *b* is being used as a surrogate for overall divergence rate. Preliminary data support the hypothesis. Divergence between sister taxa that may have been generated by sympatric speciation is lower (and thus, speciation is faster) than that between sister taxa that probably were generated by allopatric speciation – divergence between possible sympatric sister pairs is lower (and thus speciation faster) than the divergence between likely allopatric pairs (McCune and Lovejoy 1997). However, the most definitive results for this test would not fully explain the basis for rapid speciation in endemic lacustrine fishes, but may focus future studies on the efficacy of sympatric speciation in such fishes on the ecological, behavioral, and genetic conditions required for sympatric speciation.

Conclusions

Speciation in lacustrine fishes appears to be faster than speciation in animals on islands. For fishes, the time required for speciation (TFS_{ln}) is generally in the range of 1,500 to 300,000 years. Comparable estimates for TFS_{ln} for endemic birds and arthropods on islands vary from about 0.6 to 1.0 My. While estimates of speciation rate depend critically on dating the origin of the clade, this difference in TFS exists with any combination of dates that have been suggested for individual case studies. Detailed consideration of possible reasons for rapid speciation in fishes is beyond the scope of this paper, but I discuss several possible explanations that range from artifact, to ecology and geological history, to mode of speciation. I suggest a specific test of the latter hypothesis, that higher rates of speciation in endemic lacustrine fishes may be due to a predominant mode of sympatric speciation in contrast to a predominantly allopatric mode for animal endemics on islands.

Traditional biostratigraphic dates and electrophoretic data are not sufficiently precise to be very useful for estimating speciation rates, but modern methods of dating lacustrine sediments and dates derived from DNA sequence divergence have produced remarkable agreement in several cases. It is clearly useful to obtain both kinds of dates. While speciation rates are presented here on both linear and logarithmic scales, I argue that the actual speciation rate and time required for speciation in any particular group will be somewhere between these two estimates, and that the most accurate estimate will incorporate the number of divergence events for each lineage derived from a clade's phylogenetic topology.

Acknowledgments

I thank W. Allmon, T. Givnish, N. Hairston, R. Harrison, N. Lovejoy, M. McClure, A. Mooers, and D. Winkler for valuable discussion. I am particularly grateful to S. Heard for sharing an unpublished manuscript, as well as his thoughts on speciation rates and tree balance; and to A. Cohen for extensive discussion about current geological research in the African Great Lakes. T. Johnson and C. Scholz generously allowed me to cite unpublished paleolimnological and geological data on Lake Victoria and Lake Malawi, respectively. U. Strecker and A. Parker generously allowed me to cite the results of unpublished studies of sequence divergence in Lago Chichancanab and Lake Titicacan cyprinodontids, respectively. The manuscript benefited substantially from critical reviews by T. Givnish, S. Heard, N. Lovejoy, P. Wimberger, D. Winkler, and anonymous reviewers. This study was supported in part by funding from the National Science Foundation and Hatch project 183–421.

References

Asquith, A. 1995. Evolution of *Sarona* (Heteroptera, Miridae): speciation on geographic and ecological islands. Pp. 90–120 in *Hawaiian biogeography: evolution on a hot spot archipelago*, W. L. Wagner and V. A. Funk, eds. Washington, DC: Smithsonian Institution Press.
Avise, J. C. 1994. *Molecular markers, natural history and evolution*. New York, NY: Chapman and Hall.
Avise, J. C., and Aquadro, C. F. 1982. A comparative summary of genetic distances in the vertebrates: patterns and correlations. Pp. 151–185 in *Evolutionary biology, vol. 15*, M. K. Hecht, B. Wallace, and G. T. Prance eds. New York, NY: Appleton-Century-Crofts.
Bernatchez, L., and Danzmann, R. G. 1993. Congruence in control–region sequence and restriction site variation in mitochondrial DNA of brook char, *Salvelinus fontinalis* Mitchell. Molecular Biology and Evolution 10:1002–1014.
Bishop, W. W. 1969. *Pleistocene stratigraphy in Uganda*. Entebbe, Uganda: Government Printer.
Brooks, J. L. 1950. Speciation in ancient lakes. Quarterly Review of Biology 25:30–176.

Carlquist, S. 1974. *Island biology.* New York, NY: Columbia University Press.
Carson, H. L., and Clagne, D. A. 1995. Geology and biogeography of the Hawaiian Islands. Pp. 14–29 in *Hawaiian biogeography: evolution on a hot spot archipelago,* W. L. Wagner and V. A. Funk, eds. Washington, DC: Smithsonian Institution Press.
Christie, D. M., Duncan, R. A., McBirney, A. R., Richards, M. A., White, W. M., Harpp, K. W., and Fox, C. G. 1992. Drowned islands downstream from the Galapagós hotspot imply extended speciation times. Nature 355:246–248.
Cisne, J. L., Chandlee, G. O., Rabe, B. D., and Cohen, J. A. 1980. Geographic variation and episodic evolution in an Ordivician trilobite. Science 209:925–927.
Cohen, A. S., Lessar, K.-E., Tiercelin, J.-J., and Soreghan, M. 1997. New palaeogeographic and lake level reconstructions of Lake Tanganyika: implications for tectonic, climatic and biologic evolution in a rift lake. Basin Research (in press).
Cohen, A., Soreghan, M. J., and Scholz, C. A. 1993. Estimating the age of formation of lakes: an example from Lake Tanganyika, East African rift system. Geology 21:511–514.
Covich, A., and Stuiver, M. 1974. Changes in oxygen 18 as a measure of long-term fluctuations in tropical lake levels and molluscan populations. Limnology and Oceanography 19:682–691.
Dejoux, C. 1994. Lake Titicaca. Pp. 35–42 in *Speciation in ancient lakes,* K. Martens, B. Goddeeris, and G. Coulter, eds. Ergebnisse der Limnologie 44:35–42. Stuttgart, E. Schweizerbart'sche Verlagsbuchhandlung
Dixey, F. 1941. The Nyasa Rift valley. South African Geology Journal 23:21–45.
Dominey, W. 1984. Effects of sexual selection and life history on speciation: species flocks in African cichlids and Hawaiian *Drosophila.* Pp. 231–249 in *Evolution of fish species flocks,* A. A. Echelle and I. Kornfield, eds. Orono, ME: University of Maine Press.
Echelle, A. A., and Kornfield, I. 1984. *Evolution of fish species flocks.* Orono, ME: University of Maine Press.
Fryer, G. 1959. The tropic interrelationships and ecology of some littoral communities of Lake Nyasa with especial reference to the fishes, and a discussion of the evolution of a group of rock-frequenting Cichlidae. Proceedings of the Zoological Society of London 132:153–281.
Fryer, G., and Iles, T. D. 1972. *The cichlid fishes of the Great Lakes of Africa.* Neptune City, NJ: TFH Publications.
Futuyma, D. 1986. *Evolutionary biology, 2nd ed.* Sunderland, MA: Sinauer Associates.
Gillespie, R. G., and Croom, H. B. 1995. Comparison of speciation mechanisms in web-building and non-web-building groups within a lineage of spiders. Pp. 121–146 in *Hawaiian biogeography: evolution on a hot spot archipelago,* W. L. Wagner and V. A. Funk, eds. Washington, DC: Smithsonian Institution Press.
Gingerich, P. D. 1983. Rates of evolution: effects of time and temporal scaling. Science 222:159–161.
Givnish, T. J., Sytsma, K. J., Smith, J. F., and Hahn, W. J. 1995. Molecular evolution, adaptive radiation, and geographic speciation in *Cyanea* (Campanulaceae, Lobelioideae). Pp. 288–337 in *Hawaiian biogeography: evolution on a hot spot archipelago,* W. L. Wagner and V. A. Funk, eds. Washington, DC: Smithsonian Institution Press.
Gould, S. J. 1984. Smooth curve of evolutionary rate: a psychological and mathematical artifact. Science 226:994–995.
Grachev, M A., Slobodyanyuk, S. J., Kholodilov, N. G., Fyodorov, S. P., Belikov, S. I., Sherbakov, D. Y., Sideleva, V. G., Zubin, A. A., and Kharchenko, V. V. 1992. Comparative study of two protein-coding regions of mitochondrial DNA from three endemic sculpins (Cottoidei) of Lake Baikal. Journal of Molecular Evolution 34:85–90.
Grant, P. B. 1986. *Ecology and evolution of Darwin's finches.* Princeton, NJ: Princeton University Press.
Greenwood, P. H. 1965. The cichlid fishes of Lake Nabugabo, Uganda. Bulletin of the British Museum, Zoology 12:315–357.
Greenwood, P. H. 1981. *The Haplochromine fishes of the East African lakes.* Munich, Germany: Kraus International Publications.
Guyer, C., and Slowinski, J. B. 1991. Comparisons of observed phylogenetic topologies with null expectations among three monophyletic lineages. Evolution 45:340–350.
Harland, W. B., Armstrong, R. L., Cox, A. V., Craig, L. E., Smith, A. G., and Smith, D. G. 1990. *A geologic time scale.* Cambridge, England: Cambridge University Press.
Heard, S. 1992. Patterns in tree balance among cladistic, phenetic, and randomly generated phylogenetic trees. Evolution 46:1818–1826.
Heard, S. 1996. Patterns in phylogenetic tree balance with variable and evolving speciation rates. Evolution 50:2141–2148.

Hodell, D. A., Curtis, J. H., and Brenner, M. 1995. Possible role of climate in the collapse of classic Maya civilization. Nature 375:391–394.

Hulbert, R. C., Jr. 1993. The rise and fall of an adaptive radiation. Paleobiology 19:216–234.

Humphries, J. M. 1984. Genetics of speciation in pupfishes from Laguna Chichancanab, Mexico. Pp. 129–140 in *Evolution of fish species flocks*, A. A. Echelle and I. Kornfield, eds. Orono, ME: University of Maine Press.

Irwin, D. M., Kocher, T. D., and Wilson, A. C. 1991. Evolution of the cytochrome *b* gene of mammals. Journal of Molecular Evolution 32:128–144.

Kendall, R. L. 1969. The ecological history of the Lake Victoria basin. Ecological Monographs 39:121–176.

Kirkpatrick, M., and Selander, R. K. 1979. Genetics of speciation in lake whitefishes in the Allegash Basin. Evolution 33:478–485.

Kling, G. W., Clark, M. A., Compton, H. R., Devine, J. D., Evans, W. C., Humphrey, A. M., Koenigsberg, E. J., Lockwood, J. P., Tuttle, M. L., and Wagner, G. N. 1987. The 1986 Lake Nyos gas disaster in Cameroon, West Africa. Science 236:169–175.

Kondrashov, A., and Mina, M. V. 1986. Sympatric speciation: when is it possible? Biological Journal of the Linnean Society 7:201–233.

Kornfield, I., and Carpenter, K. E. 1984. Cyprinids of Lake Lanao, Philippines: taxonomic validity, evolutionary rates and speciation scenarios. Pp. 69–84 in *Evolution of fish species flocks*, A. A. Echelle and I. Kornfield, eds. Orono, ME: University of Maine Press.

Lavenu, A. 1992. Formation and geological evolution. Pp. 3–15 in *Lake Titicaca*, C. Dejoux and A. Iltis, eds. Rotterdam, The Netherlands: Kluwer Academic Publishers.

Lewis, W. M., Jr. 1978. A compositional, phytogeographical and elementary structural analysis of the phytoplankton in a tropical lake: Lake Lanao, Philippines. Journal of Ecology 66:213–266.

Lezzar, K. E., Tiercelin, J. J., DeBatist, M., Cohen, A. S., Bandora, T., VanRensbergen, P., LeTurdu, C., Mifundu, W., and Klerkx, J. 1996. New seismic stratigraphy and Late Tertiary history of the North Tanganyika Basin, East African Rift system, deduced from multichannel and high-resolution reflection seismic data and piston core evidence. Basin Research 8:1–28.

Li, W. -H., and Graur, D. 1991. *Fundamentals of molecular evolution*. Sunderland, MA: Sinauer Associates.

Livingstone, D. A. 1980. Environmental changes in the Nile headwaters. Pp. 339–359 in *The Sahara and the Nile*, M. A. J. Williams and H. Gaure, eds. Rotterdam, The Netherlands: A. A. Balkema.

Maley, J. D., Livingstone, A., Giresse, P., Thouveny, N., Brenac, P., Kelts, K., Kling, G., Stager, C., Haag, M., Fournier, M., Bandet, Y., Williamson, D., and Zogning, A. 1990. Lithostratigraphy, volcanism, paleomagnetism and palynology of Quaternary lacustrine deposits from Brombi Mbo (West Cameroon): preliminary results. Journal of Volcanology and Geothermal Research 42:319–335.

Martens, K., Goddeeris, B., and Coulter, G. eds. 1994. *Speciation in ancient lakes*. Ergebnisse der Limnologie 44:1–508. Stuttgart, E. Schweizerbart'sche Verlagsbuchhandlung.

Martin, A. P., and Palumbi, S. R. 1993. Body size, metabolic rate, generation time, and the molecular clock. Proceedings of the National Academy of Science, USA 90:4087–4091.

Mazepova, G. 1994. On comparative aspects of ostracod diversity in the Baikalian fauna. Pp. 197–202 in *Speciation in ancient lakes*, K. Martens, B. Goddeeris, and G. Coulter eds. Ergebnisse der Limnologie 44:197-202. Stuttgart, E. Schweizerbartüche Verlagsbuchhandlung..

McCune, A. R. 1987a. Toward the phylogeny of a fossil species flock: semionotid fishes from a lake deposit in the early Jurassic Towaco Formation, Newark Basin. Bulletin of the Yale Peabody Museum of Natural History 43:1–108.

McCune, A. R. 1987b. Lakes as laboratories of evolution: endemic fishes and environmental cyclicity. Palaios 2:446–454.

McCune, A. R. 1990. Evolutionary novelty and atavism in the *Semionotus* complex: relaxed selection during colonization of an expanding lake. Evolution 44:71–85.

McCune, A. R. 1996. Biogeographic and stratigraphic evidence for rapid speciation in semionotid fishes. Paleobiology 22:34–48.

McCune, A. R., and Lovejoy, N. R. 1997. The relative rate of sympatric and allopatric speciation in fishes: tests using DNA sequence divergence between sister species and among clades. *Endless forms: species and speciation*, D. Howard and S. Berlocher, eds. London, England: Oxford University Press. (In Press.)

McCune, A. R., Thomson, K. S., and Olsen, P. E. 1984. Semionotid fishes from the Mesozoic great lakes of North America. Pp. 27–44 in *Evolution of fish species flocks*, A. A. Echelle and I. Kornfield, eds. Orono, ME: University of Maine Press.

McVeigh, H. P., Bartlett, S. E., and Davidson, W. D. 1991. Polymerase chain reaction direct sequence analysis of the cytochrome *b* gene in *Salmo salar*. Aquaculture 95:225–234.

Meyer, A. 1993. Phylogenetic relationships and evolutionary processes in East African cichlid fishes. Trends in Ecology and Evolution 8:267–305.

Meyer, A., Kocher, T. D., Basasibwaki, P., and Wilson, A. C. 1990. Monophyletic origin of Lake Victoria cichlid fishes suggested by mitochondrial DNA sequences. Nature 347:550–553.

Moran, P., Kornfield, I., and Reinthal, P. N. 1994. Molecular systematics and radiation of the haplochromine cichlids (Teleostei: Perciformes) of Lake Malawi. Copeia 2:274–288.

Olsen, P. E. 1980. Fossil great lakes of the Newark Supergroup in New Jersey. Pp. 352–398 in *Field studies of New Jersey geology and guide to field trips*, W. Manspeizer, ed. Newark, NJ: Newark College of Arts and Sciences, Rutgers University.

Olsen, P. E. 1984. Comparative paleolimnology of the Newark Supergroup: a study of ecosystem evolution. Ph.D. Dissertation, Yale University.

Olsen, P. E. 1986. A 40-million year record of early Mesozoic orbital climatic forcing. Science 234:842–848.

Olsen, P. E. 1988. Continuity of strata in the Newark and Hartford Basins. United States Geological Survey Bulletin 1776:6–18.

Otte, D. 1994. *The crickets of Hawaii: origin, systematics and evolution*. Philadelphia, PA: Academy of Natural Sciences of Philadelphia.

Owen, R. B., Crossley, R., Johnson, T. C., Tweddle, D., Kornfield, I., Davison, S., Eccles, D. H., and Engstrom, D. E. 1990. Major low levels of Lake Malawi and their implications for speciation rates in cichlid fishes. Proceedings of the Royal Society of London, Biological Sciences 240:519–553.

Parenti, L. R. 1984. Biogeography of the Andean killifish genus *Orestias* with comments on the species flock concept. Pp. 85–92 in *Evolution of fish species flocks*, A. A. Echelle and I. Kornfield, eds. Orono, ME: University of Maine Press.

Parker, A., and Kornfield, I. 1995. A molecular perspective on evaluation and zoogeography of cyprinodontid killifishes. Copeia 1995:8–21

Press, F., and Siever, R. 1986. *Earth, 4th ed*. New York, NY: W. H. Freeman and Co.

Sage, R. D., Loiselle, P. V., Basasibwaki, P., and Wilson, A. C. 1984. Molecular versus morphological change among cichlid fishes of Lake Victoria. Pp. 185–202 in *Evolution of fish species flocks*, A. A. Echelle and I. Kornfield, eds. Orono, ME: University of Maine Press.

Schliewen, U. K., Tautz, D., and Paabo, S. 1994. Sympatric speciation suggested by monophyly of crater lake cichlids. Nature 368:629–632.

Schluter, D., and McPhail, J. D. 1992. Ecological displacement and speciation in sticklebacks. American Naturalist 140:85–108.

Scholz, C., and Finney, B. 1994. Late Quaternary sequence stratigraphy of Lake Malawi (Nyasa), Africa. Sedimentology 41:163–179.

Scholz, C., and Rosendahl, B. 1988. Low lake stands in Lakes Malawi and Tanganyika, East Africa, delineated with multifold seismic data. Science 240:1645–1648.

Shaw, K. L. 1995. Biogeographic patterns of two independent Hawaiian cricket radiations (*Laupala* and *Prognathogryllus*). Pp. 39–56 in *Hawaiian biogeography: evolution on a hot spot archipelago*, W. L. Wagner and V. A. Funk, eds. Washington, DC: Smithsonian Institution Press.

Sigurdsson, H. 1987. A dead chief's revenge? Natural History 96:44–49.

Smith, G. R. 1992. Introgression in fishes: significance for paleontology, cladistics, and evolutionary rates. Systematic Biology 41: 41–57.

Stager, J. C., Reinthal, P. N., and Livingstone, D. A. 1986. A 25,000 year history for Lake Victoria, East Africa and some comments on its significance for the evolution of cichlid fishes. Freshwater Biology 16:15–19.

Stanley, S. M. 1979. *Macroevolution: pattern and process*. San Francisco, CA: W. H. Freeman.

Strecker, U., Meyer, C. G., Sturmbauer, C., and Wilkens, H. 1996. Genetic divergence and speciation in an extremely young species flock in Mexico formed by the genus *Cyprinodon* (Cyprinodontidae, Teleostei). Molecular Phylogenetics and Evolution 6:143–149.

Sturmbauer, C., and Meyer, A. 1993. Mitochondrial phylogeny of the endemic mouthbrooding lineages of cichlid fishes from Lake Tanganyika in Eastern Africa. Molecular Biology and Evolution 10:751–768.

Sturmbauer, C., Verheyen, E., and Meyer, A. 1994. Mitochondrial phylogeny of the Lamprologini, the major substrate spawning lineage of cichlid fishes from Lake Tanganyika in Eastern Africa. Molecular Biology and Evolution 11:691–703.

Tarr, C. L., and Fleischer, R. C. 1995. Evolutionary relationships of the Hawaiian honeycreepers (Aves, Drepanidinae). Pp. 147–159 in *Hawaiian biogeography: evolution on a hot spot archipelago*, W. L. Wagner and V. A. Funk, eds. Washington, DC: Smithsonian Institution Press.

Temple, P. H. 1969a. Some biological implications of a revised geological history for Lake Victoria. Biological Journal of the Linnean Society 1:363–371.

Temple, P. H. 1969b. Raised strandline and shoreline evolution in the area of Lake Nabugabo, Masaka district Uganda. Pp. 119–129 in *Quaternary geology and climate*, H. E. Wright, ed. Washington, DC: National Academy of Sciences Publication 1701.

Tiercelin, J-J., and Mondeguer, A. 1991. The geology of the Tanganyika Trough. Pp. 7–48 in *Lake Tanganyika and its life*, G. W. Coulter, ed. London, England: Oxford University Press.

Vrba, E. S. 1980. Evolution, species and fossils: how does life evolve? South African Journal of Science 76:61–84.

Williamson, P. G. 1981. Paleontological documentation of speciation in Cenozoic molluscs from Turkana Basin. Nature 293:437–443.

Wirrmann, D., Ybert, J.-P., and Mourguiart, P. 1992. A 20,000 year paleohydrological record from Lake Titicaca. Pp. 40–48 in *Lake Titicaca*, C. Dejoux and A. Iltis, eds. Rotterdam, The Netherlands: Kluwer Academic Publishers.

Ybert, J.-P. 1992. Ancient lake environments as deduced from pollen analysis. Pp. 49–62 in *Lake Titicaca*, C. Dejoux and A. Iltis, eds. Rotterdam, The Netherlands: Kluwer Academic Publishers.

INDEX

12S rRNA, 140-141, 143, 147, 165, 167-168, 170
16S ribosomal gene, 192
absorptive trichomes, 261, 268, 293, 297-299, 302
accelerated transformation, 23, 173-174, 238, 251, 253, 274, 339, 341, 359, 361-364, 369, 440, 443, 446-447, 460, 467, 483, 485-486, 488, 490, 495-496, 578
acquired woodiness, 114
Acrobates, 131, 134, 328
Acrocanthosaurus, 32
active epibiotic, 414
Ada, 350, 416
adaptation, 1-2, 7-9, 11-13, 17-20, 25-29, 36-37, 130, 140, 150-151, 171, 184, 197, 213, 225-227, 231-232, 234-235, 237, 245, 247, 259-260, 272, 284, 288-289, 294-297, 299-300, 302, 313, 326, 328, 332-333, 345-346, 353-355, 364-365, 368-369, 371, 380-381, 387, 414, 438, 476-477, 479, 492, 498, 502, 504-506, 528, 536, 540, 545-547, 605
adaptationist, 18
adapted to fire, 271
adaptive divergence, 9-11, 32, 37, 501
adaptive radiation
  along the lines of least genetic resistance, 31
  and biogeography, 32-33, 107-116, 149-151, 239-240, 265-267, 286-297, 355-370, 141-420, 469-471, 511-531, 543-553
  and community structure, 550-553
  and endangered species, 37
  and hybridization, 24, 278-280, 426-427
  and homoplasy, 56, 92, 561-575
  and non-adaptive radiation, 10-12, 303-304
  and preadaptation, 20, 297-298

classic studies, 3-6, 13-16, 21, 34, 313-341, 535
definitions, 2, 8-13, 129, 225, 259, 511, 535
experimental studies, 522-526, 528-531
genetical aspects, 28
in aquatic organisms, 163-185, 225-253, 355-370, 585-606
in ecological venue or habitat, 144-147, 174-175, 359-365, 367-369, 386, 546-548, 575-578
in feeding strategies, 130-131, 196-204, 207-213, 313-328, 375-388, 475-506
in growth form, 103-125, 229-231, 243-247, 259-304, 343-346, 407-428
in reproductive stratagies, 12, 27-28, 179-181, 231-232, 234-237, 247-253, 343-346, 391-403, 431-451, 455-471, 475-506
predictability, 35-36, 550-553
role of development, 31-39, 349-364
sympatric patterns, 23-24, 279-281, 512-517, 527-528, 597-600
tempo, 33-35, 108-114, 147-149, 165-169, 240, 420, 425, 559, 578-581, 585-606
adaptive zone, 2, 8-11, 13, 21, 25, 33, 36, 129, 131, 149-152, 261, 303, 334, 337, 548
adelphoparasitic red algae, 24
Adenothamnus, 105, 110, 114
adult flight season, 358
adult vs. larval morphology, 560-561, 563-564, 566, 568-571, 574-575, 577, 579, 581
adventitious root, 269, 295
Aegyptopithecus, 192
Aeonium, 407-411, 419-420, 423
Aepyprymnus, 134, 140
aerenchyma, 282-283, 289, 295-296, 298, 302
Agrobacterium, 57
algae, 14, 24, 29, 176, 234, 375,

383, 577
algal plug, 270, 289
allopatric speciation, 11, 23, 386-388, 527-528, 530, 604-605
Alouatta, 195-196, 200, 206
Alsinodendron, 28
Amazon basin, 197, 199
amphibious flower, 235
Amphilochia, 401-402
Anacyclus, 416
anadromous fish, 512-513, 515, 517-522, 527
anaerobic environment, 501
analysis of covariance, 65
anastomosis, 29
ancient hybridization, 111, 119-120, 125, 238
ANCOVA, 62-63, 65-70, 78-79
Andinodelphys, 132
aneuploidy, 28
angiosperms, 22, 30, 63, 119, 225, 234, 261, 303, 392, 440
annual habit, 24, 113-114, 226, 231-232, 234, 245, 247, 253
Anolis, 10, 17, 36, 356, 535-537, 540-541, 543-544, 546
ants, 32, 264, 267, 269-270, 285, 485
ant nests, 198
ant-fed myrmecophyte, 261, 264, 267, 269, 278, 285, 287, 295, 297, 301-302
Antechinus, 134
antelope, 35, 130-131
Anthemideae, 407, 414-416, 418, 420
anthropoids, 192, 202
Aotus, 193, 196, 198, 200, 209, 212-213
aphid, 29
Apidium, 192
aquatic environment, 225-226, 229, 231, 234, 242
aquatic plants, 225-227, 229, 231-234, 253
Aquilegia, 391-403
Araliaceae, 487, 490, 505
Arbacia, 562-563, 566, 569-571, 574
arboreal organisms, 131, 133, 136, 146, 150-152, 204-205,

538, 546-547, 549
arborescent plants, 35-36, 264, 271, 278, 288-289, 292, 294, 301-303
archaebacteria, 29
Archaeopteropus, 314, 316
archipelagoes, 2, 112, 313, 407, 409, 419, 421, 423, 427, 455, 475, 588, 591
Argyranthemum, 28, 407, 409-414, 416, 419-428
argyrolagids, 133
Argyroxiphium, 103-106, 110-112, 115-117, 121, 124-125, 427
arms race, 11-12
asexual propagation, 163
Aspasia, 350
assault-type mating, 482, 504
assortative mating, 24, 27, 33, 529, 605
Astatoreochromis, 14, 378
Astatotilapia, 14, 378
Asteraceae, 28, 31, 36, 67, 103-104, 124, 300, 303, 344, 407, 419, 425-428
asteroids, 560
Asthenosoma, 562
Ateles, 194-196, 200, 203-204, 206, 208, 212
atmospheric oxygen, 11
atomization, 79, 83, 87, 91

backwards-opening pouch, 147
bacteria, 27, 29, 57, 480-481, 492, 498-499
bacterial symbiont, 28
basal metabolic rate, 206
bats, 26, 129, 316, 318, 326-327
Bauplan, 11, 18
beak depth, 3-4
bearded saki, 198-200, 207-208
bees, 235, 253, 267, 334, 343, 346-347, 395, 433, 445
beetles, 353, 355, 357-358, 365, 370, 424, 436, 438, 445, 522
benthic, 21, 30, 176-177, 185, 512-517, 520-531, 559-560, 581
Bettongia, 134, 146
Bidens, 425, 427
bill morphology, 18, 21
bimanual locomotion, 197, 205
biogeography, 1-2, 4, 8, 13, 31-32, 36-37, 149, 227, 239, 259-260, 265, 275, 300-303, 326, 355, 359-361, 376, 469, 511
birth-and-death model, 113
birthday paradox, 74

biting force, 209-210
body size, 10, 26-27, 164, 172-173, 189, 204-205, 209-210, 213, 479, 502-503, 529, 537, 546, 548, 589
bootstrap, 35, 89, 105, 107, 110-111, 137-141, 143, 151, 168, 192-194, 238, 274, 277-280, 290, 317, 339, 341-342, 359-360, 382, 384, 393, 399, 402, 414-416, 418-419, 421, 424, 440-442, 460-465, 481, 483, 485, 518-520, 542, 561, 565-569, 574-575
Boraginaceae, 411
borhyaenids, 131-132, 135, 151-152
Bovichidae, 21
brachiation, 205
Brachyteles, 195, 212
branch use, 206-207
Brassia, 350
breeding system, 6, 19-20, 163, 166, 169, 178-180, 183, 413, 419, 455-461, 465-467, 469-471
Brewcaria, 273, 287, 290-291, 301, 303
Brighamia, 87
Brissopsis, 562, 575
Brocchinia, 10, 25, 33, 63, 71-75, 84, 259-261, 264-304
bromeliads, 10-11, 15, 21, 33, 38, 201, 203, 261, 265, 267, 270-272, 284, 293-294, 298, 301
Bromelioideae, 261, 272
brown tree snake, 37
brushy tongues, 313
bumblebees, 395, 436, 438, 445
Burgess Shale, 11
Burramys, 134
burrowing, 11, 146-147, 151, 577
butterflies, 27, 235, 346-347, 436, 438, 445, 451

Caenolestes, 134, 149
caenolestids, 133, 135-137, 141, 143, 146, 151
Caenorhabditis, 31
Calliandra, 332
Callicebus, 198, 203, 208, 212
Callimico, 200, 202-203, 212
callitrichins, 189, 193-194, 198, 203-204, 209, 212-213
Calothrix, 270, 383
Caluromys, 133-134, 149
Camin-Sokal coding, 359, 362
Campanulaceae, 10, 26, 80, 87,

487, 490, 505
Canis, 130
Capanemia, 334, 347, 350
capuchins, 189, 193, 197, 200-201, 203-204, 206-212
carapace setation, 169, 176
Carcharodontosaurus, 32
Carlquist, Sherwin, 4, 8, 11, 27, 31, 35, 104, 114, 116-118, 247, 259, 284, 410, 414, 424, 455, 585
carnivore, 129, 131, 151, 261, 264, 267-270, 278, 285, 287, 293-294, 297, 301-303
carnivorous, 27, 32-33, 130-134, 151, 267, 270, 290-291, 293-294, 300, 302, 328
carnivorous plant, 267, 293
Caryophyllaceae, 419, 455-456, 458
Cassidulus, 562
Catopsis, 267, 293
Cebidae, 194-195, 200, 203-204, 209, 212
Cebuella, 202-203, 211-212
Cebupithecia, 192, 198-199, 207
Cebus, 193, 196, 200-201, 206, 209, 212
Centrostephanus, 562, 575
Cercartetus, 134
cetacean, 129
Chamaedorea, 26
Chamaeleolis, 537, 540, 543, 545
Chamaelinorops, 537, 540, 543
Channichthyidae, 21
Chaoborus, 174-176, 184
character displacement, 3-4, 10, 21, 24, 32, 354-356, 364-365, 370, 501, 513-517, 522-525, 528-529, 531
character diversification, 225-226, 240-241, 245, 253
character misclassification, 56, 58, 86-89, 91
character reconstructions, 239, 241, 483
Cheirodendron, 487, 489-491
Cheirolophus, 419, 426, 428
Chilotilapia, 375, 377, 382, 385
Chironectes, 134, 147
Chiropotes, 198-200, 208
chiropterans, 129, 318, 327
chloroplast DNA (cpDNA), 7, 28, 58-60, 62-63, 65-71, 76, 78-79, 81-83, 86-88, 91-92, 106-107, 110-111, 113-125, 238, 244, 246, 248-250, 260, 272-280, 282, 286, 289, 299, 301,

332, 339-346, 392, 401, 411, 414-416, 419-421, 423-428, 459-460, 463-465, 469-470
chorion, 493-495, 499, 502, 505
chromosomal speciation, 23
chromosome evolution, 110, 120-121, 123
chromosome number, 28, 110, 113, 121, 123, 125, 165, 169, 181-183, 331, 344-345
Chrysanthemum, 413-414, 416, 419-420, 425
Chrysochloris, 130
cichlids, 4, 11, 14-16, 20, 25, 32, 37, 259, 375-381, 384-387, 391-392, 401, 527, 535, 585-586, 589, 591-599, 602, 604-605
Cicindela, 355, 357-366, 368-370
Cidaris, 565
circular reasoning, 6, 71, 82, 90, 92, 259, 433, 575
Cischweinfia, 350
clade origin, 586-587
clade-level phenomena, 25
cladogenesis, 9, 85, 346, 353, 356-357, 364, 391
Cladophora, 383
Clarkia, 63, 71-75, 84, 89
claws, 135, 202, 211-213, 223
Clermontia, 487, 490
clonality, 179, 226-227, 229-232, 242, 247-248
cloud forest, 189, 261, 264, 266, 271, 283-284, 289-291, 302
coalescence, 57
coelacanth, 22
coevolutionary radiation, 12-13
Coleoptera, 353, 357
Colobocentrotus, 562
color, 11, 87, 178, 198, 208, 294, 392, 394-395, 411-412, 433, 436, 440, 445-446, 448, 451, 459, 486, 538, 550
coloration, 164, 291, 293, 380, 385, 445
columbine, 12, 21, 391-398
combined analysis, 241, 416, 419, 467, 470, 545
combined data, 83, 90, 111, 193-194, 458-460, 465, 467, 469-470, 567, 569-572, 576-577
Commelinaceae, 239-246, 250
Commelinanae, 239
community ecology, 357, 537
community evolution, 536
comparative approach, 18-19
comparative technique, 17, 19

comparative method, 316-317, 321, 325-326
Comparettia, 334, 347, 350
competition, 1, 4, 8, 11, 13, 21, 23-24, 27-28, 32, 35-36, 133, 206, 229, 232, 247, 284, 303, 337, 354, 356, 364-365, 368, 380, 387, 476-477, 479, 490-492, 498, 500-501, 504, 511, 514, 516-517, 522-528, 531, 537-539
competitive release, 4, 21, 164, 184, 208
competitive speciation, 515-516, 529
concerted convergence, 6, 58, 86, 90, 298
congruence, 6, 13, 87, 90, 103, 123, 125, 141, 300, 303, 536, 552, 561, 563, 567, 570
consensus, 55, 105, 110-111, 121, 140-141, 143, 146, 151, 167-168, 183, 238, 274, 276-278, 290, 316, 342, 385, 402, 416, 419, 440, 461, 464-465, 481, 536, 563, 566-567, 569-570, 586
conservative macroevolutionary pattern, 226
consistency index (CI, CI'), 7, 58-59, 66, 74, 78, 84, 92, 105, 110, 193, 276, 338, 341-342, 542, 564
constraintist, 19
constraints on secondary structure, 83, 87
continuous lower ankle-joint pattern (CLAJP), 135-136
convergence, 1, 6, 14, 17-19, 23, 27, 35, 55-58, 74-75, 79-83, 86-92, 130-131, 135, 141, 151, 184, 195, 197, 253, 259, 289, 298, 302, 326, 345, 375, 379, 381, 383, 386, 436, 440, 444, 450, 574
convergent adaptive radiation, 535
coprophagy, 147, 151
Coregonis, 528
cottoids, 4, 586
Crambe, 423, 425, 428
Crassulaceae, 411, 419
crinoids, 36
crown-giant, 538-539, 550-551, 553
cryopelagic specialist, 21
Ctenodaphnia, 164-167, 173, 176-178, 181, 183

Cuba, 536-537, 539, 541, 543-545, 547, 550, 552-553
cursorial predator, 358
cushion plant, 103, 118, 124
cuticular melanization and ultraviolet radiation, 169
Cyanea, 10, 26-28, 37, 80, 250, 426, 490
cyclic parthenogenesis, 165-166, 179
Cymbopappus, 413-414
Cyprinodon, 591
cyprinodontids, 585-586, 594, 596, 600, 606
Cyrtandra, 28, 427
cytochrome b, 141, 143, 589, 605

Dactyladenia, 402
Dactylopsila, 134, 139, 143
Daphne Island, 3, 21, 163
Daphnia, 163-177, 179-182, 184
Darwin, Charles, 1-6, 10, 13, 24, 26-27, 31, 56, 88, 114, 235, 259, 265, 391, 433
Darwin's finches, 1-6, 13, 24, 26, 31, 56, 259, 265, 391
dasyurids, 131, 141, 146-147, 151-152
Dasyuromorphia, 133-134, 139, 141
Dasyurus, 133-134
decay analysis, 274, 278, 342, 440-441, 462, 481, 542
decaying flower, 497, 502
deceit pollination, 334
delayed transformation (DELTRAN), 23, 173, 175, 252, 274, 282, 339, 359, 361-362, 364, 460, 467, 483 486, 490, 495, 578
Delissea, 87
Dendrobiinae, 339, 344
Dendrolagus, 133-134, 139, 146
deposit feeder, 14, 559, 562, 577
depth of water, 229
development, 2, 8, 13, 22, 29-32, 37, 79, 86, 109, 150, 160, 163, 172-173, 184, 224, 227, 232-234, 313, 333, 349, 358, 445, 475, 492-495, 498, 500-502, 515, 537, 560, 562, 575, 577-578
developmental correlation, 17, 79, 83
developmental lability, 29, 31
developmental program, 18, 35-36, 57

developmental radiation, 11-12
dewlaps, 544, 547, 549-550
Diascia, 395, 401
diaspore, 227
dicotyledons, 225
didactyly, 135, 146
Didelphimorphia, 133-134, 141
diet, 3-4, 22, 24, 28, 32, 34-35, 133, 189-190, 195, 197-201, 203-204, 206-209, 314-315, 318, 321, 324-326, 328, 383, 513-514, 523, 525-526, 531
dioecy, 6, 19-20, 455-456, 458, 460, 465, 467
Diprotodontia, 133-134, 137, 148
diprotodonty, 33, 134-135, 143, 150-151
Dipsis, 26
dipteran pollination, 445
Dischidia, 295
dispersal, 10-13, 19-20, 25-27, 31-32, 37, 103, 109, 114-116, 124-125, 132, 147, 150, 163-165, 169, 178, 201, 227, 229-230, 232, 235, 238-240, 260, 287-288, 299-303, 368-369, 379-380, 386, 391, 413, 420, 423-425, 479, 485, 517, 543, 545, 550
Distoechurus, 134, 140, 328
diurnal hawkmoth, 445, 451
Dobzhansky, Theodosius, 2, 23, 394, 511
Dorcopsulus, 134, 139
double-invasion hypothesis, 516-517, 521-522, 527-528
Drepanidinae, 14
Dromiciops, 133-137, 139-141, 146, 151
Drosophila, 30, 475-483, 485-490, 492-493, 497-498, 500-506
drosophilid, 12, 27-28, 476-477, 479-481, 483, 485-486, 488, 490, 492-496, 498, 501-505, 586, 596
Drosophilidae, 475, 478, 481-483, 485, 492, 502, 506
dry forest, 103, 189, 457
Dubautia, 103-105, 107-112, 115-116, 118-125, 427

Echinocardium, 562, 575
Echinocyamus, 562
echinoderms, 7, 560, 565, 567
Echinodiscus, 562, 581
echinoids, 559-564, 567, 569, 574-578, 580-581

Echinolampas, 562
Echinus, 562, 566
Echium, 28, 408, 411
Echymipera, 134, 143, 146
ecological innovations, 560, 575-576
ecological selection, 20, 24, 511
ecological shifts, 103, 121-122, 125, 356, 423, 458, 476, 478, 483, 486, 505
ecological specialization, 203, 225, 304, 368, 380-381, 385, 387-388, 492
ecological speciation, 24, 35
ecological venue, 146
ecological vicariance, 420-421
ecology, 1-2, 4, 6, 8, 10, 13-14, 18-19, 22, 28, 32, 36-38, 90-91, 119, 131, 133, 146, 185, 189, 196, 208, 212, 231, 234, 259-260, 264-265, 267, 275, 284-285, 289-290, 298, 300-303, 357-358, 371, 375, 382-383, 386-387, 407, 421, 425-426, 437, 440, 450-451, 477-478, 497, 502-503, 511, 522, 537-538, 540, 553, 562, 578, 606
ecology and phylogeny, 285, 298, 425
ecomorphs, 538-541, 543-544, 546, 548-553
ecomorphology, 315
egg morphology, 492
egg production, 477, 493
El Niño, 3, 24
Eleutherodactylus, 536
emergent aquatic plants, 198, 201, 226, 229-230, 232, 234, 242, 245, 247, 588
enantiostyly, 235-236, 243, 251-252
Encope, 562, 581
endangered species, 494
endoskeleton, 559
Engiscaptomyza, 478, 482-483, 496
Enos Lake, 512-513, 515, 517, 519-522, 527-528
Eonycteris, 313-314, 318, 327
ephippial morphology, 169, 178-179
Epidendrum, 402
epifaunal, 559, 562, 575
epiphytism, 33, 265, 269, 285, 289, 292, 298-299, 302, 331-334, 343, 345
epistasis, 6, 58, 79, 83, 87, 90, 164

erythrocytes, 21
ethereal scent, 433
Eucidaris, 562, 565
euglossine bees, 334, 346
Euphorbia, 411
Euphorbiaceae, 411
Euphractus, 328
Euphronia, 401-402
Eurycea, 31
Eurystemon, 227-228, 234
Eutheria, 129, 148
evolution of community structure, 550
evolution of flight, 20, 391
evolution of habitat association, 353, 355-357, 361
evolution of mating barriers, 475
evolutionary "success," 354
Exalloscaptomyza, 492, 497, 502
exaptation, 17, 20, 546
explosive diversification, 8
exposure to selection, 79-80
extinction, 9, 21, 23, 26, 32-33, 36-37, 114, 129, 133, 150, 226, 240, 304, 354, 356, 368-370, 391, 394, 400, 403, 411, 419, 428, 517, 527, 545, 549, 559, 590, 601
extracellular mucilage, 104
extreme environment, 275, 331

Fabaceae, 332, 411, 421
feeding mode, 206
Fellaster, 562
female reproductive strategies, 477, 483, 495, 495, 497, 501, 503, 505
figs, 210, 601
Fisher, Sir Ronald, 2, 17, 23, 35, 62, 66-67, 288, 476, 511
fitness, 17, 235, 353-354, 396, 523, 529
FL1G, 321, 323-326
flattened seedlings, 343, 345
flightlessness, 35
floating-leaved aquatic plants, 229-230, 245, 247
flooded forest, 198, 202, 209
floral biology, 226, 237
floral color, 440, 445-446, 448, 451
floral ecology, 234
floral form, 227, 236, 243, 251-253, 393, 399, 433, 436
floral monomorphism, 236, 251-252

floral rewards, 337
flying squirrels, 35, 130
Ford, E. B., 2, 17, 189, 192, 197, 202, 204, 211
fossil evidence, 1, 23, 33-34, 80, 84, 129, 132-133, 141, 147-150, 165, 192, 194, 198-199, 207, 239-240, 297, 314, 414, 515, 559-561, 563-564, 567, 571, 574, 579, 587-589, 594, 598, 600, 602, 604
Fosterella, 265, 271, 286, 291, 299
founder events, 90, 122, 380, 475-476, 479, 504, 511
free-floating aquatic plants, 229-230, 232, 245
frugivores, 12, 26, 199, 207-208, 314
fruits, 4, 10, 12, 20, 26, 57, 87, 109, 133, 189, 197-201, 203, 205-210, 213, 270, 317-319, 327, 332, 394, 411, 426, 462, 480, 487, 491, 497, 505
fruit bats, 313-328
fruit color, 87
Fuchsia, 89
functional analyses, 17, 575
functional constraints, 18, 448
functional radiations, 8, 204
fungus-breeders, 502-505

γ-globin, 195
Galápagos Islands, 2-3, 18, 24, 31, 354
gammarids, 4
gap, 132, 192-193, 205, 264, 284, 289-290, 292, 295, 301, 439-441
Gasterosteus, 511, 513, 518-519, 527
geckoes, 21, 546
Gekkonidae, 549
gene duplication, 29, 57
gene flow, 10, 13, 23, 25, 31, 163, 379, 387, 521, 527-531
gene sequence, order, 29-30
generalist colonization, 378, 381
genetic correlation, 18
genetic divergence, 2, 28, 84, 92, 185, 287, 300, 302, 479, 586-587, 599, 604
genetic drift, 23, 380, 479, 511
geographic distribution, 22, 134, 180, 197, 208, 211, 260, 264-265, 280, 285, 299-302, 355, 357, 361, 365, 375, 379-380, 416, 545
geographic speciation, 25

geological dating, 586-588
Geonoma, 26
Geospiza, 3, 18, 24
Geospizidae, 2
Geraniaceae, 401
Gesneriaceae, 427
gibbons, 192
Gigantosaurus, 32
gill rakers, 513-516, 524-525
Glaucomys, 130
gliding evolution, 130-131, 146
Glironia, 134
globin amino-acid sequence, 132, 149
glucosinolate families, 75
Glycine, 274, 459
Glyptocidaris, 562
Goeldi's monkey, 189, 200, 202-203, 211-212
Gomesa, 342, 350
Gondwana, 132
Goniochilus, 334, 350
Goodeniaceae, 427
Gould, John, 2
Gould, Stephen Jay, 2, 11, 17-20, 27, 29-30, 33, 35, 546, 553, 602
Gouldia, 427
Gracilinanus, 134, 136, 146-147
grass-bush, 538-540, 548, 551, 553
Greater Antilles, 36, 535-536, 538, 540, 543, 550, 553
groundhog, 35
growth form, 6, 22, 28, 35, 103, 117-118, 229-230, 232, 247, 264, 269, 278, 284, 288-289, 292, 294-295, 335, 411, 457
Guam avifauna, 37
Guayana Shield, 38, 259-262, 265-266, 270, 273, 290-292, 295-296, 299-302
Gymnobelideus, 134, 143
gymnosperms, 19-20
gynodioecy, 456-457, 460, 465, 470

Habenaria, 401, 436-437, 441
habitat partitioning, 383
habitat shift, 122, 164, 173-175, 357, 362, 364, 467, 469-470, 538
Haemodoraceae, 227, 239-243, 245, 250
Haldane, J. B. S., 2
halictids, 334, 346
halophiles, 29
haplochromine cichlids, 14, 378-379, 382, 385
haplotype, 518-521, 527-528, 530
hard-husked fruits, 207, 209, 213
Harpyionycteris, 314, 317
Hawaiian Drosophila Project, 478
Hawaiian lobelioids, 109, 114
Hawai`i, 10, 26-28, 31, 37-38, 103, 115-116, 121-125, 455, 457, 462, 478, 480, 486, 490, 492, 498-500
hawkmoths, 395-396, 436, 438, 445, 450-451
headshield morphology and predation, 171
Helcia, 350
Heliamphora, 33, 272, 293, 297, 300
Heliocidaris, 562
helmet and neckteeth formation, 173-175
Hemaris, 445
Hemibelideus, 134
Hemignathus, 14
Hemizonia, 114
Hennig, Willi, 55-56, 192-193
herbivorous diet, 203
hermaphroditism, 455-457, 460, 462-463, 465, 467, 469-471
Herpesvirus, 23
Heteranthemis, 413-414, 416, 420, 425
Heteranthera, 227-231, 233-237, 240-241, 245, 247, 252-253
heterochrony, 2, 30, 333, 335, 337, 346-348
heterocystous cyanobacteria, 270
heterophylly, 232-233
Heteropyxidaceae, 402
heterostyly, 237, 253
Himatione, 14
hindlimb length, 538, 546-549
Hispaniola, 536-537, 539, 541, 543, 550, 552-553
historical biogeography, 32, 149
historical contingency, 35-36, 553
historical ecology, 2, 13, 18, 22
homeotic genes, 11, 29, 113
Homo, 192
homology, 57-58, 79-80, 83, 86-87, 242, 245, 272-273, 275, 343-344
homoplastic genomic arrangement, 122
homoplasy, 6-7, 15, 55-59, 62-63,

65, 67-68, 70-71, 74-76, 79, 81-89, 91-92, 107, 120, 131, 195, 238-239, 253, 260, 276, 302, 337-338, 340, 342, 344, 376, 386, 441, 541, 561, 571
homoplasy, new measure of H*, 74
homoplasy slope ratio, 59
Homoptera, 28
honey-possum, 129, 133, 137, 139
honeycreepers, 4, 14-15, 37, 109, 259, 596, 601
hooked seeds, 342-343, 345
Hooker, Joseph, 2
horizontal gene transfer, 29, 57
horses, 1, 33-34, 150, 585
host plant association, 486
hot spot, 455
howler monkey, 195-197, 200, 206
hummingbirds, 313, 315, 334, 346-347, 395-396
Huxley, Julian, 1-2, 4, 8-10, 511, 522
Hyalodaphnia, 164-169, 173-174, 176-178, 181, 183
hybrid, 21, 23, 104, 118, 122-124, 164, 167, 271, 285, 395-396, 412-413, 419, 514, 521, 526, 529
hybrid, high viability and fecundity of, 24
hybrid swarm, 24, 413, 419
hybridization, DNA, 131, 139-141, 143, 147-149, 151-152, 314-318, 320-328
hybridization, filter, 337, 518
hybridization, interspecific, 7, 24, 28, 57-58, 85, 90, 111, 117, 119-120, 122, 124-125, 163, 165, 169-170, 180, 183-184, 238, 260, 274, 278, 280, 380, 413, 420-421, 424, 426-428, 479, 514, 524-525, 527-528
Hydrothrix, 227-228, 230-231, 234-235, 241, 252
Hylobates, 192
Hymenostemma, 414
Hypsiprymnodon, 134, 136

icefish, 21
immunological distance, 535
independent characters, 7, 84-85, 91-92, 260
independent contrasts, 315, 321, 323-326, 549

infaunal biota, 559, 562, 575, 577
infectious agents, 23
informative characters, 7, 65, 70, 76-77, 82, 88, 107, 168, 192, 195, 274, 397, 439, 542, 545, 559, 567, 574
insect consumption, 204, 206
inter-island dispersal, 10, 103, 125, 543
interphotoreceptor retinol-binding protein (IRBP) gene, 192
interspecific competition, 21, 35, 206, 229, 501, 523-525, 537
intralacustrine speciation, 25, 387, 604
introgression, 7, 57-58, 85, 90, 124, 164, 184, 260, 274, 278, 280, 396, 426-427, 527
Ionopsis, 333-334, 336, 342-343, 347, 350-351
Ipomoea, 489, 492
IRBP gene, 195
Ismelia, 413-414, 416, 420, 425
Isoodon, 134, 149
Isopyrum, 392-393
isozyme data, 13, 119, 125, 411, 416, 420-421, 423, 426, 514, 545
ITS spacer region, 1-3, 8, 10-11, 21-22, 58, 63, 78, 82, 105-125, 133, 139, 143, 146, 163, 176-177, 183-184, 190, 195, 203-204, 210-212, 226, 234, 237, 240, 243-244, 246-250, 259, 261, 270-271, 275-276, 280, 284, 286, 288, 300, 303, 313, 328, 332, 338, 344, 354-355, 365, 377, 385, 392-394, 401, 411, 413-414, 416, 418-421, 424, 427, 437, 439-440, 442-443, 446-447, 449, 477, 486, 498, 500, 502, 506, 513, 521, 523, 526-527, 535, 538, 549, 551, 587-588, 595, 598-602
jackknife, 140, 339, 341-342
Jamaica, 536-537, 539, 543, 551-553
jaw morphology, 11, 16-17, 22
jerboa marsupial, 35
Jessenia, 206
kangaroos, 35, 129-130, 133, 135-137, 139-140, 146-147, 150
karyotype, 478, 535
Kaua`i, 14, 103, 107-109, 111-112, 115-125, 455, 457, 462, 469, 482, 486, 491

key adaptation, 150, 171, 354, 370
key innovation, 2, 9, 14, 21, 24-25, 28, 33, 192, 204, 207, 209, 212-213, 225, 260-261, 265, 275, 285, 289, 293, 297-298, 302, 354-355, 369-371, 391-394, 396-398, 400-401, 536, 548-549
key landscape, 2, 33, 285, 300, 302
kipuka, 475

labellum, 436-438, 448, 450, 476
Labeotropheus, 382, 385
Labidochromis, 382, 385
Lack, David, 2
Lago Chichancanab, 587, 590, 594, 596, 600, 606
Lagothrix, 195, 197, 203, 206, 212
Lake Baikal, 4, 586
Lake Barombi Mbo, 591, 593, 596-597, 599, 605
Lake Lanao, 593-594, 599
Lake Nabugabo, 585, 592, 596-598
Lake Paxton, 512-515, 517, 519-525, 527-528
Lake Tanganyika, 14, 16, 25, 375, 377-379, 386, 588, 592-594, 596-599
Lake Titicaca, 585, 594, 596, 600
Lake Victoria, 14, 25, 375-376, 378-379, 384, 527, 587, 589, 591-593, 595-599, 605-606
lake-level fluctuation, 379-380, 387, 599
Lamiaceae, 411
lamprologine cichlids, 377
Lana`i, 103, 106, 457, 462, 486
land snail, 11, 37
land vertebrate, 22
larval environment, 28
larval substrate, 477, 489, 497, 501, 503, 505
Lasiorhinus, 134
Laupala, 10
Lavatera, 428
leaf character, 459-460
leaf developmental pathway, 242, 247, 249, 252
leaf shape, 28, 232, 293, 412
leaf size, 17, 22, 205
leaf turgor, 103
leaf vasculature, 104
leapfrog radiation, 331, 337
lecithotrophic development,

560, 562, 577-578
Lecythidaceae, 201
Leguminosae, 401
leks, 12, 28, 479, 503-504
Lemboglossum, 351
Leochilus, 332, 334, 336, 347, 351
Leontopithecus, 200, 202-203, 211-212
Lepidophorum, 414
lepidote inflorescence, 282
Lestodelphys, 133-134, 147
Leucohyle, 343
lianas, 103-104, 106, 119, 124, 199, 205
liana density, 205
life history, 113, 183, 226-227, 229, 245, 247, 253, 331, 333, 342, 346, 357, 364, 500, 560, 581
life-cycle duration, 227, 231, 242, 245-246
life-form, 106, 114-115, 118-119, 124, 226-232, 242, 244-245
Likoma Island, 377-386
Liliaceae, 486
limited dispersal, 12, 25-26, 37, 303, 379-380, 425
limnetic, 21, 512-517, 520-531
Limonium, 426
Lindmania, 273, 287, 291, 301, 303
lineage sorting, 57, 90, 120, 426, 527
lion tamarin, 202-203, 211
litopterns, 36
Lobelioideae, 26, 36
locomotion, 131, 136, 147, 189, 195, 197, 199, 201-205, 546, 577
long-distance dispersal, 10, 31, 114, 165, 178, 227, 238, 300, 302, 413
long-lived perennial, 231, 247
Lophiaris, 343, 346
lophodont molars, 130
loss of dispersal mechanism, 424
Lotus, 411
lowland rain forest, 189
Loxops, 14
LSU rRNA, 561, 563-566, 569-571, 574, 579, 581
Lutreolina, 133-134, 149
Lycaenoidea, 27
Lytechinus, 562

Macaca (macaque), 192

Macaronesia, 407-408, 410-411, 414, 419-420, 423
MacClade, 106, 110, 114, 143, 146, 167, 241, 275, 319-323, 325-326, 359, 382, 440, 443, 446-447, 481, 543
Macradenia, 334, 351
Macroclinium, 333-334, 336, 346-347, 351
macroevolution, 2
macroevolutionary hypothesis, 33
macroevolutionary trend, 9
macroglossine bats, 313-322, 326-328
Macroglossus, 313, 318
Macropleurodus, 375, 377
Macropus, 134, 139, 146
Macrotis, 134
Madia, 105, 108, 110-114, 116, 124
MADS box family, 30
Malawi, 14, 16, 25, 375-385, 387-388, 588, 591-593, 596-599, 604-606
Malvaceae, 428
mammals, 1, 7-8, 11, 26, 31, 36, 132-133, 135, 139, 141, 143, 178, 204-205, 209, 314, 316, 328, 354, 563, 585, 589, 592
manipulative foraging, 209
Marmosa, 134, 143, 146
marmoset, 189, 200, 202-203, 210-213
Marmosops, 134, 143, 146
marsupials, 4, 8, 15, 33, 35-37, 129-137, 139-141, 143, 146-152, 158-159, 161, 259, 328, 354, 535
marsupial lion, 134
marsupial mole, 130, 133, 141, 146-147, 151
mass extinction, 36, 150, 559
mat plant, 106, 124
mating system, 179, 226, 234-235, 387, 476, 503-505, 605
*mat*K, 68
Maui, 14, 103, 105-106, 110-111, 115-117, 121-125, 425, 457, 462-463, 482, 486, 490, 492, 498
Maui Nui, 103, 106, 115-116, 121-124, 462, 486, 490
Maxillaria, 333, 338, 352
maximum likelihood, 8, 22-23, 112, 326, 398, 541
maximum parsimony, 8, 22, 137, 339, 359, 382, 393, 399, 481,

542
Mayr, Ernst, 1-2, 4, 8-10, 12, 20-21, 23, 25, 163, 192, 259-262, 303, 375, 388, 394, 511
mbuna, 16, 25, 379-382, 385, 598
mealy-bug, 28
Megachiroptera, 148, 313
Megaloglossus, 313-314, 318-319, 327
Melanesian ant fauna, 32
Melonycteris, 313, 318, 328
Meoma, 562, 575
Mespilia, 562
Metachirus, 134
metamorphosis, 266, 560, 577
Metatheria, 129, 134, 148
metazoa, 11, 29
Metrosideros, 31
Micoureus, 134, 146
Microbiotheria, 133-134, 139, 141
Microchiroptera, 148, 327
microclimate, 537, 540
Microperoryctes, 134, 146, 149
Microtus, 585
Minuartia, 458
mitochondria, 29
mitochondrial DNA (mtDNA), 7, 13-14, 16, 165, 192-194, 212, 359-360, 362, 377, 381-382, 384, 388, 483, 485, 505, 513, 517-520, 522, 527-528, 530-531, 589, 592-594, 597-600, 604
mockingbirds, 31
Modern Synthesis, 1-2, 4, 37
modified canines and incisors, 212
modified-mouthparts clade, 27
Moehringia, 458
molecular clock, 33, 420, 548, 598
molecular systematics, 1, 6-8, 37, 87, 91-92, 131, 259-260, 272, 301-302, 337, 370, 379, 383, 413-414, 433, 438, 481, 518-519, 535-536, 540
Moloka`i, 103, 106, 457, 462-463, 486
monocarpic, 103, 116-118, 271
Monochoria, 227-230, 232, 235-237, 239-240, 247, 250, 252-253
monocotyledons, 71, 83, 225, 227, 234, 243
Monodelphis, 134, 147
monoecy, 19
monomorphic flower, 243

monophagy, 483, 486, 488, 505
monophyly, 104, 106-107, 117-119, 121, 137, 139, 141, 143, 168, 199-200, 212, 237, 274, 377, 391-392, 414, 416, 440-441, 463, 527, 592, 594, 596-597
morphological stasis, 163
mosquito, 445
mouth brooding, 25, 380
Murdannia, 242
Murexia, 134, 149
mutualism, 13, 27, 270
Myoporaceae, 498
Myrmecobius, 130, 133-134
Myrmecophaga, 130
Myrtaceae, 402

Nasalis, 192
native cats, 35
natural selection, 4, 8, 17-19, 21, 23, 33, 35, 91, 353-354, 476-477, 479, 501, 503-505, 511, 513, 521-525, 528, 531, 546
Navia, 10, 25, 273, 287, 290-291, 301, 303
*ndh*F, 67-68, 82, 237-239, 241, 244, 246, 248-250, 420
Neblinaria, 272, 297
nectar spur, 346-347, 392-403, 436-437, 450-451
nectarivores, 14, 133, 139, 313-315, 318-319, 323, 327-328
neighbor-joining, 8, 167, 422, 518-519, 521-522
Neoescobaria, 351
neoteny, 233-234, 247, 294, 333, 347
Nepenthes, 267
neuron cell size, 31
New World monkeys, 189-190, 192, 195, 199, 201, 204-205, 209, 212-213, 222
Newark Basin, 588, 594, 600-601, 604
niche differentiation, 24
Nigella, 401-402
Nile perch, 37
Nipponanthemum, 413
nitrogen fixation, 261 270, 285, 289, 292, 298, 302
noctuids, 436, 438, 441, 445
nocturnal habit, 209
Noisettia, 401-402
non-impounding habit, 261, 264, 269, 271, 276, 278, 283-285, 292, 295-296, 301-303

non-random species distribution, 32
Nothofagus, 32, 150
notonectids, 173
Notopteris, 313, 318, 328
Notoryctemorphia, 133-134, 139, 141
Notoryctes, 130, 133-134, 141
notothenioid fish, 21, 30-31
notoungulate, 36
Notylia, 334, 347, 351
nuclear ribosomal DNA (nrDNA), 28, 82, 105-125, 265, 274, 278-280, 289, 294, 439
nucleotide divergence, 518-519, 521-522
numbat, 130, 133
nutrient loading, 229

ocelots, 35
Odontoglossum, 351
oil-collecting bees, 334, 343, 433
olfactory lobes, 209
oligophagy, 483, 486, 488, 490, 492
Oncidiinae orchids, 89, 331-339, 343-350, 451
Oncidium, 272, 274-275, 334-335, 337-338, 342-343, 351-352
ophiuroids, 560
Ophrys, 440
optimality model, 17
oral jaw, 20, 375, 386
orchids (Orchidaceae), 15, 89, 272, 275, 331, 333-334, 336-337, 339, 344-346, 348-349, 394, 401, 433, 436, 440, 448, 451
Orchis, 440
Oreomystis, 14
origin of heterostyly, 253
Ornithocephalus, 333
Osborn, Henry Fairfield, 1, 8-10, 259, 354
Otoglossum, 352
Otter Point, 377-386
ovarian development, 492, 495
ovarian type, 483, 493, 495-498, 501
oviposition site, 27, 477, 479, 486, 497-498, 502
ovipositor, 480, 493, 501-502, 504-505
owl monkey, 189, 193, 198, 200-201, 204, 209, 212
oyster shells as hammer, 210

O'ahu, 103, 115-116, 121-122, 457, 462-463, 469, 486

paedomorphic, 233, 252, 333
paleontologic evidence, 132
paleontology, 131, 511
palms, 198, 205, 209-210
Palmeria, 14
Pangaea, 32
Panglossian, 18-19
Paracentrotus, 562, 566
parallel evolution, 129, 375, 384-386, 426, 460, 483, 546
parallel specialist, 380
parallelism, 14, 55-57, 129, 334, 375, 386, 440, 551
Parapithecus, 192
parasitic social insect, 24
Paroreomyza, 14
patagium, 130
Paucituberculata, 133-134, 141, 143
PAUP, 63, 105, 110-112, 137, 168, 192, 274, 338-340, 359, 382, 385, 439-440, 459-460, 481, 564, 578
Pelargonium, 401-402
Perameles, 134
Peramelina, 133-135, 137, 141
peramorphosis, 333
Pericallis, 428
Peroryctes, 134
Petauroides, 131, 134
Petaurus, 130-131, 134, 136, 139, 143, 147
Petrogale, 134
Phalacrocarpum, 416
Phalanger, 148
pharyngeal jaw, 14, 20, 401
Phascogale, 134, 149
Phascolarctos, 134
Phenacosaurus, 540, 543
phenogram, 423, 426
phenology, 355, 361, 364-366, 369-370
phenotypic stasis, 164
Philander, 134
philopatry, 10, 13, 25, 380
Philydraceae, 227, 239-243, 245-246, 250
Philydrella, 242
Philydrum, 242-243
phloxes, 4
Phormosoma, 562
phyllostomid bat, 314
phylogenetic constraints, 18-20, 35, 122, 370

phylogenetic indicator, 335
phylogenetic signal, 83, 195, 359, 384, 441, 545, 563, 574-575
phylogeny, 1-2, 5-6, 8, 13-16, 19-23, 36, 55, 59, 62, 65, 80-87, 90, 92, 105-106, 110-113, 115, 119, 123, 135, 137, 139, 143, 150-151, 165, 169-170, 176, 179, 183, 189-190, 192, 195-196, 204, 222, 240-241, 259-260, 265, 273, 275-280, 282, 285-286, 288-290, 292, 298-303, 314-316, 319, 321, 323-328, 332, 337, 342, 346, 362, 366, 369-371, 378, 386-387, 403, 414, 423-427, 433, 436, 440, 443, 446-447, 450-451, 464, 466-468, 470-471, 481-483, 485-486, 490-492, 495, 497-498, 500, 504-506, 517, 521, 527, 530, 535-536, 538, 540, 543-545, 548, 551-552, 561, 563, 567, 569-570, 574-575, 577-578, 590, 601-603
picture-winged clade, 28
Pipturus, 427
Pitcairnioideae, 259-261, 265, 270, 272, 276, 286, 295
Pithecia, 196
placental mammal, 36, 129, 133, 139, 141, 316
placental mole, 130
Planigale, 133-134, 149
planktivores, 14, 513, 516, 523, 526, 528, 530-531
planktotrophic development, 562, 577, 581
plant exudate, 189, 203, 211-213
plasmid, 29, 57, 459
Platanthera, 433, 436-438, 440-451
platyrrhine, 190, 194, 203, 208-209
pleiotropy, 6, 58, 79, 83, 87, 90, 164
Plumbaginaceae, 426
pollination biology, 6, 37, 226, 234-235, 237, 253, 334, 343, 345, 392, 394, 424, 433, 436, 438, 440-441, 443-451, 456, 458, 470-471
pollination syndrome, 234, 343, 345, 392, 436, 438, 440-441, 443-444
pollinator morphology, 395, 433
pollinator partitioning, 448, 451
pollinia, 394, 433, 436-437, 445, 447-451

pollinia placement, 436, 447-449, 451
Polychrus, 546, 549
polyphagous, 486, 490, 492
polyphagy, 483, 486, 488, 490, 505
polyphyly, 139, 314, 375, 424, 426-427
polyploidy, 23, 183
polyprotodonty, 135-136, 146
Pontederia, 230, 252-253
Pontederiaceae, 225-233, 235-250, 253
populational studies, 17-18
porcupines, 204
positional behavior, 190, 195, 197-199, 201-202, 206-207, 213
post-zygotic barrier, 413
pouch type, 146
preadaptation, 17, 20-21, 445
Precambrian, 11, 261
prehensile tail, 195, 197, 201, 204-205, 207, 213
premating barrier, 10, 12, 479
Priest Lake, 512-513, 515, 517, 519, 521-522, 527-528
primate, 23, 189-190, 192, 200, 204, 209-211, 213
proboscis monkey, 192
progenesis, 233-234, 333, 347
Prolongoa, 414
protein polymorphism, 13, 79-80
Psammechinus, 562, 566
pseudo-cleistogamy, 235
pseudobulb, 333-334, 336, 347
Pseudocheirus, 134, 140
Pseudochirops, 134, 140
Pseudomonas, 270
Pseudotropheus, 379-384, 387
Psiloxylaceae, 402
Psychopsis, 352
Psychotria, 26, 87, 114
Psygmorchis, 332-335, 342-343, 345-347, 352
Pteralopex, 314, 317-319, 328
pteropodids, 314, 317-319, 325-328
Pteropodidae, 313-314
Puerto Rico, 334, 536-540, 543, 550-553
pull-back rule, 84
pungent leaf tip, 282-283, 288-289, 295, 298
pygmy marmoset, 202
pyralid moths, 436, 438, 441, 445

rail (Rallidae), 31

Raillardella, 105, 110
Raillardiopsis, 105, 108, 110-114, 116, 124
rapid cycling, 343, 345-346
*rbc*L, 63, 67-68, 71, 82, 88, 237-239, 241, 244, 246, 248-250, 392, 401-402, 440
reciprocal translocations, 28, 104
recurrence, 6, 14, 56-57, 63, 70-71, 74-75, 83, 87-88, 91-92, 131, 259
Rediviva, 395
reduced dentition, 313
reproductive isolation, 23-24, 28, 32, 337, 386-387, 394-397, 413, 479, 523, 529, 550, 605
resampling analyses, 64, 92
resource partitioning, 4, 10, 12, 364, 511, 538-539, 575
respiratory filaments, 477, 492-493, 498-502, 505
restriction sites, 7, 14, 28, 56-60, 62-71, 76, 78-88, 91-92, 106-107, 110-111, 117, 124, 236, 238, 244, 246, 248-250, 258, 260, 265, 272-279, 282, 299, 332, 334, 337-341, 344, 350, 400, 411, 414-416, 419-421, 426-427, 458-459, 463-465, 468, 470, 513, 518-519
retention index, 59, 63, 73, 92, 105, 110, 321, 341-342, 564
reticulate speciation, 164
Reussia, 227-228, 230
reversal, 55-56, 276, 300, 450, 460, 467, 470, 495, 551
RFLP, 314, 517-518, 589, 592-593
Rhyncholestes, 134, 149
ringtail possum, 131, 133, 151-152
ritualized fighting, 479
rodents, 26, 129
Rodriguezia, 332, 334, 337, 342, 345-347, 352
rosette tree, 261
Rossioglossum, 352
Rubiaceae, 87, 269, 332, 427
Runchomyia, 271

saber-toothed tiger, 35
Saimiri, 196, 198
saki, 189, 198-200, 203, 207-208
salamanders, 13, 31
Salmacis, 562
saprophage, 477
Sarcophilus, 133-134
savanna, 1, 189, 197, 261, 264,

266, 268, 271, 273, 276, 288, 290-291, 295-296, 301, 358, 437
Scaevola, 427
Scaptomyza, 478-479, 481-482, 485, 494
scaptomyzoid flies, 478-479, 481, 483, 485, 489, 491, 493, 495, 497, 501, 504-506
Scelochilus, 333-334, 347-348, 352
Schiedea, 28, 419, 455-458, 460, 462-463, 465, 469-471
Schiekia, 243
Scholleropsis, 227, 230, 240, 245
Scincidae, 549
Scytonema, 270
sea urchin, 11, 559
secondary contact, 23, 32, 394, 397, 521, 527-528, 530
secondary sexual characteristic, 35, 476, 503
seed morphology, 332, 345
seed predator, 199, 201
seedling habit, 332-333, 345-346
selenodont grinder, 130
self-compatibility, 237, 243, 251, 411, 413
self-compatible, 237, 243, 253, 411
self-incompatibility, 227, 235, 237, 243, 251-253, 413, 420
Semiaquilegia, 397-399, 402
semionotid fish, 586, 596, 600-602, 604
sensitivity analyses, 22
separate lake basin, 25, 375
separate lower ankle-joint pattern, 135-136, 160
sequence
 colonization, 115
 developmental, 31
 evolutionary, 36, 174-175, 231, 233, 253, 294, 302, 316, 321, 326, 359, 476, 511
sequence evolution, 112, 581, 589
sequences, amino acid, 132, 149; DNA, 7, 13-14, 28, 56-63, 65-71, 78-83, 85-88, 91-92, 105-108, 110-113, 117-118, 120, 123-125, 131, 140-141, 143, 147-148, 151, 165, 167-168, 170, 179, 189, 192-195, 200, 202, 212, 224, 237-238, 244, 246, 248-250, 259-260, 274, 300, 376, 381, 392-393, 401-402, 411, 414, 418-421, 427,
440-442, 450, 483, 485-486, 505, 519, 535, 541-542, 551-553, 561, 563, 565-566, 568, 570-575, 579-581, 586, 588-589, 591-594, 596-598, 600-601, 604-606
Serranochromis, 378
sexual radiation, 12, 35, 475
sexual selection, 12, 19, 23-24, 27-28, 35, 380, 385, 387, 448, 456, 475-477, 479, 503-505, 511, 515, 517, 529
sharp canines, 207, 209, 213
short-lived perennials, 231, 247
Sideritis, 411
Silene, 458
silversword alliance, 15, 22, 24, 28, 103-122, 124-125, 391-392, 413, 424, 427
Simpson, George Gaylord, 1-2, 4, 8-10, 20-21, 24-25, 60, 129, 132, 135, 151, 225, 227, 240-243, 259, 314, 428, 511, 522, 528, 535, 545
single-invasion hypothesis, 515-516, 521
SLAJP, 135-137, 146
Slowinski and Guyer, 9, 25, 113, 397-398, 400
Smilodon, 130
Sminthopsis, 133-134, 136
snail, 11, 20, 37, 585
Solanum, 26
Sonchus, 419, 423, 428
Sophophora, 478, 481
Spatangus, 562
specialist colonization, 380-381
specialized canine, 199
speciation, 1-2, 9-14, 23-28, 31-33, 35-37, 90, 120, 124, 164-165, 169, 174-175, 225-226, 253, 260, 275, 289, 299, 301, 303-304, 353-357, 365, 370, 375, 378-381, 385-388, 391, 394, 398, 400, 403, 421, 424-425, 427, 448, 451, 455-456, 469-470, 475-477, 479, 483, 503, 506, 511-512, 515-516, 521, 527-531, 535, 539, 549-550, 585-586, 590-591, 596-598, 601-606
 models, 375, 378, 386-387, 601
 rate (SR), 591, 596
 sympatric, 23-24, 33, 164, 379, 386, 515-516, 521, 528, 530, 604-606
time required for (TFS), 585, 590-591, 596, 602, 605-606
species flock, 14, 375-379, 384, 387, 585-586
species origination and differentiation, 515
species-level phenomena, 25
Sphaerechinus, 562
spider monkey, 23, 189, 197, 199-200, 203-208
Spilocuscus, 134
squirrel monkey, 189, 193, 200-201, 203-204, 209, 211-212
SSU rRNA, 561, 563, 565-566, 568, 570-571, 574-575, 577, 579, 581
Staphylococcus, 29
stature, 26-27, 271, 281, 284, 289, 291, 294, 296, 301, 303
Stebbins, G. Ledyard, 2, 4, 12, 23, 234, 247, 253, 392
Stegolepis, 272, 297
stenotopy, 380
sticklebacks, 15-17, 21-22, 31, 511, 513-515, 517-518, 520-521, 523, 527-531
Stigonema, 270
Stomopneustes, 562, 574
stonefly wing, 17
Streptococcus, 29
Strongylocentrotus, 562, 571
structural habitat, 537, 540, 587, 600
structuralist, 29
subdioecy, 457, 460, 465, 470
successive weighting, 338-342, 344
sugar glider, 35, 130
suspended particulate, 176
suspensory habit, 195
swim-bladder, 21
swimming structure, 31
Syconycteris, 313, 318
Symphyglossum, 352
syndactyly, 135, 143, 146, 151

Taeckholmia, 425
tamarins, 31, 189, 200, 202-203, 211-212
tank epiphyte, 261, 264, 267, 269-270, 278, 282, 285, 287, 294, 301-303
tank fluid, 267, 269, 293
tank-forming habit, 261, 265
tarsier, 192
Tarsipes, 129, 134, 139, 328
Tarsius, 192

tarweed, 105-106, 108, 111-112, 114, 124, 259
Tasmanian devil, 35
taxon sampling, 22, 358
taxon-cycle hypothesis, 32, 353, 355-357, 364
taxon-pulse hypothesis, 32, 353, 355-357, 362, 364-365, 369-370
Telespiza, 14
Temnopleurus, 562, 581
Templeton's Test, 563, 567
tepui, 10, 33, 259-262, 264-268, 270-273, 276, 280, 283, 285-289, 292, 296-297, 299-304
terete leaves, 295-296, 333, 343
terra firme, 199, 202
terrestrial, 15, 33, 131, 133, 146, 150, 163, 225, 230-231, 234, 242-243, 247, 261, 264, 269-270, 276, 278, 291-294, 433, 436, 451, 538, 547, 587-588, 591-592, 594-595, 600, 604
tests of differential diversification, 397
Tetragnatha, 391, 596
tetrapods, 22
Thalarctos, 585
Thalictrum, 392-394
thermophile, 29
thylacine, 130-131, 133, 135, 151-152
Thylacinus, 130, 134
Thylacosmilus, 130, 133
Thylamys, 134, 136, 147
Thylogale, 134
Ticoglossum, 352
tiger beetles, 353, 355, 357-358, 365, 370
Tillandsia, 269, 271, 293, 295
Tillandsioideae, 261, 272, 288
time of origin, 109, 420, 586, 589
timing of the marsupial radiation, 131, 147
TIPS correlations, 323-324
tissue elastic properties, 122
tissue osmotic and elastic properties, 103
titi monkey, 189, 198, 200, 203, 208
toepad characteristic, 546
Tolumnia, 333-336, 338, 346, 352
tool use, 210
total evidence tree, 193-195, 197-198, 570
transference, 56-57, 85, 88, 91
transition
  character-state, 179, 184, 230, 242, 286, 362, 395-396, 459-460, 467, 577
  habitat, 150, 177, 225, 231
  substitutions, 112, 168, 383-384, 439, 441, 541
transversion substitutions, 112, 141, 147-149, 168, 439, 441, 541, 545
tree flux, 489, 491, 506
tree randomization, 323
Tribonanthes, 242
Trichocentrum, 336, 346, 352
trichome, 21, 261, 265, 267-271, 275, 285-286, 289-290, 292-293, 295-299, 302
Trichosurus, 134, 148
trilobite, 585
Tripneustes, 562
tristyly, 235-237, 243, 251-253
Trizeuxis, 334, 342, 352
trophic morphology, 385-386
trophic specialization, 325, 605
trophic structure, 21, 34, 386
trunk, 195, 197-199, 202-203, 207, 210-211, 333, 499, 502, 537-541, 551
trunk-crown, 538, 551
trunk-ground, 538, 540-541
tubular tongue, 14
Tulotis, 438, 441, 448, 450
twig epiphytism, 331-334, 342-343, 345-348
tympanic bulla, 130

uakari, 189, 198-199, 203, 207, 209
ultraviolet, 169, 184
unbiased sampling, 91
ungulates, 1, 129, 150, 589
universe of characters, 7, 91
unripe fruit, 199
Urticaceae, 427, 486
Utricularia, 270, 293

varillal, 198
Vavilovian series, 332, 345
vegetable debris, 261, 270-271, 285, 289
vegetative morphology and anatomy, 104
vegetative propagule, 227
velamen structure, 332
Vellozia, 297
venation, 22, 104, 122, 459
Vestiaria, 14
Viburnum, 87
vicariance, 32, 147, 260, 288, 299-303, 369, 420-421, 545
Viola, 402, 450
Violaceae, 401, 450
Vochysiaceae, 401-402
Vombatus, 134, 136

Wachendorfia, 236, 243
Wallabia, 134, 146
Warmingia, 334, 352
water hyacinth, 11, 15, 229
water-level fluctuations, 229
weighted parsimony, 274, 278, 416, 442
white-flies, 28-29
Wilkesia, 103-105, 107, 110-112, 115-119, 121, 124-125
wind pollination, 6, 394, 470-471
wolf, 61, 130, 169, 192, 195, 563
wolverine, 35
wombat, 35, 133, 137, 139, 146-147
woodiness, 26, 114, 414
woody habit, 35, 414
woolly monkey, 197, 206
woolly spider monkey, 189, 197, 203-207
Wright, Sewell, 2, 4, 23, 198, 200-201, 204, 208-209, 511, 559
Wyeomyia, 271

xenology, 57
xeric ancestry, 122

Yalkaparidontia, 133
Yp1, 481, 483, 485

Zimbawe Rock, 382
Zingiberanae, 239
zooplankton, 163, 184
Zosterella, 227-228